PROBLEMS IN THERMODYNAMICS & STATISTICAL PHYSICS

PROBLEMS IN THERMODYNAMICS & STATISTICAL PHYSICS

Edited by

PETER T. LANDSBERG

DOVER PUBLICATIONS, INC.
Mineola, New York

Bibliographical Note

This Dover edition, first published in 2014, is an unabridged republication of the work originally published in 1971 by Pion Limited, London. It is published by special arrangement with Pion Limited, 207 Brondesbury Park, London NW2 5JN England.

International Standard Book Number

ISBN-13: 978-0-486-78075-7
ISBN-10: 0-486-78075-9

Manufactured in the United States by Courier Corporation
78075901 2014
www.doverpublications.com

Contents

Authors

J.S.Blakemore	*Department of Physics, Florida Atlantic University, Boca Raton, Florida*
A.J.B.Cruickshank	*School of Chemistry, University of Bristol, Bristol*
D.J.Griffiths	*Department of Physics, University of Exeter, Exeter*
J.M.Haynes	*School of Chemistry, University of Bristol, Bristol*
P.C.Hemmer	*Institutt for Teoretisk Fysikk, NTH, Trondheim*
J.M.Honig	*Department of Chemistry, Purdue University, Lafayette, Indiana*
P.T.Landsberg	*Department of Applied Mathematics and Mathematical Physics, University College, Cardiff*
C.W.McCombie	*J.J.Thompson Physical Laboratory, University of Reading, Reading*
I.Oppenheim	*Department of Chemistry, Massachusetts Institute of Technology, Cambridge, Massachusetts*
K.E.Shuler	*Department of Chemistry, University of California, San Diego, California*
S.Simons	*Queen Mary College, London*
D.ter Haar	*Department of Theoretical Physics, University of Oxford, Oxford*
U.M.Titulaer	*Institut voor Theoretische Fysica der Rijksuniversiteit, Utrecht*
G.H.Weiss	*National Institute of Health, Bethesda, Maryland*
C.J.Wormald	*School of Chemistry, University of Bristol, Bristol*

Preface

Problems and solutions! Their production is an annual ritual—feared and unpopular—for the university teacher. Nor are examination questions particularly liked by students. Yet, take away the examination aura, and the technique appears in a new light. The problem-and-solution style of writing presents to the author new difficulties and constraints, and thus offers him a novel challenge. Perhaps it is like teasing a sculptor with a new and promising type of stone, or a poet with a new rhythm. Viewed in this light, the potential author has the opportunities which go with a new medium: for example, a novel way of arranging important results and of setting them up to be seen more clearly than is possible in a uniformly flowing and elegant exposition. As to the reader, he finds before him a series of hurdles. They are easy enough at first to bewitch him, beguile him and persuade him to join into the fun—until he gradually participates with the author in a neo-Socratic dialogue. The existence of these opportunities must surely be one of the considerations which explain how it was possible to find so select and experienced a group of persons to help in the construction of this book.

Each author contributed in his own special area of research with the result that the book presents a penetrating view of statistical physics and its uses which, as regards the width of its sweep is, I suspect, beyond any one of today's experts. Thus, in spite of the age of our subject and the love and care which has been bestowed on it by successive generations, this book presents in some sense a new unit.

I have planned its outline and have discussed various points with the authors, but have imposed only a limited uniformity of style. I have also attempted to ensure that overlap of material is not extensive, that cross references are adequate, and that the early problems in each chapter are reasonably easy. In this way I wanted the reader of each chapter to feel drawn into a fascinating world which opens up for him once he realises that he can actually derive results for himself. Here then is a book for teachers, undergraduates or graduates who want to know what can be done with reasonably simple models in thermodynamics and statistical physics. It is also suitable for self-study. It should appeal to a wide range of numerate readers: mathematicians, physicists, chemists, engineers and perhaps even economists and biologists.

The foundation of the general theory of statistical physics is involved and difficult, and its discussion takes a great deal of time in a lecture course. One can turn to this book for ideas and for inspiration if one wishes to pass on to areas of application while condensing the time spent on the foundations. This procedure will appeal to those who feel, as I do, that the foundations of the subject are best discussed briefly at first, and then again later from time to time as the effectiveness and the power of the methods are being appreciated.

It will come as no surprise to the experienced teachers that the need for care in the setting of a problem can stimulate original work. There are new presentations in various places (e.g. in Problems 1.24 to 1.26, 1.29–1.31, and 3.19 to 3.22), and in several sections of this book previously unpublished ideas will be found. For example, in Section 6 some looseness of logic inherent in previous work has been eliminated, through discussions between the author and the editor, and there are new approaches to equations of state, to the law of corresponding states and to the relationship between reduced equations of state (a) for chain molecules and (b) for their components. Problems 17.7 to 17.9 have also some novelty in the use made there of the random walk approach. There are several other places where recent research work has here been incorporated in book form for the first time (for example, in Problem 5.5).

In conclusion, I wish to thank all contributors and the publishers for the cooperation which made this venture possible.

P. T. Landsberg

PROBLEMS IN THERMODYNAMICS & STATISTICAL PHYSICS

1
The laws of thermodynamics

P.T.LANDSBERG
(University College, Cardiff)

MATHEMATICAL PRELIMINARIES

1.1 If each of the three variables A, B, C is a differentiable function of the other two, regarded as independent, prove that

$$\text{(a)} \quad \left(\frac{\partial A}{\partial B}\right)_C \left(\frac{\partial B}{\partial C}\right)_A \left(\frac{\partial C}{\partial A}\right)_B = -1,$$

$$\text{(b)} \quad \left(\frac{\partial A}{\partial C}\right)_B = 1 \bigg/ \left(\frac{\partial C}{\partial A}\right)_B.$$

Solution

Let the functional dependence of A on B and C be expressed by $f(A, B, C) = 0$, then

$$\left(\frac{\partial f}{\partial A}\right)_{B,C} dA + \left(\frac{\partial f}{\partial B}\right)_{A,C} dB + \left(\frac{\partial f}{\partial C}\right)_{A,B} dC = 0.$$

If A is constant this becomes

$$\left(\frac{\partial f}{\partial B}\right)_{A,C} \left(\frac{\partial B}{\partial C}\right)_A = -\left(\frac{\partial f}{\partial C}\right)_{A,B},$$

i.e.

$$\left(\frac{\partial B}{\partial C}\right)_A = -\left(\frac{\partial f}{\partial C}\right)_{A,B} \bigg/ \left(\frac{\partial f}{\partial B}\right)_{A,C}.$$

Similarly

$$\left(\frac{\partial C}{\partial A}\right)_B = -\left(\frac{\partial f}{\partial A}\right)_{B,C} \bigg/ \left(\frac{\partial f}{\partial C}\right)_{A,B},$$

$$\left(\frac{\partial A}{\partial B}\right)_C = -\left(\frac{\partial f}{\partial B}\right)_{A,C} \bigg/ \left(\frac{\partial f}{\partial A}\right)_{B,C}.$$

Multiplication of these three equations yields

$$\left(\frac{\partial A}{\partial C}\right)_C \left(\frac{\partial B}{\partial C}\right)_A \left(\frac{\partial C}{\partial A}\right)_B = -1.$$

Interchange of A and C in the second of these equations yields

$$\left(\frac{\partial A}{\partial C}\right)_B = 1\Big/\left(\frac{\partial C}{\partial A}\right)_B.$$

1.2 (a) Integrations over the following two paths in a plane are to be performed:

(i) the straight lines $(x_1, y_1) \to (x_2, y_1) \to (x_2, y_2)$;

(ii) the straight lines $(x_1, y_1) \to (x_1, y_2) \to (x_2, y_2)$.

$P(x_1, y_1)$, $Q(x_2, y_2)$ are two points, and $x_1 \neq x_2, y_1 \neq y_2$. The differential forms to be integrated are

$$du \equiv dx + dy,$$

$$dv \equiv x(dx + dy).$$

Show that

$$\int_{(i)} du = \int_{(ii)} du = u(Q) - u(P),$$

where $u = x + y$, and

$$\int_{(i)} dv \neq \int_{(ii)} dv,$$

and discuss the result.

[We shall denote differential forms with this property by $đv$ instead of dv, and call them **inexact**, while du is an **exact** differential. In relations of the type $du(x, y, ...) = g(x, y, ...)đv(x, y, ...)$, $g(x, y, ...)$ will be called **integrating factors**.]

(b) If

$$dF = X(x, y)dx + Y(x, y)dy$$

is an exact differential, show that

$$\left(\frac{\partial X}{\partial y}\right)_x = \left(\frac{\partial Y}{\partial x}\right)_y.$$

(c) **Pfaffian forms** have the general form

$$dv \text{ or } đv = \sum_{j=1}^{n} X_j(x_1, x_2, \dots x_n)dx_j.$$

(dv may be exact or inexact).

Show that for $n = 2$, if the X_j are single-valued, continuous and differentiable functions, $đv$ has always an integrating factor provided X_2 is non-zero in the domain of variation considered.

(d) Verify that a Pfaffian form with $n = 3$ need not have an integrating factor by considering $đv = x\,dy + k\,dz$, where k is a non-zero constant.

Solution

(a) We have

$$\int_{(i)} (dx+dy) = (x_2-x_1)+(y_2-y_1),$$

$$\int_{(ii)} (dx+dy) = (y_2-y_1)+(x_2-x_1).$$

These two line integrals have the same value, namely

$$u(Q)-u(P) = (x_2+y_2)-(x_1+y_1)\,.$$

On the other hand

$$\int_{(i)} dv = \tfrac{1}{2}(x_2^2-x_1^2)+x_2(y_2-y_1),$$

$$\int_{(ii)} dv = x_1(y_2-y_1)+\tfrac{1}{2}(x_2^2-x_1^2).$$

Since the two results are unequal, there does not exist any function $v(x, y)$ of which the differential form dv can be considered as an exact differential, for otherwise both integrals would yield $v(Q)-v(P)$.

In the present case $đv$, although inexact, has an integrating factor, i.e. a function $g(x, y)$ exists which converts $đv$ to an exact differential

$$dF = g\,đv\,.$$

If one integrating factor exists, then there exists an infinity of them, and a simple example is furnished in the present case by

$$g = 1/x, \quad dv = du\,.$$

(b) For an exact differential

$$dF = \left(\frac{\partial F}{\partial x}\right)_y dx+\left(\frac{\partial F}{\partial y}\right)_x dy \equiv X\,dx+Y\,dy,$$

so that

$$\frac{\partial^2 F}{\partial x \partial y} = \left(\frac{\partial X}{\partial y}\right)_x = \left(\frac{\partial Y}{\partial x}\right)_y.$$

(c) Let

$$đf = X\,dx+Y\,dy\,,$$

where X, Y are continuous differentiable and single-valued functions of the independent variables x, y. This restriction on X and Y together with $Y \neq 0$ means that the equation

$$\frac{dy}{dx} = -\frac{X}{Y}$$

has a solution of the form

$$F(x,y) = C, \quad \text{i.e. } dF = \left(\frac{\partial F}{\partial x}\right)_y dx+\left(\frac{\partial F}{\partial y}\right)_x dy = 0,$$

where C is a constant.

Comparing coefficients in $đf = 0$ and $dF = 0$,

$$\frac{1}{X}\left(\frac{\partial F}{\partial x}\right)_y = \frac{1}{Y}\left(\frac{\partial F}{\partial y}\right)_x \equiv g(x, y).$$

Hence

$$g\,đf = gX\,dx + gY\,dy = dF$$

and $đf$ has the integrating factor $g(x, y)$.

(d) Suppose $đv(x, y, z) \equiv x\,dy + k\,dz = g(x, y, z)\,dF$. Then

$$\left(\frac{\partial F}{\partial x}\right)_{y,z} = 0, \quad \left(\frac{\partial F}{\partial y}\right)_{x,z} = \frac{x}{g}, \quad \left(\frac{\partial F}{\partial z}\right)_{x,y} = \frac{k}{g}.$$

It follows from part (b) that

$$\frac{\partial^2 F}{\partial x \partial y} = 0 = \frac{1}{g} - \frac{1}{g^2}\left(\frac{\partial g}{\partial x}\right)_{y,z},$$

$$\frac{\partial^2 F}{\partial x \partial z} = 0 = -\frac{k}{g^2}\left(\frac{\partial g}{\partial x}\right)_{y,z},$$

$$\frac{\partial^2 F}{\partial y \partial z} = -\frac{x}{g^2}\left(\frac{\partial g}{\partial z}\right)_{x,y} = -\frac{k}{g^2}\left(\frac{\partial g}{\partial y}\right)_{x,z}.$$

These equations cannot be satisfied by a finite function $g(x, y, z)$.

QUASISTATIC CHANGES[1]

1.3 For a fluid and other simple materials any three of pressure p, volume v, and empirical temperature t are possible variables. They are connected by an equation of state so that only two of the three variables are independent. An increment of heat added quasistatically may then be expressed in the alternative ways

$$đQ = C_v\,dt + l_v\,dv = C_p\,dt + l_p\,dp = m_v\,dv + m_p\,dp,$$

where the coefficients are themselves functions and are characteristic of the fluid. The term **empirical temperature** refers to an arbitrary scale and is used to distinguish it from the **absolute temperature**, denoted by T.

Prove that the following relations hold:

(a) $$m_v = \frac{l_v C_p}{C_p - C_v}, \quad m_p = -\frac{l_p C_v}{C_p - C_v}, \quad \frac{m_v}{l_v} + \frac{m_p}{l_p} = 1;$$

(b) $$\left(\frac{\partial p}{\partial t}\right)_v = -\frac{C_p - C_v}{l_p}, \quad \left(\frac{\partial v}{\partial t}\right)_p = \frac{C_p - C_v}{l_v}.$$

(c) Express in words the physical meanings of the coefficients in the expressions for $đQ$.

[1] Changes consisting of a continuum of equilibrium states.

Solution

(a) From the first two expressions for $đQ$ an equation for $\mathrm{d}t$ is found:

$$(C_p - C_v)\mathrm{d}t = l_v\,\mathrm{d}v - l_p\,\mathrm{d}p\,.$$

Substitution for $\mathrm{d}t$ in the second expression for $đQ$

$$đQ = \frac{l_v C_p}{C_p - C_v}\mathrm{d}v + \left(l_p - \frac{l_p C_p}{C_p - C_v}\right)\mathrm{d}p.$$

Comparison with the last form for $đQ$ yields the required result.

(b) The first equation under (a) yields this result at once.

(c) C_v is the heat required per unit rise of empirical temperature at constant volume, also called the heat capacity at constant volume; l_p is the heat required per unit rise of pressure at constant temperature, sometimes called the latent heat of pressure increase. The other coefficients can be described analogously.

1.4 Let

$$\alpha_p \equiv \frac{1}{v}\left(\frac{\partial v}{\partial t}\right)_p$$

be the **coefficient of volume expansion** at constant pressure, and let

$$K_t \equiv -\frac{1}{v}\left(\frac{\partial v}{\partial p}\right)_t$$

be the **isothermal compressibility** of a fluid. The notation of Problem 1.3 will be used.

(a) Show that the **Grüneisen ratio** $\Gamma \equiv \alpha_p v / K_T C_v$ of the material satisfies

$$\Gamma = v\left(\frac{\partial p}{\partial t}\right)_v \Big/ C_v.$$

(b) Show that the ratio of heat capacities $\gamma \equiv C_p/C_v$ satisfies

$$\gamma = \left(\frac{\partial p}{\partial v}\right)_\mathrm{a} \Big/ \left(\frac{\partial p}{\partial v}\right)_t,$$

$$-\frac{1}{\gamma - 1} = \left(\frac{\partial v}{\partial t}\right)_\mathrm{a} \Big/ \left(\frac{\partial v}{\partial t}\right)_p,$$

$$\frac{\gamma}{\gamma - 1} = \left(\frac{\partial p}{\partial t}\right)_\mathrm{a} \Big/ \left(\frac{\partial p}{\partial t}\right)_v,$$

where $(\ )_\mathrm{a}$ denotes a quantity evaluated under quasistatic adiabatic conditions, i.e. for $đQ = 0$.

(c) If K_a is the adiabatic compressibility, prove that

$$\frac{K_t}{K_\mathrm{a}} = \gamma.$$

(d) Show from a consideration of d(ln$v$) that

$$\left(\frac{\partial \alpha_p}{\partial p}\right)_t = -\left(\frac{\partial K_t}{\partial t}\right)_p.$$

Solution

(a) We have

$$\Gamma = -v\left(\frac{\partial v}{\partial t}\right)_p\left(\frac{\partial p}{\partial v}\right)_t \Big/ C_v = v\left(\frac{\partial p}{\partial t}\right)_v \Big/ C_v,$$

where Problem 1.1(a) has been used.

(b) From the first equation in Problem 1.3:

$$\left(\frac{\partial p}{\partial v}\right)_a = -\frac{m_v}{m_p} = \frac{C_p l_v}{C_v l_p} \qquad \left(\frac{\partial p}{\partial v}\right)_t = \frac{l_v}{l_p},$$

[from Problem 1.3(a)]

$$\left(\frac{\partial v}{\partial t}\right)_a = -\frac{C_v}{l_v}, \qquad \left(\frac{\partial v}{\partial t}\right)_p = \frac{C_p}{m_v},$$

$$\left(\frac{\partial p}{\partial t}\right)_a = -\frac{C_p}{l_p}, \qquad \left(\frac{\partial p}{\partial t}\right)_v = -\frac{C_p - C_v}{l_p}.$$

We now take the ratios of terms in the first column to corresponding terms in the second column. The first pair yields γ at once. The second pair yields, in conjunction with Problem 1.3(a),

$$-\frac{C_v}{C_p}\frac{m_v}{l_v} = -\frac{C_v}{C_p}\frac{C_p}{C_p - C_v} = -\frac{1}{\gamma - 1}.$$

The third pair yields $\gamma/(\gamma-1)$.

(c) This follows from (b).

(d) We observe that t and p are the independent variables needed. The form of the equation to be proved suggests that the procedure of Problem 1.2(b) may be involved. Fortunately we also have the hint to consider d(ln$v$). Hence

$$\mathrm{d}(\ln v) = \frac{1}{v}\mathrm{d}v = \frac{1}{v}\left[\left(\frac{\partial v}{\partial t}\right)_p \mathrm{d}t + \left(\frac{\partial v}{\partial p}\right)_t \mathrm{d}p\right].$$

This is $\alpha_p\,\mathrm{d}t - K_t\,\mathrm{d}p$, and we can indeed apply Problem 1.2(b).

THE FIRST LAW

1.5 The first law states that an **internal energy function** U exists such that for a fluid or similar material we have, in addition to the first equation of Problem 1.3,

$$đQ = \mathrm{d}U + p\,\mathrm{d}v,$$

where p is the pressure, and $p\,\mathrm{d}v$ is the mechanical work done by the system.

(a) Show that the increment of work $\mathrm{d}W = p\,\mathrm{d}v$ is inexact.

(b) Prove that the Grüneisen ratio of Problem 1.4 is

$$\Gamma = v\bigg/\left(\frac{\partial U}{\partial p}\right)_v = \frac{v}{m_p}.$$

(c) Find the most general equation of state of a fluid whose Grüneisen ratio is independent of pressure.

Solution

(a) This is obvious since $dU = đQ - đW$. If $đW$ were an exact differential, one could integrate to find $Q = U + W + \text{constant}$, in contradiction with the inexact nature of $đQ$. Alternatively, note that we have two independent variables, so that one could write

$$đW = p\,dv + x\,dy$$

where $x = 0$ and $y = T$ or p. If $đW$ were exact we would have

$$\left(\frac{\partial p}{\partial y}\right)_v = \left(\frac{\partial x}{\partial v}\right)_y = 0.$$

This yields $1 = 0$ if y is chosen to be p, and hence $đW$ is inexact.

(b) We start with

$$đQ = dU + p\,dv = C_v\,dt + l_v\,dv$$

whence

$$C_v = \left(\frac{\partial U}{\partial t}\right)_v = \left(\frac{\partial U}{\partial p}\right)_v\left(\frac{\partial p}{\partial t}\right)_v.$$

Substitute in Problem 1.4(a) to find

$$\Gamma = v\left(\frac{\partial p}{\partial t}\right)_v\bigg/C_v = v\bigg/\left(\frac{\partial U}{\partial p}\right)_v.$$

(c) Integrate the result of (b) to find

$$U = \frac{1}{\Gamma(v)}[pv + f(v)],$$

where Γ and f can be functions of volume, and f/Γ is a constant of the integration with respect to p. The equation of state of a solid is sometimes taken in this form:

$$pv = \Gamma(v)\,U(v, T) - f(v).$$

1.6 Show that

(a)
$$m_p = \left(\frac{\partial U}{\partial p}\right)_v,$$

$$m_v = \left(\frac{\partial U}{\partial v}\right)_p + p,$$

$$\left(\frac{\partial m_p}{\partial v}\right)_p = \left(\frac{\partial m_v}{\partial p}\right)_v - 1.$$

(b)
$$\left(\frac{\partial U}{\partial t}\right)_p - \left(\frac{\partial U}{\partial t}\right)_v = \left(\frac{\partial U}{\partial v}\right)_t \left(\frac{\partial v}{\partial t}\right)_p .$$

(c)
$$C_p - C_v = \left[\left(\frac{\partial U}{\partial v}\right)_t + p\right]\left(\frac{\partial v}{\partial t}\right)_p .$$

Solution

(a) From

$$\text{đ}Q = \mathrm{d}U + p\,\mathrm{d}v = \left[\left(\frac{\partial U}{\partial v}\right)_p + p\right]\mathrm{d}v + \left(\frac{\partial U}{\partial p}\right)_v \mathrm{d}p$$

one finds

$$m_p = \left(\frac{\partial U}{\partial p}\right)_v ,$$

$$m_v = \left(\frac{\partial U}{\partial v}\right)_p + p.$$

Also, differentiating, we obtain

$$\frac{\partial^2 U}{\partial v \partial p} = \left(\frac{\partial m_p}{\partial v}\right)_p$$

$$= \left(\frac{\partial m_v}{\partial p}\right)_v - 1.$$

(b) The independent variables (v, t) and (p, t) are involved, suggesting that a substitution for one in terms of the other two is required somewhere. We have

$$\mathrm{d}U = \left(\frac{\partial U}{\partial v}\right)_t \mathrm{d}v + \left(\frac{\partial U}{\partial t}\right)_v \mathrm{d}t$$

$$= \left(\frac{\partial U}{\partial v}\right)_t \left[\left(\frac{\partial v}{\partial p}\right)_t \mathrm{d}p + \left(\frac{\partial v}{\partial t}\right)_p \mathrm{d}t\right] + \left(\frac{\partial U}{\partial t}\right)_v \mathrm{d}t$$

$$= \left(\frac{\partial U}{\partial p}\right)_t \mathrm{d}p + \left[\left(\frac{\partial U}{\partial v}\right)_t \left(\frac{\partial v}{\partial t}\right)_p + \left(\frac{\partial U}{\partial t}\right)_v\right]\mathrm{d}t.$$

Hence

$$\left(\frac{\partial U}{\partial t}\right)_p - \left(\frac{\partial U}{\partial t}\right)_v = \left(\frac{\partial U}{\partial v}\right)_t \left(\frac{\partial v}{\partial t}\right)_p .$$

(c) The first equations of Problems 1.3 and 1.5 yield

$$\text{đ}Q = C_p\,\mathrm{d}t + l_p\,\mathrm{d}p = \mathrm{d}U + p\,\mathrm{d}v ,$$

whence

$$C_p = \left(\frac{\partial U}{\partial t}\right)_p + p\left(\frac{\partial v}{\partial t}\right)_p .$$

This, combined with the relation proved in (b) and with $C_v = (\partial U/\partial t)_v$, yields the required result.

THE SECOND LAW

1.7 The second law asserts that the reciprocal **absolute temperature** $1/T$ is an integrating factor of $đQ$, the resulting function of state being called the **entropy** S, so that for quasistatic changes $dS = đQ/T$. This adds a further relation to those given at the beginning of Problems 1.3 and 1.5. Returning to these, adopt the absolute temperature as the most convenient empirical temperature scale t to establish the following results. [Note that in a quasistatic adiabatic change the entropy is constant.]

(a) By considering $dU = TdS - pdv$, $dF \equiv d(U-TS)$, $dH \equiv d(U+pv)$ and $dG \equiv d(U+pv-TS)$ establish **Maxwell's relations**

$$\left(\frac{\partial T}{\partial v}\right)_S = -\left(\frac{\partial p}{\partial S}\right)_v, \quad \left(\frac{\partial T}{\partial p}\right)_S = \left(\frac{\partial v}{\partial S}\right)_p,$$
$$\left(\frac{\partial T}{\partial v}\right)_p = -\left(\frac{\partial p}{\partial S}\right)_T, \quad \left(\frac{\partial T}{\partial p}\right)_v = \left(\frac{\partial v}{\partial S}\right)_T.$$

[F is the **Helmholtz free energy**, H is the **enthalpy**, and G is the **Gibbs free energy**.]

(b)
$$C_v = T\left(\frac{\partial S}{\partial T}\right)_v, \quad C_p = T\left(\frac{\partial S}{\partial T}\right)_p,$$
$$l_v = T\left(\frac{\partial p}{\partial T}\right)_v, \quad l_p = -T\left(\frac{\partial v}{\partial T}\right)_p,$$
$$m_p = T\left(\frac{\partial S}{\partial p}\right)_v, \quad m_v = T\left(\frac{\partial S}{\partial v}\right)_p.$$

(c) $$l_p m_v = -TC_p, \quad l_v m_p = TC_v, \quad l_v l_p = -T(C_p - C_v).$$

(d) In view of these relations, and earlier ones, which of the six functions C, l, m give a set of independent quantities in terms of which the others can be expressed?

Solution

(a) The stated equation for dU clearly yields the first required result by the process indicated in Problem 1.2(b). This can be applied next to

$$d(U-TS) = dU - TdS - SdT = -p\,dv - SdT.$$

This yields the last equation. The other two results are established analogously.

(b) These results follow immediately from the equation at the beginning of Problem 1.3, except that in the case of l_v and l_p a Maxwell relation is also needed. For example, $TdS = C_v\,dT + l_v\,dv$ implies

$$l_v = T\left(\frac{\partial S}{\partial v}\right)_T = T\left(\frac{\partial p}{\partial T}\right)_v.$$

(c) From part (b) we have

$$l_p m_v = -T^2\left(\frac{\partial S}{\partial T}\right)_p = -TC_p,$$

$$l_v m_p = T^2\left(\frac{\partial S}{\partial T}\right)_v = TC_v.$$

It was shown in Problem 1.3(a) that

$$\frac{m_v}{l_v}+\frac{m_p}{l_p} = 1.$$

Substituting the new relations for m_v and m_p which arise from the second law, we obtain

$$-\frac{TC_p}{l_p l_v}+\frac{TC_v}{l_v l_p} = 1,$$

which is the required result.

(d) Retain l_v and the heat capacities. The other quantities are

$$m_p = \frac{TC_v}{l_v} \qquad \text{from (c) above,}$$

$$m_v = \frac{l_v C_p}{C_p - C_v} \qquad \text{from a result of Problem 1.3,}$$

$$l_p = -\frac{C_p - C_v}{l_v} T \qquad \text{from (c) above.}$$

1.8 With the notation of Problem 1.4 establish the following results:

(a) $$C_p - C_v = T\left(\frac{\partial v}{\partial T}\right)_p \left(\frac{\partial p}{\partial T}\right)_v = Tv\frac{\alpha_p^2}{K_T}.$$

(b) $$K_T - K_a = Tv\frac{\alpha_p^2}{C_p}.$$

(c) Using Problem 1.5 show that in a quasistatic adiabatic change of a simple fluid

$$T\exp\left[\int_{v_0}^{v} \Gamma \,\mathrm{d}(\ln v)\right]$$

remains constant; here Γ is the Grüneisen ratio introduced in Problem 1.4, and v_0 is a standard volume.

Solution

(a) From $C_v\,\mathrm{d}T + l_v\,\mathrm{d}v = C_p\,\mathrm{d}T + l_p\,\mathrm{d}p$ it follows that

$$C_p - C_v = l_v\left(\frac{\partial v}{\partial T}\right)_p = T\left(\frac{\partial p}{\partial T}\right)_v\left(\frac{\partial v}{\partial T}\right)_p,$$

where a result of Problem 1.7(b) has been used. This is the first equation. Now use the identity of Problem 1.1(b) to write

$$C_p - C_v = -T\left(\frac{\partial v}{\partial T}\right)_p \left(\frac{\partial p}{\partial v}\right)_T \left(\frac{\partial v}{\partial T}\right)_p = -T(v\alpha_p)^2\left(-\frac{1}{vK_T}\right).$$

(b) We have from Problem 1.4(c) that $K_T/K_a = \gamma = C_p/C_v$. Hence, from part (a),

$$(K_T - K_a)C_p = K_T\left(1 - \frac{C_v}{C_p}\right)C_p = Tv\alpha_p^2.$$

(c) It was shown in Problem 1.5 that $\Gamma = v/m_p$. But in conjunction with the result of Problem 1.7(b) the second law enables us to put

$$m_p = T\left(\frac{\partial S}{\partial p}\right)_v = -T\left(\frac{\partial v}{\partial T}\right)_S.$$

Hence for an adiabatic process,

$$\Gamma \mathrm{d}(\ln v) + \mathrm{d}(\ln T) = 0.$$

It follows that

$$T\exp\left[\int \Gamma \mathrm{d}(\ln v)\right]$$

is constant.

SIMPLE IDEAL FLUIDS

1.9 (a) Establish the result

$$p + \left(\frac{\partial U}{\partial v}\right)_T = T\left(\frac{\partial S}{\partial v}\right)_T = T\left(\frac{\partial p}{\partial T}\right)_v;$$

$$p\left(\frac{\partial v}{\partial p}\right)_T + \left(\frac{\partial U}{\partial p}\right)_T = -T\left(\frac{\partial v}{\partial T}\right)_p.$$

(b) **Joule's law** for a fluid states that $(\partial U/\partial v)_T \doteq 0$. Show from (a) that it implies the existence of a function $\mathrm{f}(v)$ of the volume and that

$$p\mathrm{f}(v) = T.$$

(c) A fluid which satisfies the equation $pv = AT$, where A is a constant, is called an **ideal classical gas.** Show from (a) that it satisfies Joule's law. Is the relation $pv = At$, in which an empirical temperature t is used, adequate to infer Joule's law?

(d) Verify that for an ideal classical gas and with the notation of Problem 1.4

$$\alpha_p = \frac{1}{T}, \qquad C_p - C_v = A,$$

$$K_T = \frac{1}{p}, \qquad \Gamma = \gamma - 1.$$

(e) A fluid which satisfies the equation $pv = gU$, where g is a constant, is called an **ideal quantum gas**[(2)]. Show from Problem 1.5(b) that for such a system the **Grüneisen ratio** is $\Gamma = g$, and so is a constant. [This is called **Grüneisen's law.**]

(f) Prove that an ideal classical gas for which C_v is a constant is an ideal quantum gas with $g = \gamma - 1$ and $C_v = A/g$, provided its internal energy vanishes at the absolute zero of temperature.

Solution

(a) The independent variables are v and T. Hence the first and second laws combined can be used in the form

$$\mathrm{d}U = T\mathrm{d}S - p\,\mathrm{d}v\ ,$$

whence

$$\left(\frac{\partial U}{\partial v}\right)_T \mathrm{d}v + \left(\frac{\partial U}{\partial T}\right)_v \mathrm{d}T = T\left(\frac{\partial S}{\partial v}\right)_T \mathrm{d}v + T\left(\frac{\partial S}{\partial T}\right)_v \mathrm{d}T - p\,\mathrm{d}v.$$

Equating coefficients of $\mathrm{d}v$ we find the required result with the aid of a Maxwell relation. The second result is found similarly, using p and T as independent variables and equating coefficients of $\mathrm{d}p$.

(b) Joule's law reduces the result of (a) to $p = T(\partial p/\partial T)_v$. For constant volume this integrates to yield $\ln p + \ln \mathrm{f} = \ln T$, where $\ln \mathrm{f}$ is the constant of integration. This can of course depend on the volume.

(c) The result (a) depends on the use of the absolute temperature T and yields for a system satisfying $pv = At$

$$\left(\frac{\partial U}{\partial v}\right)_T = T\left(\frac{\partial p}{\partial t}\right)_v \frac{\mathrm{d}t}{\mathrm{d}T} - p = T\frac{p}{t}\frac{\mathrm{d}t}{\mathrm{d}T} - p = p\left(\frac{\mathrm{d}\ln t}{\mathrm{d}\ln T} - 1\right).$$

Thus if $t = T^m$,

$$\left(\frac{\partial U}{\partial v}\right)_T = (m-1)p$$

and Joule's law results only if $t = T$.

(d) α_p and K_T follow at once from $pv = AT$. For $C_p - C_v$, Problem 1.6(c) may be used with $(\partial U/\partial T)_v = 0$ from part (b) above. This yields

$$C_p - C_v = p\left(\frac{\partial v}{\partial T}\right)_p = A.$$

Substitution of these results in the definition for Γ, given in Problem 1.4, gives

$$\Gamma = \frac{\alpha_p v}{K_T C_v} = \frac{v\,p}{T C_v} = \frac{A}{C_v} = \frac{C_p - C_v}{C_v} = \gamma - 1.$$

(2) This macroscopic definition was given in P.T.Landsberg, *Am.J.Phys.*, **29**, 695 (1961). See also G.E.Uhlenbeck and E.A.Uehling, *Phys.Rev.*, **39**, 1014 (1932) and H.Einbinder, ibid. **74**, 805 (1948); G.Süssmann and E.Hilf, Proc.Intern.Conf.on Thermodynamics, Cardiff 1970. *Pure Appl.Chem.*, **22**, 243 (1970).

(e) Problem 1.5(b) yields, if $pv = gU$,

$$\Gamma = v \bigg/ \left(\frac{\partial U}{\partial p}\right)_v = g.$$

(f) If C_p is constant, then it follows from part (d) that $C_v = C_p - A$ and $\gamma = C_p/C_v$ are also constants. From part (c) we note that Joule's law holds, so that $U(v, T)$ depends on T only and $\mathrm{d}U = C_v\,\mathrm{d}T$; therefore $U = C_v T + \text{constant}$. The constant becomes zero given that U vanishes with T. Now, from part (d) we have

$$C_v T = \frac{C_v A T}{C_p - C_v} = \frac{AT}{\gamma - 1} = \frac{pv}{\gamma - 1}.$$

It follows that

$$pv = (\gamma - 1)U\,,$$

which is characteristic of an ideal quantum gas. Also,

$$pv = gU = AT$$

yields

$$C_v = \frac{\mathrm{d}U}{\mathrm{d}T} = \frac{A}{g}.$$

1.10 (a) A fluid whose equation of state is $pv = At$, where A is a constant, undergoes quasistatic adiabatic changes. Use the result of Problem 1.4(b) to establish the laws

$$pv^\gamma = B_1^\gamma, \quad tv^{\gamma-1} = B_2^{\gamma-1}, \quad \frac{t}{p^{(\gamma-1)/\gamma}} = B_3,$$

where the B_i are constants, by assuming that γ is a constant for the fluid. Show that

$$B_1^\gamma = AB_2^{\gamma-1} = A^\gamma B_3^\gamma.$$

(b) Assume that the fluid is an ideal classical gas of constant γ, i.e. that the absolute temperature scale is chosen in (a). Show that each B_i can be expressed in terms of the entropy S of the fluid through

$$B_2 = A \exp\left(\frac{S}{A} - i\right),$$

where i is a thermodynamically unidentifiable constant (not $\sqrt{-1}$!).

(c) If the fluid envisaged in part (b) expands from an initial state (T_1, v_1) into a vacuum, so that its volume increases to v_2, obtain an expression for the increase ΔS in its entropy, and show that the work done by it is $T_1 \Delta S$.

[It is desirable to choose the constants in part (a) as B_1^γ and $B_2^{\gamma-1}$, and not simply as B_1 and B_2; see Problem 1.23.]

Solution

(a) Problem 1.4(b) together with $pv = At$ yields

$$\left(\frac{\partial p}{\partial v}\right)_a = \gamma\left(\frac{\partial p}{\partial v}\right)_t = -\frac{\gamma A}{v^2} = -\frac{\gamma p}{v} .$$

Hence, for quasistatic adiabatic processes with γ constant, we have

$$\frac{dp}{p} = -\gamma\frac{dv}{p} , \quad \text{i.e. } pv^\gamma = B_1^\gamma ,$$

where B_1^γ is a constant of integration.

Under the same conditions we have also

$$tv^{\gamma-1} = \frac{t}{pv}\cdot pv^\gamma = \frac{B_1^\gamma}{A} \equiv B_2^{\gamma-1} .$$

Lastly

$$\frac{T}{p^{(\gamma-1)/\gamma}} = \frac{Tv^{\gamma-1}}{(pv^\gamma)^{(\gamma-1)/\gamma}} = \frac{B_1^\gamma/A}{B_1^{\gamma-1}} = \frac{B_1}{A} \equiv B_3 .$$

Alternatively one can start from one of the other relations established in Problem 1.4(b).

(b) The first and second laws yield

$$dS = \frac{1}{T}(dU+p\,dv) = \frac{1}{T}\left[C_v\,dT+\left(\frac{\partial U}{\partial v}\right)_T dv+p\,dv\right] .$$

Since $(\partial U/\partial v)_T = 0$ [see Problem 1.9(c)] and $p/T = A/v$, we have

$$\begin{aligned} dS &= C_v\,d\ln T+A\,d\ln v = C_v\,d\ln(Tv^{\gamma-1}) \\ &= C_v(\gamma-1)\,d\ln(T^{1/(\gamma-1)}v) = A\,d\ln(T^{1/(\gamma-1)}v) . \end{aligned}$$

Now, if γ is a constant, integration yields

$$S = A\ln(T^{1/(\gamma-1)}v)-\ln(A^{-A}/e^i) ,$$

where the last term is a constant of integration. It follows that

$$B_2 = vT^{1/(\gamma-1)} = A\exp\left(\frac{S}{A}-i\right) = A\exp\left(\frac{S}{C_p-C_v}-i\right) ,$$

$$B_1 = A^{1/\gamma}B_2^{(\gamma-1)/\gamma} = A\exp\left(\frac{S}{C_p}-i\frac{\gamma-1}{\gamma}\right) ,$$

$$B_3 = B_1/A = \exp\left(\frac{S}{C_p}-i\frac{\gamma-1}{\gamma}\right) .$$

(c) From parts (a) and (b), $S = C_v\ln(pv^\gamma)+\text{constant}$. Hence

$$\Delta S = C_v\ln\left[\frac{p_2}{p_1}\left(\frac{v_2}{v_1}\right)^\gamma\right] = C_v\ln\left[\frac{T_2}{T_1}\left(\frac{v_2}{v_1}\right)^{\gamma-1}\right] .$$

Since Joule's law holds [see Problem 1.9(b)], $U_1 = U_2$, i.e.

$$\int_0^{T_1} C_v \, dT = \int_0^{T_2} C_v \, dT ,$$

so that $T_1 = T_2$. It follows that

$$\Delta S = C_v(\gamma - 1)\ln\frac{v_2}{v_1}$$

$$= (C_p - C_v)\ln\frac{v_2}{v_1}$$

$$= A\ln\frac{v_2}{v_1} \qquad \text{[Problem 1.9(d)].}$$

The work done by the gas in the expansion is

$$\int_1^2 p\,dv = AT_1\int_1^2 \frac{1}{v}dv = AT_1\ln\frac{v_2}{v_1} = T_1\Delta S .$$

1.11 (a) The **van der Waals equation of state** is

$$\left(p + \frac{a}{v^2}\right)(v - b) = At,$$

where a, b, A are constants. On a (p, v) diagram the extrema of this equation, which are obtained by choosing various values of t, lie on a curve. Find the equation of this curve.

(b) Show that the maximum of the curve found in part (a) is given by

$$v_c = 3b, \quad p_c = \frac{a}{27b^2}, \quad t_c = \frac{8a}{27Ab},$$

so that $At_c/p_c v_c = 2{\cdot}667$.

[This point is called **the critical point.**]

(c) Show that the van der Waals equation can be written

$$\left(\pi + \frac{3}{\phi^2}\right)(3\phi - 1) = 8\tau,$$

where

$$\phi \equiv v/v_c, \quad \pi \equiv p/p_c, \quad \tau \equiv t/t_c .$$

(d) One sometimes writes a general equation of state in the form

$$pv = E_1 + E_2 p + E_3 p^2 + \ldots,$$

where $E_1, E_2, E_3, \ldots$ are functions of t and are called first, second, third, ... **virial coefficients.**

If a, b are small enough, show that a van der Waals gas has an approximate second virial coefficient

$$E_2 = b - \frac{a}{At} .$$

(e) The **Boyle temperature** t_B of a fluid is defined by $[\partial(pv)/\partial p]_{p=0} = 0$, so that $E_2 = 0$ and the fluid approximates a Boyle's law gas (for small E_n, $n \geqslant 3$) in the neighbourhood of this temperature.

Show that for a van der Waals gas

$$\frac{t_B}{t_c} = 3{\cdot}375.$$

(f) Show that C_v for a van der Waals gas is independent of volume.

(g) Show that at $v = v_c$

$$K_t = \frac{4b}{3A(t-t_c)}, \quad \alpha_p = \frac{2}{3(t-t_c)}$$

so that these quantities diverge at $t = t_c$.

[The equation of state is regarded here as given. A closely related equation is *derived* from statistical mechanics in Problem 9.7. Different definitions of virial coefficients are in use; the most common is perhaps that given in Problem 9.8. See also Problem 10.1.]

Solution

(a) From the equation of state

$$p = \frac{At}{v-b} - \frac{a}{v^2},$$

$$\left(\frac{\partial p}{\partial v}\right)_t = -\frac{At}{(v-b)^2} + \frac{2a}{v^3},$$

Thus for extrema, labelled by 'i',

$$\frac{v_i^3}{2a} = \frac{(v_i-b)^2}{At} = \frac{(v_i-b)^2}{(p_i+a/v_i^2)(v_i-b)}.$$

Hence the required curve on the (p, v)-diagram is given by

$$p_i = \frac{a}{v_i^2} - \frac{2ab}{v_i^3}.$$

To check that these extrema are maxima, note that

$$\left(\frac{\partial^2 p}{\partial v^2}\right)_t = \frac{2At}{(v-b)^3} - \frac{6a}{v^4},$$

which at an extremum becomes

$$\frac{2}{v_i-b}\frac{2a}{v_i^3} - \frac{6a}{v_i^4} \approx -\frac{2a}{v_i^4}.$$

(b) This last curve has a maximum given by

$$-\frac{2a}{v_i^3} + \frac{6ab}{v_i^4} = \frac{2a}{v_i^4}(3b-v_i) = 0.$$

To check that this is a maximum, observe that

$$\frac{d^2p_i}{dv_i^2} = \frac{6a}{v_i^4} - \frac{24ab}{v_i^5} = \frac{6a}{v_i^4}\left(1 - \frac{4b}{v_i}\right) .$$

At the point at which dp_i/dv_i is zero, this quantity is negative as required. Hence, at this maximum, $v_c = 3b$. Also $p_i = p_c$ at $v_i = 3b$, so that

$$p_c = \frac{a}{9b^2} - \frac{2ab}{27b^3} = \frac{a}{27b^2} .$$

The van der Waals equation at the critical point furnishes now a value of t_c:

$$At_c = \left(\frac{a}{27b^2} + \frac{a}{9b^2}\right)(3b-b) = \frac{8a}{27b} .$$

(c) An easy algebraic result.

(d) We write

$$pv = At\left(1 + \frac{a}{pv^2}\right)^{-1}\left(1 - \frac{b}{v}\right)^{-1} \approx At - \frac{aAt}{pv^2} + \frac{bAt}{v} .$$

For small a, b the approximate equation is $pv = At$. Using this in the two correction terms, we have

$$pv \approx At - \frac{ap}{At} + bp, \qquad \text{i.e. } E_2 = b - \frac{a}{At} .$$

(e) $E_2 = 0$ if $t = t_B = a/Ab$. Hence

$$\frac{t_B}{t_c} = \frac{a}{Ab} \cdot \frac{27Ab}{8a} = \frac{27}{8} = 3{\cdot}375 .$$

(f) Recall from Problem 1.9(a) that

$$\left(\frac{\partial U}{\partial v}\right)_T + p = T\left(\frac{\partial p}{\partial T}\right)_v$$

Differentiating with respect to temperature at constant volume yields

$$\frac{\partial^2 U}{\partial v \partial T} + \left(\frac{\partial p}{\partial T}\right)_v = T\left(\frac{\partial^2 p}{\partial T^2}\right)_v + \left(\frac{\partial p}{\partial T}\right)_v ,$$

i.e.

$$\left(\frac{\partial C_v}{\partial v}\right)_v = T\left(\frac{\partial^2 p}{\partial T^2}\right)_v .$$

The right-hand side vanishes for a van der Waals gas.

(g) From the solution of (a), and by differentiating we obtain

$$K_t = \frac{1}{\dfrac{Avt}{(v-b)^2} - \dfrac{2a}{v^2}}, \quad \alpha_p = \frac{AK_t}{v-b} .$$

On using the value of t_c and putting $v = v_c = 3b$ the required result follows.

1.12 (a) The ideal quantum gas, introduced in Problem 1.9(c) through $pv = gU$, has an internal energy which obeys the equation

$$U = T\left(\frac{\partial U}{\partial T}\right)_v - \frac{v}{g}\left(\frac{\partial U}{\partial v}\right)_T .$$

Establish this result [the equation of Problem 1.9(a) may be used], and verify that the condition is satisfied by

$$U = v^{-g}\,\mathrm{f}(Tv^g),$$

where $\mathrm{f}(x)$ is some function of its argument.

(b) Show that if this gas undergoes a quasistatic adiabatic change, then the following quantities are constants:

$$pv^{1+g}, \quad Tv^g, \quad \frac{T}{p^{g/(1+g)}} .$$

Solution

(a) The result in Problem 1.9(a) is

$$p = T\left(\frac{\partial p}{\partial T}\right)_v - \left(\frac{\partial U}{\partial v}\right)_T .$$

Multiplying it by v/g and using $pv/g = U$, one finds

$$U = T\left(\frac{\partial U}{\partial T}\right)_v - \frac{v}{g}\left(\frac{\partial U}{\partial v}\right)_T .$$

Suppose $U = v^{-g}\mathrm{f}(Tv^g)$; then, if $\mathrm{f}'(z) \equiv \mathrm{df}/\mathrm{d}z$, one has

$$\left(\frac{\partial U}{\partial T}\right)_v = \mathrm{f}',$$

$$\left(\frac{\partial U}{\partial v}\right)_T = -\frac{gU}{v} + \frac{gT}{v}\mathrm{f}',$$

so that the equation for U is satisfied.

(b) Multiply $\mathrm{d}S = T^{-1}(\mathrm{d}U + p\,\mathrm{d}v)$ by Tv^g to find

$$Tv^g\,\mathrm{d}S = v^g\left(\mathrm{d}U + \frac{gU}{v}\mathrm{d}v\right) = \mathrm{d}(Uv^g) .$$

From part (a) the right-hand side is $\mathrm{d}[\mathrm{f}(Tv^g)]$, so that the entropy is seen to be a function of Tv^g only. Hence in a change in which the entropy is constant, Tv^g is also constant. From part (a), the constancy of Tv^g implies the constancy of

$$Uv^g = \frac{1}{g}pv^{1+g} .$$

Dividing the constant Tv^g by the constant pv^{1+g} yields as a further constant during the change the quantity $T^{1/g}/p^{1/(1+g)}$, so that $Tp^{g/(1+g)}$ is also constant.

Note that the γ, in the analogous adiabatic laws for an ideal classical gas of constant heat capacity ratio γ, is replaced here by $1+g$. This might have been expected from Problem 1.9(f).

1.13 A **polytropic fluid** of index n is one for which pv^n is a constant[3].

(a) An ideal classical gas of constant heat capacities C_p, C_v undergoes a quasistatic change for which $đQ = C\mathrm{d}T$, where C is some constant. Show that it is then a polytropic fluid of index

$$n = \frac{C_p - C}{C_v - C}.$$

Further, establish that during this change the following quantities remain constant

$$pv^n = B_1^n, \quad Tv^{n-1} = \frac{B_1^n}{A}, \quad \frac{T}{p^{(n-1)/n}} = \frac{B_1}{A}.$$

(b) In what sense does part (a) generalise the result of Problem 1.10(a)?

(c) Explain why the heat capacity C for the polytropic change can be negative.

(d) An ideal quantum gas undergoes a change such that $đQ = bC_v\mathrm{d}T$, where b is a constant, but C_v need not be a constant. Show that $Tv^{g/(1-b)}$ is a constant for this change.

Solution

(a) From the basic equations of Problems 1.3 and 1.5 we have

$$đQ = \mathrm{d}U + p\,\mathrm{d}v = C_v\,\mathrm{d}T + l_v\,\mathrm{d}v\,.$$

It is easily seen that

$$l_v = p + \left(\frac{\partial U}{\partial v}\right)_T.$$

For the change envisaged, therefore,

$$C\mathrm{d}T = C_v\,\mathrm{d}T + \left[p + \left(\frac{\partial U}{\partial v}\right)_T\right]\mathrm{d}v\,.$$

Using the result of Problem 1.9(c) to neglect the differential coefficient, dividing by T, and noting that $p/T = A/v = (C_p - C_v)/v$, one finds

$$(C_v - C)\frac{\mathrm{d}T}{T} + (C_p - C_v)\frac{\mathrm{d}v}{v} = 0\,.$$

[3] A detailed discussion of polytropics is due to G.Zeuner [*Grundzüge der mechanischen Wärmetheorie*, 2nd edn., p.143 (A.Felix, Leipzig), 1866]. The theory was developed notably by R.Emden in the first quarter of this century. Its use in stellar problems was reviewed by E.A.Milne [*Handbuch der Astrophysik*, Vol.3 (1930)]. This review is reprinted in *Selected Papers on the Transfer of Radiation* (Ed. D.H.Menzel) (Dover Publications, New York), 1966.

On introducing the stated value of n this becomes

$$\frac{\mathrm{d}T}{T}+(n-1)\frac{\mathrm{d}v}{v} = 0 ,$$

so that, if B_2 is a constant,

$$Tv^{n-1} = B_2^{n-1} .$$

It follows that another constant is

$$B_1^n \equiv Atv^{n-1} = \frac{pv}{T}Tv^{n-1} = pv^n ,$$

and this makes the fluid isotropic of index n. The last result required follows from

$$\frac{T}{p^{(n-1)/n}} = \frac{Tv^{n-1}}{(pv^n)^{(n-1)/n}} .$$

(b) For $C = 0$ the change is adiabatic, whilst for $C = \infty$ the change is isothermal, and $n = \gamma$ and 1 respectively. Other possibilities exist since $-\infty \leqslant C \leqslant \infty$.

(c) From the solution for part (a)

$$đQ = \mathrm{d}U+p\,\mathrm{d}v = C_v\,\mathrm{d}T+p\,\mathrm{d}v ,$$

whence $\mathrm{d}U = C_v\,\mathrm{d}T$. Also, from $pv^n = B_1^n$, $pv = AT$ it follows that

$$v^{1-n} = \frac{AT}{B_1^n} .$$

Hence an increment of work done by the ideal classical gas in the polytropic change is

$$đW \equiv p\,\mathrm{d}v = -p\frac{v^n}{n-1}\cdot\frac{A}{pv^n}\mathrm{d}T = -\frac{A}{n-1}\mathrm{d}T = -\frac{C_p-C_v}{n-1}\mathrm{d}T .$$

Hence

$$đW = -\frac{\gamma-1}{n-1}C_v\,\mathrm{d}T = -\frac{\gamma-1}{n-1}\mathrm{d}U .$$

Thus if $1 < n < \gamma$ the work done in an increment exceeds the *drop* in the internal energy of the fluid. Hence $\mathrm{d}Q = \mathrm{d}U+\mathrm{d}W$ is positive while $\mathrm{d}T$ is negative. Now

$$C = \frac{n-\gamma}{n-1}C_v$$

and one observes that, for $1 < n < \gamma$, C is in fact negative.

(d) In the solution of Problem 1.12(b) it was noted that if $z \equiv Tv^g$ then $Uv^g = \mathrm{f}(z)$, and also

$$\mathrm{d}S = z^{-1}\mathrm{d}\mathrm{f}(z) = z^{-1}\mathrm{f}'(z)\mathrm{d}z = z^{-1}C_v\,\mathrm{d}z .$$

The last form of writing arises from the observation that

$$C_v = T\left(\frac{\partial S}{\partial T}\right)_v = Tv^g\left(\frac{\partial S}{\partial z}\right)_v = z\left(\frac{\partial S}{\partial z}\right)_v = z\cdot z^{-1}\mathrm{f}'(z) = \mathrm{f}'(z)\,.$$

The change taking place is defined by $T\mathrm{d}S = bC_v\,\mathrm{d}T$, whence

$$b\frac{\mathrm{d}T}{T} = \frac{\mathrm{d}S}{C_v} = \frac{\mathrm{d}z}{z}\,.$$

Integration shows that $zT^{-b} = T^{1-b}v^g$ is a constant.

1.14 (a) If a fluid is in equilibrium with its saturated vapour pressure, then the pressure in the system is independent of volume and depends only on temperature. Show from the Maxwell equation given in Problem 1.9(a) that under these conditions

$$T\frac{\mathrm{d}p}{\mathrm{d}T} = \frac{\Lambda_T^{1\to 2}}{v_2 - v_1} \quad \text{(Clausius-Clapeyron equation)},$$

where $\Lambda_T^{1\to 2}$ is the heat required to take a mass m of material isothermally from the state of aggregation 1 to the stage of aggregation 2, and v_1 and v_2 are the volumes occupied by this material in the two states of aggregation. [If m is taken as the unit mass, then Λ is the latent heat, and the v's are specific volumes.]

(b) Show that increase of pressure raises the boiling point and the freezing point of a normal liquid, but that, in the case of water, pressure lowers the freezing point.

Solution

(a) Observe that under the conditions stated

$$\frac{\Delta p}{\Delta T} = \frac{\Delta S}{\Delta v} = \frac{\Lambda_T^{1\to 2}}{v_2 - v_1}\,.$$

(b) Take $\Lambda > 0$; then we have

	state 1	state 2		$\mathrm{d}p/\mathrm{d}T$
Vaporization	liquid	vapour	$v_2 > v_1$	positive
Melting (normal)	solid	liquid	$v_2 > v_1$	positive
Melting of ice	solid	liquid	$v_2 < v_1$	negative

JOULE-THOMSON EFFECT

1.15 If gas is allowed to flow slowly through a porous plug between two containers, which are otherwise isolated from each other and from their surroundings, the enthalpy $H_1 \equiv U_1 + p_1v_1$ before the process is equal to H_2, the value after the process. The temperature change is measured by the **Joule–Thomson coefficient** $j \equiv (\partial T/\partial p)_H$.

(a) Show that

$$dH = T\,dS + v\,dp,$$

and hence that

$$j = \frac{v}{C_p}(T\alpha_p - 1).$$

(b) Show that, alternatively,

$$j = -\left[T\left(\frac{\partial p}{\partial T}\right)_v + v\left(\frac{\partial p}{\partial v}\right)_T\right] \Big/ C_p\left(\frac{\partial p}{\partial v}\right)_T .$$

(c) Establish that $j = 0$ for an ideal classical gas.
(d) Show that for a van der Waals gas introduced in Problem 1.11

$$j = \left(\frac{2a}{v} - \frac{3ab}{v^2} - bp\right) \Big/ \left(p - \frac{a}{v^2} + \frac{2ab}{v^3}\right) C_p .$$

(e) Obtain the curve on a (p, v)-diagram separating the region $j > 0$ from the region $j < 0$ for a van der Waals gas. This is the **inversion curve.**

[Expansion leads to cooling only for states lying in the $j > 0$ region.]

(f) If p_{io} is the pressure at the maximum of the inversion curve, show that

$$p_{io} = 9p_c .$$

for a van der Waals fluid. Show also that

$$T_{io} = 3T_c .$$

[The maximum of the inversion curve lies at a pressure and temperature which are above their critical values.]

Solution

(a) Using $dU = T\,dS - p\,dv$, we obtain

$$dH = dU + p\,dv + v\,dp = T\,dS + v\,dp ,$$

as required. It follows that

$$dH = \left[v + T\left(\frac{\partial S}{\partial p}\right)_T\right] dp + T\left(\frac{\partial S}{\partial T}\right)_p dT ,$$

so that

$$j \equiv \left(\frac{\partial T}{\partial p}\right)_H = -\left[T\left(\frac{\partial S}{\partial p}\right)_T + v\right] \Big/ T\left(\frac{\partial S}{\partial T}\right)_p = \frac{1}{C_p}\left[T\left(\frac{\partial v}{\partial T}\right)_p - v\right],$$

where one of the Maxwell relations of Problem 1.7(a) has been used. The required result follows.

(b) From part (a) and a Maxwell relation we obtain

$$j = \left[T\left(\frac{\partial v}{\partial T}\right)_p \left(\frac{\partial p}{\partial v}\right)_T - v\left(\frac{\partial p}{\partial v}\right)_T\right] \Big/ C_p\left(\frac{\partial p}{\partial v}\right)_T$$

$$= -\left[T\left(\frac{\partial p}{\partial T}\right)_v + v\left(\frac{\partial p}{\partial v}\right)_T\right] \Big/ C_p\left(\frac{\partial p}{\partial v}\right)_T .$$

This is more convenient than the result in (a) if $p = p(v, T)$ is known.

(c) From Problem 1.9(d) we have $T\alpha_p = 1$.

(d) The expression is obtained by a direct calculation of

$$j = \frac{1}{C_p}\left[T\left(\frac{\partial v}{\partial T}\right)_p - v\right].$$

(e) From the result in part (d) the curve is

$$p_{\mathrm{i}} = \frac{2a}{bv_{\mathrm{i}}} - \frac{3a}{v_{\mathrm{i}}^2} .$$

(f) Its maximum occurs at $\mathrm{d}p_{\mathrm{i}}/\mathrm{d}v_{\mathrm{i}} = 0$, i.e. at $v_{\mathrm{i}} = v_{\mathrm{i}0} = 3b$. The corresponding value of p_{i} is

$$p_{\mathrm{i}0} = \frac{2a}{3b^2} - \frac{3a}{9b^2} = \frac{a}{3b^2} .$$

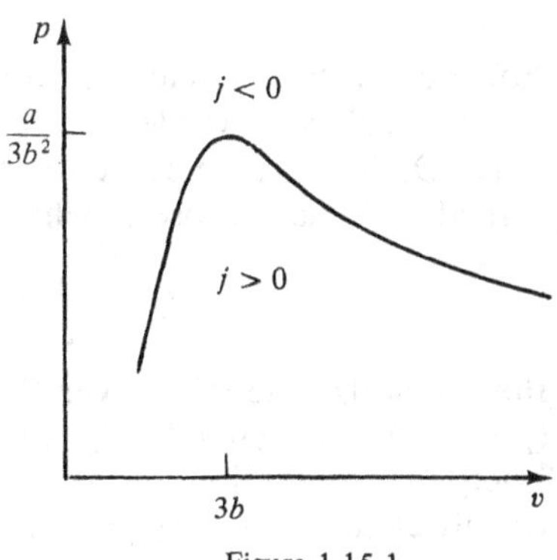

Figure 1.15.1

It follows that

$$\frac{p_{\mathrm{i}0}}{p_c} = \frac{a}{3b^2} \cdot \frac{27b^2}{a} = 9 .$$

Also

$$AT_{\mathrm{i}0} = \left(\frac{a}{3b^2} + \frac{a}{9b^2}\right)(3b - b) = \frac{8a}{9b} .$$

and

$$\frac{T_{\mathrm{i}0}}{T_{\mathrm{c}}} = \frac{8a}{9Ab} \cdot \frac{27Ab}{8a} = 3 .$$

THERMODYNAMIC CYCLES

1.16 (a) In an incremental process a fluid gains heat energy $đQ$ at temperature T. Establish that in a closed cycle

$$\oint \frac{đQ}{T} \leqslant 0 \quad \textbf{(Clausius inequality)},$$

where the equality holds for a quasistatic cycle.

(b) In a general quasistatic cycle a working fluid receives heat at various temperatures and gives up heat at various temperatures. The efficiency η of the cycle is defined as the mechanical work done divided by the sum of all the (positive) increments of heat gained. Prove, using the Clausius inequality, that

$$\eta \leqslant \frac{T_1 - T_2}{T_1} \quad (\equiv \eta_C, \text{ the \textbf{Carnot efficiency}}),$$

where T_1 is the highest temperature at which heat is gained, and T_2 is the lowest temperature at which it is given up.

(c) Show that the maximum efficiency η_C can be attained by using as a working fluid an ideal classical gas of constant heat capacities, which is working between isothermals at temperatures T_1 and T_2 separated by an adiabatic expansion and an adiabatic compression. [This is the **Carnot cycle.**]

(d) If you have used the properties of the ideal classical gas in (c), generalise the argument so that it applies to any fluid describing a Carnot cycle quasistatically.

Solution

(a) We have $đQ/T \leqslant dS$, where dS is the incremental change of the entropy of the fluid. The entropy is a function of the state of the fluid and in a cyclic process its final value equals its initial value, whence

$$\oint \frac{đQ}{T} \leqslant 0 .$$

For a quasistatic cycle the equality holds for each increment and hence for the cycle as a whole.

(b) Divide the increments of the cycle into those in which heat is gained, $đQ_+$, and those in which heat is lost, $đQ_-$. Then define

$$Q_1 \equiv \int_+ đQ_+, \quad Q_2 \equiv \int_- đQ_-,$$

the integrals extending over the appropriate increments. Q_1 is the total (positive) heat gained; Q_2 is the total heat rejected, counted positively. Energy conservation gives the mechanical work done as $W = Q_1 - Q_2$; the efficiency is $\eta = (Q_1 - Q_2)/Q_1$. Now

$$0 = \oint \frac{đQ}{T} = \int_+ \frac{đQ}{T} - \int_- \frac{đQ}{T} \geqslant \int_+ \frac{đQ}{T_1} - \int_- \frac{đQ}{T_2} = \frac{Q_1}{T_1} - \frac{Q_2}{T_2} .$$

Hence $Q_2/Q_1 \geqslant T_2/T_1$, so that $\eta \equiv 1 - Q_2/Q_1 \geqslant 1 - T_2/T_1$.

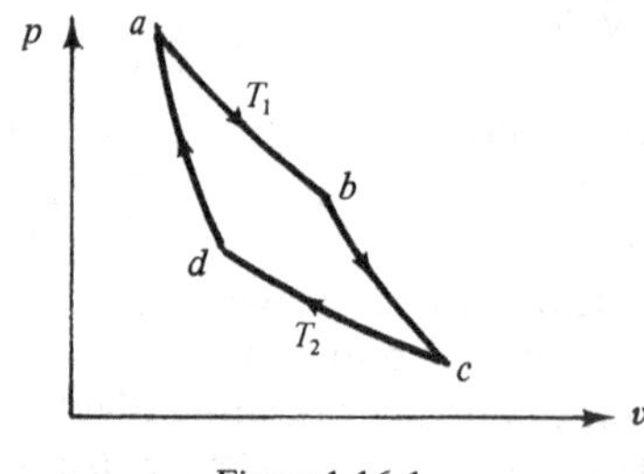

Figure 1.16.1

(c) With the notation shown in Figure 1.16.1, $Tv^{\gamma-1}$ is constant on the adiabatics bc and da [Problem 1.10(a)]. Hence

$$T_1 v_a^{\gamma-1} = T_2 v_d^{\gamma-1}, \; T_1 v_b^{\gamma-1} = T_2 v_c^{\gamma-1},$$

i.e.

$$\frac{v_b}{v_a} = \frac{v_c}{v_d} .$$

It follows from Joule's law [Problem 1.9(b)] that the internal energy remains constant on each isothermal, so that $đQ = dU + đW$ implies that the heat gained is equal to the work done. Hence

$$Q_1 = \int_a^b p\,dv = AT_1 \int_a^b \frac{dv}{v} = AT_1 \ln\frac{v_b}{v_a} ,$$

$$Q_2 = -\int_c^d p\,dv = -AT_2 \ln\frac{v_d}{v_c} = AT_2 \ln\frac{v_b}{v_a} .$$

Hence $\eta = \eta_C$.

Note that the work done on the adiabatics cancels out:

$$W_b^c = \int_b^c p\,dv = -\int_b^c dU = C_v(T_1 - T_2) ,$$

$$W_d^a = -\int_d^a dU = C_v(T_2 - T_1) .$$

(d) From the first law work done by the engine is equal to the heat gained by the fluid in a cycle: $W = Q_1 - Q_2$. Also from part (a) we have

$$\frac{Q_1}{T_1} = \frac{Q_2}{T_2} .$$

Hence

$$\eta = \frac{W}{Q_1} = \frac{Q_1 - Q_2}{Q_1} = 1 - \frac{Q_2}{Q_1} = 1 - \frac{T_2}{T_1} .$$

1.17 The working fluid of a thermodynamic engine is an ideal classical gas of constant heat capacity C_v. It works quasistatically in a cyclic process as follows:

(i) isothermal expansion at temperature T_1 from volume v_1 to v_2;
(ii) cooling at constant volume from temperature T_1 to T_2;
(iii) isothermal compression at temperature T_2 from volume v_2 to v_1;
(iv) heating at constant volume from temperature T_2 to T_1.

Obtain expressions for the amount of heat (Q_1) supplied to the gas in steps (i) and (iv) and the amount (Q_2) rejected in steps (ii) and (iii).

Show that for the above cycle the efficiency $\eta < (T_1 - T_2)/T_1$. How is the efficiency affected if (ii) and (iv) are replaced by adiabatic expansion to volume v_2 and by adiabatic compression to volume v_1, respectively.

Solution

As in the solution of Problem 1.16, the amount of heat absorbed from a reservoir during the isothermal change ab is

$$Q_{ab} = AT_1 \ln\frac{v_b}{v_a} .$$

The heat rejected on cd is

$$Q_{cd} = AT_2 \ln\frac{v_d}{v_c} = AT_2 \ln\frac{v_b}{v_a}$$

since $v_a = v_d$ and $v_b = v_c$. During the cooling bd the heat rejected is

$$Q_{bc} = C_v(T_1 - T_2) .$$

During the heating da the heat supplied is

$$Q_{da} = C_v(T_1 - T_2) .$$

Hence

$$Q_1 = Q_{da} + Q_{ab} = AT_1 \ln(v_b/v_a) + C_v(T_1 - T_2)$$

$$Q_1 - Q_2 = A(T_1 - T_2)\ln(v_b/v_a).$$

$$\eta = \frac{Q_1 - Q_2}{Q_1} = \frac{T_1 - T_2}{T_1 + \dfrac{C_v(T_1 - T_2)}{A\ln(v_2/v_1)}} < \frac{T_1 - T_2}{T_1} .$$

If steps (b) and (d) are changed as suggested, a Carnot cycle results, and the conclusions of Problem 1.16(c), (d) apply.

1.18 (a) Two fluids F_1, F_2 of fixed volumes and constant heat capacities C_1, C_2 are initially at temperatures T_1, T_2 ($T_1 > T_2$), respectively. They are adiabatically insulated from each other. A quasistatically acting Carnot engine E uses F_1 as heat source and F_2 as heat sink, and acts between the systems until they reach a common temperature, T_0 say. Obtain an expression for T_0 and for the work done by the Carnot engine.

(b) If a common temperature is established by allowing direct heat flow between F_1 and F_2, what is the final temperature and what is the change in entropy?

(c) Show that for all positive C_1, C_2 this change is an increase.

Solution

(a) Since $C_1 = T(\partial S/\partial T)_v$, the entropy lost by F_1 is

$$S_1 \equiv -C_1 \int_{T_1}^{T_0} \frac{dT}{T} = C_1 \ln\frac{T_1}{T_0} .$$

The entropy gained by F_2 is $S_2 \equiv C_2 \ln(T_0/T_2)$. If the working substance of the Carnot engine is in the same state finally as it was initially, the entropy gain of the whole system is

$$S_2 - S_1 = \ln\left[\left(\frac{T_0}{T_2}\right)^{C_2}\left(\frac{T_0}{T_1}\right)^{C_1}\right] = 0 .$$

Hence

$$T_0 = T_1^a T_2^b ,$$

where

$$a \equiv \frac{C_1}{C_1 + C_2} , \quad b \equiv \frac{C_2}{C_1 + C_2} .$$

The internal energy lost by F_1 is in the form of heat

$$Q_1 = -C_1 \int_{T_1}^{T_0} dT = C_1(T_1 - T_0) .$$

The internal energy gained by F_2 is

$$Q_2 = C_2(T_0 - T_2)\,.$$

The overall loss of internal energy is

$$Q_1 - Q_2 = C_1T_1 - C_2T_2 - (C_1 + C_2)T_0\,.$$

It follows from energy conservation that this must be equal to the total amount of work done by the Carnot engine.

(b) In the absence of the performance of work, energy conservation yields

$$C_1(T_1 - T_0) = C_2(T_0 - T_2)$$

so that

$$T_0 = aT_1 + bT_2 \quad (a+b = 1)\,.$$

The entropy lost by F_1 is

$$S_1 \equiv -C_1\int_{T_1}^{T_0}\frac{\mathrm{d}T}{T} = C_1\ln\frac{T_1}{T_0}\,.$$

The entropy gained by F_2 is

$$S_2 \equiv C_2\int_{T_2}^{T_0}\frac{\mathrm{d}T}{T} = C_2\ln\frac{T_0}{T_2}\,.$$

The gain of entropy for the whole system is

$$S_2 - S_1 = (C_1 + C_2)\ln\left[\left(\frac{T_0}{T_2}\right)^b\left(\frac{T_0}{T_1}\right)^a\right] = (C_1 + C_2)\ln\frac{T_0}{T_1^a T_2^b}\,.$$

Hence

$$S_2 - S_1 = (C_1 + C_2)\ln\frac{aT_1 + bT_2}{T_1^a T_2^b}\,.$$

(c) For all positive a, b such that $a+b = 1$, consider

$$y = aT_1 + bT_2 - T_1^a T_2^b\,.$$

This quantity has a minimum when $T_1 = T_2$ and is therefore never negative. Hence $S_2 - S_1 \geqslant 0$.

1.19 (a) Show that the **Jacobian**

$$\frac{\partial(p, v)}{\partial(T, S)} \equiv \left(\frac{\partial p}{\partial T}\right)_S\left(\frac{\partial v}{\partial S}\right)_T - \left(\frac{\partial p}{\partial S}\right)_T\left(\frac{\partial v}{\partial T}\right)_S = 1.$$

(b) A thermodynamic engine uses a cycle which is represented by closed curves on a (p, v)- and on a (T, S)-diagram. Use the fact that the work W done by the working fluid is equal to the heat Q adsorbed, expressed in the same units of energy, to obtain the result of part (a).

Solution

(a) We have

$$dp = \left(\frac{\partial p}{\partial T}\right)_S dT + \left(\frac{\partial p}{\partial S}\right)_T dS\,.$$

Hence, using two Maxwell relations of Problem 1.7(a)

$$1 = \left(\frac{\partial p}{\partial p}\right)_v = \left(\frac{\partial p}{\partial T}\right)_S \left(\frac{\partial T}{\partial p}\right)_v + \left(\frac{\partial p}{\partial S}\right)_T \left(\frac{\partial S}{\partial p}\right)_v$$

$$= \left(\frac{\partial p}{\partial T}\right)_S \left(\frac{\partial v}{\partial S}\right)_T - \left(\frac{\partial p}{\partial S}\right)_T \left(\frac{\partial v}{\partial T}\right)_S\,.$$

(b) We have for corresponding domains of integration in the (p, v)- and (T, S)-planes

$$W = \iint dp\, dv = \iint \frac{\partial(p,v)}{\partial(T,S)} dT\, dS\,,$$

where the Jacobian arises because of the change of variables. Also $Q = \iint dT\, dS$. Hence, since $W = Q$ for any domain, the result follows.

CHEMICAL THERMODYNAMICS

1.20 If two identical systems are joined together and considered as one system, the quantities whose values double are called **extensive**, and those whose values remain the same are called **intensive**. A **phase** is a thermodynamically homogeneous region of space, thermodynamically uniquely specified by its internal energy U, its volume v, and the numbers of molecules $n_1, n_2, \ldots$ of the chemical species contained in it.

(a) Verify that the equation for all $a > 0$

$$S(aU, av, an_1, an_2, \ldots) = aS(U, v, n_1, n_2, \ldots)$$

is consistent with the extensive nature of S, U, v, and the n_i.

(b) Assuming the entropy to be a function of the state of a phase, and that it satisfies

$$\left(\frac{\partial S}{\partial U}\right)_{v,n_i} = \frac{1}{T}, \quad \left(\frac{\partial S}{\partial v}\right)_{U,n_i} = \frac{p}{T},$$

even if the n_i are variable, show that

$$T dS = dU + p dv - \sum_i \mu_i dn_i, \quad \mu_i \equiv -T\left(\frac{\partial S}{\partial n_i}\right)_{U,\,v,\,\text{other } n\text{'s}}.$$

[The μ_i defined here is called the **chemical potential** of species i.]

(c) Show by differentiating with respect to a, that

$$G \equiv U - TS + pv = \sum_i \mu_i n_i\,.$$

[G is called the **Gibbs free energy**.]

(d) Establish the **Gibbs–Duhem equation**

$$S\mathrm{d}T - v\mathrm{d}p + \sum_i n_i \mathrm{d}\mu_i = 0.$$

among the intensive variables.

Solution

(a) The equation states that, if the independent variables (which can be chosen to be all extensive variables) are multiplied by a factor a, then the entropy is multiplied by the same factor, and is therefore also extensive.

A similar equation for U can be deduced from this equation, as is necessary since U is also extensive. For suppose $S(U, v, n_1, ...) = S_0$ can be solved for U; then if g is some function, the solution is, say,

$$U = g(S_0, v, n_1, ...)\,.$$

It follows that the solution of $S(aU, av, an_1, ...) = aS_0$ is

$$aU = g(aS_0, av, an_1, ...)\,.$$

But $aU = ag(S, v, n_1, ...)$, so that we have

$$g(aS_0, av, an_1, ...) = ag(S_0, v, n_1, ...)\,.$$

This is the analogue of the given equation for S. Similar equations hold for v and the n_i.

(b)

$$T\mathrm{d}S = T\left(\frac{\partial S}{\partial U}\right)_{v,\,n_i} \mathrm{d}U + T\left(\frac{\partial S}{\partial v}\right)_{U,\,n_i} \mathrm{d}v + T\sum_i \frac{\partial S}{\partial n_i}\,\mathrm{d}n_i$$

$$= \mathrm{d}U + p\,\mathrm{d}v - \sum_i \mu_i\,\mathrm{d}n_i\,.$$

(c) If $S_a \equiv S(aU, av, an_1, ...)$, then

$$\frac{\mathrm{d}S_a}{\mathrm{d}a} = \frac{\partial S_a}{\partial(aU)}\frac{\mathrm{d}(aU)}{\mathrm{d}a} + \frac{\partial S_a}{\partial(av)}\frac{\mathrm{d}(av)}{\mathrm{d}a} + \sum_i \frac{\partial S_a}{\partial(an_i)}\frac{\mathrm{d}(an_i)}{\mathrm{d}a}\,.$$

Hence

$$S = \frac{1}{T}(U + pv - \sum_i \mu_i n_i)\,.$$

(d) From part (c) we have

$$\mathrm{d}U - T\mathrm{d}S - S\mathrm{d}T + p\,\mathrm{d}v + v\,\mathrm{d}p - \sum_i(\mu_i\,\mathrm{d}n_i + n_i\,\mathrm{d}\mu_i) = 0\,.$$

From part (b),

$$\mathrm{d}U - T\mathrm{d}S + p\,\mathrm{d}v - \sum_i \mu_i\,\mathrm{d}n_i = 0\,.$$

Hence

$$SdT - v\,dp + \sum_i n_i d\mu_i = 0\,.$$

1.21 Each of n phases 1, 2, ... n consists of one and the same chemical species. A new system without inhibiting constraints is formed of these n phases by combining them so that the internal energy U, volume v, and particle number n of the resulting system are each the sum of the original quantities for the phases:

$$U = \sum_{i=1}^{n} U_i, \quad v = \sum_{i=1}^{n} v_i, \quad n = \sum_{i=1}^{n} n_i.$$

Show from the result in Problem 1.20(b) that for equilibrium in the new system

$$T_1 = T_2 = \ldots = T_n, \qquad p_1 = p_2 = \ldots = p_n, \qquad \mu_1 = \mu_2 = \ldots = \mu_n.$$

Solution

The entropy change for phase i is

$$\delta S_i = \frac{1}{T_i}(\delta U_i + p_i \delta v_i - \mu_i \delta n_i)\,.$$

We seek to maximise $\sum_i \delta S_i$ subject to

$$\sum_i \delta U_i = \sum_i \delta v_i = \sum_i \delta n_i = 0\,.$$

Using undetermined multipliers, consider

$$f = \sum_i \left[\left(\frac{1}{T_i} - \alpha\right) U_i + \left(\frac{p_i}{T_i} - \beta\right) v_i - \left(\frac{\mu_i}{T_i} - \gamma\right) n_i\right]\,.$$

Here T_i, μ_i, p_i, α, β, γ are constants. The quantity f has to be maximised for **arbitrary and independent** variations of U_i, v_i, and n_i. One finds

$$\sum_i \left[\left(\frac{1}{T_i} - \alpha\right) \delta U_i + \left(\frac{p_i}{T_i} - \beta\right) \delta v_i - \left(\frac{\mu_i}{T_i} - \gamma\right) \delta n_i\right] = 0\,.$$

This implies

$$T_1 = T_2 = \ldots = T_n = 1/\alpha\,,$$

$$p_1 = p_1 = \ldots = p_n = \beta T_i = \beta/\alpha\,,$$

$$\mu_1 = \mu_2 = \ldots = \mu_n = \gamma T_i = \gamma/\alpha\,.$$

1.22 A form of the second law suggested by Problems 1.5 and 1.16 is $\delta U - p\delta v - T\delta S \leqslant 0$, where δ denotes an incremental change the end-points of which are equilibrium states of a **closed** system (i.e. a system whose mass is fixed). Suppose that, with a generalised interpretation of δ, one has

$$\delta U + p\delta v - T\delta S = \sum \mu_i \delta n_i + \sum_j p_j \delta v_j \leqslant 0,$$

where the two sums involve pressures and volumes which can change owing to **internal** processes in the system[(4)].

(a) A system consists of two phases, labelled 1 and 2, which are initially separated mechanically at pressures p_1 and p_2. If they are then coupled mechanically, but their compositions are fixed, show that the phase whose pressure is greater will expand.

(b) If the volumes of the phases are fixed, but the molecules of a certain chemical species can suddenly be transported from one phase to the other, show that the molecules will migrate from a region of higher to one of lower chemical potential.

[The tendency to equalisation of intensive variables in equilibrium is illustrated by both Problems 1.21 and 1.22.]

Solution

(a) If one phase expands, it does so at the expense of the other, so that $\delta v_1 = -\delta v_2$. Hence

$$-p_1 \delta v_1 - p_2 \delta v_2 = -(p_1 - p_2)\delta v_1 \leqslant 0 .$$

Thus if $\delta v_1 > 0$, then $p_1 > p_2$; if $\delta v_1 < 0$, then $p_1 < p_2$. In either case the result stated in the problem follows.

(b) Let the phases be denoted by 1 and 2, so that $\delta n_1 = -\delta n_2$, since the particles lost by one phase are gained by the other. Hence

$$\mu_1 \delta n_1 + \mu_2 \delta n_2 = (\mu_1 - \mu_2)\delta n_1 \leqslant 0 .$$

If $\delta n_1 > 0$, then $\mu_2 > \mu_1$; if $\delta n_1 < 0$, then $\mu_2 < \mu_1$. In either case the result stated in the problem follows.

1.23 An ideal classical gas of constant heat capacities is considered in this problem and it is recalled that for such a system

$$\left.\begin{aligned} pv &= gU = AT, \\ C_v &= \frac{A}{g}, \quad C_p = \left(1 + \frac{1}{g}\right)A, \quad g = \gamma - 1, \end{aligned}\right\} \quad \text{(Problem 1.9)}$$

$$\frac{T^\gamma}{p^{\gamma-1}} = \exp\left[\frac{S}{C_v} - (\gamma - 1)i\right], \qquad \text{(Problem 1.10)}$$

where i is a constant (and not $\sqrt{-1}$!).

(4) For a discussion of this generalisation see P.T.Landsberg, *Thermodynamics with Quantum Statistical Illustrations* (Interscience, New York), 1961, p.156.

(a) Two such systems having identical heat capacities satisfy $pv = AT$, $pv = A'T$. Given only that A is an extensive variable, prove from Problem 1.10 that B_1 and B_2 are extensive while B_3 and i are intensive.

[Because A is extensive, it is written as kN, where k is **Boltzmann's constant** and N is the number of molecules.]

(b) Show that for one such gas

$$S = A\left[\left(1+\frac{1}{g}\right)\ln T - \ln p + i\right] = A\left(\frac{1}{g}\ln T + \ln v + i - \ln A\right).$$

[The above is a generalised **Sackur–Tetrode vapour pressure equation**, and i is here called the **chemical constant.**]

(c) Show that the chemical potential of the gas satisfies the condition

$$\frac{\mu}{kT} = \ln p + \left(1+\frac{1}{g}\right)(1-\ln T) - i.$$

[It is not possible to identify the chemical constant thermodynamically. But it is possible to do so from statistical mechanics. See Problem 3.13.]

Solution

(a) We have

$$B_1^\gamma = AB_2^{\gamma-1} = A^\gamma B_3^\gamma .$$

Hence, if $A'/A = \lambda$, we have

$$\left(\frac{B_1'}{B_1}\right)^\gamma = \lambda\left(\frac{B_2'}{B_2}\right)^{\gamma-1} = \lambda^\gamma\left(\frac{B_3'}{B_3}\right) .$$

The adiabatic through a given point (p_0, T_0) satisfies

$$\frac{t}{p^{(\gamma-1)/\gamma}} = \frac{t_0}{p_0^{(\gamma-1)/\gamma}} = B_3$$

for both gases so that $B_3 = B_3'$. Hence $B_1'/B_1 = B_2'/B_2 = \lambda$, and they are therefore extensive. The substitution of i for B_2 through $B_2 = Ae^{S/A-i}$ was therefore reasonable, and i is intensive.

(b) In Problem 1.10 it was shown that

$$B_3^\gamma = \frac{T^\gamma}{p^{\gamma-1}} = \left(\frac{B_2}{A}\right)^{\gamma-1} = \exp\left[\frac{S}{C_v} - (\gamma-1)i\right] .$$

It follows that

$$\gamma\ln T - (\gamma-1)\ln p = \frac{S}{C_v} - (\gamma-1)i .$$

Since $\gamma = g+1$ and $C_v = A/g$ (Problem 1.9), it follows further, on multiplying by A/g, that

$$S = A\left(\frac{g+1}{g}\ln T - \ln p + i\right) .$$

(c) Since $pv = gU = AT$, the Gibbs free energy of the system is

$$G = U+pv-TS = \left(1+\frac{1}{g}\right)pv-TS = AT\left[1+\frac{1}{g}-\left(1+\frac{1}{g}\right)\ln T+\ln p-i\right].$$

Since $G = \mu N$ (Problem 1.20) and $A = Nk$ [part (a)], this yields

$$\frac{\mu}{kT} = \ln p-\left(1+\frac{1}{g}\right)\ln T-i+1+\frac{1}{g}\ .$$

THIRD LAW

1.24 Suppose the entropy remains finite and continuous as $T \to 0$. [This is from reference 4, p.112 equivalent to **Nernst's heat theorem.**] Show that then $X = C_v, C_p, l_v, l_p, m_v, m_p$ each tends to zero in this limit.

Solution

(i) Recall from Problem 1.7(a) that

$$\mathrm{d}F = \mathrm{d}U-T\mathrm{d}S-S\mathrm{d}T = -S\mathrm{d}T-p\,\mathrm{d}v\ ,$$

so that

$$\left(\frac{\partial F}{\partial T}\right)_v = -S = \frac{F-U}{T}\ .$$

Now, as $T \to 0$, $F \to U$ for finite S, so that the right-hand side becomes indeterminate. Differentiating numerator and denominator and substituting the values appropriate to $T = 0$, we obtain

$$\lim_{T\to 0}\left(\frac{\partial F}{\partial T}\right)_v = \lim_{T\to 0}\left[\left(\frac{\partial F}{\partial T}\right)_v-\left(\frac{\partial U}{\partial T}\right)_v\right]\ ,$$

i.e. $S_{T=0} = S_{T=0}+C_{v,\,T=0}$. Hence $C_v \to 0$.

(ii) From $H = U+pv$, we have

$$\mathrm{d}H = T\mathrm{d}S+v\,\mathrm{d}p = C_p\,\mathrm{d}T+(l_p+v)\mathrm{d}p\ ,$$

whence $C_p = (\partial H/\partial T)_p$. Also

$$\mathrm{d}G = -S\mathrm{d}T+v\,\mathrm{d}p \quad \text{[Problem 1.7(a)]}\ .$$

Hence the following quantity is indeterminate as $T \to 0$:

$$-\left(\frac{\partial G}{\partial T}\right)_p = S = \frac{H-G}{T}\ .$$

Hence

$$\lim_{T\to 0}\left[-\left(\frac{\partial G}{\partial T}\right)_p\right] = \lim_{T\to 0}\left[C_p-\left(\frac{\partial G}{\partial T}\right)_p\right]\ ,$$

so that $C_p \to 0$.

(iii) From Problem 1.3 we have

$$T\mathrm{d}S = C_v\,\mathrm{d}T+l_v\,\mathrm{d}v = C_p\,\mathrm{d}T+l_p\,\mathrm{d}p = m_v\,\mathrm{d}v+m_p\,\mathrm{d}p\ .$$

If S remains finite and continuous

$$T\left(\frac{\partial S}{\partial x}\right)_y = C_v\left(\frac{\partial T}{\partial x}\right)_y + l_v\left(\frac{\partial v}{\partial x}\right)_y = \ldots = \ldots \to 0 .$$

Consider various cases as $T \to 0$. One finds

(x,y)	(T,v)	(T,p)	(v,T)	(p,T)	(v,p)	(p,v)
Result	$C_v \to 0$	$C_p \to 0$	$l_v \to 0$	$l_p \to 0$	$m_v \to 0$	$m_p \to 0$

This argument incorporates independent proofs of (i) and (ii), above.

1.25 Take a **strong heat theorem** in the form $(\partial S/\partial x)_y \to 0$ as $T \to 0$, where x, y are any two independent variables (other than S). Show that the result of Problem 1.24 can then be strengthened to yield the vanishing of X/T.

Solution

We have now $(\partial S/\partial x)_y \to 0$ in the solution (iii) of Problem 1.24, so that it is clear that the quantities (C_v/T), (C_p/T), etc. will vanish.

1.26 Show that an ideal classical gas of constant heat capacities (defined in Problem 1.9) cannot exist indefinitely close to the absolute zero, even if the third law is taken in the weak form given to it in Problem 1.24.

Show that an ideal quantum gas (defined in Problem 1.9) does not violate the third law even if it is taken in the strong form given to it in Problem 1.25, provided one takes $(x, y) = (v, T)$.

Solution

Consider an ideal classical gas. The result $C_p - C_v = A$ of Problem 1.9 will be violated near $T = 0$, since from Problem 1.24 we have $C_p \to 0$, $C_v \to 0$.

If $pv = gU$ (ideal quantum gas), then

$$\left(\frac{\partial p}{\partial T}\right)_v = \frac{g}{v} C_v \to 0$$

from Problem 1.24. Hence a Maxwell relation yields $(\partial S/\partial v)_T \to 0$ as $T \to 0$. Note that for $(x, y) = (T, v)$ the strong theorem fails already for an ideal electron gas for which $(\partial S/\partial T)_v$ approaches a non-zero value as $T \to 0$. To show this, use Problem 3.15(d).

PHASE CHANGES

1.27 Let $T_L(p)$ be a line L in the phase space of a fluid with two independent variables which assigns a temperature T_L to any pressure p. Let temperatures T be measured from this line: $t = T - T_L(p)$.

(a) Show that the slope $p'_{\rm L}$ of the line L satisfies

$$\left(\frac{\partial p}{\partial T}\right)_t = \left(\frac{\partial p}{\partial T}\right)_{\rm L} \equiv p'_{\rm L}\,; \quad \left(\frac{\partial p}{\partial t}\right)_T = -p'_{\rm L}\,.$$

(b) Show also that for any function of state y

$$\left(\frac{\partial y}{\partial T}\right)_p = \left(\frac{\partial y}{\partial t}\right)_p; \quad \left(\frac{\partial y}{\partial T}\right)_t = p'_{\rm L}\left(\frac{\partial y}{\partial p}\right)_t; \quad \left(\frac{\partial y}{\partial t}\right)_T = -p'_{\rm L}\left(\frac{\partial y}{\partial p}\right)_T.$$

(c) If W and X are two functions of state for the fluid, establish that

$$\left(\frac{\partial W}{\partial T}\right)_X = \left(\frac{\partial W}{\partial T}\right)_t - p'_{\rm L}\left(\frac{\partial W}{\partial X}\right)_T\left(\frac{\partial X}{\partial p}\right)_t,$$

Write down the equations resulting from $(W, X) = (S, p), (S, v), (v, p)$ and from the first and last of these deduce that

$$\frac{C_p}{T} = \left(\frac{\partial S}{\partial T}\right)_t + p'_{\rm L} v\alpha_p,$$

$$\alpha_p = \frac{1}{v}\left(\frac{\partial v}{\partial T}\right)_t + K_T p'_{\rm L},$$

where α_p and K_T have the meanings given to them in Problem 1.4. Interpret these results in terms of curves which might be plotted.

(d) Recover the result

$$C_p - C_v = T\left(\frac{\partial p}{\partial T}\right)_v\left(\frac{\partial v}{\partial T}\right)_p \qquad \text{[Problem 1.8(a)]}$$

by choosing the line $t = 0$ to be a line of constant volume.

Solution

(a) Taking the two independent variables to be T and p,

$$\mathrm{d}t = \mathrm{d}T - \frac{1}{p'_{\rm L}}\,\mathrm{d}p\,.$$

The stated results follow.

(b) We have

$$\left(\frac{\partial y}{\partial t}\right)_p = \left(\frac{\partial y}{\partial T}\frac{\partial T}{\partial t}\right)_p = \left(\frac{\partial y}{\partial T}\right)_p,$$

and from part (a)

$$\left(\frac{\partial y}{\partial t}\right)_T = \left(\frac{\partial y}{\partial p}\frac{\partial p}{\partial t}\right)_T = -p'_{\rm L}\left(\frac{\partial y}{\partial p}\right)_T,$$

$$\left(\frac{\partial y}{\partial T}\right)_t = \left(\frac{\partial y}{\partial p}\frac{\partial p}{\partial T}\right)_t = p'_{\rm L}\left(\frac{\partial y}{\partial p}\right)_t.$$

(c) For two independent variables the following result is general:

$$\left(\frac{\partial W}{\partial Y}\right)_X = \left(\frac{\partial W}{\partial Y}\right)_Z - \left(\frac{\partial W}{\partial X}\right)_Y\left(\frac{\partial X}{\partial Y}\right)_Z.$$

Putting $Y \to T, Z \to t$

$$\left(\frac{\partial W}{\partial T}\right)_X = \left(\frac{\partial W}{\partial T}\right)_t - \left(\frac{\partial p}{\partial T}\frac{\partial X}{\partial p}\right)_t \left(\frac{\partial W}{\partial X}\right)_T = \left(\frac{\partial W}{\partial T}\right)_t - p_L' \left(\frac{\partial W}{\partial X}\right)_T \left(\frac{\partial X}{\partial p}\right)_t .$$

The three special results are, using Maxwell's relations,

$$\left(\frac{\partial S}{\partial T}\right)_p = \frac{C_p}{T} = \left(\frac{\partial S}{\partial T}\right)_t + p_L' \left(\frac{\partial v}{\partial T}\right)_p , \tag{1.27.1}$$

$$\left(\frac{\partial S}{\partial T}\right)_v = \frac{C_v}{T} = \left(\frac{\partial S}{\partial T}\right)_t - p_L' \left(\frac{\partial p}{\partial T}\right)_v \left(\frac{\partial v}{\partial p}\right)_t , \tag{1.27.2}$$

$$\left(\frac{\partial v}{\partial T}\right)_p = \left(\frac{\partial v}{\partial T}\right)_t - p_L' \left(\frac{\partial v}{\partial p}\right)_T . \tag{1.27.3}$$

The last equation yields, on multiplying by v^{-1}, the connection between α_p and K_T.

Equation (1.27.1) shows that when C_p/T is plotted against $(\partial v/\partial T)_p$ one obtains a curve which close to the line L approaches a straight line of slope p_L' and intercept $(\partial S/\partial T)_L$. The notation $(\ )_L$ means: evaluated at $t = 0$, i.e. at the line L.

Equation (1.27.2) shows that the same line is an asymptote of the curve obtained by plotting C_v/T against $-(\partial p/\partial T)_v(\partial v/\partial p)_T$. The last relation under (c) shows that the curve of α_p versus K_T has an asymptote of slope p_L' and intercept $v^{-1}(\partial v/\partial T)_L$.

(d) Choose $W = S, X = p$ in the first equation of part (c), noting that

$$-p_L' \left(\frac{\partial W}{\partial X}\right)_T \left(\frac{\partial X}{\partial p}\right)_t = -\left(\frac{\partial X}{\partial T}\right)_t \left(\frac{\partial W}{\partial X}\right)_T \to \left(\frac{\partial p}{\partial T}\right)_v \left(\frac{\partial v}{\partial T}\right)_p .$$

1.28 Suppose the line L in the preceding problem marks a transition from a phase 1 to a phase 2.

(a) Assuming $(\partial S/\partial T)_t$ to be continuous across L, recover the Clausius–Clapeyron equation of Problem 1.14 by integrating the C_p/T equation of Problem 1.27(c).

(b) Show that, if p_L' is finite and non-zero, on the line L

$$\left(\frac{\partial p}{\partial T}\right)_S = \left(\frac{\partial p}{\partial T}\right)_{C_p} = \left(\frac{\partial p}{\partial T}\right)_v = p_L' ,$$

$$\left(\frac{\partial p}{\partial v}\right)_T = \left(\frac{\partial T}{\partial S}\right)_p = \left(\frac{\partial T}{\partial v}\right)_p = 0.$$

[In spite of the divergence of C_p, α_p, and K_T on L, which follows from part (b), the equations between them given in Problem 1.27(c) remain valid. These are generalisations of relations due to Pippard[(5)].]

[(5)] M.J.Buckingham and W.M.Fairbanks in *Progress in Low Temperature Physics* (Ed. C.J.Gorter), 3, 89 (North-Holland, Amsterdam), 1961, and A.B.Pippard, *Phil.Mag.*, 1, 473 (1956). See also J.Wilks, *The Properties of Liquid and Solid Helium* (Oxford University Press, Oxford), 1967, p.302.

Solution

(a) Integrate

$$\frac{C_p}{T} = \left(\frac{\partial S}{\partial T}\right)_p = \left(\frac{\partial S}{\partial T}\right)_t + p'_{\mathrm{L}}\left(\frac{\partial v}{\partial T}\right)_v$$

at constant pressure from a point just inside phase 1 to a point just inside phase 2. Then

$$\frac{\Lambda_T^{1\to 2}}{T} = \left(\frac{\mathrm{d}p}{\mathrm{d}t}\right)_L (v_2 - v_1)\,.$$

The first term in $(\partial S/\partial T)_t$ does not contribute.

(b) As in Problem 1.27(c), we have

$$\left(\frac{\partial W}{\partial p}\right)_X = \left(\frac{\partial W}{\partial p}\right)_{C_p} - \left(\frac{\partial W}{\partial X}\right)_p \left(\frac{\partial X}{\partial p}\right)_{C_p}\,.$$

Since p is the only variable on L, the last term vanishes, and

$$\left(\frac{\partial W}{\partial p}\right)_X = \left(\frac{\partial W}{\partial p}\right)_{C_p}\,.$$

Choosing $W = T$ and $X = S$ or V, we obtain

$$\left(\frac{\partial T}{\partial p}\right)_{C_p} = \left(\frac{\partial T}{\partial p}\right)_S = \left(\frac{\partial T}{\partial p}\right)_v \text{ on L}\,.$$

This ratio is $1/p'_{\mathrm{L}}$ on L. Also

$$\left(\frac{\partial T}{\partial v}\right)_p = -\left(\frac{\partial T}{\partial p}\right)_v \left(\frac{\partial p}{\partial v}\right)_T = -\frac{1}{p'_{\mathrm{L}}}\left(\frac{\partial p}{\partial v}\right)_T\,.$$

As p'_{L} is finite and non-zero and the left-hand side vanishes on L, $(\partial p/\partial v)_T = 0$ on L. Also

$$\left(\frac{\partial T}{\partial S}\right)_p = \left(\frac{\partial T}{\partial v}\right)_p = 0 \text{ on L}\,.$$

THERMAL AND MECHANICAL STABILITY

1.29 Two macroscopic thermodynamic fluids 1 and 2, free from external forces, are in thermal and mechanical contact and the total system is isolated. The four independent variables, say the volumes and entropies v_1, v_2, S_1, S_2, are constrained by the given total volume v and the given internal energy U:

$$v = v_1 + v_2\,, \tag{1.29.1}$$

$$U = U_1 + U_2\,, \tag{1.29.2}$$

since for each fluid $U = U(S, v, n)$ there are no other independent variables.

(a) Taking the remaining two independent variables as v_1 and S_1 establish the useful results

$$\left(\frac{\partial v_2}{\partial v_1}\right)_{S_1} = -1\,, \tag{1.29.3}$$

$$\left(\frac{\partial S_2}{\partial v_1}\right)_{S_1} = \left[\left(\frac{\partial U_2}{\partial v_2}\right)_{S_2} - \left(\frac{\partial U_1}{\partial v_1}\right)_{S_1}\right] \Big/ \left(\frac{\partial U_2}{\partial S_2}\right)_{v_2}, \tag{1.29.4}$$

$$\left(\frac{\partial v_2}{\partial S_1}\right)_{v_1} = 0\,, \tag{1.29.5}$$

$$\left(\frac{\partial S_2}{\partial S_1}\right)_{v_1} = -\left(\frac{\partial U_1}{\partial S_1}\right)_{v_1} \Big/ \left(\frac{\partial U_2}{\partial S_2}\right)_{v_2}. \tag{1.29.6}$$

(b) The appropriate form of the second law of thermodynamics is, from Problem 1.22, $\delta U + p\delta v - T\delta S \geqslant 0$, i.e. δS at constant U and v is to be a minimum. Show that for equilibrium between fluids 1 and 2, i.e. for $S \equiv S_1 + S_2$ to be an extremum, the necessary conditions are

$$\left(\frac{\partial U_1}{\partial v_1}\right)_{S_1} = \left(\frac{\partial U_2}{\partial v_2}\right)_{S_2}, \tag{1.29.7}$$

i.e. $p_1 = p_2$ ($= p$ say) (mechanical equilibrium) and

$$\left(\frac{\partial U_1}{\partial S_1}\right)_{v_1} = \left(\frac{\partial U_2}{\partial S_2}\right)_{v_2}, \tag{1.29.8}$$

i.e. $T_1 = T_2$ ($= T$ say) (thermal equilibrium).

(c) Show that the conditions for mechanical and thermal stability, i.e. for S to be a maximum rather than only an extremum, include

$$-\left[\left(\frac{\partial p}{\partial v_1}\right)_{S_1} + \left(\frac{\partial p}{\partial v_2}\right)_{S}\right]_0 \equiv \left[\left(\frac{1}{vK_S}\right)_1 + \left(\frac{1}{vK_S}\right)_2\right]_0 > 0\,, \tag{1.29.9}$$

$$\left[\left(\frac{\partial T}{\partial S_1}\right)_v + \left(\frac{\partial T}{\partial S_2}\right)\right]_0 \equiv \left[\left(\frac{T}{Cv}\right)_1 + \left(\frac{T}{Cv}\right)_2\right]_0 > 0\,, \tag{1.29.10}$$

where the suffix 0 indicates evaluation at equilibrium conditions.

Solution

(a) Write Equations (1.29.1) and (1.29.2) as

$$v_1 + v_2[v_1, S_1] = v, \tag{1.29.11}$$

$$U_1[v_1, S_1] + U_2[v_2(v_1, S_1),\ S_2(v_1, S_1)] = U\,. \tag{1.29.12}$$

Since v and U are constant, and S_1 and v_1 are independent, the partial derivatives with respect to S_1 and v_1 are independently zero. Differentiating Equations (1.29.11) and (1.29.12) with respect to v_1 at constant

S_1 yields

$$1 + \left(\frac{\partial v_2}{\partial v_1}\right)_{S_1} = 0\,,$$

$$\left(\frac{\partial U_1}{\partial v_1}\right)_{S_1} + \left(\frac{\partial U_2}{\partial v_2}\right)_{S_2}\left(\frac{\partial v_2}{\partial v_1}\right)_{S_1} + \left(\frac{\partial U_2}{\partial S_2}\right)_{v_2}\left(\frac{\partial S_2}{\partial v_1}\right)_{S_1} = 0\,. \quad (1.29.13)$$

The first of these is already Equation (1.29.3). Using it in the second one yields Equation (1.29.4).

Next, differentiate Equations (1.29.11) and (1.29.12) with respect to S_1 at constant v_1:

$$\left(\frac{\partial v_1}{\partial S_1}\right)_{v_1} + \left(\frac{\partial v_2}{\partial S_1}\right)_{v_1} = 0\,, \quad \text{i.e.} \left(\frac{\partial v_2}{\partial S_1}\right)_{v_1} = 0\,,$$

$$\left(\frac{\partial U_1}{\partial S_1}\right)_{v_1} + \left(\frac{\partial U_2}{\partial v_2}\right)_{S_2}\left(\frac{\partial v_2}{\partial S_1}\right)_{v_1} + \left(\frac{\partial U_2}{\partial S_2}\right)_{v_2}\left(\frac{\partial S_2}{\partial S_1}\right)_{v_1} = 0\,. \quad (1.29.14)$$

The first of these is already Equation (1.12.5). Using it in the second one yields Equation (1.29.6).

(b) Rewrite Equations (1.29.4) and (1.29.6) in the forms

$$\left(\frac{\partial S}{\partial v_1}\right)_{S_1} = \left[\left(\frac{\partial U_2}{\partial v_2}\right)_{S_2} - \left(\frac{\partial U_1}{\partial v_1}\right)_{S_1}\right] \Big/ \left(\frac{\partial U_2}{\partial S_2}\right)_{v_2} = -\frac{p_1 - p_2}{T_2}\,, \quad (1.29.15)$$

$$\left(\frac{\partial S}{\partial S_1}\right)_{v_1} = 1 - \left[\left(\frac{\partial U_1}{\partial S_1}\right)_{v_1} \Big/ \left(\frac{\partial U_2}{\partial S_2}\right)_{v_2}\right] = 1 - \frac{T_1}{T_2}\,;$$

both quantities must vanish for equilibrium.

(c) Necessary conditions to ensure $\delta S < 0$ include that the second derivatives of S with respect to S_1 and v_1 be negative. Differentiating Equation (1.29.13) with respect to v_1, at constant S_1, we obtain

$$\left(\frac{\partial^2 U_1}{\partial v_1^2}\right)_{S_1} + \left[\left(\frac{\partial^2 U_2}{\partial v_2^2}\right)_{S_2}\left(\frac{\partial v_2}{\partial v_1}\right)_{S_1} + \left(\frac{\partial^2 U_2}{\partial v_2 \partial S_2}\right)\left(\frac{\partial S_2}{\partial v_1}\right)_{S_1}\right]\left(\frac{\partial v_2}{\partial v_1}\right)_{S_1}$$
$$+ \left(\frac{\partial U_2}{\partial v_2}\right)_{S_2}\left(\frac{\partial^2 v_2}{\partial v_1^2}\right)_{S_1} + \left[\left(\frac{\partial^2 U_2}{\partial S_2^2}\right)_{v_2}\left(\frac{\partial S_2}{\partial v_1}\right)_{S_1} + \left(\frac{\partial^2 U_2}{\partial S_2 \partial v_2}\right)\left(\frac{\partial v_2}{\partial v_1}\right)_{S_1}\right]\left(\frac{\partial S_2}{\partial v_1}\right)_{S_1}$$
$$+ \left(\frac{\partial U_2}{\partial S_2}\right)_{v_2}\left(\frac{\partial^2 S_2}{\partial v_1^2}\right)_{S_1} = 0\,.$$

On using

$$\left(\frac{\partial v_2}{\partial v_1}\right)_{S_1} = -1\,, \quad \left(\frac{\partial^2 v_2}{\partial v_1^2}\right)_{S_1} = 0\,,$$

and Equation (1.29.15), the results of parts (a) and (b) yield the simplification, valid at equilibrium,

$$\left[\left(\frac{\partial^2 U_1}{\partial v_1^2}\right)_{S_1} + \left(\frac{\partial^2 U_2}{\partial v_2^2}\right)_{S_2} + \left(\frac{\partial U_2}{\partial S_2}\right)_{v_2}\left(\frac{\partial^2 S_2}{\partial v_1^2}\right)_{S_1}\right]_0 = 0\,,$$

whence

$$\left\{\left(\frac{\partial^2 S}{\partial v_1^2}\right)_{S_1} = \left(\frac{\partial^2 S_2}{\partial v_1^2}\right)_{S_1} = -\frac{1}{T}\left[\left(\frac{\partial^2 U_1}{\partial v_1^2}\right)_{S_1} + \left(\frac{\partial^2 U_2}{\partial v_2^2}\right)_{S_2}\right]\right\}_0 . \quad (1.29.16)$$

Next, on differentiating Equation (1.29.14) with respect to S_1, and omitting now terms which will not contribute, we obtain

$$\frac{\partial^2 U_1}{\partial S_1^2} + \left(\frac{\partial^2 U_2}{\partial S_2^2}\frac{\partial S_2}{\partial S_1} + \frac{\partial^2 U_2}{\partial S_2 \partial v_2}\frac{\partial v_2}{\partial S_1}\right)\frac{\partial S_2}{\partial S_1} + \frac{\partial U_2}{\partial S_2}\frac{\partial^2 S_2}{\partial S_1^2} = 0 .$$

Using the results of parts (a) and (b) we find

$$\left(\frac{\partial^2 U_1}{\partial S_1^2} + \frac{\partial^2 U_2}{\partial S_2^2} + \frac{\partial U_2}{\partial S_2}\frac{\partial^2 S_2}{\partial S_1^2}\right)_0 = 0 ,$$

whence

$$\left\{\left(\frac{\partial^2 S}{\partial S_1^2}\right)_{v_1} = \left(\frac{\partial^2 S_2}{\partial S_1^2}\right)_{v_1} = -\frac{1}{T}\left[\left(\frac{\partial^2 U_1}{\partial S_1^2}\right)_{v_1} + \left(\frac{\partial^2 U_2}{\partial S_2^2}\right)_{v_2}\right]\right\}_0 . \quad (1.29.17)$$

Now the condition for S to be a maximum, rather than just an extremum, is that the quantities (1.29.16) and (1.29.17) be negative, whence identities (1.29.9) and (1.29.10) follow.

[**Notes.**

1. We have shown that the sum of two terms must be positive from both Equation (1.29.16) and Equation (1.29.17). Each of the four terms depends on the variables of one system only and it therefore is reasonable to infer that the following inequalities hold for each system individually:

$$\left(\frac{1}{vK_S}\right)_0 = \left(-\frac{\partial p}{\partial v}\right)_{S,0} = \left(\frac{\partial^2 U}{\partial v^2}\right)_{S,0} > 0 ,$$
$$\left(\frac{T}{C_v}\right)_0 = \left(\frac{\partial T}{\partial S}\right)_{v,0} = \left(\frac{\partial^2 U}{\partial S^2}\right)_{v,0} > 0 . \quad (1.29.18)$$

2. For completeness the sign of the cross term $\partial^2 S/\partial v_1 \partial S_1$ must also be considered (see Problem 1.31).

3. The important case of stability in a two-fluid system in the presence of mass transfer is treated in Chapter 7. Note particularly Problem 7.3.]

1.30 Reconsider the derivation of the stability conditions in Problem 1.29 on the assumption that, instead of the whole system being isolated energetically, each part system is ins contact with a heat reservoir at temperature T.

[Hint: The appropriate form of the second law is $\delta F + S\delta T + p\delta v \geqslant 0$.]

Solution(6)

This problem differs from Problem 1.29 in that one condition $U_1 + U_2 = U$ on an extensive variable has been replaced by two

(6) J.T.Łopuszański, *Acta Phys. Polonica*, **33**, 953 (1968).

constraints T_1 = constant, T_2 = constant, on intensive variables. This reduces the number of independent thermodynamic variables by one. Whereas in Problem 1.29 two independent variables lead to two stability conditions, one independent variable in this problem may be expected to lead to only one stability condition.

Of the four variables, v_1, v_2, T_1, T_2 only one is independent, and we chose v_1. Condition (1.29.3) becomes

$$\frac{dv_2}{dv_1} = -1 .$$

For equilibrium at constant temperature the free energy may be minimised. So, in analogy to $S = S_1 + S_2$, we have

$$F = F_1(v_1, T_1) + F_2(v_2, T_2) ,$$

and Equation (1.29.15) is replaced by

$$\left[\left(\frac{\partial F_1}{\partial v_1}\right)_{T_1} + \left(\frac{\partial F_2}{\partial v_2}\right)_{T_2} \frac{dv_2}{dv_1}\right]_0 = \left[\left(\frac{\partial F_1}{\partial v_1}\right)_{T_1} - \left(\frac{\partial F_2}{\partial v_2}\right)_{T_2}\right]_0 = 0 .$$

Thus the variability of v_1 and v_2 yields again

$$p_{1,0} = p_{2,0}$$

as an equilibrium condition (free energy an extremum).

For a minimum rather than an extremum, the stability condition yields

$$\left(\frac{\partial^2 F}{\partial v_1^2}\right)_{T_1} = \left(\frac{\partial^2 F_1}{\partial v_1^2}\right)_{T_1} - \left(\frac{\partial^2 F_2}{\partial v_2^2}\right)_{T_2} \frac{dv_2}{dv_1} > 0 .$$

Thus

$$-\left[\left(\frac{\partial p_1}{\partial v_1}\right)_{T_1} + \left(\frac{\partial p_2}{\partial v_2}\right)_{T_2}\right]_0 > 0 . \tag{1.30.1}$$

1.31 The relation between the conditions (1.29.9), (1.29.10), and (1.30.1) is elucidated by considering one of the fluids in the preceding two problems.

(a) Let a, b, h be real constants, let x, y be variables, and let

$$I(x, y) \equiv ax^2 + 2hxy + by^2 .$$

Show that $I(x, y)$ has one and the same sign for all real x, y if $ab > h^2$.

(b) If a function f(x, y) is subjected to a Taylor expansion about its value at (x_0, y_0), say, show that the condition (a) applied to the second order terms is equivalent to a condition on the Jacobian

$$J \equiv \left|\frac{\partial(\mathrm{f}_x, \mathrm{f}_y)}{\partial(x, y)}\right| \equiv \begin{vmatrix} \left(\dfrac{\partial^2 \mathrm{f}}{\partial x^2}\right)_0 & \left(\dfrac{\partial^2 \mathrm{f}}{\partial x \partial y}\right)_0 \\ \left(\dfrac{\partial^2 \mathrm{f}}{\partial x \partial y}\right)_0 & \left(\dfrac{\partial^2 \mathrm{f}}{\partial y^2}\right)_0 \end{vmatrix} > 0 .$$

Here $I(x, y)$ is to be reinterpreted as $f(x, y) - f(x_0, y_0)$ and the suffix 0 denotes an evaluation of derivatives at the point (x_0, y_0).

(c) Show that the conditions of Problems 1.29 and 1.30 for thermodynamic stability applied to a single phase are met by the appropriate form of the condition (b) above, together with the condition that one of the diagonal terms of the determinant is positive.

Solution

(a) Write

$$I(x, y) = \frac{1}{a}[(ax + hy)^2 + (ab - h^2)y^2] .$$

Thus I has the sign of a if $ab > h^2$. This implies that a and b have like signs.

(b) We have, if x and y are measured from x_0 and y_0 respectively, and first order terms are zero,

$$f(x, y) - f(0, 0) = \left(\frac{\partial^2 f}{\partial x^2}\right)_{y,\,0} x^2 + 2\left(\frac{\partial^2 f}{\partial x \partial y}\right)_0 xy + \left(\frac{\partial^2 f}{\partial y^2}\right)_{x,\,0} y^2 .$$

This gives an interpretation of a, b, and h, and the result follows.

(c) The condition of parts (a) and (b) above is $f_{xx} f_{yy} > (f_{xy})^2$, i.e.

$$J \equiv \begin{vmatrix} f_{xx} & f_{xy} \\ f_{yx} & f_{yy} \end{vmatrix} > 0 .$$

Then f, f_{xx}, and f_{yy} have like signs. For thermodynamic stability the sign itself is prescribed. For example, with the interpretation

$$F \to U , \quad x \to v , \quad y \to S$$

one must have $U(v, S) \geqslant U_0$; the stability conditions can therefore be written as

$$J > 0 , \quad f_{yy} = \frac{T}{C_v} > 0 \tag{1.31.1}$$

from which it follows that $f_{xx} > 0$, i.e.

$$[J > 0 , \ f_{xx} > 0] \quad \Rightarrow \quad f_{yy} = \frac{T}{C_v} > 0 , \tag{1.31.2}$$

$$[J > 0 , \ f_{yy} > 0] \quad \Rightarrow \quad f_{xx} = -\left(\frac{\partial p}{\partial v}\right)_S = \frac{1}{vK_S} > 0 . \tag{1.31.3}$$

In Equation (1.29.18), on the other hand, Problem 1.29, interpreted for a single phase, yielded

$$f_{xx} > 0 , \quad f_{yy} > 0 . \tag{1.31.4}$$

It is now seen that these conditions are to be supplemented by $J > 0$, which implies a condition on the cross term $\partial^2 U/\partial v \partial S$.

We conclude with some remarks concerning the relationship between the results (1.31.1) to (1.31.3) of this problem and the result (1.31.4) of Problem 1.29. An interpretation of $J > 0$ is obtained as follows:

$$\frac{\partial(U_v, U_S)}{\partial(v, S)} = \frac{\partial(-p, T)}{\partial(v, S)} = \frac{\partial(-p, T)}{\partial(v, T)} \Big/ \frac{\partial(v, S)}{\partial(v, T)} = -\left(\frac{\partial p}{\partial v}\right)_T \Big/ \left(\frac{\partial S}{\partial T}\right)_v$$

$$= -\frac{T}{C_v} \Big/ \left(\frac{\partial v}{\partial p}\right)_T = \frac{T}{C_v} v K_T > 0 \,. \tag{1.31.5}$$

The implications (1.31.2), (1.31.3) become

$$\frac{T/C_v}{vK_T} > 0 \,, \quad vK_S > 0 \quad \Rightarrow \quad \frac{T}{C_v} > 0 \,, \quad vK_T > 0 \tag{1.31.6}$$

$$\frac{T/C_v}{vK_T} > 0 \,, \quad \frac{T}{C_v} > 0 \quad \Rightarrow \quad vK_S > 0 \,, \quad vK_T > 0 \tag{1.31.7}$$

where the last inequalities in (1.31.6) and (1.31.7) are clearly fulfilled. Indeed, Problem 1.4(c) enables one to infer also that in both cases

$$\frac{vK_T}{vK_S} = \frac{C_p}{C_v} = \frac{T/C_v}{T/C_p} > 0 \,, \qquad \text{i.e. } T/C_p > 0 \,.$$

The result (1.31.4) does not go quite as far. In this case one would like to argue

$$vK_S > 0 \,, \qquad T/C_v > 0 \quad \Rightarrow \quad \frac{T/C_v}{vK_T} > 0 \,,$$

but because of

$$K_T = K_S + Tv\frac{\alpha_p^2}{C_p} \qquad \text{[Problem 1.8(b)]}$$

this requires a condition for the right-hand side of this equation (e.g. $Tv/C_p > 0$).

The case of two fluids is considered in Problems 7.3 and 7.4.

GENERAL REFERENCES

M.W.Zemanski, *Heat and Thermodynamics*, 5th Edn. (McGraw-Hill, New York), 1968.

R.Becker, *Theory of Heat*, 2nd Edn. (Springer-Verlag, Berlin), 1967.

M.E.Fisher, *Rep.Progr.Phys.*, **30**, 615 (1967).

J.S.Rowlinson, *Liquids and Liquid Mixtures*, 2nd Edn. (Butterworths, London), 1969.

2
Statistical theory of information and of ensembles

P.T.LANDSBERG
(*University College, Cardiff*)

ENTROPY MAXIMISATION: ENSEMBLES

In these problems a **statistical** $\mathcal{S}$**ntropy** is defined, and is distinguished from the thermodynamic entropy by the symbol for its initial letter. Though k can here be any constant with the dimension of entropy, it is usually taken to be Boltzmann's constant.

2.1 A system can be in any one of N states. The probability of it being in its ith state is p_i $(i = 1, 2, ..., N)$, where $\sum_{i=1}^{N} p_i = 1$. Use the method of undetermined multipliers to show that for the maximum $\mathcal{S}$ntropy $S \equiv -k\sum_i p_i \ln p_i$ of the probability distribution,

$$p_1 = p_2 = ... = p_N = \frac{1}{N}$$

and

$$S = S_1 \equiv k \ln N.$$

Solution

We need to maximise $S = -k\sum p_i \ln p_i$ subject to $\sum p_i = 1$. Let α be an undetermined multiplier; then consider the maximisation of

$$\mathrm{f} = -k\sum_i (p_i \ln p_i - \alpha p_i)$$

with respect to each p_i. One finds the condition

$$\frac{\partial \mathrm{f}}{\partial p_j} = -k(\ln p_j + 1 - \alpha) = 0.$$

Hence $\ln p_j = \alpha - 1$ for all j. Therefore all p_i are equal. Normalisation yields the required result for p_j. The maximum $\mathcal{S}$ntropy is

$$S_1 \equiv S_{\max} = -k\sum_i \left(-\frac{1}{N}\ln\frac{1}{N}\right) = k \ln N\,.$$

2.2 Suppose that an extensive variable x takes on the value x_i when the system considered in the preceding problem is in its ith state. Reconsider the search for the maximum Entropy distribution as carried out in the preceding problem if the average value of x $\left(= \sum_{i=1}^{N} p_i x_i\right)$ is known to have the given value x_0. Show that for $j = 1, 2, \ldots, N$ and an undetermined multiplier β, one finds

$$p_j = \frac{1}{Z(x)} \exp(-\beta x_j)$$

where $Z(x) \equiv \sum \exp(-\beta x_j)$.

Also show that

$$x_0 = -\left[\frac{\partial \ln Z(x)}{\partial \beta}\right]_{x_1, x_2, \ldots}$$

and

$$S = S_2 \equiv k\beta x_0 + k \ln Z.$$

Solution

The preceding solution is generalised by considering

$$\mathrm{f} = -k \sum_i (p_i \ln p_i - \alpha p_i + \beta p_i x_i)$$

where β is the second multiplier. We have

$$\frac{\partial \mathrm{f}}{\partial p_j} = -k(\ln p_j + 1 - \alpha + \beta x_j) .$$

It follows that

$$p_j = \frac{1}{Z(x)} \exp(-\beta x_j) \qquad (j = 1, 2, \ldots)$$

where

$$Z(x) \equiv \sum_j \exp(-\beta x_j) .$$

Also

$$x_0 = \sum p_j x_j = \frac{1}{Z(x)} \sum_j x_j \exp(-\beta x_j) = -\left[\frac{\partial \ln Z(x)}{\partial \beta}\right]_{x_1, x_2, \ldots}$$

Lastly

$$S_2 \equiv S_{\max} = +k \sum_i p_j [\beta x_j + \ln Z(x)]$$

$$= k\beta x_0 + k \ln Z .$$

2.3 The extensive variables x and y take on values x_i and y_i, respectively, when the system considered in the preceding two problems is in state i. If the average values of x and y are fixed at x_0 and y_0 for the system, show that for the maximum entropy of the distribution function

$$p_j = \frac{1}{\Xi}\exp(-\beta x_j - \gamma y_j) \qquad (j = 1, 2, \dots N),$$

where $\Xi = \sum_i \exp(-\beta x_i - \gamma y_i)$, and β and γ are undetermined multipliers. Show also that

$$x_0 = -\left(\frac{\partial \ln \Xi}{\partial \beta}\right)_{x_1, \dots, y_N, \gamma}, \qquad y_0 = -\left(\frac{\partial \ln \Xi}{\partial \gamma}\right)_{x_1, \dots, y_N, \beta}$$

and

$$S = S_3 \equiv k\beta x_0 + k\gamma y_0 + k \ln \Xi .$$

Solution

Proceeding as in the preceding problems, we find

$$\mathrm{f} = -k \sum_i (p_i \ln p_i - \alpha p_i + \beta p_i x_i + \gamma p_i y_i),$$

$$\frac{\partial \mathrm{f}}{\partial p_j} = -k(\ln p_j + 1 - \alpha + \beta x_j + \gamma y_j).$$

Hence

$$p_j = \frac{1}{\Xi}\exp(-\beta x_j - \gamma y_j)$$

where

$$\Xi \equiv \sum_i \exp(-\beta x_i - \gamma y_i).$$

Also

$$x_0 = \sum_i p_i x_i = -\left(\frac{\partial \ln \Xi}{\partial \beta}\right)_{x_i, y_i, \gamma}, \qquad y_0 = -\left(\frac{\partial \ln \Xi}{\partial \gamma}\right)_{x_i, y_i, \beta}.$$

Lastly

$$S_3 \equiv S_{\max} = k \sum_i p_i(\beta x_i + \gamma y_i + \ln \Xi) = k\beta x_0 + k\gamma y_0 + k \ln \Xi .$$

2.4 A collection of copies of the system which are at any given time distributed over their states in proportion to the probabilities p_j obtained in the preceding three problems is called an **ensemble**.

Problem 2.1 describes a **microcanonical** ensemble, Problem 2.2 a **canonical** ensemble if x is the internal energy of the system, and Problem 2.3 a **grand canonical** ensemble if x is the internal energy and y the number of (identical) particles in the system. Z and Ξ are called **partition functions.**

The term **grand ensembles** is sometimes used if only the mean total number of particles is fixed. In **petit ensembles** the total number of particles itself is fixed.

With this interpretation, verify that the following identifications are consistent with thermodynamics:

$$\beta \to \frac{1}{kT}, \quad \gamma \to -\frac{\mu}{kT}, \quad -kT\ln Z \to F, \quad kT\ln\Xi \to pv,$$

where F is the Helmholtz free energy, provided only that the statistical &ntropy and the thermodynamic entropy can be identified.

Solution

In the case of S_2, x_0 is now the internal energy U, and we write

$$S_2 = k\beta U + k\ln Z = \frac{U-F}{T} \qquad \text{[cf Problem 1.7(a)]}.$$

In the case of S_3, y_0 is the mean number of particles n, and we write

$$S_3 = k\beta U + k\gamma n + k\ln\Xi = \frac{U-\mu n + pv}{T} \qquad \text{(cf Problem 1.20)}.$$

The identifications proposed in the problem follow.

2.5 (a) A system is specified by the values of $U = U_1, v = v_1, n = n_1$. Suppose its entropy is then S_1. Under different conditions it is specified by $v = v_1$, $n = n_1$, and a temperature T_2. Suppose its average internal energy is then $U = U_2(T_2)$ and its entropy $S_2(T_2)$. Choosing the appropriate ensembles, prove that if $U_2(T_2) \geqslant U_1$ then $S_2(T_2) > S_1$.

(b) Discuss this result qualitatively in terms of probability distributions[1].

Solution

(a) We have, using a canonical ensemble, for the second condition of the system

$$S_2 = \frac{U_2 - F_2}{T_2} = \frac{U_2}{T_2} + k\ln Z_2 .$$

Now, instead of summing over states in Z_2 (as in Problem 2.2), one can sum over energies E_j by inserting the degeneracies g_j of these energy levels. Then, if the suffix '0' refers to a particular energy,

$$Z_2 = \sum_j g_j \exp\left(-\frac{E_j}{kT_2}\right) > g_0 \exp\left(-\frac{E_0}{kT_2}\right).$$

Choose now E_0 as the energy U_1 of the first description of the system.

[1] See also E.A.Guggenheim, *Research*, **2**, 450 (1949).

Then $g_0 = g_1$ becomes the N value of Problem 2.1 for the microcanonical ensemble appropriate to the first condition, and $S_1 = k\ln g_1$. Hence

$$S_2 > \frac{U_2}{T_2} + k\ln g_1 - \frac{U_1}{T_2} \geqslant S_1 .$$

(b) In the second condition the energy of the system can fluctuate and the probability distribution is spread over many more states than is possible when the energy is fixed. Hence $S_2(T_2) > S_1$.

PARTITION FUNCTIONS IN GENERAL

2.6 (a) If the number n of identical particles in a system, its internal energy U, and its volume v are given, the microcanonical ensemble of Problem 2.1 is appropriate. If these quantities are varied one has to consider a set of ensembles with neighbouring values of n, U, v, which become the independent variables. Show from Problem 1.20(b) that the partition function $k\ln N$ satisfies

$$\left(\frac{\partial \ln N}{\partial v}\right)_{U,n} = \frac{p}{kT}, \quad \left(\frac{\partial \ln N}{\partial n}\right)_{U,v} = -\frac{\mu}{kT}, \quad \left(\frac{\partial \ln N}{\partial U}\right)_{v,n} = \frac{1}{kT}.$$

Systems specified in this way can be regarded as isolated systems.

(b) For a canonical ensemble the independent variables are v, n, and T, only an average value of U being now given. Systems specified in this way can be regarded as isolated apart from a large heat reservoir at temperature T with which they are supposed to be in equilibrium. Show that the partition function satisfies

$$\left(\frac{\partial \ln Z}{\partial v}\right)_{T,n} = \frac{p}{kT}, \quad \left(\frac{\partial \ln Z}{\partial n}\right)_{v,T} = -\frac{\mu}{kT}, \quad \left(\frac{\partial \ln Z}{\partial T}\right)_{v,n} = \frac{U}{kT^2}.$$

(c) For a grand canonical ensemble the independent variables are v, T, and μ, only average values of U and n being now given. Systems specified in this way are supposed to be in equilibrium with a large heat reservoir at temperature T and a large particle reservoir at chemical potential μ. Show that the partition function satisfies

$$\left(\frac{\partial \ln \Xi}{\partial v}\right)_{T,\mu} = \frac{p}{kT}, \quad \left(\frac{\partial \ln \Xi}{\partial \mu}\right)_{v,T} = \frac{n}{kT}, \quad \left(\frac{\partial \ln \Xi}{\partial T}\right)_{v,\mu} = \frac{U-\mu n}{kT^2}.$$

(d) Establish

$$\left(\frac{\partial U}{\partial \mu}\right)_{T,v} - \mu\left(\frac{\partial n}{\partial \mu}\right)_{T,v} = T\left(\frac{\partial n}{\partial T}\right)_{\mu,v}.$$

Solution

(a) The results follow from Problems 1.20(b) and 2.4 which yield

$$T\mathrm{d}S = kT\ln N = \mathrm{d}U + p\,\mathrm{d}v - \mu\,\mathrm{d}n .$$

(b) From Problem 1.7(a) we have

$$dF = d(U-TS) = -S\,dT - p\,dv + \mu\,dn$$

and from Problem 2.4

$$dF = -d(kT\ln Z) = -k\ln Z\,dT - kT\,d(\ln Z)\,.$$

Hence

$$d\ln Z = \frac{U}{kT^2}dT + \frac{p}{kT}dv - \frac{\mu}{kT}dn\,,$$

and the results follow.

(c) From Problems 2.4 and 1.20(c) we need

$$d\left(\frac{pv}{kT}\right) = d\left(\frac{\mu n}{kT} - \frac{U}{kT} + \frac{S}{k}\right).$$

Now write down the derivative on the right and use Problem 1.20(b) to simplify the result to

$$d\left(\frac{pv}{kT}\right) = \frac{n}{kT}d\mu + \frac{1}{kT^2}(U-\mu n)\,dT + \frac{p}{kT}dv\,.$$

(d) This follows since the last result is an exact differential. See also Problem 20.4.

2.7 A system is kept at fixed chemical potential and temperature. Show that the logarithm of its grand partition function is proportional to the volume.

[Hint: The first result of Problem 2.6(c) is useful.]

Solution

From Problem 2.6(c) $x \equiv \ln\Xi$ satisfies

$$\left(\frac{\partial x}{\partial v}\right)_{\mu,\,T} = \frac{p}{kT} = \frac{x}{v}\,.$$

Hence for systems at constant μ and T

$$\frac{x}{v} = \text{const.}$$

Alternatively, if μ and T are constants, the Gibbs-Duhem relation of Problem 1.20(d) shows that p is constant. Hence

$$\ln\Xi = \frac{pv}{kT} \propto v\,.$$

2.8 The replacement of the partition functions Z and Ξ in Problem 2.6 by their thermodynamic equivalents, given in Problem 2.4, leads to thermodynamic results of some interest.

(a) Infer from

$$\left(\frac{\partial \ln \Xi}{\partial \mu}\right)_{v,\,T} = \frac{n}{kT}$$

that

$$n = v\left(\frac{\partial p}{\partial \mu}\right)_{v,\,T}.$$

(b) From

$$\left(\frac{\partial \ln \Xi}{\partial T}\right)_{v,\,\mu} = \frac{U-\mu n}{kT^2},$$

show that

$$S = k\left[\ln \Xi + T\left(\frac{\partial \ln \Xi}{\partial T}\right)_{v,\,\mu}\right].$$

Hence establish the relation

$$S = v\left(\frac{\partial p}{\partial T}\right)_{v,\,\mu}.$$

(c) Establish the thermodynamic relations as found in parts (a) and (b) by purely thermodynamic methods.

Solution

(a) Replacing $\ln \Xi$ by pv/kT we obtain

$$\frac{n}{kT} = \left[\frac{\partial(pv/kT)}{\partial \mu}\right]_{v,\,T} = \frac{v}{kT}\left(\frac{\partial p}{\partial \mu}\right)_{v,\,T}$$

as required.

(b) From the equation stated

$$\left(\frac{\partial \ln \Xi}{\partial T}\right)_{v,\,\mu} = \frac{TS-pv}{kT^2} = \frac{S}{kT} - \frac{pv}{kT^2} = \frac{S}{kT} - \frac{1}{T}\ln \Xi,$$

which is the first result stated under (b). Replacing $\ln \Xi$ as in part (a), we obtain

$$\begin{aligned} S &= \frac{pv}{T} + kT\left[\frac{\partial(pv/kT)}{\partial T}\right]_{v,\,\mu} \\ &= \frac{pv}{T} + vT\left[\frac{1}{T}\left(\frac{\partial p}{\partial T}\right)_{v,\,\mu} - \frac{p}{T^2}\right] = v\left(\frac{\partial p}{\partial T}\right)_{v,\,\mu}. \end{aligned}$$

(c) From the Gibbs-Duhem relation of Problem 1.20(d)

$$S\,\mathrm{d}T - v\,\mathrm{d}p + n\,\mathrm{d}\mu = 0$$

one finds

$$n = v\left(\frac{\partial p}{\partial \mu}\right)_T$$

and

$$S = v\left(\frac{\partial p}{\partial T}\right)_\mu.$$

ENTROPY MAXIMISATION. PROBABILITY DISTRIBUTIONS

2.9 A one-dimensional normal distribution of zero mean and standard deviation σ is given by

$$p(x) = (2\pi\sigma^2)^{-1/2} \exp\left(-\frac{x^2}{2\sigma^2}\right) \qquad -\infty < x < \infty .$$

(a) Show that its &ntropy is $\frac{1}{2}k\ln(2\pi e\sigma^2)$, where e is the base of the natural logarithms.

(b) Show that for given $\int_{-\infty}^{\infty} x^2 p(x)\,dx \equiv \sigma^2$ the normalised probability distribution having the largest &ntropy is the one-dimensional normal distribution.

Solution

(a) The &ntropy is

$$\begin{aligned} S &= -k\int p(x)\ln p(x)\,dx \\ &= -k\int p(x)\left(-\frac{\ln 2\pi\sigma^2}{2} - \frac{x^2}{2\sigma^2}\right)dx \\ &= \frac{k}{2}\ln(2\pi\sigma^2) + \frac{k}{2\sigma^2}\int x^2 p(x)\,dx \\ &= \tfrac{1}{2}k\ln(2\pi e\sigma^2) . \end{aligned}$$

(b) We must maximise with respect to arbitrary variations of $p(x)$ in the integrand of

$$f[p(x)] \equiv -k\int_{-\infty}^{\infty} p(x)\ln p(x)\,dx - \alpha\int_{-\infty}^{\infty} p(x)\,dx - \beta\int_{-\infty}^{\infty} x^2 p(x)\,dx$$

where α, β are Lagrangian multipliers. Hence

$$-k - k\ln p(x) - \alpha - \beta x^2 = 0 .$$

It follows that

$$p(x) = a\exp(-\beta x^2)$$

for the maximum. The normalisation yields[2]

$$a\int_{-\infty}^{\infty} \exp(-bx^2)\,dx = a\left(\frac{\pi}{\beta}\right)^{1/2} = 1 .$$

(2) $$\int_{-\infty}^{\infty} x^{2r}\exp(-\beta x^2)\,dx = \frac{1 \times 3 \times 5 \times \ldots (2r-1)}{2r}\left(\frac{\pi}{\beta^{2r+1}}\right)^{1/2} \quad \text{for } r = 1, 2, \ldots .$$

For $r = 0$ the result is $(\pi/\beta)^{1/2}$.

Also

$$\sigma^2 = a\int_{-\infty}^{\infty} x^2 \exp(-bx^2)\mathrm{d}x = \frac{a}{2}\left(\frac{\pi}{\beta^3}\right)^{1/2} = \frac{1}{2\beta}\,.$$

Hence

$$p(x) = \frac{1}{(2\pi\sigma^2)^{1/2}}\exp\left(-\frac{x^2}{2\sigma^2}\right).$$

under the conditions stated.

2.10 Let

$$\frac{1}{B_r} \equiv 2r^{1/r}\Gamma\left(1+\frac{1}{r}\right), \text{ where } \Gamma \text{ is the Gamma function,}$$

$$p(x) \equiv p(-x), \text{ a probability distribution,}$$

$$M_r \equiv \left[\int_{-\infty}^{\infty} |x|^r p(x)\mathrm{d}x\right]^{1/r} \quad (r > 0), \text{ the } r\text{th moment, and}$$

$$S[p(x)] = -k\int p\ln p\,\mathrm{d}x, \text{ the Entropy of probability distribution.}$$

Show that

$$M_r \geqslant B_r \exp\left\{\frac{S[p(x)]}{k} - \frac{1}{r}\right\}$$

and that the equality sign holds when

$$p(x) = \frac{B_r}{M_r}\exp\left(\frac{|x|^r}{rM_r^r}\right).$$

Solution

The following integral will be needed:

$$I_{rs} \equiv \int_0^{\infty} x^s \exp(-\beta x^r)\mathrm{d}x\,.$$

Put

$$y = x^r,\ \mathrm{d}x = \frac{1}{r}x^{-(r-1)}\mathrm{d}y = \frac{1}{r}y^{-(r-1)/r}\mathrm{d}y\,.$$

Then

$$\begin{aligned} I_{r,s} &= \frac{1}{r}\int_0^{\infty} y^p \exp(-\beta y)\mathrm{d}y \quad \left(p \equiv \frac{s+1}{r} - 1\right) \\ &= \frac{1}{r}\frac{\Gamma(p+1)}{\beta^{p+1}} \\ &= \frac{1}{r}\beta^{-(s+1)/r}\Gamma\left(\frac{s+1}{r}\right). \end{aligned}$$

(a) Maximise as in Problem 2.5 the integrand in

$$\mathrm{f} = \int p(x)[-k\ln p(x) - \alpha - \beta|x|^r]\,\mathrm{d}x\,,$$

whence

$$-k - k\ln p(x) - \alpha - \beta|x|^r = 0\,.$$

The distribution of largest Entropy compatible with the specifications is

$$p_0(x) = a\exp(-\beta|x|^r)\,.$$

The constants a, β can be identified by

$$1 = \int_{-\infty}^{\infty} p(x)\,\mathrm{d}x = 2a\int_0^{\infty}\exp(-\beta x^r)\,\mathrm{d}x = 2a\Gamma\left(1+\frac{1}{r}\right)\beta^{-1/r}\,,$$

$$M_r^r = 2a\int_0^{\infty} x^r\exp(-\beta x^r)\,\mathrm{d}x = \frac{2a}{r}\beta^{-(r+1)/r}\Gamma\left(\frac{r+1}{r}\right) = \frac{1}{r\beta}\,.$$

It follows that

$$a = \frac{1}{2\Gamma\left(\frac{r+1}{r}\right)}\frac{1}{M_r}r^{-1/r} = \frac{B_r}{M_r}\,,$$

and

$$p_0(x) = \frac{B_r}{M_r}\exp\left(-\frac{|x|^r}{rM_r^r}\right),$$

as required.

The Entropy of the distribution p_0 satisfies the relation

$$\frac{S_0}{k} - \frac{1}{r} = -\int_{-\infty}^{\infty} p_0(x)\left(\ln\frac{B_r}{M_r} - \frac{|x|^r}{rM_r^r}\right)\mathrm{d}x - \frac{1}{r} = \ln\frac{M_r}{B_r}\,.$$

It follows that for given M_r all other distributions have a smaller Entropy, i.e.

$$\frac{M_r}{B_r} = \exp\left(\frac{S_0}{k} - \frac{1}{r}\right) \geqslant \exp\left\{\frac{S[p(x)]}{k} - \frac{1}{r}\right\},$$

which is the required result.

MOST PROBABLE DISTRIBUTION METHOD

2.11 The states of a quantum-mechanical system are labelled by a complete set of quantum numbers. Suppose one of these determines its energy E_j and that the corresponding energy degeneracy is g_j $(j = 1, 2, \ldots)$. Consider an ensemble of N copies of this system in the sense of Problem 2.4. Let one State of such an ensemble be specified by the numbers $(n_1, n_2, \ldots)$, where n_j is the number of systems with energy E_j. The capital S is a reminder that a State of an ensemble is considered.

(a) Prove that the number of ways of realising this State is

$$G = \frac{N! g_1^{n_1} g_2^{n_2} \ldots}{n_1! n_2! \ldots} .$$

(b) If n is large enough, check from a book on special functions that **Stirling's approximation** holds:

$$n! = \Gamma(1+n) = n^n e^{-n} (2\pi n)^{1/2}$$

where Γ is the Gamma function and n is a positive integer.

(c) Make the **continuity assumption**(3) with regard to n, and define Gauss' Ψ-function by

$$\Psi(n) \equiv \frac{d}{dn} \Gamma(1+n) .$$

Prove from (b) that

$$\Psi(n) \approx \ln(n+\tfrac{1}{2}) .$$

Verify, by the use of tables of $\Psi(n)$, that for $n \geqslant 3$ this approximation holds with an error of less than 0·26%.

(d) Make an assumption of equal *a priori* probabilities of different ways of realising a State, and assume that the energy of the ensemble is given. Hence show, using part (c), that for the most probable State of the ensemble the probabilities n_j/N are given by

$$\frac{n_j}{N} = \frac{n_j^*}{N} \approx \frac{g_j e^{-\beta E_j}}{\sum_j g_j e^{-\beta E_j} + O(1/N)} ,$$

where β is a Lagrangian multiplier.

(e) Discuss the relation between this result and that of Problem 2.2.

Solution

(a) In any State of the ensemble, let us give systems with the same energy the same letter, and systems with different energies different letters. Hence $G(n_1, n_2, \ldots)$ gives the number of distinguishable arrangements of N letters, n_1 of one type, n_2 of another type, etc., and is equal to $N!/n_1!n_2!\ldots$ if the degeneracies are neglected. As a result of the degeneracy g_j, each of the G arrangements gives rise to a number of further arrangements equal to the number of ways of assigning n_j systems to g_j states, i.e. to $g_j^{n_j}$ arrangements. The result of part (a) is thus obtained.

(b) This matter is discussed in many books.

(3) The importance of this assumption and the usefulness of $\Psi(n)$ was pointed out in *Proc. Natl. Acad. Sci., U.S.*, **40**, 149 (1954).

(c) From part (b) and the definition of $\Psi(n)$, we have

$$\Psi(n) \approx \frac{d}{dn}[n\ln n - n + \tfrac{1}{2}\ln(2\pi n)] = \ln n + 1 - 1 + \tfrac{1}{2}n$$
$$\approx \ln n + \ln(1+\tfrac{1}{2}n) = \ln(n+\tfrac{1}{2}) .$$

From tables we find $\Psi(3) = 1{\cdot}2561$, $\ln(3{\cdot}5) = 1{\cdot}2528$, so that the error in $\Psi(3) \approx \ln(3{\cdot}5)$ is $0{\cdot}0033/1{\cdot}2561 = 0{\cdot}26\%$. It is less for $n > 3$.

(d) For the most probable State of the ensemble, we have to consider

$$f \equiv \ln G(n_1, n_2, ...) - \alpha \sum_j n_j - \beta \sum_j E_j n_j ,$$

where α and β are undetermined multipliers to take account of the conditions that N and E_0 are given, i.e.

$$\sum_j n_j = N , \qquad \sum_j E_j n_j = E_0 .$$

Instead of maximising G with respect to each n_j, it is more convenient to maximise $\ln G$. We have

$$\frac{\partial f}{\partial n_j} = \frac{\partial}{\partial n_j}[-\ln n_j! + n_j \ln g_j - \alpha n_j - \beta E_j n_j]$$
$$= -\Psi(n_j) + \ln g_j - \alpha - \beta E_j = 0 .$$

Hence, if $(n_1^*, n_2^*, ...)$ is the most probable State,

$$\Psi(n_j^*) = \ln[g_j \exp(-\alpha - \beta E_j] ,$$

i.e.

$$n_j^* = g_j \exp(-\alpha - \beta E_j) - \tfrac{1}{2} .$$

If these are M energy levels, α may be identified by summing over the energy levels to find

$$N + \tfrac{1}{2}M = \exp(-\alpha) \sum_j g_j \exp(-\beta E_j) = Z \exp(-\alpha) ,$$

where the canonical partition function has been introduced. It follows that

$$\frac{n_j^*}{N} = \left(1 + \frac{M}{2N}\right) \frac{g_j \exp(-\beta E_j)}{Z} - \frac{1}{2N} .$$

When the terms in $1/N$ are collected together, the desired result is obtained.

(e) The most probable State of the ensemble leads to the most probable values n_j^*. The canonical ensemble of Problem 2.2 yields the mean values of the n_j as averaged over all States of the ensemble. In this average the most probable State makes a dominant contribution, but less probable States will also contribute, and hence the different results which are, strictly speaking, obtained by the two methods.

The present method requires the n_j^* to be large enough for the assumption of continuity to be justifiable. This condition is difficult to fulfil unless $N \to \infty$. Although M is often also infinitely large the usual result follows only if $M/N \to 0$. These difficulties are often overlooked. It is particularly easy to overlook them if the coarser approximation $\Psi(n) \sim \ln n$ is used, which yields at once

$$\frac{n_j^*}{N} = \frac{g_j \exp(-\beta E_j)}{Z} .$$

This is the standard result of Problems 2.2 and 2.4 if we rewrite it for quantum states instead of for energy levels. It then becomes

$$\frac{n_i^*}{N} = \frac{\exp(-\beta E_i)}{Z} .$$

2.12[4] Develop the ideas of Problem 2.11 for an ensemble of N identical systems each of which has two non-degenerate states. Take N to be even, and specify a State of the ensemble by the number n (which is even) giving the difference between the number of systems in the lower energy state and the number in the upper energy state. Assume that N, but not the total energy, of the ensemble is given.

(a) For $n \ll N$ show that the number of distinct ways of realising a State n of the ensemble is, within the Stirling approximation,

$$G(n) = 2^N \left(\frac{2}{\pi N}\right)^{1/2} \exp\left(-\frac{n^2}{2N}\right) .$$

(b) Assuming that distinct ways of realising a State of the ensemble are equiprobable, show that the probability of a state n ($\ll N$) is

$$P(n) = A\left(\frac{2}{\pi N}\right)^{1/2} \exp\left(-\frac{n^2}{2N}\right)$$

where A is a normalisation constant. Determine its value, and hence verify that $P(n)$ is a one-dimensional normal distribution of zero mean and standard deviation $\sqrt{N}$ (defined in Problem 2.9).

(c) Compare the mean value of n and the most probable value of n.

(d) Compare the total number of States of the ensemble, G_T, with the number of ways $G(0)$ of realising the most probable State, as N becomes very large. Repeat this process for the corresponding entropies, and use the result to discuss the key properties of the method of the most probable distribution for this example.

[4] Problems 2.12 and 2.13 are of interest in various contexts. See, for example, J.E.Mayer and M.Goeppert Mayer, *Statistical Mechanics* (John Wiley, New York), 1940, p.75, and C.Kittel, *Elementary Statistical Physics* (John Wiley, New York), 1958, p.22.

Solution

(a) Let u refer to the upper level, and l to the lower level. Then

$$n_l + n_u = N, \qquad n_l - n_u = n,$$

whence

$$n_l = \tfrac{1}{2}(N+n), \qquad n_u = \tfrac{1}{2}(N-n).$$

Since N is even, it follows that n is even.

The number of ways of realising a state n is, from Problem 2.11(a) with $g_1 = g_2 = 1$,

$$G(n) = \frac{N!}{[\frac{1}{2}(N+n)]![\frac{1}{2}(N-n)]!}.$$

Hence using Problem 2.11(b)

$$\begin{aligned}\ln G(n) &= N\ln N - N + \tfrac{1}{2}\ln(2\pi N) - \tfrac{1}{2}(N+n)\ln[\tfrac{1}{2}(N+n)] + \tfrac{1}{2}(N+n) \\ &\quad - \tfrac{1}{2}\ln[\pi(N+n)] - \tfrac{1}{2}(N-n)\ln[\tfrac{1}{2}(N-n)] + \tfrac{1}{2}(N-n) - \tfrac{1}{2}\ln[\pi(N-n)] \\ &= N\ln 2 + \tfrac{1}{2}\ln\left(\frac{2}{\pi N}\right) - \tfrac{1}{2}(N+n+1)\ln\left(1+\frac{n}{N}\right) - \tfrac{1}{2}(N-n+1)\ln\left(1-\frac{n}{N}\right)\end{aligned}$$

Using $\ln(1+x) = x - \frac{1}{2}x$ for $x \ll 1$, we find

$$\ln G(n) = N\ln 2 + \tfrac{1}{2}\ln\left(\frac{2}{\pi N}\right) - \frac{n^2}{N} + \frac{n^2(N+1)}{2N^2},$$

whence the result follows.

(b) The number of States of the ensemble is 2^N since each of N systems can be in one of two states. Hence

$$P(n) = \frac{G(n)}{2^N}.$$

To test the normalisation, put $y = \frac{1}{2}(N+n)$, when $\frac{1}{2}(N-n) = N-y$. Then $y = 0$ at $n = -N$ and $y = N$ at $n = N$, and the corresponding values are given below:

y	0	1	...	$\frac{1}{2}N$	$\frac{1}{2}N+1$	...	N
n	$-N$	$-N+2$		0	2		N

Observe also that

$$(a+b)^N = \sum_{y=0}^{N} \frac{N!}{y!(N-y)!} a^y b^{N-y},$$

whence

$$\sum_{y=0}^{N} \frac{N!}{y!(N-y)!} = 2^N.$$

Using all these results, we can verify the correct normalisation of $P(n)$:

$$\sum_{\substack{n=-N \\ (\text{even } n)}}^{N} P(n) = 2^{-N} \sum_{\substack{n=-N \\ (\text{even } n)}}^{N} G(n) = 2^{-N} \sum_{y=0}^{N} \frac{N!}{y!(N-y)!} = 1\,.$$

Hence the value of A introduced in the problem is unity.

One may be tempted to check the normalisation by evaluating the integral

$$\int_{-\infty}^{\infty} P(n)\,\mathrm{d}n = \left(\frac{2}{\pi N}\right)^{\frac{1}{2}} \int_{-\infty}^{\infty} \exp\left(-\frac{n^2}{2N}\right) \mathrm{d}n = \left(\frac{2}{\pi N}\right)^{\frac{1}{2}} (2N\pi)^{\frac{1}{2}} = 2\,.$$

This suggests that the normalisation is incorrect. However, an approximate formula for $n \ll N$ has here been used for $-N \leqslant n \leqslant N$ as n and $N \to \infty$, and the fact that n takes only every other integral value has been ignored. This supplies a correction factor of $\frac{1}{2}$.

(c) The symmetry between the two states of the system gives $G(n) = G(-n)$, whence the mean value of n is

$$\sum_{\substack{n=-N \\ (\text{even } n)}}^{N} nP(n) = 2^{-N} \sum nG(n) = 0\,.$$

Direct algebraic study of the original expression for $G(n)$ in terms of factorials gives the most probable value of n as $n = 0$. This is also the value obtained from the expression given in part (b). Hence the mean and the most probable values coincide in this case.

(d)

$$G_{\mathrm{T}} = 2^N\,, \quad G(0) = \frac{N!}{[(\frac{1}{2}N)!]^2}$$

and one has

$$\mathrm{f}_1 \equiv \frac{G_{\mathrm{T}} - G(0)}{G_{\mathrm{T}}} = 1 - \frac{2}{(\pi N)^{\frac{1}{2}}} \to 1\,.$$

Also

$$S_{\mathrm{T}} = kN\ln 2, \quad S(0) \equiv k \ln G(0)\,.$$

Hence

$$\mathrm{f}_2 \equiv \frac{S_{\mathrm{T}} - S(0)}{S_{\mathrm{T}}} = \frac{\ln(\frac{1}{2}\pi N)}{2N\ln 2} \to 0\,.$$

Thus the value of f, which gives the fractional error in replacing all States by the most probable States in a calculation, is large in a calculation of the number of States, but small for the &ntropy. This circumstance is typical and shows that the &ntropy is a very *insensitive* function.

The probability $P(n)$ has a maximum at $n = 0$. This is not, however, a sufficient condition for an average over the ensemble to be approximated well by the most probable States. One must show that this maximum is also very steep. As a measure of the 'steepness' one could take $P(0)$ divided by the standard deviation. This ratio is

$$\frac{2^N\left(\frac{2}{\pi N}\right)^{1/2}}{N^{1/2}} = \left(\frac{2}{\pi}\right)^{1/2}\frac{2^N}{N}$$

and it indicates infinite 'steepness' as $N \to \infty$. The replacement of all States by the most probable States is therefore justifiable in this case.

2.13 Each of a system of N ($\gg 1$) weakly interacting indistinguishable atoms align themselves either parallel or antiparallel to the applied magnetic field H. Let n be the difference between the number of atoms in the lower level and the number of atoms in the upper level, and let μ be the magnetic moment of an atom, then a state of the system can be specified by the integer n. The energy of the system in a typical state n, referred to an energy zero at $n = 0$, is $E_n = -n\mu H$ and the magnetic moment is μn.

(a) Show by using Problem 2.12 that in the limit of zero magnetic field the number of arrangements which can give rise to a state $n \ll N$ is

$$G(n) = 2^N\left(\frac{2}{\pi N}\right)^{1/2}\exp\left(-\frac{n^2}{2N}\right).$$

(b) If the system is in equilibrium at temperature T show that the probability of finding a state n in an external magnetic field H is Gaussian with the mean value $\langle n\rangle$ and the standard deviation σ given by

$$\langle n\rangle = \frac{\mu HN}{kT}, \qquad \sigma = \sqrt{N}.$$

Solution

(a) The value of $G(n)$ is derived in Problem 2.12(a). However, there is a change in the point of view since only one system is contemplated. The number N is now the number of particles in this system. But the formula

$$G(n) = \frac{N!}{[\frac{1}{2}(N+n)]!\,[\frac{1}{2}(N-n)]!} \equiv \frac{N!}{n_1!\,n_2!}$$

still applies. In Problem 2.12(a) one divides by the factorial factors since it clearly does not matter to the specification of the State of the ensemble which of the systems are in state 1 and which are in state 2, so long as the number in each state is definite. In the present problem the division arises again because the atoms are assumed indistinguishable.

(b) We have for a canonical ensemble that the probability of state n is

$$P(n) = \frac{G(n)}{Z}\exp\left(\frac{-E_n}{kT}\right) = \frac{2^N}{Z}\left(\frac{2}{\pi N}\right)^{1/2}\exp[\mathrm{f}(n)],$$

where Z is the partition function and

$$\mathrm{f}(n) = \frac{n\mu H}{kT} - \frac{n^2}{2N}\,.$$

Writing $n_1 \equiv \mu HN/kT$, we now have

$$\mathrm{f}(n) = \frac{2nn_1 - n^2}{2N} = \frac{n_1^2}{2N} - \frac{(n-n_1)^2}{2N}\,,$$

so that

$$P(n) = A\exp\left[-\frac{(n-n_1)^2}{2N}\right].$$

Here A is a normalisation constant which depends on N.

SOME GENERAL PRINCIPLES

2.14 The states of a system form W groups labelled by the suffixes $i = 1, 2, \ldots, W$, the ith group having G_i equiprobable states. The probability that the system is in *any* of these states is P_i. Let the probability per unit time that the system makes a transition from a particular state of group i to a particular state of group j be a constant which we shall denote by A_{ij} (with $A_{ii} \equiv 0$).

(a) Prove that the transition rate from group i to group j is

$$R_{ij} = P_i A_{ij} G_j\,.$$

(The **principle of detailed balance** asserts $\mathbf{D}: R_{ij} = R_{ji}$ for all i, j.)

(b) Prove that the rate of change of a typical P_i can be written as

$$\dot{P}_i = \sum_j (R_{ji} - R_{ij}) = G_i \sum_j G_j A_{ji}\left(\frac{P_j}{G_j} - \frac{P_i}{G_i}\right) - P_i F_i\,,$$

where

$$F_i \equiv \sum_j (A_{ij} - A_{ji}) G_j\,.$$

(c) A system is in a **steady state** if the following principle holds: $\mathbf{S}: \dot{P}_i = 0$ for all i. Prove that $\mathbf{S}$ is satisfied if the following two principles hold: $\mathbf{X}$: $F_i = 0$; and $\mathbf{P}$: the suffixes i fall into classes $\alpha, \beta, \ldots$ such that within each class there holds $A_{ij} \neq 0$ if and only if $P_i/G_i = P_j/G_j = K_\alpha$, say ($i, j$ within αth group). Interpret this result.

(d) Show that the &ntropy is

$$S = -k\sum_i P_i \ln\frac{P_i}{G_i}$$

and prove that the &ntropy production rate satisfies $\mathbf{H}: \dot{S} \geqslant 0$ if the **principle of microscopic reversibility** $\mathbf{M}$ ($A_{ij} = A_{ji}$ for all i, j) holds.

(e) Verify that the implications proved under (c) and (d) belong to a wider scheme of implications given by

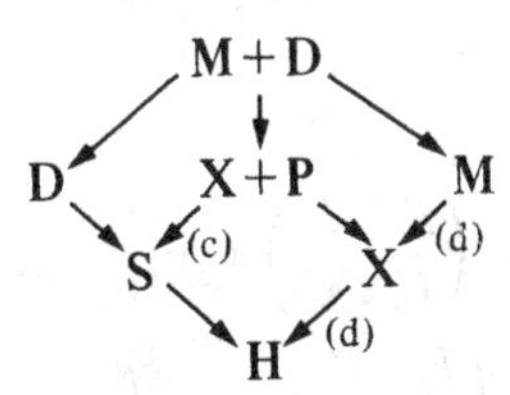

[The equation for $\dot{P}_i$ is sometimes called the **master equation** and was first discussed by W.Pauli in 1928. Its generalisations are currently topics of research. The principle **X** generalises the principle **M**, the principle **P** generalises the principle of equiprobability of states. Note that **S** is weaker than **D**, and these therefore are not equivalent principles. The subjects of master equations and detailed balance are pursued further in Chapter 26.]

Solution

(a) Multiply the probability of finding any state in group i by A_{ij}. This gives only the transition probability per unit time into a particular state of group j, and has to be multiplied by the factor G_j.

(b) We have

$$\dot{P}_i = \sum_j P_j A_{ji} G_i - \sum_j P_i A_{ij} G_j$$

$$= \sum_j P_j A_{ji} G_i - P_i F_i - P_i \sum_j A_{ji} G_j$$

$$= \sum_j G_j A_{ji} G_i \left(\frac{P_j}{G_j} - \frac{P_i}{G_i}\right) - P_i F_i \,.$$

(c) The principles **X** and **P** clearly imply $\dot{P}_i = 0$ if we take into account the results of part (b). If the principle **P** holds, the W groups of states decompose into a smaller number of classes of states between which transitions are not possible. The principle **S** is fulfilled if the probability per state has the same value *for each class of states*. If $F_i = 0$, the equiprobability *of all states* leads to detailed balance only if one assumes also that all states are *interconnected.*

(d) The entropy is

$$-k \sum_{\text{all states}} \frac{P_i}{G_i} \ln \frac{P_i}{G_i} \,.$$

Sum over all states in group i first. Their probabilities are all equal to P_i/G_i, so that the expression given in the problem is found.

Next we have that all F_i vanish, and

$$\dot{S} = -k \sum_i \left[\dot{P}_i \ln\left(\frac{P_i}{G_i}\right) + P_i \frac{G_i}{P_i} \cdot \frac{\dot{P}_i}{G_i} \right] .$$

Since $\sum_i \dot{P}_i = 0$ by normalisation, the last sum vanishes. Using part (b) we find

$$\dot{S} = -k\sum_{i,j} G_i G_j \left(\frac{P_j}{G_j} - \frac{P_i}{G_i}\right) A_{ji} \ln\frac{P_i}{G_i}$$
$$= -\tfrac{1}{2}k\sum_{i,j}\left[G_i G_j\left(\frac{P_j}{G_j} - \frac{P_i}{G_i}\right)A_{ji}\ln\frac{P_i}{G_i} + G_i G_j\left(\frac{P_i}{G_i} - \frac{P_j}{G_j}\right)A_{ji}\ln\frac{P_j}{G_j}\right]$$
$$= -\tfrac{1}{2}k\sum_{i,j} G_i G_j\left(\frac{P_j}{G_j} - \frac{P_i}{G_i}\right)A_{ji}\ln\frac{P_i G_j}{P_j G_i} \ .$$

If $P_i/G_i = P_j/G_j$ the (i, j) contribution to the double sum vanishes. If $P_i/G_i \neq P_j/G_j$ then the contribution is positive. Hence $\dot{S} \geqslant 0$.

(e) The proofs are simple[5].

2.15 (a) Show that, for all positive x, $\ln x \geqslant 1 - 1/x$, where the equality holds if, and only if, $x = 1$.

(b) Show from (a) that if $\{p_i\}$, $\{p_i^0\}$ are two probability distributions for the same set of states and such that $p_i^0 > 0$ for all i, then

$$K(p, p^0) \equiv k\sum_i p_i \ln\frac{p_i}{p_i^0}$$

is positive unless the distributions are identical ($p_i = p_i^0$ for all i).

[Suppose the p_i^0 are the equilibrium probabilities of a possibly non-isolated system, for example it could be the canonical distribution if the system is in contact with a heat reservoir. Then the knowledge that the actual distribution is (say) p_i under these conditions, represents an 'information gain' $K(p, p^0)$. Conversely, the passage from p_i to p_i^0 corresponds to an internally produced entropy $K(p, p^0)$. The quantity $K(p, p^0)$ was introduced by A. Rényi[6].]

Solution

(a) Let $y \equiv \ln x - 1 + 1/x$. Then

$$\frac{dy}{dx} = -\frac{1-x}{x^2} \ .$$

This is negative for $0 < x < 1$ and positive for $1 < x$. The least value of y occurs therefore for $x = 1$, so that $y \geqslant 0$.

(b) With $x_i \equiv p_i/p_i^0$,

$$K(p, p^0) = k\sum_i p_{i0} x_i \ln x_i \geqslant k\sum_i p_{i0} x_i\left(1 - \frac{1}{x_i}\right) = k\sum_i (p_i - p_i^0) = 0\ .$$

Hence K is non-negative. If all x_i's are unity, $K = 0$. If one $x_i \neq 1$ this term will contribute a positive quantity to K and $K > 0$.

[5] P.T.Landsberg, *Phys.Rev.*, **96**, 1420 (1954) and Section 34 of P.T.Landsberg, *Thermodynamics with Quantum Statistical Illustrations* (Interscience, New York), 1961.

[6] For recent discussions, see F.Schlögl, *J.Phys.Soc.Jap.*, **26** Supplement, 215 (1969).

3
Statistical mechanics of ideal systems

P.T.LANDSBERG
(*University College, Cardiff*)

The statistical &ntropy of Chapter 2 and the thermodynamic entropy of Chapter 1 are regarded as identical in the rest of this book.

MAXWELL DISTRIBUTION

3.1 (a) A system is specified by variables $x_1, x_2, ..., x_N$ which have independent normalised probability distributions $p_1(x_1), p_2(x_2), ..., p_N(x_N)$. Show that the entropy may be written as

$$S = -k\sum_{i=1}^{N}\left[\int p_i(x_i)\ln p_i(x_i)\,\mathrm{d}x_i\right].$$

(b) The x_i are interpreted as the three Cartesian velocity components V_i of a particle in a gas with point interactions (i.e. the particles are points and interact only if they are at the same point) in equilibrium at temperature T. Assuming that the mean kinetic energy associated with each component is $\frac{1}{2}kT$, show that for a state of maximum entropy

$$p_i(V_i) = \left(\frac{m}{2\pi kT}\right)^{\frac{1}{2}} \exp\left(-\frac{mV_i^2}{2kT}\right),$$

where m is the mass of a molecule.

(c) Derive from (b) the probability that a molecule in this gas has a speed in the range $(V, V+\mathrm{d}V)$ is $p(V)\mathrm{d}V$, where V can have any value from 0 to ∞ and

$$p(V) = 4\pi V^2\left(\frac{m}{2\pi kT}\right)^{\frac{3}{2}} \exp\left(-\frac{mV^2}{2kT}\right).$$

This is the **Maxwell velocity distribution.**

(d) Discuss the range of validity of these results by inspecting the assumptions needed to obtain them.

Solution

(a) The independence of the probability distributions $p_i(x_i)$ implies that the probability of finding a state $x_1, x_2, ..., x_n$ of the system is

$$P(x_1, x_2, ..., x_N) = \prod_{i=1}^{N} p_i(x_i).$$

Hence the entropy is

$$S = -k\int\ldots\int P\ln P\,\mathrm{d}\tau = -k\int\ldots\int p_1\ldots p_N \ln(p_1\ldots p_N)\,\mathrm{d}x_1\ldots\mathrm{d}x_N$$

$$= -k\sum_i\left(\int p_i \ln p_i\,\mathrm{d}x_i\right).$$

(b) The following constraints apply to the p_i's:

$$\int_{-\infty}^{\infty} p_i(V_i)\,\mathrm{d}V_i = 1, \qquad \int_{-\infty}^{\infty} p_i(\tfrac{1}{2}mV_i^2)\,\mathrm{d}V_i = \tfrac{1}{2}kT \quad (i = 1, 2, 3).$$

Maximising the entropy subject to these constraints as in Problem 2.2, consider

$$\mathrm{f} = -k\sum_{i=1}^{3}\left(\int p_i \ln p_i\,\mathrm{d}V_i + \alpha_i'\int p_i\,\mathrm{d}V_i + \beta_i\int p_i V_i^2\,\mathrm{d}V_i\right).$$

Hence

$$\frac{\partial \mathrm{f}}{\partial p_j} = -k\int\left(\ln p_j + \alpha_j' + \beta_j V_j^2 + 1\right)\mathrm{d}V_j = 0,$$

whence

$$p_j = \exp(-\beta_j V_j^2 - \alpha_j), \qquad \alpha_j \equiv \alpha_j' + 1.$$

To identify the Lagrangian multipliers observe, using the integral given in the solution of Problem 2.9, that

$$\int_{-\infty}^{\infty} \exp(-\beta_j V_j^2 - \alpha_j)\,\mathrm{d}V_j = 1, \qquad \text{i.e.}\ \left(\frac{\pi}{\beta_j}\right)^{1/2}\exp(-\alpha_j) = 1,$$

$$\int_{-\infty}^{\infty} V_j^2 \exp(-\beta_j V_j^2 - \alpha_j)\,\mathrm{d}V_j = \frac{kT}{m}.$$

This yields

$$\left(\frac{\pi}{\beta_j}\right)^{1/2}\exp(-\alpha_j) = 1 = \frac{m}{2kT\beta_j}\left(\frac{\pi}{\beta_j}\right)^{1/2}\exp(-\alpha_j),$$

so that

$$\exp(-\alpha_j) = \left(\frac{m}{2\pi kT}\right)^{1/2}, \qquad \beta_j = \frac{m}{2kT},$$

and the stated expression for p_j follows.

(c) On integrating the probability $p_1 p_2 p_3$ of a velocity with components in the ranges $(V_1, V_1 + \mathrm{d}V_1)$, $(V_2, V_2 + \mathrm{d}V_2)$, $(V_3, V_3 + \mathrm{d}V_3)$ over a shell of essentially positive radius $V = (V_1^2 + V_2^2 + V_3^2)^{1/2}$, we find the required probability

$$p(V)\,\mathrm{d}V = \left(\frac{m}{2\pi kT}\right)^{3/2}\exp\left(-\frac{mV^2}{2kT}\right)4\pi V^2\,\mathrm{d}V,$$

since

$$\iiint_{\text{shell}} \mathrm{d}V_1\,\mathrm{d}V_2\,\mathrm{d}V_3 = 4\pi V^2\,\mathrm{d}V.$$

(d) The main assumption is that $p_1(V_1)$, $p_2(V_2)$, $p_3(V_3)$ are independent probability distributions and that the point particles have only point interactions. In a dense gas the interactions cannot be approximated in this way. It has also been assumed that one particle can be treated separately from the rest. This is invalid for a system of indistinguishable particles when exchange effects must be expected.

3.2 Obtain the following quantities for a Maxwell distribution of velocities:

(a) The average of the nth power of the velocity is

$$\langle V^n \rangle = \frac{2}{\sqrt{\pi}} \left(\frac{2kT}{m} \right)^{n/2} \Gamma\left(\frac{n+3}{2} \right),$$

where n is real, $n > -1$, and Γ is the gamma function.

(b) The average speed is

$$\langle V \rangle = \left(\frac{8kT}{\pi m} \right)^{\frac{1}{2}}.$$

(c) The 'fluctuation' in speed is

$$\langle (V - \langle V \rangle)^2 \rangle = \frac{kT}{m}\left(3 - \frac{8}{\pi}\right).$$

(d) The 'fluctuation' in the kinetic energy is

$$(\tfrac{1}{2}m)^2 \langle (V^2 - \langle V^2 \rangle)^2 \rangle = \tfrac{3}{2}(kT)^2.$$

(e) The most probable speed is

$$V_0 = \left(\frac{2kT}{m} \right)^{\frac{1}{2}}.$$

Solution

(a)

$$\langle V^n \rangle = 4\pi \left(\frac{m}{2\pi kT} \right)^{3/2} \int_0^\infty V^{n+2} \exp\left(-\frac{mV^2}{2kT} \right) \mathrm{d}V$$

$$= 4\pi \left(\frac{m}{2\pi kT} \right)^{3/2} \frac{kT}{m} \left(\frac{2kT}{m} \right)^{(n+1)/2} \int_0^\infty x^{(n+1)/2} \mathrm{e}^{-x} \,\mathrm{d}x.$$

The integral is $\Gamma\left(\dfrac{n+3}{2}\right)$ and the result follows.

(b) Put $n = 1$ in (a).

(c) $$\langle V^2 \rangle = \frac{3kT}{m}$$

$$\langle (V - \langle V \rangle)^2 \rangle = \langle V^2 - 2V\langle V \rangle + (\langle V \rangle)^2 \rangle = \langle V^2 \rangle - (\langle V \rangle)^2 = \frac{3kT}{m} - \frac{8kT}{\pi m}.$$

(d) The 'fluctuation' is

$$\langle(\tfrac{1}{2}mV^2-\langle\tfrac{1}{2}mV^2\rangle)^2\rangle = (\tfrac{1}{2}m)^2\langle V^4-2V^2\langle V^2\rangle+(\langle V^2\rangle)^2\rangle$$
$$= (\tfrac{1}{2}m)^2[\langle V^4\rangle-(\langle V^2\rangle)^2]$$
$$= \tfrac{15}{4}k^2T^2-(\tfrac{3}{2}kT)^2 = \tfrac{3}{2}(kT)^2.$$

(e) The probability has the form

$$p = AV^2\exp\left(-\frac{mV^2}{2kT}\right).$$

Hence $dp/dV = 0$ yields

$$A\left(2V-V^2\frac{mV}{kT}\right)\exp\left(-\frac{mV^2}{2kT}\right) = 0.$$

The required result follows.

3.3 **Maxwell's original argument.** Let $4\pi V^2 g(V^2)dV$ be the probability of finding a molecule in the gas with velocity magnitude in the range $(V, V+dV)$. Here $g(V^2)$ is an unidentified differentiable function. Obtain the Maxwell velocity distribution on the assumption that the probability distribution for the three Cartesian components of a velocity vector are (a) independent, and (b) identical.

Solution

Let $f(V_x^2)$ be the probability distribution function for the x-component of the velocity. Then one can put

$$g(V^2) = g(V_x^2+V_y^2+V_z^2) = f(V_x^2)f(V_y^2)f(V_z^2).$$

It follows from these two expressions for $g(V^2)$ that

$$\left(\frac{\partial g}{\partial V_x^2}\right)_{V_y,V_z} = \frac{dg}{dV^2}\left(\frac{\partial V^2}{\partial V_x^2}\right)_{V_y,V_z} = \frac{dg}{dV^2}$$

and that

$$\left(\frac{\partial g}{\partial V_x^2}\right)_{V_y,V_z} = \frac{df(V_x^2)}{dV_x^2}f(V_y^2)f(V_z^2).$$

Hence

$$\frac{1}{g}\frac{dg}{dV^2} = \frac{1}{f(V_x^2)}\frac{df(V_x^2)}{dV_x^2} \quad (\equiv -\beta).$$

The quantity β can depend only on V_x, but similar results hold for the y- and z-components. Hence β is independent of the velocity components. It follows that

$$f(V_x^2) = \exp(\alpha-\beta V_x^2),$$

where α is a constant of integration. Finally

$$g(V^2) = \exp(3\alpha-\beta V^2).$$

Normalisation of the distribution according to

$$\int_0^\infty 4\pi V^2 g(V^2)\mathrm{d}V = 1$$

leads back to the Maxwell distribution.

CLASSICAL STATISTICAL MECHANICS

3.4 The canonical partition function in **classical statistical mechanics** is given by an integral over a phase space instead of a sum over quantum states. In fact, each state is specified by a point in the space in which generalised coordinates $q_1, ..., q_f$ and generalised momenta $p_1, ..., p_f$ are the axes. A factor gh^{-f}, where h is a constant with the dimension of action, reduces Z again to a dimensionless number and allows each state to be g-fold degenerate. If H is the Hamiltonian of the system, the probability of finding the system in an element $\mathrm{d}\tau$ of phase space is proportional to $\exp(-H/kT)\mathrm{d}\tau$ and

$$Z = gh^{-f}\int ... \int \exp\left(-\frac{H}{kT}\right)\mathrm{d}p_1...\mathrm{d}q_f.$$

(a) Show that for one particle of mass m moving classically and non-relativistically in a field-free container of volume v (with $g = 1$)

$$Z_1 = h^{-3}v(2\pi mkT)^{3/2},$$

where T is the temperature of this system.

(b) For n distinguishable particles moving independently but with point interactions as in (a), show that one would expect classically

$$Z_n = Z_1^n.$$

Explain the qualitative effect on Z_n if the particles are indistinguishable.

(c) Writing $Z_n = (Z_1 f_n)^n$ for generality, where f_n depends only on n, consider the restrictions on f_n if the Helmholtz free energy of the system is to be an extensive quantity (cf Problem 1.20). In particular, show that for large n the classical partition function $(Z_n)_{\mathrm{cl}}$ satisfies

$$Z_n = (Z_n)_{\mathrm{cl}}/n!$$

(d) Obtain Boyle's law for the system specified above and show that it does not depend on the values of g, h, and f_n.

[The correction factor f_n is needed because particles are **indistinguishable** in a simple gas. It has an interesting history. The **quantum-statistical approach** has no need for such corrections. See Problem 3.12.]

Solution

(a) Take the potential energy of the particle as zero. Then

$$H = \frac{p^2}{2m}.$$

The integrations over the three coordinates, conveniently taken to be Cartesian coordinates, yields v. Hence

$$Z_1 = h^{-3}v\int_{-\infty}^{\infty}\int_{-\infty}^{\infty}\int_{-\infty}^{\infty}\exp\left(-\frac{p_1^2+p_2^2+p_3^2}{2mkT}\right)\mathrm{d}p_1\mathrm{d}p_2\mathrm{d}p_3$$

$$= h^{-3}v(2mkT)^{3/2}\left[\int_{-\infty}^{\infty}\exp(-x^2)\mathrm{d}x\right]^3$$

$$= h^{-3}v(2\pi mkT)^{3/2}.$$

(b) In this case if K_i is the kinetic energy of the ith particle

$$Z_n = h^{-3n}v^n\int\exp\left(-\frac{K_1+K_2+\cdots+K_n}{kT}\right)\mathrm{d}p_1 \dots \mathrm{d}p_{3n}$$

$$= h^{-3n}v^n\left[\int\exp\left(-\frac{K_1}{kT}\right)\mathrm{d}p_1\,\mathrm{d}p_2\,\mathrm{d}p_3\right]^n = Z_1^n.$$

For indistinguishable particles the value of Z_n must be smaller. For example, if $n = 2$, the above integral treats $K_1 = 1$ eV and $K_2 = 2$ eV as contributing equally with $K_1 = 2$ eV and $K_2 = 1$ eV. For indistinguishable particles there can be only one such contribution.

(c) From Problem 2.4

$$F_n = -kT\ln Z_n = -nkT\ln(Z_1 f_n)$$

$$= -kTn[\ln(vf_n)+\tfrac{3}{2}\ln T+\tfrac{3}{2}\ln(2\pi mk/h^2)].$$

F_n, n, and v are extensive; T and also constant terms can be regarded as intensive. It follows that f_n cannot be unity; instead it must make vf_n intensive. So, if D is a constant,

$$f_n = \frac{D}{n}.$$

It follows that the correction factor for Z_n is

$$f_n^n = \left(\frac{D}{n}\right)^n,$$

and this is for large n approximately $1/n!$ if D is interpreted as the base of the natural logarithms.

(d) It follows from the solution of Problem 1.7(a), which dealt with a closed system $n =$ constant, that the pressure is

$$p = -\left(\frac{\partial F}{\partial v}\right)_{T,n} = \frac{kTn}{v}.$$

3.5 We are retaining the notation of Problem 3.4. A classical statistical mechanical system is at temperature T and has Hamiltonian

$$H = K(p_1, \dots, p_f)+M(q_1, \dots, q_f),$$

where K is the kinetic energy and M is the potential energy. Let $P_p\,\mathrm{d}p_1...\mathrm{d}p_f$ be the probability of finding the momenta in the ranges $(p_i, p_i+\mathrm{d}p_i)$ and let $P_q\,\mathrm{d}q_1...\mathrm{d}q_f$ be corresponding probabilities for the coordinates.

(a) If B and C are normalisation constants, prove that

$$P_p = B\exp(-K/kT),$$
$$P_q = C\exp(-M/kT).$$

(b) Check the P_p-formula by applying it to a simple gas of n distinguishable particles of mass m whose momenta can range from $-\infty$ to $+\infty$ and show that

$$B = (2\pi mkT)^{-3/2}.$$

Also check that the Maxwell velocity distribution of Problems 3.1, 3.2, and 3.3 is obtainable by this method.

(c) Using Problem 3.4, show that the canonical partition function is for $g = 1, f = 3n$

$$Z_n = \left(\frac{2\pi mkT}{h^2}\right)^{3n/2}\frac{Q_n}{n!},$$

provided that $K = \sum_{i=1}^{n} p_i^2/2m$ and that the so-called **configurational integral or partition function** is

$$Q_n \equiv \int\!...\!\int \exp\left(-\frac{M}{kT}\right)\mathrm{d}q_1...\mathrm{d}q_{3n},$$

where the integration is over the volume of the fluid.

Show also that the grand partition function (Problem 2.4) satisfies

$$\Xi = \exp\frac{pv}{kT} = \sum_{n=0}^{\infty}\frac{Q_n z^n}{n!},$$

where the so-called fugacity is

$$z \equiv \left(\frac{2\pi mkT}{h^2}\right)^{3/2}\exp\frac{\mu}{kT}.$$

(d) A right circular cylinder of base area A, of great height, and at uniform temperature T contains n particles each of mass m. They are acted upon by a gravitational acceleration g, independent of height. Use the P_q-formula to show that the particle concentration at height z is given by the **barometer-formula**

$$\rho(z) = \frac{mgn}{AkT}\exp\left(-\frac{mgz}{kT}\right).$$

If the law $pv = nkT$ holds at all levels show that the pressure variation is

$$p(z) = \frac{mgn}{A}\exp\left(-\frac{mgz}{kT}\right).$$

[The configurational integrals for real fluids are studied in Problems 9.2 to 9.7.]

Solution

(a) To obtain P_p integrate the probability distribution

$$A \exp\left(-\frac{K+M}{kT}\right) \mathrm{d}q_1 \ldots \mathrm{d}p_f$$

(A is a normalisation constant) over all q's to find the stated result with a new normalisation constant B. A similar argument leads to P_q.

(b) Put $f = 3n$ and $K = \sum_{i=1}^{3n} \frac{p_i^2}{2m}$. It follows that

$$B^{-1} = \left[\int_{-\infty}^{\infty} \exp(-ap^2)\,\mathrm{d}p\right]^{3n} = (2mkT)^{3n/2}\left[\int_{-\infty}^{\infty} \exp(-x^2)\,\mathrm{d}x\right]^{3n}$$

where $a = 1/2mkT$. The integral is $\sqrt{\pi}$ and this yields the required result. Next take $n = 1$ to find

$$P_p\,\mathrm{d}p_1\mathrm{d}p_2\mathrm{d}p_3 = \frac{m^3}{(2\pi mkT)^{3/2}} \exp\left[-\frac{m(V_1^2+V_2^2+V_3^2)}{2kT}\right] \mathrm{d}V_1\mathrm{d}V_2\mathrm{d}V_3$$

$$\equiv P_{V_1V_2V_3}\,\mathrm{d}V_1\mathrm{d}V_2\mathrm{d}V_3 \equiv P_V\,\mathrm{d}V.$$

Since $\mathrm{d}V_1\mathrm{d}V_2\mathrm{d}V_3 = 4\pi V^2\mathrm{d}V$, one finds

$$P_V = 4\pi V^2\left(\frac{m}{2\pi kT}\right)^{3/2} \exp\left(-\frac{mV^2}{2kT}\right).$$

(c) From Problem 3.4(c)

$$Z_n = \frac{1}{n!}\frac{1}{h^{3n}} Q_n \int \ldots \int \exp\left(-\frac{K}{kT}\right) \mathrm{d}p_1 \ldots \mathrm{d}p_f.$$

This is integrated as in part (b). From Problem 2.4

$$\Xi = \sum_{n=0}^{\infty} Z_n \exp\frac{\mu n}{kT} = \sum_{n=0}^{\infty} \frac{1}{n!}\left(\frac{2\pi mkT}{h^2}\right)^{3n/2} Q_n \exp\frac{\mu n}{kT}.$$

This is the required result.

(d) The probability that particle 1 is in the range $(q_1, q_1+\mathrm{d}q_1)$, particle 2 in $\mathrm{d}q_2$, etc., is

$$P_{q_1q_2}\ldots\mathrm{d}q_1\mathrm{d}q_2\ldots\mathrm{d}q_n = C\exp\left(-\frac{mg}{kT}\sum_{i=1}^{n} q_i\right) \mathrm{d}q_1\ldots\mathrm{d}q_n$$

$$= \prod_{i=1}^{n}\left[\frac{\exp(-mgq_i/kT)\,\mathrm{d}q_i}{\int_0^{\infty} \exp(-mgq_i/kT)\,\mathrm{d}q_i}\right] = \prod_{i=1}^{n}\left[\frac{mg}{kT}\exp\left(-\frac{mgq_i}{kT}\right)\mathrm{d}q_i\right].$$

Hence the probability of being in the range $(z, z+\mathrm{d}z)$ for any one particle is

$$P_z \,\mathrm{d}z = \frac{mg}{kT}\exp\left(-\frac{mgz}{kT}\right)\mathrm{d}z.$$

To find the number of particles in range $\mathrm{d}z$, we must multiply by n, and to obtain the concentration we must divide the result by $A\,\mathrm{d}z$. The stated result is then found. To obtain the pressure at level z, put

$$p(z) = \frac{n(z)}{v}kT = \rho(z)kT\,.$$

3.6 Let r_i be any of the generalised momenta $p_1, \ldots, p_f$ or coordinates $q_1, \ldots, q_f$. Suppose that it can range between the values a and b in a certain system and that $a = 0$ or $H(a) = \infty$ or both, and $b = 0$ or $H(b) = \infty$ or both. Let $\langle\,\rangle$ denote an average over the classical canonical distribution discussed in Problems 3.4 and 3.5. This yields a number of **equipartition theorems.**

(a) Prove that[1]

$$\left\langle r_i \frac{\partial H}{\partial r_i}\right\rangle = kT.$$

(b) If

$$H = \sum_{i=1}^{\alpha} a_i p_i^r + \sum_{j=1}^{\beta} b_j q_j^r\,,$$

show that

$$\langle H\rangle = \frac{(\alpha+\beta)kT}{r}$$

(c) For a classical relativistic gas of particles with point interactions, $H = c(p_1^2+p_2^2+p_3^2+m_0^2c^2)^{1/2}$ applies to one particle with rest mass m_0. Prove that

$$\left\langle\frac{p_j^2}{2m}\right\rangle = \left\langle\frac{mV_j^2}{2}\right\rangle = \tfrac{1}{2}kT,$$

where $m = \beta m_0$ and $\beta \equiv (1 - V^2/c^2)^{-1/2}$.

[In quantum statistics these theorems hold only in the **classical limit**; see Problem 3.9.]

Solution

(a) The normalisation integral for the canonical distribution is

$$1 = A\int\!\ldots\!\int \exp\left(-\frac{H}{kT}\right)\mathrm{d}q_1\ldots\mathrm{d}p_f.$$

[1] In this general form the result is due to R.C.Tolman.

Integrate partially with respect to q_1:

$$1 = A\int\ldots\int \left| q_1 \exp\left(-\frac{H}{kT}\right)\right|_a^b \mathrm{d}q_2 \ldots \mathrm{d}p_f$$

$$+\frac{A}{kT}\int\ldots\int q_1 \frac{\partial H}{\partial q_1}\exp\left(-\frac{H}{kT}\right)\mathrm{d}q_1\ldots\mathrm{d}p_f.$$

The first integral does not contribute by hypothesis and the result follows.

(b) It follows from part (a) that each quadratic term in the energy contributes an amount kT/r.

(c) In this case

$$kT = p_j \frac{\partial H}{\partial p_j} = p_j \frac{c}{2}\frac{c}{H} 2p_j = \frac{c^2 p_j^2}{H}.$$

The right-hand side is, from the principles of relativistic mechanics,

$$\frac{\beta^2 m_0^2 V_j^2 c^2}{c(\beta^2 m_0^2 V^2 + m_0^2 c^2)^{½}} = \frac{\beta m_0^2 V_j^2 c}{(m_0^2 V^2 + m_0^2 c^2/\beta^2)^{½}} = \frac{\beta m_0^2 V_j^2 c}{(m_0^2 V^2 + m_0^2 c^2 - m_0^2 V^2)^{½}}$$

$$= \beta m_0 V_j^2 = m V_j^2 = \frac{p_j^2}{m}.$$

VIRIAL THEOREM

3.7 A particle i has coordinate $\mathbf{r}_i = (q_{ix}, q_{iy}, q_{iz})$ when the force acting on it is $\mathbf{F}_i = \mathrm{d}\mathbf{p}_i/\mathrm{d}t$, where $\mathbf{p}_i = (p_{ix}, p_{iy}, p_{iz})$. The **virial** of a system of n particles is $C \equiv -\frac{1}{2}\sum_{i=1}^{n} \overline{\mathbf{F}_i \cdot \mathbf{r}_i}$, where the bar denotes a time average.

(a) Assuming (i) the **Hamiltonian equations of motion** ($\mathrm{d}q_{ij}/\mathrm{d}t = \partial H/\partial p_{ij}$, $\mathrm{d}p_{ij}/\mathrm{d}t = -\partial H/\partial q_{ij}$; $i = 1, 2, \ldots, n$; $j = x, y, z$), and (ii) the **ergodic hypothesis** that ensemble and time averages yield identical results, prove that $C = \frac{3}{2}nkT$.

(b) If the forces are derivable from a potential W, $F_{ij} = -\partial W/\partial q_{ij}$, and the momenta are involved only in a kinetic energy of the form $K = \sum_{i=1}^{n} p_i^2/2m$, prove that

$$\overline{K} = \frac{1}{2}\sum_i \overline{\nabla W \cdot \mathbf{r}_i} = \frac{3}{2}nkT.$$

(c) For a single particle ($n = 1$) under a central force of potential energy $W = ar^u$ prove that

$$\overline{K} = \frac{u}{2}\overline{W} = \frac{u}{2+u}\overline{E},$$

where E denotes the total energy.

[The virial theorem $\overline{K} = C$ is due to Clausius, and can be derived independently of the equipartition theorem. This is demonstrated in Problem 9.16, which deals with closely related questions.]

Solution

(a) From Problem 3.6

$$\left\langle q_{ij}\frac{\partial H}{\partial q_{ij}}\right\rangle = \left\langle -q_{ij}F_{ij}\right\rangle = kT.$$

This becomes $C = \frac{3}{2}nkT$.

(b) The above result is

$$\frac{1}{2}\sum_{i,j}\overline{\left(\frac{\partial W}{\partial q_{ij}}q_{ij}\right)} = \frac{3}{2}nkT.$$

Also

$$\overline{K} = \sum_i \overline{\left(\frac{p_i^2}{2m}\right)} = \frac{1}{2}\sum_{i,j}\overline{\left(p_{ij}\frac{\partial H}{\partial p_{ij}}\right)} = \frac{3}{2}nkT.$$

(c) Use the fact that

$$\nabla W = r\frac{\mathrm{d}W}{\mathrm{d}r} = uW.$$

Also

$$\overline{E} = \overline{K}+\overline{W} = \left(1+\frac{2}{u}\right)\overline{K}.$$

3.8 The particles of a gas interact with forces $f(|\mathbf{r}_j-\mathbf{r}_k|) \equiv f(r_{jk})$ which depend only on the distance between the particles. For the results to be established, use the assumptions and inferences given in Problem 3.7.

(a) Show that the interaction forces contribute $-\frac{1}{2}\sum\overline{r_{jk}f(r_{jk})}$ to the virial, where the sum extends over all pairs of particles.

(b) Show that the force exerted by a container of volume v on a gas at pressure p contributes $\frac{3}{2}pv$ to the virial.

(c) Show that for a classical imperfect gas of n particles at temperature T

$$pv = nkT+\frac{1}{3}\sum_{\text{(pairs)}} r_{jk}f(r_{jk}).$$

(d) Establish the virial theorem

$$(u+2)\overline{K} = u\overline{E}+3pv$$

for a gas whose interaction forces are derived from a potential energy which is a homogeneous function of order u in the coordinates.

Solution

(a)

$$\mathbf{F} \leftarrow \text{- - - - -} \rightarrow \mathbf{G}$$

$\mathbf{r}_j$ $\mathbf{r}_k$

We have, taking $f(r_{jk})$ as positive for repulsive forces,

$$\mathbf{F} = \frac{\mathbf{r}_j - \mathbf{r}_k}{r_{jk}} f(r_{jk}), \qquad \mathbf{G} = -\mathbf{F}.$$

The contribution to the virial due to the pair (j, k) is

$$-\tfrac{1}{2}(\mathbf{r}_j \cdot \mathbf{F} - \mathbf{r}_k \cdot \mathbf{F}) = -\tfrac{1}{2}(\mathbf{r}_j - \mathbf{r}_k) \cdot \mathbf{F} = -\tfrac{1}{2} r_{jk} f(r_{jk}) \, .$$

Hence the result.

(b) The force exerted by the container on an element of area da is $-p\mathbf{n}\,da$, where $\mathbf{n}$ is the unit outwards drawn normal. The contribution to the virial is

$$\tfrac{1}{2}p \int_s \mathbf{n} \cdot \mathbf{r}\, da = \tfrac{1}{2} \int_v \operatorname{div}\mathbf{r}\, dv = \tfrac{3}{2}pv,$$

where Gauss's theorem and $\operatorname{div}\mathbf{r} = 3$ has been used.

(c) We have from Problem 3.7 and the above results

$$C = \bar{K} = \tfrac{3}{2}nkT = \tfrac{3}{2}pv - \tfrac{1}{2} \sum_{\text{(pairs)}} \overline{[r_{jk} f(r_{jk})]} \, .$$

(d) In this case $r_{jk} f(r_{jk}) = -r_{jk}(\partial W / \partial r_{jk}) = -uW$. It follows that

$$\bar{K} = \tfrac{3}{2}pv + \tfrac{1}{2}u\bar{W}.$$

Multiply by 2 and add $u\bar{K}$ to both sides to find

$$(u+2)K = 3pv + u\bar{E}.$$

OSCILLATORS AND PHONONS

3.9 The state of a $[w]$-**oscillator** of angular frequency $\omega = 2\pi\nu$ is specified by w quantum numbers $\mathbf{n} = (n_1, \ldots, n_w)$, its energy in this state being

$$E(\mathbf{n}) = [(n_1+\tfrac{1}{2}) + (n_2+\tfrac{1}{2}) + \ldots + (n_w+\tfrac{1}{2})]\hbar\omega.$$

The n_i are positive integers or zero and $\hbar$ is **Planck's constant** divided by 2π.

(a) Show that the energies may be expressed in the form

$$E_j = (\tfrac{1}{2}w + j)\hbar\omega, \qquad j = 0, 1, 2, \ldots$$

the degeneracy of this level being

$$g_j = \frac{(w+j-1)!}{j!(w-1)!} \, .$$

Discuss some simple special cases.

(b) Prove that the canonical partition function of one such quantum oscillator at temperature T is

$$(Z_1)_q = \left(\frac{\exp(-x)}{1-\exp(-2x)}\right)^w = \left(\frac{1}{2\sinh x}\right)^w,$$

where $x \equiv \hbar\omega/2kT$. Hence show that its mean energy is

$$(U_1)_q = \hbar\omega w[\tfrac{1}{2}+(\exp 2x-1)^{-1}].$$

(c) Obtain expressions for the classical partition function and mean energy of a one-dimensional oscillator of mass m, angular frequency ω, and temperature T, according to the prescription of Problem 3.4, given the Hamiltonian

$$H(p, q) = ap^2+bq^2 \qquad (-\infty < p, q < \infty),$$

where $a \equiv 1/2m$, $b \equiv \frac{1}{2}m\omega^2$. Check that $(Z_1)_q$, $(U_1)_q$ go over into these expressions in the limit $T \to \infty$ (**classical limit**).

(d) Prove that N identical and distinguishable $[w]$-oscillators are equivalent to Nw [1]-oscillators.

(e) Show that the heat capacity of N identical, distinguishable quantum [1]-oscillators is $(C_v)_q = NkE(2x)$, where

$$E(y) \equiv \frac{y^2 \exp y}{(\exp y-1)^2} \qquad \text{(Einstein function)}.$$

[Part (c) is an independent check on the equipartition theorem in a special case.]

Solution

(a) By rewriting $E(\mathbf{n})$ as E_j we make the substitution $\sum_{i=1}^{w} n_i = j$. Clearly j can assume all positive integral values or zero. Given one of these, the degeneracy g_j of this level is given by the following number theory problem: g_j is the number of ways of expressing an integer j as a sum of w integers, zero and repetitions being allowed and order being important. This in turn is clearly the same as the number of ways of placing j indistinguishable balls into w distinguishable boxes. This number is $(w+j-1)!/j!(w-1)!$.[2]

Note that the ground state $j = 0$ is always non-degenerate, $g_0 = 1$. Also all states of the [1]-oscillator (i.e. the one-dimensional oscillator) are non-degenerate.

(b) The quantum-statistical partition function is

$$(Z_1)_q = \exp(-wx)\sum_{j=0}^{\infty} g_j \exp(-2jx) = \exp(-wx)\sum_{j=0}^{\infty}\frac{(w+j-1)!}{j!(w-1)!}\exp(-2jx).$$

[2] P.T.Landsberg, *Thermodynamics with Quantum Statistical Illustrations* (Interscience, New York), 1961, p.447.

The binomial theorem

$$(1+a)^q = 1+qa+\frac{q(q-1)a^2}{2!}+\cdots+\frac{q(q-1)\ldots(q-j+1)a^j}{j!}+\cdots$$

yields

$$(1-a)^{-w} = 1+wa+\frac{w(w-1)a^2}{2!}+\cdots+\frac{w(w+1)\ldots(w+j-1)a^j}{j!}+\cdots$$

$$= \sum_{j=0}^{\infty}\frac{(w+j-1)!}{j!(w-1)!}a^j.$$

Hence

$$(Z_1)_q = \exp(-wx)[1-\exp(-2x)]^{-w}.$$

It follows from Problem 2.6(b) that the internal energy is

$$(U_1)_q = kT^2\left(\frac{\partial \ln Z_1}{\partial T}\right)_v = -\frac{\hbar\omega}{2}\frac{\partial \ln Z_1}{\partial x},$$

and this gives the stated results.

(c) We have

$$(Z_1)_{cl} = h^{-1}\iint_{-\infty}^{\infty}\exp\left(-\frac{a}{kT}p^2-\frac{b}{kT}q^2\right)\mathrm{d}p\,\mathrm{d}q$$

$$= h^{-1}\left(2\pi mkT\frac{2\pi kT}{m\omega^2}\right)^{1/2} = \frac{kT}{h\nu}.$$

The limit $T\to\infty$ or $x\to 0$ of $(Z_1)_q$ is for $w = 1$

$$(Z_1)_q \to \frac{1}{2x} = (Z_1)_{cl}.$$

Also classically

$$(U_1)_{cl} = kT^2\frac{\partial \ln Z}{\partial T} = kT^2Z^{-1}\frac{\partial Z}{\partial T}$$

$$= kT^2\frac{\hbar\omega}{kT}\frac{k}{\hbar\omega} = kT.$$

From part (b) we have

$$(U_1)_q \to \hbar\omega\frac{1}{2x} = kT = (U_1)_{cl}.$$

(d) The correction for indistinguishability (Problem 3.4) does not apply and

$$Z_N = Z_1^N = \left(\frac{1}{2\sinh x}\right)^{Nw}$$

is the partition function of N $[w]$-oscillators. It is equal to the partition function of Nw [1]-oscillators.

(e) This result follows from

$$C_v = \frac{\partial U}{\partial T} = -\frac{N\hbar\omega}{2kT^2}\frac{\partial U}{\partial x}.$$

3.10 Consider a basic set of r points in space called a **unit cell**. If it is repeated N times it yields a periodic set of points called a **lattice**. If atoms are placed on the rN points, one has an idealised representation of a **crystal**. If each atom is a [3]-oscillator and if its internal structure is disregarded, one has an ideal crystal with $3rN$ degrees of freedom. Each degree of freedom gives rise to a so-called **mode of oscillation** $(\mathbf{q}, s)$. The **wave vector** $\mathbf{q}$ assumes N values; the **polarisation index** s takes on $3r$ values.

(a) Prove that the (canonical) average for temperature T of the quantum number $n(\mathbf{q}, s)$ of an oscillator of mode $(\mathbf{q}, s)$ is $[\exp(\hbar\omega/kT)-1]^{-1}$, ω being its angular frequency.

(b) Show that the energy of the ideal crystal can be expressed as

$$E_z + E_{th} \equiv \sum_{\mathbf{q}, s} \tfrac{1}{2}\hbar\omega(\mathbf{q}, s) + \sum_{\mathbf{q}, s} \hbar\omega(\mathbf{q}, s)\langle n(\mathbf{q}, s)\rangle,$$

where E_z occurs at all temperatures (zero-point energy) and E_{th} is the thermal energy and vanishes at $T = 0$.

(c) Let θ, ϕ specify the direction of $\mathbf{q}$, let $d\Omega = d(\cos\theta)\,d\phi$ be an element of solid angle in $\mathbf{q}$-space, and note that $v/8\pi^3$ is the number of wave vectors per unit volume of $\mathbf{q}$-space in a crystal of volume v for each polarisation. Replacing the $\mathbf{q}$-sum by an integration obtain the formal expression

$$E_{th} = \frac{vkT}{8\pi^3}\sum_s \iiint \frac{y}{\exp y - 1} q^2\,dq\,d\Omega \qquad \left(y \equiv \frac{\hbar\omega}{kT}\right),$$

where q is the length of $\mathbf{q}$, and the surface $q = q_0(\theta, \phi)$ up to which the integrations are carried out satisfies

$$\frac{v}{8\pi^3}\sum_s \int q^2\,dq\,d\Omega = 3rN.$$

[Instead of an oscillator in state $n(\mathbf{q}, s)$ one can think of $n(\mathbf{q}, s)$ **quanta** of excitation with wave number $\mathbf{q}$ and polarisation s. They are related to sound waves as photons are to light waves, and are called **phonons**. More generally, excitations or particles whose mean occupation numbers per quantum state satisfy (a) are called **bosons** of zero chemical potential. But see Problem 3.12 for a generalisation to arbitrary chemical potential.]

Solution

(a) Put

$$\begin{aligned} Z &= \sum_{n=0}^{\infty} \exp[-(n+\tfrac{1}{2})\beta\hbar\omega] \\ &= (1+a+a^2+\cdots)\exp(-\tfrac{1}{2}\beta\hbar\omega) \qquad a \equiv \exp(-\beta\hbar\omega) \\ &= \frac{\exp(-\tfrac{1}{2}\beta\hbar\omega)}{1-a}. \end{aligned}$$

Then

$$\ln Z = -\tfrac{1}{2}\beta\hbar\omega - \ln[1-\exp(-\beta\hbar\omega)].$$

Now

$$\langle n\rangle = Z^{-1}\sum_{n=0}^{\infty} n\exp[-(n+\tfrac{1}{2})\beta\hbar\omega] = -\frac{1}{\hbar\omega}\frac{\partial\ln Z}{\partial\beta} - \frac{1}{2}$$

$$= \frac{1}{[\exp(\beta\hbar\omega)-1]}\,.$$

(b) This follows from Problem 3.9.

(c) The angular frequency depends on **q** and s, i.e. on (q, θ, ϕ, s). With this understanding and replacing $\sum_{\mathbf{q}} f(\mathbf{q}, s)$ by

$$\frac{v}{8\pi^3}\int f(\mathbf{q}, s)q^2 \mathrm{d}q\,\mathrm{d}\Omega,$$

one finds

$$E_{\text{th}} = \sum_{\mathbf{q},s}\frac{\hbar\omega}{\exp y - 1} = kT\sum_{\mathbf{q},s}\frac{y}{\exp y - 1}$$

$$= \frac{vkT}{8\pi^3}\sum_s\iiint\frac{y}{\exp y - 1}q^2\mathrm{d}q\,\mathrm{d}\Omega.$$

The limiting surface $q_0(\theta, \phi)$ in **q**-space must be such as to give the correct number of modes of oscillation.

3.11 A system of non-interacting bosons is confined to a volume v. The state of a boson is specified by labels **q** and s, $v/8\pi^3$ being again the number of wave vectors per unit volume in **q**-space. The results and notation of Problem 3.10 must be used.

(a) If the **dispersion relation** between ω and q is $\omega = a(s, \theta, \phi)q^t$, where $t > 0$ is a constant, establish the law

$$C_v \propto T^{3/t}$$

for the low-temperature heat capacity of the system.

(b) If a is a constant a_s, and the surface $q = q_0(\theta, \phi)$ is a sphere of radius q_s for polarisation s, show that

$$E_{\text{th}} = \frac{kTv}{6\pi^2}\sum_{s=1}^{3r}\left(\frac{kTx_s}{\hbar a_s}\right)^{3/t} D\left(\frac{3}{t}, x_s\right), \qquad x_s \equiv a_s q_s^t,$$

where

$$D(m, x) \equiv \frac{m}{x^m}\int_0^x \frac{y^m\,\mathrm{d}y}{\exp y - 1}$$

is a generalised Debye function. The values of x_s are subject to

$$\frac{18\pi^2 rN}{v} = \sum_s q_s^3 = \sum_s\left(\frac{kTx_s}{\hbar a_s}\right)^{3/t}.$$

(c) Establish the high- and low-temperature approximations for the result of part (b), verifying that the former is in agreement with the equipartition theorem, while the latter yields

$$E_{\text{th}} = \Gamma\left(1+\frac{3}{t}\right)\zeta\left(1+\frac{3}{t}\right)\frac{kTv}{2\pi^2 t}\sum_{s=1}^{3r}\left(\frac{kT}{\hbar a_s}\right)^{3/t}$$

in agreement with part (a). Here $\zeta(a) \equiv \sum_{j=1}^{\infty} j^{-a}$ $(a \gg 1)$ is the Riemann zeta function, and you may assume that $D(m,x) \sim mx^{-m}\Gamma(m+1)\zeta(m+1)$ for large x.

(d) For $t = 1$ and low temperatures putting $\zeta(4) = \pi^4/90$, verify that the above model yields

$$E_{\text{th}} = \frac{4}{15}\frac{\pi^5 v k^4 T^4}{h^3}\sum_{s=1}^{3r}\frac{1}{a_s^3}.$$

[In the simple **Debye theory of specific heats (phonons)** one has $r = t = 1$, $a_1 = a_2$ is the transverse sound velocity, and a_3 is the longitudinal sound velocity. For **black body radiation (photons)** the low-temperature theory applies, with $t = 1, 3r = 2, a_1 = a_2$ being the velocity of light. For spin waves in a ferromagnetic solid (magnons) one has $t = 2$. The laws $C_v \propto T^3$ (Debye solid at low temperatures, and black body radiation) and $C_v \propto T^{3/2}$ (magnons) are implied.]

Solution

(a) Write, with $\beta = 1/kT, x = \beta\hbar\omega = \beta\hbar a q^t$, so that

$$q^2 \mathrm{d}q = \left(\frac{kT}{\hbar a}\right)^{3/t}\left(\frac{1}{t}\right)x^{3/t-1}\,\mathrm{d}x.$$

Hence

$$E_{\text{th}} = \frac{kTv}{8\pi^3 t}\left(\frac{kT}{\hbar}\right)^{3/t}\sum_s\int\frac{\mathrm{d}\Omega}{a^{3/t}}\int\frac{x^{3/t}\,\mathrm{d}x}{\exp x - 1}.$$

For fixed θ, ϕ, i.e. for fixed solid angle, the upper limit of the x-integration is given by the limiting surface in **q**-space. However, at low enough temperatures the integration extends to $x = \infty$ so that the last integral is replaced by a constant numerical value. The temperature dependence is therefore given by the factor in front of the summation: $E_{\text{th}} \propto T^{3/t+1}$. This leads to the stated form of the heat capacity.

(b) We now have, with $x_s = \hbar\omega_s/kT = \hbar a q_s^t/kT$,

$$E_{\text{th}} = \frac{kTv}{8\pi^3 t}\left(\frac{kT}{\hbar}\right)^{3/t}\sum_s 4\pi(a_s)^{-3/t}\int_0^{x_s}\frac{x^{3/t}\,\mathrm{d}x}{\exp x - 1}.$$

If

$$D(m,x) \equiv \frac{m}{x^m}\int_0^x\frac{y^m\,\mathrm{d}y}{\exp y - 1},$$

then, in the classical limit (see Problem 3.9), $T \to \infty$, $x \to 0$ and so $D \to 1$.

This is a useful way of normalising the new function D. The stated equation for E_{th} now follows. Also the values of q_s are subject to

$$3rN = \frac{v}{6\pi^2}\sum_s q_s^3.$$

(c) At high temperature the two results of part (b) are easily combined to yield $E_{\text{th}} = 3rNkT$. This is the equipartition result for $3rN$ one-dimensional oscillators (Problem 3.9).

At low temperature $x_s \to \infty$; and also $D(m, x) \to mx^{-m}\Gamma(m+1)\zeta(m+1)$ is required, where ζ is the Riemann zeta function (ref. 2, p.263). One finds

$$E_{\text{th}} = \frac{kTv}{2\pi^2 t}\Gamma\left(1+\frac{3}{t}\right)\zeta\left(1+\frac{3}{t}\right)\sum_{s=1}^{3r}\left(\frac{kT}{\hbar a_s}\right)^{3/t}.$$

(d) The result is immediate.

THE IDEAL QUANTUM GAS[3]

3.12 The grand canonical mean occupation number $\langle n_j \rangle$ of the quantum state j of energy E_j of a particle in the volume v of a system at temperature T is $\{\exp[(E_j - \mu)/kT \pm 1]\}^{-1}$ for fermions and bosons respectively, μ being the chemical potential. A system of non-interacting fermions or non-interacting bosons will be considered.

(a) From Problem 2.4 establish the general result

$$\langle n_j \rangle = -\frac{\partial \ln \Xi}{\partial (E_j/kT)},$$

and hence prove from the given information that

$$\frac{pv}{kT} = \ln \Xi = \pm \sum_j \ln(1 \pm t_j),$$

where $t_j \equiv \exp[(\mu - E_j)/kT]$.

(b) Assume that the number of single-particle quantum states in the energy range $(E, E+\mathrm{d}E)$ is in a continuous spectrum approximation $AE^s\,\mathrm{d}E$, where $s > -1$ is a constant and A is independent of energy. Show that then

$$\ln \Xi = A(kT)^{s+1}\Gamma(s+1)I(\mu/kT, s+1, \pm),$$

where

$$I(a, s, \pm) \equiv \frac{1}{\Gamma(s+1)}\int_0^\infty \frac{x^s\,\mathrm{d}x}{\exp(x-a) \pm 1}.$$

Hence establish from Problem 2.7 that A is proportional to volume.

(c) Show that the mean total number of particles, the Helmholtz free energy, the entropy, the internal energy, and the pressure of the system

[3] Problems 3.12 to 3.17 follow the exposition introduced in Section 28 of reference 2.

are given by

$$\langle n\rangle \equiv \left\langle \sum_j n_j \right\rangle = A(kT)^{s+1}\Gamma(s+1)I(\mu/kT, s, \pm),$$

$$F = \mu\langle n\rangle - A(kT)^{s+2}\Gamma(s+1)I(\mu/kT, s+1, \pm),$$

$$U = (s+1)\langle n\rangle kT\frac{I(\mu/kT, s+1, \pm)}{I(\mu/kT, s, \pm)},$$

$$S = k\langle n\rangle \left[(s+2)\frac{I(\mu/kT, s+1, \pm)}{I(\mu/kT, s, \pm)} - \frac{\mu}{kT}\right],$$

$$(s+1)pv = U.$$

(d) With

$$x \equiv \frac{I(\mu/kT, s+1, \pm)}{I(\mu/kT, s, \pm)}$$

check that

$$U = (s+1)pv = (s+1)\langle n\rangle kTx,$$

$$F = \mu\langle n\rangle - kT\langle n\rangle x,$$

$$S = k\langle n\rangle[(s+2)x - \mu/kT].$$

(e) Check that the expressions for $\langle n\rangle$, F, S, and U in part (c) are **extensive**, that the system is an ideal quantum gas in the sense of Problem 1.9 with $g = 1/(s+1)$, and that $U - TS + pv$ is equal to $\mu\langle n\rangle$.

(f) If $s = \frac{1}{2}$ and $A = 4\pi v g m h^{-3}\sqrt{(2m)}$ (as in Problem 3.10), where g is the spin degeneracy of each state and m the mass of a particle, show that

$$p = kTg\left(\frac{2\pi mkT}{h^2}\right)^{3/2} I(\mu/kT, \tfrac{3}{2}, \pm).$$

[Non-interacting particles means particles with point interactions. This term implies that the particles are pictured as geometrical spheres of radius r, and that they do not interact except when they collide. The limit is then taken as $r \to 0$.]

Solution

(a) From Problem 2.4

$$\ln\Xi = \frac{pv}{kT}.$$

Also the probability of a set of occupation numbers $n_1, n_2, \ldots$ is

$$\Xi^{-1}\exp\left(-\sum_i \frac{n_i E_i}{kT} + \sum_i \frac{\mu n_i}{kT}\right),$$

since $\sum_i n_i E_i$ is the energy of the system and $\sum_j n_j$ the number of particles in a general state. Hence

$$\langle n_j\rangle = \sum_{n_1, n_2, \ldots} \ldots \sum n_j \Xi^{-1} \exp\left[\frac{n_i}{kT}(\mu - E_j)\right]$$
$$= -\frac{\partial \ln \Xi}{\partial(E_j/kT)}.$$

Substituting for $\langle n_j\rangle$, the equation for Ξ is

$$\frac{1}{t_j^{-1} \pm 1} = -\frac{\partial \ln \Xi}{\partial t_j}\bigg/\frac{\partial(E_j/kT)}{\partial t_j} = t_j\frac{\partial \ln \Xi}{\partial t_j}.$$

Hence, if all t's are kept fixed, except for t_j,

$$\ln \Xi = \int \frac{\mathrm{d}t_j}{1 \pm t_j} = \pm \ln(1 \pm t_j) + \text{constant}.$$

The general expression is therefore that stated in the problem.

(b) We have

$$\ln \Xi = \pm A \int E^s \ln(1 \pm t_j)\,\mathrm{d}E$$
$$= \pm A(kT)^{s+1} \int_0^\infty x^s \ln[1 \pm \exp(\mu/kT)]\,\mathrm{d}x$$
$$= \pm A(kT)^{s+1}\left\{\left|\frac{x^{s+1}}{s+1}\ln[1 \pm \exp(\mu/kT - x)]\right|_0^\infty \pm \frac{1}{s+1}\int_0^\infty \frac{x^{s+1}\,\mathrm{d}x}{\exp(x-\mu/kT) \pm 1}\right\}$$
$$= A(kT)^{s+1}\Gamma(s+1)I(\mu/kT, s+1, \pm).$$

(c) As in part (b)

$$\left\langle \sum_j n_j \right\rangle = A(kT)^{s+1}\int \frac{x^s\,\mathrm{d}x}{\exp(x-\mu/kT)}.$$

F is obtained from

$$F = U - TS = \mu\langle n\rangle - pv \qquad \text{(Problem 1.20)}$$
$$= \mu\langle n\rangle - kT\ln \Xi.$$

U can be obtained from

$$\sum \langle n_j\rangle E_j = A(kT)^{s+2}\int_0^\infty \frac{x^{s+1}\,\mathrm{d}x}{\exp(x-\mu/kT) \pm 1}.$$

S can be obtained from $(U-F)/T$. The pressure is obtainable from

$$pv = kT\ln \Xi.$$

(d) These results are immediate.

(e) We have

$$U - TS + pv = F + kT \ln \Xi = \langle n \rangle (\mu - xkT + xkT)$$

as required.

(f) This result follows from a simple substitution, using $\Gamma(\frac{3}{2}) = \frac{1}{2}\sqrt{\pi}$.

3.13 Develop the classical implications of the model of Problem 3.12 as follows. The distributions are then referred to as **non-degenerate**.

(a) Show that in the **classical approximation** ($\mu \to -\infty$)

$$I(a, s, \pm) \to e^{a} \qquad \text{and hence } x \to 1.$$

(b) By using the classical approximation and Problem 1.23(c) obtain the expression

$$i = s + 2 + \ln[\Gamma(s+1)k^{s+2}A/v]$$

for the chemical constant.

(c) Show that in the simple case of a gas of structureless particles in a box of volume v with $v/8\pi^3$ wave vectors $\mathbf{k} = \mathbf{p}/\hbar$ per unit volume of $\mathbf{k}$-space [as in Problem 3.10(c)]

$$s = \tfrac{1}{2}, \qquad A = 4\pi v m g h^{-3}\sqrt{(2m)},$$

where g is the spin degeneracy of each level. Hence find an expression for i.

(d) Show that in the classical approximation

$$S = k\langle n \rangle [(s+2)\ln T - \ln p + i].$$

[This problem supplements the thermodynamic results of Problem 1.23. These are seen to be valid only in the classical approximation. Note that the chemical constant is sometimes defined differently.]

Solution

(a) As $\mu \to -\infty$

$$I(a, s, \pm) \approx \frac{1}{\Gamma(s+1)} \int_0^\infty x^s e^{a-x} dx = e^{a},$$

where the definition of the gamma function

$$\Gamma(s) = \int_0^\infty x^s e^{-x} dx$$

has been used.

(b) We have

$$pv = \frac{U}{s+1} = A(kT)^{s+2}\Gamma(s+1)\exp\frac{\mu}{kT}.$$

Hence

$$\frac{\mu}{kT} = \ln p - (s+2)\ln T + (s+2) - i$$

provided

$$i = s + 2 + \ln\frac{\Gamma(s+1)k^{s+2}A}{v}.$$

(c) We have

$$AE^s\,\mathrm{d}E = g\frac{v}{8\pi^3}4\pi k^2\,\mathrm{d}k = \frac{4\pi vgp^2\,\mathrm{d}p}{h^3}.$$

Since $p^2 = 2mE$, $p^2\,\mathrm{d}p = \sqrt{(2Em)}m\,\mathrm{d}E$ and the quoted result follows.

$$i = \tfrac{5}{2} + \ln[(2\pi m)^{3/2}k^{5/2}gh^{-3}].$$

(d) Substitute the expression for μ from part (b) into that for the entropy:

$$S = k\langle n\rangle\left(s+2-\frac{\mu}{kT}\right)$$

$$= k\langle n\rangle[s+2-\ln p+(s+2)\ln T-(s+2)+i].$$

3.14 Develop the implications of a zero chemical potential for the model of Problem 3.12.

(a) Show from Problem 3.11(c) that

$$I(0, s, -) = \zeta(s+1).$$

(b) Putting $D \equiv \Gamma(s+1)I(0, s+1, \pm)k^{s+2}\,A/v$, show that if $\mu = 0$

$$U = (s+1)pv = (s+1)vDT^{s+2} = \frac{s+1}{s+2}TS,$$

$$\langle n\rangle = \frac{I(0, s, \pm)S}{(s+2)kI(0, s+1, \pm)}.$$

(c) Show by an argument analogous to that used in Problem 3.13(c) that for a gas of photons (i.e. black body radiation)

$$s = 2, \qquad A = 8\pi vh^{-3}c^{-3},$$

where c is the velocity of light. Given that $\zeta(4) = \pi^4/90$, verify that

$$D = \tfrac{8}{45}\pi^5k^4h^{-3}c^{-3},$$

$$U = \tfrac{8}{15}\pi^5vh^{-3}c^{-3}(kT)^4.$$

(d) For the model of Problem 3.11 establish the dispersion relation

$$\omega = ck$$

for photons and phonons, and show that for a sound field or phonon gas at low temperature

$$E_{\text{th}} = \tfrac{4}{15}\pi^5vh^{-3}(kT)^4\left(\frac{2}{c_{\text{t}}^3}+\frac{1}{c_{\text{l}}^3}\right)$$

where c_{t} and c_{l} are the velocities of sound for transverse and longitudinal directions.

[For black body radiation based on bosons the characteristic laws $U = 3pv = \frac{3}{4}TS = 3vDT^4$ hold. If one puts $U = (4\sigma/c)vT^4$, then

$\sigma = \pi^2 k^4/60\hbar^3 c^2 \sim 5\cdot 7 \times 10^{-5}$ erg cm^{-2} sec^{-1} deg^{-4} is Stefan's constant. These laws apply also to low-temperature phonons in a Debye-type theory. The number of phonons or photons $\langle n \rangle$ is not generally conserved, but it does remain constant in a quasistatic adiabatic change, since S is then constant. The generalised black body relations of part (b) do not presuppose a boson gas and the question arises if they can be illustrated also by a simple model of a fermion gas; see Problem 5.2.]

Solution

(a), (b) Use Problem 3.12 to find

$$\langle n \rangle = A(kT)^{s+1}\Gamma(s+1)I(0, s, \pm),$$

$$TS = (s+2)A(kT)^{s+2}\Gamma(s+1)I(0, s+1, \pm),$$

$$U = A(kT)^{s+2}\Gamma(s+2)I(0, s+1, \pm).$$

The stated results follow.

(c) As before in Problem 3.13(c)

$$AE^s\,\mathrm{d}E = 4\pi vgh^{-3}p^2\,\mathrm{d}p.$$

For a photon

$$E = h\nu, \qquad p = \frac{h}{\lambda} = \frac{E}{\nu\lambda} = \frac{E}{c}.$$

The required result follows noting that $g = 2$ for the two directions of polarisation.

(d) As in part (c), since the wave vector $k = 2\pi/\lambda$, we have

$$\hbar\omega = h\nu = \frac{hc}{\lambda} = \hbar ck,$$

so that $t = 1$ in Problem 3.11.

3.15 Thermodynamic properties of the model of Problem 3.12.

(a) Writing for simplicity $I(s)$ for $I(\mu/kT, s, \pm)$ whenever this causes no confusion, and $\gamma \equiv \mu/kT$, prove

$$\left(\frac{\partial \gamma}{\partial T}\right)_{p,\langle n\rangle} = -\left[\frac{s+1}{T} + \frac{1}{v}\left(\frac{\partial v}{\partial T}\right)_{p,\langle n\rangle}\right]\frac{I(s)}{I(s-1)},$$

$$\left(\frac{\partial \gamma}{\partial v}\right)_{T,\langle n\rangle} = -\frac{1}{v}\frac{I(s)}{I(s-1)},$$

$$\left(\frac{\partial \gamma}{\partial T}\right)_{v,\langle n\rangle} = -\frac{s+1}{v}\frac{I(s)}{I(s-1)},$$

(b) The coefficient of volume expansion

$$\frac{1}{v}\left(\frac{\partial v}{\partial T}\right)_{p,\langle n\rangle}$$

is

$$\alpha_p = \frac{1}{T}\left[(s+2)\frac{I(s+1)I(s-1)}{I^2(s)} - s - 1\right].$$

(c) The isothermal compressibility

$$-\frac{1}{v}\left(\frac{\partial v}{\partial p}\right)_T$$

is

$$K_T = \frac{I(s+1)I(s-1)}{pI^2(s)}.$$

(d) The heat capacity is

$$C_v = (s+1)k\langle n\rangle\frac{I(s+1)}{I(s)}\left[s+2-(s+1)\frac{I^2(s)}{I(s+1)I(s-1)}\right].$$

(e) Check that the Grüneisen ratio $\Gamma = \alpha_p v/K_t C_v$ [cf Problems 1.4(a), 1.9(e), 3.12(e)] is $1/(s+1)$.

Solution

(a) Use

$$\langle n\rangle = A(kT)^{s+1}\Gamma(s+1)I(s).$$

Hence $\left(\frac{\partial\langle n\rangle}{\partial T}\right)_{p,\langle n\rangle}$ is zero and this yields

$$0 = \frac{s+1}{T}\langle n\rangle + \frac{\langle n\rangle}{v}\left(\frac{\partial v}{\partial T}\right)_{p,\langle n\rangle} + A(kT)^{s+1}\Gamma(s+1)I(s-1)\left(\frac{\partial\gamma}{\partial T}\right)_{p,\langle n\rangle}.$$

This gives the first result. Also

$$\left(\frac{\partial\langle n\rangle}{\partial v}\right)_{T,\langle n\rangle} = 0$$

$$= \frac{\langle n\rangle}{v} + \langle n\rangle\frac{I(s-1)}{I(s)}\left(\frac{\partial\gamma}{\partial v}\right)_{T,\langle n\rangle}.$$

This gives the second result. The third result is obtained similarly.

(b) From

$$v = \frac{\langle n\rangle kT}{p}\frac{I(s+1)}{I(s)},$$

using the first result of part (a), we have

$$\left(\frac{\partial v}{\partial T}\right)_{p,\langle n\rangle} = \frac{v}{T} - \frac{\langle n\rangle kT}{p}\left[1 - \frac{I(s+1)I(s-1)}{I^2(s)}\right]\left[\frac{s+1}{T} + \frac{1}{v}\left(\frac{\partial v}{\partial T}\right)_{p,\langle n\rangle}\right]\frac{I(s)}{I(s-1)},$$

$$\alpha_p = \frac{1}{T} - \frac{I(s)}{I(s+1)}\left[\frac{I(s)}{I(s-1)} - \frac{I(s+1)}{I(s)}\right]\left[\frac{s+1}{T} + \frac{1}{v}\left(\frac{\partial v}{\partial T}\right)_{p,\langle n\rangle}\right],$$

$$\alpha_p\left[1 + \frac{I^2(s)}{I(s+1)I(s-1)} - 1\right] = \frac{1}{T}\left[1 - (s+1)\frac{I^2(s)}{I(s-1)I(s+1)} + s + 1\right].$$

Hence the required result.

(c) The second equation of part (a) is now needed. The argument is

$$p = \frac{\langle n\rangle kT}{v}\frac{I(s+1)}{I(s)}$$

$$\left(\frac{\partial p}{\partial v}\right)_{T,\langle n\rangle} = -\frac{p}{v}-\frac{\langle n\rangle kT}{v}\left[1-\frac{I(s+1)I(s-1)}{I^2(s)}\right]\frac{1}{v}\frac{I(s)}{I(s-1)}$$

$$= -\frac{p}{v}\frac{I^2(s)}{I(s+1)I(s-1)}.$$

(d) The third result of part (a) is now needed. We argue that

$$U = (s+1)\langle n\rangle kT\frac{I(s+1)}{I(s)},$$

$$C_v = \frac{U}{T}-(s+1)\langle n\rangle kT\left[1-\frac{I(s+1)I(s-1)}{I^2(s)}\right]\frac{s+1}{T}\frac{I(s)}{I(s-1)}.$$

The stated equation follows.

(e) The proof is immediate.

3.16 The model of Problem 3.12 will be applied to non-interacting electrons of mass m in a volume v at low temperature T.

(a) Show that for this case

$$I(c, s, +) \approx \frac{c^{s+1}}{\Gamma(s+2)}.$$

(b) Show also that for the values of A and s given in Problem 3.13

$$\mu = \left(\frac{3\langle n\rangle}{4\pi vg}\right)^{2/3}\frac{h^2}{2m}$$

$$U = \frac{3}{5}\langle n\rangle\left(\frac{3\langle n\rangle}{4\pi vg}\right)^{2/3}\frac{h^2}{2m}$$

$$p = \frac{2}{5}\frac{\langle n\rangle}{v}\left(\frac{3\langle n\rangle}{4\pi vg}\right)^{2/3}\frac{h^2}{2m}.$$

(c) Obtain an expression for K_T. Using $\langle n\rangle/v = 2{\cdot}56 \times 10^{22}\ \text{cm}^{-3}$ (sodium), show that

$$K_T = 11{\cdot}6 \times 10^{-12}\ \text{cm}^2\ \text{dyn}^{-1}.$$

Solution

(a) We have

$$I(c, s, +) = \frac{1}{\Gamma(s+1)}\int_0^\infty \frac{x^s\,\mathrm{d}x}{\exp(x-c)+1}$$

$$\approx \frac{1}{\Gamma(s+1)}\int_0^c x^s\,\mathrm{d}x$$

$$\approx \frac{c^{s+1}}{\Gamma(s+2)}.$$

(b) From Problem 3.13 we have

$$\langle n\rangle \approx vg\left(\frac{2\pi mkT}{h^2}\right)^{3/2}\frac{(\mu/kT)^{3/2}}{\Gamma(\frac{5}{2})}$$

$$= \frac{4gv}{3\sqrt{\pi}}\left(\frac{2\pi m}{h^2}\right)^{3/2}\mu^{3/2},$$

$$U \approx (s+1)kT\langle n\rangle\frac{\mu}{kT}\frac{\Gamma(s+2)}{\Gamma(s+3)} = \frac{s+1}{s+2}\mu\langle n\rangle,$$

$$p \approx \frac{\langle n\rangle kT}{v}\frac{I(s+1)}{I(s)} \approx \frac{\langle n\rangle kT}{v}\frac{\mu}{kT}\frac{1}{s+2} = \frac{2}{5v}\langle n\rangle\mu.$$

(c) It follows from Problem 3.15, putting $g = 2$, that

$$K_T = \frac{3}{5}p = \frac{4m}{h^2}\left(\frac{\langle n\rangle}{v}\right)^{-5/3}(3\pi^2)^{1/3}.$$

Inserting the values $m = 9\cdot11 \times 10^{-28}$ g, $h = 6\cdot62 \times 10^{-27}$ erg sec, and $\langle n\rangle/v = 2\cdot56 \times 10^{22}$ cm^{-3}, we obtain

$$K_T \sim 11\cdot6 \times 10^{-12}\ \text{cm}^2\ \text{dyn}^{-1}.$$

3.17 The expressions for the Fermi and Bose mean occupation numbers according to the grand canonical ensemble have been used in Problem 3.12 without being proved. They can be obtained as part of a more general procedure.

Each quantum state of a particle in a system of weakly interacting indistinguishable particles can accommodate 0, 1, 2, ..., a particles, where a is a constant. Let

$$t_j \equiv \exp\left(\gamma - \frac{E_j}{kT}\right), \qquad \gamma \equiv \frac{\mu}{kT},$$

$$I(\gamma, s, a) = \frac{1}{\Gamma(s+1)}\int_0^\infty \left[\frac{1}{\exp(x-\gamma)-1} - \frac{a+1}{\exp[(a+1)(x-\gamma)]-1}\right]x^s\,\mathrm{d}x.$$

(a) Show that for $s > -1$

$$\frac{\mathrm{d}}{\mathrm{d}\gamma}[I(\gamma, s+1, a)] = I(\gamma, s, a).$$

(b) Show that the mean occupation number per quantum state is

$$\langle n_j\rangle = \frac{1}{\exp(x-\gamma)-1} - \frac{a+1}{\exp[(a+1)(x-\gamma)]-1}.$$

(c) Show that the mean total number of particles in the system is for a continuous spectrum approximation [$\rho(E) = AE^s$ as investigated in Problem 3.12]

$$\langle n\rangle = A(kT)^{s+1}\Gamma(s+1)I(\gamma, s, a).$$

(d) Prove that

$$\left(\frac{d\gamma}{dT}\right)_{v,\langle n\rangle} = -\frac{s+1}{T}\frac{I(\gamma, s, a)}{I(\gamma, s-1, a)}.$$

(e) Also show that

$$F = \mu\langle n\rangle - A(kT)^{s+2}\Gamma(s+1)I(\gamma, s+1, a)$$

$$pv = \langle n\rangle kT\frac{I(\gamma, s+1, a)}{I(\gamma, s, a)} = \frac{1}{s+1}U$$

$$S = k\langle n\rangle\left[(s+2)\frac{I(\gamma, s+1, a)}{I(\gamma, s, a)} - \gamma\right].$$

(f) Verify that for $a = 1$ and ∞ the results for fermions and bosons respectively are obtained. These are the expressions for bosons and fermions given in Problem 3.12(a).[4]

Solution

(a)

$$\frac{\mathrm{d}}{\mathrm{d}\gamma}I(\gamma, s+1, a)$$

$$= \frac{1}{\Gamma(s+2)}\int_0^\infty\left[\frac{\exp(x-\gamma)}{[\exp(x-\gamma)-1]^2} - \frac{(a+1)^2\exp[(a+1)(x-\gamma)]}{\{\exp[(a+1)(x-\gamma)]-1\}^2}\right]x^{s+1}\mathrm{d}x.$$

A partial integration on the first term with

$$u = x^{s+1},\ v' = \frac{\exp(x-\gamma)}{[\exp(x-\gamma)-1]^2};$$

$$u' = (s+1)x^s,\ v = -\frac{1}{\exp(x-\gamma)-1},$$

yields

$$\frac{1}{\Gamma(s+2)}\left[-\left|\frac{x^{s+1}}{\exp(x-\gamma)-1}\right|_0^\infty + (s+1)\int_0^\infty\frac{x^s}{\exp(x-\gamma)-1}\mathrm{d}x\right]$$

$$= \frac{1}{\Gamma(s+1)}\int_0^\infty\frac{x^s}{\exp(x-\gamma)-1}\mathrm{d}x.$$

In the second term we use

$$u = x^{s+1},\ v' = \frac{(a+1)^2\exp[(a+1)(x-\gamma)]}{\{\exp[(a+1)(x-\gamma)]-1\}^2};$$

$$u' = (s+1)x^s,\ v = -\frac{a+1}{\exp[(a+1)(x-\gamma)]-1}.$$

[4] P.T.Landsberg, *Molec.Phys.*, **6**, 341 (1963).

One finds a vanishing uv-term, and the surviving term is

$$-\frac{1}{\Gamma(s+1)}\int_0^\infty \frac{(a+1)x^s}{\exp[(a+1)(x-\gamma)]-1}\,dx,$$

whence the result follows.

(b) For indistinguishable and non-interacting particles

$$\Xi = \sum \exp\frac{\mu N - E}{kT} = \sum \exp\left[\frac{1}{kT}\sum_j(\mu n_j - E_j n_j)\right] = \sum t_1^{n_1} t_2^{n_2} \ldots,$$

where the sum is over all single-particle quantum states. Here E_j and n_j are respectively the energies and occupation numbers of quantum states j. Also $t_j = \exp[(\mu - E_j)/kT]$. Specifying the system quantum states by the n_j, we have

$$\Xi = \sum_{n_1=0}^{a}\sum_{n_2=0}^{a} \ldots t_1^{n_1} t_2^{n_2} \ldots$$

$$= \prod_j\left(\sum_{n_j=0}^{a} t_j^{n_j}\right) = \prod_j \frac{1-t_j^{a+1}}{1-t_j},$$

$$\ln\Xi = \sum_j[\ln(1-t_j^{a+1}) - \ln(1-t_j)].$$

Hence from Problem 3.12(a)

$$\langle n_j\rangle = -\frac{\partial \ln\Xi}{\partial(E_j/kT)} = t_j\frac{\partial\ln\Xi}{\partial t_j}$$

$$= t_j\left[-\frac{(a+1)}{1-t_j^{a+1}}t_j^a + \frac{1}{1-t_j}\right] = \frac{1}{t_j^{-1}-1} - \frac{a+1}{t_j^{-(a+1)}-1}.$$

(c) This follows from

$$\langle n\rangle = \sum_j\langle n_j\rangle$$

with the continuous spectrum approximation.

(d) From

$$\langle n\rangle = A(kT)^{s+1}\Gamma(s+1)I(\gamma, s, a)$$

it follows that

$$0 = \left(\frac{\partial\langle n\rangle}{\partial T}\right)_{v,\langle n\rangle} = \frac{(s+1)\langle n\rangle}{T} + \frac{I(\gamma, s-1, a)}{I(\gamma, s, a)}\left(\frac{\partial\gamma}{\partial T}\right)_{v,\langle n\rangle}.$$

The stated result follows.

(e) From Problem 2.4, we find an expression for the Helmholtz free energy

$$F = \mu\langle n\rangle - pv = \mu\langle n\rangle - kT\ln\Xi.$$

Using next the solution of part (b), we obtain

$$F-\mu\langle n\rangle = -kT\sum_j \ln(1-t_j^{a+1})+kT\sum_j \ln(1-t_j)$$
$$= -(kT)^{s+2}A\left\{\int_0^\infty x^s[\ln(1-t^{a+1})-\ln(1-t)]\,\mathrm{d}x\right\}.$$

The integral is

$$\left|\frac{x^{s+1}}{s+1}\ln\{-\exp[(a+1)(\gamma-x)]\}\right|_0^\infty-\frac{1}{s+1}\int_0^\infty x^{s+1}\frac{(a+1)}{t^{-(a+1)}-1}\,\mathrm{d}x$$
$$-\left|\frac{x^{s+1}}{s+1}\ln[1-\exp(\gamma-x)]\right|_0^\infty+\frac{1}{s+1}\int_0^\infty x^{s+1}\frac{1}{t^{-1}-1}\,\mathrm{d}x.$$

This gives the required expression

$$F-\mu\langle n\rangle = -(kT)^{s+2}A\Gamma(s+1)I(\gamma, s+1, a).$$

Next, on the basis of the argument just given, we can write

$$pv = kT\ln\Xi = (kT)^{s+2}A\Gamma(s+1)I(\gamma, s+1, a)$$
$$= kT\langle n\rangle\frac{I(\gamma, s+1, a)}{I(\gamma, s, a)}.$$

Also

$$TS = -T\left(\frac{\partial F}{\partial T}\right)_{v,\langle n\rangle}$$
$$= -T\langle n\rangle\left(\frac{\partial \mu}{\partial T}\right)_{v,\langle n\rangle}+(s+2)A(kT)^{s+2}\Gamma(s+1)I(\gamma, s+1, a)$$
$$+A(kT)^{s+2}\Gamma(s+1)I(\gamma, s, a)T\left(\frac{\partial \gamma}{\partial T}\right)_{v,\langle n\rangle}.$$

The first and last term add to yield

$$kT\langle n\rangle T\frac{\partial\gamma}{\partial T}-T\langle n\rangle\left(kT\frac{\partial\gamma}{\partial T}+\frac{\mu}{T}\right) = -\mu\langle n\rangle.$$

Hence

$$TS = kT\langle n\rangle\left[(s+2)\frac{I(\gamma, s+1, a)}{I(\gamma, s, a)}-\gamma\right].$$

Lastly observe that

$$U = F+TS = \mu\langle n\rangle-kT\langle n\rangle\frac{I(\gamma, s+1, a)}{I(\gamma, s, a)}-\mu\langle n\rangle+kT\langle n\rangle(s+2)\frac{I(\gamma, s+1, a)}{I(\gamma, s, a)}$$
$$= (s+1)kT\langle n\rangle\frac{I(\gamma, s+1, a)}{I(\gamma, s, a)}$$
$$= (s+1)pv.$$

(f) For $a = 1$ the solution of part (b) yields

$$\Xi = \prod_j (1+t_j).$$

For $a = \infty$,

$$\Xi = \prod_j (1-t_j)^{-1}.$$

These are the partition functions for fermions and bosons.

CONSTANT PRESSURE ENSEMBLES

3.18 The **(petit) pressure ensemble** is one for which p, T, and the number of particles n are given, together with the **mean** energy E_0 and the mean volume v_0.

(a) Show from Problem 2.3 that the probability of a state (E, v) is

$$P(E, v) = \frac{\exp[-(E+pv)/kT]}{Z_p(p, T, n)},$$

where Z_p is a normalising constant related to the Gibbs free energy by

$$Z_p(p, T, n) = \exp(-G/kT).$$

(b) Discuss the difficulties associated with the partition function

$$Z_p(p, T, n) = \sum_{m,l} \exp\left[-\frac{E_m(v_l)+pv_l}{kT}\right].$$

(c) Show that these difficulties do not arise if one takes

$$Z_p(p, T, n) = \frac{1}{v_0}\int \sum_m \exp\left[-\frac{E_m(v)+pv}{kT}\right] dv$$
$$= \frac{1}{v_0}\int_0^\infty Z(v, T, n) \exp\left(-\frac{pv}{kT}\right) dv,$$

where Z is the canonical partition function, and v_0 is a suitably chosen constant volume, the definition of which presents difficulty.

(d) Find an expression analogous to that under (b) for the classical partition function Z_p, and verify that there are no difficulties in this case. Show also that the ensemble average of the volume is

$$v = -kT\left[\frac{\partial \ln Z_p(p, T, n)}{\partial p}\right]_{T,n}$$

in analogy with the result

$$p = kT\left[\frac{\partial \ln Z(v, T, n)}{\partial v}\right]_{T,n}$$

of Problem 2.6(b).

[The classical ensemble is used in Problem 9.4. The difficulties under

(b) and (c) are discussed by Lloyd and O'Dwyer [5] who give references to earlier work.]

Solution

(a) In Problem 2.3 identify x with the energy E and y with the volume v. A state i of the system in the ensemble is specified by the volume v_l and one of the energies $E_m(v_l)$ available for that volume. Thus i implies *two* specifications. The result of Problem 2.3

$$S = k\beta E_0 + k\gamma v_0 + k \ln \Xi$$

and the thermodynamic result

$$TS = E_0 + pv_0 - G$$

must be equivalent for all E_0 and v_0 and this leads to

$$P(E, v) = \exp \frac{-(E+pv)/kT}{Z_p(p, T, n)}.$$

(b) In general the 'eigenvolumes' v_l form a *continuous* set so that the l-summation diverges.

(c) By integration over v one avoids the divergence, but shifts the difficulty to the question of how v_0 should be defined. However, the precise value of v_0 will not affect those thermodynamic quantities which depend on a differentiation of $\ln Z_p$. The reference cited gives an introduction to the current situation.

(d) In the classical case E may be replaced by the classical Hamiltonian H expressed in terms of the f generalised coordinates and momenta. Following the procedure in Problem 3.4,

$$Z_p(p, T, n) = h^{-f} \int \dots \int \exp \left[-\frac{H(p_1 \dots q_f) + pv}{kT} \right] \mathrm{d}p_1 \dots \mathrm{d}q_f.$$

There is no divergence and no dimensional difficulty here. The differentiation indicated in the problem now yields the average volume.

RADIATIVE EMISSION AND ABSORPTION

3.19 In a certain fermion system transitions occur from a group i of single-particle states $I(\epsilon i)$ to a group j of states $J(\epsilon j)$ at a rate

$$u_{ij} = \sum_{\substack{I(\epsilon i) \\ J(\epsilon j)}} [p_I S_{IJ} q_J - p_J S_{JI} q_I],$$

where p_I is the mean occupation number of a quantum state I, $(1 - q_J)$ is the mean occupation number of a quantum state J, and S_{IJ} is a transition probability per unit time.

[5] P.Lloyd and J.J.O'Dwyer, *Molec.Phys.*, **6**, 573 (1963). See also D.R.Cruise, *J.Phys.Chem.*, **74**, 405 (1970).

(a) Show that for free fermions in equilibrium at temperature T, to be denoted by $(...)_0$,

$$\left(\frac{q_I}{p_I}\right)_0 = \exp\left(\frac{E_I - \mu_I}{kT}\right)$$

where μ_I is the chemical potential appropriate to state I.

(b) Let N_ν be the photon occupation number of a mode of frequency ν where $h\nu = E_I - E_J$. Assuming (a) valid, show that the quantity $Y_{\nu IJ}$ defined by

$$Y_{\nu IJ} \equiv \left[N_\nu\left(\frac{q_I p_J}{p_I q_J} - 1\right)\right]_0$$

has the value unity.

(c) Assuming the result (a) and that $S_{IJ} = (S_{IJ})_0$ for all transition probabilities, show that

$$\frac{S_{JI}}{S_{IJ}} = \exp\left(\frac{E_J - E_I}{kT}\right),$$

where E_I is the energy of the state I.

(d) Show with the assumptions of (c) that the ratio of the reverse rate p_{ji} to forward rate p_{ij} is

$$\frac{p_{ji}}{p_{ij}} = \exp\left(\frac{\mu_j - \mu_i}{kT}\right).$$

[They are equal in thermal equilibrium in accordance with the principle of detailed balance.]

Solution

(a) Writing $x_I \equiv \exp[(E_I - \mu_I)/kT]$ one has essentially from Problem 3.12

$$(p_I)_0 = \frac{1}{1 + x_I}\ .$$

Since one fermion is the maximum mean occupation number of any state I,

$$(q_I)_0 = 1 - (p_I)_0 = \frac{x_I}{1 + x_I}\ .$$

The result follows.

(b) Radiation in thermal equilibrium corresponds to black body radiation (see Problem 3.14). Hence, using also part (a),

$$Y_{\nu IJ} = \frac{1}{\exp(h\nu/kT) - 1}\left[\exp\left(\frac{E_I - \mu_I + \mu_J - E_J}{kT}\right)_0 - 1\right].$$

In thermal equilibrium all chemical potentials are equal, as follows from Problem 1.21. This establishes the result.

(c) In thermal equilibrium the net rate $I \to J$ is zero. Hence, since in equilibrium $\mu_i = \mu_j$,

$$\frac{S_{JI}}{S_{IJ}} = \left(\frac{S_{JI}}{S_{IJ}}\right)_0 = \left(\frac{p_I q_J}{p_J q_I}\right)_0 = \exp\frac{E_J - E_I}{kT} .$$

(d) Reverse to forward rates are in the ratio

$$\frac{S_{JI}}{S_{IJ}} = \frac{p_J}{q_J}\frac{q_I}{p_I} = \exp\frac{E_J - E_I + \mu_j - E_J + E_I - \mu_i}{kT} .$$

It follows that

$$u_{ij} = \sum_{I,J} p_I S_{IJ} q_J \left(1 - \exp\frac{\mu_j - \mu_i}{kT}\right) .$$

Hence

$$u_{ij} = p_{ij} - p_{ji} ,$$

where

$$\frac{p_{ji}}{p_{ij}} = \exp\frac{\mu_j - \mu_i}{kT} .$$

These results are of use in the recombination theory of semiconductors where the μ_i's are driving forces for transitions. They are then referred to as quasi-Fermi levels, see Chapter 16.

3.20 Stimulated and **spontaneous emission** rates and **absorption rates** of photons due to single particle transitions between states I and J are, respectively, in the notation of Problem 3.19:

$$u^{\text{st}} \equiv B_{IJ} N_\nu p_I q_J , \quad u^{\text{sp}} \equiv A_{IJ} p_I q_J , \quad u^{\text{abs}} \equiv B_{JI} N_\nu p_J q_I .$$

Here J corresponds to a lower energy, and it will be assumed that

$$B_{IJ} = (B_{IJ})_0, \quad A_{IJ} = (A_{IJ})_0 .$$

The quantum-mechanical result that $B_{IJ} = B_{JI}$ may be used in these equations. Note that the spontaneous rate does not depend on N_ν.

(a) Show from Problem 3.19 that $A_{IJ} = B_{IJ}$.

(b) Show that

$$\frac{u^{\text{st}} - u^{\text{abs}}}{u^{\text{sp}}} = N_\nu \left(1 - \exp\frac{\mu_J - \mu_I + h\nu}{kT}\right).$$

(c) Show that the ratio of reverse to forward rate is

$$\frac{N_\nu}{N_\nu + 1} \exp\frac{\mu_J - \mu_I + h\nu}{kT} .$$

(d) Show that the result (c) satisfies the equation of Problem 3.19(d) only if the radiation field is that of a black body at temperature T.

[All these results are important and are used in semiconductor theory, Problem 16.3, and in the theory of the laser, Problems 15.6 to 15.8.]

Solution

(a) In this case the rate is zero in thermal equilibrium, i.e.

$$(B_{IJ}N_{\nu 0}+A_{IJ})(p_I q_J)_0 = B_{IJ}N_{\nu 0}(p_J q_I)_0 \, .$$

It follows that

$$\left(\frac{N_\nu q_I p_J}{p_I q_J}\right)_0 = N_{\nu 0}+\frac{A_{IJ}}{B_{IJ}} \, .$$

Applying the result of Problem 3.19(b), we find $B_{IJ} = A_{IJ}$.

(b) The ratio of the net stimulated emission rate to the spontaneous emission rate is

$$\frac{B_{IJ}N_\nu p_I q_J(1-p_J q_I/p_I q_J)}{B_{IJ}p_I q_J} = N_\nu\left(1-\exp\frac{E_I-E_J-\mu_I+\mu_J}{kT}\right)$$
$$= N_\nu\left(1-\exp\frac{\mu_J-\mu_I+h\nu}{kT}\right) .$$

(c) The forward rate is $B_{IJ}(N_\nu+1)p_I q_J$. The reverse rate is $B_{IJ}N_\nu p_J q_I$. Hence the required ratio is

$$\frac{\text{reverse rate}}{\text{forward rate}} = \frac{N_\nu}{N_\nu+1}\exp\frac{\mu_J-\mu_I+h\nu}{kT} \, .$$

(d) For black body radiation

$$\frac{N_\nu}{N_\nu+1} = \exp\left(-\frac{h\nu}{kT}\right) .$$

The results of Problem 3.19 apply in this special case with

$$S_{IJ} = B_{IJ}(N_{\nu 0}+1) \, ,$$
$$S_{JI} = B_{IJ}N_{\nu 0} \, .$$

3.21 In Problem 3.20 $(N_\nu)_0$ was assumed known because the result of Problem 3.19(b) was used. Assume now the existence of the three rates, with unknown $(N_\nu)_0$ and unknown temperature-independent transition probabilities B_{IJ}, A_{IJ}, B_{JI}; assume also that fermions make the transitions with $E_I-E_J = h\nu$ and that the system is in equilibrium at temperature T.

(a) Show that $B_{IJ} = B_{JI}$ by supposing that $(N_\nu)_0 \to \infty$ as $T \to \infty$.

(b) Assuming Wien's law $(N_\nu)_0 \propto \nu^3\exp(-h\nu/kT)$ for $\nu \to \infty$ show that

$$\frac{A_{IJ}}{B_{IJ}} \propto \nu^3 \, .$$

[This is Einstein's original argument[6] which established the existence of spontaneous emission. The coefficients introduced here are called **Einstein's** A **and** B **coefficients.**]

[6] A.Einstein, *Ann.Physik*, **18**, 121 (1917).

Solution

(a) Let $x_{IJ} \equiv p_J q_I / p_I q_J$. Then

$$N_\nu = \frac{A_{IJ}/B_{IJ}}{(x_{IJ}B_{JI}/B_{IJ}) - 1} .$$

It follows from Problem 3.19(a) that in equilibrium

$$(x_{IJ})_0 = \exp \frac{h\nu}{kT} .$$

Hence as $T \to \infty$, $(x_{IJ})_0 \to 1$; so that the result follows.

(b) One finds

$$(N_\nu)_0 = \frac{A_{IJ}/B_{IJ}}{\exp(h\nu/kT) - 1}$$

so that the result follows.

GENERAL REFERENCES

G.H.Wannier, *Statistical Physics* (John Wiley, New York), 1966.

4

Ideal classical gases of polyatomic molecules

C.J.WORMALD
(*University of Bristol, Bristol*)

THE TRANSLATIONAL PARTITION FUNCTION

4.1 From Problem 2.2 we have that the total energy E of an assembly of N systems is given by

$$E = -N\left(\frac{\partial \ln Z}{\partial \beta}\right)_{V,N} = NkT^2\left(\frac{\partial \ln Z}{\partial T}\right)_{V,N},$$

where the partition function Z of a single system is given by

$$Z = \sum_j \exp(-\epsilon_j/kT)\ ,$$

in which ϵ_j is the energy of the jth quantum state.

(a) For an assembly of systems each of which possesses energy levels of the form $\epsilon_{(1)i}+\epsilon_{(2)j}+\epsilon_{(3)k}$, where i, j, k label the levels of essentially three independent spectra, show that the total partition function Z_{tot} can be written as the product $Z_{tot} = Z_{(1)}Z_{(2)}Z_{(3)}$.

(b) For one mole of gaseous molecules confined within a container of fixed volume V we have from Problem 3.5 the translational partition function for one mole of an ideal classical gas in the form

$$Z_N(T, V) = \left(\frac{2\pi mkT}{h^2}\right)^{3N/2}\frac{V^N}{N!}\,i(T)^N\ ,$$

where i is the internal partition function for a single molecule and $N = N_0$. Show that the molar Gibbs free energy and the entropy of the gas are given by the equations

$$\tilde{G} = -N_0kT[\tfrac{5}{2}\ln T+\tfrac{3}{2}\ln M-\ln P-3\cdot 6605]-N_0kT\ln i(T)\ ,$$

$$\tilde{S} = N_0k[\tfrac{5}{2}\ln T+\tfrac{3}{2}\ln M-\ln P-1\cdot 1605]+N_0k\left[\ln i(T)+T\frac{\partial \ln i(T)}{\partial T}\right],$$

where P is in atmospheres and M is the molecular weight of the gas.

[Note: 1 atmosphere = $1\cdot 01325 \times 10^6$ dyne cm^{-2}.]

(c) Would you expect these formulae to be valid at all temperatures and pressures?

Solution

(a) The total energy E of any system can be written as the sum of the different types of energy, $E = \epsilon_{(1)} + \epsilon_{(2)} + \epsilon_{(3)}$. The total partition function is therefore

$$Z_{\text{tot}} = \sum_{ijk} \exp[(-\epsilon_{(1)i} - \epsilon_{(2)j} - \epsilon_{(3)k})/kT] ,$$

where i, j, and k are the numbers of energy levels associated with the energies of types (1), (2), and (3). We can write Z_{tot} in the form

$$Z_{\text{tot}} = \sum_i \exp(-\epsilon_{(1)i}/kT) \sum_j \exp(-\epsilon_{(2)j}/kT) \sum_k \exp(-\epsilon_{(3)k}/kT) ,$$

so that

$$Z_{\text{tot}} = Z_{(1)} Z_{(2)} Z_{(3)} .$$

(b) We calculate first the Helmholtz free energy $F = -kT \ln Z_N$.

$$\ln Z_N = \ln \frac{1}{N!} + \tfrac{3}{2} N \ln \frac{2\pi mkT}{h^2} + N \ln V + N \ln i(T) .$$

Using Stirling's approximation that $\ln N! = N \ln N - N$ [Problem 2.11(b)], putting $M = mN$ and $V = NkT/P$ we obtain

$$\frac{\ln Z_N}{N} = 1 + \tfrac{3}{2} \ln\left(\frac{2\pi}{Nh^2}\right) + \tfrac{5}{2} \ln k - \ln(1{\cdot}01325 \times 10^6)$$

$$+ \tfrac{5}{2} \ln T + \tfrac{3}{2} \ln M - \ln P + \ln i(T) .$$

Using $N = N_0 = 6{\cdot}0225 \times 10^{23}$ mole^{-1}, $h = 6{\cdot}6256 \times 10^{-27}$ erg s, and $k = 1{\cdot}38054 \times 10^{-16}$ erg deg^{-1} we obtain

$$\tilde{F} = -N_0 kT[\tfrac{5}{2} \ln T + \tfrac{3}{2} \ln M - \ln P - 2{\cdot}6605] - N_0 kT \ln i(T) .$$

Finally $\tilde{G} = \tilde{F} + N_0 kT$, so that

$$\tilde{G} = -N_0 kT[\tfrac{5}{2} \ln T + \tfrac{3}{2} \ln M - \ln P - 3{\cdot}6605] - N_0 kT \ln i(T) .$$

The entropy $\tilde{S}$ is obtained using $\tilde{S} = -(\mathrm{d}\tilde{G}/\mathrm{d}T)_{P, N}$. Noting that the term $T \times \tfrac{5}{2} \ln T$ yields $\tfrac{5}{2} \ln T + \tfrac{5}{2}$, we have

$$\tilde{S} = N_0 k[\tfrac{5}{2} \ln T + \tfrac{3}{2} \ln M - \ln P - 1{\cdot}1605] + N_0 k \left[\ln i(T) + T \frac{\partial \ln i(T)}{\partial T}\right] .$$

(c) We note that as T approaches zero the entropy approaches $-\infty$, whereas it should approach the value zero, which corresponds to the system being in a single quantum state. However, if we have only one quantum state we cannot write a partition function for the canonical ensemble of the form $Z_N = Z^N/N!$ as this formula is conditional upon the number of available quantum states being much greater than the number of molecules. For gases at low temperatures and high densities it is necessary to replace Boltzmann statistics by Bose-Einstein or Fermi-Dirac statistics, and when this is done the paradox is resolved.

THERMODYNAMIC PROPERTIES AND THE THEORY OF FLUCTUATIONS

4.2 (a) For a single molecule the distribution function for the energy is broad, whereas for an assembly of many (N) molecules the distribution function for the total energy is sharp. This contrast arises because in the assembly the relative fluctuations in the energies of individual molecules almost cancel each other out, and only a small relative fluctuation in the energy remains. For an assembly, in which there is a Boltzmann distribution, show that the mean square deviation of the energy E of the individual molecules, which is fluctuating about the average energy $\langle E \rangle$ of all the molecules, is given by

$$\langle E - \langle E \rangle \rangle^2 = N(\langle \epsilon^2 \rangle - \langle \epsilon \rangle^2) ,$$

where ϵ is the energy of an individual molecule and $\langle \epsilon \rangle$ is its mean value.

(b) Show that the heat capacity C_v of an assembly can be written in the form

$$C_v = \frac{Nk}{T^2}\left[\frac{Z''}{Z'} - \left(\frac{Z'}{Z}\right)^2\right],$$

where Z is the partition function of a single system, $Z' = \partial Z/\partial(1/T)$, and $Z'' = \partial^2 Z/\partial(1/T)^2$, both taken at constant volume.

(c) Use the formula obtained in (b) above to show that the heat capacity C_v of an assembly is proportional to the rate of change of the average energy $\langle E \rangle$ with temperature so that (see also Problem 22.3)

$$C_v = \frac{N}{kT^2}(\langle \epsilon^2 \rangle - \langle \epsilon \rangle^2) .$$

(d) The fluctuation at equilibrium in any thermodynamic quantity x from its equilibrium value x_0, at constant y, is given by the general formula[(1)]

$$\langle x - x_0 \rangle^2 = kT \bigg/ \left(\frac{\partial^2 E}{\partial x^2} - T\frac{\partial^2 S}{\partial x^2}\right)_y .$$

Show that the fluctuation in the total energy E of an assembly from its equilibrium value U is the heat capacity C_v given in part (c).

Solution

(a) We have

$$E = \sum_{i=1}^{N} \epsilon_i, \quad \langle E \rangle = \sum_{i=1}^{N} \langle \epsilon_i \rangle$$

(note that E is the internal energy U), where ϵ_i is the energy of the ith molecule and $\langle \epsilon_i \rangle$ is the energy of the ith molecule averaged over the whole assembly.

[(1)] This holds for a system in contact with a heat reservoir at temperature T. It can be obtained from the modification of the result $p \propto \exp[S(X)/k]$ (Problem 22.4) to this situation, when it reads $p \propto \exp[-F(X)/kT]$, where F is the free energy.

We can write

$$E - \langle E \rangle = \sum_{i=1}^{N} (\epsilon_i - \langle \epsilon_i \rangle) ,$$

so that

$$\langle E - \langle E \rangle \rangle^2 = \sum_{i=1}^{N} \sum_{j=1}^{N} \langle (\epsilon_i - \langle \epsilon_i \rangle)(\epsilon_j - \langle \epsilon_j \rangle) \rangle .$$

In a Boltzmann distribution the molecules i and j are statistically independent, and we can readily show that terms for which $i \neq j$ average to zero.

Using the Boltzmann equation to express the fraction of the molecules i and j in the assembly which are in the states p_i and p_j we have

$$\langle (\epsilon_i - \langle \epsilon_i \rangle)(\epsilon_j - \langle \epsilon_j \rangle) \rangle = \left[\frac{1}{Z_{p_i}} \sum_{p_i} (\epsilon_i - \langle \epsilon_i \rangle) \exp(-\epsilon_{i,p_i}/kT) \right.$$

$$\left. \times \frac{1}{Z_{p_j}} \sum_{p_j} (\epsilon_j - \langle \epsilon_j \rangle) \exp(-\epsilon_{j,p_j}/kT) \right]$$

$$= \langle \epsilon_i - \langle \epsilon_i \rangle \rangle \langle \epsilon_j - \langle \epsilon_j \rangle \rangle = (\langle \epsilon_i \rangle - \langle \epsilon_i \rangle)(\langle \epsilon_j \rangle - \langle \epsilon_j \rangle) = 0 .$$

Retaining only terms for which $i = j$ we have

$$\langle E - \langle E \rangle \rangle^2 = \sum_{i=1}^{N} \langle \epsilon_i - \langle \epsilon_i \rangle \rangle^2 = N \langle \epsilon - \langle \epsilon \rangle \rangle^2$$

$$= N(\langle \epsilon^2 \rangle - 2\langle \epsilon \rangle \langle \epsilon \rangle + \langle \epsilon \rangle^2) = N(\langle \epsilon^2 \rangle - \langle \epsilon \rangle^2) .$$

(b) Alternative expressions for the energy E can readily be obtained from $E = F + TS$, which we can write in terms of Z as

$$E = -kT \ln \frac{Z^N}{N!} + T \left[k \ln \frac{Z^N}{N!} + NkT \left(\frac{\partial \ln Z}{\partial T} \right)_V \right] ;$$

hence

$$E = NkT^2 \left(\frac{\partial \ln Z}{\partial T} \right)_V = -\frac{Nk}{Z} \left[\frac{\partial Z}{\partial (1/T)} \right]_V = -Nk \frac{Z'}{Z} .$$

From this we have

$$C_v = \left(\frac{\partial E}{\partial T} \right)_V = -\frac{1}{T^2} \left(\frac{\partial E}{\partial (1/T)} \right)_V = \frac{Nk}{T^2} \frac{\partial (Z'/Z)}{\partial (1/T)} .$$

Differentiation using the quotient rule immediately yields

$$C_v = \frac{Nk}{T^2} \left[\frac{Z''}{Z'} - \left(\frac{Z'}{Z} \right)^2 \right] .$$

(c) The average energy of a molecule is given by

$$\langle \epsilon \rangle = \sum_i n_i \epsilon_i$$

where

$$n_i = \frac{\exp(-\epsilon_i/kT)}{\sum\limits_i \exp(-\epsilon_i/kT)}$$

so that

$$\langle\epsilon\rangle = \frac{\sum\limits_i \epsilon_i \exp(-\epsilon_i/kT)}{\sum\limits_i \exp(-\epsilon_i/kT)} \ .$$

From $Z = \sum\limits_i \exp(-\epsilon_i/kT)$ we have

$$\frac{dZ}{d(1/T)} = -\frac{1}{k}\sum \epsilon_i \exp(-\epsilon_i/kT) = Z' \ ,$$

so that

$$k\langle\epsilon\rangle = \frac{1}{Z}\frac{dZ}{d(1/T)} = \frac{Z'}{Z}$$

and

$$k\left[\frac{\partial\langle\epsilon\rangle}{\partial(1/T)}\right]_V = \left[\frac{\partial(Z'/Z)}{\partial(1/T)}\right]_V = \frac{Z''}{Z} - \left(\frac{Z'}{Z}\right)^2 .$$

Here

$$Z'' = \frac{\partial Z'}{\partial(1/T)} = \frac{1}{k^2}\sum_i \epsilon_i^2 \exp(-\epsilon_i/kT) \ ,$$

the ratio

$$\frac{Z''}{Z} = \frac{\sum\limits_i \epsilon_i^2 \exp(-\epsilon_i/kT)}{k^2 \sum\limits_i \exp(-\epsilon_i/kT)} = \frac{\langle\epsilon^2\rangle}{k^2} \ ,$$

and the ratio

$$\left(\frac{Z'}{Z}\right)^2 = \frac{\langle\epsilon\rangle^2}{k^2} \ .$$

Now

$$C_v = \frac{Nk}{T^2}\left(\frac{Z''}{Z} - \frac{Z'}{Z}\right) = \frac{Nk}{T^2}\left(\frac{\langle\epsilon^2\rangle}{k^2} - \frac{\langle\epsilon\rangle^2}{k^2}\right) = \frac{N}{kT^2}(\langle\epsilon^2\rangle - \langle\epsilon\rangle^2) \ .$$

(d) Putting $x = E$ and $y = V$ and noting that the average value of the total energy $\langle E\rangle$ is the internal energy U, we have

$$\langle E - U\rangle^2 = kT \bigg/ \left(\frac{\partial^2 E}{\partial E^2} - T\frac{\partial^2 S}{\partial U^2}\right)_V \ .$$

Using

$$\left(\frac{\partial S}{\partial U}\right)_V = \frac{1}{T} \ ; \quad \left(\frac{\partial^2 S}{\partial U^2}\right)_V = -\frac{1}{T^2}\left(\frac{\partial T}{\partial U}\right)_V = -\frac{1}{T^2 C_v}$$

we obtain

$$\langle E - U\rangle^2 = kT^2 C_v \ .$$

Using the result obtained in part (a) we find

$$C_v = \frac{N}{kT^2}(\langle\epsilon^2\rangle - \langle\epsilon\rangle^2) \ .$$

THE CLASSICAL ROTATIONAL PARTITION FUNCTION

4.3 (a) For most polyatomic molecules at temperatures above the normal boiling point of the fluid, the rotational energy levels are so closely spaced that they approximate to a continuum, and rotational energy may be calculated accurately on the basis that the molecule is a rigid body obeying the laws of classical mechanics.

Show that the rotational kinetic energy of a heteronuclear diatomic molecule about its centre of mass is given by

$$\epsilon = \frac{1}{2I}\left(P_\theta^2 + \frac{P_\phi^2}{\sin^2\theta}\right),$$

where I is the moment of inertia, and P_θ and P_ϕ are the moments conjugate to θ and ϕ.

(b) Obtain the rotational partition function for the rotation of a heteronuclear diatomic molecule by evaluating the phase integral introduced in Problem 3.4.

$$Z_{\text{rot}} = \frac{1}{h^2}\int_0^{2\pi}\int_0^{\pi}\int_{-\infty}^{\infty}\int_{-\infty}^{\infty}\exp\left[-\frac{1}{2IkT}\left(P_\theta^2 + \frac{P_\phi^2}{\sin^2\theta}\right)\right] \mathrm{d}P_\theta\,\mathrm{d}P_\phi\,\mathrm{d}\theta\,\mathrm{d}\phi\,.$$

(c) The general quantum mechanical solution for the rotational coordinates of a rigid body with three degrees of rotational freedom is complicated but, since the moments of inertia of all but the very lightest of molecules are large, and as the quantum levels are closely spaced compared with kT at temperatures above the normal boiling point of the substance, the classical partition function may be safely used.

Evaluation of the classical phase integral yields

$$Z_{\text{rot}} = \frac{\sqrt{\pi}}{\sigma h^2}(8\pi^2 kT)^{3/2}(I_a I_b I_c)^{1/2},$$

where σ is the symmetry number of the molecule and is defined as the number of values of the rotational coordinates which all correspond to one orientation of the molecule. The quantity $I_a I_b I_c$ is the product of the three principal moments of inertia of the molecule.

Obtain expressions for the rotational energy, heat capacity, free energy, and entropy from the above partition function.

Solution

(a) The molecule rotates about the centre of gravity c. The total energy is the sum of the precessional energy, in which the atoms of mass m_A and m_B follow the circular orbits shown in Figure 4.3.1 with velocity $\mathrm{d}\phi/\mathrm{d}t$, and the energy of end-over-end rotation with velocity $\mathrm{d}\theta/\mathrm{d}t$. The moment of inertia is

$$I = m_A r_A^2 + m_B r_B^2 = \frac{m_A m_B}{m_A + m_B}(r_A + r_B)^2\,.$$

The precessional inertia is

$$I_\alpha = m_A(\mathrm{Aa})^2 + m_B(\mathrm{Bb})^2 = m_A(r_A \sin\theta)^2 + m_B(r_B \sin\theta)^2 = I\sin^2\theta \; .$$

The precessional kinetic energy E_α is

$$E_\alpha = \tfrac{1}{2}[m_A(r_A \sin\theta)^2 + m_B(r_B \sin\theta)^2]\frac{\mathrm{d}\phi}{\mathrm{d}t} = \tfrac{1}{2}I\sin^2\theta\left(\frac{\mathrm{d}\phi}{\mathrm{d}t}\right)^2 .$$

For the end-over-end rotation energy E_β we have simply

$$E_\beta = \tfrac{1}{2}m_A\left(r_A\frac{\mathrm{d}\theta}{\mathrm{d}t}\right)^2 + \tfrac{1}{2}m_B\left(r_B\frac{\mathrm{d}\theta}{\mathrm{d}t}\right)^2 = \tfrac{1}{2}I\left(\frac{\mathrm{d}\theta}{\mathrm{d}t}\right)^2 .$$

The total energy ϵ is therefore

$$\epsilon = E_\alpha + E_\beta = \tfrac{1}{2}I(\dot{\theta}^2 + \dot{\phi}^2\sin^2\theta) \; .$$

Differentiating with respect to $\dot{\theta}$ and $\dot{\phi}$ we obtain the angular momenta P_θ and P_ϕ.

$$P_\theta = \frac{\partial\epsilon}{\partial\dot{\theta}} = I\dot{\theta} \; , \quad P_\phi = \frac{\partial\epsilon}{\partial\dot{\phi}} = I\dot{\phi}\sin^2\theta$$

so that

$$\epsilon = \frac{1}{2I}\left(P_\theta^2 + \frac{P_\phi^2}{\sin^2\theta}\right) .$$

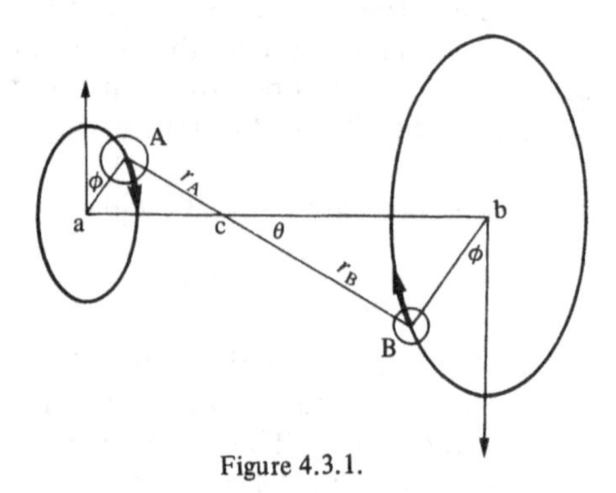

Figure 4.3.1.

(b) The fourfold integration is straightforward.

1.
$$\int_0^{2\pi} \mathrm{d}\phi = 2\pi$$

2.
$$\int_{-\infty}^{\infty} \exp\left(-\frac{P_\theta^2}{2IkT}\right)\mathrm{d}P_\theta \; .$$

This has the form of the Gaussian error integral

$$\int_0^{\infty} \exp(-\mathrm{d}x^2)\,\mathrm{d}x = \tfrac{1}{2}\left(\frac{\pi}{\alpha}\right)^{1/2}$$

so that

$$2\int_0^\infty \exp\left(-\frac{P_\theta^2}{2IkT}\right) dP_\theta = (2\pi IkT)^{1/2} .$$

Using the same standard integral again we have

3. $$Z_{\rm rot} = \frac{2\pi}{h^2}(2\pi IkT)^{1/2} \int_0^\infty \int_{-\infty}^\infty \exp\left(-\frac{P_\phi^2}{2IkT\sin^2\theta}\right) dP_\phi \, d\theta .$$

4. $$Z_{\rm rot} = \frac{2\pi}{h^2}(2\pi IkT)^{1/2}(2\pi IkT)^{1/2} \int_0^\pi \sin\theta \, d\theta = \frac{8\pi^2 IkT}{h^2} .$$

(c) $$Z_{\rm rot} = \frac{8\pi^2}{\sigma h^3}(8\pi^2 I_a I_b I_c)^{1/2}(kT)^{3/2} .$$

We at once obtain the molar rotational thermodynamic functions

$$\tilde{E}_{\rm rot} = N_0 kT^2 \frac{\partial \ln Z}{\partial T} = N_0 kT^2 \frac{d}{dT}\left[\tfrac{1}{2}\ln\frac{8\pi^2}{\sigma h^3}(I_a I_b I_c)^{1/2} + \tfrac{3}{2}\ln kT\right] ,$$

whence $\tilde{E} = \frac{3}{2}N_0 kT$ and $\tilde{C}_v = \frac{3}{2}N_0 k$.

$$\tilde{F}_{\rm rot} = -N_0 kT \ln Z = -N_0 kT\left(\tfrac{3}{2}\ln T + \tfrac{1}{2}\ln I_a I_b I_c - \ln\sigma + \tfrac{1}{2}\ln\pi + \tfrac{3}{2}\ln\frac{8\pi^2 k}{h^2}\right)$$

and

$$\tilde{S}_{\rm rot} = \frac{\tilde{E}}{T} + N_0 k \ln Z$$

$$= N_0 k\left(\tfrac{3}{2} + \tfrac{3}{2}\ln T + \tfrac{1}{2}\ln I_a I_b I_c - \ln\sigma + \tfrac{1}{2}\ln\pi + \tfrac{3}{2}\ln\frac{8\pi^2 k}{h^2}\right) .$$

Collecting together the numerical constants we obtain

$$\tilde{S}_{\rm rot} = N_0 k(\tfrac{3}{2}\ln T + \tfrac{1}{2}\ln I_a I_b I_c - \ln\sigma + 134{\cdot}684) .$$

THE QUANTUM MECHANICAL ROTATIONAL PARTITION FUNCTION

4.4 Polyatomic molecules in the gaseous phase are free to rotate about the centre of mass of the molecule. This rotational motion is strictly independent of the translational motion, and at moderate temperatures in which the vibrational modes of the molecule are excited only to a negligible extent, it can be assumed to be independent of vibrational motion within the molecule.

A heteronuclear diatomic molecule has two degrees of rotational freedom, and the rotational energy is quantized such that the energy ϵ_J in the Jth rotational state is given by

$$\epsilon_J = J(J+1)\frac{h^2}{8\pi^2 I} = J(J+1)k\theta_r ,$$

where $\theta_r = h^2/8\pi^2 Ik$ and has the dimensions of temperature. Each rotational state except the first ($J = 0$) is doubly degenerate, as the rotation can be either left or right handed, and the degeneracy ω_J of the Jth level is in this case $(2J+1)$.

(a) Assuming that the rotational energy levels are close enough together to be approximated to a continuum (i.e. when $\theta_r/T < 1$) show that the quantum mechanical rotational partition function is the same as the classical rotational partition function obtained in Problem 4.3(b). From this partition function obtain expressions for the rotational energy, heat capacity, and entropy.

(b) At low temperatures when $\theta/T \gg 1$ the rotational energy cannot be approximated to a continuum and a summation must be used. Investigate the form of the rotational energy and heat capacity in the low temperature limit. Sketch the temperature dependence of the rotational energy, and deduce the form of the rotational heat capacity curve.

Solution

(a) The rotational energy ϵ_J and the degeneracy ω_J are given by

$$\epsilon_J = J(J+1)\frac{h^2}{8\pi^2 I}, \qquad \omega_J = (2J+1)\,.$$

Hence

$$Z_{\text{rot}} = \sum_0^\infty (2J+1)\exp\left[-\frac{J(J+1)\theta_r}{T}\right].$$

When $\theta/T < 1$ we can replace the sum by an integral

$$Z_{\text{rot}} = \int_0^\infty \exp\left[-\frac{J(J+1)\theta_r}{T}\right](2J+1)\,\mathrm{d}J = \int_0^\infty \exp\left[-\frac{(J^2+J)\theta_r}{T}\right]\mathrm{d}(J^2+J)$$

We have this integral in the form $\int e^{-\alpha}\,\mathrm{d}(-\alpha) = \int \mathrm{d}(e^{-\alpha}) = e^{-\alpha}$ so that

$$Z_{\text{rot}} = -\frac{T}{\theta_r}\int_0^\infty \exp\left[-\frac{(J^2+J)\theta_r}{T}\right]\mathrm{d}\left[-\frac{(J^2+J)\theta_r}{T}\right]$$

$$= -\frac{T}{\theta_r}\left\{\exp\left[-\frac{(J^2+J)\theta_r}{T}\right]\right\}_{J=0}^{J=\infty} = \frac{T}{\theta_r}\,.$$

But $\theta_r = h^2/8\pi^2 Ik$ so that

$$Z_{\text{rot}} = \frac{8\pi^2 IkT}{h^2}\,.$$

From the rotational partition function we obtain

$$\tilde{E}_{\text{rot}} = N_0 kT^2\left(\frac{\partial \ln Z_{\text{rot}}}{\partial T}\right) = N_0 kT^2\frac{\mathrm{d}}{\mathrm{d}T}\left(\ln\frac{8\pi^2 IkT}{h^2}\right) = N_0 kT$$

so that

$$\tilde{C}_{\text{rot}} = N_0 k\,.$$

The rotational entropy is given by

$$\tilde{S}_{\text{rot}} = \frac{E_{\text{rot}}}{T} + N_0 k \ln Z_{\text{rot}} = N_0 k + N_0 k \ln\left(\frac{8\pi^2 IkT}{h^2}\right)$$

$$= N_0 k\left(1 + \ln IT + \ln\frac{8\pi^2 k}{h^2}\right),$$

which can be written in the alternative form

$$\tilde{S}_{\text{rot}} = N_0 k\left(1 + \ln\frac{T}{\theta_{\text{r}}}\right).$$

(b) The rotational energy is given by

$$E_{\text{rot}} = NkT^2\frac{\mathrm{d}\ln Z_{\text{rot}}}{\mathrm{d}T} = -\frac{Nk\theta}{Z}\frac{\mathrm{d}Z}{\mathrm{d}(\theta/T)}.$$

Using the rotational partition function in the form

$$Z_{\text{rot}} = \sum(2J+1)\exp\left[-\frac{J(J+1)\theta_{\text{r}}}{T}\right],$$

we obtain

$$E_{\text{rot}} = \frac{Nk\theta}{Z}\sum(2J+1)(J^2+J)\exp\left[-\frac{J(J+1)\theta_{\text{r}}}{T}\right]$$

and

$$C_{\text{rot}} = \frac{Nk}{Z^2}\left(\frac{\theta}{T}\right)^2\sum(2J+1)(J^2+J)^2\exp\left[-\frac{J(J+1)\theta_{\text{r}}}{T}\right].$$

In the limit of low temperature the partition function approaches unity, and only the first term of the summations is important. When $J = 1$ we have

$$\tilde{E}_{\text{rot}} = 6N_0 k\left(\frac{\theta}{T}\right)\exp\left(-2\frac{\theta}{T}\right),$$

$$\tilde{C}_{\text{rot}} = 12N_0 k\left(\frac{\theta}{T}\right)\exp\left(-2\frac{\theta}{T}\right).$$

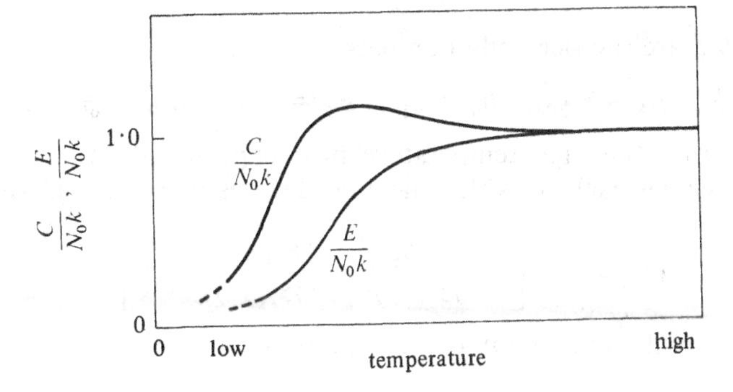

Figure 4.4.1.

As we know that the high temperature limits are $E_{rot} = N_0kT$ and $\bar{C}_{rot} = N_0k$, we can sketch the rotational energy and heat capacity curves (see Figure 4.4.1). The point of inflection in the energy curve results in a maximum in the corresponding heat capacity curve.

A CONVENIENT FORMULA FOR THE HIGH TEMPERATURE ROTATIONAL PARTITION FUNCTION

4.5 The partition function for a molecule in which two degrees of rotational freedom are possible is given by the equation

$$Z_{rot} = \sum_0^\infty (2J+1)\exp\left[-\frac{J(J+1)\theta_r}{T}\right].$$

If θ_r/T is large, Z_{rot} can be evaluated only by direct summation. If $\theta_r/T < 1$, it is possible to construct a simple and convenient expression for Z_{rot} which does not contain a summation of exponential terms. Mulholland[2] used the Euler–Maclaurin expansion, which expresses the difference between the (unknown) sum and the corresponding (known) integral in polynomial form, to obtain a simple expression for Z_{rot}.

(a) Obtain an expression for the rotational partition function of a heteronuclear diatomic molecule as a polynomial in θ_r/T using the Euler-Maclaurin summation formula in the form

$$\sum_{n=0}^\infty f(n) = \int_0^\infty f^0(J)\,dJ + \tfrac{1}{2}f^0(0) - \frac{1}{12}f^1(0) + \frac{1}{720}f^3(0) - \frac{1}{30\,240}f^5(0),$$

where $f^k(0)$ is the kth derivative of the (J) function with $J = 0$.

(b) Show that the polynomial obtained in part (a) is given by the general formula

$$Z_{rot} = \frac{T}{\theta_r}\exp\left(\frac{\theta_r}{4T}\right)\left[1 + \sum_{n=0}^\infty a_n\left(\frac{\theta_r}{T}\right)^{n+1}\right],$$

where

$$a_n = \frac{(-1)^n(1-2^{-2n-1})}{(n+1)!}B_{2n+2}$$

and where B_{2n} are the Bernoulli numbers

$$B_2 = \tfrac{1}{6},\quad B_4 = -\tfrac{1}{30},\quad B_6 = \tfrac{1}{42},\quad B_8 = -\tfrac{1}{30},\quad B_{10} = \tfrac{5}{66},\quad \ldots.$$

(c) Show that the high temperature heat capacity of an assembly of $N = N_0$ molecules each of which has two degrees of rotational freedom is given by

$$C_{rot} = Nk\left[1 + \frac{1}{45}\left(\frac{\theta_r}{T}\right)^2 + \frac{16}{945}\left(\frac{\theta_r}{T}\right)^3 + \ldots\right]$$

and obtain a similar expression for the rotational entropy.

(2) H.P.Mulholland, *Proc.Cambridge Phil.Soc.*, **24**, 280 (1928).

Solution

(a) Putting

$$f^0(J) = (2J+1)\exp\left[-\frac{J(J+1)\theta_r}{T}\right]$$

we have that $\int_0^\infty f(J)\,dJ$ is of the form $\frac{T}{\theta_r}\int_0^\infty e^{-\alpha}\,d\alpha = \frac{T}{\theta_r}$ and when $J = 0$, $f^0(0) = 1$.

The higher derivatives are readily obtained

$$f^1(0) = \left[2-\frac{(2J+1)^2\theta_r}{T}\right]\exp\left[-\frac{J(J+1)\theta_r}{T}\right],$$

so that when $J = 0$

$$f^1(0) = 2-\frac{\theta_r}{T},$$

$$f^3(0) = -12\frac{\theta_r}{T} + 12\left(\frac{\theta_r}{T}\right)^2 - \left(\frac{\theta_r}{T}\right)^3,$$

$$f^5(0) = 120\left(\frac{\theta_r}{T}\right)^2 - 180\left(\frac{\theta_r}{T}\right)^3 + 30\left(\frac{\theta_r}{T}\right)^4 - \left(\frac{\theta_r}{T}\right)^5.$$

Substituting into the Euler-Maclaurin formula and collecting terms we obtain

$$Z_{rot} = \frac{T}{\theta_r}\left[1+\frac{1}{3}\frac{\theta_r}{T}+\frac{1}{15}\left(\frac{\theta_r}{T}\right)^2+\frac{4}{315}\left(\frac{\theta_r}{T}\right)^3+\cdots\right].$$

(b) The general formula gives

$$Z_{rot} = \frac{T}{\theta_r}\exp\frac{\theta_r}{4T}\left[1+\frac{1}{12}\frac{\theta_r}{T}+\frac{7}{480}\left(\frac{\theta_r}{T}\right)^2+\frac{31}{8064}\left(\frac{\theta_r}{T}\right)^3+\cdots\right];$$

expanding the exponential we have

$$\exp\frac{\theta_r}{4T} = 1+\frac{1}{4}\frac{\theta_r}{T}+\frac{1}{32}\left(\frac{\theta_r}{T}\right)^2+\frac{1}{384}\left(\frac{\theta_r}{T}\right)^3+\cdots.$$

On multiplying out and collecting terms we obtain the formula derived in part (a).

(c) To obtain the heat capacity we make use of the formulae

$$F = -NkT\ln Z_{rot}, \qquad E = -T^2\frac{d(F/T)}{dT}, \qquad C_v = \left(\frac{\partial E}{\partial T}\right)_v.$$

Now

$$\ln Z(T) = -\ln\frac{\theta_r}{T} - \ln(1-x) = -\left(\ln\frac{\theta_r}{T}+x+\frac{x^2}{2}+\frac{x^3}{3}+\frac{x^4}{4}+\cdots\right)$$

where

$$x = -\left[\frac{1}{3}\frac{\theta_r}{T}+\frac{1}{15}\left(\frac{\theta_r}{T}\right)^2+\frac{4}{315}\left(\frac{\theta_r}{T}\right)^3+\cdots\right].$$

Multiplying out and collecting terms, we find

$$\ln Z(T) = -\left[\ln\frac{\theta_r}{T} - \frac{1}{3}\frac{\theta_r}{T} - \frac{1}{90}\left(\frac{\theta_r}{T}\right)^2 - \frac{8}{2835}\left(\frac{\theta_r}{T}\right)^3 + \cdots\right]$$

from which we obtain

$$E = NkT\left[1 - \frac{1}{3}\frac{\theta_r}{T} - \frac{1}{45}\left(\frac{\theta_r}{T}\right)^2 - \frac{8}{945}\left(\frac{\theta_r}{T}\right)^3 + \cdots\right].$$

Note that as $T \to \infty$ the energy approaches the limiting value of $Nk(T - \frac{1}{3}\theta_r)$ rather than the classical value NkT. Terms in $\frac{1}{3}\theta_r$ do not occur in either the entropy or the heat capacity:

$$C_v = Nk\left[1 + \frac{1}{45}\left(\frac{\theta_r}{T}\right)^2 + \frac{16}{945}\left(\frac{\theta_r}{T}\right)^3 + \cdots\right].$$

Using $S = E/T + Nk\ln Z$ we obtain

$$S = Nk\left[1 - \frac{1}{3}\frac{\theta_r}{T} - \frac{1}{45}\left(\frac{\theta_r}{T}\right)^2 - \frac{8}{945}\left(\frac{\theta_r}{T}\right)^3\right] - Nk\left[\ln\frac{\theta_r}{T} - \frac{1}{3}\frac{\theta_r}{T} - \frac{1}{90}\left(\frac{\theta_r}{T}\right)^2 - \frac{8}{2835}\left(\frac{\theta_r}{T}\right)^3\right]$$

$$= Nk\left[1 - \ln\frac{\theta_r}{T} - \frac{1}{90}\left(\frac{\theta_r}{T}\right)^2 - \frac{16}{2835}\left(\frac{\theta_r}{T}\right)^3 + \cdots\right].$$

The limiting value Nk of the heat capacity as $T \to \infty$ is approached from above. This is in accordance with the ideas of Problem 4.4.

THERMODYNAMIC PROPERTIES ARISING FROM SIMPLE HARMONIC MODES OF VIBRATION

4.6 (a) In Problem 3.9 it was shown that the partition function for a [1]-oscillator of angular frequency $\omega = 2\pi\nu$ is given by

$$Z_{\text{vib}} = \frac{\exp(-\frac{1}{2}\hbar\omega/kT)}{1 - \exp(-\hbar\omega/kT)}.$$

Show that the heat capacity and entropy of a [1]-oscillator are independent of the amount of energy which may be possessed by the oscillator in its lowest vibrational state.

(b) The vibrational partition function for a polyatomic molecule with i vibrational modes may be written as the product of the partition functions of i distinguishable [1]-oscillators [see Problem 3.9(d)].

$$Z_{\text{vib}} = \prod_i \frac{1}{1 - \exp(-\hbar\omega_i/kT)}.$$

A non-linear polyatomic molecule has $3n-6$ vibrational modes, and a linear molecule has $3n-5$. Of the $3n-6$ modes $n-1$ are stretching vibrations and these generally have frequencies $>1000\ \text{cm}^{-1}$. The remaining $2n-5$ are bending modes, and generally have frequencies considerably lower. Vibrational modes are excited at high temperatures, and an error in the assignment of these modes does not usually introduce an appreciable error into the calculated thermodynamic quantities.

It was shown in Problem 3.9(e) that the heat capacity of N identical distinguishable quantum [1]-oscillators is $C_{\text{vib}} = NkE(2x)$ where $x = \hbar\omega/2kT$ and

$$E(y) = \frac{y^2 e^y}{(e^y - 1)^2}.$$

By graphical interpolation of the Einstein function $E(y)$ tabulated below calculate the molar vibrational heat capacity and entropy at $298{\cdot}15°\text{K}$ of O_2, Cl_2, and Br_2 for which the fundamental vibrational frequencies are 1580, 565, and 323 cm^{-1} respectively.

y	$\tilde{C}_{\text{vib}}/N_0k$	$\tilde{S}_{\text{vib}}/N_0k$	y	$\tilde{C}_{\text{vib}}/N_0k$	$\tilde{S}_{\text{vib}}/N_0k$	y	$\tilde{C}_{\text{vib}}/N_0k$	$\tilde{S}_{\text{vib}}/N_0k$	y	$\tilde{C}_{\text{vib}}/N_0k$	$\tilde{S}_{\text{vib}}/N_0k$
0	1·000	∞	3·0	0·496	0·208	5·5	0·125	0·028	8·0	0·021	0·003
0·5	0·979	1·704	3·5	0·393	0·139	6·0	0·090	0·017	8·5	0·015	0·002
1·0	0·921	1·041	4·0	0·304	0·093	6·5	0·064	0·011	9·0	0·010	0·001
1·5	0·832	0·683	4·5	0·230	0·066	7·0	0·045	0·007	9·5	0·007	0·001
2·0	0·724	0·458	5·0	0·171	0·041	7·5	0·031	0·005	10·0	0·005	0·000
2·5	0·609	0·309									

(c) By expanding the exponential term in the vibrational partition function and retaining only terms up to $(\theta_v/T)^2$, obtain simple approximate formulae for the heat capacity and entropy of a [1]-oscillator which are valid in the region of temperature for which $(\theta_v/T) < 1$.

Use these formulae to calculate the vibrational heat capacity of diatomic iodine vapour at 100°C and compare with the values obtained using the Einstein functions. (The fundamental vibrational frequency of I_2 is $214{\cdot}6\ \text{cm}^{-1}$.)

Solution

(a) Rather than rewrite the numerator of the partition function in the form $\exp(E_{\text{min}}/kT)$, we will carry the calculation through using $E_{\text{min}} = \frac{1}{2}h\nu$.

The energy of the [1]-oscillator is given by

$$E_{\text{vib}} = NkT^2\left(\frac{\partial \ln Z_{\text{vib}}}{\partial T}\right)_v$$

so that

$$E_{\text{vib}} = NkT^2\frac{d}{dT}\left\{\ln\frac{1}{1-\exp(-h\nu/kT)} + \ln[\exp(-\tfrac{1}{2}h\nu/kT)]\right\}$$

$$= Nk\left(\frac{h\nu}{k}\right)\frac{1}{\exp(h\nu/kT)-1} + \tfrac{1}{2}Nh\nu\,.$$

The right hand term, which is the minimum (zero point) energy of the oscillator, disappears on subsequent differentiation. For the heat capacity we have

$$C_{vib} = \left(\frac{\partial E}{\partial T}\right)_v = Nk\left(\frac{h\nu}{k}\right)\frac{d}{dT}\frac{1}{\exp(h\nu/kT)-1}$$

$$= Nk\left(\frac{h\nu}{kT}\right)^2\frac{\exp(h\nu/kT)}{[\exp(h\nu/kT)-1]^2} .$$

For the entropy of the [1]-oscillator we have

$$S_{vib} = NkT\left(\frac{\partial \ln Z_{vib}}{\partial T}\right) + Nk\ln Z_{vib} ,$$

so that

$$S_{vib} = NkT\frac{h\nu}{kT^2}\frac{1}{\exp(h\nu/kT)-1} + NkT\frac{d}{dT}\ln[\exp(-\tfrac{1}{2}h\nu/kT)]$$

$$-Nk\ln[1-\exp(-h\nu/kT)] + Nk\ln[\exp(-\tfrac{1}{2}h\nu/kT)] ,$$

which yields

$$S_{vib} = Nk\left(\frac{h\nu}{kT}\right)\frac{1}{\exp(h\nu/kT)-1} + \tfrac{1}{2}\frac{Nh\nu}{T}$$

$$-Nk\ln[1-\exp(-h\nu/kT)] - \tfrac{1}{2}\frac{Nh\nu}{T} .$$

Terms arising from the zero point energy cancel out.

(b) We will denote the spectroscopic units, which are wave numbers, by w (cm^{-1}). Multiplication by the velocity of light c converts wave numbers into frequency ν. Using the numerical values of the fundamental constants we have that

$$2x = \frac{\hbar\omega}{kT} = \frac{h\nu}{kT} = \frac{hcw}{kT} = 1{\cdot}4387\frac{w}{T} = \frac{\theta_v}{T} ,$$

where θ_v is the vibrational characteristic temperature of the [1]-oscillator and θ_v/T is y in the table given.

For oxygen at $298{\cdot}15°K$, $\theta_v/T = 1{\cdot}4387 \times 1580/298{\cdot}15 = 7{\cdot}62$. Graphical interpolation of the Einstein functions gives $\tilde{C}_{vib}/N_0k = 0{\cdot}027$ and $\tilde{S}_{vib}/N_0k = 0{\cdot}0044$, so that $\tilde{C}_{vib} = 0{\cdot}054$ cal mole^{-1} deg^{-1} and $\tilde{S}_{vib} = 0{\cdot}0088$ cal mole^{-1} deg^{-1}. Similarly, for chlorine and bromine we obtain $\tilde{C}_{vib} = 1{\cdot}095$ and $1{\cdot}64$ cal mole^{-1} deg^{-1}, and $\tilde{S}_{vib} = 0{\cdot}52$ and $1{\cdot}32$ cal mole^{-1} deg^{-1}, respectively. These values agree well with experimental values. However, at temperatures for which $\theta_v/T \ll 1$, thermodynamic functions calculated from the harmonic oscillator model are not in good agreement with experiment.

(c) Expanding $Z_{vib} = [1-\exp(-\theta_v/T]^{-1}$ we obtain

$$Z^{-1} = \frac{\theta_v}{T}\left[1 - \tfrac{1}{2}\left(\frac{\theta_v}{T}\right) + \tfrac{1}{6}\left(\frac{\theta_v}{T}\right)^2 - \cdots\right].$$

We use the logarithmic series expansion (which is valid only when $\theta_v/T < 1$) to obtain $\ln Z^{-1} = -\ln Z$;

$$\ln(1-x) = -x - \frac{x^2}{2} - \frac{x^3}{3} = \ln\left[1 - \left\{\tfrac{1}{2}\left(\frac{\theta_v}{T}\right) - \tfrac{1}{6}\left(\frac{\theta_v}{T}\right)^2 + \cdots\right\}\right]$$

$$= -\tfrac{1}{2}\left(\frac{\theta_v}{T}\right) + \tfrac{1}{6}\left(\frac{\theta_v}{T}\right)^2 - \tfrac{1}{8}\left(\frac{\theta_v}{T}\right)^2 \cdots$$

The free energy F, energy E, and vibrational heat capacity C_{vib}, and the entropy S_{vib} are now readily obtained.

$$F = -NkT\ln Z = NkT\left[\ln\left(\frac{\theta_v}{T}\right) - \tfrac{1}{2}\left(\frac{\theta_v}{T}\right) + \tfrac{1}{24}\left(\frac{\theta_v}{T}\right)^2 - \cdots\right],$$

$$E = NkT^2\frac{\mathrm{d}}{\mathrm{d}T}\ln Z = NkT\left[1 - \tfrac{1}{2}\left(\frac{\theta_v}{T}\right) + \tfrac{1}{12}\left(\frac{\theta_v}{T}\right)^2 - \cdots\right],$$

$$C_{vib}\left(\frac{\partial E}{\partial T}\right)_v = Nk\left[1 - \tfrac{1}{12}\left(\frac{\theta_v}{T}\right)^2 + \cdots\right],$$

$$S_{vib} = NkT\left(\frac{\partial \ln Z}{\partial T}\right) + Nk\ln Z = Nk\left[1 - \ln\left(\frac{\theta_v}{T}\right) + \tfrac{1}{24}\left(\frac{\theta_v}{T}\right)^2 - \cdots\right].$$

For iodine vapour (I_2) we have

$$\theta_v/T = 214{\cdot}6 \times 1{\cdot}4387/373{\cdot}15 = 0{\cdot}827\,.$$

From the approximate equation we have

$$\tilde{C}_{vib} = N_0k[1 - \tfrac{1}{12}(0{\cdot}827)^2] = 0{\cdot}943N_0k = 1{\cdot}873 \text{ cal mole}^{-1}\text{ deg}^{-1}\,,$$

and by graphical interpolation of the Einstein function we have

$$\tilde{C}_{vib} = 0{\cdot}945N_0k = 1{\cdot}877 \text{ cal mole}^{-1}\text{ deg}^{-1}\,.$$

At $\theta_v/T = 1$ the approximate equation yields $\tilde{C}_{vib}/N_0k = 0{\cdot}917$, which differs from the value given by the Einstein function (0·921) by only about a quarter of one percent.

Similarly, for the entropy the approximate equation yields $\tilde{S}_{vib}/N_0k = 1{\cdot}218$ whereas the value obtained by graphical interpolation of the Einstein function is 1·215.

CORRECTIONS TO THE RIGID ROTATOR-HARMONIC OSCILLATOR MODEL

4.7 (a) To treat the vibration of a diatomic molecule as simple harmonic motion is unrealistic because on this model dissociation of the molecule can never occur. A more realistic potential function was given by Morse, who proposed the equation

$$u(r) = D_e\{1 - \exp[-\beta(r - r_e)]^2\}\,,$$

where D_e is the dissociation energy of the molecule, β is an empirical constant, r is the separation of the nuclei of the two atoms, and r_e is the equilibrium separation.

Show that at low amplitudes of vibration this model allows simple harmonic oscillation, and show that the constant β is then given by

$$\beta = w_e \left(\frac{2\pi^2 \mu c^2}{D_e} \right)^{1/2} ,$$

where w_e is the frequency of vibration of the molecule, expressed in wave numbers, about its equilibrium position, and where μ is the reduced mass of the molecule.

(b) When the amplitude of vibration of a diatomic molecule becomes large the simple harmonic oscillator model is inadequate for precise work and a contribution to the thermodynamic properties due to anharmonic vibration must be considered. The vibrational spectra of diatomic molecules are often represented empirically by energy levels which are described by the equation

$$\epsilon_V = h\nu_e(V+\tfrac{1}{2}) - x_e h\nu_e(V+\tfrac{1}{2})^2 + y_e h\nu_e(V+\tfrac{1}{2})^3 ,$$

where V is the vibrational quantum number and where x_e, y_e, etc. are called anharmonicity constants. When the Morse potential is put into the Schrödinger equation, the following approximate expression for the allowed energy levels is obtained

$$\epsilon_V = h\nu_e(V+\tfrac{1}{2}) - \frac{(h\nu_e)^2(V+\tfrac{1}{2})^2}{4D_e} .$$

Given that the fundamental vibration frequency of $^{35}Cl_2$ is $564{\cdot}9\ \mathrm{cm}^{-1}$ and the coefficient β of the Morse potential is $2{\cdot}05 \times 10^8\ \mathrm{cm}^{-1}$, calculate the molar dissociation energy $\tilde{D}_e$ in kilocalories and the anharmonicity constant x_e of chlorine gas.

(c) Show that for an anharmonic oscillator, whose energy levels are described by the equation

$$\epsilon_V = h\nu(V+\tfrac{1}{2}) - x_e h\nu(V+\tfrac{1}{2})^2 ,$$

the corresponding partition function is

$$Z_{\mathrm{anh}} = \frac{\exp(\frac{1}{4}x_e u - \frac{1}{2}u)}{1 - e^{-u}} \left[1 + \frac{2x_e u}{(e^u - 1)^2} \right] ,$$

where $u \equiv h\nu/kT \equiv \theta/T$.

[Hint: Expand the exponential expression containing x_e and retain only the leading term. Simplify the summation over this term by utilizing the first and second derivatives with respect to u of the simple harmonic oscillator function.]

(d) It can be shown that terms which allow for the fact that the moment of inertia of a vibrating rotator is greater than that of a nonvibrating rotator (δ term), and which allow for the centrifugal stretching of the molecule (γ term), must also be added to the logarithm of the partition function:

$$\ln Z_\delta = \frac{\delta}{e^{\theta/T}-1}, \quad \text{where } \delta \equiv 6\frac{B_e}{w_e}\left[\left(\frac{w_e x_e}{B_e}\right)^{1/2} - 1\right];$$

$$\ln Z_\gamma = \frac{8\gamma}{\theta/T}, \quad \text{where } \gamma \equiv \frac{B_e}{w_e}.$$

B_e is the rotational constant $(h/8\pi^2 I)$ calculated with the atoms at their equilibrium separation r_e.

For the above terms in x_e, δ, and γ obtain expressions for the corresponding terms which must be added to the heat capacity of the molecule.

(e) The equilibrium separation of the nuclei in a chlorine molecule is $1{\cdot}988 \times 10^{-8}$ cm and the corresponding value of B_e is $0{\cdot}2438\ \text{cm}^{-1}$. Calculate the contribution to the heat capacity of chlorine gas at $\theta/T = 2$ (i.e. $133{\cdot}2$°C) arising from the above terms, and express these contributions as a percentage of the heat capacity obtained from the simple harmonic oscillator model.

Solution

(a) At low amplitudes of vibration we may expand the exponential of the Morse potential and examine the leading term

$$u(r) = D_e\{1 - \exp[-\beta(r-r_e)]^2\}.$$

Using $e^{-x} = 1 - x + \frac{1}{2}x^2 - \frac{1}{6}x^3$ and putting $x = \beta(r-r_e)$ we obtain

$$u(r) = -D_e[\beta^2(r-r_e)^2 - \beta^3(r-r_e)^3 + \tfrac{5}{12}\beta^4(r-r_e)^4 - \cdots].$$

From the leading term for the potential energy we obtain the force F:

$$F = -2D_e\beta^2(r-r_e).$$

The equations of motion for the two atoms of mass m_1 and m_2 at distances r_1 and r_2 from the centre of mass are given by

$$m_1\frac{d^2 r_1}{dt^2} = -2D_e\beta^2(r_1+r_2-r_e),$$

$$m_2\frac{d^2 r_2}{dt^2} = -2D_e\beta^2(r_1+r_2-r_e),$$

so that

$$\frac{d^2(r_1+r_2)}{dt^2} = -2D_e\beta^2(r_1+r_2-r_e)\left(\frac{1}{m_1}+\frac{1}{m_2}\right)$$

which can be written

$$\mu\frac{d^2 r}{dt^2} = -2D_e\beta^2(r-r_e),$$

where $r_1+r_2 = r$, and $1/m_1 + 1/m_2 = 1/\mu$.

When a mass μ is constrained by a force constant of $-2D_e\beta^2$ the frequency of vibration ν is

$$\nu = \frac{1}{2\pi}\left(\frac{2\beta^2 D_e}{\mu}\right)^{1/2}.$$

Using $\nu = cw_e$ we obtain

$$\beta = w_e\left(\frac{2\pi^2\mu c^2}{D_e}\right)^{1/2}.$$

(b) The reduced mass μ of $^{35}Cl_2$ is given by

$$\mu = \frac{m_1 m_2}{m_1 + m_2},$$

where $m_1 = m_2 = 35/N_0$, so that $\mu = 17{\cdot}5/N_0$.

We note that if in the numerator of our equation for β we use $17{\cdot}5/N_0$, then in the denominator we use the dissociation energy per molecule $D_e = \tilde{D}_e/N_0$, and the N_0 cancels out. Hence we obtain for the molar dissociation energy

$$\tilde{D}_e = \frac{w^2}{\beta^2}(2\pi^2\mu c^2)$$

$$= \left(\frac{564{\cdot}9}{2{\cdot}05 \times 10^8}\right)^2 (2 \times 3{\cdot}1416^2 \times 17{\cdot}5 \times 2{\cdot}9979^2 \times 10^{20})$$

$$= 2{\cdot}3574 \times 10^{12} \text{ erg mole}^{-1} = 56{\cdot}34 \text{ kcal mole}^{-1}.$$

Comparison of the empirical equation for the vibrational energy levels of a diatomic molecule with the form obtained from the Morse potential gives

$$x_e = \frac{h\nu_e}{4D_e} = \frac{hcw_e}{4\tilde{D}_e/N}$$

$$= \frac{6{\cdot}6256 \times 10^{-27} \times 2{\cdot}9979 \times 10^{10} \times 564{\cdot}9 \times 6{\cdot}0225 \times 10^{23}}{4 \times 2{\cdot}3574 \times 10^{12}}$$

$$= 7{\cdot}166 \times 10^{-3}.$$

(c) The partition function for the anharmonic oscillator is

$$Z_{anh} = \sum_{V=0}^{\infty} \exp[-u(V+\tfrac{1}{2}) + ux_e(V+\tfrac{1}{2})^2]$$

$$= \sum_{V=0}^{\infty} \exp\{(\tfrac{1}{4}x_e u - \tfrac{1}{2}u)[-uV + x_e uV(V+1)]\}$$

$$= \exp(\tfrac{1}{4}x_e u - \tfrac{1}{2}u) \sum_{V=0}^{\infty} e^{-uV}\{1 + \exp[x_e(V+1)]\}.$$

Expanding the exponential term in x_e we obtain

$$Z_{anh} = \text{const}\left[\sum_{V=0}^{\infty} e^{-uV} + \sum_{V=0}^{\infty} x_e uV(V+1)e^{-uV}\right].$$

The first term is the simple harmonic oscillator (SHO) partition function equal to $1/(1-e^{-u})$. We write the anharmonicity term in the form

$$x_e u \sum V e^{-uV} + V^2 e^{-uV} .$$

Now

$$\sum e^{-uV} = \frac{1}{1-e^{-u}}$$

and

$$\frac{d}{du}\sum e^{-uV} = \sum V e^{-uV} = -\frac{e^{-u}}{(1-e^{-u})^2}$$

and

$$\frac{d^2}{du^2}\sum e^{-uV} = \sum V^2 e^{-uV} = e^{-u}(1-e^{-u})^2 + \frac{2e^{-2u}}{(1-e^{-u})^3} .$$

Using these expressions the anharmonicity term becomes

$$x_e u \frac{2e^{-2u}}{(1-e^{-u})^3} ,$$

which can be written

$$\frac{2x_e u}{(1-e^{-u})(e^u-1)^2} ,$$

so that finally we obtain

$$Z_{\text{anh}} = \frac{\exp(\frac{1}{4}x_e u - \frac{1}{2}u)}{1-e^{-u}}\left[1 + \frac{2x_e u}{(e^u-1)^2}\right] .$$

(d) The correction term which must be added to the SHO partition function is obtained by quite straightforward differentiation:

$$\ln Z_{\text{corr}} = \tfrac{1}{4}x_e u - \tfrac{1}{2}u + 2x_e u(e^u-1)^2 + \frac{\delta}{e^u-1} + \frac{8\gamma}{u} .$$

Using

$$E_{\text{corr}} = NkT^2\frac{d\ln Z_{\text{corr}}}{dT}$$

we obtain for the four terms

$$E_{\text{corr}} = Nk\theta(-\tfrac{1}{4}x_e + \tfrac{1}{2}) + Nk \times 2x_e\theta\frac{e^{\theta/T}(2\theta/T-1)+1}{(e^{\theta/T}-1)^3}$$
$$+ Nk\delta\theta\, e^{\theta/T}(e^{\theta/T}-1)^2 + NkT^2 \times \frac{8\gamma}{\vartheta}$$

and finally

$$C_{\text{corr}} = \left(\frac{dE_{\text{corr}}}{dT}\right)_V = Nk \times 4x_e(\theta/T)^2 e^{\theta/T}\frac{2(\theta/T)e^{\theta/T} - 2e^{\theta/T} + \theta/T + 2}{(e^{\theta/T}-1)^4}$$
$$+ Nk\delta(\theta/T)^2 e^{\theta/T}\frac{e^{\theta/T}+1}{(e^{\theta/T}-1)^3} + \frac{Nk \times 16\gamma}{\theta/T} .$$

(e) The constants γ and δ for chlorine are

$$\gamma = \frac{B_e}{w_e} = \frac{0\cdot 2438}{564\cdot 9} = 4\cdot 316 \times 10^{-4} ,$$

$$\delta = 6\frac{B_e}{w_e}\left[\left(\frac{w_e x_e}{B_e}\right)^{\frac{1}{2}} - 1\right] = 7\cdot 979 \times 10^{-3} .$$

Using $N_0 k = 1\cdot 987$ cal mole^{-1} deg^{-1}, $\theta/T = 2\cdot 0$, $e^{\theta/T} = 7\cdot 389$, we obtain the following values for the contributions to the heat capacity arising from the terms in x_e, δ, and γ.

$$C_{x_e} = 0\cdot 0190 \text{ cal mole}^{-1} \text{ deg}^{-1} ,$$

$$C_\delta = 0\cdot 0151 \text{ cal mole}^{-1} \text{ deg}^{-1} ,$$

$$C_\gamma = 0\cdot 0071 \text{ cal mole}^{-1} \text{ deg}^{-1} .$$

The heat capacity of a simple harmonic oscillator with $\theta/T = 2\cdot 0$ (from Problem 4.6) is $1\cdot 438$ cal mole^{-1} deg^{-1}, so the above terms contribute $1\cdot 32\%$, $1\cdot 05\%$, and $0\cdot 49\%$ respectively, a total of $2\cdot 86\%$.

CONTRIBUTIONS TO THE THERMODYNAMIC PROPERTIES ARISING FROM LOW LYING ELECTRONIC ENERGY LEVELS

4.8 (a) For certain molecules, notably oxygen and nitric oxide, there is an additional contribution to the thermodynamic properties, which arises from the presence of two electronic energy levels separated by an energy which is small compared with kT, so that thermal excitation is sufficient to populate appreciably the upper level. Construct the partition function for this two-level system and show that the electronic contribution to the heat capacity is given by

$$C_{el} = \omega N k e^{\epsilon/kT}\left(\frac{\epsilon/kT}{e^{\epsilon/kT} + \omega}\right)^2 ,$$

where ω is the ratio ω_1/ω_0 of the degeneracies, ϵ is the energy of the upper level, and the energy of the lower level is taken as zero.

(b) Investigate the high- and the low-temperature limits of the electronic heat capacity, and show that $\ln C_{el}$ has a maximum value when

$$\frac{\epsilon}{kT} = \ln\omega + \ln\left(\frac{\epsilon/kT + 2}{\epsilon/kT - 2}\right).$$

(c) Obtain an expression for the electronic heat capacity of a molecule with three equally-spaced singly-degenerate energy levels. By taking numerical values of ϵ/kT between $1\cdot 0$ and $4\cdot 0$ compare the magnitude of the maximum in the electronic heat capacity of this molecule with that of a molecule in which the lower energy level is singly degenerate and the upper level has threefold degeneracy.

(d) The nitric oxide molecule has two doubly-degenerate energy levels which are separated by the unusually small gap of $\epsilon/k = 174°K$. The oxygen molecule has two levels which are separated by an energy gap of $\epsilon/k = 11\,300°K$, and at high temperatures the lower level is triply degenerate and the upper level is doubly degenerate. Estimate the temperature at which the electronic heat capacity has its maximum and calculate the corresponding molar electronic heat capacity for these molecules.

[Slide rule accuracy is adequate for parts (c) and (d) of this problem.]

Solution

(a) The partition function for a system with two energy levels is

$$Z = \omega_0 e^{-\epsilon_0/kT} + \omega_1 e^{-\epsilon_1/kT}.$$

Taking the zero of energy $\epsilon_0 = 0$ we have

$$Z = \omega_0(1 + \omega e^{-\epsilon/kT}),$$

where $\omega = \omega_1/\omega_0$ and $\epsilon = \epsilon_1 - \epsilon_0$.

The energy and heat capacity follow directly

$$E = NkT^2\left[\frac{d}{dT}\ln\omega_0 + \frac{d}{dT}\ln(1+\omega e^{-\epsilon/kT})\right] = \omega N\epsilon\frac{e^{-\epsilon/kT}}{\omega e^{-\epsilon/kT}+1} = \frac{\omega N\epsilon}{e^{\epsilon/kT}+\omega}$$

and

$$C = \left(\frac{\partial E}{\partial T}\right)_v = \omega Nk e^{\epsilon/kT}\left(\frac{\epsilon/kT}{e^{\epsilon/kT}+\omega}\right)^2.$$

(b) At high temperatures $\epsilon/kT \ll 1$, and

$$C \approx \omega Nk\left(\frac{\epsilon/kT}{1+\omega}\right)^2 \to 0.$$

At low temperatures $\epsilon/kT \gg 1$, and

$$C \approx \omega Nk(\epsilon/kT)^2 e^{\epsilon/kT} \to 0.$$

We locate the maximum in the heat capacity by putting the temperature derivative of $\ln C$ equal to zero:

$$\frac{C}{\omega Nk} = e^{\epsilon/kT}\frac{(\epsilon/kT)^2}{(e^{\epsilon/kT}+\omega)^2},$$

$$\ln\left(\frac{C}{\omega Nk}\right) = \frac{\epsilon}{kT} + 2\ln\left(\frac{\epsilon}{kT}\right) - 2\ln(e^{\epsilon/kT}+\omega),$$

$$\frac{d}{dT}\ln\left(\frac{C}{\omega Nk}\right) = -\frac{\epsilon}{kT^2} - \frac{2}{T} + \frac{2(\epsilon/kT^2)e^{\epsilon/kT}}{e^{\epsilon/kT}+\omega} = 0;$$

multiplying by T we have

$$\left(\frac{\epsilon}{kT}+2\right)(e^{\epsilon/kT}+\omega) = 2\frac{\epsilon}{kT}e^{\epsilon/kT}$$

and

$$e^{\epsilon/kT} + \omega = \frac{2(\epsilon/kT)e^{\epsilon/kT}}{\epsilon/kT+2},$$

$$1 + \frac{\omega}{e^{\epsilon/kT}} = \frac{2(\epsilon/kT)}{(\epsilon/kT+2)},$$

so that

$$\omega e^{-\epsilon/kT} = \frac{2(\epsilon/kT)}{\epsilon/kT+2} - 1 = \frac{(\epsilon/kT-2)}{(\epsilon/kT+2)};$$

taking logs we obtain as the condition for the maximum

$$\frac{\epsilon}{kT} = \ln\omega + \ln\left(\frac{\epsilon/kT+2}{\epsilon/kT-2}\right).$$

(c) The energies of the three level system are 0, ϵ, and 2ϵ, and the partition function is

$$Z = 1 + e^{-\epsilon/kT} + e^{-2\epsilon/kT}.$$

The energy and heat capacity follow directly

$$\tilde{E} = N_0\epsilon \frac{e^{-\epsilon/kT} + 2e^{-2\epsilon/kT}}{1 + e^{-\epsilon/kT} + e^{-2\epsilon/kT}},$$

from which we obtain

$$\tilde{C} = N_0 k\left(\frac{\epsilon}{kT}\right)^2 \frac{e^{-\epsilon/kT}(1 + e^{-2\epsilon/kT} + 4e^{-\epsilon/kT})}{(1 + e^{-\epsilon/kT} + e^{-2\epsilon/kT})^2}.$$

Working to three significant figures we calculate for the two-level system with $\omega = 3$ and for the three-level system with $\omega = 1$ the following values:

ϵ/kT	1·0	1·5	2·0	2·5	3·0	3·5	4·0
$\tilde{C}/N_0k$							
(two-level)	0·249	0·540	0·821	0·991	1·02	0·933	0·790
(three-level)	0·424	0·602	0·634	0·578	0·486	0·390	0·303

The maximum for the two-level system is 1·023 at $\epsilon/kT = 2{\cdot}8$ and the maximum for the three-level system is 0·637 at $\epsilon/kT = 1{\cdot}9$.

(d) We can locate the maximum in the electronic heat capacity using the equation

$$\frac{\epsilon}{kT} = \ln\omega + \ln\left(\frac{\epsilon/kT+2}{\epsilon/kT-2}\right).$$

For nitric oxide $\omega = 1$, and for oxygen $\omega = 2/2$. By trial-and-error substitution we find that the electronic heat capacity of nitric oxide has its maximum value at 72°K and the corresponding maximum heat capacity is 0·875 cal mole^{-1} deg^{-1}. The maximum for oxygen is at 4900°K and it has the value 0·616 cal mole^{-1} deg^{-1}.

CALCULATION OF THE THERMODYNAMIC PROPERTIES OF HCl FROM SPECTROSCOPIC DATA

4.9 The fundamental vibration-rotation spectrum of isotopically pure $H^{35}Cl$ is shown in the lower part of Figure 4.9.1. The relative intensity I_{rel} and frequency w of each line, expressed in wave numbers, is indicated below the spectrum. This type of spectrum arises in the following way. When the vibrational quantum number V increases by unity ($\Delta V = +1$) as a result of the absorption of radiation, the rotational energy levels may be affected in two ways. In the transition to a higher vibrational level, the increase in bond length of the molecule is accompanied by an increase in the moment of inertia. If the molecule absorbs only a small amount of energy, then it may fall into a lower rotational energy level ($\Delta J = -1$) and give rise to the P branch of the spectrum. If a large amount of energy is absorbed, there may be an increase in the rotational energy, sufficient to raise the molecule into a higher rotational level ($\Delta J = +1$). This gives rise to the R branch of the spectrum. The upper part of Figure 4.9.1 shows the first eight rotational energy levels of the molecule and the double headed arrows indicate the transitions which give rise to the line in the spectrum below the arrow. For $\Delta V = +1$ the rotational transitions which we have just discussed are indicated by upward pointing arrows, whereas for $\Delta V = -1$ the downward pointing arrows will pertain.

The selection rule for vibration-rotation spectra is $\Delta J = 0, \pm 1$. Now the vibration-rotation energy of a diatomic molecule on the rigid rotator-harmonic oscillator model is given by

$$E_{V,R} = (V+\tfrac{1}{2})hc\overline{w}+J(J+1)hcB ,$$

where $\overline{w}$ is the fundamental vibrational frequency expressed in wave numbers and B is the rotational constant $h/8\pi^2 Ic$ for a particular vibrational state. The constant c is the velocity of light so that B is also in wave numbers. The above equation represents just one energy level, so that the total vibrational-rotational energy of the molecule is given by the sum over all V of such terms. We now use the equation for $E_{V,R}$ together with the Bohr frequency condition

$$E'_{V,R}-E''_{V,R} = hcw ,$$

where the single prime and double primes indicate upper and lower vibrational states, and w is the frequency expressed in wave numbers, in conjunction with the selection rule ($\Delta V = +1$ and $\Delta J = -1$) for the P branch, to obtain a general expression for the difference of energy between any two levels. The set of frequencies $P(J)$ in the P branch is then given by

$$P(J) = \overline{w}-(B'-B'')J+(B'-B'')J^2 ,$$

where J may have any integral values 1, 2, 3 etc. other than zero.

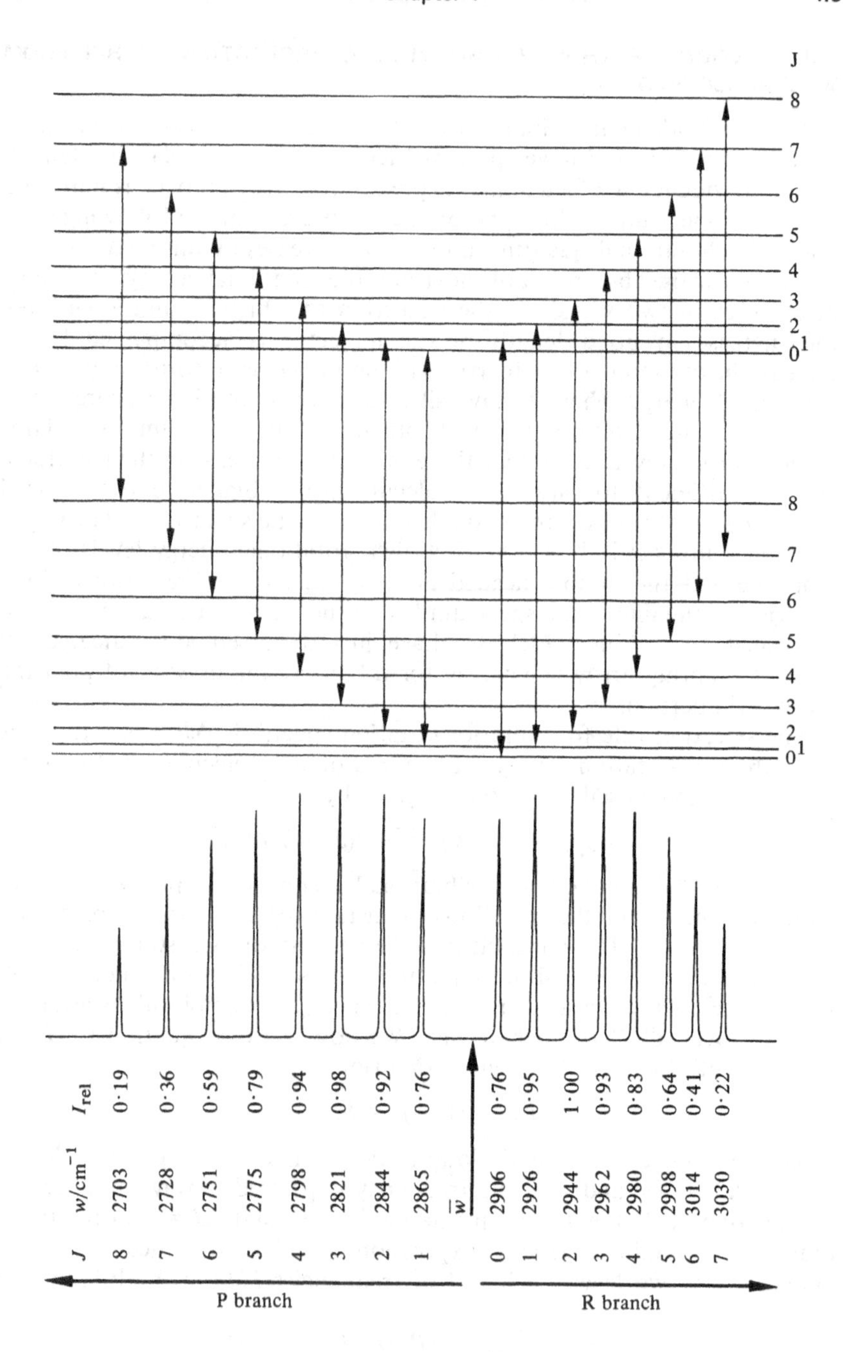

Figure 4.9.1. The vibration-rotation spectrum of $H^{35}Cl$ showing the energy levels and transitions which give rise to the P and R branches.

In the same way the set of frequencies $R(J)$ in the R branch is given by

$$R(J) = \overline{w} + 2B' + (3B' - B'')J + (B' + B'')J^2 ,$$

where J may now have any integral value including zero.

(a) Analyse the spectrum of $H^{35}Cl$ in the following way. Calculate the combination terms $R(J) - P(J)$ and $R(J-1) - P(J+1)$, and by considering them as a function of $2J+1$ determine B' and B'' respectively, hence obtain the fundamental vibrational frequency $\overline{w}$.

(b) The rotational constant in a vibrational state may be related to the equilibrium value B_e by the equation $B_V = B_e - (V + \frac{1}{2})\alpha$ where α is a vibration-rotation constant. Calculate B_e and hence obtain the equilibrium moment of inertia for $H^{35}Cl$. [$BI = h/8\pi^2 = 5{\cdot}0553 \times 10^{11}$ atomic mass units Å^2 Hz.]

(c) Given that the degeneracy of the Jth rotational energy level is $2J+1$ (cf Problem 4.4), derive an expression, based on the Boltzmann distribution law, for the population density of the rotational energy levels. Show that the most densely populated level is that for which

$$J_{\text{max}} = \left(\frac{kT}{2Bh}\right)^{1/2} - \frac{1}{2} .$$

(d) The area beneath an observed peak is proportional to the intensity of the line, and this is in turn proportional to the population density of the energy level. Absolute intensities are difficult to measure, but relative intensities are readily available from the observed spectrum. Show that a plot of $\ln(I_{\text{rel}}/2J+1)$ against $J(J+1)$ has a slope of $-hB/kT$. Analyse the R branch of the spectrum and determine the temperature of the gas.

(e) Calculate the entropy, Helmholtz free energy, and heat capacity of $H^{35}Cl$ at 300°K and 1 atm pressure. [The molecular weight of $H^{35}Cl$ is $35{\cdot}9877$.]

Solution

(a) Using the equations for $R(J)$ and $P(J)$ we readily obtain the combination terms

$$R(J) - P(J) = 2B' + (3B' + B'')J + (B' + B'')J + (B' - B'')J^2 - (B' - B'')J^2$$
$$= 2B'(J+1) ,$$

and

$$R(J-1) = \overline{w} + J(B' + B'') + J^2(B' - B'') ,$$
$$P(J+1) = \overline{w} + J(B' - 3B'') + J^2(B' - B'') - 2B'' ;$$

on subtraction we obtain

$$R(J-1) - P(J+1) = 2B''(2J+1) .$$

We can obtain the combination terms from the spectrum in the following manner:

J	0	1	2	3	4	5	6	7	8
$2J+1$		3	5	7	9	11	13	15	17
$P(J)$		2865	2844	2821	2798	2775	2751	2728	2703
$P(J+1)$	2865	2844	2821	2798	2775	2751	2728	2703	
$R(J)$	2906	2926	2944	2962	2980	2998	3014	3030	
$R(J-1)$		2906	2926	2944	2962	2980	2998	3014	3030
$R(J)-P(J)$		61	100	141	182	223	263	302	
$R(J-1)-P(J+1)$		62	105	146	187	229	270	311	

We now plot $R(J)-P(J)$ to $2J+1$ and obtain from the slope $B' = 10\cdot0558\ \text{cm}^{-1}$, and from a graph of $R(J-1)-P(J+1)$ to $2J+1$ we obtain $B'' = 10\cdot3448\ \text{cm}^{-1}$. Using these constants and the value of $P(J) = 2865\ \text{cm}^{-1}$ at $J = 1$, we obtain $\overline{w} = 2885\cdot7\ \text{cm}^{-1}$.

(b) For $V = 0$ we have $B_{V0} = 10\cdot3448\ \text{cm}^{-1}$, and for $V = 1$ we have $B_{V1} = 10\cdot0558\ \text{cm}^{-1}$.

We obtain the vibration-rotation constant α by eliminating B_e between the equations

$$10\cdot3448 = B_e - (0+\tfrac{1}{2})\alpha\,,$$

$$-10\cdot0558 = -B_e + (1+\tfrac{1}{2})\alpha\,;$$

addition gives $\alpha = 0\cdot289$, and back substitution of α gives $B_e = 10\cdot4888\ \text{cm}^{-1}$.

Using

$$B = \frac{h}{8\pi^2 Ic} = \frac{5\cdot0553\times10^{11}}{2\cdot9979\times10^{10}} = 16\cdot8628 \text{ atomic mass units Å}^2\ \text{cm}^{-1}$$

we calculate the equilibrium moment of inertia to be

$$I_e = \frac{16\cdot8628}{10\cdot4888} = 1\cdot60770 \text{ atomic mass units Å}^2$$

$$= \frac{1\cdot60770\times10^{-16}}{6\cdot0225\times10^{23}} = 2\cdot6695\times10^{-40}\ \text{g cm}^{-2}\,.$$

(c) The relative population of the rotational energy levels is given by the Boltzmann distribution law

$$\frac{n_J}{n} = g_J \exp(E_r/kT)\,.$$

For HCl the statistical weight g_J is simply the degeneracy of the level $2J+1$, and the energy E_r is given by

$$E_r = \frac{J(J+1)h^2}{\pi^2 I}$$

so that

$$\frac{E_r}{h} = J(J+1)B\,.$$

The total number of molecules occupying all the rotational levels is simply the rotational partition function, which can be approximated by the integral

$$\int_0^\infty (2J+1)\exp\left[-J(J+1)\frac{hB}{kT}\right]\mathrm{d}J = \frac{kT}{hB}\,.$$

The fraction of molecules in the Jth rotational state is therefore

$$\frac{n_J}{n} = (2J+1)\frac{hB}{kT}\exp\left[-J(J+1)\frac{hB}{kT}\right].$$

The value of J for which n_J/n is a maximum is given by

$$\frac{\mathrm{d}}{\mathrm{d}J}\left(\frac{n_J}{n}\right) = -(2J+1)(2J+1)\frac{hB}{kT}\exp\left[-J(J+1)\frac{hB}{kT}\right] + 2\exp\left[-J(J+1)\frac{hB}{kT}\right]$$

$$= 0$$

so that

$$J^2+J+\left(\tfrac{1}{4}-\frac{kT}{2hB}\right) = 0$$

and

$$J_{\max} = \left(\frac{kT}{2hB}\right)^{\frac{1}{2}} - \tfrac{1}{2}\,.$$

(d) For convenience we select the line $J(I_{\max})$ of greatest intensity $I_{\max}$ in the R branch, and we scale all other intensities relative to it

$$I_{\mathrm{rel}} = \frac{I_J}{I_{\max}} = \frac{hB/kT(2J+1)}{hB/kT(2J_{\mathrm{m}}+1)}\,\frac{\exp[-J(J+1)hB/kT]}{\exp[-J_{\mathrm{m}}(J_{\mathrm{m}}+1)hB/kT]}$$

so that

$$\ln I_{\mathrm{rel}} = \ln(2J+1) - \ln(2J_{\mathrm{m}}+1) + [J_{\mathrm{m}}(J_{\mathrm{m}}+1) - J(J+1)]\frac{hB}{kT}\,,$$

where the integer J_{m} is a constant.

Analysing the lines of the R branch we have:

$J(J+1)$	0	2	6	12	20	30	42	56
$2J+1$	1	3	5	7	9	11	13	15
I_{rel}	0·76	0·95	1·00	0·93	0·83	0·64	0·41	0·22
$\ln 100I_{\mathrm{rel}}/2J+1$	4·33	3·94	2·99	2·58	2·22	1·76	1·15	0·40

Plotting $\ln 100I_{\mathrm{rel}}/2J+1$ against $J(J+1)$ for values of $J > J(I_{\max})$ we obtain a straight line of slope $0{\cdot}05 \pm 0{\cdot}001$, so that

$$\frac{Bhc}{k\times 0{\cdot}05} = \frac{10{\cdot}488\times 6{\cdot}6256\times 10^{-27}\times 2{\cdot}9979\times 10^{10}}{1{\cdot}38054\times 10^{-16}\times 0{\cdot}05} = 302 \pm 15^\circ\mathrm{K}\,.$$

The intensity of the spectral lines is not a simple exponential function as the Boltzmann distribution law alone suggests, as there is a term which involves the dependence of the transition probability upon J which we have not considered. However when $J > J(I_{max})$ this term is small compared with the Boltzmann term[3].

(e) Formulae for the contributions to thermodynamic properties of a gas which arise from translational, rotational, and vibrational motion of the molecule have been obtained in Problems 4.1, 4.4 and 4.6. We shall consider first the vibrational contributions. From the fundamental vibrational frequency $\overline{w} = 2885{\cdot}7\ \text{cm}^{-1}$ we obtain the ratio $\theta_v/T = 1{\cdot}4387 \times 2885{\cdot}7/300 = 13{\cdot}839$. The value of $\exp(\theta_v/T)$ which occurs in the formulae for the vibrational thermodynamic functions is then approximately 1×10^6, so that the magnitude of these thermodynamic functions is exceedingly small, and they can be completely neglected. For the translational and rotational contributions we have for the entropy:

$$\begin{aligned}
\tilde{S}_{trans} &= 1{\cdot}98717[\tfrac{5}{2}\ln 300 - \ln 1 + \tfrac{3}{2}\ln 35{\cdot}9877 - 1{\cdot}1605] \\
&= 36{\cdot}71\ \text{cal mole}^{-1}\ \text{deg}^{-1}, \\
\tilde{S}_{rot} &= 1{\cdot}98717[1 + \ln 2{\cdot}66948 \times 10^{-40} + \ln 300 + \ln(8\pi^2 k/h^2)] \\
&= 1{\cdot}98717[1 - 91{\cdot}1215 + 5{\cdot}70378 + 88{\cdot}4077] \\
&= 7{\cdot}9288\ \text{cal mole}^{-1}\ \text{deg}^{-1}, \\
\tilde{S}_{trans} + \tilde{S}_{rot} &= 44{\cdot}6391\ \text{cal mole}^{-1}\ \text{deg}^{-1}.
\end{aligned}$$

This value is in excellent agreement with the calorimetric value of $44{\cdot}5$ cal mole^{-1} deg^{-1} obtained at $298{\cdot}1$°K for a mixture of $H^{35}Cl$ and $H^{37}Cl$ in their natural abundancies of 75·4% and 24·6% respectively.

Similarly for the Helmholtz free energy we have:

$$\begin{aligned}
\tilde{F}_{trans} &= -1{\cdot}98717 \times 300[\tfrac{5}{2}\ln 300 + \tfrac{3}{2}\ln 35{\cdot}9877 - \ln 1 - 2{\cdot}6605] \\
&= 10118{\cdot}8\ \text{cal mole}^{-1}, \\
\tilde{F}_{rot} &= -1{\cdot}98717 \times 300[\ln 300 + \ln 2{\cdot}66948 \times 10^{-40} + \ln(8\pi^2 k/h^2)] \\
&= -1782{\cdot}5\ \text{cal mole}^{-1}, \\
\tilde{F}_{trans} + \tilde{F}_{rot} &= -11901\ \text{cal mole}^{-1}.
\end{aligned}$$

Finally for the heat capacity C_p we have:

$$\begin{aligned}
\tilde{C}_p &= \tilde{C}_v + R = \tfrac{3}{2}N_0 k_{trans} + \tfrac{2}{2}N_0 k_{rot} + N_0 k = 3{\cdot}5 \times 1{\cdot}98717 \\
&= 6{\cdot}956\ \text{cal mole}^{-1}\ \text{deg}^{-1}.
\end{aligned}$$

The calorimetric value of the heat capacity is $6{\cdot}96$ cal mole^{-1} deg^{-1}.

[3] G.Herzberg, *Molecular Spectra and Molecular Structure,* 2nd Edn. Vol.I, *Spectra of Diatomic Molecules* (Van Nostrand, Princeton), 1950, p.125.

THERMODYNAMIC PROPERTIES OF ETHANE

4.10 At 500°K and a pressure of 1 atm the calorimetric heat capacity C_p of ethane is 18·66 cal mole^{-1} deg^{-1} and the calorimetric entropy is 62·79 cal mole^{-1} deg^{-1}. The molecular weight of ethane is 30·047, and the principal moments of inertia of the molecule are 42·23, 42·23 and 10·81 x 10^{-40} g cm^2. The following vibrational frequencies (cm^{-1}) have been assigned from the spectrum:

C–H stretching	2955	2954	2996_d	2963_d
C–C stretching	993			
C–C bending			821_d	1190_d
CH_3 group deformation	1375	1375	1472_d	1460_d

Frequencies with the subscript d are doubly degenerate.

(a) The only assignment which has not been made is that of the twisting mode about the C–C bond. Calculate the contribution to the heat capacity and to the entropy which arises from this mode.

Investigate the following three models (denoted b, c, and d below) which purport to describe the nature of the motion in this mode by choosing (where possible) the parameters of the model to fit the heat capacity, and using these parameters to calculate the entropy:

(b) The free rotational model in which the methyl groups rotate freely about the C–C bond. The reduced moment of inertia I_r of the rotating group is given by

$$I_r = I_g \frac{I_m}{I_g} + I_m \,,$$

where I_g is the moment of inertia of the rotating group about the axis of rotation and I_m is the moment of inertia of the rest of the molecule about the same axis. The kinetic energy ϵ of the rotating group is given by

$$\epsilon = \frac{1}{2I_r} p_\theta^2 \,,$$

where p_θ is the momentum conjugate to rotation about this axis at an angle θ. [Note: Include a symmetry number in the derived thermodynamic formulae.]

(c) The harmonic oscillator model in which the two methyl groups undergo torsional oscillation.

(d) The restricted rotator model in which at low temperatures only torsional oscillations take place, but as the temperature is increased the amplitude of these oscillations increases until a certain energy barrier is overcome, and thereafter rotation of the hindered group becomes possible. The potential barrier V is given by

$$V = \tfrac{1}{2} V_0 (1 - \cos n\theta) \,,$$

where $n = 3$ for ethane and V_0 is the maximum energy of the barrier.

Pitzer and Gwinn[4] tabulated thermodynamic functions for the hindered rotator model, and a section of their tables is reproduced below. In these tables Z is the numerical value of the classical partition function for the group with reduced moment of inertia I_r, and the tabulated functions have the dimensions of cal mole^{-1} deg^{-1}.

$\frac{V_0}{RT}$	Entropy Z^{-1}				Heat capacity Z^{-1}			
	0·25	0·30	0·35	0·40	0·25	0·30	0·35	0·40
2·0	3·355	3·004	2·709	2·458	1·632	1·606	1·574	1·541
2·5	3·180	2·836	2·548	2·303	1·840	1·801	1·756	1·717
3·0	3·008	2·667	2·380	2·138	1·996	1·952	1·900	1·846
3·5	2·838	2·500	2·218	1·978	2·106	2·054	1·995	1·934
4·0	2·678	2·343	2·069	1·834	2·168	2·110	2·048	1·980

Solution

(a) We calculate first the contributions to the heat capacity and the entropy which arise from motion within the ethane molecule about the C–C bond. This is done by subtracting the statistical thermodynamic quantity from the calorimetric quantity.

For translational motion we have [cf.Problem 4.1(b)]:

$$\tilde{S}_{\text{trans}} = N_0 k[\tfrac{5}{2}\ln T - \ln P + \tfrac{3}{2}\ln M - 1\cdot164]$$
$$= 1\cdot9872 \times 2\cdot3026[\tfrac{5}{2}\lg 500 - \lg 1 + \tfrac{3}{2}\lg 30\cdot047 - 1\cdot164]$$
$$= 38\cdot75 \text{ cal mole}^{-1}\text{ deg}^{-1},$$

$$C_{\text{trans}} = \tfrac{3}{2}N_0 k = 2\cdot980 \text{ cal mole}^{-1}\text{ deg}^{-1}.$$

For rotational motion we have [cf Problem 4.3(c)]:

$$\tilde{S}_{\text{rot}} = N_0 k[\tfrac{3}{2}\ln T + \tfrac{1}{2}\ln I_a I_b I_c - \ln\sigma + 134\cdot68]$$
$$= 1\cdot9872 \times 2\cdot3026[\tfrac{3}{2}\lg 500 + \tfrac{1}{2}\lg(42\cdot23 \times 42\cdot23 \times 10\cdot81 \times 10^{-120}) - \lg 6] + 1\cdot9872 \times 134\cdot68$$
$$= 17\cdot83 \text{ cal mole}^{-1}\text{ deg}^{-1},$$

$$\tilde{C}_{\text{rot}} = \tfrac{3}{2}N_0 k = 2\cdot980 \text{ cal mole}^{-1}\text{ deg}^{-1}.$$

For vibrational motion [cf. Problem 4.6(a), (b)] we obtain the entropy and heat capacity of each vibrational mode by graphical interpolation of the harmonic oscillator functions tabulated in Problem 4.6. Now $\theta_v/T = 1\cdot4387w/500$ so that we have

w (cm^{-1})	993	1375	1375	2954	2955
θ_v/T	2·86	3·95	3·95	8·50	8·50
$\tilde{C}_{\text{vib}}/N_0k$	0·525	0·315	0·315	0·013	0·013
$\tilde{S}_{\text{vib}}/N_0k$	0·232	0·097	0·097	0·002	0·002

(4) K.S.Pitzer and W.D.Gwinn, *J. Chem. Phys.*, **10**, 428 (1942).

Total for singly degenerate vibrations:

$$\tilde{C}_{\text{vib}}/N_0k = 1\cdot181\,, \quad \tilde{S}/N_0k = 0\cdot430\,.$$

$w\,(\text{cm}^{-1})$	821·5	1190	1460	1472	2963	2996
θ_y/T	2·36	3·42	4·20	4·24	8·51	8·63
$\tilde{C}_{\text{vib}}/N_0k$	0·645	0·410	0·275	0·270	0·013	0·010
$\tilde{S}_{\text{vib}}/N_0k$	0·346	0·148	0·079	0·074	0·002	0·002

Total for doubly degenerate vibrations:

$$\tilde{C}_{\text{vib}}/N_0k = 3\cdot246\,, \quad \tilde{S}/N_0k = 1\cdot302\,,$$

so that

$$\tilde{C}_{\text{vib}} = 1\cdot9872\,(1\cdot181+3\cdot246) = 8\cdot796 \text{ cal mole}^{-1}\text{ deg}^{-1}\,,$$

$$\tilde{S}_{\text{vib}} = 1\cdot9872\,(0\cdot430+1\cdot302) = 3\cdot45 \text{ cal mole}^{-1}\text{ deg}^{-1}\,.$$

For motion about the C-C bond we therefore have,

$$\tilde{C}_{\text{vib}} = (\tilde{C}_p - N_0k)_{\text{expt}} - \tilde{C}_{v\,\text{calc}} = 16\cdot67-(2\cdot980+2\cdot980+8\cdot796)$$

$$= 1\cdot91 \text{ cal mole}^{-1}\text{ deg}^{-1}\,,$$

$$\tilde{S}_{\text{vib}} = 62\cdot79-(38\cdot75+17\cdot83+3\cdot45)$$

$$= 2\cdot76 \text{ cal mole}^{-1}\text{ deg}^{-1}\,.$$

We can now investigate the proposed models.

(b) The free rotational model. We must first obtain expressions for the thermodynamic functions of a rotor with one degree of freedom. We follow the methods of Problem 4.3.

The kinetic energy of a part of a molecule which is rotating about a single axis with respect to the rest of the molecule is given by

$$\epsilon = \frac{1}{2I_r}p_\theta^2\,.$$

The classical partition function is therefore

$$Z = \frac{1}{\sigma h}\int_0^{2\pi}\int_{-\infty}^{\infty}\exp\left(-\frac{p_\theta^2}{2I_rkT}\right)\mathrm{d}\theta\,\mathrm{d}P_\theta = \frac{2\pi}{\sigma h}(2\pi I_r kT)^{1/2} = \frac{(8\pi^3 I_r kT)^{1/2}}{\sigma h}\,.$$

The relevant thermodynamic formulae follow at once:

$$\tilde{E} = N_0kT^2\frac{\partial \ln Z}{\partial T} = \tfrac{1}{2}N_0kT\,,$$

$$\tilde{C}_v = \tfrac{1}{2}N_0k\,,$$

$$\tilde{F} = -N_0kT\ln Z = -N_0kT\left(\tfrac{1}{2}\ln T+\tfrac{1}{2}\ln I_r - \ln\sigma + \tfrac{1}{2}\ln\frac{8\pi^2k}{h^2}\right),$$

$$\tilde{S} = \frac{\tilde{E}}{T} + N_0k\ln Z = N_0k\left(\tfrac{1}{2}\ln T+\tfrac{1}{2}\ln I_r - \ln\sigma+\tfrac{1}{2}\ln\frac{8\pi^2k}{h^2}+\tfrac{1}{2}\right).$$

The smallest moment of inertia listed above for the ethane molecule is $10{\cdot}81 \times 10^{-40}$ g cm^2, and this clearly corresponds to joint rotation of the two methyl groups along the axis formed by the C-C bond. The reduced moment for one methyl group is consequently

$$I_r = \frac{(5{\cdot}405 \times 10^{-40})^2}{2 \times 5{\cdot}405 \times 10^{-40}} = 2{\cdot}702 \times 10^{-40} \text{ g cm}^2 .$$

In the course of the complete rotation of one methyl group with respect to the other we see that there are three identical configurations; hence $\sigma = 3$. From the above equations we obtain

$$\tilde{C}_{\text{rot}} = 0{\cdot}99 \text{ cal mole}^{-1} \text{ deg}^{-1} ,$$

$$\tilde{S}_{\text{rot}} = 2{\cdot}46 \text{ cal mole}^{-1} \text{ deg}^{-1} .$$

For this model there is no parameter which we can adjust to improve agreement with the experimental values.

(c) The harmonic oscillator model. Using the Einstein functions (Problem 4.6) we find that $\tilde{C}_{\text{vib}}/N_0k = 1{\cdot}91/1{\cdot}987 = 0{\cdot}96$ corresponds to $\theta/T = 0{\cdot}70$, and this in turn is equivalent to assigning a frequency of $243{\cdot}3$ cm^{-1} to the torsional oscillation mode.

We obtain an accurate value of the vibrational entropy using the formula obtained in Problem 4.6(a) rather than by interpolation of the tabulated Einstein functions, so that for $\theta_v/T = 0{\cdot}70$ we have,

$$\frac{\tilde{S}_{\text{vib}}}{N_0k} = \frac{\theta_v/T}{\exp(\theta_v/T) - 1} - \ln[1 - \exp(-\theta_v/T)]$$

$$= \frac{0{\cdot}7}{1{\cdot}01375} - \ln 0{\cdot}49659 = 1{\cdot}377 .$$

Hence

$$\tilde{S}_{\text{vib}} = 2{\cdot}736 \text{ cal mole}^{-1} \text{ deg}^{-1} .$$

(d) The restricted rotator model. We calculate first the numerical value of the classical rotational partition function obtained in part (b). Putting $I_r = 2{\cdot}702 \times 10^{-40}$ g cm^2, $T = 500°$K, and $\sigma = 3$, we obtain $1/Z = 0{\cdot}292$. Plotting and cross plotting the tabulated thermodynamic functions we find that a heat capacity of $1{\cdot}91$ corresponds to a value of $V/RT = 2{\cdot}80$, and that $2{\cdot}66$ cal mole^{-1} deg^{-1} is the corresponding entropy.

On comparison of the calorimetric entropy, $2{\cdot}76$ cal mole^{-1} deg^{-1}, with that obtained for the three models: (b) $2{\cdot}46$, (c) $2{\cdot}74$, (d) $2{\cdot}66$ cal mole^{-1} deg^{-1}, it would appear that the harmonic oscillator model is marginally better than the restricted rotator model. However when the above comparison is made over a wide temperature range it becomes evident that the restricted rotator model is superior.

In Problems 4.9 and 4.10 we have calculated some thermodynamic properties of ideal classical gases of polyatomic molecules from statistical mechanical formulae based on simple molecular models. We have used in these formulae constants derived from experimental spectroscopic data, and have made a comparison of the calculated values with those obtained from direct calorimetric measurements. Agreement between the values obtained via these two routes validates the molecular models upon which the statistical mechanical formulae are based.

5

Ideal relativistic classical and quantum gases

P.T.LANDSBERG
(University College, Cardiff)

5.1 A system of non-interacting identical particles of rest mass m_0 is in a cubic box of side L. In a single-particle quantum state (j_1, j_2, j_3, σ) the energy is $\epsilon(j_1, j_2, j_3)$ and the momentum components are $p_r = (h/L)j_r$ $(r = 1, 2, 3)$, where the j's are integers which cannot all be zero. The chemical potential μ is written as γkT, T being the temperature of the system. The spin label σ assumes g values if each single-particle state has spin degeneracy g.

(a) Prove that

$$pv = \eta kTg\sum_{j_1}\sum_{j_2}\sum_{j_3} \ln[1+\eta t(j_1,j_2,j_3)]$$

where $\eta = +1$ for fermions and $\eta = -1$ for bosons, and

$$t(j_1,j_2,j_3) \equiv \exp\left[\gamma - \frac{\epsilon(j_1,j_2,j_3)}{kT}\right].$$

(b) Using a continuous spectrum approximation, assuming the energy to depend only on the magnitude of the momentum, and treating the magnitude of the momentum, $p(\epsilon)$, as a function of the energy ϵ, show that the pressure is

$$p = \frac{4\pi g}{3h^3}\int_{p=0}^{p=\infty} \frac{[p(\epsilon)]^3\,d\epsilon}{\exp(\epsilon/kT-\gamma)+\eta}.$$

(c) Show that in the non-relativistic case

$$p = kTg\left(\frac{2\pi m_0 kT}{h^2}\right)^{3/2} I(\gamma, \tfrac{3}{2}, \pm)$$

in agreement with Problem 3.12(f).

(d) Given that in the relativistic case $\epsilon^2 = p^2c^2+\epsilon_0^2$, where $\epsilon_0 = m_0c^2$, show that

$$p = \frac{4\pi g}{3h^3c^3}\int_0^\infty \frac{(E^2+2E\epsilon_0)^{3/2}dE}{\exp[(E+\epsilon_0)/kT-\gamma]+\eta}$$

where $E \equiv \epsilon-\epsilon_0$.

(e) Show from $n = v(\partial p/\partial\mu)_T$ (Problem 2.8) that the mean number of particles in the above case is

$$\langle n\rangle = \frac{4\pi vg}{h^3c^3}\int_0^\infty \frac{(E+\epsilon_0)(E^2+2E\epsilon_0)^{1/2}\,dE}{\exp[(E+\epsilon_0)/kT-\gamma]+\eta}.$$

Solution

(a) The grand partition function is, from Problems 2.4 and 3.12,

$$\Xi = \sum_i \exp\{(\mu N_i - E_i)/kT\}$$

where the sum extends over all **many-particle states** of the system. If the system is in a typical state i, it contains a total number of particles and a total energy given respectively by ($j_1 = j_2 = j_3 = 0$ excluded)

$$N_i = \sum_{\sigma=1}^{g} \sum_{j_1=-\infty}^{\infty} \sum_{j_2=-\infty}^{\infty} \sum_{j_3=-\infty}^{\infty} n_i(j_1, j_2, j_3, \sigma)$$

$$E_i = \sum_{\sigma} \sum_{j_1, j_2, j_3} \epsilon(j_1, j_2, j_3) n_i(j_1, j_2, j_3, \sigma) .$$

Hence

$$\Xi = \sum_i \prod_{\sigma} \prod_{j_1} \prod_{j_2} \prod_{j_3} [t(j_1, j_2, j_3)]^{n_i(j_1, j_2, j_3, \sigma)}$$

A many-particle state i can for indistinguishable particles be specified by the set of occupation numbers

$$n_i(1, 0, 0, \sigma_1), \quad n_i(0, 0, 1, \sigma_1), \quad n_i(0, 1, 0, \sigma_1), \ldots . \qquad (5.1.1)$$

These $n_i(j_1, j_2, j_3, \sigma)$ can have values 0 or 1 for fermions and $0, 1, 2, \ldots, \infty$ for bosons. A summation over i is then equivalent to summing over all admissible values of the numbers (5.1.1). Hence

$$\Xi = \sum_{n(1,0,0,1)=0}^{1 \text{ or } \infty} \; \sum_{n(0,1,0,1)=0}^{1 \text{ or } \infty} \cdots \prod_{\sigma} \prod_{j_1} \prod_{j_2} \prod_{j_3} [t(j_1, j_2, j_3)]^{n(j_1, j_2, j_3, \sigma)} .$$

The product goes over all quantum numbers. Carrying out the summations first

$$\Xi = \prod_{\sigma} \prod_{j_1} \prod_{j_2} \prod_{j_3} [1 + \eta t(j_1, j_2, j_3)]^{\eta} = \prod_{j_1, j_2, j_3} [1 + \eta t(j_1, j_2, j_3)]^{\eta g} .$$

Hence

$$pv = kT \ln \Xi = \eta k T g \sum_{j_1} \sum_{j_2} \sum_{j_3} \ln[1 + \eta t(j_1, j_2, j_3)] . \qquad (5.1.2)$$

(b) We can put

$$\sum_{j_1} \sum_{j_2} \sum_{j_3} \ldots \to \int_{-\infty}^{\infty} \ldots \mathrm{d}j_1 \, \mathrm{d}j_2 \, \mathrm{d}j_3 = \frac{4\pi L^3}{h^3} \int_0^{\infty} \ldots p^2 \, \mathrm{d}p .$$

Applying this to Equation (5.1.2), with $v = L^3$, we obtain

$$p = \frac{4\pi\eta k T g}{h^3} \int_0^{\infty} \ln[1 + \eta t(p)] p^2 \, \mathrm{d}p$$

where t is regarded to be a function of p through its dependence on the j_r. Also

$$t(p) = \exp[\gamma - \epsilon(p)/kT] .$$

Integrating partially one finds the stated relation.

(c) We have $p(\epsilon) = (2m_0\epsilon)^{1/2}$ so that the pressure is

$$p = \frac{4\pi g}{3h^3}\int_0^\infty \frac{(2m_0)^{3/2}\epsilon^{3/2}\mathrm{d}\epsilon}{\exp(\epsilon/kT-\gamma)+\eta} ,$$

whence the result follows.

(d) We have $cp(\epsilon) = (\epsilon^2 - \epsilon_0^2)^{1/2}$, so that

$$\begin{aligned} c^3[p(\epsilon)]^3 &= [(\epsilon - m_0c^2)(\epsilon + m_0c^2)]^{3/2} \\ &= E^{3/2}(E+2\epsilon_0)^{3/2} = (E^2+2E\epsilon_0)^{3/2} . \end{aligned}$$

This gives the result.

(e) The stated result is found after a partial integration.

5.2 Let

$$L_r \equiv \int_0^\infty \frac{E^r\,\mathrm{d}E}{(E^2+2E\epsilon_0)^{1/2}[\exp(E/kT-\alpha)]+\eta} \qquad (r = 0, 1, 2, 3, 4)$$

where $\alpha \equiv (\mu - m_0c^2)/kT$, and $B = 4\pi vg/3h^3c^3$.

(a) Establish

$$\begin{aligned} \langle n\rangle &= 3B[L_3+3\epsilon_0L_2+2\epsilon_0^2L_1] , \\ pv &= B[L_4+4\epsilon_0L_3+4\epsilon_0^2L_2] , \\ TS &= B[4L_4+(13\epsilon_0-3\alpha kT)L_3+(10\epsilon_0-9\alpha kT)\epsilon_0L_2-6\alpha kT\epsilon_0^2L_1] , \\ U &= 3B[L_4+3\epsilon_0L_3+2\epsilon_0^2L_2] , \end{aligned}$$

where U is the internal energy excluding the energy due to the rest mass of the particles.

(b) Verify that

$$U - TS + pv = (\mu - \epsilon_0)\langle n\rangle ,$$

and discuss this result.

(c) Verify that this system is not an ideal quantum gas according to the definition in Problem 1.9(e).

(d) A system is said to be **ultrabaric** if its energy (including its rest mass) per unit volume is exceeded by its pressure. Show that this does not apply here.

[If interactions are taken into account then a system may become ultrabaric.[1]]

Solution

(a) and (b) If one multiplies the integrand in the expression for $\langle n\rangle$ in Problem 5.1(e) by $(E^2+2\epsilon_0)^{1/2}/(E^2+2\epsilon_0)^{1/2}$ one has the denominator in a

[1] S.A.Bludman and M.Ruderman, *Phys.Rev.*, 170, 1176 (1968).

form suitable for substituting the L_r's. In this way one finds $\langle n \rangle$ and pv. The entropy is found, by using $S = v(\partial p/\partial T)_{v,\mu}$ (see Problem 2.8) from the expression for pv,

$$TS = \frac{B}{kT}\left\{\int_0^\infty \frac{(E^2+2E\epsilon_0)^{3/2}E\exp(-\alpha+E/kT)\mathrm{d}E}{[\exp(E/kT-\alpha)+\eta]^2} - \alpha kT\int_0^\infty \frac{(E^2+2E\epsilon_0)^{3/2}\exp(-\alpha+E/kT)\mathrm{d}E}{[\exp(E/kT-\alpha)+\eta]^2}\right\}$$

$$= B[4L_4+13\epsilon_0 L_3+10\epsilon_0 L_2-\alpha kT(3L_3+9\epsilon_0 L_2+6\epsilon_0^2 L_1)]\,.$$

It is convenient to make a table of coefficients as follows:

	L_4	L_3	L_2	L_1
$\frac{TS}{B}$	4	$13\epsilon_0-3\alpha kT$	$10\epsilon_0^2-9\alpha kT\epsilon_0$	$-6\alpha kT\epsilon_0^2$
$\frac{\mu\langle n\rangle}{B}$	0	3μ	$9\epsilon_0\mu$	$6\epsilon_0^2\mu$
$-\frac{pv}{B}$	-1	$-4\epsilon_0$	$-4\epsilon_0^2$	0
$-\frac{\langle n\rangle\epsilon_0}{B}$	0	$-3\epsilon_0$	$-9\epsilon_0^2$	$-6\epsilon_0^3$
$\frac{(\mu-\epsilon_0)\langle n\rangle+TS-pv}{B}$	3	$9\epsilon_0$	$6\epsilon_0^2$	0

The last line gives U/B. This shows that the relativistic theory leads to a renormalisation of the chemical potential from μ to $\mu' = \mu - m_0c^2$. Alternatively one can keep the μ of non-relativistic theory and renormalise the internal energy from U to $U' = U+\epsilon_0\langle n\rangle$ by the energy due to the rest mass of the particles.

(c) Neither the relation $pv = gU$ nor the relation $pv = gU'$ is satisfied with a constant g.

(d) A calculation of $U+\epsilon_0\langle n\rangle - pv$ yields the following sum of positive terms:

$$B(2L_4+8\epsilon_0 L_3+11\epsilon_0^2 L_2+6\epsilon_0^3 L_1)\,.$$

5.3 Using equations of the preceding problem, establish the following results:

(a) In the non-relativistic limit, $u \equiv m_0c^2/kT \gg 1$,

$$L_r \simeq (kT)^r\left(\frac{kT}{2m_0c^2}\right)^{1/2}\Gamma(r+\tfrac{1}{2})I(\alpha, r-\tfrac{1}{2}, \pm)\,.$$

In the extreme relativistic limit, $u \ll 1$,

$$L_r \simeq (kT)^r \Gamma(r) I(\alpha, r-1, \pm).$$

Discuss the meaning of these approximations.

(b) Use these approximations to establish the results given in the following table.

	$u \gg 1$	$u \ll 1$
$\langle n \rangle$	$vg\left(\frac{2\pi m_0 kT}{h^2}\right)^{3/2} I(\alpha, \frac{1}{2}, \pm)$	$8\pi vg\left(\frac{kT}{hc}\right)^3 I(\alpha, 2, \pm)$
U	$\frac{3}{2}vg\left(\frac{2\pi m_0 kT}{h^2}\right)^{3/2} kTI(\alpha, \frac{3}{2}, \pm)$	$24\pi vgkT\left(\frac{kT}{hc}\right)^3 I(\alpha, 3, \pm)$
pv	$\frac{2}{3}U$	$\frac{1}{3}U$

(c) Discuss the approximation $e^\alpha \ll 1$ when made in conjunction with $u \gg 1$, and show that it leads to

$$L_r = (kT)^r \left(\frac{kT}{2m_0c^2}\right)^{1/2} \Gamma(r+\tfrac{1}{2})\, e^\alpha .$$

[The results of Problem 3.12 for an ideal quantum gas have here been recovered with $s = \frac{1}{2}$ and $s = 2$ respectively. The extreme relativistic thermodynamic properties of a Fermi gas are seen to be rather similar to the thermodynamic properties of black-body radiation. An important difference is, however, that $\alpha = 0$ in the latter case. The theory can be taken further by obtaining expressions for other quantities.[(2)]]

Solution

(a) The integral is

$$L_r \equiv (kT)^r \int_0^\infty \frac{x^r \, dx}{(x^2+2xu)^{1/2}[\exp(x-\alpha)+\eta]} .$$

The square root in the denominator is $(2xu)^{-1/2}$ or x^{-1} in the two approximations.

The approximation $u \equiv m_0c^2/kT \gg 1$ implies relatively low temperatures, or relatively heavy particles, or both. The thermal velocities of these particles will therefore be small enough for a non-relativistic treatment to be valid. In the limit $m_0 \to 0$, however, any temperature above absolute zero is capable of imparting high thermal velocities to the particles, and such a situation may be called 'extreme relativistic'.

(b) The results stated are simple algebraic consequences of part (a) and the results of Problem 5.2.

(2) P.T.Landsberg and J.Dunning-Davies, in *Proceedings of the International Symposium on Statistical Mechanics and Thermodynamics* (Ed. J.Meixner) (North-Holland, Amsterdam), 1965.

(c) The approximation implies 'non-degeneracy' as used in connection with gases in statistical mechanics. It means that $m_0c^2 - \mu \gg kT$ or $u - \mu/kT \gg 1$. Since $u \gg 1$ is given, the restriction on μ is that it can be negative, but if it is positive it must satisfy $\mu/kT \ll u$.

5.4 A four-vector $(c\mathbf{b}, a)$ in an inertial frame I and the corresponding one $(c\mathbf{b}', a')$ in an inertial frame I′ are related by

$$a' = \beta(a + \mathbf{v} \cdot \mathbf{b})\,, \tag{5.4.1}$$

$$\left.\begin{aligned} \mathbf{v} \cdot \mathbf{b}' &= \beta\left(\mathbf{v} \cdot \mathbf{b} + \frac{v^2}{c^2}a\right)\,, \\ \mathbf{v} \times \mathbf{b}' &= \mathbf{v} \times \mathbf{b}\,, \end{aligned}\right. \tag{5.4.2}$$

where $\beta = (1 - v^2/c^2)^{-1/2}$, $\mathbf{v}$ is the velocity of the origin of I in I′, and c is the velocity of light.

(a) Prove the reciprocal relations

$$a = \beta(a' - \mathbf{v} \cdot \mathbf{b}')\,, \tag{5.4.3}$$

$$\mathbf{v} \cdot \mathbf{b} = \beta\left(\mathbf{v} \cdot \mathbf{b}' - \frac{v^2}{c^2}a'\right)\,. \tag{5.4.4}$$

(b) An inertial frame I_0 can be defined by specifying that the vector $\mathbf{b}$ shall be zero, i.e. $\mathbf{b}_0 = 0$. If the origin of I_0 has velocity $\mathbf{w}$ in I′ and the corresponding value of β is denoted by β_w, prove that

$$a' = \beta_w a_0\,, \tag{5.4.5}$$

$$\mathbf{b}' = \frac{a'}{c^2}\mathbf{w}\,, \tag{5.4.6}$$

$$a' = \frac{a_0}{\beta_w} + \mathbf{w} \cdot \mathbf{b}'\,. \tag{5.4.7}$$

(c) If I_0' be a frame with velocity $\mathbf{w} + \mathrm{d}\mathbf{w}$ in I so that $(a, \mathbf{b})$ in I corresponds to $(a_0 + \mathrm{d}a_0,\ \mathbf{b}_0 + \mathrm{d}\mathbf{b}_0)$ in I_0' then, assuming an equation of the type (5.4.7) to apply to I and I_0', establish

$$-a_0\,\mathrm{d}\left(\frac{1}{\beta_w}\right) = \mathbf{b}' \cdot \mathrm{d}\mathbf{w} \tag{5.4.8}$$

and hence that

$$\mathrm{d}a' = \frac{\mathrm{d}a_0}{\beta_w} + \mathbf{w} \cdot \mathrm{d}\mathbf{b}'\,. \tag{5.4.9}$$

[A physical application of these results is given in the following problem.]

Solution

(a) From Equation (5.4.1) substitute for $\mathbf{v} \cdot \mathbf{b}$ in Equation (5.4.2). Similarly from Equation (5.4.2) substitute for a in Equation (5.4.1).

(b) Equations (5.4.5) and (5.4.6) follow immediately from Equations (5.4.1) and (5.4.2). Equation (5.4.7) is then a useful consequence of these results.

(c) Use the relation

$$-\mathrm{d}\left(\frac{1}{\beta_w}\right) = \frac{\beta_w}{c^2}\mathbf{w}\cdot\mathrm{d}\mathbf{w}\,. \tag{5.4.10}$$

Now note that

$$\mathbf{b}'\cdot\mathrm{d}\mathbf{w} = \frac{a'}{c^2}\mathbf{w}\cdot\mathrm{d}\mathbf{w} = \frac{\beta_w a_0}{c^2}\mathbf{w}\cdot\mathrm{d}\mathbf{w} = -a_0\,\mathrm{d}\left(\frac{1}{\beta_w}\right),$$

which is Equation (5.4.8). Now differentiate Equation (5.4.7) to find

$$\mathrm{d}a' = \frac{\mathrm{d}a_0}{\beta_w} + a_0\,\mathrm{d}\left(\frac{1}{\beta_w}\right) + \mathbf{b}'\cdot\mathrm{d}\mathbf{w} + \mathbf{w}\cdot\mathrm{d}\mathbf{b}'\,.$$

The second and third terms cancel as a consequence of Equation (5.4.8), yielding Equation (5.4.9).

5.5 The quantity $a = E + \lambda pv$ and linear momentum $\mathbf{b} = \mathbf{P}$ of a system transform like a four-vector $(c\mathbf{b}, a)$ of Problem 5.4. The suffix '0' refers in this problem to the inertial frame in which the centre of mass of the system is *initially* at rest, $\beta \equiv (1 - w^2/c^2)^{-1/2}$, and $\mathbf{w}$ is the initial velocity of the centre of mass in an inertial frame I. For a fluid in a container $\lambda = 1$, and $\lambda = 0$ for a solid, a box, or a fluid with the energy and momentum (but not the rest mass) due to the stresses in the container included with the system. Assume $v = v_0/\beta$, $p = p_0$ for the transformation of volume and pressure.

(a) Establish the result

$$E - \mathbf{w}\cdot\mathbf{P} = \frac{E_0}{\beta}\,.$$

(b) Establish the relation

$$\mathrm{d}E - \mathbf{w}\cdot\mathrm{d}\mathbf{P} + \lambda p_0\,\mathrm{d}v + \frac{(1-\lambda)p_0}{\beta}\mathrm{d}v_0 = \frac{1}{\beta}(\mathrm{d}E_0 + p_0\,\mathrm{d}v_0)\,.$$

(c) Justify physically the following expression for the compressional work done on the system:

$$\begin{aligned}\mathrm{d}W_c &= -p_0\left[\mathrm{d}v - (1-\lambda)v_0\,\mathrm{d}\left(\frac{1}{\beta}\right)\right]\\ &= -p_0\left(\lambda\,\mathrm{d}v + \frac{1-\lambda}{\beta}\mathrm{d}v_0\right).\end{aligned}$$

(d) Justify the expression for the translational work $\mathrm{d}W_{tr} = \mathbf{w}\cdot\mathrm{d}\mathbf{P}$ if the non-relativistic definitions of force and work are adopted.

(e) From the first law in the form $\mathrm{d}Q = \mathrm{d}E - \mathrm{d}W = \mathrm{d}E - \mathrm{d}W_c - \mathrm{d}W_{tr}$ show that the heat transforms as $\mathrm{d}Q = \mathrm{d}Q_0/\beta$.

[These considerations, basic to relativistic thermodynamics, have been recently a subject of controversy.[3] Note that the transformation of dQ is independent of the value of λ.]

Solution

(a) From Equation (5.4.7) it follows that

$$E' + \lambda p' v' = \frac{E_0 + \lambda p_0 v_0}{\beta_w} + \mathbf{w} \cdot \mathbf{P}' .$$

Hence the transformation of p and v makes the terms involving λ cancel out. On changing the notation to the present problem, this yields $E = E_0/\beta_w + \mathbf{w} \cdot \mathbf{P}$.

(b) We use Equation (5.4.9) and then evaluate $da' - da_0/\beta = \mathbf{w} \cdot d\mathbf{b}'$:

$$\begin{aligned}\mathbf{w} \cdot d\mathbf{P} &= dE + \lambda p_0\, dv + \lambda v\, dp_0 - \frac{1}{\beta} dE_0 - \frac{\lambda p_0}{\beta} dv_0 - \frac{\lambda v_0}{\beta} dp_0 \\ &= dE - \frac{1}{\beta} dE_0 + \lambda p_0\, dv - \frac{\lambda p_0}{\beta} dv_0 \\ &= dE - \frac{1}{\beta} dE_0 - \frac{p}{\beta} dv_0 + \lambda p_0\, dv + \frac{(1-\lambda) p_0}{\beta} dv_0\end{aligned}$$

as required.

(c) An increment of volume transforms as

$$dv = \frac{1}{\beta} dv_0 + v_0\, d\left(\frac{1}{\beta}\right)$$

so that there are two causes of compressional work in frame I for $\lambda = 1$. One is compression of the system; the other is due to Lorentz contraction of the volume when it suffers acceleration. As the system is 'bare' and the walls are not included, the only way of producing an acceleration is by regarding the pressures on the system as doing the work. If $\lambda = 0$ the system behaves like an accelerated box and compressional work arises only if $dv_0 \neq 0$. This work is $-p_0 dv_0$ in I_0, but $-p_0 dv_0/\beta$ in I. So we must combine $-p_0 dv$ for $\lambda = 1$ with $-p_0 dv_0/\beta$ for $\lambda = 0$.

(d)

$$dW_{tr} = \mathbf{f} \cdot d\mathbf{s} = \frac{d\mathbf{P}}{dt} \cdot d\mathbf{s} = \mathbf{w} \cdot d\mathbf{P} .$$

(e) The right-hand side of part (b) is

$$\frac{1}{\beta}(dE_0 + p_0 dv_0) = \frac{1}{\beta}(dE_0 - dW_0) = \frac{1}{\beta} dQ_0 .$$

The left-hand side is

$$dE - dW_{tr} - dW_c = dQ .$$

[3] P.T.Landsberg and K.A.Johns, *J.Phys.Soc.Jap.*, **26** Supplement, 310–312 (1969) and *Annals of Physics*, **56**, 299 (1970). P.T.Landsberg, *Essays in Physics*, **2**, 93 (1970) and papers in *A Critical Review of Thermodynamics*, Eds. E.B.Stuart, B.Gal-Or, and A.J.Brainard (Mono Press, Baltimore), 1970.

6
Non-electrolyte liquids and solutions

A.J.B.CRUICKSHANK
(*University of Bristol, Bristol*)

6.0 The condensed fluid is generally discussed in terms of so-called configurational thermodynamic properties, this designation implying the parts of the corresponding total properties which may be attributed to the inter-relations between the molecules. Unfortunately, the conventional definitions[(1)] of the configurational properties give rise to difficulties which are inimical to the purposes of this book.

For indistinguishable molecules possessing neither rotational nor vibrational degrees of freedom, the canonical partition function takes the form stated for proof in Problem 3.5(c), namely,

$$Z_N = \left(\frac{2\pi mkT}{h^2}\right)^{3N/2}\frac{Q_N}{N!},$$

where Q_N is the configurational integral defined by

$$Q_N = \int_v \!\!\ldots\!\int \exp\left(-\frac{U^*}{kT}\right)\mathrm{d}\mathbf{r}_1 \ldots \mathrm{d}\mathbf{r}_N ;$$

here the potential energy M of Problem 3.5(c) is identified as the configurational intrinsic energy U^*, and the $\mathbf{r}_i$ are three-dimensional position vectors. Z_N is dimensionless, and it may be used to study solutions either as it stands, or by factorizing it in either of two ways:

(a) $$Z_N = \left(\frac{2\pi mkT}{h^2}\right)^{3N/2}\frac{v^N}{N!}\frac{Q_N}{v^N} \equiv Z_{\mathrm{trs}}Z^* ;$$

(b) $$Z_N = \left(\frac{2\pi mkT}{h^2}\right)^{3N/2}\frac{Q_N}{N!} \equiv Z_{\mathrm{mol}}Z_{\mathrm{conf}} .$$

The first factors represent the **translational** or **molecular part**, and the second factors the **configurational part.** The Helmholtz free energy F ($= -kT\ln Z_N$, cf Problem 2.4) can accordingly be exhibited as a sum in either of two ways:

(a) $F = F_{\mathrm{trs}} + F^*$, $\quad F_{\mathrm{trs}} \equiv -kT\ln Z_{\mathrm{trs}}$, $\quad F^* \equiv -kT\ln Z^*$;

(b) $F = F_{\mathrm{mol}} + F_{\mathrm{conf}}$, $\quad F_{\mathrm{mol}} \equiv -kT\ln Z_{\mathrm{mol}}$, $\quad F_{\mathrm{conf}} \equiv -kT\ln Z_{\mathrm{conf}}$.

(1) R.H.Fowler and E.A.Guggenheim, *Statistical Thermodynamics* (Cambridge University Press, Cambridge), 1952, pp.700-701. J.S.Rowlinson, *Liquids and Liquid Mixtures*, 2nd Edn. (Butterworths, London), 1969, pp.250-252.

If necessary we may define, in addition, **rotational** and **vibrational Helmholtz free energies.** The corresponding partition functions may be assumed independent of molecular environment in the fluid state, at least in the absence of hydrogen bonding etc., so they include no volume-dependent terms, and do not contribute to the change in the free energy on mixing. The procedure usual in solution theory is based on (b), the justification being that $(2\pi mkT/h^2)^{3N/2}$ is independent of volume, and so cannot change in an isothermal mixing process; its demerit is that the factors of Z_N are not dimensionless, but have the dimensions of $[v]^{-N}$ and $[v]^N$ respectively. The corresponding 'free energies' are, however, individually extensive, cf solution to Problem 3.4(c). The consequent logical difficulties can be resolved by using molar quantities and referring volumes to unit molar volume, but since the procedure based on (a) avoids these difficulties altogether, it will be adopted here. The two procedures necessarily give equivalent results for the thermodynamic functions of mixing, as does using Z_N itself, see solution to Problem 6.11.

To describe liquid-vapour equilibria, i.e. the possibility of two phases of different densities at the same p, T, an equation of state $\xi(p,v,T)=0$ must have three real roots in v over some region of p and T (see Chapter 7). It is consequently convenient to use T, v rather than T, p as independent variables, and we therefore separate p (rather than v) into configurational and translational parts. Thus we define

$$p^* \equiv -\left(\frac{\partial F^*}{\partial v}\right)_T, \quad p_{\text{trs}} \equiv -\left(\frac{\partial F_{\text{trs}}}{\partial v}\right)_T,$$

respectively, as the **configurational** or **internal**[2] **pressure** and the **translational**[3] or **kinetic pressure.** The latter is the quantity described by the kinetic theory of gases, i.e. $p_{\text{trs}} \equiv NkT/v$.

F^* may be evaluated for the ideal gas by putting $U^*=0$ (the necessary and sufficient condition to define the ideal gas) in the configurational integral, giving $Q_N = v^N$, whence $Z^* = 1$. Thus the configurational Helmholtz free energy for the ideal gas, $F^*_{\text{id}} = 0$, whence $p^*_{\text{id}} \equiv -(\partial F^*_{\text{id}}/\partial v)_T = 0$, and the total pressure for the ideal gas is $p_{\text{id}} \equiv p_{\text{trs}} + p^*_{\text{id}} = NkT/v$, see Problem 6.1(a). It follows that all the configurational thermodynamic potentials for the ideal gas are zero; since 'configurational' is always taken to mean 'pertaining to or resulting from the forces between molecules', this conclusion has at least the virtue of semantic consistency.

For the general case of the non-ideal fluid, F^* may be evaluated (i) by proceeding in the same general way as above, but using a molecular model to specify U^* in the configurational integral; or (ii) for a specific

[2] J.O.Hirschfelder, C.F.Curtiss, and R.B.Bird, *Molecular Theory of Gases and Liquids* (Wiley, New York), 1954, Chapter 4, Section 2, p.255.

[3] J.O.Hirschfelder, C.F.Curtiss, and R.B.Bird, *Molecular Theory of Gases and Liquids* (Wiley, New York), 1954, Chapter 2, Section 6, p.116.

equation of state, $\xi(p,v,T) = 0$ which yields a total pressure $p_\xi(v,T)$, by integrating the corresponding internal pressure, $p_\xi^* \equiv p_\xi - NkT/v$, with respect to v, as

$$F_\xi^*(v) - F_\xi^*(\infty) = -\int_\infty^v p_\xi^* \, dv \, .$$

Since all equations of state necessarily approach ideality as $v \to \infty$, it follows that $F_\xi^*(\infty) = F_{id}^*(\infty) = 0$. The other configurational thermodynamic potentials are then defined as, for example,

$$G_\xi^* \equiv F_\xi^* + p_\xi^* v \; ; \quad U_\xi^* \equiv F_\xi^* + TS_\xi^* \, .$$

The configurational properties X_ξ^* as defined above are identical to the **'residual' properties**[4] obtained by comparing the volume derivative of the total property to that for the ideal gas over the range ∞ to v,

$$X_\xi^* \equiv \int_\infty^v \left[\left(\frac{\partial X_\xi}{\partial v} \right)_T - \left(\frac{\partial X_{id}}{\partial v} \right)_T \right] dv \, ,$$

as may be seen on subtracting $(\partial X_{trs}/\partial v)_T$ from both terms of the integrand, since $(\partial X_{id}^*/\partial v)_T$ is zero. For example, in the case of $G \equiv F + pv$,

$$\left(\frac{\partial G}{\partial v} \right)_T = \left(\frac{\partial F}{\partial v} \right)_T + p + v\left(\frac{\partial p}{\partial v} \right)_T = v\left(\frac{\partial p}{\partial v} \right)_T \, ,$$

and

$$G_\xi^* = \int_\infty^v \left[v\left(\frac{\partial p_\xi}{\partial v} \right)_T - v\left(\frac{\partial p_{id}}{\partial v} \right) \right] dv = \int_\infty^v v\left(\frac{\partial p_\xi^*}{\partial v} \right) dv \, ,$$

implying $G_\xi^* = F_\xi^* + p_\xi^* v$. In this argument, however, it is assumed that $(\partial G_{trs}/\partial v)_T = v(\partial p_{trs}/\partial v)_T$, i.e. that $G_{trs} \equiv F_{trs} + p_{trs} v$, which is equivalent to defining $G^* \equiv F^* + p^* v$ at the outset.

CELL THEORIES OF THE LIQUID STATE

6.1 Simple cell theory. The cell theories all assume that each molecule is, during most of the time, confined to a cell whose boundaries are determined by the potential due to the neighbouring molecules. For a hard-sphere fluid, this implies that $U^* = \infty$ whenever a molecule overlaps its cell boundary, and $U^* = 0$ otherwise. Suppose the cell boundary to be defined by planes which perpendicularly bisect the lines joining the centre of the molecule to the centres of its neighbours when all are at rest at the centres of their cells, so that the cells completely fill the volume without overlap.

(a) If the volume of each molecule is negligibly small, and the molecules are identical, prove that the configurational Helmholtz free energy,

$$F^* = -kT \ln \frac{Q_N}{v^N} \, ,$$

[4] J.S.Rowlinson, *Liquids and Liquid Mixtures*, 2nd Edn. (Butterworths, London), 1969, pp.57-58.

is, for this model,

$$F^* = NkT\,,$$

and compare the model to the ideal gas. Remember that there are $N!$ ways of arranging N particles among N cells.

(b) Assuming that the polyhedral cell may be approximated by the sphere of the same volume, i.e. of radius b, $\frac{4}{3}\pi b^3 \equiv v/N$, and that the molecule, diameter σ, cannot approach within $\frac{1}{2}\sigma$ of the boundary, prove that

$$F^* = NkT - 3NkT\ln\left[1 - \eta\left(\frac{v_\sigma}{v}\right)^{1/3}\right], \tag{6.1.1}$$

where $\eta \equiv (\sqrt{2}\pi/6)^{1/3}$, and v_σ is the volume of the system at maximum density (close packing) $v_\sigma \equiv N\sigma^3/\sqrt{2}$.

(c) Derive the equation of state, $p = p(v, T, N)$ corresponding to the configurational Helmholtz free energy expression in part (b) and prove that the isothermal bulk modulus $B_T \equiv -v(\partial p/\partial v)_T$ is

$$B_T = p\left[1 + \frac{1}{3}\eta\left(\frac{v_\sigma}{v}\right)^{1/3}\frac{pv}{NkT}\right].$$

Solution

(a) Assume each cell to be of average size v/N. Consider first a particular arrangement, λ, of the molecules among the cells, and evaluate the contribution $Q_N^{(\lambda)}$ of this arrangement to Q_N. The N independently ranging vectors $\mathbf{r}_1, ..., \mathbf{r}_N$ are weighted by $\exp(-U^*/kT)$, and this is zero whenever $\mathbf{r}_i$ is outside the cell to which molecule i has been assigned. Thus $Q_N^{(\lambda)}$ breaks up into separate integrals each over a cell of volume v/N. Then

$$Q_N^{(\lambda)} = \left(\int_{v/N} \mathrm{d}\mathbf{r}_1\right)\left(\int_{v/N} \mathrm{d}\mathbf{r}_2\right)\ldots\left(\int_{v/N} \mathrm{d}\mathbf{r}_N\right) = \left(\frac{v}{N}\right)^N.$$

There are $N!$ such arrangements avoiding multiple occupancy, so

$$Q_N = \sum_\lambda Q_N^{(\lambda)} = N!\left(\frac{v}{N}\right)^N,$$

whence

$$\frac{Q_N}{v^N} = \frac{N!}{N^N}\,;$$

using Stirling's approximation,

$$\frac{Q_N}{v^N} = \mathrm{e}^{-N},$$

which leads directly to the required result.

In the case of the ideal gas, the total volume is accessible to every molecule, i.e. $\exp(-U^*/kT)$ is unity for all values of each $\mathbf{r}_i$. Then $Q_N = v^N$, and $Z^* \equiv Q_N/v^N = 1$, so $F^* = U^* = 0$. The difference in F^* between this cell model and the ideal gas corresponds to $S^* = -Nk$ in the former case. This is the so-called **communal entropy**[5], which would be gained by the assembly in the cell model if the molecules were allowed freely to interchange between cells, i.e. if multiple occupancy were allowed.

(b) Since $U^* = \infty$ whenever a molecule overlaps its cell boundary, its centre cannot approach within $\frac{1}{2}\sigma$ of the boundary. The average volume accessible to the centre of the molecule—the so-called free volume per molecule—v_f is

$$v_f = \tfrac{4}{3}\pi(b-\tfrac{1}{2}\sigma)^3 .$$

Now $\frac{4}{3}\pi b^3 \equiv v/N$, and $\sigma^3 \equiv \sqrt{2}(v_\sigma/N)$, so

$$v_f = \frac{1}{N}[v^{1/3} - (\tfrac{4}{3}\pi)^{1/3}(\tfrac{1}{8}\sqrt{2})^{1/3}(v_\sigma)^{1/3}]^3 = \frac{1}{N}(v^{1/3} - \eta v_\sigma^{1/3})^3 = \frac{v}{N}\left[1-\eta\left(\frac{v_\sigma}{v}\right)^{1/3}\right]^3 .$$

The argument of part (a) gives

$$Q_N = N!\,v_f^N = N!\left(\frac{v}{N}\right)^N\left[1-\eta\left(\frac{v_\sigma}{v}\right)^{1/3}\right]^{3N} ,$$

whence

$$\frac{Q_N}{v^N} = N!N^{-N}\left[1-\eta\left(\frac{v_\sigma}{v}\right)^{1/3}\right]^{3N} ,$$

and the given result follows.

(c) From the result of part (b) we have

$$\left(\frac{\partial F^*}{\partial v}\right)_T = -3NkT\frac{\eta}{3v}\left(\frac{v_\sigma}{v}\right)^{1/3}\left[1-\eta\left(\frac{v_\sigma}{v}\right)^{1/3}\right]^{-1} ;$$

hence

$$p^* = \frac{NkT}{v}\eta\left(\frac{v_\sigma}{v}\right)^{1/3}\left[1-\eta\left(\frac{v_\sigma}{v}\right)^{1/3}\right]^{-1} ;$$

$$p = p^* + p_{trs} = \frac{NkT}{v}\left\{\eta\left(\frac{v_\sigma}{v}\right)^{1/3}\left[1-\eta\left(\frac{v_\sigma}{v}\right)^{1/3}\right]^{-1}+1\right\},$$

and

$$pv = NkT\left[1-\eta\left(\frac{v_\sigma}{v}\right)^{1/3}\right]^{-1} .$$

To obtain the isothermal bulk modulus, differentiate both sides with respect to v at constant T:

$$p+v\left(\frac{\partial p}{\partial v}\right)_T = -NkT\left[1-\eta\left(\frac{v_\sigma}{v}\right)^{1/3}\right]^{-2}\frac{\partial}{\partial v}\left[1-\eta\left(\frac{v_\sigma}{v}\right)^{1/3}\right]$$

$$= -pv\left[1-\eta\left(\frac{v_\sigma}{v}\right)^{1/3}\right]^{-1}[-\eta v_\sigma^{1/3}(-\tfrac{1}{3}v^{-4/3})]$$

$$= -p\left[\tfrac{1}{3}\eta\left(\frac{v_\sigma}{v}\right)^{1/3}\right]\left[1-\eta\left(\frac{v_\sigma}{v}\right)^{1/3}\right]^{-1} .$$

(5) J.O.Hirschfelder, C.F.Curtiss, and R.B.Bird, *Molecular Theory of Gases and Liquids* (Wiley, New York), 1954, pp.273-276.

Hence

$$-v\left(\frac{\partial p}{\partial v}\right)_T = p\left\{1+\left[\tfrac{1}{3}\eta\left(\frac{v_\sigma}{v}\right)^{1/3}\right]\left[1-\eta\left(\frac{v_\sigma}{v}\right)^{1/3}\right]^{-1}\right\},$$

and the given result follows. It is clear that for this model B_T approaches (from above) the ideal gas value in the limit of large v. As (v/v_σ) decreases towards unity, B_T values lie in the range found experimentally for solids rather than in that for liquids.

6.2 Hirschfelder's cell theory for hard spheres assumes that the cell is bounded by the polyhedron whose apices are the centres of the neighbouring molecules when the latter are at rest at the centres of their own cells, mutually distant a, in close-packed array; it is further assumed that this cell may be approximated by the sphere of radius a, and that the centre of the molecule may not approach within σ of the cell boundary [6].

(a) Taking $a^3 \equiv \sqrt{2}(v/N)$, $\sigma^3 \equiv \sqrt{2}(v_\sigma/N)$, as for close packing, prove that the configurational entropy is

$$S^* = -Nk + 3Nk\ln\left[1-\left(\frac{v_\sigma}{v}\right)^{1/3}\right].$$

Compare this to the configurational entropy according to Problem 6.1(b). [Hint: it is necessary to take account of the fact that this way of defining the cells leads to a multiple overlap, so that each element of volume is counted more than once.]

(b) Express the isobaric thermal expansivity and the isothermal compressibility

$$\alpha_p \equiv \frac{1}{v}\left(\frac{\partial v}{\partial T}\right)_p, \qquad K_T \equiv -\frac{1}{v}\left(\frac{\partial v}{\partial p}\right)_T$$

(cf Problem 1.4) as functions of (T,v), (p,v), respectively, and compare these to the corresponding properties derived from the equation of state obtained in Problem 6.1(c).

Solution

(a) $v_f = (\tfrac{4}{3}\pi)(a-\sigma)^3$, i.e. the volume over which the centre of the molecule can move. With

$$a \equiv \sqrt[6]{2}\left(\frac{v}{N}\right)^{1/3}, \quad \sigma \equiv \sqrt[6]{2}\left(\frac{v_\sigma}{N}\right)^{1/3},$$

$$Nv_f = \sqrt{32}\times\tfrac{1}{3}\pi(v^{1/3}-v_\sigma^{1/3})^3\,.$$

As in Problem 6.1,

$$Q_N = N!v_f^N\,, \quad \frac{Q_N}{v^N} = N!N^{-N}\left(\frac{Nv_f}{v}\right)^N,$$

[6] R.J.Buehler, R.H.Wentorf, J.O.Hirschfelder, and C.F.Curtiss, *J.Chem.Phys.*, 19, 61 (1951).

so

$$F^* = NkT[1 - \ln(v^{1/3} - v_\sigma^{1/3})^3 + \ln v - \ln(\sqrt{32} \times \tfrac{1}{3}\pi)] .$$

To obtain the given result, we have to remove the term

$$-NkT\ln(\sqrt{32} \times \tfrac{1}{3}\pi) .$$

In close-packed array, the polyhedral cells defined by the twelve neighbouring centres clearly have a fourfold overlap. The actual figure here is 5·92; the difference arises from approximating the cells by spheres, increasing the overlap. Thus the normalized configurational free energy is

$$F^* = NkT\left\{1 - 3\ln\left[1 - \left(\frac{v_\sigma}{v}\right)^{1/3}\right]\right\},$$

and the required result follows directly from $S^* = -(\partial F^*/\partial T)_v$.

From Equation (6.1.1)

$$S^* = -Nk\left\{1 - 3\ln\left[1 - \eta\left(\frac{v_\sigma}{v}\right)^{1/3}\right]\right\}.$$

The only difference between these equations is in the value of the coefficient η, which increases from 0·905 to 1·0 as the amount of overlap between neighbouring cells is increased from zero.

(b) It is simplest to consider the general equation of state for both models as

$$p = \frac{NkT}{v}\left[1 - \eta\left(\frac{v_\sigma}{v}\right)^{1/3}\right]^{-1},$$

or

$$v - \eta v_\sigma^{1/3} v^{2/3} = \frac{NkT}{p} .$$

Differentiating with respect to v we have

$$1 - \tfrac{2}{3}\eta\left(\frac{v_\sigma}{v}\right)^{1/3} = \frac{Nk}{p}\left(\frac{\partial T}{\partial v}\right)_p ,$$

whence

$$\alpha_p \equiv \frac{1}{v}\left(\frac{\partial v}{\partial T}\right)_p = \frac{Nk}{pv}\left[1 - \tfrac{2}{3}\eta\left(\frac{v_\sigma}{v}\right)^{1/3}\right]^{-1},$$

and substituting for pv from the equation of state,

$$\alpha_p = \frac{1}{T}\left[1 - \eta\left(\frac{v_\sigma}{v}\right)^{1/3}\right]\left[1 - \tfrac{2}{3}\eta\left(\frac{v_\sigma}{v}\right)^{1/3}\right]^{-1}.$$

Similarly,

$$K_T = \frac{NkT}{vp^2}\left[1 - \tfrac{2}{3}\eta\left(\frac{v_\sigma}{v}\right)^{1/3}\right] = \frac{1}{p}\left[1 - \eta\left(\frac{v_\sigma}{v}\right)^{1/3}\right]\left[1 - \tfrac{2}{3}\eta\left(\frac{v_\sigma}{v}\right)^{1/3}\right]^{-1}.$$

It is clear that at given T and v, p increases and α_p and K_T decrease as η increases.

6.3 The tunnel theories for the hard sphere fluid replace the regularly-packed polyhedral cells of the preceding problems by hexagonal prisms or tunnels, stacked parallel in two-dimensional array. The **longitudinal motions** of the spheres in neighbouring tunnels are taken to be mutually independent, the **transverse motion** of each sphere being limited only by the geometry of the tunnel. Barker's tunnel theory[7] is analogous to the cell theory of Problem 6.2: the tunnel cross-section is defined by the hexagon whose apices are the centres of neighbouring tunnels; it is assumed that the cross-section may be approximated by the circle of radius b, b being the centre-to-centre distance in the two-dimensional array, and that $U^* = \infty$ whenever the centre of a molecule approaches within σ of its tunnel boundary.

(a) Formulate the configurational integral for transverse motion in terms of a free area a_f per molecule. For motion parallel to the axis of the tunnel (longitudinal motion), the problem is that of M hard spheres, diameter σ, arranged with their centres on a straight line (the tunnel axis) so that they lie always within a distance cM, where c, the mean distance per molecule, is greater than σ. If the positions of two spheres, i, j, are defined by x_i, x_j, then $U^* = \infty$ whenever $|x_i - x_j| < \sigma$. The configurational integral, Q_M, for one tunnel is

$$Q_M = M^M(c-\sigma)^M .$$

Use this expression to obtain Q_N for a K-tunnel model, where $N \equiv MK$. For the derivation of Q_M see Problem 9.3.

(b) For given v/N, only one of b and c can vary independently, but the ratio b/c can take any positive value consistent with b, $c > \sigma$; thus Q_N must be maximized with respect to b/c. Show that this maximization gives

$$\ln \frac{Q_N}{v^N} = 3N \ln \left[1 - \sqrt[6]{\tfrac{3}{2}}\left(\frac{v_\sigma}{v}\right)^{1/3}\right]$$

where, as in preceding problems, $v_\sigma \equiv N\sigma^3/\sqrt{2}$.

(c) Comment on the general features of the equations of state of Problems 6.1 to 6.3.

Solution

(a) There are clearly $N!/(M!)^K$ ways of arranging N molecules into K sets of M; thus, by an argument exactly parallel to that of Problem 6.2(a) we have

$$Q_N(\text{transverse}) = a_f^N \frac{N!}{(M!)^K} ,$$

whence

$$Q_N = [a_f N(c-\sigma)]^N . \tag{6.3.1}$$

[7] J.A.Barker, *Australian J.Chem.*, **13**, 187 (1960).

(b) The area of the hexagon of side b is $\frac{1}{2}\sqrt{3^3}b^2$, so the volume per tunnel is $\frac{1}{2}\sqrt{3^3}b^2cM$, and the volume of N/M tunnels is $\frac{1}{2}\sqrt{3^3}b^2cN$; there is a threefold overlap, so

$$\frac{v}{N} = \tfrac{1}{2}\sqrt{3}b^2c \equiv \alpha b^2 c\,. \tag{6.3.2}$$

Given $a_f = \pi(b-\sigma)^2$, we have

$$Q_N = [\pi N(b-\sigma)^2(c-\sigma)]^N\,. \tag{6.3.3}$$

Substituting for c according to Equation (6.3.2) into Equation (6.3.3) we obtain

$$Q_N = \left[\pi N(b-\sigma)^2\left(\frac{v}{N\alpha b^2}-\sigma\right)\right]^N.$$

It is convenient to maximize $\ln Q_N$ rather than Q_N:

$$\left(\frac{\partial \ln Q_N}{\partial b}\right)_v = \frac{2N}{b-\sigma} - N\left(\frac{v}{N\alpha b^2}-\sigma\right)^{-1}\frac{2v}{N\alpha b^3}\,.$$

This is zero when

$$b-\sigma = b - \frac{N\alpha b^3\sigma}{v}\,,$$

or $N\alpha b^3\sigma/v = \sigma$; $\alpha b^3 = v/N$. But $v/N = \alpha b^2 c$, so the extremum of $\ln Q_N$ is at $b = c$. The second derivative confirms that this is a maximum. Then

$$Q_N = (\pi N)^N(b^*-\sigma)^{3N} = \left[\pi N b^{*3}\left(1-\frac{\sigma}{b^*}\right)^3\right]^N.$$

Now

$$b^* = \left(\frac{v}{N}\right)^{1/3}\left(\frac{1}{\alpha}\right)^{1/3}, \quad \sigma \equiv \left(\frac{v_\sigma}{N}\right)^{1/3}\sqrt[6]{2}\,,$$

so

$$\frac{\sigma}{b^*} = \left(\frac{v_\sigma}{v}\right)^{1/3}\frac{\sqrt[6]{2}}{\sqrt[3]{2}}\sqrt[6]{3} = \left(\frac{v_\sigma}{v}\right)^{1/3}\sqrt[6]{\tfrac{3}{2}}$$

and

$$b^{*3} = \frac{v}{N}\frac{1}{\alpha}\,,$$

whence

$$\ln\frac{Q_N}{v^N} = N\ln\pi + N\ln\frac{2}{\sqrt{3}} + 3N\ln\left[1-\sqrt[6]{\tfrac{3}{2}}\left(\frac{v_\sigma}{v}\right)^{1/3}\right],$$

and the required result follows on elimination of the term $N\ln(2\pi/\sqrt{3})$, which is due to incomplete normalization, cf.solution to Problem 6.2(a). Note that this model includes the communal entropy; this is because the summation of Q_M allows a molecule to be anywhere within its tunnel.

(c) Differentiating $\ln(Q_N/v^N)$ with respect to v at constant T, we find

$$p^* = \frac{NkT}{v}\eta\left(\frac{v_\sigma}{v}\right)^{1/3}\left[1-\eta\left(\frac{v_\sigma}{v}\right)^{1/3}\right]^{-1},$$

and so, for the total pressure,

$$pv = NkT\left[1-\sqrt[6]{\tfrac{3}{2}}\left(\frac{v_\sigma}{v}\right)^{1/3}\right]^{-1}.$$

This is clearly a member of the class

$$pv = NkT\left[1-\eta\left(\frac{v_\sigma}{v}\right)^{1/3}\right]^{-1}.$$

The three equations of state of Problems 6.1-6.3 differ only in the value of η, which is determined by the degree of overlap of neighbouring cells or tunnels. The index of (v_σ/v) is $\frac{1}{3}$ because the restriction on molecular movement is expressed by a radial quantity, i.e. in dimensions of length.

6.4 Smoothed-potential theory. Suppose that the molecules of Problem 6.2 interact according to a spherically symmetrical potential $\phi(r)$,

$$\phi(r) \to \infty \text{ as } r \to 0\,,$$

$$\phi(r) \to 0 \text{ as } r \to \infty\,,$$

$$\phi(r) < 0 \text{ for } \sigma < r < \infty$$

and that the configurational intrinsic energy is the sum of the interaction potentials of all pairs of molecules without extra contributions from groups of more than two, i.e.

$$U^* = \tfrac{1}{2}\sum_{i,j}\phi(r_{ij})\,.$$

The configurational intrinsic energy $\omega(s)$ of a molecule in its cell obviously is negative when the molecule is at the cell centre, increasing toward $+\infty$ as the radial displacement from the cell centre approaches the nearest-neighbour distance a. The smoothed potential, or square well, theory approximates $\omega(s)$ by

$$\omega = \omega_0\,, \quad 0 < s < (a-\sigma)\,;$$

$$\omega = \infty\,, \quad (a-\sigma) < s\,,$$

where $\omega_0 = \omega_0(a)$ is the configurational intrinsic energy of a molecule at rest at its cell centre, and σ is defined by $\phi(\sigma) \equiv 0$.

(a) Let z be the **coordination number** of the cell lattice, i.e. the number of neighbours of a particular molecule with which it is in contact when $a = \sigma$. Then z has the maximum value 12 for face-centred cubic and hexagonal close-packed lattices.

If

$$\phi(r) = 4\epsilon^*\left[\left(\frac{\sigma}{r}\right)^{12} - \left(\frac{\sigma}{r}\right)^{6}\right],$$

where $-\epsilon^*$ is the minimum value of the pair potential corresponding to

$r = r^* = \sqrt[6]{2}\sigma$, prove that if only nearest neighbour interactions are counted, either

$$U^* = 2zN\epsilon^*\left[\left(\frac{v_\sigma}{v}\right)^4 - \left(\frac{v_\sigma}{v}\right)^2\right],$$

or

$$U^* = +\infty,$$

according to whether or not any molecule has $s_i > (a-\sigma)$; v_σ has the same meaning as in the preceding problems.

(b) Derive the explicit form of $Q_N = Q_N(v, T)$ for this theory, and thence prove that the equation of state is

$$pv = NkT\left[1 - \left(\frac{v_\sigma}{v}\right)^{1/3}\right]^{-1} + 4zN\epsilon^*\left[2\left(\frac{v_\sigma}{v}\right)^4 - \left(\frac{v_\sigma}{v}\right)^2\right].$$

(c) Prove that equation of state of part (b) predicts 'condensation', i.e. the possibility of co-existence of stable high and low density phases at the same pressure and temperature. Note that it is a sufficient, though not a necessary, condition for two-phase equilibrium that the equation of state has two real positive roots in v for $p = 0$ over some range of T.

Solution

(a) It follows from the postulates of the model that U^* has only two values open to it, $U^* = N\omega_0$ or $U^* = \infty$. In the first case, corresponding to

$$s_i < (a-\sigma), \quad i = 1, 2, ..., N,$$

U^* has the same value as when every molecule is at its cell centre, i.e. all pair-interaction distances are $r_{ij} = a$. Then $\omega_0 = \frac{1}{2}z\phi(a)$, with

$$\phi(a) = 4\epsilon^*\left[\left(\frac{\sigma}{a}\right)^{12} - \left(\frac{\sigma}{a}\right)^6\right] = 4\epsilon^*\left[\left(\frac{v_\sigma}{v}\right)^4 - \left(\frac{v_\sigma}{v}\right)^2\right],$$

whence

$$\omega_0 = 2z\epsilon^*\left[\left(\frac{v_\sigma}{v}\right)^4 - \left(\frac{v_\sigma}{v}\right)^2\right].$$

(b) Since the cell is bounded by an infinite potential barrier at $s_i = a-\sigma$, the free volume is, as in Problem 6.2(a), $v_f = \frac{4}{3}\pi(a-\sigma)^3$, with $a \equiv \sqrt[6]{2}(v/N)^{1/3}$, $\sigma \equiv \sqrt[6]{2}(v_\sigma/N)^{1/3}$. Then

$$Nv_f = \frac{4\sqrt{2}\pi}{3N}(v^{1/3} - v_\sigma^{1/3})^3.$$

$$Q_N = N!(v_f)^N \exp\left(-\frac{N\omega_0}{kT}\right),$$

and

$$\ln\frac{Q_N}{v^N} = -N + 3N\ln\left[1 - \left(\frac{v_\sigma}{v}\right)^{1/3}\right] - \frac{2zN\epsilon^*}{kT}\left[\left(\frac{v_\sigma}{v}\right)^4 - \left(\frac{v_\sigma}{v}\right)^2\right],$$

when the numerical term $\frac{1}{3}\sqrt{32\pi}$, due to overlap, is dropped. Differentiating with respect to v at constant T, and adding p_{trs} gives, cf.Problem 6.3(c),

$$p = \frac{NkT}{v}\left[1-\left(\frac{v_\sigma}{v}\right)^{1/3}\right]^{-1} + 2zN\epsilon^*\left(4\frac{v_\sigma^4}{v^5} - 2\frac{v_\sigma^2}{v^3}\right),$$

whence the required result follows.

(c) Putting $p = 0$ in this equation of state gives

$$\left[1-\left(\frac{v_\sigma}{v}\right)^{1/3}\right]^{-1} = \frac{4z\epsilon^*}{kT}\left[\left(\frac{v_\sigma}{v}\right)^2 - 2\left(\frac{v_\sigma}{v}\right)^4\right].$$

Setting $y = (v_\sigma/v)$ transforms this into

$$y^2(1-2y^2)(1-y^{1/3}) = \frac{kT}{4z\epsilon^*}\ .$$

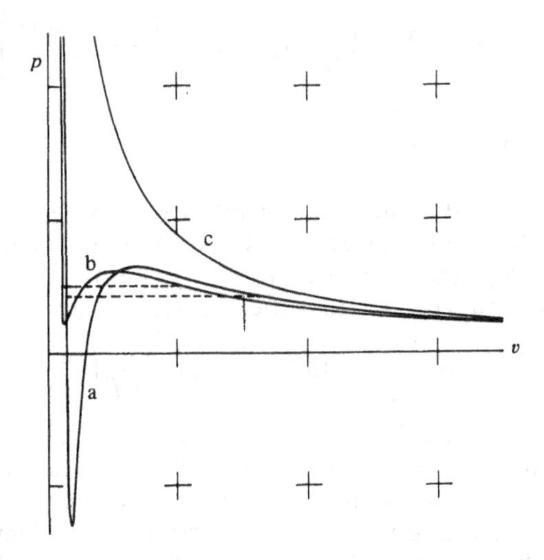

Figure 6.4.1. Comparison of van der Waals and Dieterici equations of state for two substances having similar vapour pressure (indicated by the broken lines): a, van der Waals; b, Dieterici; c, ideal gas. The vapour pressure has been located by the *equal area* rule, see Problem 7.5(b), and Figure 11.14.1.

There is clearly a range of T, $0 < T < T'$, for which there are two real roots, $0 < y < 1/\sqrt{2}$, while for $T > T'$ there are no real roots. Note that the van der Waals equation, see Problem 1.11(a), also shows this behaviour; the Dieterici equation of state,

$$p = \frac{RT}{v-b}\exp\left(-\frac{a}{RTv}\right),$$

has no real roots in v for $p = 0$, but nevertheless predicts condensation. The difference is illustrated by Figure 6.4.1.

6.5 Hole theory. Instead of accounting for variation of total volume by variation of the cell radius, we consider an assembly of fixed-volume cells which may be either occupied or vacant. With N molecules distributed among $(N+N_0)$ cells in close-packing, the total volume is, cf Problem 6.1,

$$v = \frac{(N+N_0)a^3}{\sqrt{2}} = v_a \frac{N+N_0}{N}, \quad \text{with} \quad v_a \equiv \frac{Na^3}{\sqrt{2}},$$

where, as in the preceding problems, a is the parameter of the cell lattice. It is assumed that only nearest-neighbour attractions are significant, and that the configurational energy per cell has the value corresponding to the molecule being at the cell centre [i.e. $\frac{1}{2}\phi(a)$ per neighbour] everywhere within a free volume v_f, and the value $+\infty$ outside v_f.

(a) Show that for this theory the general form of the configurational integral is

$$Q_N = \sum_{\lambda} \left[\prod_{i=1}^{N} v_f(i, \lambda) \right] \exp\left[-\frac{Nz_\lambda \phi(a)}{2kT} \right],$$

where z_λ is the average number of occupied cells neighbouring an occupied cell in the configuration λ (a member of the set appropriate to $N, N+N_0$).

(b) Assuming that z_λ may be replaced by its average over all configurations, that $v_f(i, \lambda)$ may be treated likewise, and that the average, v_f, so obtained is independent of overall density, i.e. is a function only of the lattice parameter a, confirm that the equation of state is

$$p = \frac{NkT}{v_a} \left[-\ln\left(1 - \frac{v_a}{v}\right) + \frac{z\phi(a)}{2kT} \left(\frac{v_a}{v}\right)^2 \right]$$

where z is the coordination number of the cell lattice.

(c) Establish the approximate form of this equation of state appropriate to $v/v_a \gg 1$, and compare it to the van der Waals equation, see Problem 1.11(a), and to the equation of state of Problem 6.4(b).

(d) Show that the equation of state of part (b) predicts a two-phase region; the argument of Problem 6.4(c) is not appropriate because of the logarithmic term; use instead the condition $(\partial p/\partial v)_T = 0$, when two real roots ensure phase separation.

Solution

(a) The assumption of uniform potential over each $v_f(i, \lambda)$ for the ith molecule in the configuration λ allows the configurational energy $U_\lambda^* = z_\lambda \phi(a) N/2kT$ to be taken outside the integral. As in the solution to Problem 6.1(a), each integration $\mathrm{d}\mathbf{r}_i$ yields $v_f(i)$, within the configuration λ. Then

$$Q_N^{(\lambda)} = \exp\left[-\frac{Nz_\lambda \phi(a)}{2kT} \right] \prod_{i=1}^{N} v_f(i, \lambda),$$

and the required result follows.

(b) In the expression for Q_N, put $v_f(i, \lambda) = v_f$, and take the average coordination number for every configuration to be that for all configurations, $z_\lambda = z = zN/(N+N_0)$. Then

$$Q_N = v_f^N \exp\left[-\frac{zN^2\phi(a)}{2(N+N_0)kT}\right]\sum_\lambda 1 \ .$$

The number of arrangements of N molecules among $N+N_0$ cells is clearly $(N+N_0)!/(N_0!)$, so, since $N+N_0 = Nv/v_a$,

$$Q_N = \frac{(Nv/v_a)!}{(Nv/v_a - N)!} v_f^N \exp\left[-\frac{zNv_a\phi(a)}{2kTv}\right] .$$

Using Stirling's approximation in the logarithmic form and simplifying, we obtain

$$\ln\frac{Q_N}{v^N} = \frac{Nv}{v_a}\ln\frac{Nv}{v_a} - \left(\frac{Nv}{v_a} - N\right)\ln\left(\frac{Nv}{v_a} - N\right) + N\ln\frac{v_f}{v} - \frac{zNv_a\phi(a)}{2kTv} - N ,$$

$$kT\ln\frac{Q_N}{v^N} = NkT\left[\frac{v}{v_a}\ln\frac{v}{v_a} - \left(\frac{v-v_a}{v_a}\right)\ln\left(\frac{v-v_a}{v_a}\right) + \ln\frac{Nv_f}{v} - \frac{zv_a\phi(a)}{2kTv} - 1\right] .$$

Differentiating with respect to v at constant T, we find

$$p^* = \frac{NkT}{v_a}\left[\ln\frac{v}{v-v_a} - \frac{v_a}{v} + \frac{z\phi(a)}{2kT}\left(\frac{v_a}{v}\right)^2\right]$$

$$= \frac{NkT}{v_a}\left[-\ln\left(1 - \frac{v_a}{v}\right) + \frac{z\phi(a)}{2kT}\left(\frac{v_a}{v}\right)^2 - \frac{NkT}{v}\right].$$

Adding p_{trs} gives the required result.

(c) Expand the logarithmic term in the equation of state, retaining the second power in (v_a/v), since this occurs in the second main term; the result reduces to

$$pv = NkT\left(1 + \frac{v_a}{2v}\right) + \frac{Nz\phi(a)}{2}\frac{v_a}{v}$$

$$\approx NkT\left(1 - \frac{v_a}{2v}\right)^{-1} + \frac{Nz\phi(a)}{2}\frac{v_a}{v} .$$

Recall that $\phi(a)$ is necessarily negative for a pair potential like that of Problem 6.4. The corresponding form of the van der Waals equation is

$$pv = RT\left(1 - \frac{b}{v}\right)^{-1} - \frac{a}{v} .$$

Thus the statistical theory produces a result formally the same as that obtained by empirical modification of the ideal gas law.

If we compare this equation of state with that obtained in Problem 6.4(b), we find that the salient points are that the first term in the former suggests a second class of equations of state similar to those of the cell theories, but having the index of v_a/v equal to unity rather than

$\frac{1}{3}$. In fact, one may construct models for which this index is $\frac{2}{3}$, so that defining v_f by a *radius* gives $(v_\sigma/v)^{1/3}$, by a *cross-section* gives $(v_\sigma/v)^{2/3}$, and by a *volume* gives $(v_\sigma/v)^1$. In the low density limit the cohesive energy term from the equation of state of Problem 6.4(b) takes the form $-a/v^2$ instead of $-a/v$ as in the hole theory. Thus a family of equations of state, having the same sort of 'theoretical' justification as the van der Waals equation, may be set up; cf Problems 9.7, 9.17, and 11.12 to 11.15.

(d) Differentiating the equation of state of part (b) with respect to volume, we obtain

$$\frac{\partial p}{\partial v} = \frac{NkT}{v_a}\left[\frac{z\phi(a)v_a^2}{kT\ v^3} - \frac{v_a}{v^2 - v_a v}\right].$$

This is zero when

$$v^2 + \frac{v_a z\phi(a)}{kT}v + \frac{v_a^2 z\phi(a)}{kT} = 0\,.$$

This has two real roots for $T < -z\phi(a)/4k$, otherwise two imaginary roots; the two real roots are coincident for $T = -z\phi(a)/4k$. The locus of the real roots is the boundary of the region within which a single phase is unstable, and $-z\phi(a)/4k$ defines the 'critical temperature' above which phase instability does not occur.

EQUATION OF STATE TREATMENT OF LIQUIDS

6.6 The van der Waals liquid (i). The van der Waals equation of state is usually written for one mole, as

$$\left(p + \frac{a}{\bar{v}^2}\right)(\bar{v} - b) = RT$$

with $R \equiv N_0 k$, N_0 being the Avogadro number, and $\bar{v}$ the molar volume, $\bar{v} \equiv N_0 v/N$. It is the simplest explicit equation of state capable of describing qualitatively the observed phase behaviour of fluids and their mixtures[8].

(a) Use the methods suggested in the Introduction to this chapter to prove that for one mole of fluid obeying this equation of state,

$$U^* = -\frac{a}{\bar{v}}\ ;$$

$$H^* = -\frac{2a}{\bar{v}} + RT\frac{b}{\bar{v} - b}\ ;$$

$$F^* = -\frac{a}{\bar{v}} + RT\ln\frac{\bar{v}}{\bar{v} - b}\ ;$$

$$G^* = -\frac{2a}{\bar{v}} + RT\frac{b}{\bar{v} - b} + RT\ln\frac{\bar{v}}{\bar{v} - b}\,.$$

[8] R.L.Scott and P.H.van Konynenberg, *Disc.Faraday Soc.*, **49**, 87 (1970).

What are the corresponding extensive quantities?

(b) Problem 1.8(a) states for proof

$$C_p - C_v = \frac{Tv\alpha_p^2}{K_T} ,$$

where

$$\alpha_p \equiv \frac{1}{v}\left(\frac{\partial v}{\partial T}\right)_p ; \quad K_T \equiv -\frac{1}{v}\left(\frac{\partial v}{\partial p}\right)_T .$$

Thus $(C_p - C_v)$ is specified completely by the equation of state and derivatives. Use a simple adaptation of the method of part (a) to prove that for one mole of the van der Waals fluid

$$C_p^* = (C_p^* - C_v^*) = \frac{2aR(\bar{v}-b)^2}{\bar{v}^3 RT - 2a(\bar{v}-b)^2} .$$

Note that while C_v^* is defined by $(\partial U^*/\partial T)_v$, $C_p^* \equiv (\partial H^*/\partial T)_p$. The identity given in Problem 1.1(a),

$$\left(\frac{\partial p}{\partial v}\right)_T \left(\frac{\partial v}{\partial T}\right)_p \left(\frac{\partial T}{\partial p}\right)_v = -1 ,$$

is useful in avoiding the difficulties due to p being treated as the dependent variable.

Solution

(a) It is simplest first to establish F^*, and then to use $G^* \equiv F^* + p^*\bar{v}$, $U^* = F^* + TS^*$, etc., for the others; but all four may be established independently. From the equation of state,

$$p^*(\text{vdW}) = RT\left(\frac{1}{\bar{v}-b} - \frac{1}{\bar{v}}\right) - \frac{a}{\bar{v}^2}$$

and

$$F^*(\text{vdW}) = -RT\int_\infty^{\bar{v}}\left(\frac{1}{\bar{v}-b} - \frac{1}{\bar{v}}\right)\mathrm{d}\bar{v} - \frac{a}{\bar{v}} ,$$

since $F^*(\bar{v} = \infty) = F^*_{\text{id}} = 0$. Hence

$$F^*(\text{vdW}) = RT\ln\frac{\bar{v}}{\bar{v}-b} - \frac{a}{\bar{v}} .$$

Since $S^* \equiv -(\partial F^*/\partial T)_v$, we have

$$S^*(\text{vdW}) = -R\ln\frac{\bar{v}}{\bar{v}-b}, \qquad U^*(\text{vdW}) = (F^* + TS^*)_{\text{vdW}} = -\frac{a}{\bar{v}} .$$

Similarly

$$G^*(\text{vdW}) = RT\ln\frac{\bar{v}}{\bar{v}-b} - \frac{a}{\bar{v}} + RT\frac{b}{\bar{v}-b} - \frac{a}{\bar{v}} .$$

$$H^*(\text{vdW}) = -\frac{a}{\bar{v}} + RT\frac{b}{\bar{v}-b} - \frac{a}{\bar{v}} .$$

Alternatively, for U^*, since Problem 1.9(a) gives

$$\left(\frac{\partial U}{\partial v}\right)_T = -p + T\left(\frac{\partial p}{\partial T}\right)_v ,$$

we have

$$U^* = \int_\infty^{\bar{v}} \left[-p_\xi^* + T\left(\frac{\partial p_\xi^*}{\partial T}\right)_v\right] \mathrm{d}\bar{v} .$$

For the van der Waals fluid,

$$-p^* + T\left(\frac{\partial p^*}{\partial T}\right)_v = -RT\left(\frac{1}{\bar{v}-b} - \frac{1}{\bar{v}}\right) + RT\left(\frac{1}{\bar{v}-b} - \frac{1}{\bar{v}}\right) + \frac{a}{\bar{v}^2}$$

and

$$U^*(\mathrm{vdW}) = -\frac{a}{\bar{v}} .$$

The expression for $(\partial H/\partial v)_T$ is obtained from the partial differential identities,

$$\left(\frac{\partial H}{\partial v}\right)_T = \left(\frac{\partial H}{\partial p}\right)_T \left(\frac{\partial p}{\partial v}\right)_T ; \quad \left(\frac{\partial H}{\partial p}\right)_T = \left(\frac{\partial H}{\partial p}\right)_S + \left(\frac{\partial H}{\partial S}\right)_p \left(\frac{\partial S}{\partial p}\right)_T$$

which give

$$\left(\frac{\partial H}{\partial v}\right)_T = \left[v - T\left(\frac{\partial v}{\partial T}\right)_p\right]\left(\frac{\partial p}{\partial v}\right)_T ;$$

thence, using the identity

$$\left(\frac{\partial p}{\partial v}\right)_T \left(\frac{\partial v}{\partial T}\right)_p \left(\frac{\partial T}{\partial p}\right)_v = -1 ,$$

$$\left(\frac{\partial H}{\partial v}\right)_T = v\left(\frac{\partial p}{\partial v}\right)_T + T\left(\frac{\partial p}{\partial T}\right)_v ,$$

and so

$$H_\xi^* = \int_\infty^{\bar{v}} \left[\bar{v}\left(\frac{\partial p_\xi^*}{\partial \bar{v}}\right)_T + T\left(\frac{\partial p_\xi^*}{\partial T}\right)_v\right] \mathrm{d}\bar{v} .$$

For the van der Waals fluid,

$$\bar{v}\left(\frac{\partial p^*}{\partial \bar{v}}\right)_T = \frac{2a}{\bar{v}^2} - RT\left[\frac{\bar{v}}{(\bar{v}-b)^2} - \frac{1}{\bar{v}}\right],$$

$$\bar{v}\left(\frac{\partial p^*}{\partial \bar{v}}\right)_T + T\left(\frac{\partial p^*}{\partial T}\right)_v = \frac{2a}{\bar{v}^2} - RT\left[\frac{\bar{v}}{(\bar{v}-b)^2} - \frac{1}{\bar{v}}\right] + RT\left(\frac{1}{\bar{v}-b} - \frac{1}{\bar{v}}\right)$$

$$= \frac{2a}{\bar{v}^2} - RT\frac{b}{(\bar{v}-b)^2} ,$$

and

$$H^*(\mathrm{vdW}) = -\frac{2a}{\bar{v}} + RT\frac{b}{\bar{v}-b} .$$

From the solution to Problem 1.24,

$$\left(\frac{\partial G}{\partial v}\right)_T = v\left(\frac{\partial p}{\partial v}\right)_T ,$$

so

$$G_\xi^* = \int_\infty^v \bar{v}\frac{\partial p^*}{\partial \bar{v}}\mathrm{d}\bar{v} .$$

For the van der Waals fluid

$$\begin{aligned} G^*(\mathrm{vdW}) &= -\frac{2a}{\bar{v}} - RT\int_\infty^{\bar{v}}\left[\frac{\bar{v}}{(\bar{v}-b)^2} - \frac{1}{\bar{v}}\right]\mathrm{d}\bar{v} \\ &= -\frac{2a}{\bar{v}} - RT\int_\infty^{\bar{v}}\left[\frac{1}{\bar{v}-b} - \frac{1}{\bar{v}} + \frac{b}{(\bar{v}-b)^2}\right]\mathrm{d}\bar{v} \\ &= -\frac{2a}{\bar{v}} + RT\ln\frac{\bar{v}}{\bar{v}-b} + RT\frac{b}{\bar{v}-b} . \end{aligned}$$

The corresponding extensive properties may be discussed in terms of G^*, since it includes all three types of term: $2a/\bar{v} = 2an/v$; the corresponding extensive term must be linear in n, i.e. $2an^2/v$.

$$\ln\frac{\bar{v}}{\bar{v}-b} = \ln\frac{v}{v-nb} \; ; \quad RT\frac{b}{\bar{v}-b} = RT\frac{nb}{v-nb} .$$

Thus

$$G^*(\mathrm{vdW}) = -\frac{2an^2}{v} + nRT\ln\frac{v}{v-nb} + nRT\frac{nb}{v-nb} .$$

(b) The quantity $\int_\infty^v \left(\frac{\partial X}{\partial v}\right)_T \mathrm{d}v$ used in the argument of the introduction gives precisely the volume-dependent part of X. If, as with $(C_p - C_v)$, there is no other part, the definition

$$X_\xi^* \equiv \int_\infty \left[\left(\frac{\partial X}{\partial v}\right)_T - \left(\frac{\partial X_{\mathrm{id}}}{\partial v}\right)_T\right]\mathrm{d}v$$

reduces to

$$(C_p - C_v)_\xi^* = (C_p - C_v)_\xi - (C_p - C_v)_{\mathrm{id}} .$$

Now

$$(C_p - C_v) = \frac{vT\alpha_p^2}{K_T} = -T\left(\frac{\partial v}{\partial T}\right)_p^2\left(\frac{\partial p}{\partial v}\right)_T ,$$

but this involves differentiation at constant pressure, as with C_p itself. This is avoided by substituting for $(\partial v/\partial T)_p$ according to

$$\left(\frac{\partial v}{\partial T}\right)_p = -\left(\frac{\partial p}{\partial T}\right)_v \bigg/ \left(\frac{\partial p}{\partial v}\right)_T :$$

yielding

$$C_p - C_v = -T\left(\frac{\partial p}{\partial T}\right)_v^2 \bigg/ \left(\frac{\partial p}{\partial v}\right)_T .$$

For the van der Waals fluid,

$$\left(\frac{\partial p}{\partial T}\right)_v = \frac{R}{\bar{v}-b} \; ; \quad \left(\frac{\partial p}{\partial \bar{v}}\right)_T = -\frac{RT}{(\bar{v}-b)^2} + \frac{2a}{\bar{v}^3} \; ,$$

yielding

$$C_p - C_v = \frac{R^2 T}{(\bar{v}-b)^2}\left[\frac{RT}{(\bar{v}-b)^2} - \frac{2a}{\bar{v}^3}\right]^{-1} = R\left[1 - \frac{2a(\bar{v}-b)^2}{\bar{v}^3 RT}\right]^{-1} .$$

For the ideal gas,

$$\left(\frac{\partial p}{\partial T}\right)_v = \frac{R}{\bar{v}} \; ; \quad \left(\frac{\partial p}{\partial v}\right)_T = -\frac{RT}{\bar{v}^2}$$

yielding

$$(C_p - C_v) = R$$

and

$$(C_p^* - C_v^*)(\text{vdW}) = R\left[1 - \frac{2a(\bar{v}-b)^2}{\bar{v}^3 RT}\right]^{-1} - R$$

$$= R\left[\frac{2a(\bar{v}-b)^2}{\bar{v}^3 RT}\right]\left[1 - \frac{2a(\bar{v}-b)^2}{\bar{v}^3 RT}\right]^{-1} .$$

It is obvious that

$$C_v^*(\text{vdW}) = -\left[\frac{\partial(a/\bar{v})}{\partial T}\right]_v = 0 \, .$$

[Note: The methods explored in this problem are applicable to any explicit equation of state $\xi(p, v, T) = 0$.]

6.7 The van der Waals liquid (ii). The 'liquid' phase of the van der Waals fluid at a temperature below the boiling point may conveniently be specified by the condition $p = 0$, taking the smaller of the two positive roots in v. In this problem, since we shall deal in molar quantities throughout, we shall omit the bar signifying this.

(a) Use the equation given in Problem 6.6(a) to express G^*, the molar configurational Gibbs free energy for the van der Waals liquid at $p = 0$ as a function of T, the liquid molar volume v_l, and the parameter a. Thence use the reduced variables

$$\tau = \frac{T}{T_c} \; , \quad \phi = \frac{v}{v_c} \; ,$$

where the critical temperature, T_c, and critical molar volume v_c for the van der Waals fluid, cf.Problem 1.11(b), are

$$T_c = \frac{8a}{27Rb} \; , \quad v_c = 3b \; ,$$

to show that for the van der Waals liquid

$$\frac{G^*}{RT} = -\frac{2}{Y(\tau)} - \ln Y(\tau) + \ln 2 - 1 \; ,$$

where

$$Y(\tau) = 1-[1-(\tfrac{32}{27})\tau]^{1/2} .$$

(b) Apply this expression for G^*/RT to two van der Waals liquids, 0, 1, at the same reduced temperature τ, and thence prove

$$G_1^*(T) = f_1 G_0^* \frac{T}{f_1} ,$$

where

$$f_1 \equiv \frac{T_{c1}}{T_{c0}} ; \quad h_1 \equiv \frac{v_{c1}}{v_{c0}} .$$

(c) Show that the analogous relations for the configurational enthalpy and the molar volume (at $p = 0$) are

$$H_1^*(T) = f_1 H_0^* \left(\frac{T}{f_1}\right),$$

$$v_{l1}(T) = h_1 v_{l0}\left(\frac{T}{f_1}\right).$$

Solution

(a) When $p = 0, v/(v-b) = a/RTv$. This equation is clearly quadratic in v, having two real positive roots for T less than some determinate value: the smaller represents the stable phase, for which $(\partial p/\partial v)_T > 0$. Substituting into the expression for G^* given in Problem 6.6(a), for $v/(v-b)$—since a/RTv is the simpler function—we obtain

$$G^*(v_l, T) = RT\left(-\frac{2a}{RTv_l} + \frac{a}{RTv_l} - 1 + \ln\frac{a}{RTv_l}\right)$$

where v_l is the smaller root of the quadratic:

$$v_l = \frac{a}{2RT}\left[1-\left(1-\frac{4RTb}{a}\right)^{1/2}\right] .$$

On transforming into reduced variables, this becomes

$$\frac{v_l}{3b} = \frac{a}{6RTb}\left[1-\left(1-\frac{\frac{32}{37}T}{8a/27Rb}\right)^{1/2}\right]$$

or

$$\phi = \frac{9}{16\tau}[1-(1-\tfrac{32}{27}\tau)^{1/2}] ,$$

or

$$\frac{a}{v_l RT} = \frac{2}{1-(1-\frac{32}{27}\tau)^{1/2}} \equiv \frac{2}{Y(\tau)} .$$

Considering the other terms in $G^*(v_l, T)$ we have

$$\ln\frac{a}{v_l RT} = \ln 2 - \ln[1-(1-\tfrac{32}{27}\tau)^{1/2}] \equiv \ln 2 - \ln Y(\tau) .$$

Then, substitution into the expression for $G^*(v_1, T)$ gives the required result.

(b) At temperature T, for substance 1

$$\frac{G_1^*}{RT} = \frac{2}{Y(\tau)} - \ln Y(\tau) + \ln 2 - 1 \,.$$

For the reduced temperature to be the same, $\tau = T/T_{c1} = T'/T_{c0}$, so we must consider the reference substance, 0, at temperature $T' = T/f_1$. Then

$$\frac{f_1 G_0^*}{RT} = \frac{2}{Y(\tau)} - \ln Y(\tau) + \ln 2 - 1 = \frac{G_1^*}{RT} \,,$$

and the required result follows.

(c) From Problem 6.6(a), for the van der Waals liquid

$$H^* = -\frac{2a}{v} + \frac{RTb}{v-b} = -\frac{2a}{v} + \frac{RTv}{v-b} - RT \,,$$

and at $p = 0$, this reduces to

$$\frac{H^*}{RT} = -\frac{a}{v_l RT} - 1 \,.$$

Now, from part (a),

$$\frac{a}{v_l RT} = \frac{2}{Y(\tau)}$$

whence

$$H_1^*(T) = \frac{2RT}{Y(\tau)} - RT \,,$$

$$H_0^*(T') = \frac{2RT'}{Y(\tau)} - RT'$$

and

$$f_1 H_0^*\left(\frac{T}{f_1}\right) = \frac{2RT}{Y(\tau)} - RT = H_1^*(T) \,.$$

Similarly, since

$$\phi_l = \tfrac{9}{16} Y(\tau)$$

we have

$$v_{l1}(T) = v_{c1} \frac{9}{16\tau} Y(\tau) \,,$$

$$v_{l0}\left(\frac{T}{f_1}\right) = v_{c0} \frac{9}{16\tau} Y(\tau) \,,$$

and

$$h_1 v_{l0}\left(\frac{T}{f_1}\right) = v_{c1} \frac{9}{16\tau} Y(\tau) = v_{l1}(T) \,.$$

These two relations, together with that proven in part (b), are valid for any pair of liquids which follow the same reduced equation of state, irrespective of its explicit form. The second provides a powerful method of testing whether two liquids do in fact follow the same reduced equation of state, at least in the region $p \sim 0$ [(9)].

6.8 The principle of corresponding states (i). The relations proven in Problem 6.7(b), (c) can be derived also by considering the configurational integral itself[(10)]. Problems 6.4(c) and 6.5(d) exemplify the proposition that if $\phi(r)$ takes the form

$$\phi(r) = 4\epsilon^*\left[\left(\frac{\sigma}{r}\right)^m - \left(\frac{\sigma}{r}\right)^n\right], \quad m > n,$$

then the critical temperature T_c correlates directly[(11)] with ϵ^*/k, and the critical volume v_c with σ^3.

(a) By considering the general form of the configurational integral, Q_N, show that if the form of $\phi(r)$ for two substances, 1 and 0, is as above, then at $p = 0$,

$$Q_N^{(1)}(T,N) = h_1^N Q_N^{(0)}\left(\frac{T}{f_1}, N\right),$$

where

$$f_1 \equiv \frac{T_{c1}}{T_{c0}} = \frac{\epsilon_1^*}{\epsilon_0^*}\ ; \quad h_1 \equiv \frac{v_{c1}}{v_{c0}} = \left(\frac{\sigma_1}{\sigma_0}\right)^3 .$$

[Hint: consider a particular configuration of the N molecules of species i, $\mathbf{r}_1, \ldots, \mathbf{r}_N$, in volume v_i, and take the configurational intrinsic energy in this configuration to be

$$U^*(\lambda) = \tfrac{1}{2}\sum_{i,j}\phi(r_{ij}),$$

cf Problem 6.4.]

(b) Show that, at $p = 0$, the result of part (a) implies, for one mole, $N = N_0$,

$$F_1^*(T,N_0) = f_1 F_0^*\left(\frac{T}{f_1}, N_0\right)\ ; \quad G_1^*(T,N_0) = f_1 G_0^*\left(\frac{T}{f_1}, N_0\right)\ ;$$

just as when the comparison is made in terms of a specific equation of state, cf Problem 6.7(b), (c).

(c) Show that the corresponding relation for the $F^\dagger$, defined as the sum of the translational and configurational parts of the molar Helmholtz free energy, is

$$F_1^\dagger(T) = f_1 F_0^\dagger\left(\frac{T}{f_1}\right) - RT\ln h_1 - \tfrac{3}{2}RT\ln f_1 - \tfrac{3}{2}RT\ln\frac{M_1}{M_0},$$

(9) A.J.B.Cruickshank and C.P.Hicks, *Disc.Faraday Soc.*, **49**, 106 (1970).

(10) K.S.Pitzer, *J.Chem.Phys.*, **7**, 583 (1939).

(11) J.S.Rowlinson, *Liquids and Liquid Mixtures*, 2nd Edn. (Butterworths, London), 1960, pp.265-266.

where M_1 and M_0 are the two molecular weights; thence prove that at $p = 0$,

$$H_1^\dagger(T) = f_1 H_0^\dagger\left(\frac{T}{f_1}\right).$$

(d) Generalize these relations to the case $p \neq 0$.

Solution

(a) Any pair of molecules of species 0 mutually distant r contribute

$$\phi_0(r) = 4\epsilon_0^*\left[\left(\frac{\sigma_0}{r}\right)^m - \left(\frac{\sigma_0}{r}\right)^n\right]$$

to U_0^*. Similarly, a pair of molecules of species 1 mutually distant $r\sigma_1/\sigma_0$ contribute

$$\phi_1\left(\frac{r\sigma_1}{\sigma_0}\right) = 4\epsilon_1^*\left[\left(\frac{\sigma_0}{r}\right)^m - \left(\frac{\sigma_0}{r}\right)^n\right]$$

to U_1^*. Then $\phi_1(r\sigma_1/\sigma_0) = f_1\phi_0(r)$.

Thus if the volumes occupied by N_0 molecules of species 0 and species 1, respectively, are so related that in geometrically similar configurations the separations of *all* pairs of species 1 are σ_1/σ_0 times the corresponding separations of all pairs of species 0, then for that configuration (λ),

$$U_1^*(\lambda) = f_1 U_0^*(\lambda)\ ;$$

this condition clearly implies that every dimension of the container of 1 is σ_1/σ_0 times the corresponding dimension of the container of 0. This in turn implies $v_1 = h_1 v_0$, or $v_0 = v_1/h_1$. Thus, if the assembly of species 1 is defined by T, v, N, and that of species 0 by $T/f_1, v/h_1, N$, then

$$\frac{U_1^*(v,\lambda,N)}{kT} = \frac{U_0^*(v/h_1,\lambda,N)}{kT/f_1},$$

and this will be true of every configuration, as long as we may assume a 1:1 correspondence between configurations in the two assemblies. Since Q_N has the dimensions of v^N, it follows that in this case

$$Q_N^{(1)}(v,T,N) = h_1^N Q_N^{(0)}\left(\frac{v}{h_1}, \frac{T}{f_1}, N\right).$$

At $p = 0$, $\phi = \phi(\tau)$ and $\tau_1 = T/T_{c1} = \tau_2 = T/f_1 T_{c2}$, so

$$Q_N^{(1)}(T,N) = h_1^N Q_N^{(0)}\left(\frac{T}{f_1}, N\right).$$

(b) It follows directly that

$$\frac{Q_N^{(1)}}{v^N} = \frac{h_1^N Q_N^{(0)}}{v^N} = \frac{Q_N^{(0)}}{(v/h_1)^N}$$

whence

$$F_1^*(T,N) = -kT\ln\frac{Q_N^{(1)}}{v^N} = -\frac{f_1 kT}{f_1}\ln\frac{Q_N^{(0)}}{(v/h)^N} = f_1 F_0^*\left(\frac{T}{f_1}, N\right).$$

Changing N to N_0, with $N_0 kT = RT$, the required result follows.

(c) From the translational partition function,

$$F_{\text{trs}} = -NkT\ln\left[\left(\frac{2\pi mkT}{h^2}\right)^{3/2}\frac{v}{N}\right] - NkT,$$

when Stirling's approximation is used for $N!$. Then, for one mole, denoting F_{trs} per mole by F',

$$F' = -RT\ln\left[\left(\frac{2\pi mkT}{h^2}\right)^{3/2}\frac{v}{N_0}\right] - RT = -RT\left\{\ln\left[\left(\frac{2\pi mkT}{h^2}\right)^{3/2}\frac{v_c}{N_0}\right] + \ln\phi + 1\right\}.$$

Then, for substance 1 at T, $p = 0$,

$$F_1'(T) = -RT\left\{\ln\phi + 1 + \ln\left[\left(\frac{2\pi mkT}{h^2}\right)^{3/2}\frac{v_{c1}}{N_0}\right]\right\},$$

and for substance 0 at the same reduced temperature, i.e. at T/f_1,

$$F_0'\left(\frac{T}{f_1}\right) = -\frac{RT}{f_1}\left\{\ln\phi + 1 + \ln\left[\left(\frac{2\pi m_0 kT}{fh^2}\right)^{3/2}\frac{v_{c0}}{N_0}\right]\right\},$$

whence

$$f_1 F_0'\left(\frac{T}{f_1}\right) = -RT\left\{\ln\phi + 1 + \ln\left[\left(\frac{2\pi m_1 kT}{h^2}\right)^{3/2}\frac{v_{c1}}{N_0}\right]\right\} + \tfrac{3}{2}RT\ln\frac{m_1 f_1}{m_0} + RT\ln\frac{v_{c1}}{v_{c0}}$$

and

$$F_1'(T) = f_1 F_0'\left(\frac{T}{f_1}\right) - \tfrac{3}{2}RT\ln\frac{m_1 f_1}{m_0} - RT\ln h_1 .$$

Now, from part (b),

$$F_1^*(T) = f_1 F_0^*\left(\frac{T}{f_1}\right),$$

and simple addition gives the required result.

Differentiating, $(\partial F'/\partial v)_T$, gives $p_{\text{trs}} = NkT/v$, so

$$G' = -RT\ln\left[\left(\frac{2\pi mkT}{h^2}\right)^{3/2}\frac{v}{N_0}\right]$$

and the same argument gives

$$G_1'(T) = f_1 G_0'\left(\frac{T}{f_1}\right) - \tfrac{3}{2}RT\ln\frac{m_1 f_1}{m_0} - RT\ln h_1 .$$

We can now either differentiate the given expression with respect to T, $(\partial F^\dagger/\partial T)_v = -S^\dagger$ and obtain $U^\dagger \equiv F^\dagger + TS^\dagger$, or obtain $S' \equiv -(\partial F'/\partial T)_v$ as

$$S' = R\ln\left[\left(\frac{2\pi mkT}{h^2}\right)^{3/2}\frac{v}{N_0}\right] + RT(T^{-3/2} \times \tfrac{3}{2}T^{1/2}) + R\,,$$

whence

$$TS' = RT\ln\left[\left(\frac{2\pi mkT}{h^2}\right)^{3/2}\frac{v}{N_0}\right] + \tfrac{5}{2}RT$$

$$U'(T) = \tfrac{3}{2}RT\,;$$

as

$$H'(T) \equiv U'(T) + p'v = U'(T) + RT = \tfrac{5}{2}RT$$

we have

$$U_1'(T) = \tfrac{3}{2}RT = \tfrac{3}{2}f_1\frac{RT}{f_1}\,U_0'\!\left(\frac{T}{f_1}\right),$$

and similarly for $H'(T)$. Adding $U_1^*(T)$ and $H_1^*(T)$, respectively, gives the required result.

(d) When $p \neq 0$, v is no longer a function only of T, but is either itself an independent variable, or a function of T, p, according to $\xi(p, v, T) = 0$. Then, to ensure 1 and 0 are at the same reduced volume, compare $F_1^\dagger(T, v)$ and $F_0^\dagger(T/f_1, v/h_1)$. Similarly compare $G_1^\dagger(T, p)$ and $G_0^\dagger(T/f_1, p/k_1)$, where $k_1 \equiv p_{c1}/p_{c0}$. The procedure is exemplified by extending the van der Waals case, cf Problem 6.7(b), as

$$F^*(T, v) = RT\left(-\frac{a}{vRT} + \ln\frac{v}{v-b}\right) = RT\left(-\frac{9}{8\phi\tau} + \ln\frac{\phi}{\phi - \frac{1}{3}}\right)$$

whence

$$F_1^*(T, v) = RT\left(-\frac{9}{8\phi\tau} + \ln\frac{\phi}{\phi - \frac{1}{3}}\right)$$

$$F_0^*\left(\frac{T}{f_1}, \frac{v}{h_1}\right) = \frac{RT}{f_1}\left(-\frac{9}{8\phi\tau} + \ln\frac{\phi}{\phi - \frac{1}{3}}\right).$$

Thus

$$F_1^\dagger(T, v) = f_1F_0^\dagger\left(\frac{T}{f_1}, \frac{v}{h_1}\right) - RT\ln h_1 - \tfrac{3}{2}RT\ln f_1 - \tfrac{3}{2}RT\ln\frac{m_1}{m_0}$$

$$G_1^\dagger(T, p) = f_1G_0^\dagger\left(\frac{T}{f_1}, \frac{p}{k_1}\right) - RT\ln h_1 - \tfrac{3}{2}RT\ln f_1 - \tfrac{3}{2}RT\ln\frac{m_1}{m_0}$$

$$H_1^\dagger(T, p) = f_1H_0^\dagger\left(\frac{T}{f_1}, \frac{p}{k_1}\right)$$

$$U_1^\dagger(T, v) = f_1U_0^\dagger\left(\frac{T}{f_1}, \frac{v}{h_1}\right).$$

The rigorous derivation for $G^\dagger$ is as follows:

$$G_1^\dagger = F_1^\dagger + v_1 p = F_1^\dagger + \phi v_{c1} p = F_1^\dagger + h_1 \phi v_{c0} p \,,$$

$$G_0^\dagger = F_0^\dagger + \frac{v_0 p}{k_1} = F_0^\dagger + \frac{\phi v_{c0} p}{k_1} \,,$$

whence

$$f_1 G_0^\dagger = f_1 F_0^\dagger + \frac{\phi v_{c0} p f_1}{k_1} = f_1 F_0^\dagger + h_1 \phi v_{c0} p \,.$$

Now

$$F_1^\dagger = f_1 F_0^\dagger - RT \ln h_1 - \tfrac{3}{2} RT \ln f_1 - \tfrac{3}{2} RT \ln \frac{m_1}{m_0}$$

and addition of $h_1 \phi v_{c0} p$ to both sides converts this to the required result for $G^\dagger$.

6.9 The principle of corresponding states (ii). The necessary condition for the relations proven in Problem 6.8 to be useful is that there exists (over the relevant range of T and v) an approximation to $Q_N^{(0)}$ or F_0^*, which can be expressed by a Taylor series expansion about a datum T, v. Because $Q_N^{(0)}$ determines F_0^*, but not *vice versa*, we examine first the expansions for the configurational thermodynamic potentials. For simplicity, we shall consider only the case $p = 0$.

(a) Expand $H_0^*(T/f)$ about $T = \theta$ to obtain

$$H_0^*\left(\frac{T}{f}\right) = \sum_{n=0}^{\infty} \frac{(-\theta)^n}{n!} \left(\frac{\partial^n H_0^*}{\partial T^n}\right)_{p,\,T=\theta} t^n \,,$$

where $t \equiv 1 - T/\theta f$; and thence prove the alternative expansions:

$$H_0^*\left(\frac{T}{f}\right) = \sum_{n=0}^{\infty} \frac{T^n}{f^n n!} \sum_{r=n}^{\infty} \frac{(-\theta)^{r-n}}{(r-n)!} \left(\frac{\partial^r H_0^*}{\partial T^r}\right)_{p,\,T=\theta} ;$$

$$H_0^*\left(\frac{T}{f}\right) = \sum_{n=0}^{\infty} \frac{T^n}{f^n n!} \left(\frac{\partial^n H_0^*}{\partial T^n}\right)_{p,\,T=0} .$$

(b) Use the Gibbs-Helmholtz relation, $G = H + T(\partial G/\partial T)_p$, to show that for $n > 2$,

$$\left(\frac{\partial^n G}{\partial T^n}\right)_{p,\,T=\theta} = \left(-\frac{1}{\theta}\right)^{n-1} (n-2)! \sum_{r=1}^{n-1} \frac{(-\theta)^{r-1}}{(r-1)!} \left(\frac{\partial^r H}{\partial T^r}\right)_{p,\,T=\theta} ;$$

and thence prove

$$G_0^*\left(\frac{T}{f}\right) = H_0^*(\theta) + \frac{T}{f}\left(\frac{\partial G_0^*}{\partial T}\right)_{p,\,T=\theta} + \sum_{n=1}^{\infty} \frac{(-\theta)^n}{(n-1)!} \left(\frac{\partial^n H_0^*}{\partial T^n}\right)_{p,\,T=\theta} \times \sum_{r=n-1}^{\infty} \frac{t^{r+2}}{(r+1)(r+2)} \,.$$

(c) Show that the leading term ($n = 1$) in the summation above is

$$-\theta\left(\frac{\partial H_0^*}{\partial T}\right)_{T=\theta}\left(1-\frac{T}{\theta f}+\frac{T}{\theta f}\ln\frac{T}{\theta f}\right).$$

Solution

(a) The first series is obtained by expanding in powers of $[(T/f)-\theta]$ and then factorizing as $[(T/f)-\theta]^n = (-\theta)^n t^n$. This, rather than $\theta^n(-t)^n$, is chosen to conform to the series in part (b) below. The most direct way of converting the series in t^n into a series in $(T/f)^n$ is to multiply out the t^n according to the binomial series:

$$\left(1-\frac{T}{\theta f}\right)^n = \sum_{r=0}^{n}\frac{n!}{(n-r)!r!}\left(-\frac{T}{\theta f}\right)^r.$$

This gives the following array, where all $(\partial^n H^*/\partial T^n)_p$ are at $T = \theta$:

$$\begin{aligned}H^*\left(\frac{T}{f}\right) = {} & H^*(\theta)-\theta\left(\frac{\partial H^*}{\partial T}\right)_p+\frac{\theta^2}{2!}\left(\frac{\partial^2 H^*}{\partial T^2}\right)_p-\frac{\theta^3}{3!}\left(\frac{\partial^3 H^*}{\partial T^3}\right)_p+\cdots\\ &+\frac{T}{f}\left[\left(\frac{\partial H^*}{\partial T}\right)_p-\theta\left(\frac{\partial^2 H^*}{\partial T^2}\right)_p+\frac{\theta^2}{2!}\left(\frac{\partial^3 H^*}{\partial T^3}\right)_p-\frac{\theta^3}{3!}\left(\frac{\partial^4 H^*}{\partial T^4}\right)_p+\cdots\right]\\ &+\frac{T^2}{2!f^2}\left[\left(\frac{\partial^2 H^*}{\partial T^2}\right)_p-\theta\left(\frac{\partial^3 H^*}{\partial T^3}\right)_p+\frac{\theta^2}{2!}\left(\frac{\partial^4 H^*}{\partial T^4}\right)_p-\frac{\theta^3}{3!}\left(\frac{\partial^5 H^*}{\partial T^5}\right)_p+\cdots\right]\\ &+\frac{T^3}{3!f^2}\left[\left(\frac{\partial^3 H^*}{\partial T^3}\right)_p-\theta\left(\frac{\partial^4 H^*}{\partial T^4}\right)_p+\frac{\theta^2}{2!}\left(\frac{\partial^5 H^*}{\partial T^5}\right)_p-\frac{\theta^3}{3!}\left(\frac{\partial^6 H^*}{\partial T^6}\right)_p+\cdots\right]\\ &+\cdots,\end{aligned}$$

which is the required form. By inspection, each of the constituent series gives directly the value of its leading term at $T = 0$, whence the third alternative form of the expansion may be written down.

Measured heat quantities, especially those pertaining to solution processes, chemical reactions, and ionisation processes, have most often been analysed using truncations of series belonging to the same family as the third series established here, namely that for which the reference state is $T = 0$. This form has no special advantage, however, and the choice between the first and third forms should be made in each situation according to their relative rapidity of convergence.

(b)

$$G = H + T\left(\frac{\partial G}{\partial T}\right)_p;$$

it is clear that $(\partial G/\partial T)_p$ cannot be functionally related to H alone. Differentiate both sides with respect to T:

$$\left(\frac{\partial G}{\partial T}\right)_p = \left(\frac{\partial H}{\partial T}\right)_p + \left(\frac{\partial G}{\partial T}\right)_p + T\left(\frac{\partial^2 G}{\partial T^2}\right)_p.$$

Thence

$$\left(\frac{\partial^2 G}{\partial T^2}\right)_p = -\frac{1}{T}\left(\frac{\partial H}{\partial T}\right)_p,$$

$$\left(\frac{\partial^3 G}{\partial T^3}\right)_p = +\frac{1}{T^2}\left(\frac{\partial H}{\partial T}\right)_p - \frac{1}{T}\left(\frac{\partial^2 H}{\partial T^2}\right),$$

$$\left(\frac{\partial^4 G}{\partial T^4}\right)_p = -\frac{2}{T^3}\left(\frac{\partial H}{\partial T}\right)_p + \frac{1+1}{T^2}\left(\frac{\partial^2 H}{\partial T^2}\right) - \frac{1}{T}\left(\frac{\partial^3 H}{\partial T^3}\right)_p,$$

$$\left(\frac{\partial^5 G}{\partial T^5}\right)_p = +\frac{3!}{T^4}\left(\frac{\partial H}{\partial T}\right)_p - \frac{2+4}{T^3}\left(\frac{\partial^2 H}{\partial T^2}\right)_p + \frac{2+1}{T^2}\left(\frac{\partial^3 H}{\partial T^3}\right)_p - \frac{1}{T}\left(\frac{\partial^4 H}{\partial T^4}\right)_p,$$

$$\left(\frac{\partial^6 G}{\partial T^6}\right)_p = -\frac{4!}{T^5}\left(\frac{\partial H}{\partial T}\right)_p + \frac{4!}{T^4}\left(\frac{\partial^2 H}{\partial T^2}\right)_p - \frac{2\times 3!}{T^3}\left(\frac{\partial^3 H}{\partial T^3}\right)_p + \frac{3+1}{T^2}\left(\frac{\partial^4 H}{\partial T^4}\right)_p - \frac{1}{T}\left(\frac{\partial^5 H}{\partial T^5}\right)_p,$$

$$\left(\frac{\partial^7 G}{\partial T^7}\right)_p = +\frac{5!}{0!T^6}\left(\frac{\partial H}{\partial T}\right)_p - \frac{5!}{1!T^5}\left(\frac{\partial^2 H}{\partial T^2}\right)_p + \frac{5!}{2!T^4}\left(\frac{\partial^3 H}{\partial T^3}\right)_p - \frac{5!}{3!T^3}\left(\frac{\partial^4 H}{\partial T^4}\right)_p + \cdots,$$

etc.

On putting $T = \theta$, inspection shows the stated result to be correct.

Expanding $G_0^*(T/f)$ about $T = \theta$ and factorizing as in part (a), we find

$$G_0^*\left(\frac{T}{f}\right) = G_0^*(\theta) - \theta\left(\frac{\partial G_0^*}{\partial T}\right)_p + \frac{T}{f}\left(\frac{\partial G_0^*}{\partial T}\right)_p + \frac{\theta^2}{2!}\left(\frac{\partial^2 G_0^*}{\partial T^2}\right)_p t^2 - \frac{\theta^3}{3!}\left(\frac{\partial^3 G_0^*}{\partial T^3}\right)_p t^3 + \cdots,$$

and putting $H_0^*(\theta) = G_0^*(\theta) - \theta(\partial G_0^*/\partial T)_p$ gives

$$G_0^*\left(\frac{T}{f}\right) = H_0^*(\theta) + \frac{T}{f}\left(\frac{\partial G_0^*}{\partial T}\right)_p + \sum_{n=2}^{\infty}\frac{(-\theta)^n}{n!}\left(\frac{\partial^n G_0^*}{\partial T^n}\right)_p t^n.$$

Substituting for $(\partial^n G_0^*/\partial T^n)_p$, $n > 2$, as above, gives

$$G_0^*\left(\frac{T}{f}\right) = H_0^*(\theta) + \frac{T}{f}\left(\frac{\partial G_0^*}{\partial T}\right)_{p,T=\theta} + \sum_{n=2}^{\infty}(-\theta)t^n\frac{(n-2)!}{n!}\sum_{r=1}^{n-1}\frac{(-\theta)^{r-1}}{(r-1)!}\left(\frac{\partial^r H_0^*}{\partial T^r}\right)_{p,T=\theta}.$$

The double summation transforms directly into the required result, as is clear on writing it as the array:

$$-\frac{\theta t^2}{2\times 1}\left(\frac{\partial H_0^*}{\partial T}\right)_p$$

$$-\frac{\theta t^3}{3\times 2}\left(\frac{\partial H_0^*}{\partial T}\right)_p+\frac{\theta^2 t^3}{3\times 2}\left(\frac{\partial^2 H_0^*}{\partial T^2}\right)_p$$

$$-\frac{\theta t^4}{4\times 3}\left(\frac{\partial H_0^*}{\partial T}\right)_p+\frac{\theta^2 t^4}{4\times 3}\left(\frac{\partial^2 H_0^*}{\partial T^2}\right)_p-\frac{\theta^3 t^4}{4\times 3}\left(\frac{\partial^3 H_0^*}{\partial T^3}\right)_p$$

$$-\frac{\theta t^5}{5\times 4}\left(\frac{\partial H_0^*}{\partial T}\right)_p+\frac{\theta^2 t^5}{5\times 4}\left(\frac{\partial^2 H_0^*}{\partial T^2}\right)_p-\frac{\theta^3 t^5}{5\times 4}\left(\frac{\partial^3 H_0^*}{\partial T^3}\right)_p+\frac{\theta^4 t^5}{5\times 4}\left(\frac{\partial^4 H_0^*}{\partial T^4}\right)_p$$

$$-\cdots,$$

and collecting columns.

(c) For $n = 1$, the summation over t^{r+2} becomes

$$\begin{aligned}\sum_{r=0}^{\infty}\frac{t^{r+2}}{(r+1)(r+2)} &= \frac{t^2}{1\times 2}+\frac{t^3}{2\times 3}+\frac{t^4}{3\times 4}+\frac{t^5}{4\times 5}+\cdots\\ &= \frac{t^2}{1}-\frac{t^2}{2}+\frac{t^3}{2}-\frac{t^3}{3}+\frac{t^4}{3}-\frac{t^4}{4}+\frac{t^5}{4}-\frac{t^5}{5}+\cdots\\ &= t-t+t^2-\tfrac{1}{2}t^2(1-t)-\tfrac{1}{3}t^3(1-t)-\tfrac{1}{4}t^4(1-t)\ldots\\ &= t+(1-t)[-t-\tfrac{1}{2}t^2-\tfrac{1}{3}t^3-\tfrac{1}{4}t^4\ldots]\\ &= t+(1-t)\ln(1-t)\,.\end{aligned}$$

Replacing t by $(1-T/\theta f)$, $1-t = T\theta/f$, gives the required result. Since $T/\theta f > 0$, the series is always absolutely or conditionally convergent.

It is interesting to note that the expansion for $G_0^*(T/f)$ of part (b) cannot be transformed into terms of $(\partial^n H_0^*/\partial T^n)_{p,T=0}$ by using the binomial series for t^n; this procedure produces an array whose lines alternate in sign, but (except for the top two) are divergent series. The reason is that, in general, G^* is non-differentiable at $T = 0$, as is obvious from the general form for $(\partial^n G/\partial T^n)_p$, see solution to part (b). The difficulty is avoided, however, by proceeding as in part (c) for all values of n in the general series for $G_0^*(T/f)$, e.g. writing the second term as

$$\theta^2\left(\frac{\partial^2 H_0^*}{\partial T^2}\right)_{p,T=\theta}\left[1-\frac{T}{\theta f}+\frac{T}{\theta f}\ln\frac{T}{\theta f}-\tfrac{1}{2}\left(1-\frac{T}{\theta f}\right)^2\right].$$

BINARY SOLUTIONS

Theories of solutions deal with the **molar functions of mixing** defined by

$$X_{\mathrm{M}} \equiv X_x - \sum_i x_i X_i\,,$$

where X is U, H, F, G, or v; the subscript x denotes the molar function for the solution, and x_i is the mole fraction of species i:

$$x_i \equiv \frac{N_i}{\sum_i N_i} .$$

X_M specifies the change in the value of the function when one mole of solution is formed from x_i mole of each species. The mixing process is taken to be (i) **isothermal** and (ii) either **isobaric** or **isochoric.**

6.10 Ideal solutions. Ideal solutions may be defined as those for which the functions of mixing are the same as those for a mixture of ideal gases.

(a) Commencing from the partition function for one ideal gas molecule in volume v at temperature T,

$$Z_1 = \left(\frac{2\pi mkT}{h^2}\right)^{3/2} v ,$$

use the general argument of Problem 3.4(c) to prove that for a mixture of N_1 molecules of ideal gas species 1, and N_2 molecules of ideal gas species 2, in volume v at temperature T,

$$Z_N = \left(\frac{2\pi m_1 kT}{h^2}\right)^{3xN/2}\left(\frac{2\pi m_2 kT}{h^2}\right)^{3(1-x)N/2} \frac{v^N}{(xN)![(1-x)N]!} ,$$

where $N \equiv N_1 + N_2$, $x \equiv N_1/N$, and thence prove that for such a mixture:

$$F_M = -kT\ln\frac{Q_N^{(\text{id})}}{Q_{xN}^{(\text{id})}Q_{(1-x)N}^{(\text{id})}} = NkT[x\ln x + (1-x)\ln(1-x)] ,$$

where $Q_N^{(\text{id})} = v^N$ [see Problem 6.1(a)] and $Q_{xN}^{(\text{id})} \equiv (xv)^{xN}$; remember that Q_N is *not* an extensive quantity.

(b) Show that if the molecules of species 1 and species 2 are now taken to be identical, the procedure of part (a) gives the necessary result, $F_M = 0$.

(c) Show that the above results for F_M obtain also for the cell model of Problem 6.1(a).

(d) Deduce from the above expression for F_M that the chemical potentials of the two species in an ideal solution [see Problem 1.20(b)]

$$\mu_i \equiv \left(\frac{\partial F}{\partial n_i}\right)_{T,\,v,\,\text{other } n} \equiv \left(\frac{\partial G}{\partial n_i}\right)_{T,\,p,\,\text{other } n} ,$$

(where n_i is the number of moles of i, $n_i \equiv N_i/N_0$) follow

$$\mu_i = \mu_i^0 + RT\ln x_i ;$$

here μ_i^0 refers to pure i at the temperature and pressure of the solution. Thence show that the equivalent expression in terms of the partial pressure, $p_i \equiv x_i p$, is

$$\mu_i = \mu_i^\dagger + RT\ln\frac{p_i}{p^\dagger} ,$$

$p^\dagger$ being the arbitrarily chosen standard pressure at which, for pure i, $\mu_i^0(T, p) = \mu_i^\dagger(T)$.

Solution

(a) Considering at first all the N molecules to be distinguishable,

$$(Z_N)_{\text{Cl}} = Z_{1(1)}^{xN} Z_{1(2)}^{(1-x)N}$$

$$= \left(\frac{2\pi m_1 kT}{h^2}\right)^{3xN/2} v^{xN} \left(\frac{2\pi m_2 kT}{h^2}\right)^{3(1-x)N/2} v^{(1-x)N}.$$

The requirement that the Helmholtz free energy shall be extensive is met by writing

$$Z_N = [Z_{1(1)}\alpha_1]^{xN}[Z_{1(2)}\alpha_2]^{(1-x)N}.$$

One might proceed either by putting $\alpha_1 = D/xN$, $\alpha_2 = D'/(1-x)N$, whence, taking $D = D' = \text{e}$ gives the required result; or by putting $\alpha_1 = \alpha_2 = \text{e}/N$, giving

$$Z_N = \left(\frac{2\pi m_1 kT}{h^2}\right)^{3xN/2} \left(\frac{2\pi m_2 kT}{h^2}\right)^{3(1-x)N/2} \frac{v^N}{N!}.$$

That this second result is incorrect follows from its giving an absurd result for part (d) below, i.e. that the chemical potential of species i in a binary mixture is independent of x, remaining finite as $x \to 0$. Thus the extensivity of F is, alone, an inadequate criterion. If we interpret the factors α_i as expressing the fact, [cf solution to Problem 3.4(b)] that molecules of the same species are indistinguishable in the sense of having the same set of allowed energies (for quantized translational motion) then—since interchanging energies between two molecules of different species changes, in general, the list of momenta—the result is a distinct arrangement. The correct result for Z_N is therefore the one stated in the problem.

The required result for F_M is most simply obtained by writing, for xN molecules of species 1 in volume $xv = xNkT/p$,

$$Z_{xN} = \left(\frac{2\pi m_1 kT}{h^2}\right)^{3xN/2} \frac{(xv)^{xN}}{(xN)!},$$

and similarly for $(1-x)N$ molecules of species 2 in volume $(1-x)v$. Then

$$F_\text{M} = -kT\ln\frac{Z_N}{Z_{xN}Z_{(1-x)N}} = -kT\ln\frac{v^N}{[xv]^{xN}[(1-x)v]^{(1-x)N}}$$

$$= -kT\ln\frac{Q_N^{(\text{id})}}{Q_{xN}^{(\text{id})}Q_{(1-x)N}^{(\text{id})}},$$

since $Q_N^{(\text{id})} = v^N$, etc., see Problem 6.1(a).

Putting $v = NkT/p$, $xv = xNkT/p$, etc., we obtain

$$F_{\mathrm{M}} = -kT\ln\frac{N^N}{[xN]^{xN}[(1-x)N]^{(1-x)N}} = NkT\ln[x^x(1-x)^{(1-x)}] .$$

(b) If the two species are identical, then Z_{xN} and $Z_{(1-x)N}$ are formally unchanged, whereas Z_N is clearly

$$Z_N = \left(\frac{2\pi mkT}{h^2}\right)^{3xN/Z}\left(\frac{2\pi mkT}{h^2}\right)^{3(1-x)N/Z}\frac{(NkT/p)^N}{N!} .$$

Then

$$F_{\mathrm{M}} = NkT\ln[x^x(1-x)^{(1-x)}] - kT\ln\frac{[xN]![(1-x)N]!}{N!} ,$$

and, applying Stirling's approximation to the factorials, we see that the two contributions to F_{M} cancel; but recall that $Q_N = v^N$, so

$$-kT\ln\frac{Q_N^{(\mathrm{id})}}{Q_{xN}^{(\mathrm{id})}Q_{(1-x)N}^{(\mathrm{id})}} = NkT\ln[x^x(1-x)^{(1-x)}] ,$$

irrespective of whether or not the two species are indistinguishable.

(c) For the simple cell model, $Q_N = N!(v/N)^N$, $Q_{xN} = [xN]!(v/N)^{xN}$, $Q_{(1-x)N} = [(1-x)N]!(v/N)^{(1-x)N}$, since for the isochoric process, (v/N) is the same for both pure species and for the mixture. Then, for two distinguishable species

$$Z_N = \left(\frac{2\pi m_1 kT}{h^2}\right)^{3xN/2}\left(\frac{2\pi m_2 kT}{h^2}\right)^{3(1-x)N/2}\frac{N!(v/N)^N}{[xN]![(1-x)N]!} ,$$

$$Z_{xN} = \left(\frac{2\pi m_1 kT}{h^2}\right)^{3xN/2}\frac{[xN]!(v/N)^{xN}}{[xN]!} ,$$

$$Z_{(1-x)N} = \left(\frac{2\pi m_2 kT}{h^2}\right)^{3(1-x)N/2}\frac{[(1-x)N]!(v/N)^{(1-x)N}}{[(1-x)N]!} ,$$

and

$$\frac{Z_N}{Z_{xN}Z_{(1-x)N}} = \frac{Q_N}{Q_{xN}Q_{(1-x)N}} = \frac{N!}{[xN]![(1-x)N]!} ,$$

$$F_{\mathrm{M}} = -kT\ln\frac{N!}{[xN]![(1-x)N]!} ,$$

and Stirling's approximation leads to the required result.

If the two species are now taken to be indistinguishable,

$$Z_N = \left(\frac{2\pi mkT}{h^2}\right)^{3xN/2}\left(\frac{2\pi mkT}{h^2}\right)^{3(1-x)N/2}\frac{N!(v/N)^N}{N!} ,$$

and

$$\frac{Z_N}{Z_{xN}Z_{(1-x)N}} = \frac{(v/N)^N}{(v/N)^{xN}(v/N)^{(1-x)N}} = 1 .$$

(d) Now, from the definition,

$$\mu_i = N_0\left(\frac{\partial F}{\partial N_i}\right)_{T,\, v,\, \text{other } N} .$$

Since

$$F_{\text{M}} = kT\left(N_1 \ln\frac{N_1}{N_1+N_2} + N_2 \ln\frac{N_2}{N_1+N_2}\right),$$

we have

$$\left(\frac{\partial F_{\text{M}}}{\partial N_1}\right)_{T,\, v,\, N_2} = kT\left[\ln N_1 + 1 - \ln(N_1+N_2) - \frac{N_1}{N_1+N_2} - \frac{N_2}{N_1+N_2}\right],$$

$$\mu_{1\text{M}} = N_0 kT \ln x = RT\ln x \ ,$$

$$\mu_1 = \mu_1^0 + RT\ln x \ ,$$

where μ_1^0 is the value for pure 1 at the same T and p. The same result is obtained from $G = G_{\text{M}} + x_1\mu_1^0 + x_2\mu_2^0$.

Since, for a one-component system, cf.Problem 1.20(d),

$$\mathrm{d}\mu = -S\mathrm{d}T + \bar{v}\mathrm{d}p \ ,$$

for an ideal gas

$$\left(\frac{\partial \mu}{\partial p}\right)_T = \bar{v} = \frac{RT}{p} \ ,$$

so

$$\mu^0(p) = \mu^\dagger(p = p^\dagger) + RT\ln\left(\frac{p}{p^\dagger}\right).$$

Thus, in a mixture at p,

$$\mu_i = \mu_i^\dagger + RT\ln\left(\frac{p}{p^\dagger}\right) + RT\ln x_i = \mu_i^\dagger(T) + RT\ln\frac{p_i}{p^\dagger} \ .$$

This result is widely used in approximate treatments of gas-phase chemical equilibria.

The ideal solution result, $F_{\text{M}} = NkT[x\ln x + (1-x)\ln(1-x)]$ generalizes for multi-component systems to

$$F_{\text{M}} = RT\sum_i x_i \ln x_i \ .$$

It is clear from $G \equiv F + pv$ that, similarly,

$$G_{\text{M}} = RT\sum_i x_i \ln x_i$$

for mixing at constant pressure.

6.11 Non-ideal solutions. One general way of constructing a theory for non-ideal solutions is to start with the result of Problem 6.10(c),

$$F_{\text{M}} = -kT\ln\frac{Q_N^{(x)}}{Q_{xN}^{(1)}Q_{(1-x)N}^{(2)}} \ ,$$

whose generality may be inferred from the general form of Z_N [cf Problem 3.5(c)]. Alternatively, F_M may be expressed in terms of the configurational quantities F_i^*, together with the translational quantities F_i' [see solution to Problem 6.8(c)]; or F_M may be expressed directly in terms of the configurational-translational properties $F_i^\dagger$. In all three cases the principle of corresponding states plays an essential role in the derivation of usable explicit forms for F_M. Because the direct outcome of experiment is always G_M, H_M for the isothermal, isobaric process, we concentrate on this, rather than the isothermal, isochoric process. It is convenient to use molar quantities throughout.

(a) Starting from the above expression for F_M, prove that for the isothermal, isobaric process,

$$G_M = -kT\ln\left\{\frac{Q_N^{(0)}[T/f_x, p/k_x, N]}{Q_{xN}^{(0)}[T/f_1, p/k_1, xN]Q_{(1-x)N}^{(0)}[T/f_2, p/k_2, (1-x)N]}\right\} - RT\ln{}^{E}h + pv_M ,$$

where, as before, superscript (0) identifies the reference substance, and

$$\ln{}^{E}h \equiv \ln h_x - x\ln h_1 - (1-x)\ln h_2 ;$$

and derive an explicit form for v_M in terms of the properties of the reference substance (a first-order Taylor expansion in T, p suffices).

(b) Show that, in terms of G^* and G' [see Problem 6.8(c)], G_M may be written

$$G_M = G_M^* - RT\ln{}^{E}v + G_M^{(id)} ,$$

where

$$G_M^*(T,p) \equiv f_x G_0^*\left(\frac{T}{f_x}, \frac{p}{k_x}\right) - xf_1 G_0^*\left(\frac{T}{f_1}, \frac{p}{k_1}\right) - (1-x)f_2 G_0^*\left(\frac{T}{f_2}, \frac{p}{k_2}\right)$$

$$\ln{}^{E}v \equiv \ln v_x(T,p) - x\ln v_1(T,p) - (1-x)\ln v_2(T,p) ,$$

$$G_M^{(id)} = RT[x\ln x + (1-x)\ln(1-x)] ;$$

verify that this is equivalent to the result of part (a).

(c) Similarly, show that in terms of $G^\dagger$ [see Problem 6.8(d)]

$$G_M = G_M^\dagger + G_M^{(id)} ,$$

where

$$G_M^\dagger(T,p) \equiv f_x G_0^\dagger\left(\frac{T}{f_x}, \frac{p}{k_x}\right) - xf_1 G_0^\dagger\left(\frac{T}{f_1}, \frac{p}{k_1}\right) - (1-x)f_2 G_0^\dagger\left(\frac{T}{f_2}, \frac{p}{k_2}\right) - RT\ln{}^{E}h - \tfrac{3}{2}RT\ln{}^{E}f ;$$

verify that this is equivalent to the result of part (b).

(d) Suggest a procedure for calculating F_M for the isochoric and isobaric processes, applicable to the model of Problem 6.4, for the case

$v_{\sigma 1} = v_{\sigma 2} = v_{\sigma x}$; start with the general expression for F_M quoted in the preamble to this problem.

Solution

(a) The appropriate form of the principle of corresponding states is obviously that of Problem 6.8(a):

$$Q_N^{(i)}(T, v) = h_i^N Q_N^{(0)}\left(\frac{T}{f_i}, \frac{v}{h_i}\right).$$

Substituting into the expression for F_M, we obtain

$$\begin{aligned} F_M &= -kT\ln\frac{h_x Q_N^{(0)}(T/f_x, v_x/h_x)}{h_1^{xN} Q_{xN}^{(0)}(T/f_1, xv_1/h_1) h_2^{(1-x)N} Q_{(1-x)N}^{(0)}(T/f_2, (1-x)v_2/h_2)}, \\ &= -kT\ln\left[\frac{Q_N^{(0)}(T/f_x, v_x/h_x)}{Q_{xN}^{(0)}(T/f_1, xv_1/h_1) Q_{(1-x)N}^{(0)}(T/f_2, (1-x)v_2/h_2)}\right] - NkT\ln^E h. \end{aligned}$$

Note that the v retain their identifying subscripts because, even if $h_1 = h_2 = h_x$,

$$v_i(T, p.N) = h_i v_0\left(\frac{T}{f_i}, \frac{p}{k_i}, N\right) \neq h_j v_0\left(\frac{T}{f_j}, \frac{p}{k_j}, N\right) = v_j(T, p, N),$$

cf. Problems 6.7(c), 6.8(d). But since $v_i(T, p, N)/h_i = v_0(T/f_i, p/k_i, N)$, $Q_N^{(0)}(T/f_x, v_x/h_x)$ may equivalently be written $Q_N^{(0)}(T/f_x, p/k_x, N)$. Now $G \equiv F + pv$, and adding the appropriate terms to F_M for the isobaric process gives $p[v_x - xv_1 - (1-x)v_2]$ for N_0 molecules of solution. These arguments lead to the required result, with $N_0 kT = RT$.

It is convenient to expand $v_0(T/f_i, p/k_i)$ about the standard temperature and pressure, $T = \theta$, $p = 1$ (e.g. $\theta = 298°\mathrm{K}$, $p = 1$ bar); to the first order:

$$v_0\left(\frac{T}{f}, \frac{p}{k}\right) = v_0(\theta, 1) + \left(\frac{\partial v_0}{\partial T}\right)_{p,\,T=\theta}\left(\frac{T}{f} - \theta\right) + \left(\frac{\partial v_0}{\partial p}\right)_{T,\,p=1}(p-1),$$

and

$$h_i v_0\left(\frac{T}{f_i}, \frac{p}{k_i}\right) = h_i v_0(\theta, 1)\left(1 + \frac{T\alpha_p}{f_i} - \theta\alpha_p - \frac{p\beta_T}{k_i} + \beta_T\right);$$

since for corresponding-states substances $hk = f$,

$$v_M(T, p) = v_0(1 - \theta\alpha_p + \beta_T)\left(\frac{f}{k}\right)^E + v_0 T\alpha_p\left(\frac{1}{k}\right)^E - v_0 p\beta_T\left(\frac{f}{k^2}\right)^E,$$

where v_0, α_p, β_T all refer to the reference substance at $\theta, 1, N$, this implies only that the equation of state $\xi_0(p, v, T)$ is known in the neighbourhood of θ, 1, and that it may be derived from $Q_N^{(0)}(T, v, N)$.

(b) From Problem 6.8(b)

$$G_x^*(T, p) = f_x G_0^*\left(\frac{T}{f_x}, \frac{p}{k_x}\right), \qquad xG_1^*(T, p) = xf_1 G_0^*\left(\frac{T}{f_1}, \frac{p}{k_1}\right),$$

etc., and the result quoted for G^*_{M} follows. From Problems 6.8(c), 6.10(b), for one mole of solution,

$$G'_x(T,p) = -N_0kT\ln\left\{\left(\frac{2\pi m_1kT}{h^2}\right)^{3x/2}\left(\frac{2\pi m_2kT}{h^2}\right)^{3(1-x)/2} \times \frac{v_x}{[xN_0]^x[(1-x)N_0]^{(1-x)}}\right\},$$

and, since G' is extensive, $xG'_1(T,p,N_0) = G'_1(T,p,xN_0)$ so

$$xG'_1(T,p) = -N_0kT\ln\left[\left(\frac{2\pi m_1kT}{h^2}\right)^{3x/2}\left(\frac{v_1}{N_0}\right)^x\right]$$

etc.; thus the contributions to G'_{M} are comparable directly at T, p, so we may write:

$$G'_{\mathrm{M}} \equiv G'_x - xG'_1 - (1-x)G'_2 = -N_0kT[\ln v_x - x\ln v_1(1-x)\ln v_2] + N_0kT\ln[x^x(1-x)^{(1-x)}],$$

and the required result follows from $G_{\mathrm{M}} = G^*_{\mathrm{M}} + G'_{\mathrm{M}}$.

By definition,

$$F^*_x \equiv -kT\ln\frac{Q^{(x)}_N}{v^N_x}, \quad G^*_x = -kT\ln\frac{Q^{(x)}_N}{v^N_x} + pv_x - N_0kT,$$

since $p^*v_x = pv_x - N_0kT$. Now G^* is an extensive quantity, so

$$xG^*_1(T,p,N_0) = G^*_1(T,p,xN_0) = kT\ln\frac{Q^{(1)}_{xN}}{(xv_1)^{xN}} + xpv_1 - xN_0kT;$$

whence

$$G^*_{\mathrm{M}} = -kT\ln\frac{Q^{(x)}_N}{Q^{(1)}_{xN}Q^{(2)}_{(1-x)N}} + kT\ln\frac{v^N_x}{[xv_1]^{xN}[(1-x)v_2]^{(1-x)N}} + pv_{\mathrm{M}}$$

$$= -kT\ln\frac{Q^{(x)}_N}{Q^{(1)}_{xN}Q^{(2)}_{(1-x)N}} + N_0kT\ln{}^{\mathrm{E}}v - N_0kT\ln[x^x(1-x)^{(1-x)}] + pv_{\mathrm{M}}$$

and

$$G^*_{\mathrm{M}} - RT\ln{}^{\mathrm{E}}v + G^{(\mathrm{id})}_{\mathrm{M}} = -kT\ln\frac{Q^{(x)}_N}{Q^{(1)}_{xN}Q^{(2)}_{(1-x)N}} + pv_{\mathrm{M}}.$$

This particular relation suggests that a pseudo-ideal solution having $G^*_{\mathrm{M}} = RT\ln{}^{\mathrm{E}}v$, would not have $v_{\mathrm{M}} = 0$; the only ideal solution is one for which $f_x = f_1 = f_2$, $h_x = h_1 = h_2$.

(c) From the solution to Problem 6.8(d) we have

$$xG^\dagger_1(T,p) = xf_1G^\dagger_0\left(\frac{T}{f_1}, \frac{p}{k_1}\right) - RTx\ln h_1 - \tfrac{3}{2}RTx\ln f_1 - \tfrac{3}{2}RT\ln\left(\frac{m_1}{m_0}\right)^x,$$

$$(1-x)G^\dagger_2(T,p) = (1-x)f_2G^\dagger_0\left(\frac{T}{f_2}, \frac{p}{k_2}\right) - RT(1-x)\ln h_2 - \tfrac{3}{2}RT(1-x)\ln f_2 - \tfrac{3}{2}RT\ln\left(\frac{m_2}{m_0}\right)^{(1-x)}.$$

On comparing $G_x^\dagger(T,p)$ to $f_x G_0^\dagger(T/f_x, p/k_x)$, as with $G_x'(T,p)$ and $f_x G_0'(T/f_x, p/k_x)$ in part (b), it becomes clear that $G_x^\dagger(T,p)$ also includes the extra term $G_M^{(id)}$. The remainder is said to describe the *equivalent substance*, which differs from the solution only by $G_M^{(id)}$. Then

$$G_x^\dagger(T,p) = f_x G_0^\dagger\left(\frac{T}{f_x}, \frac{p}{k_x}\right) - RT\ln h_x - \tfrac{3}{2}RT\left(\ln f_x - \ln\frac{m_1^x m_2^{1-x}}{m_0}\right) + G_M^{(id)},$$

whence the required result follows. To establish the equivalence of this result to that of part (b), it suffices to prove

$$G_M^* - RT\ln {}^E v = G_M^\dagger .$$

This requires that $G_0^\dagger$ be expressed in terms of G_0^*. From the definitions [see solution to Problem 6.8(c)],

$$G^\dagger \equiv G^* + G',$$

$$G' \equiv -RT\ln\left[\left(\frac{2\pi mkT}{h^2}\right)^{3/2}\frac{v}{N_0}\right],$$

$$f_x G_0^\dagger\left(\frac{T}{f_x}, \frac{p}{k_x}\right) = f_x G_0^*\left(\frac{T}{f_x}, \frac{p}{k_x}\right) + \tfrac{3}{2}RT\ln f_x - RT\ln\left[\left(\frac{2\pi m_0 kT}{h^2}\right)^{3/2}\frac{v_0(T/f_x, p/k_x)}{N_0}\right],$$

whence

$$G_M^\dagger = G_M^* + \tfrac{3}{2}RT\ln {}^E f - RT\ln\frac{v_0(T/f_x, p/k_x)}{v_0^x(T/f_1, p/k_1)v_0^{1-x}(T/f_2, p/k_2)} - \tfrac{3}{2}RT\ln {}^E f - RT\ln {}^E h .$$

On separating $-RT\ln {}^E h$ into its components and incorporating these into the v_0 term, this equation reduces to

$$G_M^\dagger = G_M^* - RT\ln {}^E v(T,p) .$$

While these equivalence proofs are essentially tautologous, they emphasise the importance in corresponding states arguments of (i) the precise definitions, and (ii) the temperature and pressure to which each extensive quantity relates.

(d) Consider first F_M for the general process. The solution to Problem 6.4(b) gives

$$\ln\frac{Q_N}{v^N} = -N + 3N\ln\left[1 - \left(\frac{v_\sigma}{v}\right)^{1/3}\right] - \frac{2zN\epsilon}{kT}\left[\left(\frac{v_\sigma}{v}\right)^4 - \left(\frac{v_\sigma}{v}\right)^2\right],$$

whence

$$-kT\ln\frac{Q_N^{(x)}}{Q_{xN}^{(1)}Q_{(1-x)N}^{(2)}} = -3NkT\frac{\ln[1-(v_\sigma/v_x)^{1/3}]}{[1-(v_\sigma/v_1)^{1/3}]^x[1-(v_\sigma/v_2)^{1/3}]^{1-x}}$$

$$+2zN\left\{\epsilon_x\left[\left(\frac{v_\sigma}{v_x}\right)^4-\left(\frac{v_\sigma}{v_x}\right)^2\right]-x\epsilon_1\left[\left(\frac{v_\sigma}{v_1}\right)^4-\left(\frac{v_\sigma}{v_1}\right)^2\right]\right.$$

$$\left.-(1-x)\epsilon_2\left[\left(\frac{v_\sigma}{v_2}\right)^4-\left(\frac{v_\sigma}{v_2}\right)^2\right]\right\}-NkT\ln{}^{\mathrm{E}}v(T,p)+F_{\mathrm{M}}^{(\mathrm{id})}\,.$$

This relation is formally the same for both the isochoric and the isobaric processes. In the former case, however, $v_x = x_1v_1+(1-x)v_2$, while in the latter, $v_x = h_xv_0(T/f_x, p/k_x)$. Thus F_{M} (isochoric) clearly differs from F_{M} (isobaric) in all three terms. It is obvious from the equation of state [Problem 6.4(b)] that for the isobaric process at $p = 0$,

$$-\ln\left[1-\left(\frac{v_\sigma}{v}\right)^{1/3}\right] = \ln\left\{\frac{4z\epsilon}{kT}\left[\left(\frac{v_\sigma}{v}\right)^2-\left(\frac{v_\sigma}{v}\right)^4\right]\right\},$$

and this simplifies the expression for F_{M}. A possible line of progress is to use $v_i(T,0) = v_0(T/f_i,0)$, $f_i = \epsilon_i/\epsilon_0$ to express the v_i in terms of $v_0(T,p)$ and the f_i. Some tedious algebra and drastic approximation yield expressions for F_{M} in terms of ϵ_1, ϵ_2, when a recipe is assumed for $\epsilon_x(\epsilon_1,\epsilon_2,x)$ (see *Molecular Theory of Solutions*[12] for detailed expositions).

Alternatively, one might use Taylor expansions for $Q_N^{(0)}(T/f, v/h)$ about $f = h = 1$, but this requires an explicit form for $v_0(T,p)$, as a solution to the equation of state. Probably the simplest procedure is to solve the equation of state for each v_i and thence to evaluate $Q_N^{(x)}$, $Q_{xN}^{(1)}$ and $Q_{(1-x)N}^{(2)}$ using particular recipes for ϵ_x, $v_{\sigma x}$ in terms of ϵ_1, ϵ_2; $v_{\sigma 1}$, $v_{\sigma 2}$.

A second type of solution theory follows the method of Problem 6.6. Expressions are derived for the X^* (or $X^\dagger$) in terms of the parameters of the chosen equation of state, $\xi(p,v,T) = 0$, e.g. the parameters a and b of the van der Waals equation. The values of the parameters which best reproduce the observed p, v, T behaviour of the reference liquid having been decided, the principle of corresponding states is used to estimate the f_i, h_i values characterizing the components of the solution[13]. Given a recipe for $f_x = f_x(f_1,f_2,h_1,h_2,x)$ and another for h_x, either (i) X^* are calculated for 1, 2, and x as in Problem 6.6, and thence the X_{M} are calculated directly, or (ii) the corresponding-states formulae for X_i^* in terms of X_0^*, f_i, h_i, cf Problem 6.9, are used, and the X_{M} calculated directly as in parts (b) and (c) above. A variation on method (ii) is to

(12) I.Prigogine, *Molecular Theory of Solutions* (North Holland, Amsterdam), 1957.

(13) A.J.B.Cruickshank and C.P.Hicks, *Disc.Faraday Soc.*, **49**, 106 (1940); D.Patterson and J.M.Bardin, *Trans.Faraday Soc.*, **66**, 321 (1970).

treat the X_0^* as adjustable parameters to fit experimental H_M, G_M, and v_M values for one system; the X_0^* thus specified are then used to predict X_M for other systems in terms of their f_i and h_i. This procedure does not assume an explicit equation of state. Note that the X_0^* are calculable from measured p, v, T data, although not very precisely (cf.Introduction to this section). Detailed exposition of these various procedures is a matter of computation rather than algebra.

6.12 The regular solution (i). This theory[14] is related to a simplified version of the smoothed potential cell model (see Problem 6.4), in which the variation of U^* with volume is ignored, i.e. it takes the form

$$U^* = -\tfrac{1}{2}\sum_{i,j} \epsilon_{ij} \text{ (nearest neighbours only)} .$$

While obviously invalid as a model for the liquid state, this provides a way of describing solutions which, although itself of limited applicability, is of interest as a basis for introducing the concept of incompletely random mixing (see Problem 6.13). The three pairwise interaction energies are denoted by $-\epsilon_1, -\epsilon_2, -\epsilon_{12}$, and we define the **interchange energy** w by

$$w \equiv -z\epsilon_{12} + \tfrac{1}{2}z(\epsilon_1 + \epsilon_2) ,$$

where, as in preceding problems, z is the co-ordination number of the (implied) cell lattice. Thus, if a molecule of species 1 from pure liquid 1 is interchanged with a molecule of species 2 from pure liquid 2, the net increase in the energy of the two systems is $2w$.

(a) Write down the configurational intrinsic energy U^* for an assembly of N_1 molecules of species 1 and N_2 molecules of species 2, $N_1 + N_2 \equiv N$, such that there are zY 1, 2 pairwise interactions, remembering that there are $\tfrac{1}{2}zN$ pairwise interactions in total, and thence prove that, for the binary solution, U_M is related to the average value, $\overline{Y}$, of Y by

$$U_M = \overline{Y}w .$$

(b) Assuming the free volume per cell to be v_f for all cells, and considering each possible configuration in turn, as in Problems 6.1(a), 6.5(a), show that

$$\frac{Q_N}{Q_{xN}Q_{(1-x)N}} = \sum_{\lambda} \frac{\exp(-Y_\lambda w/kT)}{N_1!N_2!} .$$

By extracting the mean, $\exp(-\overline{\overline{Y}}w/kT)$, show that

$$F_M = F_M^{(\text{id})} + \overline{\overline{Y}}w ,$$

and use the relation

$$U_M = F_M - T\left(\frac{\partial F_M}{\partial T}\right)_v$$

[14] E.A.Guggenheim, *Mixtures* (Clarendon Press, Oxford), 1952, pp.30-32.

to prove

$$\overline{Y} = \overline{\overline{Y}} - T\frac{\mathrm{d}\overline{\overline{Y}}}{\mathrm{d}T} .$$

(c) Consider, as a first approximation, the case of completely random mixing, corresponding formally to the limiting case as $w/kT \to 0$. Show that to this approximation

$$\overline{Y} = \frac{N_1 N_2}{N} = x(1-x)N ,$$

$$U_{\mathrm{M}} = x(1-x)Nw ,$$

$$F_{\mathrm{M}} = F_{\mathrm{M}}^{(\mathrm{id})} + x(1-x)Nw .$$

(d) Show that these results are obtained also by taking only the first term of each of the corresponding-states expansions for F_{M} and U_{M} obtained from the expansions in $F_0^*(0)$ and $U_0^*(0)$ analogous to those in $G_0^*(0)$ and $H_0^*(0)$ in Problem 6.9(a), (b), provided that f_x is taken as

$$f_x = x^2 f_1 + 2x(1-x)f_{12} + (1-x)^2 f_2 ,$$

which is appropriate to random mixing of molecules of equal size.

Solution

(a) A molecule of species 1, with z nearest neighbours is engaged in z pairwise interactions. Let αz of these, on average, be 1, 2 interactions and $(1-\alpha)z$ be 1, 1 interactions. Then there are $\alpha z N_1$ 1, 2 interactions in all, i.e. $\alpha N_1 = Y$. The total number of 1, 1 interactions is clearly

$$\tfrac{1}{2}(1-\alpha)zN_1 = \tfrac{1}{2}z(N_1 - Y) .$$

Similarly, if the average number of 2, 1 interactions per molecule of 2 is βz, the total is $\beta z N_2 = zY$, and the total number of 2, 2 interactions is $\frac{1}{2}z(N_2 - Y)$. The total number of pairwise interactions is then $\frac{1}{2}z(N_1 + N_2)$ as required. The total configurational intrinsic energy is

$$U_\lambda^* = -\tfrac{1}{2}z\epsilon_1(N_1 - Y) - \tfrac{1}{2}z\epsilon_2(N_2 - Y) - z\epsilon_{12}Y ,$$

and for the average of all configurations

$$U^* = -\tfrac{1}{2}z\epsilon_1(N_1 - \overline{Y}) - \tfrac{1}{2}z\epsilon_2(N_2 - \overline{Y}) - z\epsilon_{12}\overline{Y} .$$

Since $U_1^* = -\frac{1}{2}z\epsilon_1 N_1$, $U_2^* = \frac{1}{2}z\epsilon_2 N_2$,

$$U_{\mathrm{M}} = \tfrac{1}{2}z\epsilon_1\overline{Y} + \tfrac{1}{2}z\epsilon_2\overline{Y} - z\epsilon_{12}\overline{Y} = \overline{Y}w .$$

(b) Since U_λ^* is constant for each arrangement of the molecules in their cells, as in Problem 6.1(b), Q_N breaks up into separate integrals each over

v_f, so

$$Q_N = v_f^N \sum_\lambda \exp\left(-\frac{U_\lambda^*}{kT}\right) = v_f^N \sum_\lambda \exp\frac{\frac{1}{2}z\epsilon_1 N_1 + \frac{1}{2}z\epsilon_2 N_2 - Y_\lambda w}{kT},$$

$$Q_{xN} = N_1! v_f^{xN} \exp\frac{\frac{1}{2}z\epsilon_1 N_1}{kT},$$

$$Q_{(1-x)N} = N_2! v_f^{(1-x)N} \exp\frac{\frac{1}{2}z\epsilon_2 N_2}{kT}$$

and so

$$\frac{Q_N}{Q_{xN}Q_{(1-x)N}} = \sum_\lambda \frac{\exp(-Y_\lambda w/kT)}{N_1!N_2!};$$

extracting the mean, and remembering that there are $N!$ arrangements, we obtain

$$\frac{Q_N}{Q_{xN}Q_{(1-x)N}} = \frac{\exp(-\overline{\overline{Y}}w/kT)N!}{N_1!N_2!},$$

and the required result for F_M follows at once.

$$F_M^{(id)} = RT[x\ln x + (1-x)\ln(1-x)],$$

whence

$$-T\left(\frac{\partial F_M^{(id)}}{\partial T}\right)_v = -F_M^{(id)},$$

$$U_M = \overline{\overline{Y}}w - T\left(\frac{\partial \overline{\overline{Y}}w}{\partial T}\right)_v.$$

But $U_M = \overline{Y}w$, so that

$$\overline{Y}w = \overline{\overline{Y}}w - T\left(\frac{d\overline{\overline{Y}}w}{dT}\right)$$

$$\overline{Y} = \overline{\overline{Y}} - T\left(\frac{d\overline{\overline{Y}}}{dT}\right).$$

(c) In the case of random mixing $\alpha = (1-x)$, and the mean number of 1, 2 interactions per molecule of 1 is $z(1-x)$, so the total number is $N_1 z(1-x) = x(1-x)Nz$. The number of 1, 1 interactions is $\frac{1}{2}zxN_1$, and the number of 2, 2 interactions is $\frac{1}{2}z(1-x)N_2$. The total number is then

$$(x-x^2)Nz + \tfrac{1}{2}x^2Nz + \tfrac{1}{2}(1-x)^2Nz = (x-x^2)Nz + \tfrac{1}{2}x^2Nz + \tfrac{1}{2}Nz - xNz + \tfrac{1}{2}x^2Nz = \tfrac{1}{2}Nz,$$

and

$$\overline{Y} = x(1-x)N.$$

For $\overline{\overline{Y}}$ independent of temperature, the differential equation of part (b) is obviously satisfied by

$$\overline{\overline{Y}} = \overline{Y} = x(1-x)N$$

and the required results follow immediately.

The assumption that the free volume is the same for species 1 and 2, and for the solution, when the reduced temperatures all differ, implies a curious mixing process in which the components are initially at different pressures, while the final solution is at a third pressure; it is obviously closer to the isochoric process than to the isobaric process.

(d) Considering only the leading term in the $U_0^*(T/f)$ expansion analogous to that of Problem 6.9(a),

$$f_x U_0^*\left(\frac{T}{f_x}\right) = f_x U_0^*(0)\,,$$

and, since $U_M = U_M^*$,

$$U_M = f_x U_0^*\left(\frac{T}{f_x}\right) - x f_1 U_0^*\left(\frac{T}{f_1}\right) - (1-x) f_2 U_0^*\left(\frac{T}{f_2}\right),$$

$$= U_0^*(0)[f_x - x f_1 - (1-x) f_2]\,.$$

If f_x is taken as

$$f_x = x^2 f_1 + 2x(1-x) f_{12} + (1-x)^2 f_2\,,$$

then

$$U_M = U_0^*(0)[x^2 f_1 + 2x(1-x) f_{12} + (1-x)^2 f_2 - x f_1 - (1-x) f_2]$$

$$= U_0^*(0)[-f_1 x(1-x) - f_2(1-x)x + 2 f_{12} x(1-x)]$$

$$= U_0^*(0) x(1-x) w / \tfrac{1}{2} z \epsilon_0\,,$$

since $f_i \equiv \epsilon_i/\epsilon_0$; but $U_0^*(0) = \frac{1}{2} z N_0 \epsilon_0$ in the regular solution model, so to this approximation

$$U_M = x(1-x) N_0 w$$

for one mole of solution.

For the Helmholtz free energy, isochoric process, it is clear from Problem 6.9(c) that

$$F_M = U_0^*(0) f^E + T C_{v0}^*(0) \ln {}^E f + F_M^{(id)}\,,$$

since $RT \ln {}^E v = 0$ for this process. Neglecting the second term we obtain

$$F_M = x(1-x) N_0 w + F_M^{(id)}\,.$$

6.13 The regular solution (ii). In the model of Problem 6.12, consider, instead of random mixing, the **quasi-chemical equilibrium**:

$$(1, 1) + (2, 2) \rightleftharpoons 2(1, 2)\,,$$

for which the energy increase per molecular unit of reaction is clearly $\Delta u = 2w/z$. The equilibrium 'concentrations' of the three types of interaction are assumed to accord with a quasi-chemical equilibrium constant K, defined by

$$\Delta F \equiv -N_0 k T \ln K\,,$$

where ΔF is the Helmholtz free energy change per molar unit of reaction, or, equivalently, by

$$K \equiv \Omega \exp\left(\frac{-\Delta U}{N_0 kT}\right) = \Omega \exp\left(\frac{-\Delta u}{kT}\right),$$

where ΔU, the intrinsic energy change per molar unit of reaction, is given by the usual relation

$$\Delta U = \Delta F - T\left(\frac{\mathrm{d}\Delta F}{\mathrm{d}T}\right),$$

and Ω is the change in the degeneracy factor such that

$$kT \ln \Omega \equiv T\Delta S\,.$$

K is equal to the ratio of the product of the concentrations of the reaction-product species (each raised to the power equal to its numerical coefficient in the reaction equation) to the corresponding product of the concentrations of the reactants. Thus, denoting the equilibrium concentration C_{ij}, we have for the above reaction

$$K = \frac{[C_{12}]^2}{[C_{11}][C_{22}]}\,.$$

(a) Express the concentrations of the different types of pair interactions in terms of zY [see Problem 6.12(a)], the number of 1, 2 interactions, and show that the equilibrium value of Y obeys

$$(N_1 - \overline{Y})(N_2 - \overline{Y}) - \eta^2 \overline{Y}^2 = 0$$

where $\eta \equiv \exp(w/zkT)$. If U_{M} be written

$$U_{\mathrm{M}} \equiv \frac{2}{(\beta+1)} x(1-x)Nw\,,$$

so that $U_{\mathrm{M}} \to x(1-x)Nw$ as $\beta \to 1$, prove that

$$\beta = [1 + 4(\eta^2 - 1)x(1-x)]^{1/2}\,.$$

(b) Since β is a function of T, $\overline{Y}$ is also a function of T, and the relation $\overline{Y} = \overline{\overline{Y}} - T(\mathrm{d}\overline{\overline{Y}}/\mathrm{d}T)$ cannot be solved for $\overline{\overline{Y}}$ directly. Verify the alternative[15] form $\overline{Y} = \mathrm{d}(\overline{\overline{Y}}/T)/\mathrm{d}(1/T)$ and show that substituting for $\overline{Y}$ and for $\mathrm{d}(1/T)$ in terms of β, and integrating between $\beta = 1$ and $\beta = \beta$ gives for F_{M}

$$F_{\mathrm{M}} = F_{\mathrm{M}}^{(\mathrm{id})} + \tfrac{1}{2} zN_0 kT(1-x)\ln\left[\frac{\beta+1-2x}{(1-x)(\beta+1)} + x\ln\frac{\beta-1+2x}{x(\beta+1)}\right],$$

when $\overline{\overline{Y}}(\beta = 1)$ is put equal to zero.

(c) Show that $\overline{\overline{Y}}(\beta = 1) = 0$ is the necessary condition that

$$\overline{Y} = \overline{\overline{Y}} - T\frac{\mathrm{d}\overline{\overline{Y}}}{\mathrm{d}T}$$

[15] E.A.Guggenheim, *Mixtures* (Clarendon Press, Oxford), 1952, pp.38, 39.

holds at $T = \infty$. Is this last assertion necessarily valid? Suggest an alternative way of evaluating F_M which avoids the logical difficulties associated with the reference temperature $T = \infty$.

Solution

(a) From Problem 6.12(a), the numbers of interactions are

$$1, 1: \quad \tfrac{1}{2}z(N_1 - Y);$$

$$2, 2: \quad \tfrac{1}{2}z(N_2 - Y);$$

$$1,2: \quad zY,$$

and the concentrations are given by dividing each number by v. Then

$$\frac{4\overline{Y}^2}{(N_1 - \overline{Y})(N_2 - \overline{Y})} = \Omega \exp\left(-\frac{2w}{zkT}\right).$$

For a 1, 1 interaction, there is clearly only one distinguishable arrangement, and likewise for a 2, 2 interaction; but two 1, 2 interactions may be achieved in four ways. This is confirmed by the fact that for $w/zkT = 0$ we must have $\overline{Y}^2 = (N_1 - \overline{Y})(N_2 - \overline{Y})$ to conform to random mixing, $\overline{Y} = N_1N_2/(N_1 + N_2) = x(1-x)N$.

On putting $\exp(-2w/zkT) = 1/\eta^2$, the equilibrium relation becomes

$$\frac{\overline{Y}^2}{(N_1 - \overline{Y})(N_2 - \overline{Y})} = \frac{1}{\eta^2}$$

giving

$$(N_1 - \overline{Y})(N_2 - \overline{Y}) = \eta^2 \overline{Y}^2 .$$

Putting $\overline{Y} = 2x(1-x)N/(\beta + 1)$, so that,

$$N_1 - \overline{Y} = N\left[x - \frac{2x(1-x)}{\beta + 1}\right],$$

$$N_2 - \overline{Y} = N\left[1 - x - \frac{2x(1-x)}{\beta + 1}\right],$$

$$\overline{Y}^2 = N^2 \frac{4x^2(1-x)^2}{(\beta + 1)^2},$$

the equilibrium relation becomes

$$\frac{N^2}{(\beta + 1)^2}[(\beta + 1)x - 2x(1-x)][(\beta + 1)(1-x) - 2x(1-x)]$$

$$= \frac{4\eta^2 x^2 (1-x)^2 N^2}{(\beta + 1)^2},$$

which reduces to

$$\beta^2 - 4(\eta^2 - 1)x(1-x) - 1 = 0$$

and the required result follows.

(b) The required form follows exactly the standard alternative form of the Gibbs-Helmholtz relation: $H = [\partial(G/T)/\partial(1/T)]_p$. Thus

$$\overline{Y} = \overline{\overline{Y}} - T\frac{d\overline{\overline{Y}}}{dT} = \overline{\overline{Y}} - T\frac{d\overline{\overline{Y}}}{d(1/T)}\Big/\frac{d(1/T)}{dT} = \overline{\overline{Y}} + \frac{1}{T}\frac{d\overline{\overline{Y}}}{d(1/T)} = \frac{d(\overline{\overline{Y}}/T)}{d(1/T)} \cdot$$

$$\overline{Y} \equiv \frac{2x(1-x)N}{\beta+1} ,$$

and, from part (a),

$$\beta^2 - 4(\eta^2 - 1)x(1-x) - 1 = 0$$

giving

$$\eta^2 = \frac{\beta^2 - (1-2x)^2}{4x(1-x)} ,$$

$$\frac{2w}{zkT} = \ln[\beta^2 - (1-2x)^2] - \ln 4x(1-x) ,$$

whence

$$\frac{2w}{zk}d(1/T) = \frac{2\beta}{\beta^2 - (1-2x)^2}d\beta .$$

Substituting for $\overline{Y}$ and for $d(1/T)$,

$$d\frac{\overline{\overline{Y}}}{T} = \frac{zNkx(1-x)}{2w}\left\{\frac{4\beta\, d\beta}{[\beta+1][\beta^2 - (1-2x)^2]}\right\} .$$

Separating the integrand on the right-hand side into partial fractions, we find

$$\frac{4\beta}{[\beta+1][\beta+(1-2x)][\beta-(1-2x)]}$$
$$= \frac{1}{x(1-x)}\left(\frac{1-x}{\beta+1-2x} + \frac{x}{\beta-1+2x} - \frac{1}{\beta+1}\right)$$

whence

$$d\frac{\overline{\overline{Y}}}{T} = \frac{zNk}{2w}\left(\frac{1-x}{\beta+1-2x} + \frac{x}{\beta-1+2x} - \frac{1}{\beta+1}\right)d\beta .$$

Integrating between $\beta = 1$ and $\beta = \beta$ we obtain

$$\overline{\overline{Y}} = \frac{zNkT}{2w}[(1-x)\ln(\beta+1-2x) + x\ln(\beta-1+2x) - \ln(\beta+1)]_1^\beta$$
$$+ \overline{\overline{Y}}(\beta = 1) .$$

$$= \frac{zNkT}{2w}\left[(1-x)\ln\frac{\beta+1-2x}{1-x} + x\ln\frac{\beta-1+2x}{x} - \ln(\beta+1)\right] + \overline{\overline{Y}}(\beta = 1) .$$

The given result follows at once when $\overline{\overline{Y}}(\beta = 1)$ is put equal to zero.

(c) $\beta = 1$ corresponds to $\eta^2 = 1$, $2w/zkT = 0$, i.e. $T = \infty$, in the general case $w \neq 0$. The simplest way to proceed is to evaluate

$T(\mathrm{d}\overline{\overline{Y}}/\mathrm{d}T)$ at $T = \infty$. From the result of part (b)

$$T\frac{\mathrm{d}\overline{\overline{Y}}}{\mathrm{d}T} = \overline{\overline{Y}} - \overline{\overline{Y}}(\beta = 1) + \frac{zNkT^2}{2w}\left(\frac{1-x}{\beta+1-2x} + \frac{x}{\beta-1+2x} - \frac{1}{\beta+1}\right)\frac{\mathrm{d}\beta}{\mathrm{d}T}\,.$$

From the result of part (a)

$$\beta^2 = 4(\eta^2 - 1)x(1-x) + 1\,.$$

Hence

$$\frac{\mathrm{d}\beta}{\mathrm{d}T} = \frac{2x(1-x)}{\beta}\left(\frac{\mathrm{d}\eta^2}{\mathrm{d}T}\right) = \frac{4x(1-x)}{\beta}\left(-\frac{w}{zkT^2}\right)\exp\frac{2w}{zkT}\,,$$

$$T\frac{\mathrm{d}\overline{\overline{Y}}}{\mathrm{d}T} = \overline{\overline{Y}} - \overline{\overline{Y}}(\beta = 1) - \frac{2x(1-x)N}{\beta}\left(\frac{1-x}{\beta+1-2x} + \frac{x}{\beta-1+2x} - \frac{1}{\beta+1}\right)\exp\frac{2w}{zkT}\,,$$

When $\beta = 1$, then $T = \infty$, $\exp(2w/zkT) = 1$, and

$$\left(T\frac{\mathrm{d}\overline{\overline{Y}}}{\mathrm{d}T}\right)_{\beta=1} = \overline{\overline{Y}}(\beta = 1) - \overline{\overline{Y}}(\beta = 1) - \overline{Y}(\beta = 1)\,.$$

Thus at $T = \infty$, $\overline{\overline{Y}}(\beta = 1) = 0$ is the only assumption which satisfies $\overline{Y} = \overline{\overline{Y}} - T(\mathrm{d}\overline{\overline{Y}}/\mathrm{d}T)$. In fact, to assume that the general relation $[\partial(\Delta G/T)/\partial(1/T)]_p = \Delta H$ holds at $T = \infty$ implies categorically that $\Delta G(T = \infty) = 0$. This is to assert that $\Delta G/T$ remains differentiable with respect to $1/T$ at $1/T = 0$. There is no simple way to establish the validity of this assertion. It can be avoided, in principle, by choosing a reference temperature $T = \theta$, other than $T = \infty$. Then $\overline{\overline{Y}}$ must be evaluated at $T = \theta$ through the ratio $Q_N/Q_{xN}Q_{(1-x)N}$. Various approximations have been used on this problem, and it has been proven(16) that the result of part (b) is obtained by assuming that all pairwise interactions are mutually independent. Higher-order approximations give slightly different results; the algebra is too tedious for inclusion here; the methods are described in the reference cited. The result of part (b) is important because it suggests a temperature dependence of $F_{\mathrm{M}} - F_{\mathrm{M}}^{(\mathrm{id})}$ which is different from that of U_{M}, as is usually observed when experimental results are converted to give F_{M} (isochoric) and U_{M} (isochoric). Note the qualitative similarity to the inclusion of the term $TC_{v0}^{*}(0)\ln{}^{\mathrm{E}}f$ in the corresponding-states formula [cf Problem 6.12(a)].

6.14 Solutions of chain molecules (i). Problem 6.10 shows that the ideal solution formulae may be derived from the simple cell model. For the case $v_{\sigma 1} = v_{\sigma 2} = v_{\sigma x}$ the same result is obtained from the models of Problems 6.2 and 6.3. There is thus some justification for using the simple cell model to derive the thermodynamic functions of mixing for

(16) E.A.Guggenheim, *Mixtures* (Clarendon Press, Oxford), 1952, pp.42-47.

solutions of non-interacting chain molecules. The essential assumption is that the molecules of both species are chains whose links, or segments, can each, interchangeably, occupy a single cell. The simplest case is a solution of 'monomer' (species 1) and 'r-mer' (species 2), the molecules of the latter each occupying a chain of r cells. The total number of cells is then $N \equiv N_1 + rN_2$. As in Problem 6.1(b) and Problem 6.2,

$$Q_N = \sum_{\lambda} v_f^N \,.$$

The present problem is then to formulate the number of ways of arranging N_2 r-mers among N cells; there are clearly $N_1!$ ways of arranging the N_1 monomers among the remaining N_1 cells. The problem is greatly simplified if we make the general assumption that the average proportion of occupied cells neighbouring any empty cell is the same as the proportion for the whole assembly of N cells. Only a trivial error is introduced by taking this average proportion as constant during the placing of one complete r-mer.

(a) Using these two assumptions, prove that the number of ways of arranging N_2 r-mers among $N \equiv N_1 + rN_2$ cells is

$$g(N_2) = \left(\frac{z}{z-1}\right)^{(1-\phi)N/r} \left(\frac{z-1}{N}\right)^{(1-\phi)N(1-1/r)} \left(\frac{[N/r]!}{[\phi N/r]!}\right),$$

where z is the co-ordination number of the cell lattice, and ϕ (the segment fraction, or volume fraction of monomer) is defined by $\phi \equiv N_1/N$. A generally valid way of proceeding is to consider the situation after n r-mers have been placed; count the number of choices open for the placing of each segment after the first, and so formulate the number of ways of placing the $(n+1)$th r-mer.

(b) Repeat the procedure of part (a) for N_2 r-mers among rN_2 cells, and thence prove that for the binary mixture, of N_2 r-mers and N_1 monomers, $N_1 + N_2 \equiv N_0$,

$$F_M = RT[x \ln\phi + (1-x)\ln(1-\phi)] \,.$$

Solution

(a) Once the first segment of the $(n+1)$th r-mer has been placed, there are $z(N-rn)/N$ possibilities for the second segment, i.e. z times the average fraction of vacant cells. For the third segment there are $(z-1)$ cells accessible, since one of the z adjacent cells is occupied by the first segment, so the number of possibilities is $(z-1)(N-rn)/N$. Thus, for every siting of the first segment, there are

$$z\left(\frac{N-rn}{N}\right)\left[(z-1)\frac{N-rn}{N}\right]^{r-2} = \frac{z}{z-1}\left[(z-1)\frac{N-rn}{N}\right]^{r-1}$$

ways of placing the complete r-mer. Since there are $N-rn$ possibilities for the first segment, there are

$$\frac{z}{z-1}(N-rn)\left[(z-1)\frac{N-rn}{N}\right]^{r-1}$$

ways of placing the $(n+1)$th r-mer. Thus the total number of ways of arranging $N_2 = (1-\phi)N/r$ r-mers among $N = rN_2+N_1$ cells is

$$g(N_2, N) = \prod_{n=0}^{N_2-1}\left(\frac{z}{z-1}\right)(N-rn)\left[(z-1)\frac{N-rn}{N}\right]^{r-1}$$

$$= \left(\frac{z}{z-1}\right)^{N_2}\left(\frac{z-1}{N}\right)^{N_2(r-1)}(r)^{rN_2}\prod_{n=0}^{N_2-1}\left(\frac{N}{r}-n\right)^{r}.$$

$$= \left(\frac{z}{z-1}\right)^{(1-\phi)N/r}\left(\frac{z-1}{N}\right)^{(1-\phi)N(1-1/r)} r^{(1-\phi)N}\left\{\frac{(N/r)!}{[(N/r)-N_2]!}\right\}^{r}.$$

Since $[(N/r)-N_2] = (N/r)-(1-\phi)N/r = \phi N/r$, the given result follows.

(b) For N_2 r-mers distributed among rN_2 cells, the result (a) evidently becomes

$$g(N_2, rN_2) = \left(\frac{z}{z-1}\right)^{N_2}\left(\frac{z-1}{rN_2}\right)^{N_2(r-1)}(r)^{rN_2}(N_2!)^{r}.$$

Then, for the solution (N cells, $N = N_1+rN_2$):

$$Q_N = g(N_2, N)N_1!v_f^N;$$

while for the pure r-mer [rN_2 cells, $rN_2 = (1-\phi)N$]:

$$Q_{(1-\phi)N} = g(N_2, rN_2)v_f^{(1-\phi)N};$$

and for the pure monomer (N_1 cells, $N_1 = \phi N$):

$$Q_{\phi N} = N_1!v_f^{\phi N}.$$

Then

$$F_M = -kT\ln\frac{Q_N}{Q_{(1-\phi)N}Q_{\phi N}} = -kT\ln\frac{g(N_2, N)}{g(N_2, rN_2)}$$

$$= -kT\ln\left\{\frac{[N/r]!}{[(N/r)-N_2]!N_2!}\right\}^{r}\left(\frac{rN_2}{N}\right)^{N_2(r-1)}.$$

Now $(rN_2/N) = 1-\phi$, $(N/r)-N_2 = \phi N/r$ [see part (a) above], and $N_2(r-1) = (1-\phi)N-N_2$. Thus

$$F_M = -kTr\ln\frac{[N/r]!}{[\phi N/r]![(1-\phi)N/r]!}-(1-\phi)NkT\ln(1-\phi)$$

$$+N_2kT\ln(1-\phi).$$

Using Stirling's approximation for the factorials, we obtain

$$F_M = -kT\left[N\ln\frac{N}{r} - \phi N\ln\frac{\phi N}{r} - (1-\phi)N\ln\frac{(1-\phi)N}{r}\right]$$

$$- kT\{1-\phi)N\ln(1-\phi) - N_2\ln(1-\phi)\},$$

$$= kT[(1-\phi)N\ln(1-\phi) + \phi N\ln\phi - (1-\phi)N\ln(1-\phi) + N_2\ln(1-\phi)]$$

$$= kT[N_1\ln\phi + N_2\ln(1-\phi)],$$

and for one mole of solution, $N_0 = N_1 + N_2$, $x = N_1/N_0$, $1-x = N_2/N_0$,

$$F_M = N_0 kT[x\ln\phi + (1-x)\ln(1-\phi)].$$

It is clear that an exactly parallel argument for N_1 r_1-mers and N_2 r_2-mers will give the same formal result if ϕ is now defined by

$$\phi \equiv \frac{r_1 N_1}{r_1 N_1 + r_2 N_2}.$$

The effect of the second assumption in the derivation vanishes as $z \to \infty$, and this is held to justify using the general result for solutions. In fact, just as with the ideal solution formula, alternative derivations, independent of the cell model, give the same general result[17].

6.15 Solutions of chain molecules (ii). For chain-molecule solutions in which the enthalpy of mixing is significantly different from zero, formulae may be derived which bear the same relation to that of Problem 6.14(b), as do the regular solution formulae to those for ideal solutions. The essentials of the method are apparent in the first approximation (random mixing) based on the cell model of Problem 6.14.

(a) For a solution of species 1 (r_1 identical segments) and species 2 (r_2 identical segments), prove that

$$U_M = \tfrac{1}{2}zq_1N_1(1-\phi)(\epsilon_1 - \epsilon_{12}) + \tfrac{1}{2}zq_2N_2\phi(\epsilon_2 - \epsilon_{12}),$$

where zq_i is the number of external contacts per molecule of species i, comprising $(z-1)$ for each end segment and $(z-2)$ for each segment other than the end ones. Thence show that taking the formal limit $z \to \infty$ gives

$$U_M = (r_1N_1 + r_2N_2)\phi(1-\phi)w,$$

where w is the interchange energy per segment, defined (cf Problem 6.12) by

$$w \equiv -z\epsilon_{12} + \tfrac{1}{2}z(\epsilon_1 + \epsilon_2).$$

(b) Defining W_{12} as the interchange energy per molecule of species 1, so that when r_2 molecules of species 1 from pure 1 are interchanged with r_1 molecules of species 2 from pure 2, the net gain in interaction energy

[17] P.J.Flory, Spiers Memorial Lecture 1970, *Disc.Faraday Soc.*, 49, 7 (1970).

is r_2 times W_{12}, i.e.

$$W_{12} \equiv -zE_{12}+\tfrac{1}{2}z\left(E_1+\frac{r_2}{r_1}E_2\right),$$

where $-\frac{1}{2}zE_1N_0$ is the configurational intrinsic energy per mole of species 1; show that

$$U_{\mathrm{M}} = \left[x+\frac{r_2}{r_1}(1-x)\right]\phi(1-\phi)W_{12}N_0 .$$

(c) The result of part (b) holds, in fact, whether or not all the segments in each species are identical; it therefore suffices to derive W_{12} for each type of solution. A case of particular interest is the solution of species 1 and species 2, each of which has two types of segment, type A and type B (e.g. middles and ends, respectively). Suppose that the molecule of species 1 has r_1 segments, comprising a_1r_1 type A and b_1r_1 type B; and that species 2 has r_2 segments, comprising a_2r_2 type A and b_2r_2 type B. Evaluate E_1 for a molecule of species 1 in pure 1, E_2 for a molecule of species 2 in pure 2, and E_{12} for a molecule of species 1 in pure 2, and thence show that

$$W_{12} = r_1(a_1-a_2)^2w_{ab},$$

where w_{ab} is the segment interchange energy defined by

$$w_{ab} \equiv -z\epsilon_{ab}+\tfrac{1}{2}z(\epsilon_{aa}+\epsilon_{bb}) .$$

(d) Verify that the result for U_{M} obtained by substituting the result of part (c) into the result of part (b) is obtained also by direct counting of all types of segment interactions in a solution of 1 and 2 characterized by the volume fractions ϕ, $1-\phi$, respectively, and of all types of segment interactions in pure 1 and in pure 2.

Solution

(a) The given definition of q_i leads to

$$zq_i = zr_i - 2r_i + 2 ,$$

$$q_i = r_i - \frac{2(r_i-1)}{z},$$

and, in the limit $z \to \infty$, $q_i = r_i$.

Consider first a molecule of 1 in the solution: the number of 1, 1 interactions per segment is simply $z\phi$, and the number per molecule of 1 is then $z\phi q_1$, so the total number is $\frac{1}{2}zq_1\phi N_1$. Similarly the number of 1, 2 interactions is $zq_1(1-\phi)N_1$. The corresponding numbers derived by considering a molecule of 2 are $\frac{1}{2}zq_2(1-\phi)N_2$ and $zq_2\phi N_2$. It is clear that adding all types thus enumerated will give twice the number of 1, 2 interactions, so take half of each count, i.e. $\frac{1}{2}zq_1(1-\phi)N_1+\frac{1}{2}zq_2\phi N_2$. The total is now $\frac{1}{2}z(q_1N_1+q_2N_2)$ as required. The corresponding numbers for pure 1 and pure 2 are clearly $\frac{1}{2}zq_1N_1$ and $\frac{1}{2}zq_2N_2$, and so

$$U_M = -\tfrac{1}{2}z[q_1N_1\phi\epsilon_1+q_2N_2(1-\phi)\epsilon_2]-\tfrac{1}{2}z[q_1N_1(1-\phi)\epsilon_{12}+q_2N_2\phi\epsilon_{12}] + \tfrac{1}{2}z[q_1N_1\epsilon_1+q_2N_2\epsilon_2]\,,$$

whence the required result follows on simple rearrangement. Replacing q_i by r_i, corresponding to the limit $z \to \infty$, we obtain

$$U_M = \tfrac{1}{2}z[r_1N_1(1-\phi)(\epsilon_1-\epsilon_{12})+r_2N_2\phi(\epsilon_2-\epsilon_{12})]\,;$$

taking out the total number of segments, $r_1N_1+r_2N_2$, we obtain

$$U_M = \tfrac{1}{2}z[r_1N_1+r_2N_2][\phi(1-\phi)(\epsilon_1-\epsilon_{12}+\epsilon_2-\epsilon_{12})]\,,$$

and then introducing the definition of w, we find

$$U_M = [r_1N_1+r_2N_2]\phi(1-\phi)w\,.$$

(b) Since $-\frac{1}{2}zE_1N_0$ is the configurational intrinsic energy per mole of 1, clearly $E_1 = r_1\epsilon_1$. Similarly $E_2 = r_2\epsilon_2$, and $E_{12} = r_1\epsilon_{12}$, since it refers to a molecule of species 1. It follows then from the definition of W_{12},

$$W_{12} = -zr_1\epsilon_{12}+\tfrac{1}{2}zr_1\epsilon_1+\tfrac{1}{2}zr_1\epsilon_2 = r_1w\,.$$

Alternatively, interchanging r_2 molecules of 1 from pure 1 with r_1 molecules of 2 from pure 2, the energy increment is clearly

$$-\tfrac{1}{2}zr_1r_2(2\epsilon_{12}-\epsilon_{11}-\epsilon_{22})\,;$$

this is stated to be equal to r_2W_{12}, so again, $W_{12} = r_1w$. The two definitions are thus consistent. Substituting W_{12}/r_1 for w in the result of part (a) yields

$$U_M = \left(N_1+\frac{r_2}{r_1}N_2\right)\phi(1-\phi)W_{12}\,,$$

and, taking $N_1+N_2 = N_0$, for one mole of solution,

$$U_M = \left(\frac{N_1}{N_0}+\frac{r_2N_2}{r_1N_0}\right)\phi(1-\phi)W_{12}N_0 = \left[x+\frac{r_2}{r_1}(1-x)\right]\phi(1-\phi)W_{12}N_0\,.$$

The importance of this apparently trivial result is that, being in terms of W_{12} (defined in terms of E_{12}, E_1, and E_2), it is independent of the segmental constitution of species 1 and species 2.

(c) E_1 may be constructed as the sum of the two types of interaction open to type A segments, and the two types of interaction open to type B segments, thus

$$E_1 = (r_1a_1\epsilon_a+r_1b_1\epsilon_{ab})za_1+(r_1a_1\epsilon_{ab}+r_1b_1\epsilon_b)zb_1\,.$$

Similarly

$$E_2 = (r_2a_2\epsilon_a+r_2b_2\epsilon_{ab})za_2+(r_2a_2\epsilon_{ab}+r_2b_2\epsilon_b)zb_2\,.$$

For a molecule of 1 in pure 2

$$E_{12} = (r_1a_1\epsilon_a+r_1b_1\epsilon_{ab})za_2+(r_1a_1\epsilon_{ab}+r_1b_1\epsilon_b)zb_1\,.$$

Then

$$-W_{12} = \tfrac{1}{2}zr_1\begin{bmatrix} 2a_1a_2\epsilon_a + 2b_1a_2\epsilon_{ab} + 2a_1b_2\epsilon_{ab} + 2b_1b_2\epsilon_b \\ -a_1^2\epsilon_a - 2a_1b_1\epsilon_{ab} - b_1^2\epsilon_b \\ -a_2^2\epsilon_a - 2a_2b_2\epsilon_{ab} - b_2^2\epsilon_b \end{bmatrix}$$

whence

$$W_{12} = -\tfrac{1}{2}zr_1[(a_1-a_2)^2\epsilon_a + 2(a_1-a_2)(b_1-b_2)\epsilon_{ab} + (b_1-b_2)^2\epsilon_b],$$

and, since $a_1+b_1 = 1$, $a_2+b_2 = 1$, $(a_1-a_2) = -(b_1-b_2)$, so that

$$W_{12} = -\tfrac{1}{2}zr_1(a_1-a_2)^2(2\epsilon_{ab} - \epsilon_a - \epsilon_b),$$

we have

$$W_{12} = r_1(a_1-a_2)^2 w_{ab}.$$

The symmetry of these equations ensures that if 2 is taken as reference species instead of 1,

$$W_{21} = r_2(a_1-a_2)^2 w_{ab} = \frac{r_2}{r_1}W_{12}.$$

The argument generalizes at once to any number of segment types on writing

$$E_{ii} = zr_i\sum_{\alpha,\beta}\alpha_i\beta_i\epsilon_\alpha; \quad E_{ij} = zr_i\sum_{\alpha,\beta}\alpha_i\beta_j\epsilon_{\alpha\beta},$$

where the summation extends over all α, β;

$$\alpha_i = a_i, b_i, c_i, \ldots; \quad \beta_i = a_i, b_i, c_i, \ldots.$$

Thus for three types of segment we obtain

$$\frac{W_{12}}{r_1} = (a_1-a_2)(b_1-b_2)w_{ab} + (b_1-b_2)(c_1-c_2)w_{bc} + (c_1-c_2)(a_1-a_2)w_{ca}.$$

(d) For an assembly of N_1 molecules of species 1 and N_2 molecules of species 2 in random mixing, list the energies associated with each type of interaction, dividing each by 2 to normalize the total number of interactions to $\tfrac{1}{2}z(r_1N_1+r_2N_2)$:

$$\begin{aligned}
aa(1)&: \quad -\tfrac{1}{2}r_1N_1a_1[\phi a_1 + (1-\phi)a_2]z\epsilon_a,\\
ab(1)&: \quad -\tfrac{1}{2}r_1N_1a_1[\phi b_1 + (1-\phi)b_2]z\epsilon_{ab},\\
ba(1)&: \quad -\tfrac{1}{2}r_1N_1b_1[\phi a_1 + (1-\phi)a_2]z\epsilon_{ab},\\
bb(1)&: \quad -\tfrac{1}{2}r_1N_1b_1[\phi b_1 + (1-\phi)b_2]z\epsilon_b,\\
aa(2)&: \quad -\tfrac{1}{2}r_2N_2a_2[(1-\phi)a_2 + \phi a_1]z\epsilon_a,\\
ab(2)&: \quad -\tfrac{1}{2}r_2N_2a_2[(1-\phi)b_2 + \phi b_1]z\epsilon_{ab},\\
ba(2)&: \quad -\tfrac{1}{2}r_2N_2b_2[(1-\phi)a_2 + \phi a_1]z\epsilon_{ab},\\
bb(2)&: \quad -\tfrac{1}{2}r_2N_2b_2[(1-\phi)b_2 + \phi b_1]z\epsilon_b.
\end{aligned}$$

Then the total energy of *aa* interactions is

$$U^*_{aa} = \{-\tfrac{1}{2}r_1N_1[\phi a_1^2 + (1-\phi)a_1a_2] - \tfrac{1}{2}r_2N_2[(1-\phi)a_2^2 + \phi a_1a_2]\}z\epsilon_a$$
$$= -\tfrac{1}{2}[r_1N_1 + r_2N_2][\phi a_1 + (1-\phi)a_2]^2 z\epsilon_a \ .$$

Similarly for the *bb* interactions

$$U^*_{bb} = -\tfrac{1}{2}[r_1N_1 + r_2N_2][\phi b_1 + (1-\phi)b_2]^2 z\epsilon_b \ ,$$

and for the *ab* interactions

$$U^*_{ab} = -[r_1N_1 + r_2N_2][\phi a_1 + (1-\phi)a_2][\phi b_1 + (1-\phi)b_2]z\epsilon_{ab} \ .$$

For N_1 molecules of pure 1 and N_2 molecules of pure 2, the total energy is

$$U^*_{1,2} = -\tfrac{1}{2}z[r_1N_1 + r_2N_2]\left\{\begin{matrix}[\phi a_1^2 + (1-\phi)a_2^2]\epsilon_a + [\phi b_1^2 + (1-\phi)b_2^2]\epsilon_b \\ + 2[\phi a_1b_1 + (1-\phi)a_2b_2]\epsilon_{ab}\end{matrix}\right\} .$$

Straightforward algebra then gives for U_{M}, as before,

$$U_{\mathrm{M}} = -\tfrac{1}{2}z(r_1N_1 + r_2N_2)\phi(1-\phi)[(a_1-a_2)^2\epsilon_a + (b_1-b_2)^2\epsilon_b + 2(a_1-a_2)(b_1-b_2)\epsilon_{ab}]$$
$$= (r_1N_1 + r_2N_2)\phi(1-\phi)(a_1-a_2)^2 w_{ab} \ .$$

GENERAL REFERENCES

In addition to the works cited in the Introduction to this chapter, the following books provide comprehensive surveys, including detailed expositions of the theories explored here in a preliminary way. They also introduce the group of theories based upon the radial distribution function (cf Chapter 9) which has been excluded here, not because they are mistakenly thought to be conceptually difficult, but because (i) an essential prerequisite is to become familiar with an extensive and specialized vocabulary, and (ii) they involve a 'surprisingly formidable' amount of algebra which is not intrinsically interesting.

THE LIQUID STATE

Barker, J. A., *Lattice Theories of the Liquid State* (Pergamon Press, New York), 1963.

Egelstaff, P. A., *An Introduction to the Liquid State* (Academic Press, New York), 1967.

Fisher, I. Z., *Statistical Theory of Liquids* (trans. T.M.Switz, with supplement by S.A.Rice and P.Gray) (Chicago University Press, Chicago), 1964.

Frisch, H. L., and Lebowitz, J. L., *The Equilibrium Theory of Classical Fluids* (Benjamin, New York), 1964.

Temperley, H. N. V., Rowlinson, J. S., and Rushbrooke, G. S., *The Physics of Simple Liquids* (North-Holland, Amsterdam), 1968.

SOLUTIONS

Guggenheim, E. A., *Mixtures* (Clarendon Press, Oxford), 1952.

Prigogine, I., *Molecular Theory of Solutions* (North-Holland, Amsterdam), 1957.

7

Phase stability, co-existence, and criticality

A.J.B.CRUICKSHANK
(*University of Bristol, Bristol*)

SINGULARY SYSTEMS

7.1 Thermodynamic stability of the single phase. Consider a uniform system (i.e. in which each intensive variable, T, p, has the same value at all points) of n moles of a single component, constrained to constant entropy and volume. Phases α and β will be identified by suffixes, and the initial (unperturbed) state by index (0). Initially the system is entirely in phase α. The perturbation is $\delta\xi$ mole going over into phase β, only slightly different from α. The condition for equilibrium in such a system at constant S, v, diffusional processes being allowed, is that [1] $(\partial U/\partial\xi)_{S,v} > 0$, i.e. that U is a minimum. That this is equivalent to $T(\partial S/\partial\xi)_{U,v} < 0$, i.e. that S is a maximum [1] is obvious also from the second law as stated in the preamble to Problem 1.22. The derivation of the particular conditions for phase stability is simplified by:

(i) using molar (intensive) rather than total quantities for each phase, since $\overline{U} = \overline{U}(\overline{S}, \overline{v})$ whereas $U = U(S, v, n)$, cf Problem 1.20(a),

(ii) expressing both $\overline{U}_\alpha$ and $\overline{U}_\beta$ by Taylor expansions about $\overline{U}_\alpha^{(0)}$, valid because phase β is similar to phase α.

Then, initially,

$$n_\alpha^{(0)} = n\,, \quad n_\beta^{(0)} = 0\,; \quad S^{(0)} = n\overline{S}_\alpha^{(0)}\,, \quad v^{(0)} = n\overline{v}_\alpha^{(0)}\,;$$

and as a result of the perturbation

$$\begin{aligned} n_\alpha &= n - \delta\xi\,, & \overline{S}_\alpha &= \overline{S}_\alpha^{(0)} + \delta S\,, & \overline{v}_\alpha &= \overline{v}_\alpha^{(0)} + \delta v\,, \\ n_\beta &= \delta\xi\,, & \overline{S}_\beta &= \overline{S}_\alpha^{(0)} + \Delta S\,, & \overline{v}_\beta &= \overline{v}_\alpha^{(0)} + \Delta v\,. \end{aligned}$$

The distinction between ΔS and δS emphasizes that as $\delta\xi \to 0$, $\delta S \to 0$, but ΔS does not, being the difference between the molar entropy of phase α and that of the hypothetical neighbouring phase β.

(a) Prove that for the complete system,

$$\left(\frac{\partial U}{\partial\xi}\right)_{S,v} = \frac{1}{2}\left\{\left(\frac{\partial^2\overline{U}}{\partial\overline{S}^2}\right)_{v,n}(\Delta S)^2 + 2\left(\frac{\partial^2\overline{U}}{\partial\overline{S}\partial\overline{v}}\right)_n \Delta S\Delta v + \left(\frac{\partial^2\overline{U}}{\partial\overline{v}^2}\right)_{S,n}(\Delta v)^2\right\},$$

where all derivatives refer to phase α in the unperturbed state.

[1] J.W.Gibbs, *Collected Works*, Vol.1 (Yale University Press), 1948, p.56, Eqns.1, 2. I.Prigogine and R.Defay, *Chemical Thermodynamics* (trs.D.H.Everett), (Longmans, London), 1954, pp.35, 36.

(b) Show that the necessary and sufficient conditions to ensure $(\partial U/\partial \xi)_{S,v} > 0$ with respect to formation of any phase β differing only slightly from α are

$$\left(\frac{\partial^2 U}{\partial S^2}\right)_{v,n}\left(\frac{\partial^2 U}{\partial v^2}\right)_{S,n} - \left(\frac{\partial^2 U}{\partial S \partial v}\right)_n^2 > 0 \tag{7.1.1}$$

together with either of

$$\left(\frac{\partial^2 U}{\partial S^2}\right)_{v,n} > 0\,, \tag{7.1.2}$$

$$\left(\frac{\partial^2 U}{\partial v^2}\right)_{S,n} > 0\,. \tag{7.1.3}$$

(c) Show that the conditions (7.1.1-7.1.3) may be expressed in the equivalent forms

$$\frac{T}{C_p}\,,\quad \frac{1}{vK_S} > 0\,, \tag{7.1.4}$$

$$\frac{T}{C_v}\,,\quad \frac{1}{vK_T} > 0\,. \tag{7.1.5}$$

Solution

(a) The Taylor expansion for $\overline{U}_\beta$ is

$$\overline{U}_\beta = \overline{U}_\alpha^{(0)} + \left(\frac{\partial \overline{U}}{\partial \overline{S}}\right)_{\overline{v}} \Delta S + \left(\frac{\partial \overline{U}}{\partial \overline{v}}\right)_{\overline{S}} \Delta v + \frac{1}{2}\left(\frac{\partial^2 \overline{U}}{\partial \overline{S}^2}\right)_{\overline{v}} (\Delta S)^2 + \left(\frac{\partial^2 \overline{U}}{\partial \overline{S}\partial \overline{v}}\right) \Delta S \Delta v + \frac{1}{2}\left(\frac{\partial^2 \overline{U}}{\partial \overline{v}^2}\right)_{\overline{S}} (\Delta v)^2\,.$$

and similarly for $\overline{U}_\alpha$, but with δS, δv, instead of ΔS, Δv. The net change in U (total) is

$$\begin{aligned}\delta U &= (n-\delta\xi)\overline{U}_\alpha + \delta\xi \overline{U}_\beta - n\overline{U}_\alpha^{(0)} \\ &= \left(\frac{\partial \overline{U}}{\partial \overline{S}}\right)_{\overline{v}} [(n-\delta\xi)\delta S + \delta\xi \Delta S] + \left(\frac{\partial \overline{U}}{\partial \overline{v}}\right)_{\overline{S}} [(n-\delta\xi)\delta v + \delta\xi\Delta v] \\ &\quad + \frac{1}{2}\left(\frac{\partial^2 \overline{U}}{\partial \overline{S}^2}\right)_{\overline{v}} [(n-\delta\xi)(\delta S)^2 + \delta\xi(\Delta S)^2] + \frac{1}{2}\left(\frac{\partial^2 \overline{U}}{\partial \overline{v}^2}\right)_{\overline{S}} [(n-\delta\xi)(\delta v)^2 + \delta\xi(\Delta v)^2] \\ &\quad + \left(\frac{\partial^2 \overline{U}}{\partial \overline{S}\partial \overline{v}}\right) [(n-\delta\xi)\delta S\delta v + \delta\xi \Delta S \Delta v]\,.\end{aligned}$$

The conditions S, v, constant relate δS, δv to ΔS, Δv, as

$$S = (n-\delta\xi)(\overline{S}_\alpha^{(0)} + \delta S) + \delta\xi(\overline{S}_\alpha^{(0)} + \Delta S) = n\overline{S}_\alpha^{(0)}$$

whence

$$(n-\delta\xi)\delta S + \delta\xi\Delta S = 0\,; \quad (n-\delta\xi)^2(\delta S)^2 = (\delta\xi)^2(\Delta S)^2\,; \tag{7.1.6}$$

similarly

$$(n-\delta\xi)\delta v+\delta\xi\Delta v = 0\,; \quad (n-\delta\xi)^2(\delta v)^2 = (\delta\xi)^2(\Delta v)^2\,, \quad (7.1.7)$$

whence

$$(n-\delta\xi)^2\delta S\delta v = (\delta\xi)^2\Delta S\Delta v\,. \quad (7.1.8)$$

Clearly the first two terms in δU are zero; using Equations (7.1.6)-(7.1.8) as appropriate on the last three terms gives

$$\delta U = \frac{1}{2}\frac{n\delta\xi}{n-\delta\xi}\left[\left(\frac{\partial^2\overline{U}}{\partial\overline{S}^2}\right)_v(\Delta S)^2+2\left(\frac{\partial^2\overline{U}}{\partial S\partial v}\right)\Delta S\Delta v+\left(\frac{\partial^2\overline{U}}{\partial\overline{v}^2}\right)_S(\Delta v)^2\right],$$

and taking the limit as $\delta\xi\to 0$ gives the required result. To ensure equilibrium, the right hand side of the result of part (a) must be positive, irrespective of the higher order terms from the Taylor expansions.

(b) The right hand side of this result is a simple quadratic in either $\Delta v/\Delta S$ or $\Delta S/\Delta v$; it is everywhere positive provided (i) there are no real roots, and (ii) it is positive as $\Delta S/\Delta v\to\infty$. The first is ensured by condition (7.1.1) expressed in molar quantities, i.e. multiplied through by n^2. Now, condition (7.1.1) implies that $(\partial^2U/\partial S^2)_v$ and $(\partial^2U/\partial v^2)_S$ have the same sign, so the second provision is ensured by either one of conditions (7.1.2) or (7.1.3) being met.

(c) The left hand side of condition (7.1.1) reduces at once to

$$-\left(\frac{\partial p}{\partial v}\right)_S\left(\frac{\partial T}{\partial S}\right)_v-\left(\frac{\partial T}{\partial v}\right)_S^2, \quad \text{or} \quad -\left(\frac{\partial p}{\partial v}\right)_S\left(\frac{\partial T}{\partial S}\right)_v-\left(\frac{\partial p}{\partial S}\right)_v^2.$$

The first may be transformed into a simple product by using

$$\left(\frac{\partial T}{\partial v}\right)_S = \left(\frac{\partial T}{\partial p}\right)_S\left(\frac{\partial p}{\partial v}\right)_S;$$

it then becomes

$$-\left(\frac{\partial p}{\partial v}\right)_S\left[\left(\frac{\partial T}{\partial S}\right)_v+\left(\frac{\partial T}{\partial v}\right)_S\left(\frac{\partial T}{\partial p}\right)_S\right] = -\left(\frac{\partial p}{\partial v}\right)_S\left[\left(\frac{\partial T}{\partial S}\right)_v+\left(\frac{\partial T}{\partial v}\right)_S\left(\frac{\partial v}{\partial S}\right)_p\right],$$

when the appropriate Maxwell relation is used, and this in turn reduces to $-(\partial p/\partial v)_S(\partial T/\partial S)_p$ in virtue of the standard change-of-constraint formula. The definition of K_S, with $(\partial S/\partial T)_p = C_p/T$, gives the required result. Similarly, with $(\partial p/\partial S)_v = (\partial p/\partial T)_v(\partial T/\partial S)_v$, the second form becomes

$$-\left(\frac{\partial T}{\partial S}\right)_v\left[\left(\frac{\partial p}{\partial v}\right)_S+\left(\frac{\partial p}{\partial S}\right)_v\left(\frac{\partial p}{\partial T}\right)_v\right] = -\left(\frac{\partial T}{\partial S}\right)_v\left[\left(\frac{\partial p}{\partial v}\right)_S+\left(\frac{\partial p}{\partial S}\right)_v\left(\frac{\partial S}{\partial v}\right)_T\right]$$

and the change-of-constraint formula again gives the required result.

An alternative procedure is apparent on writing condition (7.1.1) as

$$\begin{vmatrix} U_{2S} & U_{Sv} \\ U_{vS} & U_{2v} \end{vmatrix} > 0\,, \qquad U_{2S} \equiv \left(\frac{\partial^2 U}{\partial S^2}\right)_v\,, \quad U_{Sv} \equiv \left(\frac{\partial^2 U}{\partial S \partial v}\right).$$

Continue via the Jacobian, as

$$\begin{vmatrix} U_{2S} & U_{Sv} \\ U_{vS} & U_{2v} \end{vmatrix} = \frac{\partial(T, -p)}{\partial(S, v)} = \frac{\partial(T, -p)}{\partial(S, p)} \bigg/ \frac{\partial(S, v)}{\partial(S, p)} = -\left(\frac{\partial T}{\partial S}\right)_p \bigg/ \left(\frac{\partial v}{\partial p}\right)_S$$

or equivalently,

$$\frac{\partial(U_v, U_S)}{\partial(v, S)} = \frac{\partial(-p, T)}{\partial(v, S)} = \frac{\partial(-p, T)}{\partial(v, T)} \bigg/ \frac{\partial(v, S)}{\partial(v, T)} = -\left(\frac{\partial p}{\partial v}\right)_T \bigg/ \left(\frac{\partial S}{\partial T}\right)_v .$$

Note that the phase stability conditions (7.1.4) or (7.1.5) may be derived equivalently by considering the case where the total intrinsic energy and the total volume are kept constant[2]. The results (7.1.4) or (7.1.5), whose equivalence reflects the identity $K_S/K_T = C_v/C_p$, indicate that in a one-component system diffusional stability is secured by the conditions for thermal stability ($T/C_v > 0$) and for mechanical stability ($1/vK_T > 0$). It will be seen (Problem 7.7) that this is not so for a two-component system, an additional condition of diffusional stability being required.

7.2 Higher order stability conditions. For simplicity we consider uniform systems in which one intensive variable is, additionally, held constant. The appropriate forms of the equilibrium condition are, cf.preamble to Problem 1.22,

$$\left(\frac{\partial F}{\partial \xi}\right)_{T,\,v} > 0\,, \quad F \equiv U - TS\,, \quad \text{or} \quad \left(\frac{\partial H}{\partial \xi}\right)_{S,\,p} > 0\,, \quad H \equiv U + pv\,.$$

It is necessary also to validate the first-order stability conditions thus obtained, by comparison with conditions (7.1.4) and (7.1.5).

(a) Prove that the first-order stability condition for a one-component, one-phase system constrained to constant volume and uniform and constant temperature is

$$\left(\frac{\partial^2 F}{\partial v^2}\right)_{T,\,n} = -\left(\frac{\partial p}{\partial v}\right)_{T,n} = \frac{1}{vK_T} > 0\,, \tag{7.2.1}$$

and that the condition for stability when $1/vK_T = 0$ is

$$\left(\frac{\partial^2 p}{\partial v^2}\right)_{T,\,n} = 0\,, \quad \left(\frac{\partial^3 p}{\partial v^3}\right)_{T,\,n} < 0\,. \tag{7.2.2}$$

[2] I.Prigogine and R.Defay, *Chemical Thermodynamics* (trs.D.H.Everett), (Longmans, London), 1954, pp.209–212.

(b) Prove that the corresponding conditions when the constraints are constant total entropy and uniform and constant pressure are either

$$\frac{T}{C_p} > 0\,; \tag{7.2.3}$$

or

$$\frac{T}{C_p} = 0\,, \quad \left(\frac{\partial^2 T}{\partial S^2}\right)_p = 0\,, \quad \left(\frac{\partial^3 T}{\partial S^3}\right)_p > 0\,. \tag{7.2.4}$$

(c) Relate the condition (7.2.1) to the conditions (7.1.1) and (7.1.2). [Hint: The relation $K_S/K_T = C_v/C_p$ is useful, as is the relation of Problem 1.8(a). The general relation between $(\partial^2 F/\partial v^2)_T$ and $(\partial^2 U/\partial v^2)_S$ is also illuminating.]

(d) Given that $(\partial p/\partial T)_v$ is almost always positive and without singularities, what can you deduce about the approach to instability?

(e) Show, in the case of the van der Waals equation of state (Problems 1.11 and 6.7), that a single phase within the normal two-phase region (e.g. at $v = 3b$, $T < 8a/27Rb$) where it would clearly be unstable, has $1/K_T < 0$, $T/C_v > 0$, $\alpha_p < 0$, and hence show that T/C_p, $1/K_S$ may be either positive or negative.

Solution

(a) Proceeding as in the solution to Problem 7.1(a), write the Taylor series for $\bar{F}_\alpha$ and $\bar{F}_\beta$ about $\bar{F}_\alpha^{(0)}$, including terms up to $(\partial^4 \bar{F}/\partial \bar{v}^4)_T$. Then

$$\begin{aligned}
\delta F &= (n-\delta\xi)\bar{F}_\alpha + \delta\xi \bar{F}_\beta - n\bar{F}_\alpha^{(0)}\,,\\
&= \left(\frac{\partial F}{\partial v}\right)_T [(n-\delta\xi)\delta v + \delta\xi \Delta v] + \frac{1}{2!}\left(\frac{\partial^2 F}{\partial v^2}\right)_T [(n-\delta\xi)(\delta v)^2 + \delta\xi(\Delta v)^2]\\
&\quad + \frac{1}{3!}\left(\frac{\partial^3 F}{\partial v^3}\right)_T [(n-\delta\xi)(\delta v)^3 + \delta\xi(\Delta v)^3]\\
&\quad + \frac{1}{4!}\left(\frac{\partial^4 F}{\partial v^4}\right)_T [(n-\delta\xi)(\delta v)^4 + \delta\xi(\Delta v)^4] + \cdots\,.
\end{aligned}$$

Using the relations (7.1.7), together with the corresponding higher-order relations for increasing powers of Δv gives the coefficients of the derivatives of $\bar{F}_\alpha^{(0)}$, leaving off the redundant indices and suffixes, as

$\left(\frac{\partial \bar{F}}{\partial \bar{v}}\right)_T$	$0\,,$
$\left(\frac{\partial^2 \bar{F}}{\partial \bar{v}^2}\right)_T$	$\frac{1}{2!}(\Delta v)^2 \frac{n\delta\xi}{n-\delta\xi}$
$\left(\frac{\partial^3 \bar{F}}{\partial \bar{v}^3}\right)_T$	$\frac{1}{3!}(\Delta v)^3 \frac{n^2\delta\xi}{(n-\delta\xi)^2}\left(1-\frac{2\delta\xi}{n}\right)$
$\left(\frac{\partial^4 \bar{F}}{\partial \bar{v}^4}\right)_T$	$\frac{1}{4!}(\Delta v)^4 \frac{n^3\delta\xi}{(n-\delta\xi)^3}\left[1-\frac{3\delta\xi}{n}+3\left(\frac{\delta\xi}{n}\right)^2\right]\,.$

Thus, taking the limit as $\delta\xi \to 0$,

$$\left(\frac{\partial F}{\partial \xi}\right)_{T,\,v} = \frac{1}{2!}\left(\frac{\partial^2 \overline{F}}{\partial \bar{v}^2}\right)_T (\Delta v)^2 + \frac{1}{3!}\left(\frac{\partial^3 \overline{F}}{\partial \bar{v}^3}\right)_T (\Delta v)^3 + \frac{1}{4!}\left(\frac{\partial^4 \overline{F}}{\partial \bar{v}^4}\right)_T (\Delta v)^4 + \cdots .$$

The condition that $(\partial F/\partial \xi)_{T,\,v} > 0$ is clearly either (7.2.1) or, if $(\partial^2 F/\partial v^2)_T = 0$, since $(\Delta v)^3$ can take either sign, (7.2.2). In the general case, the condition for stability is that the first non-vanishing derivative of p must be (i) of odd order, (ii) negative.

(b) The proof is exactly analogous to that of part (a).

(c) The various sets of conditions are related as follows. That either condition (7.1.4) or condition (7.1.5) ensure all four of $1/K_T$, $1/K_S$, T/C_p, T/C_v positive is obvious from the relations cited. Note that condition (7.1.1) alone does not ensure all four parameters positive; nor does $1/K_S$, $1/K_T > 0$, etc. The relation between condition (7.2.1) and conditions (7.1.1), (7.1.2), (7.1.3) is obvious on writing

$$\left(\frac{\partial^2 F}{\partial v^2}\right)_T = \frac{1}{vK_T} = \frac{T/C_v v K_T}{T/C_v} > 0 . \tag{7.2.5}$$

This makes it obvious that while condition (7.1.5) implies condition (7.2.1), the reverse is not true; rather, condition (7.2.1) implies only that the two conditions (7.1.5) are met, or not, together. Taking conditions (7.2.1) and (7.2.3) together adds only that $1/vK_S$, T/C_v have the same sign. The algebraic relation between the $F(T, v)$ and $U(S, v)$ stability conditions gives the same information as condition (7.2.5), but deriving it establishes a general method which is useful in wider contexts. Commence with

$$\left(\frac{\partial F}{\partial v}\right)_{T,\,n} = \left(\frac{\partial U}{\partial v}\right)_{S,\,n} = -p ,$$

whence

$$\left(\frac{\partial^2 F}{\partial v^2}\right)_{T,\,n} = \left[\frac{\partial}{\partial v}\left(\frac{\partial U}{\partial v}\right)_S\right]_{T,\,n} = \left[\frac{\partial}{\partial v}\left(\frac{\partial U}{\partial v}\right)_S\right]_{S,\,n} + \left[\frac{\partial}{\partial S}\left(\frac{\partial U}{\partial v}\right)_S\right]_{v,\,n}\left(\frac{\partial S}{\partial v}\right)_{T,\,n}$$

according to the change of constraint formula. Making use of

$$\left(\frac{\partial S}{\partial v}\right)_{T,\,n}\left(\frac{\partial v}{\partial T}\right)_{S,\,n}\left(\frac{\partial T}{\partial S}\right)_{v,\,n} = -1 ,$$

and putting

$$\left(\frac{\partial T}{\partial v}\right)_{S,\,n} = \left(\frac{\partial^2 U}{\partial v\,\partial S}\right)_n , \qquad \left(\frac{\partial T}{\partial S}\right)_{v,\,n} = \left(\frac{\partial^2 U}{\partial S^2}\right)_{v,\,n} ,$$

we obtain

$$\begin{aligned}\left(\frac{\partial^2 F}{\partial v^2}\right)_{T,\,n} &= \left(\frac{\partial^2 U}{\partial v^2}\right)_{S,\,n} - \left(\frac{\partial^2 U}{\partial S\partial v}\right)_n \left(\frac{\partial T}{\partial v}\right)_{S,\,n}\left(\frac{\partial S}{\partial T}\right)_{v,\,n}\\ &= \left(\frac{\partial^2 U}{\partial v^2}\right)_{S,\,n} - \left(\frac{\partial^2 U}{\partial S\partial v}\right)_n^2 \Bigg/ \left(\frac{\partial^2 U}{\partial S^2}\right)_{v,\,n} ,\end{aligned}$$

i.e.

$$F_{2v}U_{2S} = U_{2S}U_{2v} - U_{vS}U_{vS} = \begin{vmatrix} U_{2S} & U_{Sv} \\ U_{Sv} & U_{2v} \end{vmatrix} .$$

(d) The conditions (7.1.4), or (7.1.5), equivalently imply that all four of the reciprocals of C_p, C_v, K_T, K_S are positive; $(\partial p/\partial T)_{v,n}$ being positive and non-singular implies that α_p has the same sign as K_T, and has the same zeros and/or infinite discontinuities. Then, in virtue of the relations $C_p = C_v + T\alpha_p^2 v/K_T$ and $K_T = K_S + T\alpha_p^2 v/C_p$, a stable phase has $C_p > C_v > 0$, $T/C_v > T/C_p > 0$; $K_T > K_S > 0$, $1/vK_S > 1/vK_T > 0$. This suggests that in the absence of an infinite discontinuity in T/C_v, stability is likely to fail first by T/C_p, $1/vK_T$ going through zero simultaneously. If C_p, K_T go through zero, however, then C_v, K_S approach zero faster.

(e) In the van der Waals equation, using molar quantities,

$$\left(\frac{\partial p}{\partial T}\right)_v = \frac{R}{v-b} > 0 \quad \text{for all } v > b . \qquad \left(\frac{\partial p}{\partial v}\right)_T = -\frac{RT}{(v-b)^2} + \frac{2a}{v^3} .$$

If $v = 3b$, $T < 8a/27b$, we have for the uniform-density state [3]

$$\left(\frac{\partial p}{\partial v}\right)_T > 0 , \quad \frac{1}{vK_T} < 0 , \quad \alpha_p < 0 .$$

Now $C_v^* = 0$, from Problem 6.6(b); $C_v' = \frac{3}{2}R$, $C_v(\text{internal}) > 0$; thus $\infty > C_v > 0$ for all $T, v > 0$, i.e. $T/C_v > 0$. Further, $C_p - C_v < 0$ at the point specified, since $1/vK_T < 0$, so C_p may have either sign, and K_S has the sign opposite to that of C_p. Along this isotherm, $(\partial p/\partial v)_T$ has two zeros which, with $C_v > 0$ everywhere, necessarily define stability limits. Thus the stability condition which fails, at the so-called **spinodal locus,** is $T/C_p > 0$, $1/vK_T > 0$, and it fails by passing through zero; if C_p is positive anywhere in the unstable region, the part of that region within which $C_p > 0$ is bounded by zeros of C_p, the locus coinciding with zeros of K_S.

It is the fact that phase stability normally breaks down by T/C_p, $1/vK_T$ passing through zero simultaneously, which justifies using conditions (7.2.1) or (7.2.3)—usually the former—rather than the complete conditions (7.1.4) or (7.1.5) when searching for critical points or for spinodal loci (see following problems).

7.3 Co-existence of two phases (i). A one-component system, constrained to constant total entropy, volume, and mass, is supposed to comprise two phases, α and β, in thermal, mechanical, and diffusional contact. The phases are both internally uniform, but they may differ in

[3] That this state has a higher free energy than the two-phase state may be deduced from the results of Problem 6.6(a). For an analogous problem in statistical mechanics see Problems 11.14 and 11.15.

T, p, and μ. The perturbation to be considered is the transfer of δS, δv, δn (extensive quantities) from phase β to phase α. This is equivalent to a generalization of Problem 1.29. The general condition for equilibrium, as in the preceding problems, is $(\delta U)_{S,v} > 0$.

(a) Taking S_α, v_α, and n_α as independent variables, express U_α and U_β by Taylor series expansions about $U_\alpha^{(0)}$ and $U_\beta^{(0)}$, and thence show that the conditions for equilibrium, i.e. for $U \equiv U_\alpha + U_\beta$ to be an extremum, are [4]

$$T_\alpha = T_\beta, \quad p_\alpha = p_\beta, \quad \mu_\alpha = \mu_\beta ; \tag{7.3.1}$$

and that those for stability, i.e. for U to be a minimum rather than merely an extremum, are

$$\begin{vmatrix} U_{2S}^+ & U_{Sv}^+ & U_{Sn}^+ \\ U_{vS}^+ & U_{2v}^+ & U_{vn}^+ \\ U_{nS}^+ & U_{nv}^+ & U_{2n}^+ \end{vmatrix} > 0 ; \tag{7.3.2}$$

$$\begin{vmatrix} U_{2S}^+ & U_{Sv}^+ \\ U_{vS}^+ & U_{2v}^+ \end{vmatrix} > 0 ; \tag{7.3.3}$$

$$U_{2S}^+ > 0 ; \tag{7.3.4}$$

where

$$U_{2S}^+ \equiv \left(\frac{\partial^2 U_\alpha}{\partial S_\alpha^2}\right)_{v,n} + \left(\frac{\partial^2 U_\beta}{\partial S_\beta^2}\right)_{v,n}, \quad \text{etc.}$$

(b) Prove that the conditions (7.3.4) and (7.3.3) are secured by

$$\frac{1}{v_\alpha K_{T\alpha}}, \quad \frac{T}{C_{v\alpha}} > 0 ; \quad \frac{1}{v_\beta K_{T\beta}}, \quad \frac{T}{C_{v\beta}} > 0 . \tag{7.3.5}$$

The results of Problem 7.1(c) are useful.

(c) To prove that condition (7.3.2) also is secured by condition (7.3.5) is both formidable and tedious, but the method is similar to that of part (b). Prove the following general proposition, which is a prerequisite for proving that condition (7.3.2) is secured by condition (7.3.5),

$$\begin{vmatrix} U_{2S} & U_{Sv} & U_{Sn} \\ U_{vS} & U_{2v} & U_{vn} \\ U_{nS} & U_{nv} & U_{2n} \end{vmatrix} = U_{2S} \begin{vmatrix} F_{2v} & F_{vn} \\ F_{nv} & F_{2n} \end{vmatrix} = U_{2S} F_{2v} G_{2n} = 0 , \tag{7.3.6}$$

[4] That the co-existence condition $p_\alpha = p_\beta$ may be deduced equivalently by maximizing the canonical partition function for non-uniform density at $T < T_c$ is seen from Problem 11.14. A related analysis using the grand partition function establishes also that $\mu_\alpha = \mu_\beta$.

where

$$U_{2S} \equiv \left(\frac{\partial^2 U}{\partial S^2}\right)_{v,\,n}, \quad F_{2v} \equiv \left(\frac{\partial^2 F}{\partial v^2}\right)_{T,\,n}, \quad G_{2n} \equiv \left(\frac{\partial^2 G}{\partial n^2}\right)_{T,\,p}, \quad \text{etc.}$$

(d) Use the same general method as in part (b) to prove that the conditions for stability under constant and uniform temperature,

$$\begin{vmatrix} F_{2v}^{+} & F_{vn}^{+} \\ F_{nv}^{+} & F_{2n}^{+} \end{vmatrix} > 0\,, \quad F_{2n}^{+} > 0\,,$$

are secured by

$$\frac{1}{v_\alpha K_{T\alpha}}, \quad \frac{1}{v_\beta K_{T\beta}} > 0\,.$$

Note that it follows from Problem 1.20(b) that

$$\mu \equiv \left(\frac{\partial U}{\partial n}\right)_{S,\,v} = \left(\frac{\partial H}{\partial n}\right)_{S,\,p} = \left(\frac{\partial F}{\partial n}\right)_{T,\,v} = \left(\frac{\partial G}{\partial n}\right)_{T,\,p} = \frac{G}{n}\,;$$

also, it follows from the Gibbs equation, $S\,\mathrm{d}T - v\,\mathrm{d}p + n\,\mathrm{d}\mu = 0$, that

$$\left(\frac{\partial \mu}{\partial T}\right)_p = -\bar{S}\,, \quad \left(\frac{\partial \mu}{\partial p}\right)_T = \bar{v}\,;$$

and that the molar entropy and volume may be defined also by

$$\bar{S} \equiv \left(\frac{\partial S}{\partial n}\right)_{T,\,p}, \quad \bar{v} \equiv \left(\frac{\partial v}{\partial n}\right)_{T,\,p},$$

with

$$\bar{U} \equiv \left(\frac{\partial U}{\partial n}\right)_{T,\,p}.$$

Solution

(a) Since $\delta S_\beta = -\delta S_\alpha$, $\delta v_\beta = -\delta v_\alpha$, $\delta n_\beta = -\delta n_\alpha$, the first-order terms in δU are:

$$\delta U = \left[\left(\frac{\partial U_\alpha}{\partial S_\alpha}\right)_{v,\,n} - \left(\frac{\partial U_\beta}{\partial S_\beta}\right)_{v,\,n}\right]\delta S_\alpha + \left[\left(\frac{\partial U_\alpha}{\partial v_\alpha}\right)_{S,\,n} - \left(\frac{\partial U_\beta}{\partial v_\beta}\right)_{S,\,n}\right]\delta v_\alpha$$
$$+ \left[\left(\frac{\partial U_\alpha}{\partial n_\alpha}\right)_{S,\,v} - \left(\frac{\partial U_\beta}{\partial n_\beta}\right)_{S,\,v}\right]\delta n_\alpha + \cdots.$$

Clearly, conditions (7.3.1) are the necessary and sufficient conditions to ensure $\delta U = 0$ to the first order.

When these conditions are met,

$$\delta U = \frac{1}{2}\left[\left(\frac{\partial^2 U_\alpha}{\partial S_\alpha^2}\right)_{v,n} + \left(\frac{\partial^2 U_\beta}{\partial S_\beta^2}\right)_{v,n}\right](\delta S_\alpha)^2 + \left[\left(\frac{\partial^2 U_\alpha}{\partial S_\alpha \partial v_\alpha}\right)_n + \left(\frac{\partial^2 U_\beta}{\partial S_\beta \partial v_\beta}\right)_n\right](\delta S_\alpha \delta v_\alpha)$$
$$+ \frac{1}{2}\left[\left(\frac{\partial^2 U_\alpha}{\partial v_\alpha^2}\right)_{S,n} + \left(\frac{\partial^2 U_\beta}{\partial v_\beta^2}\right)_{S,n}\right](\delta v_\alpha)^2 + \left[\left(\frac{\partial^2 U_\alpha}{\partial v_\alpha \partial n_\alpha}\right)_S + \left(\frac{\partial^2 U_\beta}{\partial v_\beta \partial n_\beta}\right)_S\right](\delta v_\alpha \delta n_\alpha)$$
$$+ \frac{1}{2}\left[\left(\frac{\partial^2 U_\alpha}{\partial n_\alpha^2}\right)_{S,v} + \left(\frac{\partial^2 U_\beta}{\partial n_\beta^2}\right)_{S,v}\right](\delta n_\alpha)^2 + \left[\left(\frac{\partial^2 U_\alpha}{\partial n_\alpha \partial S_\alpha}\right)_v + \left(\frac{\partial^2 U_\beta}{\partial n_\beta \partial S_\beta}\right)_v\right](\delta n_\alpha \delta S_\alpha)$$
$$+ \text{terms in } (\delta S_\alpha)^3, \text{ etc.} \qquad (7.3.7)$$

Arraying the coefficients in the expression (7.3.7) so that the three 'even' coefficients, i.e. those of $(\delta S_\alpha)^2$, $(\delta v_\alpha)^2$, and $(\delta n_\alpha)^2$, are diagonal, gives the determinant on the left hand side of inequality (7.3.2). Since δU is to be positive for all possible combinations of δS_α, δv_α, δn_α, it must be zero in particular for any two of them being zero, so that the three diagonal elements must be positive; when any one is zero Equation (7.3.7) reduces to a simple quadratic, and the argument of Problem 7.1(b) thus adds that the 2×2 minors of the determinant shall also be positive. Since the determinant comprises the totality of the coefficients in Equation (7.3.7), it follows that the conditions (7.3.2), (7.3.3), and (7.3.4) are both necessary and sufficient.

(b) Using the notation

$$\frac{\partial^2 U_\alpha}{\partial S_\alpha^2} = \alpha_{2S}\,, \quad \frac{\partial^2 U_\beta}{\partial S_\beta \partial v_\beta} = \beta_{Sv}\,, \quad \text{etc.,}$$

we can write the determinant (7.3.3)

$$D = (\alpha_{2S} + \beta_{2S})(\alpha_{2v} + \beta_{2v}) - (\alpha_{Sv} + \beta_{Sv})^2\,.$$

Multiplying out and collecting like terms we obtain

$$D = \alpha_{2S}\alpha_{2v} - \alpha_{Sv}^2 + \beta_{2S}\beta_{2v} - \beta_{Sv}^2 + \alpha_{2S}\beta_{2v} - \alpha_{Sv}\beta_{Sv} + \alpha_{2v}\beta_{2S} - \alpha_{Sv}\beta_{Sv}\,.$$

According to the solution to Problem 7.1(c), the first four terms become

$$\frac{T}{v_\alpha K_{T\alpha} C_{v\alpha}} + \frac{T}{v_\beta K_{T\beta} C_{v\beta}}\,.$$

The last four are partially separable as

$$\alpha_{2S}\alpha_{2v}\frac{\beta_{2v}}{\alpha_{2v}} - \alpha_{Sv}^2\frac{\beta_{Sv}}{\alpha_{Sv}} + \beta_{2S}\beta_{2v}\frac{\alpha_{2v}}{\beta_{2v}} - \beta_{Sv}^2\frac{\alpha_{Sv}}{\beta_{Sv}} =$$

$$(\alpha_{2S}\alpha_{2v} - \alpha_{Sv}^2)\frac{\beta_{2v}}{\alpha_{2v}} + (\beta_{2S}\beta_{2v} - \beta_{Sv}^2)\frac{\alpha_{2v}}{\beta_{2v}} + \alpha_{Sv}^2\left(\frac{\beta_{2v}}{\alpha_{2v}} - \frac{\beta_{Sv}}{\alpha_{Sv}}\right) + \beta_{Sv}^2\left(\frac{\alpha_{2v}}{\beta_{2v}} - \frac{\alpha_{Sv}}{\beta_{Sv}}\right).$$

The last two terms rearrange to the quadratic form

$$\left(\alpha_{Sv} - \beta_{Sv}\frac{\alpha_{2v}}{\beta_{2v}}\right)^2\frac{\beta_{2v}}{\alpha_{2v}}.$$

Thus D has the form

$$D = (1+x)A + \left(1+\frac{1}{x}\right)B + xC^2,$$

$$A = \frac{T}{v_\alpha K_{T\alpha} C_{v\alpha}}, \quad \text{etc.,} \quad x = \frac{v_\beta K_{S\beta}}{v_\alpha K_{S\alpha}}.$$

Evidently $D > 0$ is secured by $x > 0$, $[(1+x)A + (1+1/x)B] > 0$. According as $x \lessgtr 1$, however, the latter condition may be met with *either* A or $B < 0$, and this is not pre-empted by condition (7.3.4), $T/C_{v\alpha} + T/C_{v\beta} > 0$. Thus the conditions (7.3.3) and (7.3.4) are secured by the condition (7.3.5), though not by $(T/v_\alpha K_{T\alpha} C_{v\alpha} + T/v_\beta K_{T\beta} C_{v\beta})$, $(T/C_{v\alpha} + T/C_{v\beta}) > 0$; but conditions (7.3.3) and (7.3.4) do not require condition (7.3.5) as *necessary*. The further inequalities from condition (7.3.2) also are secured by condition (7.3.5). Note that condition (7.3.2) includes terms in

$$\left(\frac{\partial^2 U}{\partial v\, \partial n}\right)_S = \frac{1}{nK_S}\left(1 + \frac{S\alpha_p T}{C_p}\right),$$

$$\left(\frac{\partial^2 U}{\partial n^2}\right)_{S,\, v} = \frac{1}{n^2}\left(\frac{v}{K_T} + \frac{TS^2}{C_v}\right).$$

A two-way perturbation always leads to a condition like (7.3.3), implying through its cross-terms that some property has the same sign for α and β, whereas the diagonal terms of condition (7.3.2) being positive implies only that the sum for α and β is positive. This is because a determinant of sums is not equal to the sum of the corresponding determinants for α and β separately.

(c) The form in which the problem is stated suggests working from right to left in Equations (7.3.6). The method suggested by the solution to Problem 7.2(c) is to use the change-of-constraint formula. To prove $(\partial^2 G/\partial n^2)_{T,\,p} = 0$, start with the definition $G \equiv U - TS + pv$, whence given the Gibbs equation [Problem 1.20(d)] for one component, $S\,dT - v\,dp + n\,d\mu = 0$, we have

$$dG = -S\,dT + v\,dp + \mu\,dn$$

whence

$$\left(\frac{\partial G}{\partial n}\right)_{T,\,p} = \mu\,, \quad \left(\frac{\partial^2 G}{\partial n^2}\right)_{T,\,p} = \left(\frac{\partial \mu}{\partial n}\right)_{T,\,p}\,;$$

but

$$\mathrm{d}\mu = -\bar{S}\mathrm{d}T + \bar{v}\,\mathrm{d}p\,, \quad (\mathrm{d}\mu)_{T,\,p} = 0\,.$$

To relate $(\partial^2 G/\partial n^2)_{T,\,p}$ to second derivatives of F, write

$$\begin{aligned}
\left(\frac{\partial^2 G}{\partial n^2}\right)_{T,\,p} = \left(\frac{\partial \mu}{\partial n}\right)_{T,\,p} &= \left(\frac{\partial \mu}{\partial n}\right)_{T,\,v} + \left(\frac{\partial \mu}{\partial v}\right)_{T,\,n}\left(\frac{\partial v}{\partial n}\right)_{T,\,p}\,,\\
&= \left(\frac{\partial^2 F}{\partial n^2}\right)_{T,\,v} + \left(\frac{\partial^2 F}{\partial v\,\partial n}\right)_{T}\left(\frac{\partial \mu}{\partial p}\right)_{T,\,n}\,,\\
&= \left(\frac{\partial^2 F}{\partial n^2}\right)_{T,\,v} + \left(\frac{\partial^2 F}{\partial v\,\partial n}\right)_{T}\left(\frac{\partial \mu}{\partial v}\right)_{T,\,n}\Bigg/\left(\frac{\partial p}{\partial v}\right)_{T,\,n}\,,\\
&= \left(\frac{\partial^2 F}{\partial n^2}\right)_{T,\,v} - \left(\frac{\partial^2 F}{\partial v\,\partial n}\right)_{T}^{2}\Bigg/\left(\frac{\partial^2 F}{\partial v^2}\right)_{T,\,n}\,,
\end{aligned}$$

and the required result follows on multiplying through by $(\partial^2 F/\partial v^2)_{T,\,n}$. Alternatively, use the method of Jacobians, cf. solution to Problem 7.1(c),

$$\begin{aligned}
\begin{vmatrix} F_{2n} & F_{nv} \\ F_{vn} & F_{2v} \end{vmatrix} &= \frac{\partial(\mu,-p)_T}{\partial(n,v)_T} = \frac{\partial(\mu,-p)}{\partial(n,-p)}\Bigg/\frac{\partial(n,v)}{\partial(n,-p)} = -\left(\frac{\partial \mu}{\partial n}\right)_{T,\,p}\Bigg/\left(\frac{\partial v}{\partial p}\right)_{T,\,n}\\
&= \left(\frac{\partial^2 G}{\partial n^2}\right)_{T,\,p}\left(\frac{\partial^2 F}{\partial v^2}\right)_{T,\,n}.
\end{aligned}$$

The third step necessitates expressing the three second derivatives of $F = F(v,n)$ at constant temperature in terms of the second derivatives of $U = U(S,v,n)$. The above procedure gives

$$\left(\frac{\partial^2 F}{\partial n^2}\right)_{T,\,v} = \left(\frac{\partial^2 U}{\partial n^2}\right)_{S,\,v} - \left(\frac{\partial^2 U}{\partial S\partial n}\right)_{v}\Bigg/\left(\frac{\partial^2 U}{\partial S^2}\right)_{v,\,n}\,,$$

and from Problem 7.1(c) we obtain

$$\left(\frac{\partial^2 F}{\partial v^2}\right)_{T,\,n} = \left(\frac{\partial^2 U}{\partial v^2}\right)_{S,\,n} - \left(\frac{\partial^2 U}{\partial S\partial v}\right)_{n}^{2}\Bigg/\left(\frac{\partial^2 U}{\partial S^2}\right)_{v,\,n}.$$

Similarly

$$\begin{aligned}
\left(\frac{\partial^2 F}{\partial v\,\partial n}\right)_{T} = \left(\frac{\partial \mu}{\partial v}\right)_{T} &= \left(\frac{\partial \mu}{\partial v}\right)_{S} + \left(\frac{\partial \mu}{\partial S}\right)_{v,\,n}\left(\frac{\partial S}{\partial v}\right)_{T,\,n}\\
&= \left(\frac{\partial^2 U}{\partial v\,\partial n}\right)_{S} + \left(\frac{\partial^2 U}{\partial S\partial n}\right)_{v}\left(\frac{\partial p}{\partial T}\right)_{v,\,n}\\
&= \left(\frac{\partial^2 U}{\partial v\,\partial n}\right)_{S} + \left(\frac{\partial^2 U}{\partial S\partial n}\right)_{v}\left(\frac{\partial p}{\partial S}\right)_{v,\,n}\left(\frac{\partial S}{\partial T}\right)_{v,\,n}\\
&= \left(\frac{\partial^2 U}{\partial v\,\partial n}\right)_{S} - \left(\frac{\partial^2 U}{\partial S\partial n}\right)_{v}\left(\frac{\partial^2 U}{\partial v\,\partial S}\right)_{n}\Bigg/\left(\frac{\partial^2 U}{\partial S^2}\right)_{v,\,n}.
\end{aligned}$$

Forming the determinant in F and multiplying through by $(\partial^2 U/\partial S^2)_{v,\,n}$ gives the required result.

The result (7.3.6) has the following implication in regard to expression (7.3.7). If the latter is separated into two quadratic forms, one for each phase, then the corresponding pair of 3 x 3 determinants are each zero by (7.3.6). This means that for each phase there is a particular combination of δS, δv, δn (such that $\delta S/\delta n = \bar{S}$, $\delta v/\delta n = \bar{v}$) for which the quadratic form has two coincident real roots. Thus, provided that the principal minors of the determinant for that phase are positive, i.e.

$$\begin{vmatrix} U_{2S} & U_{Sv} \\ U_{vS} & U_{2v} \end{vmatrix} > 0\,, \quad U_{2S} > 0\,,$$

the quadratic form for that contribution to δU is everywhere positive or zero. These are the conditions (7.1.1)-(7.1.3) secured by conditions (7.1.4) or (7.1.5). The same applies to the other phase; but unless the two phases are identical, there is no combination of δS, δv, δn for which δU overall is zero, since non-identity implies $\delta S/\delta n = \bar{S}_\alpha \neq \bar{S}_\beta$, etc. Thus condition (7.1.4) or (7.1.5) for each phase secures overall stability.

(d) Using the notation

$$\left(\frac{\partial^2 F_\alpha}{\partial n_\alpha^2}\right)_{T,\,v} = \alpha_{2n}\,, \quad \left(\frac{\partial^2 F_\beta}{\partial v_\beta \partial n_\beta}\right)_T = \beta_{v,\,n}, \text{ etc.,}$$

we can write the determinant

$$(\alpha_{2n}+\beta_{2n})(\alpha_{2v}+\beta_{2v})-(\alpha_{vn}+\beta_{vn})^2 = \alpha_{2n}\alpha_{2v}+\beta_{2n}\beta_{2v}-\alpha_{vn}^2-\beta_{vn}^2$$
$$+\alpha_{2v}\beta_{2n}+\alpha_{2n}\beta_{2v}-2\alpha_{vn}\beta_{vn}\,.$$

The first four terms are zero by Equation (7.3.6). That equation enables the last three terms to be put into quadratic form as:

$$\alpha_{vn}^2\frac{\beta_{2n}}{\alpha_{2n}}+\beta_{vn}^2\frac{\alpha_{2n}}{\beta_{2n}}-2\alpha_{vn}\beta_{vn} = \frac{(\alpha_{vn}\beta_{2n}-\beta_{vn}\alpha_{2n})^2}{\alpha_{2n}\beta_{2n}}\,.$$

This is positive if, and only if $(\partial^2 F_\alpha/\partial n_\alpha^2)_{T,\,v}$ and $(\partial^2 F_\beta/\partial n_\beta^2)_{T,\,v}$ have the same sign. Now

$$\left(\frac{\partial^2 F}{\partial n^2}\right)_{T,\,v} = -\left(\frac{\partial \mu}{\partial v}\right)_{T,\,n}\left(\frac{\partial v}{\partial n}\right)_{T,\,p} = -\left(\frac{\partial \mu}{\partial p}\right)_{T,\,n}\left(\frac{\partial v}{\partial n}\right)_{T,\,p}\left(\frac{\partial p}{\partial v}\right)_{T,\,n},$$

cf part (c), so

$$\left(\frac{\partial^2 F}{\partial n^2}\right)_{T,\,v} = \frac{\bar{v}^2}{vK_T} = \frac{v}{n^2 K_T}\,.$$

Thus the determinant is positive if, and only if, $K_{T\alpha}$ and $K_{T\beta}$ have the same sign, negative volume being excluded, and the other condition

$$\left(\frac{\partial^2 F_\alpha}{\partial n_\alpha^2}\right)_{T,\,v}+\left(\frac{\partial^2 F_\beta}{\partial n_\beta^2}\right)_{T,\,v} > 0$$

requires that sign to be positive. Compare this result with that of Problem 1.30; again, a two-way perturbation requires for stability that the two phases or fluids be individually stable, whereas a one-way perturbation requires only that they be collectively stable.

7.4 Co-existence of two phases (ii). Reconsider the system of Problem 7.3, but use molar rather than extensive properties as the variables. The perturbation may then be described in terms similar to those of Problem 7.1, i.e. as $\delta\xi$ mole passing from phase α to phase β, with consequent changes in the molar entropies and volumes of the phases imposed by the constraints to constant total entropy and volume.

(a) Show that

$$(n_\alpha - \delta\xi)\delta S_\alpha + (n_\beta + \delta\xi)\delta S_\beta - (\overline{S}_\alpha^{(0)} - \overline{S}_\beta^{(0)})\delta\xi = 0\,, \qquad (7.4.1)$$

$$(n_\alpha - \delta\xi)\delta v_\alpha + (n_\beta + \delta\xi)\delta v_\beta - (\overline{v}_\alpha^{(0)} - \overline{v}_\beta^{(0)})\delta\xi = 0\,, \qquad (7.4.2)$$

$$\delta U = n_\alpha(\overline{U}_\alpha - \overline{U}_\alpha^{(0)}) + n_\beta(\overline{U}_\beta - \overline{U}_\beta^{(0)}) + \delta\xi(\overline{U}_\beta - \overline{U}_\alpha)\,, \qquad (7.4.3)$$

where the indices and suffixes have the same meanings as in Problem 7.1, with the incidental difference that neither δS_β nor δS_α necessarily tends to zero as $\delta\xi$ tends to zero (and similarly for the δv_α, δv_β).

(b) Use Taylor series expansions for $\overline{U}_\alpha$, $\overline{U}_\beta$ about $\overline{U}_\alpha^{(0)}$, $\overline{U}_\beta^{(0)}$ respectively to confirm that the conditions for equilibrium are (7.3.1).

(c) Thence show that thermodynamic stability is secured by the condition that each phase be stable by the criteria (7.1.4) or (7.1.5).

(d) Give a physical interpretation of condition (7.3.1) in terms of the tangent planes to the surfaces $\overline{U}_\alpha = \overline{U}_\alpha(\overline{S}_\alpha, \overline{v}_\alpha)$, $\overline{U}_\beta = \overline{U}_\beta(\overline{S}_\beta, \overline{v}_\beta)$. What is the maximum number of phases which can, in general, co-exist in a one-component system?

Solution

(a) Initial state: $S = n_\alpha S_\alpha^{(0)} + n_\beta S_\beta^{(0)}$; perturbed state: $\overline{S}_\alpha = \overline{S}_\alpha^{(0)} + \delta S_\alpha$ $\overline{S}_\beta = \overline{S}_\beta^{(0)} + \delta S_\beta$, $n_\alpha - \delta\xi$, etc., and this specification is complete, since δS_α, δv_α, $\delta\xi$ are independent variables. Then

$$\delta S = 0 = (n_\alpha - \delta\xi)(\overline{S}_\alpha^{(0)} + \delta S_\alpha) + (n_\beta + \delta\xi)(\overline{S}_\beta^{(0)} + \delta S_\beta) - n_\alpha \overline{S}_\alpha^{(0)} - n_\beta \overline{S}_\beta^{(0)}\,.$$

Rearrangement gives the required result, and similarly for $\delta v = 0$. Since $\overline{U}_\alpha = \overline{U}_\alpha(\overline{S}_\alpha, \overline{v}_\alpha)$,

$$\delta U = (n_\alpha - \delta\xi)\overline{U}_\alpha + (n_\beta + \delta\xi)\overline{U}_\beta - n_\alpha \overline{U}_\alpha^{(0)} - n_\beta \overline{U}_\beta^{(0)}\,.$$

Note that if $n_\beta \to 0$, and $\overline{U}_\beta^{(0)} \to \overline{U}_\alpha^{(0)}$, this reduces to the corresponding relation of Problem 7.1(a).

(b) The Taylor expansion for $\overline{U}_\alpha$ is

$$\overline{U}_\alpha = \overline{U}_\alpha^{(0)} + \left(\frac{\partial \overline{U}_\alpha}{\partial \overline{S}_\alpha}\right)_{\overline{v}} \delta S_\alpha + \left(\frac{\partial \overline{U}_\alpha}{\partial \overline{v}_\alpha}\right)_{\overline{S}} \delta v_\alpha$$
$$+ \frac{1}{2}\left(\frac{\partial^2 \overline{U}_\alpha}{\partial \overline{S}_\alpha^2}\right)_{\overline{v}} (\delta S_\alpha)^2 + \left(\frac{\partial^2 \overline{U}_\alpha}{\partial \overline{S}_\alpha \partial \overline{v}_\alpha}\right) \delta S_\alpha \delta v_\alpha + \frac{1}{2}\left(\frac{\partial^2 \overline{U}_\alpha}{\partial \overline{v}_\alpha^2}\right)_{\overline{S}} (\delta \overline{v}_\alpha)^2 + \cdots,$$

where the constraints on the partial derivatives refer to the molar quantities for phase α; and similarly for $\bar{U}_\beta$. Substituting in δU we obtain

$$\delta U = n_\alpha\left[\left(\frac{\partial \bar{U}_\alpha}{\partial \bar{S}_\alpha}\right)_{\bar{v}}\delta S_\alpha + \left(\frac{\partial \bar{U}_\alpha}{\partial \bar{v}_\alpha}\right)_{\bar{S}}\delta v_\alpha\right] + n_\beta\left[\left(\frac{\partial \bar{U}_\beta}{\partial \bar{S}_\beta}\right)_{\bar{v}}\delta S_\beta + \left(\frac{\partial \bar{U}_\beta}{\partial \bar{v}_\beta}\right)\delta v_\beta\right]$$

$$-\delta\xi\left[\bar{U}_\alpha^{(0)} + \left(\frac{\partial \bar{U}_\alpha}{\partial \bar{S}_\alpha}\right)_{\bar{v}}\delta S_\alpha + \left(\frac{\partial \bar{U}_\alpha}{\partial \bar{v}_\alpha}\right)_{\bar{S}}\delta \bar{v}_\alpha\right]$$

$$+\delta\xi\left[\bar{U}_\beta^{(0)} + \left(\frac{\partial \bar{U}_\beta}{\partial \bar{S}_\beta}\right)\delta S_\beta + \left(\frac{\partial \bar{U}_\beta}{\partial \bar{v}_\beta}\right)_{\bar{S}}\delta \bar{v}_\beta\right] + \text{second-order terms}$$

$$= [(n_\alpha - \delta\xi)\delta S_\alpha - \bar{S}_\alpha^{(0)}\delta\xi]\left(\frac{\partial \bar{U}_\alpha}{\partial \bar{S}_\alpha}\right)_{\bar{v}} + [(n_\alpha - \delta\xi)\delta v_\alpha - \bar{v}_\alpha^{(0)}\delta\xi]\left(\frac{\partial \bar{U}_\alpha}{\partial \bar{v}_\alpha}\right)_{\bar{S}}$$

$$+[(n_\beta + \delta\xi)\delta S_\beta + \bar{S}_\beta^{(0)}\delta\xi]\left(\frac{\partial \bar{U}_\beta}{\partial \bar{S}_\beta}\right)_{\bar{v}} + [(n_\beta + \delta\xi)\delta v_\beta + \bar{v}_\beta^{(0)}\delta\xi]\left(\frac{\partial \bar{U}_\beta}{\partial \bar{v}_\beta}\right)_{\bar{S}}$$

$$+\delta\xi\left[\bar{U}_\beta^{(0)} - \bar{U}_\alpha^{(0)} + \bar{S}_\alpha^{(0)}\left(\frac{\partial \bar{U}_\alpha}{\partial \bar{S}_\alpha}\right)_{\bar{v}} - S_\beta^{(0)}\left(\frac{\partial \bar{U}_\beta}{\partial \bar{S}_\beta}\right)_{\bar{v}}\right.$$

$$\left.+\bar{v}_\alpha^{(0)}\left(\frac{\partial \bar{U}_\alpha}{\partial \bar{v}_\alpha}\right)_{\bar{S}} - \bar{v}_\beta^{(0)}\left(\frac{\partial \bar{U}_\beta}{\partial \bar{v}_\beta}\right)_{\bar{S}}\right] + \text{second-order terms}.$$

Then, by virtue of Equations (7.4.1), (7.4.2),

$$\delta U = [(n_\alpha - \delta\xi)\delta S_\alpha - \bar{S}_\alpha^{(0)}\delta\xi]\left[\left(\frac{\partial \bar{U}_\alpha}{\partial \bar{S}_\alpha}\right)_{\bar{v}} - \left(\frac{\partial \bar{U}_\beta}{\partial \bar{S}_\beta}\right)_{\bar{v}}\right]$$

$$+[(n_\alpha - \delta\xi)\delta v_\alpha - \bar{v}_\alpha^{(0)}\delta\xi]\left[\left(\frac{\partial \bar{U}_\alpha}{\partial \bar{v}_\alpha}\right)_{\bar{S}} - \left(\frac{\partial \bar{U}_\beta}{\partial \bar{v}_\beta}\right)_{\bar{S}}\right]$$

$$+\delta\xi[(\bar{U}_\beta^{(0)} - T_\beta\bar{S}_\beta^{(0)} + p_\beta\bar{v}_\beta^{(0)}) - (\bar{U}_\alpha^{(0)} - T_\alpha\bar{S}_\alpha^{(0)} + p_\alpha\bar{v}_\alpha^{(0)})]$$

$$+\text{second-order terms}. \tag{7.4.4}$$

Clearly, for $\delta U = 0$ (to first order) $T_\alpha = T_\beta$ and $p_\alpha = p_\beta$. Now $G \equiv U - TS + pv$ so, on dividing by n, we obtain $\mu = \bar{U} - T\bar{S} + p\bar{v}$; Alternatively, take $(\partial U/\partial n)_{S,v} = \mu$, change variable twice to give $(\partial U/\partial n)_{T,p} = \bar{U}$, and use

$$\left(\frac{\partial p}{\partial T}\right)_v\left(\frac{\partial T}{\partial v}\right)_p\left(\frac{\partial v}{\partial p}\right)_T = -1$$

to obtain $\mu = \bar{U} - T\bar{S} + p\bar{v}$. Thus the third condition is $\mu_\alpha = \mu_\beta$.

(c) The second-order terms in δU are

$$(n_\alpha - \delta\xi)\left[\frac{1}{2}\left(\frac{\partial^2 \bar{U}_\alpha}{\partial \bar{S}_\alpha^2}\right)_{\bar{v}}(\delta S_\alpha)^2 + \left(\frac{\partial^2 \bar{U}_\alpha}{\partial \bar{S}_\alpha \partial \bar{v}_\alpha}\right)\delta S_\alpha \delta v_\alpha + \frac{1}{2}\left(\frac{\partial^2 \bar{U}_\alpha}{\partial \bar{v}_\alpha}\right)_{\bar{S}}(\delta v_\alpha)^2\right]$$

$$+(n_\beta + \delta\xi)\left[\frac{1}{2}\left(\frac{\partial^2 \bar{U}_\beta}{\partial \bar{S}_\beta^2}\right)_{\bar{v}}(\delta S_\beta)^2 + \left(\frac{\partial^2 \bar{U}_\beta}{\partial \bar{S}_\beta \partial \bar{v}_\beta}\right)\delta S_\beta \delta v_\beta + \frac{1}{2}\left(\frac{\partial^2 \bar{U}_\beta}{\partial \bar{v}_\beta}\right)_{\bar{S}}(\delta v_\beta)^2\right]. \tag{7.4.5}$$

The simplest condition ensuring that this sum is positive is that both square brackets be positive, i.e. that each phase fulfils (7.1.4). Since δS_β, δv_β, are dependent variables, it follows that while this condition is sufficient, it may not be necessary. Substituting for δS_β and δv_β according to (7.4.1) and (7.4.2) does not reduce the second-order terms to a single bracket, *except if the two phases are identical.* Thus the conditions which may be formulated in terms of sums

$$\left[\left(\frac{\partial^2 \overline{U}_\alpha}{\partial \overline{S}_\alpha^2}\right) + \left(\frac{\partial^2 \overline{U}_\beta}{\partial \overline{S}_\beta^2}\right)\right], \text{ etc.,}$$

are not sufficient.

(d) The stability conditions (7.1.4), (7.1.5) ensure that for a stable phase the $\overline{U} = \overline{U}(\overline{S}, \bar{v})$ surface is concave upwards. For two co-existing phases, the equilibrium conditions $T_\alpha = T_\beta$, $p_\alpha = p_\beta$ ensure that the tangent planes to $\overline{U}_\alpha$, $\overline{U}_\beta$ at $\overline{S}_\alpha, \bar{v}_\alpha$ and $\overline{S}_\beta, \bar{v}_\beta$ respectively are parallel; it is evident from the third term of expression (7.4.4) that $\mu_\alpha = \mu_\beta$ ensures additionally that the tangent planes are identical. Thus two phases can co-exist in a one-component system over a range of temperature, but fixing the co-existence temperature fixes the co-existence pressure. There is in general one, and only one, way of putting a common tangent plane to three uniformly curved surfaces. Thus three phases can co-exist at a determinate temperature and pressure. It might seem that four phases could co-exist fortuitously, but since the tangency must be geometrically exact, this occurrence has zero probability.

7.5 Criticality, continuous equation of state. The region of liquid-vapour co-existence of real fluids is invariably bounded in temperature, and the same is true of the van der Waals and similar continuous equations of state, $\xi(p, \bar{v}, T) = 0$. Designate this bounding temperature T_c. Then for all $T > T_c$ the fluid is stable, by the criteria of Problem 7.1(b), (c), over the accessible range of volume; and for $T < T_c$, the continuity of $\xi(p, \bar{v}, T) = 0$ implies a region of instability bounded by the so-called **spinodal locus,**

$$\begin{vmatrix} \overline{U}_{2\overline{S}} & \overline{U}_{\overline{S}v} \\ \overline{U}_{v\overline{S}} & \overline{U}_{2v} \end{vmatrix} = 0,$$

cf. Problem 7.2(e). The locus joining the molar entropies and volumes of co-existing phases is called the **binodal locus,** and defined by the co-existence conditions (7.3.1).

(a) Consider two co-existing phases, α and β. Use the three co-existence conditions in terms of the $\overline{F} = \overline{F}(T, \bar{v})$ surface to show that if the tie-line connecting $\overline{F}_\alpha(T, \bar{v}_\alpha)$ to $\overline{F}_\beta(T, \bar{v}_\beta)$ becomes vanishingly short as $T \to T_c$, then at $T = T_c$

$$\left(\frac{\partial \overline{F}}{\partial \bar{v}}\right)_T < 0, \quad \left(\frac{\partial^2 \overline{F}}{\partial \bar{v}^2}\right)_T = \left(\frac{\partial^3 \overline{F}}{\partial \bar{v}^3}\right)_T = 0, \quad \left(\frac{\partial^4 \overline{F}}{\partial \bar{v}^4}\right)_T > 0,$$

and that the binodal locus is tangential to the spinodal locus.

(b) Consider a temperature ΔT below T_c. The co-existing phases are defined by Δp, Δv_α, Δv_β; and since the equation of state is continuous, Δp is expressible as an explicit function of Δv, ΔT along this isotherm from Δv_α to Δv_β. Use a Taylor series expansion for Δp in terms of Δv and ΔT, together with the co-existence condition $\mu_\alpha = \mu_\beta$ to show that

$$[(\Delta v_\alpha)^2-(\Delta v_\beta)^2]\left\{\left(\frac{\partial^2 p}{\partial v\,\partial T}\right)_c \Delta T+\frac{1}{4}\left(\frac{\partial^3 p}{\partial v^3}\right)_c [(\Delta v_\alpha)^2+(\Delta v_\beta)^2]\right\}=0\,.$$

Derivatives with respect to T of order higher than first may be neglected.

(c) Use Taylor series expansions for $\Delta p = \Delta p(\Delta v_\alpha, \Delta T)$ and $\Delta p = \Delta p(\Delta v_\beta, \Delta T)$, together with the result of part (b) to show that

$$\Delta v_\alpha = -\Delta v_\beta = \left[-6\left(\frac{\partial^2 p}{\partial T\partial \bar{v}}\right)_c \Delta T\bigg/\left(\frac{\partial^3 p}{\partial \bar{v}^3}\right)_{T,\,c}\right]^{1/2},$$

i.e. that the locus of the mid-points of the tie-lines connecting the molar volumes of co-existing phases passes through the critical point, the binodal locus near the critical point being quadratic in volume, and that

$$\frac{\Delta p}{\Delta T} = \left(\frac{\partial p}{\partial T}\right)_{\bar{v},\,c},$$

i.e. that the vapour-pressure curve is co-linear with the isochore passing through the critical point.

(d) In real fluids, the binodal locus near the critical point is described better by

$$\Delta v_\alpha \approx -\Delta v_\beta \approx X(\Delta T)^{1/3}$$

than by the result of part (c); what does this imply about $\xi(p,\bar{v},T)=0$ in the critical region?

Solution

(a) The co-existence conditions are:

$$T_\alpha = T_\beta,\quad p_\alpha = p_\beta,\quad \bar{F}_\alpha = \bar{F}_\beta - p(\bar{v}_\beta - \bar{v}_\alpha)\,,$$

so that at each T, $\bar{F}_\alpha$ and $\bar{F}_\beta$ have a common tangent of slope $-p$. If $\bar{F}=\bar{F}(T,\bar{v})$ is continuous between $\bar{v}_\alpha$ and $\bar{v}_\beta$ there is necessarily an intervening region of instability ($\bar{F}=\bar{F}(\bar{v})_T$ convex upwards) bounded by points of inflexion $(\partial^2\bar{F}/\partial\bar{v}^2)_T = 0$, and these are necessarily both between $\bar{v}_\alpha$ and $\bar{v}_\beta$, i.e. the spinodal locus is necessarily inside the binodal locus. If the binodal locus is to close, $\bar{F}=\bar{F}(T,\bar{v})$ being everywhere analytic implies that the points $\bar{v}_\alpha$ and $\bar{v}_\beta$ and the two points of inflexion $(\partial^2\bar{F}/\partial\bar{v}^2)_T = 0$ must coincide at $T = T_c$, i.e. the spinodal locus coincides with the binodal locus at $T = T_c$, with

$$\left(\frac{\partial^2\bar{F}}{\partial\bar{v}^2}\right)_{T,\,c} = -\left(\frac{\partial p}{\partial\bar{v}}\right)_{T,\,c} = 0\,.$$

If this point is to be stable, then the given conditions on the higher derivatives of $\overline{F}$ must hold. The remaining condition, $p > 0$, follows from the requirement that all continuous equations of state necessarily converge to $p\bar{v} = RT$ as $\bar{v} \to \infty$. Thus a dilute phase (vapour) can never sustain negative pressure, and so liquid-vapour co-existence can only occur at $p > 0$; and this applies also at the liquid-vapour critical point.

The critical point is the *only* point on the spinodal locus at which $(\partial^2\overline{F}/\partial\bar{v}^2)_T$ and $(\partial^3\overline{F}/\partial\bar{v}^3)_T$ are simultaneously zero.

(b) Since

$$\left(\frac{\partial p}{\partial \bar{v}}\right)_{T,\,\mathrm{c}} = \left(\frac{\partial^2 p}{\partial \bar{v}^2}\right)_{T,\,\mathrm{c}} = 0\,,$$

the expansion for Δp is

$$\Delta p = \left(\frac{\partial p}{\partial T}\right)_{\bar{v},\,\mathrm{c}} \Delta T + \left(\frac{\partial^2 p}{\partial T \partial \bar{v}}\right)_{\mathrm{c}} \Delta v \Delta T + \frac{1}{3!}\left(\frac{\partial^3 p}{\partial \bar{v}^3}\right)_{T,\,\mathrm{c}} (\Delta v)^3 + \cdots.$$

The condition $\mu_\alpha = \mu_\beta$ has to be expressed in terms of Δv_α and Δv_β, given $\Delta p_\alpha = \Delta p_\beta$, and ΔT. Now $(\partial\mu/\partial p)_T = \bar{v}$, cf Problem 7.3(d), so $\mu_\alpha = \mu_\beta$ implies

$$\int_{p_\beta}^{p_\alpha} \bar{v}\,\mathrm{d}p = \int_{\Delta v_\beta}^{\Delta v_\alpha} \Delta v\,\mathrm{d}p = 0\,.$$

The physical meaning of this condition—known as Maxwell's equal area rule—is obvious from the diagram of Problem 6.4. Expressed in terms of Δv as independent variable it becomes

$$\int_{\Delta v_\beta}^{\Delta v_\alpha} \Delta v \left[\frac{\partial(\Delta p)}{\partial(\Delta v)}\right]_{T,\,\mathrm{c}} \mathrm{d}(\Delta v) = 0\,,$$

and $[\partial(\Delta p)/\partial(\Delta v)]_{T,\,\mathrm{c}}$ is obtained by differentiating the Taylor series as

$$\left(\frac{\partial(\Delta p)}{\partial(\Delta v)}\right)_{T,\,\mathrm{c}} = \left(\frac{\partial^2 p}{\partial T \partial \bar{v}}\right)_{T,\,\mathrm{c}} \Delta T + \frac{1}{2}\left(\frac{\partial^3 p}{\partial \bar{v}^3}\right)_{T,\,\mathrm{c}} (\Delta v)^2 + \cdots.$$

Substituting into the argument of the integral and integrating gives for the first two terms

$$\frac{1}{2}\left[(\Delta v)^2 \left(\frac{\partial^2 p}{\partial T \partial \bar{v}}\right)_{\mathrm{c}} \Delta T\right]_{v_\beta}^{v_\alpha} + \frac{1}{8}\left[\left(\frac{\partial^3 p}{\partial \bar{v}^3}\right)_{\mathrm{c}} (\Delta v)^4\right]_{v_\beta}^{v_\alpha}\,,$$

and the required result follows.

(c) The Taylor expansions are as in part (b) except that Δv is replaced by Δv_α, Δv_β, respectively. Subtracting the second series from the first eliminates Δp as

$$0 = \left(\frac{\partial^2 p}{\partial T \partial \bar{v}}\right)_{\mathrm{c}} (\Delta v_\alpha - \Delta v_\beta)\Delta T + \frac{1}{3!}\left(\frac{\partial^3 p}{\partial \bar{v}^3}\right)_{\mathrm{c}} [(\Delta v_\alpha)^3 - (\Delta v_\beta)^3]\,,$$

or, since $\Delta v_\alpha \neq \Delta v_\beta$,

$$\left(\frac{\partial^2 p}{\partial T \partial \bar{v}}\right)_c \Delta T = -\frac{1}{3!}\left(\frac{\partial^3 p}{\partial \bar{v}^3}\right)_{T,c} [(\Delta v_\alpha)^2 + \Delta v_\alpha \Delta v_\beta + (\Delta v_\beta)^2] .$$

Substituting for $(\partial^2 p/\partial T \partial \bar{v})_c \Delta T$ in the result of part (b), we obtain

$$[(\Delta v_\alpha)^2 - (\Delta v_\beta)^2]\left(\frac{\partial^3 p}{\partial \bar{v}^3}\right)_{T,c} \left\{-\frac{1}{3!}[(\Delta v_\alpha)^2 + \Delta v_\alpha \Delta v_\beta + (\Delta v_\beta)^2] + \frac{1}{4}[(\Delta v_\alpha)^2 + (\Delta v_\beta)^2]\right\} = 0 ,$$

whence

$$\frac{1}{12}\left(\frac{\partial^3 p}{\partial \bar{v}^3}\right)_{T,c} [(\Delta v_\alpha)^2 - (\Delta v_\beta)^2](\Delta v_\alpha - \Delta v_\beta)^2 = 0 ,$$

$$\left(\frac{\partial^3 p}{\partial \bar{v}^3}\right)_{T,c} (\Delta v_\alpha + \Delta v_\beta)(\Delta v_\alpha - \Delta v_\beta)^3 = 0 ,$$

and

$$\Delta v_\alpha = -\Delta v_\beta .$$

Substituting back into the equation for $(\partial^2 p/\partial T \partial \bar{v})_c \Delta T$, we obtain

$$\left(\frac{\partial^2 p}{\partial T \partial \bar{v}}\right)_c \Delta T = -\frac{1}{3!}\left(\frac{\partial^3 p}{\partial \bar{v}^3}\right)_{T,c} (\Delta v_\alpha)^2 ,$$

and the required result follows.

To obtain $\Delta p/\Delta T$, add the two Taylor series expansions for Δp, and put $\Delta v_\alpha = -\Delta v_\beta$; since $(\Delta v_\alpha)^3 = -(\Delta v_\beta)^3$, the second and third terms on the right hand side are each zero, and the required result follows.

(d) The results of part (c) demonstrate that the physical consequence of the assumption that $\bar{F}$ is a continuous differentiable function of T and $\bar{v}$ is that the binodal locus has a rounded top which is quadratic in $\bar{v}$. The order of the curve depends on the order of the first non-vanishing derivative of $\bar{F}$ with respect to $\bar{v}$ at constant T. If the fourth derivative is zero, then stability requires that the fifth be likewise, with the sixth being positive. If the first non-vanishing derivative is of the order $2n$ (since it must be even), an analysis similar to the above must show that Δv_α is proportional to $(\Delta T)^{1/(2n-2)}$. In other words, no continuous equation of state $\xi(p, \bar{v}, T) = 0$ can give Δv_α proportional to $(\Delta T)^{1/3}$ or to any similar power of ΔT. Note that if the equation of state is not continuous from liquid-like to gas-like states, the spinodal locus ceases to have any simple meaning; but the critical isotherm is necessarily continuous, even if non-analytic, and, at least

$$\left(\frac{\partial p}{\partial \bar{v}}\right)_c = \left(\frac{\partial^2 p}{\partial \bar{v}^2}\right)_c = 0 .$$

7.6 Stability at a critical point. The preceding problem examines the critical point in terms of derivatives of $\overline{F} = \overline{F}(T,\bar{v})$, i.e. in terms of the volume derivatives of pressure. To elucidate the relevance of the condition of thermal stability, $T/C_v > 0$, requires the $\overline{U} = \overline{U}(\overline{S},\bar{v})$ surface.

(a) Express the slope of the projected tie-lines (on the $\overline{S}, \bar{v}$ plane) by alternative equations derived, respectively, from the co-existence conditions $T_\alpha = T_\beta$, $p_\alpha = p_\beta$, by using Taylor series expansions for ΔT and Δp in terms of $\Delta S_\alpha, \Delta v_\alpha$; $\Delta S_\beta, \Delta v_\beta$. Thence show that the binodal locus at $T = T_c$ is characterized by $-(\partial p/\partial \bar{v})_T = 0$, irrespective of whether or not $T/C_v = 0$; i.e. that the binodal locus on the $\overline{U} = \overline{U}(\overline{S},\bar{v})$ surface coincides, at the critical point, with the spinodal locus as specified by the condition of mechanical stability alone.

(b) Use the equation of the tangent at $T = T_c$ to the spinodal locus on the $\overline{U} = \overline{U}(\overline{S},\bar{v})$ surface,

$$D = \begin{vmatrix} \overline{U}_{2\overline{S}} & \overline{U}_{\overline{S}\bar{v}} \\ \overline{U}_{\bar{v}\overline{S}} & \overline{U}_{2\bar{v}} \end{vmatrix} = 0 \,,$$

to show that at the critical point

$$\frac{T}{C_v}\left(\frac{\partial^2 p}{\partial \bar{v}^2}\right)_{T,\,c} = 0 \,.$$

Solution

(a) The expansions for ΔT and Δp are [leaving off the bar, since all extensive variables are to be read as molar (intensive)]:

$$\Delta T = \Delta\left(\frac{\partial U}{\partial S}\right)_v = \left(\frac{\partial^2 U}{\partial S^2}\right)_{v,\,c}\Delta S + \left(\frac{\partial^2 U}{\partial S \partial v}\right)_c \Delta v + \cdots \,,$$

$$-\Delta p = \Delta\left(\frac{\partial U}{\partial v}\right)_S = \left(\frac{\partial^2 U}{\partial S \partial v}\right)_c \Delta S + \left(\frac{\partial^2 U}{\partial v^2}\right)_{S,\,c}\Delta v + \cdots \,.$$

Putting $T_\alpha = T_\beta$, $p_\alpha = p_\beta$, and subtracting the series for β from that for α, and neglecting terms of order higher than the first, we obtain

$$\left(\frac{\partial^2 U}{\partial S^2}\right)_{v,\,c}(\Delta S_\alpha - \Delta S_\beta) + \left(\frac{\partial^2 U}{\partial S \partial v}\right)_c (\Delta v_\alpha - \Delta v_\beta) = 0 \,,$$

$$\left(\frac{\partial^2 U}{\partial S \partial v}\right)_c (\Delta S_\alpha - \Delta S_\beta) + \left(\frac{\partial^2 U}{\partial v^2}\right)_{S,\,c}(\Delta v_\alpha - \Delta v_\beta) = 0 \,.$$

At the critical point, the limiting forms of these equations equivalently define the tangent to the binodal locus:

$$\left(\frac{\partial^2 U}{\partial S^2}\right)_{v,\,c} \mathrm{d}S + \left(\frac{\partial^2 U}{\partial S \partial v}\right)_c \mathrm{d}v = 0 \,, \quad \left(\frac{\partial^2 U}{\partial S \partial v}\right)_c \mathrm{d}S + \left(\frac{\partial^2 U}{\partial v^2}\right)_{S,\,c} \mathrm{d}v = 0 \,.$$

$$\frac{\partial^2 U}{\partial S \partial v} = \left(\frac{\partial T}{\partial v}\right)_S = -\left(\frac{\partial S}{\partial v}\right)_T \left(\frac{\partial T}{\partial S}\right)_v = -\frac{T}{C_v}\left(\frac{\partial p}{\partial T}\right)_v ,$$

The conditions of stability above the critical point require that $(\partial^2 p/\partial v\,\partial T)_c < 0$; dividing through we obtain

$$\left(\frac{T}{C_v}\right)_c^2 dS - \left(\frac{T}{C_v}\right)_c^2 \left(\frac{\partial p}{\partial T}\right)_{v,c} dv + \left[\left(\frac{T}{C_v}\right)_c \Big/ \frac{\partial^2 p}{\partial v\,\partial T}\right] \left(\frac{\partial^2 p}{\partial v^2}\right)_{T,c} dv = 0\ . \quad (7.6.3)$$

Comparing Equation (7.6.3) with (7.6.1) and (7.6.2) evidently justifies the assertion of the problem.

It is sometimes argued[5] that the identity of (7.6.1), (7.6.2), (7.6.3) requires that the ratio $(\partial p/\partial v)_{T,c}/(T/C_v)_c = 0$, so that if C_v is infinite at the critical point, the infinity is of a lower order than that in K_T. Certainly, for $(T/C_v)_c$ finite, even if vanishingly small, the first and second isothermal derivatives of p with respect to v must be zero, and criticality is determined by the conditions of mechanical stability, i.e. in the one-phase region near T_c,

$$\frac{T}{C_v} > \frac{T}{C_p} \geqslant 0\,, \quad \frac{1}{vK_S} > \frac{1}{vK_T} \geqslant 0\,,$$

cf.solution to Problem 7.2(d). It is doubtful, however, whether the foregoing analysis, being based on a continuous analytic equation of state [for which $(T/C_v)_c$ is finite, cf Problem 7.2(e)] is applicable to real fluids in which, apparently, $(T/C_v)_c = 0$, because the critical isotherm in such systems is non-analytic at the critical point.

BINARY SYSTEMS

7.7 Diffusional stability of the single phase. In a binary system the local composition is completely specified by a single intensive variable, the mole fraction of component 1,

$$x \equiv \frac{n}{n+m} \equiv \frac{N}{N+M}\,,$$

where n is the number of moles and N the number of molecules of component 1, and m the number of moles and M the number of molecules of component 2, within an arbitrarily small element of volume. We therefore expect stability with respect to diffusion of either or both components in a binary system at uniform temperature and pressure to be ensured by a single condition.

Suppose a two-component, one-phase system to be maintained at constant and uniform temperature and pressure. The appropriate form of the general equilibrium condition of Problem 1.22 is

$$\delta G_{T,p} > 0\ .$$

Consider two regions, α and β, each internally uniform with respect to the chemical potentials μ_1 and μ_2, α containing n_α and m_α moles of the

[5] J.S.Rowlinson, *Liquids and Liquid Mixtures*, 2nd Edn. (Butterworths, London), 1969, p.83.

two components respectively, and β containing n_β and m_β moles, so that the extensive Gibbs free energies are, initially,

$$G^{(0)} = G_\alpha^{(0)} + G_\beta^{(0)} = n_\alpha \mu_{1\alpha} + m_\alpha \mu_{2\alpha} + n_\beta \mu_{1\beta} + m_\beta \mu_{2\beta} .$$

The perturbation to be considered is the passing of δn mole of component 1 from α to β and, simultaneously, the passing of δm mole of component 2 from β to α, without restriction on the ratio $\delta n/\delta m$.

(a) Express δG in terms of δn, δm, and thence show that a necessary condition for equilibrium (terms first order in δn, δm to sum to zero) is

$$\mu_{1\alpha} = \mu_{1\beta} , \quad \mu_{2\alpha} = \mu_{2\beta} . \tag{7.7.1}$$

(b) Use the definition $G \equiv n\mu_1 + m\mu_2$, together with the Gibbs-Duhem equation, cf Problem 1.20(d),

$$S \mathrm{d}T + v \,\mathrm{d}p + \sum_i n_i \,\mathrm{d}\mu_i = 0 ,$$

to prove the results

$$n\left(\frac{\partial \mu_1}{\partial n}\right)_{T,\,p,\,m} + m\left(\frac{\partial \mu_2}{\partial n}\right)_{T,\,p,\,m} = 0 , \tag{7.7.2}$$

$$\left(\frac{\partial \mu_1}{\partial m}\right)_{T,\,p,\,n} - \left(\frac{\partial \mu_2}{\partial n}\right)_{T,\,p,\,m} = 0 , \tag{7.7.3}$$

$$n\left(\frac{\partial \mu_1}{\partial n}\right)_{T,\,p,\,m} + m\left(\frac{\partial \mu_1}{\partial m}\right)_{T,\,p,\,n} = 0 . \tag{7.7.4}$$

(c) Formulate the first and second derivatives of x with respect to n and m, at constant temperature and pressure, and thence prove

$$\left(\frac{\partial \mu_1}{\partial n}\right)_m = \frac{1-x}{n+m}\frac{\partial \mu_1}{\partial x} ; \quad \left(\frac{\partial \mu_2}{\partial n}\right)_m = \frac{1-x}{n+m}\frac{\partial \mu_2}{\partial x} ;$$

$$\left(\frac{\partial \mu_1}{\partial m}\right)_n = -\frac{x}{n+m}\frac{\partial \mu_1}{\partial x} ; \quad \left(\frac{\partial \mu_2}{\partial m}\right)_n = -\frac{x}{n+m}\frac{\partial \mu_2}{\partial x} ;$$

$$\left(\frac{\partial^2 \mu_1}{\partial n^2}\right)_m = \left(\frac{1-x}{n+m}\right)^2 \frac{\partial^2 \mu_1}{\partial x^2} - \frac{2(1-x)}{(n+m)^2}\frac{\partial \mu_1}{\partial x} ;$$

$$\left(\frac{\partial^2 \mu_2}{\partial n^2}\right)_m = \left(\frac{1-x}{n+m}\right)^2 \frac{\partial^2 \mu_2}{\partial x^2} - \frac{2(1-x)}{(n+m)^2}\frac{\partial \mu_2}{\partial x} ;$$

$$\left(\frac{\partial^2 \mu_1}{\partial m^2}\right)_n = \left(\frac{x}{n+m}\right)^2 \frac{\partial^2 \mu_1}{\partial x^2} + \frac{2x}{(n+m)^2}\frac{\partial \mu_1}{\partial x} ;$$

$$\left(\frac{\partial^2 \mu_2}{\partial m^2}\right)_n = \left(\frac{x}{n+m}\right)^2 \frac{\partial^2 \mu_2}{\partial x^2} + \frac{2x}{(n+m)^2}\frac{\partial \mu_2}{\partial x} ;$$

$$\frac{\partial^2 \mu_1}{\partial m \partial n} = -\frac{x(1-x)}{(n+m)^2}\frac{\partial^2 \mu_1}{\partial x^2} + \frac{2x-1}{(n+m)^2}\frac{\partial \mu_1}{\partial x} ;$$

$$\frac{\partial^2 \mu_2}{\partial m \partial n} = -\frac{x(1-x)}{(n+m)^2}\frac{\partial^2 \mu_2}{\partial x^2} + \frac{2x-1}{(n+m)^2}\frac{\partial \mu_2}{\partial x} .$$

Thence establish that μ_1 and μ_2 are completely specified by

$$\mu_1 = \mu_1(T, p, x)\,; \quad \mu_2 = \mu_2(T, p, x)\,. \tag{7.7.5}$$

(d) Expand $\delta\mu_{1\alpha}$, $\delta\mu_{1\beta}$, $\delta\mu_{2\alpha}$, $\delta\mu_{2\beta}$ as Taylor series in powers of δn, δm, and use the results of part (b) to show that the condition (7.1.1) being met ensures that the first-order terms in $\delta G = \delta G(\delta n, \delta m)$ sum to zero.

(e) Use the results of part (c) to convert the second-order terms in $\delta G = \delta G(\delta n, \delta m)$ into derivatives with respect to x; derive a result from Equation (7.7.2) which reduces the second derivatives of μ_i with respect to x to first derivatives; and thence show that for stability (terms second-order in δn, δm, to sum to a result greater than zero)

$$\left(\frac{\partial \mu_1}{\partial x}\right)_{T,\,p} - \left(\frac{\partial \mu_2}{\partial x}\right)_{T,\,p} > 0\,. \tag{7.7.6}$$

(f) Show that condition (7.7.6) is exactly equivalent to

$$\frac{\partial^2 \overline{G}}{\partial x^2} > 0 \tag{7.7.7}$$

and express it also in extensive variables.

Solution

(a) Write down the expression for $G = G_\alpha + G_\beta$ (subsequently to the perturbation) corresponding to the given expression for $G^{(0)}$. By inspection

$$\delta G = (n_\alpha - \delta n)\delta\mu_{1\alpha} + (m_\alpha + \delta m)\delta\mu_{2\alpha} + (n_\beta + \delta n)\delta\mu_{1\beta} + (m_\beta - \delta m)\delta\mu_{2\beta}$$
$$+ (\mu_{1\beta} - \mu_{1\alpha})\delta n - (\mu_{2\beta} - \mu_{2\alpha})\delta m\,. \tag{7.7.8}$$

The last two terms include only the independent variables, so to ensure that the first-order terms sum to zero, these terms must be independently zero. The condition (7.7.1) follows. Thus, in addition to being uniform with respect to temperature and pressure, the system has to be uniform with respect to each of the chemical potentials. Note that condition (7.7.1) must apply also in the case where α and β are co-existing phases.

(b) For a binary system at constant temperature and pressure the Gibbs-Duhem equation reduces to

$$n\,\mathrm{d}\mu_1 + m\,\mathrm{d}\mu_2 = 0\,.$$

Put

$$\mathrm{d}\mu_1 = \left(\frac{\partial \mu_1}{\partial n}\right)_m \mathrm{d}n + \left(\frac{\partial \mu_1}{\partial m}\right)_n \mathrm{d}m\,,$$

$$\mathrm{d}\mu_2 = \left(\frac{\partial \mu_2}{\partial n}\right)_m \mathrm{d}n + \left(\frac{\partial \mu_2}{\partial m}\right)_n \mathrm{d}m\,;$$

then

$$\left[n\left(\frac{\partial \mu_1}{\partial n}\right)_m + m\left(\frac{\partial \mu_2}{\partial n}\right)_m\right]\mathrm{d}n + \left[n\left(\frac{\partial \mu_1}{\partial m}\right)_n + m\left(\frac{\partial \mu_2}{\partial m}\right)_n\right]\mathrm{d}m = 0\,.$$

$$\frac{\partial^2\mu_1}{\partial n\partial m} = \left[\frac{\partial}{\partial n}\left(\frac{\partial\mu_1}{\partial m}\right)_n\right]_m = \left\{\frac{\partial}{\partial n}\left[\frac{\partial\mu_1}{\partial x}\frac{-n}{(n+m)^2}\right]\right\}_m$$
$$= -\frac{n}{(n+m)^2}\left[\frac{\partial}{\partial n}\left(\frac{\partial\mu_1}{\partial x}\right)\right]_m + \frac{\partial\mu_1}{\partial x}\left[-\frac{1}{(n+m)^2}+\frac{2n}{(n+m)^3}\right]$$
$$= -\frac{x(1-x)}{(n+m)^2}\frac{\partial^2\mu_1}{\partial x^2} + \frac{2x-1}{(n+m)^2}\frac{\partial\mu_1}{\partial x};$$

etc.; and similarly for μ_2. To establish the functions (7.7.5), put

$$d\mu_1 = \left(\frac{\partial\mu_1}{\partial n}\right)_m dn + \left(\frac{\partial\mu_1}{\partial m}\right)_n dm = \frac{\partial\mu_1}{\partial x}\left[\left(\frac{\partial x}{\partial n}\right)_m dn + \left(\frac{\partial x}{\partial m}\right)_n dm\right]$$
$$= \frac{\partial\mu_1}{\partial x}dx\ .$$

This is a tautology; it reflects the possibility of segregating variables, as

$$(n+m)\left(\frac{\partial\mu_1}{\partial n}\right)_m = (1-x)\frac{\partial\mu_1}{\partial x}\ ,$$
$$(n+m)\left(\frac{\partial\mu_1}{\partial m}\right)_n = -x\frac{\partial\mu_1}{\partial x}\ ,$$

etc. Note that, consequently, condition (7.7.1) implies $x_\alpha = x_\beta$.

(d) Substituting the Taylor series expansions for the $\delta\mu$ into the residue of expression (7.7.8) (less the last two terms) gives

$$\delta G = (n_\alpha - \delta n)\left[-\left(\frac{\partial\mu_{1\alpha}}{\partial n}\right)_m \delta n + \left(\frac{\partial\mu_{1\alpha}}{\partial m}\right)_n \delta m\right] + \cdots$$
$$+ (m_\alpha + \delta m)\left[-\left(\frac{\partial\mu_{2\alpha}}{\partial n}\right)_m \delta n + \left(\frac{\partial\mu_{2\alpha}}{\partial m}\right)_n \delta m\right] + \cdots$$
$$+ (n_\beta + \delta n)\left[\left(\frac{\partial\mu_{1\beta}}{\partial n}\right)_m \delta n - \left(\frac{\partial\mu_{1\beta}}{\partial m}\right)_n \delta m\right] + \cdots$$
$$+ (m_\beta - \delta m)\left[\left(\frac{\partial\mu_{2\beta}}{\partial n}\right)_m \delta n - \left(\frac{\partial\mu_{2\beta}}{\partial m}\right)_n \delta m\right] + \cdots; \quad (7.7.9)$$

the terms which are actually first order are

$$\delta G =$$
$$-\left[n_\alpha\left(\frac{\partial\mu_{1\alpha}}{\partial n}\right)_m + m_\alpha\left(\frac{\partial\mu_{2\alpha}}{\partial n}\right)_m\right]\delta n + \left[n_\beta\left(\frac{\partial\mu_{1\beta}}{\partial n}\right)_m + m_\beta\left(\frac{\partial\mu_{2\beta}}{\partial n}\right)_m\right]\delta n$$
$$+\left[n_\alpha\left(\frac{\partial\mu_{1\alpha}}{\partial m}\right)_n + m_\alpha\left(\frac{\partial\mu_{2\alpha}}{\partial m}\right)_n\right]\delta m - \left[n_\beta\left(\frac{\partial\mu_{1\beta}}{\partial m}\right)_n + m_\beta\left(\frac{\partial\mu_{2\beta}}{\partial m}\right)_n\right]\delta m\ ,$$

and Equation (7.7.2) shows that all four brackets are zero.

so the equations to the tangent become

$$\left(\frac{T}{C_v}\right)_c dS - \left(\frac{T}{C_v}\right)_c \left(\frac{\partial p}{\partial T}\right)_{v,c} dv = 0\,, \qquad (7.6.1)$$

$$\left(\frac{T}{C_v}\right)_c \left(\frac{\partial p}{\partial T}\right)_{v,c} dS + \left(\frac{\partial p}{\partial v}\right)_{S,c} dv = 0\,,$$

or, changing constraint on the last term, and using

$$\left(\frac{\partial T}{\partial v}\right)_S \left(\frac{\partial v}{\partial S}\right)_T \left(\frac{\partial S}{\partial T}\right)_v = -1\,,$$

$$\left(\frac{T}{C_v}\right)_c \left(\frac{\partial p}{\partial T}\right)_{v,c} dS + \left(\frac{\partial p}{\partial v}\right)_{T,c} dv - \left(\frac{T}{C_v}\right)_c \left(\frac{\partial p}{\partial T}\right)^2_{v,c} dv = 0\,.$$

The result of Problem 7.5(c) establishes $(\partial p/\partial T)_{v,c} > 0$; dividing through gives

$$\left(\frac{T}{C_v}\right)_c dS - \left(\frac{T}{C_v}\right)_c \left(\frac{\partial p}{\partial T}\right)_{v,c} dv + \left[\left(\frac{\partial p}{\partial v}\right)_{T,c} \Big/ \left(\frac{\partial p}{\partial T}\right)_{v,c}\right] dv = 0\,. \quad (7.6.2)$$

If Equations (7.6.1) and (7.6.2) are to describe the same line, then $(\partial p/\partial v)_{T,c}$ must be zero irrespective of $(T/C_v)_c$.

(b) From Problem 7.1(c),

$$D = -\left(\frac{\partial T}{\partial S}\right)_v \left(\frac{\partial p}{\partial v}\right)_T\,,$$

$$\left(\frac{\partial D}{\partial S}\right)_v = -\left(\frac{\partial T}{\partial S}\right)_v \left[\frac{\partial}{\partial S}\left(\frac{\partial p}{\partial v}\right)_T\right]_v - \left(\frac{\partial^2 T}{\partial S^2}\right)_v \left(\frac{\partial p}{\partial v}\right)_T\,,$$

and since $(\partial p/\partial v)_T = 0$, changing the differentiating variable from S to T, we obtain

$$\left(\frac{\partial D}{\partial S}\right)_v = -\left(\frac{\partial T}{\partial S}\right)^2_v \frac{\partial^2 p}{\partial v\,\partial T}\,.$$

$$\left(\frac{\partial D}{\partial v}\right)_S = -\left(\frac{\partial T}{\partial S}\right)_v \left[\frac{\partial}{\partial v}\left(\frac{\partial p}{\partial v}\right)_T\right]_S - \frac{\partial^2 T}{\partial S\,\partial v}\left(\frac{\partial p}{\partial v}\right)_T\,,$$

$$= -\left(\frac{\partial T}{\partial S}\right)_v \left[\left(\frac{\partial^2 p}{\partial v^2}\right)_T + \frac{\partial^2 p}{\partial v\,\partial T}\left(\frac{\partial T}{\partial v}\right)_S\right],$$

$$= -\left(\frac{\partial T}{\partial S}\right)_v \left(\frac{\partial^2 p}{\partial v^2}\right)_T + \frac{\partial^2 p}{\partial v\,\partial T}\left(\frac{\partial S}{\partial v}\right)_T \left(\frac{\partial T}{\partial S}\right)^2_v\,.$$

Thus the equation to the tangent to the spinodal locus is

$$\left(\frac{T}{C_v}\right)^2_c \left(\frac{\partial^2 p}{\partial v\,\partial T}\right)_c dS + \left(\frac{T}{C_v}\right)_c \left(\frac{\partial^2 p}{\partial v^2}\right)_{T,c} dv - \left(\frac{T}{C_v}\right)^2_c \left(\frac{\partial p}{\partial T}\right)_{v,c} \left(\frac{\partial^2 p}{\partial v\,\partial T}\right)_c dv = 0\,.$$

Since this is to be true for all dn, dm, the two terms must be independently zero.

Differentiating $G = n\mu_1 + m\mu_2$, we obtain

$$\left(\frac{\partial G}{\partial n}\right)_m = \mu_1 + n\left(\frac{\partial \mu_1}{\partial n}\right)_m + m\left(\frac{\partial \mu_2}{\partial n}\right)_m = \mu_1 ,$$

and

$$\frac{\partial^2 G}{\partial m \partial n} = \left(\frac{\partial \mu_1}{\partial m}\right)_n .$$

Similarly,

$$\frac{\partial^2 G}{\partial n \partial m} = \left(\frac{\partial \mu_2}{\partial n}\right)_m ,$$

whence Equation (7.7.3) follows, and Equations (7.7.2) and (7.7.3) give Equation (7.7.4).

(c) The first derivatives of $x \equiv n/(m+n)$ are:

$$\left(\frac{\partial x}{\partial n}\right)_m = \frac{1}{n+m} - \frac{n}{(n+m)^2} = \frac{m}{(n+m)^2} = \frac{1-x}{n+m} ,$$

$$\left(\frac{\partial x}{\partial m}\right)_n = -\frac{n}{(n+m)^2} = -\frac{x}{n+m} .$$

The second derivatives are

$$\left(\frac{\partial^2 x}{\partial n^2}\right)_m = \left\{\frac{\partial[m/(n+m)^2]}{\partial n}\right\}_m = -\frac{2m}{(n+m)^3} = -\frac{2(1-x)}{(n+m)^2}$$

$$\frac{\partial^2 x}{\partial m \partial n} = \left\{\frac{\partial[m/(n+m)^2]}{\partial m}\right\}_n = \frac{1}{(n+m)^2} - \frac{2m}{(n+m)^3} = \frac{2x-1}{(n+m)^2} ,$$

$$\left(\frac{\partial^2 x}{\partial m^2}\right)_n = \left\{\frac{\partial[-n/(n+m)^2]}{\partial m}\right\}_n = \frac{2n}{(n+m)^3} = \frac{2x}{(n+m)^2} .$$

The first four results quoted follow at once. The simplest method for the other six is as

$$\begin{aligned}
\left(\frac{\partial^2 \mu_1}{\partial n^2}\right) &= \left[\frac{\partial}{\partial n}\left(\frac{\partial \mu_1}{\partial n}\right)_m\right]_m \\
&= \left\{\frac{\partial}{\partial n}\left[\left(\frac{\partial \mu_1}{\partial x}\right)_m \frac{m}{(n+m)^2}\right]\right\}_m \\
&= \frac{m}{(n+m)^2}\left[\frac{\partial}{\partial n}\left(\frac{\partial \mu_1}{\partial x}\right)\right]_m + \frac{\partial \mu_1}{\partial x}\left\{\frac{\partial}{\partial n}\left[\frac{m}{(n+m)^2}\right]\right\}_m \\
&= \left[\frac{m}{(n+m)^2}\right]^2 \frac{\partial^2 \mu_1}{\partial x^2} - \frac{2m}{(n+m)^3}\frac{\partial \mu_1}{\partial x} \\
&= \frac{(1-x)^2}{(n+m)^2}\frac{\partial^2 \mu_1}{\partial x^2} - \frac{2(1-x)}{(n+m)^2}\frac{\partial \mu_1}{\partial x} ;
\end{aligned}$$

(e) The second-order terms from the expression (7.7.9) are

$$\left[\left(\frac{\partial\mu_{1\alpha}}{\partial n}\right)_m+\left(\frac{\partial\mu_{1\beta}}{\partial n}\right)_m\right](\delta n)^2-\left[\left(\frac{\partial\mu_{1\alpha}}{\partial m}\right)_n+\left(\frac{\partial\mu_{1\beta}}{\partial m}\right)_n\right]\delta n\delta m$$
$$-\left[\left(\frac{\partial\mu_{2\alpha}}{\partial n}\right)_m+\left(\frac{\partial\mu_{2\beta}}{\partial n}\right)_m\right]\delta n\delta m+\left[\left(\frac{\partial\mu_{2\alpha}}{\partial m}\right)_n+\left(\frac{\partial\mu_{2\beta}}{\partial m}\right)_n\right](\delta m)^2 . \quad (7.7.10)$$

On changing variables to x, note that if T, p are uniform, and if $\mu_{1\alpha}=\mu_{1\beta}$, $\mu_{2\alpha}=\mu_{2\beta}$, it follows, since $\mu_1=\mu_1(T,p,x)$, $\mu_2=\mu_2(T,p,x)$, that $x_\alpha = x_\beta$ and thence that

$$\frac{\partial\mu_{1\alpha}}{\partial x}=\frac{\partial\mu_{1\beta}}{\partial x}, \quad \frac{\partial\mu_{2\alpha}}{\partial x}=\frac{\partial\mu_{2\beta}}{\partial x} .$$

The terms of (7.7.10) thus reduce to

$$\left(\frac{1}{n_\alpha+m_\alpha}+\frac{1}{n_\beta+m_\beta}\right)\left\{\frac{\partial\mu_1}{\partial x}[(1-x)(\delta n)^2+x\delta n\delta m]\right.$$
$$\left.-\frac{\partial\mu_2}{\partial x}[(1-x)\delta n\delta m+x(\delta m)^2]\right\} .$$

Of the second-order terms in the Taylor series for the $\delta\mu$, the following contribute second-order terms to δG:

$$n_\alpha\left[\frac{1}{2}\left(\frac{\partial^2\mu_{1\alpha}}{\partial n^2}\right)_m(\delta n)^2-\frac{\partial^2\mu_{1\alpha}}{\partial n\partial m}\delta n\delta m+\frac{1}{2}\left(\frac{\partial^2\mu_{1\alpha}}{\partial m^2}\right)_n(\delta m)^2\right]$$
$$+n_\beta\left[\frac{1}{2}\left(\frac{\partial^2\mu_{1\beta}}{\partial n^2}\right)_m(\delta n)^2-\frac{\partial^2\mu_{1\beta}}{\partial n\partial m}\delta n\delta m+\frac{1}{2}\left(\frac{\partial^2\mu_{1\beta}}{\partial m^2}\right)_n(\delta m)^2\right]$$
$$+m_\alpha\left[\frac{1}{2}\left(\frac{\partial^2\mu_{2\alpha}}{\partial n^2}\right)_m(\delta n)^2-\frac{\partial^2\mu_{2\alpha}}{\partial n\partial m}\delta n\delta m+\frac{1}{2}\left(\frac{\partial^2\mu_{2\alpha}}{\partial m^2}\right)_n(\delta m)^2\right]$$
$$+m_\beta\left[\frac{1}{2}\left(\frac{\partial^2\mu_{2\beta}}{\partial n^2}\right)_m(\delta n)^2-\frac{\partial^2\mu_{2\beta}}{\partial n\partial m}\delta n\delta m+\frac{1}{2}\left(\frac{\partial^2\mu_{2\beta}}{\partial m^2}\right)_n(\delta m)^2\right] . \quad (7.7.11)$$

On changing the differentiating variables to x and collecting terms in pairs this becomes, when the terms from (7.7.10) are added in,

$$\delta G=\left\{\left[x\frac{\partial^2\mu_1}{\partial x^2}+(1-x)\frac{\partial^2\mu_2}{\partial x^2}\right][\tfrac{1}{2}(1-x)^2(\delta n)^2+x(1-x)\delta n\delta m\right.$$
$$+\tfrac{1}{2}x^2(\delta m)^2]+\left(\frac{\partial\mu_1}{\partial x}-\frac{\partial\mu_2}{\partial x}\right)[(1-x)^2(\delta n)^2+2x(1-x)\delta n\delta m$$
$$\left.+x^2(\delta m)^2]\right\}\left(\frac{1}{n_\alpha+m_\alpha}+\frac{1}{n_\beta+m_\beta}\right) . \quad (7.7.12)$$

From Equation (7.7.2) we obtain

$$x\frac{\partial\mu_1}{\partial x}+(1-x)\frac{\partial\mu_2}{\partial x}=0\,,$$

and thence by differentiating

$$\frac{\partial\mu_1}{\partial x}+x\frac{\partial^2\mu_1}{\partial x^2}-\frac{\partial\mu_2}{\partial x}+(1-x)\frac{\partial^2\mu_2}{\partial x^2}=0\,;$$

this gives

$$x\frac{\partial^2\mu_1}{\partial x^2}+(1-x)\frac{\partial^2\mu_2}{\partial x^2}=-\left(\frac{\partial\mu_1}{\partial x}-\frac{\partial\mu_2}{\partial x}\right).$$

Thus expression (7.7.12) becomes

$$\begin{aligned}\delta G=&-\frac{1}{2}\left(\frac{\partial\mu_1}{\partial x}-\frac{\partial\mu_2}{\partial x}\right)[(1-x)\delta n+x\delta m]^2\left(\frac{1}{n_\alpha+m_\alpha}+\frac{1}{n_\beta+m_\beta}\right)\\&+\left(\frac{\partial\mu_1}{\partial x}-\frac{\partial\mu_2}{\partial x}\right)[(1-x)\delta n+x\delta m]^2\left(\frac{1}{n_\alpha+m_\alpha}+\frac{1}{n_\beta+m_\beta}\right),\end{aligned}$$

and since n_α, m_α, n_β, m_β cannot be negative, the stability condition is the inequality (7.7.6).

(f) From

$$G=n_1\mu_1+n_2\mu_2$$

we have

$$\overline{G}=x\mu_1+(1-x)\mu_2\,,$$

and hence

$$\begin{aligned}\left(\frac{\partial\overline{G}}{\partial x}\right)_{T,\,p}&=x\left(\frac{\partial\mu_1}{\partial x}\right)_{T,\,p}+\mu_1+(1-x)\left(\frac{\partial\mu_2}{\partial x}\right)_{T,\,p}-\mu_2\,,\\&=\mu_1-\mu_2\quad\text{by Equation (7.7.2)}\,,\\\left(\frac{\partial^2\overline{G}}{\partial x^2}\right)_{T,\,p}&=\left(\frac{\partial\mu_1}{\partial x}\right)_{T,\,p}-\left(\frac{\partial\mu_2}{\partial x}\right)_{T,\,p}.\end{aligned}$$

It follows that

$$(1-x)\frac{\partial^2\overline{G}}{\partial x^2}=\frac{\partial\mu_1}{\partial x}\,,\quad x\frac{\partial^2\overline{G}}{\partial x^2}=-\frac{\partial\mu_2}{\partial x}\,,\tag{7.7.13}$$

so that the stability condition may be expressed equivalently as

$$\left(\frac{\partial\mu_1}{\partial x}\right)_{T,\,p}>0\quad\text{or}\quad\left(\frac{\partial\mu_2}{\partial x}\right)_{T,\,p}<0\,.$$

By the results of part (c) these are in turn equivalent to

$$\left(\frac{\partial\mu_1}{\partial n}\right)_m>0\,,\ \text{or}\ \left(\frac{\partial\mu_1}{\partial m}\right)_n<0\,,\ \text{or}\ \left(\frac{\partial\mu_2}{\partial n}\right)_m<0\ \text{or}\ \left(\frac{\partial\mu_2}{\partial m}\right)_n>0\,.$$

Thus, in extensive variables, they become

$$\left(\frac{\partial^2 G}{\partial n^2}\right)_m > 0\,, \quad \text{or} \quad \left(\frac{\partial^2 G}{\partial m^2}\right)_n > 0\,, \quad \text{or} \quad \frac{\partial^2 G}{\partial m \partial n} < 0\,.$$

These equivalent conditions correspond to the fact, cf Equation (7.3.6), that

$$\begin{vmatrix} G_{2n} & G_{mn} \\ G_{nm} & G_{2m} \end{vmatrix} = 0\,,$$

so that, in respect of two regions, [cf solution to Problem 7.3(c)] stability is ensured by either $G_{2n} > 0$ or $G_{2m} > 0$, for each region. Note that for a homogeneous region, adding material at the same T, p, and x is a reversible process.

Note that proceeding by analogy to Problem 7.2(a), i.e. using $\overline{G} = \overline{G}(T, p, x)$, and the perturbation $\delta n_\alpha, -\delta m_\alpha$, leads directly to

$$\tfrac{1}{2}[(n_\alpha + m_\alpha)(\delta x_\alpha)^2 + (n_\beta + m_\beta)(\delta x_\beta)^2]\left(\frac{\partial^2 \overline{G}}{\partial x^2}\right)_{T,\,p} > 0\,.$$

A third procedure is to truncate the expansions for $\mu_{1\alpha}$, $\mu_{1\beta}$, etc. at the first derivative. This gives, for the second order terms, (7.7.10) rather than the full list (7.7.12). Separating the terms (7.7.10) into two inequalities, one for each region, gives simple quadratic forms for both. The corresponding determinants are zero by Equation (7.7.4), and, by an argument similar to that of the solution to Problem 7.3(c), the stability condition is

$$\left(\frac{\partial \mu_1}{\partial n}\right)_m, \quad \left(\frac{\partial \mu_2}{\partial m}\right)_n > 0$$

for each region, i.e.

$$\frac{\partial \mu_1}{\partial x}\,, \quad -\frac{\partial \mu_2}{\partial x} > 0$$

for the system.

7.8 Criticality, continuous equation of state. The region of two-phase co-existence in a binary system is frequently bounded in temperature, either by an upper critical solution temperature (UCST), as with liquid-vapour co-existence, or by a lower critical solution temperature (LCST), as when a one-phase binary system separates into two liquid phases with increasing temperature. In either case, if $\overline{G} = \overline{G}(p, T, x)$ is continuous, there is necessarily a region of instability bounded by a spinodal locus, $(\partial^2 \overline{G}/\partial x^2)_{T,\,p} = 0$, analogous to the locus $(\partial^2 \overline{F}/\partial \bar{v}^2) = 0$ of Problem 7.5; the corresponding binodal locus is defined by the co-existence conditions, cf conditions (7.7.1),

$$T_\alpha = T_\beta, \quad p_\alpha = p_\beta, \quad \mu_{1\alpha} = \mu_{1\beta}, \quad \mu_{2\alpha} = \mu_{2\beta}\,. \tag{7.8.1}$$

(a) Consider two co-existing phases, α and β. Use the conditions (7.8.1), together with the relations derived in the solution to Problem 7.7(f) to show that the tie-line connecting $\overline{G}_\alpha(T, p, x_\alpha)$ to $\overline{G}_\beta(T, p, x_\beta)$ has the equation

$$\frac{\overline{G}_\beta - \overline{G}_\alpha}{x_\beta - x_\alpha} = \left(\frac{\partial \overline{G}}{\partial x}\right)_{\alpha,\beta};$$

and that the binodal locus spans a larger range of x, $(x_\beta - x_\alpha)$ than does the spinodal locus.

(b) Thence show that, if the tie-line becomes vanishingly short as $T \to T_c$, then at $T = T_c$,

$$\left(\frac{\partial^2 \overline{G}}{\partial x^2}\right)_{T,p} = 0\,,\quad \left(\frac{\partial^3 \overline{G}}{\partial x^3}\right)_{T,p} = 0\,,\quad \left(\frac{\partial^4 \overline{G}}{\partial x^4}\right)_{T,p} > 0\,,$$

and that the binodal locus is tangential to the spinodal locus.

(c) Use a Taylor series expansion for $\overline{G}$ about the critical point, at constant pressure (omitting powers of ΔT higher than the first), with the results of part (a) to show that

$$\Delta x_\alpha = -\Delta x_\beta\,, \tag{7.8.2}$$

$$\left(\frac{\partial^4 \overline{G}}{\partial x^4}\right)_{T,p,c} (\Delta x)^2 = -6\left(\frac{\partial^3 \overline{G}}{\partial^2 x \partial T}\right)_{p,c} \Delta T\,, \tag{7.8.3}$$

i.e. that Δx is quadratic in $(T - T_c)$ near the critical point.

Solution

(a) The solution to Problem 7.7(f) gives [leaving off the bar, since all extensive variables are to be read as molar (intensive)],

$$\mu_{1\alpha} = \mu_{2\alpha} + \frac{\partial G_\alpha}{\partial x_\alpha}$$

whence

$$(1 - x_\alpha)\mu_{1\alpha} = (1 - x_\alpha)\mu_{2\alpha} + (1 - x_\alpha)\frac{\partial G_\alpha}{\partial x_\alpha}\,,$$

$$\mu_{1\alpha} = x_\alpha \mu_{1\alpha} + (1 - x_\alpha)\mu_{2\alpha} + (1 - x_\alpha)\frac{\partial G_\alpha}{\partial x_\alpha}\,,$$

$$\mu_{1\alpha} = G_\alpha + (1 - x_\alpha)\frac{\partial G_\alpha}{\partial x_\alpha}\,.$$

Thus, from the co-existence condition $\mu_{1\alpha} = \mu_{1\beta}$,

$$G_\alpha + (1 - x_\alpha)\frac{\partial G_\alpha}{\partial x_\alpha} = G_\beta + (1 - x_\beta)\frac{\partial G_\beta}{\partial x_\beta}\,. \tag{7.8.4}$$

Similarly

$$\mu_{2\alpha} = G_\alpha - x_\alpha \frac{\partial G_\alpha}{\partial x_\alpha} = G_\beta - x_\beta \frac{\partial G_\beta}{\partial x_\beta}\,. \tag{7.8.5}$$

Write Equation (7.8.4) as

$$G_\alpha - G_\beta = \frac{\partial G_\beta}{\partial x_\beta} - \frac{\partial G_\alpha}{\partial x_\alpha} - x_\beta\frac{\partial G_\beta}{\partial x_\beta} + x_\alpha\frac{\partial G_\alpha}{\partial x_\alpha} .$$

On comparing this with Equation (7.8.5) we obtain

$$\frac{\partial G_\beta}{\partial x_\beta} = \frac{\partial G_\alpha}{\partial x_\alpha} , \tag{7.8.6}$$

and the equation to the tie-line is

$$G_\alpha - G_\beta = (x_\alpha - x_\beta)\left(\frac{\partial G}{\partial x}\right)_{\alpha,\,\beta} . \tag{7.8.7}$$

Thus the tie-line is the common tangent at the points x_α, x_β to the curve $G = G(x)$, for constant temperature and pressure. Since, for stability, $(\partial^2 G/\partial x)^2$ is positive at x_α, x_β, and the continuity of $G = G(x)$ requires a range of x for which $(\partial^2 G/\partial x^2)$ is negative, this range must be between x_α and x_β, and be smaller than $x_\beta - x_\alpha$.

(b) The argument is exactly parallel to that of Problem 7.5(a), except for the last part of the latter; there is no restriction on the sign of $(\partial G/\partial x)_c = (\mu_1 - \mu_2)_c$.

(c) The co-existence conditions should be used in terms of G and its derivatives; Equations (7.8.6) and (7.8.5) or (7.8.7) are suitable forms. On using Taylor series expansions to fourth order for $(\partial G/\partial x)$, and recalling [cf Problem 7.5(c)] that $\Delta x_\alpha \neq \Delta x_\beta$, and that $(\partial^2 G/\partial x^2)_c$, $(\partial^3 G/\partial x^3)_c = 0$, we obtain from Equation (7.8.6), on dividing through by $(\Delta x_\alpha - \Delta x_\beta)$

$$\left(\frac{\partial^3 G}{\partial x^2 \partial T}\right)_c \Delta T + \frac{1}{2!}\left(\frac{\partial^4 G}{\partial x^3 \partial T}\right)_c \Delta T(\Delta x_\alpha + \Delta x_\beta)$$
$$+\frac{1}{3!}\left(\frac{\partial^4 G}{\partial x^4}\right)_c [(\Delta x_\alpha)^2 + \Delta x_\alpha \Delta x_\beta + (\Delta x_\beta)^2] = 0 . \tag{7.8.8}$$

The forms obtained using Equation (7.8.5) are more symmetrical than those obtained from Equation (7.8.7). The relation may be set up as follows

$$G_\alpha - G_\beta = \left(\frac{\partial G}{\partial x}\right)_c (\Delta x_\alpha - \Delta x_\beta) + \left(\frac{\partial^2 G}{\partial x \partial T}\right)_c \Delta T(\Delta x_\alpha - \Delta x_\beta)$$
$$+\frac{1}{2!}\left(\frac{\partial^3 G}{\partial x^2 \partial T}\right)_c \Delta T[(\Delta x_\alpha)^2 - (\Delta x_\beta)^2]$$
$$+\frac{1}{3!}\left(\frac{\partial^4 G}{\partial x^3 \partial T}\right)_c \Delta T[(\Delta x_\alpha)^3 - (\Delta x_\beta)^3]$$
$$+\frac{1}{4!}\left(\frac{\partial^4 G}{\partial x^4}\right)_c [(\Delta x_\alpha)^4 - (\Delta x_\beta)^4] + \cdots ,$$

and

$$-x_\alpha \frac{\partial G_\alpha}{\partial x_\alpha} = -\left(\frac{\partial G}{\partial x}\right)_c x_\alpha - \left(\frac{\partial^2 G}{\partial x \partial T}\right)_c \Delta T x_\alpha - \left(\frac{\partial^3 G}{\partial x^2 \partial T}\right)_c \Delta T x_\alpha \Delta x_\alpha$$
$$-\frac{1}{2!}\left(\frac{\partial^4 G}{\partial x^3 \partial T}\right)_c \Delta T x_\alpha (\Delta x_\alpha)^2 - \frac{1}{3!}\left(\frac{\partial^4 G}{\partial x^4}\right)_c x_\alpha (\Delta x_\alpha)^3 + \cdots,$$

$$+x_\beta \frac{\partial G_\beta}{\partial x_\beta} = +\left(\frac{\partial G}{\partial x}\right)_c x_\beta + \left(\frac{\partial^2 G}{\partial x \partial T}\right)_c \Delta T x_\beta + \cdots$$

and, on addition, Equation (7.8.5) becomes

$$\frac{1}{2!}\left(\frac{\partial^3 G}{\partial x^2 \partial T}\right)_c \Delta T[(\Delta x_\alpha)^2 - (\Delta x_\beta)^2 - 2x_\alpha \Delta x_\alpha + 2x_\beta \Delta x_\beta]$$
$$+\frac{1}{3!}\left(\frac{\partial^4 G}{\partial x^3 \partial T}\right)_c \Delta T[(\Delta x_\alpha)^3 - (\Delta x_\beta)^3 - 3x_\alpha (\Delta x_\alpha)^2 + 3x_\beta (\Delta x_\beta)^2]$$
$$+\frac{1}{4!}\left(\frac{\partial^4 G}{\partial x^4}\right)_c [(\Delta x_\alpha)^4 - (\Delta x_\beta)^4 - 4x_\alpha (\Delta x_\alpha)^3 + 4x_\beta (\Delta x_\beta)^3] = 0. \quad (7.8.9)$$

Substituting for $(\partial^3 G/\partial x^2 \partial T)_c$ from Equation (7.8.8) gives finally

$$2\left(\frac{\partial^4 G}{\partial x^3 \partial T}\right)_c \Delta T[(\Delta x_\alpha)^3 + (\Delta x_\beta)^3]$$
$$+\left(\frac{\partial^4 G}{\partial x^4}\right)_c [(\Delta x_\alpha)^2 - (\Delta x_\beta)^2][\Delta x_\alpha - \Delta x_\beta]^2 = 0$$

or

$$2\left(\frac{\partial^4 G}{\partial x^3 \partial T}\right)_c \Delta T[\Delta x_\alpha + \Delta x_\beta][(\Delta x_\alpha)^2 - (\Delta x_\alpha)(\Delta x_\beta) + (\Delta x_\beta)^2]$$
$$+\left(\frac{\partial^4 G}{\partial x^4}\right)_c [\Delta x_\alpha + \Delta x_\beta][\Delta x_\alpha - \Delta x_\beta]^3 = 0,$$

and relation (7.8.2) follows unambiguously. Substitution from Equation (7.8.2) in Equation (7.8.8) gives the result (7.8.3). The analogy to Problem 7.5(b) and (c) is obvious. Note that, extending Equation (7.8.8) we have

$$\left(\frac{\partial^3 G}{\partial x^2 \partial T}\right)_c \Delta T + \frac{1}{3!}\left(\frac{\partial^4 G}{\partial x^4}\right)_c (\Delta x)^2 + \frac{1}{3!}\left(\frac{\partial^5 G}{\partial x^4 \partial T}\right)_c \Delta T(\Delta x)^2$$
$$+\frac{1}{5!}\left(\frac{\partial^6 G}{\partial x^6}\right)_c (\Delta x)^4 = 0.$$

7.9 Stability at a critical point. To elucidate the relevance of the condition of mechanical stability at a critical point in a binary system evidently requires the $\bar{F} = \bar{F}(\bar{v}, x)$ surface. In terms of this surface, the phase co-existence conditions include, in addition to conditions (7.7.1), the condition $p_\alpha = p_\beta$.

(a) Prove that

$$\begin{vmatrix} \overline{F}_{2\bar{v}} & \overline{F}_{\bar{v}x} \\ \overline{F}_{x\bar{v}} & \overline{F}_{2x} \end{vmatrix} = \overline{F}_{2\bar{v}}\overline{G}_{2x},$$

where

$$\overline{F}_{2\bar{v}} \equiv \left(\frac{\partial^2 \overline{F}}{\partial \bar{v}^2}\right)_{T,\,x},$$

$$\overline{F}_{2x} \equiv \left(\frac{\partial^2 \overline{F}}{\partial x^2}\right)_{T,\,\bar{v}}, \text{ etc.,}$$

and thence deduce the stability conditions for a phase at constant temperature and volume.

(b) Proceed by analogy to Problem 7.6 to show that for the alternative equations for the $(\bar{v}, x)$ projection of the tangent to the binodal locus at the critical point to be equivalent requires that $\overline{G}_{2x}$ be zero whether or not $\overline{F}_{2\bar{v}}$ is zero.

(c) Differentiate the determinant of part (a) to obtain the equation for the tangent to the spinodal locus at the critical point, and use the relations obtained in the solution to part (b) to reduce it, for $(\partial p/\partial \bar{v})_c \neq 0$, to

$$\overline{F}_{3x} - 3\overline{F}_{2x\bar{v}}\frac{\overline{F}_{x\bar{v}}}{\overline{F}_{2\bar{v}}} + 3\overline{F}_{x2\bar{v}}\left(\frac{\overline{F}_{x\bar{v}}}{\overline{F}_{2\bar{v}}}\right)^2 - \overline{F}_{3\bar{v}}\left(\frac{\overline{F}_{x\bar{v}}}{\overline{F}_{2\bar{v}}}\right)^3 = 0\,.$$

(d) Prove that the result of part (c) is simply

$$\left(\frac{\partial^3 \overline{G}}{\partial x^3}\right)_{T,\,p,\,c} = 0\,.$$

Solution

(a) The change of constraint formula gives, in molar quantities,

$$\begin{aligned}\left(\frac{\partial G}{\partial x}\right)_{T,\,v} &= \left(\frac{\partial G}{\partial x}\right)_{T,\,p} + \left(\frac{\partial p}{\partial x}\right)_{T,\,v}\left(\frac{\partial G}{\partial p}\right)_{T,\,x} \\ &= \left(\frac{\partial G}{\partial x}\right)_{T,\,p} + v\left(\frac{\partial p}{\partial x}\right)_{T,\,v},\end{aligned}$$

and from $F \equiv G - pv$,

$$\left(\frac{\partial F}{\partial x}\right)_{T,\,v} = \left(\frac{\partial G}{\partial x}\right)_{T,\,v} - v\left(\frac{\partial p}{\partial x}\right)_{T,\,v} = \left(\frac{\partial G}{\partial x}\right)_{T,\,p}.$$

Hence

$$\left(\frac{\partial^2 G}{\partial x^2}\right)_{T,\,p} = \left(\frac{\partial^2 F}{\partial x^2}\right)_{T,\,v} + \left(\frac{\partial^2 F}{\partial v\,\partial x}\right)_T\left(\frac{\partial v}{\partial x}\right)_{T,\,p};$$

$$\begin{aligned}\left(\frac{\partial v}{\partial x}\right)_{T,\,p} &= \left(\frac{\partial^2 G}{\partial x\,\partial p}\right)_T = \left[\frac{\partial}{\partial p}\left(\frac{\partial F}{\partial x}\right)_{T,\,v}\right]_{T,\,x} \\ &= \left[\frac{\partial}{\partial v}\left(\frac{\partial F}{\partial x}\right)_{T,\,v}\right]_{T,\,x}\left(\frac{\partial v}{\partial p}\right)_{T,\,x} = -\left(\frac{\partial^2 F}{\partial v\,\partial x}\right)_T \bigg/ \left(\frac{\partial^2 F}{\partial v^2}\right)_{T,\,x}\end{aligned}$$

and
$$\left(\frac{\partial^2 G}{\partial x^2}\right)_{T,p}\left(\frac{\partial^2 F}{\partial v^2}\right)_{T,x}=\left(\frac{\partial^2 F}{\partial x^2}\right)_{T,v}\left(\frac{\partial^2 F}{\partial v^2}\right)_{T,x}-\left(\frac{\partial^2 F}{\partial v\partial x}\right)_T^2 .$$

Alternatively, use the method of Jacobians:

$$\begin{vmatrix} F_{2x} & F_{xv} \\ F_{xv} & F_{2v} \end{vmatrix}_T = \frac{\partial(F_x, F_v)}{\partial(x, v)} = \frac{\partial[(\mu_1-\mu_2), -p]}{\partial(x, v)}$$
$$= \frac{\partial[(\mu_1-\mu_2), -p]}{\partial(x, -p)}\bigg/\frac{\partial(x, v)}{\partial(x, -p)} = -\left[\frac{\partial(\mu_1-\mu_2)}{\partial x}\right]_{T,p}\bigg/\left(\frac{\partial v}{\partial p}\right)_{T,x}$$
$$= \left(\frac{\partial^2 G}{\partial x^2}\right)_{T,p}\left(\frac{\partial^2 F}{\partial v^2}\right)_{T,x} .$$

Evidently the stability conditions are,

$$\left(\frac{\partial \mu_1}{\partial x}\right)_{T,p} > 0\,, \quad \frac{1}{vK_T} > 0\,, \tag{7.9.1}$$

and, correspondingly, when fluctuations in entropy also are allowed,

$$\left(\frac{\partial \mu_1}{\partial x}\right)_{T,p} > 0\,, \quad \frac{1}{vK_T} > 0\,, \quad \frac{T}{C_v} > 0\,. \tag{7.9.2}$$

(b) In terms of F and its derivatives (molar quantities being understood) the co-existence conditions $p_\alpha = p_\beta$ and (7.7.1) are:

$$-p_\alpha = \left(\frac{\partial F_\alpha}{\partial v_\alpha}\right)_{T,x} = \left(\frac{\partial F_\beta}{\partial v_\beta}\right)_{T,x} = -p_\beta\,,$$
$$(\mu_{1\alpha}-\mu_{2\alpha}) = \left(\frac{\partial F_\alpha}{\partial x_\alpha}\right)_{T,v} = \left(\frac{\partial F_\beta}{\partial x_\beta}\right)_{T,v} = (\mu_{1\beta}-\mu_{2\beta})\,,$$
$$\mu_{2\alpha} = F_\alpha - v_\alpha\left(\frac{\partial F_\alpha}{\partial v_\alpha}\right)_{T,x} - x_\alpha\left(\frac{\partial F_\alpha}{\partial x_\alpha}\right)_{T,v}$$
$$= F_\beta - v_\beta\left(\frac{\partial F_\beta}{\partial v_\beta}\right)_{T,x} - x_\beta\left(\frac{\partial F_\beta}{\partial x_\beta}\right)_{T,v} = \mu_{2\beta}\,,$$

since $(\partial F/\partial x)_{T,v} = (\partial G/\partial x)_{T,p}$. The Taylor series expansion for $-\Delta p$, to first order in Δx, Δv, is

$$-\Delta p = \Delta\left(\frac{\partial F}{\partial v}\right)_{T,x} = \left(\frac{\partial^2 F}{\partial v\,\partial x}\right)_{T,c}\Delta x + \left(\frac{\partial^2 F}{\partial v^2}\right)_{T,x,c}\Delta v + \cdots .$$

Subtracting the series for β from that for α, and proceeding to the limit,

$$\left(\frac{\partial^2 F}{\partial v\,\partial x}\right)_{T,c}\mathrm{d}x + \left(\frac{\partial^2 F}{\partial v^2}\right)_{T,x,c}\mathrm{d}v = 0\,. \tag{7.9.3}$$

Similarly for

$$\Delta(\mu_1-\mu_2) = \Delta\left(\frac{\partial G}{\partial x}\right)_{T,p} = \Delta\left(\frac{\partial F}{\partial x}\right)_{T,v}$$

we have

$$\left(\frac{\partial^2 F}{\partial x^2}\right)_{T,\,v,\,c} \mathrm{d}x + \left(\frac{\partial^2 F}{\partial v \partial x}\right)_{T,\,c} \mathrm{d}v = 0\,. \tag{7.9.4}$$

Equations (7.9.3) and (7.9.4) may be written, respectively, cf.part (a) above:

$$-\left(\frac{\partial p}{\partial x}\right)_{T,\,v,\,c} \mathrm{d}x - \left(\frac{\partial p}{\partial v}\right)_{T,\,x,\,c} \mathrm{d}v = 0\,, \tag{7.9.5}$$

$$\left(\frac{\partial^2 G}{\partial x^2}\right)_{T,\,p,\,c} \mathrm{d}x - \left[\left(\frac{\partial p}{\partial x}\right)^2_{T,\,v,\,c} \Big/ \left(\frac{\partial p}{\partial v}\right)_{T,\,x,\,c}\right] \mathrm{d}x - \left(\frac{\partial p}{\partial x}\right)_{T,\,v,\,c} \mathrm{d}v = 0\,. \tag{7.9.6}$$

If $(\partial p/\partial v)_{T,\,x,\,c} \neq 0$, then it follows from Equation (7.9.5) that $(\partial p/\partial x) \neq 0$, and Equation (7.9.6) becomes

$$\left(\frac{\partial^2 G}{\partial x^2}\frac{\partial p}{\partial v}\Big/\frac{\partial p}{\partial x}\right) \mathrm{d}x - \frac{\partial p}{\partial x}\mathrm{d}x - \frac{\partial p}{\partial v}\mathrm{d}v = 0\,, \tag{7.9.7}$$

and equivalence of Equations (7.9.3), (7.9.5) requires $(\partial^2 G/\partial x^2) = 0$; if $(\partial p/\partial v)_{T,\,x,\,c} = 0$, then $(\partial p/\partial x)_{T,\,v,\,c} = 0$, and Equation (7.9.6) requires $(\partial^2 G/\partial x^2)_{T,\,p,\,c} = 0$. The situation differs from that in a one-component system (continuous equation of state) in that

$$\left(\frac{\partial^2 G}{\partial x^2}\right)_{T,\,p,\,c} = \left(\frac{\partial p}{\partial v}\right)_{T,\,x,\,c} = 0$$

can occur, e.g. at a liquid-vapour critical point; the critical point is, in this case, also an azeotrope, and the occurrence is unique in T, p, x, like the one-component critical point, whereas normal binary-system critical points define a continuous locus on the p, x, or T, p, projections.

(c) Denoting the determinant by D, the tangent to the projection of the spinodal locus on the v, x plane is

$$\left(\frac{\partial D}{\partial x}\right)_{T,\,v,\,c} \mathrm{d}x + \left(\frac{\partial D}{\partial v}\right)_{T,\,x,\,c} \mathrm{d}v = 0\,,$$

or, explicitly,

$$(F_{3x}F_{2v} + F_{x2v}F_{2x} - 2F_{2xv}F_{xv})\mathrm{d}x + (F_{3v}F_{2x} + F_{2xv}F_{2v} - 2F_{x2v}F_{xv})\mathrm{d}v = 0\,. \tag{7.9.8}$$

To obtain the given result, this must be reduced to terms of one variable. This may be done using Equations (7.9.3) and (7.9.4); the two possible ways are equivalent. Thus, changing terms in $\mathrm{d}v$ to terms in $\mathrm{d}x$, we obtain

$$-2[F_{x2v}F_{xv}]\mathrm{d}v = 2[F_{x2v}F_{2x}]\mathrm{d}x\,, \quad [F_{2xv}F_{2v}]\mathrm{d}v = -[F_{2xv}F_{xv}]\mathrm{d}x\,,$$

and Equation (7.9.8) becomes

$$[F_{3x}F_{2v} - 3F_{2xv}F_{xv} + 3F_{x2v}F_{2x}]\mathrm{d}x + [F_{3v}F_{2x}]\mathrm{d}v = 0\,;$$

but, $D = 0$ gives

$$F_{2x} = \frac{(F_{xv})^2}{F_{2v}},$$

and

$$F_{3v}F_{2x}\,dv = \frac{F_{3v}(F_{xv})^2}{F_{2v}}\,dv\,.$$

Hence

$$[F_{3x}F_{2v} - 3F_{2xv}F_{xv} + 3F_{x2v}F_{2x}]\,dx + \left[F_{3v}\frac{(F_{xv})^2}{F_{2v}}\right]dv = 0\,,$$

$$\left[F_{3x}F_{2v} - 3F_{2xv}F_{xv} + 3F_{x2v}\frac{(F_{xv})^2}{F_{2v}} - F_{3v}F_{xv}\frac{F_{2x}}{F_{2v}}\right]dx = 0\,,$$

$$\left[F_{3x}F_{2v} - 3F_{2xv}F_{xv} + 3F_{x2v}\frac{(F_{xv})^2}{F_{2v}} - F_{3v}\frac{(F_{xv})^3}{(F_{2v})^2}\right]dx = 0\,,$$

$$F_{2v}\left[F_{3x} - 3F_{2xv}\frac{F_{xv}}{F_{2v}} + 3F_{x2v}\left(\frac{F_{xv}}{F_{2v}}\right)^2 - 3F_{3v}\left(\frac{F_{xv}}{F_{2v}}\right)^3\right]dx = 0\,. \qquad (7.9.9)$$

As with (T/C_v) in Equation (7.6.3), the identity of Equation (7.9.8) with Equations (7.9.3), (7.9.4) implies, cf.Equation (7.9.9), that $(\partial^2\overline{F}/\partial\overline{v}^2)_{T,\,x,\,c}$ is finite; if so, Equation (7.9.9) leads at once to the required result; if not (critical azeotrope) then Equation (7.9.9) is not usable ($\overline{G}$ is singular) and behaviour in the critical region is determined by $(\partial^2 p/\partial v^2)_{T,\,x} = 0$, etc., as for a pure substance.

(d) To express $G_{3x} \equiv (\partial^3 G/\partial x^3)_{T,\,p}$ in terms of $F = F(T, v, x)$, start from the result of part (a) in the form

$$G_{2x} = F_{2x} - \frac{(F_{xv})^2}{F_{2v}}\,.$$

Now

$$\left(\frac{\partial F_{2x}}{\partial x}\right)_{T,\,p} = F_{3x} + F_{2xv}\left(\frac{\partial v}{\partial x}\right)_{T,\,p},$$

from the standard change of constraint formula. But

$$\left(\frac{\partial v}{\partial x}\right)_{T,\,p} = -\left(\frac{\partial p}{\partial x}\right)_{T,\,v}\Big/\left(\frac{\partial p}{\partial v}\right)_{T,\,x} = -\frac{F_{xv}}{F_{2v}},$$

whence

$$\left(\frac{\partial F_{2x}}{\partial x}\right)_{T,\,p} = F_{3x} - F_{2xv}\frac{F_{xv}}{F_{2v}}\,. \qquad (7.9.10)$$

Now

$$\left[\frac{\partial (F_{xv})^2}{\partial x}\right]_{T,\,p} = 2F_{xv}\left(\frac{\partial F_{xv}}{\partial x}\right)_{T,\,p}$$

whence

$$\left[\frac{\partial (F_{xv})^2}{\partial x}\right]_{T,\,p} = 2F_{xv}\left[F_{2xv} + F_{x2v}\left(\frac{\partial v}{\partial x}\right)_p\right]$$

$$\left[\frac{\partial (F_{xv})^2}{\partial x}\right]_{T,\,p} = 2F_{2xv}F_{xv} - 2F_{x2v}\frac{(F_{xv})^2}{F_{2v}} .$$

Finally,

$$\left[\frac{\partial (F_{2v})^{-1}}{\partial x}\right]_{T,\,p} = \left[\frac{\partial (F_{2v})^{-1}}{\partial x}\right]_{T,\,v} + \left[\frac{\partial (F_{2v})^{-1}}{\partial v}\right]_{T,\,x}\left(\frac{\partial v}{\partial x}\right)_{T,\,p},$$

$$= -(F_{2v})^{-2}F_{x2v} + (F_{2v})^{-2}F_{3v}\frac{F_{xv}}{F_{2v}} ,$$

whence

$$\left[\frac{\partial (F_{2v})^{-1}}{\partial x}\right]_{T,\,p} = -\frac{F_{x2v}}{(F_{2v})^2} + F_{3v}\frac{F_{xv}}{(F_{2v})^3} .$$

Thus

$$-\left[\frac{\partial (F_{xv})^2}{\partial x}\right]_{T,\,p}\Big/ F_{2v} = -2F_{2xv}\frac{F_{xv}}{F_{2v}} + 2F_{x2v}\left(\frac{F_{xv}}{F_{2v}}\right)^2 , \qquad (7.9.11)$$

$$-(F_{xv})^2\left[\frac{\partial (F_{2v})^{-1}}{\partial x}\right]_{T,\,p} = +F_{x2v}\left(\frac{F_{xv}}{F_{2v}}\right)^2 - F_{3v}\left(\frac{F_{xv}}{F_{2v}}\right)^3 . \qquad (7.9.12)$$

Adding expressions (7.9.10), (7.9.11), and (7.9.12) gives the required result.

Thus, as in Problem 7.6, the higher-order stability condition ensures that the binodal locus and the spinodal locus coincide at the critical point.

It is clear from Problems 7.1 and 7.2, and the note to Problem 7.7[6] that the method of partial molar quantities is well suited to the problem of the single phase, since it suppresses the difficulties associated with the 'reversible' perturbation, adding or subtracting matter at the same values of the intensive variables. Problem 7.4 shows, however, that this method is incompletely rigorous for co-existing phases; it always establishes the correct stability conditions as *sufficient*, but may fail to establish them as *necessary*. This is because the list of second-order terms obtained by this method may not correspond exactly to the list obtained using extensive variables, cf.the expressions (7.3.7) and (7.4.5), noting that the latter contains reducible third-order terms such as

$$\left(\frac{\partial^2 \overline{U}}{\partial \overline{S}^2}\right)_{\overline{v}} (\delta S)^2 \delta\xi ,$$

or

$$n^2\left[\frac{\partial^2}{\partial S^2}\left(\frac{\partial U}{\partial n}\right)_{T,\,p}\right]_{v,\,n} (\delta S)^2 \delta\xi .$$

[6] page 221.

8
Surfaces

J.M.HAYNES
(*University of Bristol, Bristol*)

THE GIBBS MODEL OF A SURFACE[1]

8.0 When two bulk phases are placed in contact, their various properties do not, in general, remain uniform right up to their common boundary or **interface**. Rather, the operation of long-range intermolecular forces will cause the transition at the molecular level from one phase to the other to be more or less gradual, within an **interfacial region**. This is so even in systems of only one component; in multicomponent systems, especially, the net composition of the interfacial region may differ very markedly from that of either bulk phase—the phenomenon of **adsorption**.

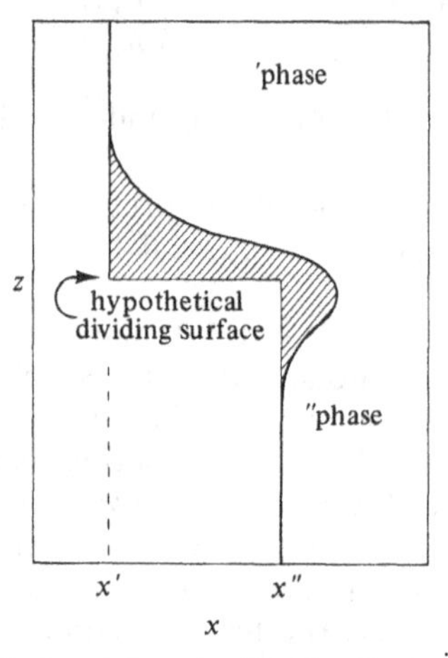

Figure 8.0.1. Real (heavy line) and hypothetical (fine line) variation of the quantity x in the z direction normal to the interface.

Development of a quantitative surface thermodynamics based on a realistic molecular-statistical picture of the interfacial region is impeded by our ignorance of its microscopic structure. For many purposes, however, it is an acceptable simplification to imagine the two homogeneous bulk phases to be separated by a surface of zero thickness, across which the thermodynamic properties change abruptly. Thus, in Figure 8.0.1

[1] J.W.Gibbs, *Scientific Papers* (Dover reprint, New York), 1961, Vol.1. R.Defay, I.Prigogine, A.Bellemans, and D.H.Everett, *Surface Tension and Adsorption* (Longmans, London), 1966.

the heavy line may represent the real variation of the volume density x of some extensive thermodynamic quantity X, in the direction z normal to the interface between two bulk phases (distinguished by single and double primes), whilst the fine line represents the model situation. The total quantity of X within the whole system of volume V is then made up by assigning any discrepancy between model and real systems to the surface itself. Thus (using a superscript s to distinguish quantities associated with the surface),

$$X = V'x' + V''x'' + Ax^{\mathrm{s}} , \tag{8.0.1}$$

where A is the area of the interface, Ax^{s} is proportional to the shaded area in Figure 8.0.1, and the volumes V' and V'' are defined by the position of the model surface of discontinuity and by the conservation equation

$$V' + V'' = V . \tag{8.0.2}$$

For example, n_i, the number of moles of the ith chemical component in the system, may be subdivided into contributions n_i' and n_i'' from the two bulk phases, and an additional number of moles n_i^{s} which can be attributed to the surface. Thus, applying Equation (8.0.1), we have

$$n_i = V'c_i' + V''c_i'' + A\Gamma_i \tag{8.0.3}$$

where c_i', c_i'' are concentrations, and Γ_i ($= n_i^{\mathrm{s}}/A$) is the surface excess or adsorption of i per unit area of interface.

Since such a dividing surface is entirely hypothetical, and its location quite arbitrary, the surface excess quantities defined with reference to it are not susceptible to experimental measurement (unless by methods involving direct observation of the interfacial region). It is therefore necessary to define **relative surface excess** quantities which are independent of the location of the dividing surface, and which may be deduced directly from observations of the two bulk phases alone.

Since the location chosen in the model for the surface dividing the two bulk phases determines the relative magnitudes of V' and V'', this location also controls the values of the surface excess quantities x^{s} (which may be energy, entropy, or the mass of any chemical component).

8.1 Show that the **relative adsorption** of i, $\Gamma_{i,1}$, defined by

$$\Gamma_{i,1} \equiv \Gamma_i - \Gamma_1 \frac{c_i' - c_i''}{c_1' - c_1''} , \tag{8.1.1}$$

is independent of the position of the dividing surface. [Proceed by using Equation (8.0.3) to express Γ_i and Γ_1 in terms of V' and V'', which are then eliminated using Equation (8.0.2).]

Solution

Γ_1 is defined [cf.Equation (8.0.3)] by

$$\Gamma_1 = \frac{1}{A}[n_1 - V'c_1' - V''c_1''] . \tag{8.1.2}$$

Using Equation (8.0.2) to eliminate V' and V'' between Equations (8.0.3) and (8.1.2) yields

$$\Gamma_i - \Gamma_1 \frac{c_i' - c_i''}{c_1' - c_1''} = \frac{1}{A}\left[n_i - Vc_i' - (n_1 - Vc_1')\frac{c_i' - c_i''}{c_1' - c_1''}\right].$$

All the quantities on the right-hand side are experimentally accessible and are independent of the position of the hypothetical dividing surface. So, therefore, is $\Gamma_{i,1}$, as defined by Equation (8.1.1).

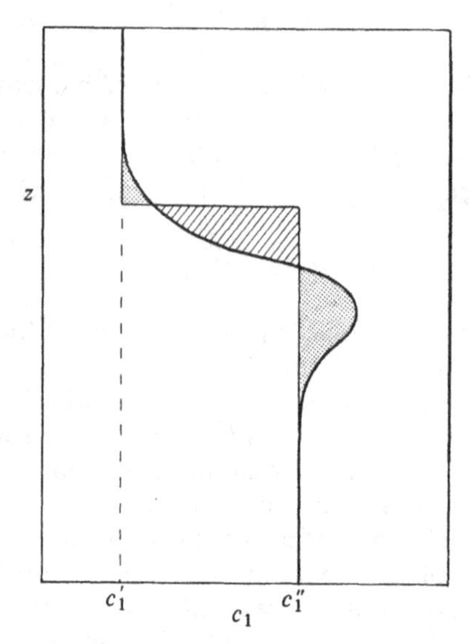

Figure 8.1.1. Choice of dividing surface which gives $\Gamma_1 = 0$.

The values of all the Γ_i (including Γ_1) depend critically on the position of the dividing surface with respect to which they are defined. In particular, there exists a dividing surface for which $\Gamma_1 = 0$; that is to say the relative magnitudes of V' and V'' may be chosen so that

$$n_1 = V'c_1' + V''c_1'' ,$$

and the shaded and stippled areas of Figure 8.1.1 then become equal. In this case, the last term of Equation (8.1.1) vanishes, thus giving the physical meaning of $\Gamma_{i,1}$ as the adsorption of the ith component relative to the dividing surface at which the adsorption of component 1 is zero.

Relative surface energy and entropy may be similarly defined.

8.2 The first law of thermodynamics applied to a single-component system of two phases separated by an interface of area A states that

$$dQ = dU + p'dV' + p''dV'' - \sigma dA \ . \tag{8.2.1}$$

The new intensive variable σ is called the **surface tension.** Show from this that a spherical liquid drop of radius r, at equilibrium with its vapour, is under an excess pressure $2\sigma/r$.

Solution

Let the liquid drop ($''$ phase) and its surrounding vapour ($'$ phase) constitute an isolated system of constant total volume, in which for any infinitesimal fluctuation about the equilibrium state,

$$dQ = dU$$

and

$$-dV' = dV'' = \tfrac{1}{2}r dA \ .$$

Thus, from the given statement of the first law,

$$0 = -p'dV'' + p''dV'' - \frac{2\sigma}{r}dV''$$

or

$$p'' - p' = \frac{2\sigma}{r} \ ,$$

the **Laplace equation.**

This result is also obtainable as a mechanical equilibrium condition of curved membranes under uniform tension. The present derivation demonstrates that the model surface of zero thickness in which the surface tension appears to act mechanically is the same as the one that correctly defines the volumes V' and V''. The distinction is only of importance when the radius r is comparable with the thickness of the actual interfacial region.

More generally, the factor $2/r$ in the Laplace equation may be replaced by the mean curvature C defined as the reciprocal of the harmonic mean of the principal radii:

$$C = \frac{1}{r_1} + \frac{1}{r_2} \ .$$

When the effects of external fields such as gravity are negligible (e.g. small systems, or nearly equal densities in the two phases), mechanical equilibrium requires that C be uniform over the interface. Mathematical studies of such surfaces of constant mean curvature are important in dealing with problems of capillarity.

It may be shown variationally [2] that for reversible fluctuations in such a system of constant curvature C,

$$\frac{dA}{dV''} = C \ ,$$

[2] C.F.Gauss, *Theorie der Gestalt von Flüssigkeiten* (Wilhelm Engelmann, Leipzig), 1903, p.46. See also J.W.Gibbs, *loc.cit.*, pp.219-229.

which leads immediately to the balance of the pV and σA work terms in the first law statement.

8.3 Consider a system involving equilibrium between liquid (l) and vapour (v) phases each in contact with an inert solid (s) (such as the case of a liquid droplet resting on a plane solid surface, as in Figure 8.3.1). Assume obedience to Young's equation

$$\sigma_{sv} = \sigma_{sl} + \sigma_{lv}\cos\theta\ , \tag{8.3.1}$$

in which θ, the **contact angle**, is the angle included by the s/l and l/v interfaces at their intersection. Show that in such cases an **effective area,** defined by

$$\Omega = A_{lv} - A_{sl}\cos\theta\ ,$$

replaces the area A in the First Law statement of Equation (8.2.1).

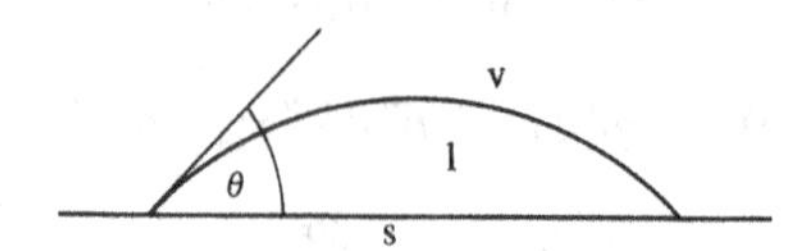

Figure 8.3.1. Three-phase system showing contact angle θ.

Solution

In this system the term $-\mathrm{d}A$ of Equation (8.2.1) must be replaced by the three terms $-(\sigma_{lv}\mathrm{d}A_{lv} + \sigma_{sl}\,\mathrm{d}A_{sl} + \sigma_{sv}\mathrm{d}A_{sv})$. These are not independent however, since σ_{lv}, σ_{sl}, and σ_{sv} are related by Young's equation (8.3.1), and, furthermore, for any variation (for example, of the radius of the drop at constant θ) it must be true that

$$\mathrm{d}A_{sv} = -\mathrm{d}A_{sl}\ .$$

The three terms above may then be reduced to the form

$$-\sigma_{lv}(\mathrm{d}A_{lv} - \mathrm{d}A_{sl}\cos\theta)\ ,$$

which, according to the definition of the effective area Ω, is equal to $-\sigma_{lv}\mathrm{d}\Omega$. This quantity therefore replaces the surface work term in the first law.

If r is the radius of curvature of the drop (assumed to be unaffected by gravity), and h is its maximum height (Figure 8.3.1), then (using single and double primes as before to denote the vapour and liquid phases)

$$V'' = \tfrac{1}{3}\pi h^2(3r-h)\ ,$$

$$A_{lv} = 2\pi rh\ ,$$

and

$$A_{sl} = \pi r^2\sin^2\theta\ ,$$

where θ is the contact angle. Since, evidently, $\cos\theta = 1-h/r$, it is readily shown that for variations in which the contact angle is unchanged

$$\mathrm{d}V'' = -\mathrm{d}V' = \pi r^2(1-\cos\theta)^2(2+\cos\theta)\mathrm{d}r\,,$$

$$\mathrm{d}A_{\mathrm{sl}} = -\mathrm{d}A_{\mathrm{sv}} = 2\pi r\sin^2\theta\,\mathrm{d}r\,,$$

and

$$\mathrm{d}A_{\mathrm{lv}} = 4\pi r(1-\cos\theta)\mathrm{d}r\,.$$

Insertion of these quantities into the first law statement (8.2.1), as in the preceding solution, yields the Laplace equation once again.

If the l/v interface in such a system has constant mean curvature, then [3]

$$\frac{\mathrm{d}\Omega}{\mathrm{d}V''} = C\,.$$

This equation is the counterpart of $\mathrm{d}A/\mathrm{d}V'' = C$ in Problem 8.2.

8.4 Derive the **Kelvin equation**, $RT\ln(p/p^0) = 2\sigma/r$, relating p, the equilibrium vapour pressure of the liquid drop of Problem 8.2, with p^0, that over a plane surface of the same liquid at the same temperature T.

Solution

The hydrostatic pressure inside the drop exceeds that at a plane interface by

$$p''_{\mathrm{d}} - p''_{\mathrm{p}} = p'_{\mathrm{d}} - p'_{\mathrm{p}} + \frac{2\sigma}{r}$$

where subscripts d and p refer to droplet and plane interfaces. The chemical potential in the liquid phase is thus enhanced by

$$\Delta\mu'' = v''(p''_{\mathrm{d}} - p''_{\mathrm{p}}) = v''\left(p'_{\mathrm{d}} - p'_{\mathrm{p}} + \frac{2\sigma}{r}\right)$$

(assuming the molar volume v'' to be independent of pressure). That in the vapour phase is changed by

$$\Delta\mu' = RT\ln(p'_{\mathrm{d}}/p'_{\mathrm{p}})$$

(assuming ideality).

The vapour pressure over the droplet, p'_{d} (written simply as p) is obtained by equating $\Delta\mu'$ and $\Delta\mu''$. The approximation $(p'_{\mathrm{d}} - p'_{\mathrm{p}}) \ll 2\sigma/r$ yields the Kelvin equation in the form

$$RT\ln(p/p^0) = \frac{2\sigma}{r}$$

in which p^0 has been written for p'_{p}, the vapour pressure over the plane interface at the temperature T. Thus, for droplets in equilibrium with their vapour, $p > p^0$. The effect is quite small for radii greater than a

[3] C.F.Gauss, *loc.cit.*

micron or so. For radii less than about 100 Å, corrections may be significant for compressibility, vapour non-ideality, and the variation of surface tension with curvature.

The same result can be obtained by considering equilibrium at the meniscus in a capillary rise experiment, in which the reduced liquid hydrostatic pressure is given by the Laplace equation, and the vapour pressure in the adjacent vapour column of equal height is given by the hypsometric equation. Thus, in a column of vapour of average density ρ', the pressure p at a height h above the level where $p = p^0$ is given by

$$p = p^0 \exp(-\rho' gh/p^0) .$$

If the same substance, existing as a liquid of density ρ'', attains a capillary rise height of h when its meniscus curvature is C, then balancing the hydrostatic pressures at the meniscus yields

$$h\rho'' g + \sigma C = p^0 - p \approx 0 .$$

(The same approximation was made in the alternative derivation.) Eliminating h we obtain

$$\ln(p/p^0) = \frac{\rho'}{p^0}\frac{\sigma C}{\rho''} .$$

Since $\rho' v' = \rho'' v''$, and $pv' = RT$ for an ideal gas, it follows that

$$\frac{RT}{v''}\ln(p/p^0) = \sigma C ,$$

as before. In this case the interfacial curvature is negative and $p < p^0$.

The Kelvin equation has been experimentally verified for moderate positive curvatures[4] (droplets in the micron range), but not for negative curvatures[5]. Despite this and other objections, it is frequently used to estimate pore sizes in porous media, from measurements of the vapour pressure of volatile liquids condensed therein[6].

Note that droplets, having $dp/dV'' < 0$, are unstable in equilibrium with their vapour. Such equilibria may be stabilized if the droplet contains an involatile solute. Systems of this kind are of meteorological interest[7].

8.5 In a system of two bulk phases and their interface, containing i components, the Helmholtz free energy

$$F = F(T, V', V'', A, n_i', n_i'', n_i^s)$$

[4] V.K.La Mer and R.Gruen, *Trans.Faraday Soc.*, **48**, 410 (1952).

[5] J.L.Shereshefsky, *J.Amer.Chem.Soc.*, **55**, 3149 (1933).

[6] S.J.Gregg and K.S.W.Sing, *Adsorption, Surface Area and Porosity* (Academic Press, London), 1967.

[7] L.Dufour and R.Defay, *The Thermodynamics of Clouds* (Academic Press, New York), 1963.

can be subdivided, according to the Gibbs model, into bulk and surface contributions:

$$F = F' + F'' + F^{s} .$$

(a) Show the relationship of the specific surface free energy f^{s} $(= F^{s}/A)$ to the surface tension.

(b) Deduce the **Gibbs adsorption equation**

$$d\sigma = -s^{s}\,dT - \sum_i \Gamma_i\,d\mu_i , \qquad (8.5.1)$$

where $s^{s} = S^{s}/A$, and the chemical potential μ_i is defined by

$$\mu_i = \left(\frac{\partial F}{\partial n_i}\right)_{T,\,V,\,n_j,\,A} .$$

(c) Rewrite this result in a form which is independent of the position of the Gibbs dividing surface.

Solution

(a) Writing

$$dF = -S\,dT - p'\,dV' - p''\,dV'' + \sigma\,dA + \sum_i \mu_i'\,dn_i' + \sum_i \mu_i''\,dn_i'' + \sum_i \mu_i^{s}\,dn_i^{s}$$

$$= dF' + dF'' + dF^{s}$$

and expressing dF', dF'' in the form

$$dF' = -S'\,dT - p'\,dV' + \sum_i \mu_i'\,dn_i' , \text{ etc.,}$$

yields

$$dF^{s} = -S^{s}\,dT + \sigma\,dA + \sum_i \mu_i^{s}\,dn_i^{s} .$$

Holding the intensive variables T, μ_i^{s}, and σ constant and integrating we obtain

$$F^{s} = \sigma A + \sum_i \mu_i^{s} n_i^{s} .$$

The constant of integration is clearly zero, since F^{s} must be zero in a system with no interface ($A = n_i^{s} = 0$). Dividing by the area gives

$$f^{s} = \sigma + \sum_i \mu_i^{s}\Gamma_i .$$

Thus, the surface tension (an experimental quantity) is not in general equal to the specific surface free energy (whose definition, like that of Γ_i, depends on the location of the Gibbs dividing surface). Only for that choice of dividing surface for which $\sum_i \mu_i^{s}\Gamma_i = 0$ is $\sigma = f^{s}$.

(b) Differentiating the last expression for F^s we obtain

$$\mathrm{d}F^s = \sigma\,\mathrm{d}A + A\,\mathrm{d}\sigma + \sum_i \mu_i\,\mathrm{d}n_i^s + \sum_i n_i^s\,\mathrm{d}\mu_i\ .$$

Comparing with the previous expression for $\mathrm{d}F^s$, we find

$$A\,\mathrm{d}\sigma = -S^s\,\mathrm{d}T - \sum_i n_i^s\,\mathrm{d}\mu_i$$

or

$$\mathrm{d}\sigma = -s^s\,\mathrm{d}T - \sum_i \Gamma_i\,\mathrm{d}\mu_i\ ,$$

which is the surface analogue of the Gibbs-Duhem equation [Problem 1.20(d)].

(c)

$$\mathrm{d}\sigma = -s_1^s\,\mathrm{d}T - \sum_i \Gamma_{i,1}\,\mathrm{d}\mu_i$$

(see Problem 8.1).

8.6 The Gibbs adsorption equation (8.5.1) may also be written in terms of the spreading pressure ϕ, which appears thermodynamically as an intensive variable equal, to within an additive constant, to $-\sigma$. (An *operational* definition for systems in which surface tensions are experimentally accessible, such as solutions, is

$$\phi = \sigma_0 - \sigma\ ,$$

where σ_0 is the surface tension of the pure solvent.)

Consider the adsorption of a single gaseous component, at pressure p, at the surface of an involatile inert solid. For **isothermal** variations, $\mathrm{d}\mu$ in the last term of Equation (8.5.1) is equal to $RT\mathrm{d}\ln p$ for the gas (assumed ideal), and is zero for the solid; the Gibbs equation then becomes

$$\mathrm{d}\phi = RT\Gamma\,\mathrm{d}\ln p\ ,$$

where Γ is the surface excess of the gas defined relative to a dividing surface essentially coincident with the solid surface.

The spreading pressure may also be expressed as a function of temperature and adsorption by means of a **surface equation of state:**

$$\phi = \phi(\Gamma, T)\ .$$

An **adsorption isotherm equation** can now be formed by elimination of ϕ between the preceding two equations:

$$\Gamma = \Gamma(T, p)\ .$$

(a) Derive by this method the adsorption isotherm equation corresponding to the surface equation of state of the van der Waals type

$$(\phi + a\Gamma^2)(1 - b\Gamma) = RT\Gamma\ .$$

In comparing this equation with its three-dimensional form (Problem 1.11), it will be seen that ϕ and $1/\Gamma$ are the surface analogues of the pressure and volume, respectively, of a bulk phase, and the temperature-dependent constants a and b have the same significance as before.

(Hint: Differentiate the equation of state to find an expression for $d\phi$ at constant temperature.)

(b) Examine the special cases of (i) an ideal two-dimensional surface film, and (ii) a film in which the adsorbate molecules are non-interacting hard spheres of finite size.

Discuss in each case the behaviour of the adsorbed phase predicted by these equations.

Solution

(a) Differentiating the two-dimensional van der Waals equation and introducing the Gibbs equation gives

$$d\ln p = \frac{d\Gamma}{\Gamma(1-b\Gamma)^2} - \frac{2a}{RT}d\Gamma .$$

Integration yields the **Hill-de Boer equation:** [8]

$$p = K\frac{\Gamma}{1-b\Gamma}\exp\left(\frac{b\Gamma}{1-b\Gamma} - \frac{2a\Gamma}{RT}\right),$$

where K is an integration constant.

Because the surface equation of state on which this isotherm is based is of the van der Waals type, there will be a two-dimensional phase transition below a critical point defined by

$$\frac{dp}{d\Gamma} = \frac{d^2p}{d\Gamma^2} = 0 .$$

Performing the differentiation we find

$$\Gamma_c = \frac{1}{3b}, \quad T_c = \frac{8a}{27R},$$

as for the van der Waals equation itself [compare Problem 1.11(b)].

(b) For ideal behaviour of the surface film, we set $a = b = 0$, yielding the linear isotherm

$$p = K\Gamma ,$$

which is the two-dimensional form of **Henry's Law.**

For non-interacting hard spheres we retain only the co-volume term which yields the **Volmer equation** [9]

$$p = K\frac{\Gamma}{1-b\Gamma}\exp\frac{b\Gamma}{1-b\Gamma} .$$

[8] J.H.de Boer, *The Dynamical Character of Adsorption* (Oxford University Press, Oxford), 1953.

[9] M.Volmer, *Z.physik.Chem.*, 115, 253 (1925).

The larger the value of b, the greater the pressure needed to reach a given surface coverage, as expected.

8.7 In the treatment of the previous problem the adsorbed phase is regarded as a continuum. By contrast, the adsorbate molecules may be regarded as occupying definite sites in the surface. (An identical formalism describes reaction occurring at fixed sites on a macromolecule.) The simplest such **lattice model** leads to the **Langmuir equation**

$$\theta = \frac{ap}{1+ap},$$

in which θ is the fraction of the surface sites occupied when the pressure of unadsorbed gas is p, and a is a temperature-dependent constant. Such an equation describes monolayer (i.e. $\theta < 1$) adsorption of non-interacting particles at identical localized sites.

It may be derived by a 'steady-state' method, in which the rate of condensation on the fraction $(1-\theta)$ of unoccupied surface, $\alpha p(1-\theta)$, is equated to the rate of evaporation from the occupied sites, $\nu\theta$. The temperature-dependent proportionality constant ν may be factorized to show a term in $\exp(-q/kT)$, where q is the heat evolved when one molecule is adsorbed from the gas phase (in some suitably defined standard state). The Langmuir model postulates q independent of θ, from which the Langmuir equation follows immediately, with $a = \alpha/\nu$.

(a) Extend this derivation to the case of adsorption from a mixture of two gaseous species.

(b) Multilayer (i.e. θ-unrestricted) adsorption may be included by permitting filled sites in the first layer to act as adsorption sites for the formation of a second layer, and so on. By assuming that the evaporation and condensation properties of the second and all higher layers are those of the bulk condensed adsorbate, show that

$$\theta = \frac{cx}{(1-x)(1+cx-x)},$$

where c is a constant and $x = p/p^0$, p^0 being the saturated vapour pressure of the bulk condensed adsorbate.

Solution

(a) Extending the Langmuir assumptions slightly, we suppose that the rate of evaporation of species A is unaffected by the presence of species B, and *vice versa*, and that each species condenses, at a rate proportional to its partial pressure, only on those parts of the surface not covered by either species.

Hence

$$\theta_A = a_A p_A(1-\theta),$$

where θ_A is the fraction of sites occupied by particles of species A,

$a_A = \alpha_A/\nu_A$, and $\theta = \theta_A + \theta_B$. Combining this with an analogous equation for species B gives

$$\theta = \frac{a_A p_A + a_B p_B}{1 + a_A p_A + a_B p_B} .$$

Note that in this model a particle of species A is supposed to cover the same area (i.e. one site) as a particle of species B.[10]

(b) The steady-state condition for the ith layer is

$$a_i p s_{i-1} = b_i s_i \exp\frac{q_i}{kT} ,$$

where s_i is the number of ith layer sites filled but not covered by occupied sites in the $(i+1)$th layer, and a_i, b_i are constants.

This set of simultaneous equations defines the s_i, which give the total number n of adsorbed particles in the expression

$$n = n_0 \sum_{i=0}^{\infty} i s_i \Big/ \sum_{i=0}^{\infty} s_i .$$

Here, n_0 is the number of particles which would completely fill any one layer (assumed independent of i).

The assumption suggested in the problem may be formulated as follows:

(i) If

$$g_i = \frac{a_i}{b_i} \exp\frac{q_i}{kT}$$

then

$$g_2 = g_3 = \dots = g_\infty ;$$

(ii)

$$\frac{1}{g_\infty} = p_0 = \frac{b_\infty}{a_\infty} \exp\left(-\frac{q_L}{kT}\right) ,$$

where p^0 is the saturated vapour pressure of the bulk adsorbate at the temperature T, and q_L is the corresponding latent heat of evaporation (per molecule). Thus, it is only the adsorbed layer nearest the solid surface that is assumed to behave differently from the bulk adsorbate. Hence,

$$s_1 = g_1 p s_0 \quad \text{(for layer 1)}$$

and

$$s_i = x s_{i-1} \quad (i > 1) ,$$

where $x = p/p^0$.

The second of these equations becomes

$$s_i = x^{i-1} s_1$$

[10] For an experimental test of this equation see F.C.Tompkins and D.M.Young, *Trans. Faraday Soc.*, 47, 88 (1951).

(since $s_{i-1} = xs_{i-2}$, etc.) and hence

$$s_i = cx^i s_0 ,$$

where

$$c = \frac{g_1}{g} = \frac{a_1 b_\infty}{a_\infty b_1} \exp\left(\frac{q_1 - q_L}{kT}\right) .$$

The summations for n can now be effected, yielding eventually

$$\frac{n}{n_0} = \theta = \frac{cx}{(1-x)(1+cx-x)} .$$

For $x > 1/(1+\sqrt{c})$, $\theta > 1$, corresponding to **multilayer adsorption,** and $\theta \to \infty$ as $x \to 1$ (i.e. $p \to p^0$).

This isotherm equation, known as the **Brunauer, Emmett, and Teller (BET) equation**[11], may be rearranged in the linear form

$$\frac{x}{n(1-x)} = \frac{1}{n_0 c} + \frac{c-1}{n_0 c} x .$$

Experimental data giving n as a function of x at constant T can thus be plotted as $x/n(1-x)$ against x and, from the slope and intercept of such a plot, the quantities n_0 and c can be calculated. If the area per site is known, n_0 may in turn be used to evaluate the surface area of the solid adsorbent.[12]

8.8 The results of the preceding problem may also be obtained by statistical methods. For example, if the surface is assumed to contain n_0 independent sites, then, for the case of monolayer adsorption, each site may contain 0 or 1 particles. Let α_0 and α_1 be the respective probabilities of zero and unit occupancy of a site. Then, if λ is the absolute activity of the adsorbed species, and q the ordinary partition function for a single particle, we have

$$\frac{\alpha_0}{\alpha_1} = \frac{1}{\lambda q} .$$

On identifying α_1 with θ, the fraction of filled sites, and using $\alpha_0 + \alpha_1 = 1$ it follows that

$$\theta = \frac{\lambda q}{1+\lambda q} .$$

Since λ is proportional to p, and q is independent of it, this is the Langmuir result.

A more general method consists of expressing Ξ, the grand partition function, in terms of λ and q for a given model, and determining n, the

[11] S.Brunauer, P.H.Emmett, and E.Teller, *J.Amer.Chem.Soc.*, 60, 309 (1938).

[12] S.J.Gregg and K.S.W.Sing, *loc.cit.*

number of filled sites, from the identity

$$n \equiv \lambda\left(\frac{\partial \ln \Xi}{\partial \lambda}\right)_{T,\,V,\,A}.$$

Re-derive the results of Problem 8.7 by this method, given the following expressions for the grand partition function:

(a) monolayer adsorption (single species):

$$\Xi = (1+\lambda q)^{n_0};$$

(b) mixed monolayer adsorption (species A and B):

$$\Xi = (1+\lambda_A q_A + \lambda_B q_B)^{n_0};$$

(c) multilayer adsorption:

$$\Xi = (1+\lambda q_1 + \lambda^2 q_1 q_2 + \lambda^3 q_1 q_2 q_3 + \ldots)^{n_0};$$

where q_i is the partition function for a particle adsorbed in the ith layer.

Solution

(a)

$$n \equiv \lambda \frac{\partial \ln \Xi}{\partial \lambda} = \lambda \frac{\partial}{\partial \lambda}[n_0 \ln(1+\lambda q)] = n_0 \frac{\lambda q}{1+\lambda q},$$

as before.

(b) For either species considered separately,

$$n_A \equiv \lambda_A\left(\frac{\partial \ln \Xi}{\partial \lambda_A}\right)_{\lambda_B,\,T,\,V,\,A}$$

leading to

$$\theta_A = \lambda_A q_A (1-\theta),$$

where

$$\theta = \theta_A + \theta_B.$$

Hence,

$$\theta = \frac{\lambda_A q_A + \lambda_B q_B}{1+\lambda_A q_A + \lambda_B q_B}.$$

(c) With the assumption (see solution to Problem 8.7) that

$$q_1 = cq,$$

$$q_2 = q_3 = \ldots = q,$$

the grand partition function becomes

$$\Xi = (1+\lambda c q + \lambda^2 c q^2 + \lambda^3 c q^3 + \ldots)^{n_0} = \left[\frac{1+(c-1)\lambda q}{1-\lambda q}\right]^{n_0}.$$

If we use

$$\theta = \frac{n}{n_0} = \frac{\lambda}{n_0}\left(\frac{\partial \ln \Xi}{\partial \lambda}\right)$$

and assume further that

$$\lambda q = p/p^0 = x$$

we obtain

$$\theta = \frac{cx}{(1-x)(1+cx-x)},$$

as before.[13]

8.9 A more realistic model of monolayer adsorption would permit a particle in a surface site to interact with particles in neighbouring sites. Introduction of an appropriate interaction potential might be expected to reproduce in the model such co-operative phenomena as surface phase transitions (cf Problem 8.5).

An exact treatment is impeded by the difficulty of calculating the configurational part of the partition function, and averaging the interaction energy, whilst allowing for the re-arrangement of the adsorbed particles under their mutual influence.

The **Bragg-Williams approximate treatment** simply assumes that the arrangement of particles is *random*, as it would be in the absence of interaction. This somewhat severe approximation, which is the same as that underlying the van der Waals treatment of gas imperfection (cf Problem 1.11) is not quantitatively satisfactory; nevertheless it succeeds in predicting the occurrence of surface phase transitions below a certain critical temperature, as does van der Waals' equation.

Write down the canonical partition function for such a system, calculate the chemical potential of the adsorbed species, and hence derive the adsorption isotherm equation and find the two-dimensional critical temperature.

Solution

Let the fractional surface occupation be $\theta = n/n_0$; then, of the z sites surrounding any adsorbed particle, an average number zn/n_0 will be occupied by other particles, yielding altogether $zn^2/2n_0$ interacting pairs. (The 2 in the denominator accounts for each pair having been counted twice.) Thus, the average total interaction energy will be $wzn^2/2n_0$, where w is the interaction energy for one pair.

The canonical ensemble partition function is then

$$Q = \frac{n_0!}{n!(n_0-n)!}q^n \exp\left(-\frac{wzn^2}{2n_0kT}\right).$$

It has been assumed that the distribution of filled sites is random, in evaluating both the pre-exponential factor and the number of interacting pairs (Bragg-Williams approximation).

Use of Stirling's approximation yields

$$\ln Q = n_0 \ln n_0 - n\ln n - (n_0-n)\ln(n_0-n) + n\ln q - \frac{wzn^2}{2n_0kT}.$$

[13] For a more detailed statistical treatment see T.L.Hill, *J.Chem.Phys.*, **14**, 263 (1946). See also E.A.Guggenheim, *Applications of Statistical Mechanics* (Oxford University Press, Oxford), 1966.

Hence

$$\frac{\mu}{kT} \equiv -\left(\frac{\partial \ln Q}{\partial n}\right)_{n_0, T} = \ln\frac{\theta}{(1-\theta)q} + \frac{wz\theta}{kT} ,$$

and

$$\lambda \equiv \exp\frac{\mu}{kT} = \frac{\theta}{(1-\theta)q}\exp\frac{wz\theta}{kT} .$$

If $q\lambda = p/p^0$ (see solution 8.8) then

$$\frac{p}{p^0} = \frac{\theta}{1-\theta}\exp\frac{wz\theta}{kT} ,$$

which reduces to the Langmuir form for $w = 0$.

From the symmetry property

$$p(\theta)p(1-\theta) = [p(\tfrac{1}{2})]^2 ,$$

where $p(\theta)$ denotes the pressure at which the fractional coverage is θ, we deduce that the critical point lies at $\theta_c = \frac{1}{2}$.

From the condition

$$\left(\frac{dp}{d\theta}\right)_{\theta = \theta_c} = 0$$

it follows that

$$T_c = -\frac{zw}{4k} .$$

Thus a phase transition, giving a vertical discontinuity in the isotherm, is to be expected for an attractive interaction (negative w).

9
The imperfect classical gas

P.C.HEMMER
(*Institutt for Teoretisk Fysikk, NTH, Trondheim*)

THE EQUATION OF STATE

9.1 The **intermolecular potential energy** $U(\mathbf{r}_1, \mathbf{r}_2, \dots \mathbf{r}_N)$ of a real gas of N particles is a homogeneous function of degree γ in the position coordinates of the particles. Show that the equation of state is of the form

$$pT^{-1+3/\gamma} = \mathrm{f}(vT^{-3/\gamma})$$

where f is an undetermined function of one variable.

[Proceed broadly as in Problem 3.5. In the partition function only the integral over the position variables,

$$Q_N = \int \dots \int \exp\left[-\frac{U(\mathbf{r}_1, \dots \mathbf{r}_N)}{kT}\right] d\mathbf{r}_1 \dots d\mathbf{r}_N ,$$

contributes to the pressure. Q_N is called the **configuration integral.**]

Solution

If we introduce new integration variables by $\mathbf{r}_n = \lambda \mathbf{r}_n^*$ in the configuration integral then we obtain

$$Q_N(v, T) = \lambda^{3N} \int_{v\lambda^{-3}} \dots \int_{v\lambda^{-3}} \exp\left[-\frac{\lambda^\gamma U(\mathbf{r}_1^*, \dots \mathbf{r}_N^*)}{kT}\right] d\mathbf{r}_1^* \dots d\mathbf{r}_N^* ,$$

using the homogeneity property $U(\lambda \mathbf{r}_1^*, \dots \lambda \mathbf{r}_N^*) = \lambda^\gamma U(\mathbf{r}_1^*, \dots \mathbf{r}_N^*)$. Consider now the following function of v and T:

$$T^{-3N/\gamma} Q_N(v, T) = (T\lambda^{-\gamma})^{-3N/\gamma} \int_{v\lambda^{-3}} \dots \int_{v\lambda^{-3}} \exp\left[-\frac{U(\mathbf{r}_1^*, \dots \mathbf{r}_N^*)}{kT\lambda^{-\gamma}}\right] d\mathbf{r}_1^* \dots d\mathbf{r}_N^* .$$

The substitution $T \to T\lambda^\gamma$, $v \to v\lambda^3$ leaves this function invariant, hence it must depend upon v and T through the combination $vT^{-3/\gamma}$:

$$T^{-3N/\gamma} Q_N(v, T) = \mathrm{g}(vT^{-3/\gamma})$$

with an unknown function g. The pressure

$$p = kT\frac{\partial}{\partial v} \ln Q_N = kT^{1-3/\gamma} \mathrm{g}'(vT^{-3/\gamma})\mathrm{g}^{-1}$$

is of the required form, with $k\mathrm{g}'(x)/\mathrm{g}(x) = \mathrm{f}(x)$.

9.2 Assume that for a class of substances the pair potential is of the form

$$\varphi(\mathbf{r}) = \epsilon\psi(\mathbf{r}/\sigma),$$

where ψ is a universal function, and where σ (of dimension length) and ϵ may differ from one substance to another.

(a) Prove from the configuration integral that in terms of appropriately scaled thermodynamic variables all these substances have the same equation of state. This is one form of the **law of corresponding states.**

(b) Prove also that if these substances have a critical point, then they have the same **critical ratio** $\kappa_c = p_c\tilde{v}_c/kT_c$.

Solution

(a) Introduce the reduced (dimensionless) variables

$$\mathbf{r}^* = \frac{\mathbf{r}}{\sigma}\,, \quad T^* = \frac{kT}{\epsilon}\,, \quad \text{and } v^* = \frac{v}{\sigma^3}\,,$$

into the configuration integral for N particles

$$Q_N = \sigma^{3N}\int_{v^*} d\mathbf{r}_1^* \ldots d\mathbf{r}_N^* \exp\left[-\sum_{i<k}\psi(r_{ik}^*)/T^*\right].$$

The dimensionless configuration integral

$$Q_N^* = Q_N\sigma^{-3N}$$

is therefore the same for all substances, cf solution to Problem 6.8(a).

For the pressure, $p = kT\partial \ln Q_N/\partial v$, introduce the dimensionless quantity,

$$p^* = k\sigma^3\epsilon^{-1}p\,.$$

The resulting expression for the reduced pressure p^*,

$$p^* = T^*\frac{\partial \ln Q_N^*}{\partial v^*} = p^*(v^*, T^*)\,,$$

contains no reference to the specific properties of the substances, and thus the law of corresponding states holds. Two substances are in corresponding states if the reduced variables are the same for the two substances.

(b) Since (with $\tilde{v}_c = v_c/N$)

$$\kappa_c = \frac{p_c\tilde{v}_c}{kT_c} = \frac{p_c^*\tilde{v}_c^*}{T_c^*}\,,$$

the critical ratio is a dimensionless number, the same for *all* these substances.

9.3 **A gas of hard rods.** N particles interacting via a hard core of length d are moving on a line segment of length v. Calculate the configuration integral and find the equation of state in the **thermodynamic limit** $N \to \infty$, $v \to \infty$, N/v fixed.

Solution

Since the integrand of the configuration integral is symmetric in the positions $x_1, \dots x_N$ of the N particles we may choose one particular ordering of the particles and multiply by $N!$:

$$Q_N = N! \int_{0 < x_1 < x_2 \dots < x_N < v} \cdots \int dx_1 \dots dx_N .$$

The existence of the hard core puts the restrictions $x_{i+1} - x_i > d$ upon the domain of integration. Introducing the new variables

$$y_n = x_n - (n-1)d ,$$

we obtain

$$Q_N = N! \int_{0 < y_1 < \dots < y_N < v-(N-1)d} \cdots \int dy_1 \dots dy_N$$

$$= \int_0^{v-(N-1)d} \cdots \int_0^{v-(N-1)d} dy_1 \dots dy_N = [v-(N-1)d]^N .$$

The equation of state:

$$p = kT \frac{\partial}{\partial v} \ln Q_N = \frac{NkT}{v-(N-1)d} .$$

In the thermodynamic limit $N \to \infty$, $v \to \infty$, $N/v = \rho$, we obtain **Tonks' equation of state,**[1]

$$p = \frac{kT\rho}{1-\rho d} .$$

9.4 A one-dimensional gas consists of N particles that interact pairwise via the nearest neighbour potential (see Figure 9.4.1).

$$\varphi(x) = \begin{cases} \infty & \text{for } x \leqslant d \\ \text{arbitrary} & \text{for } d < x < 2d \\ 0 & \text{for } 2d \leqslant x \end{cases}$$

The external pressure is p, the temperature is T, and the volume (length) v is defined as the distance between the end particle on the right hand side and a wall on the left hand side.

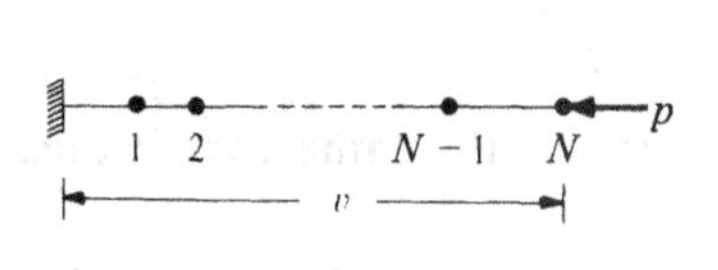

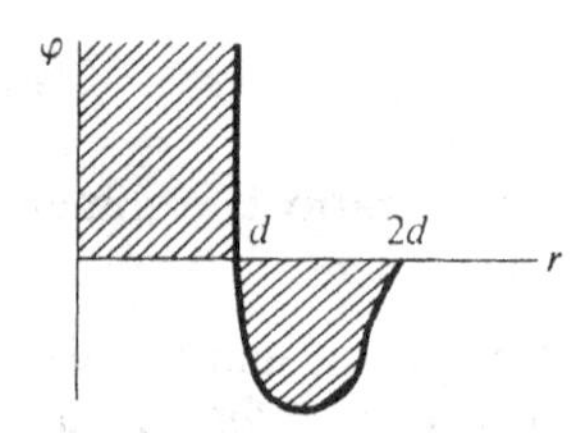

Figure 9.4.1.

[1] L. Tonks, *Phys. Rev.*, **50**, 955 (1936).

(a) Use the constant pressure ensemble of Problem 3.18 to show that the equation of state (in the limit $N \to \infty$) is given by

$$\bar{v} = \frac{\int_0^\infty \exp[-\beta pu - \beta\varphi(u)]\,du}{\int_0^\infty \exp[-\beta pu - \beta\varphi(u)]\,du},$$

where $\beta = 1/kT$.

(b) Show that this system does not exhibit a phase transition.

(c) Calculate the equation of state for the pure hard core potential, defined by

$$\varphi(x) = \begin{cases} \infty & \text{for } x \leqslant d \\ 0 & \text{for } x > d\,. \end{cases}$$

Solution

(a) The partition function Z_p of the constant pressure canonical ensemble, related to the canonical partition function Z by

$$Z_p(p, T) = \frac{1}{v_0}\int_0^\infty \exp\left(-\frac{pv}{kT}\right) Z(v, T)\,dv\,,$$

may be evaluated for this one-dimensional gas. A suitably chosen volume v_0 is inserted to make Z_p dimensionless. In $Z(v, T)$ symmetry allows us to choose one ordering of the particles, $0 < x_1 < x_2 \ldots < x_N = v$, and multiply by $N!$. The energy is then

$$E = \sum_{n=1}^{N} \frac{p_n^2}{2m} + \sum_{n=1}^{N-1} \varphi(x_{n+1} - x_n)\,,$$

Integrating over the momenta we obtain

$$Z_p = \lambda^{-N} v_0^{-1} \int_0^\infty \exp(-\beta p x_N)\,dx_N \int_0^{x_N} dx_{N-1} \ldots \int_0^{x_2} dx_1 \exp[-\beta \sum \varphi(x_{n+1} - x_n)]$$

with $\beta = 1/kT$ and $\lambda = h(2\pi mkT)^{-1/2}$. Introducing relative coordinates

$$u_1 = x_1$$
$$u_{n+1} = x_{n+1} - x_n \qquad (n = 1, 2, \ldots N-1)\,,$$

we may write

$$Z_p = \lambda^{-N} v_0^{-1} \int_0^\infty \ldots \int_0^\infty \exp\left(-\beta p \sum_{n=1}^{N} u_n\right) du_N \prod_{n=1}^{N-1} \exp[-\beta\varphi(u_n)]\,du_n$$
$$= \lambda^{-N} (v_0 p\beta)^{-1} \left\{ \int_0^\infty \exp[-\beta pu - \beta\varphi(u)]\,du \right\}^{N-1}.$$

The average volume $\overline{v}$ is given by $\overline{v} = -kT\partial \ln Z_p/\partial p$. We obtain for the volume per particle $\tilde{v} = \overline{v}/N$:

$$\tilde{v} = \frac{\int_0^\infty u \exp[-\beta p u - \beta\varphi(u)]\,du}{\int_0^\infty \exp[-\beta p u - \beta\varphi(u)]\,du},$$

neglecting terms that vanish in the thermodynamic limit $N \to \infty$.

(b) At a given temperature $\tilde{v}$ is determined uniquely by the pressure through the last formula in part (a). Consequently the isotherms in a (p, v)-diagram cannot exhibit the horizontal part characteristic of a first-order phase transition. One can also show directly by differentiation that $(\partial\tilde{v}/\partial p)_T$ is always finite and negative when $p > 0$.

(c) In the case $\varphi = \infty$ for $r < d$ and zero otherwise, we obtain

$$\tilde{v} = \frac{\int_d^\infty u \exp(-\beta p u)\,du}{\int_d^\infty \exp(-\beta p u)\,du} = d + \frac{1}{p\beta},$$

or $p = kT/(\tilde{v} - d)$. This is Tonks' equation of state obtained by a different method in Problem 9.3.

9.5 Determine the equation of state of a two-dimensional gas of N positive and N negative charges interacting via the pair potential

$$\varphi(|\mathbf{r}_i - \mathbf{r}_j|) = -q_i q_j \ln|\mathbf{r}_i - \mathbf{r}_j|.$$

Here $q_i = \pm q$ is the charge (in suitable units) of particle i. Assume the container to be the square $0 < x < L$, $0 < y < L$.

Show that the canonical partition function does not exist if $T < q^2/2k$.

Solution

In the configuration integral

$$Q = \int \ldots \int d\mathbf{r}_1 \ldots d\mathbf{r}_{2N} \exp(\beta \sum_{i<j} q_i q_j \ln r_{ij})$$

introduce the dimensionless two-dimensional vector variable $\mathbf{R}_i = \mathbf{r}_i/L$. The integration limits now become independent of L and we may write

$$Q = L^{4N} \exp(\beta \sum_{i<j} q_i q_j \ln L) f,$$

where f is the factor independent of L. We have

$$2\sum_{i<j} q_i q_j = \left(\sum_i q_i\right)^2 - \sum_i q_i^2 = -2Nq^2,$$

using the neutrality condition $\sum_i q_i = 0$. In terms of the two-dimensional 'volume' $v = L^2$ of the container,

$$Q = fv^{2N - \frac{1}{2}\beta q^2 N}$$

This yields the pressure

$$\frac{p}{kT} = \frac{\partial}{\partial v}\ln Q = \frac{2N - \frac{1}{2}\beta q^2 N}{v} ,$$

or

$$p = (kT - \tfrac{1}{4}q^2)\rho$$

in terms of the number density $\rho = 2N/v$. We observe that this imperfect gas follows the equation of state of an ideal gas with the temperature scale shifted by an amount

$$T_0 = \frac{q^2}{4k} .$$

The configuration integral may diverge for configurations in which oppositely charged particles coalesce. If $q_i = -q_j$ then the integrand contains the divergent factor $r_{ij}^{-\beta q^2}$. The (two-dimensional) integral over $\mathbf{r}_j$, say, exists only if $\beta q^2 < 2$, or

$$T > \frac{q^2}{2k} = 2T_0 .$$

For $T < 2T_0$ the gas collapses into neutral pairs.

Three isochores at densities $\rho_3 > \rho_2 > \rho_1$ are shown in Figure 9.5.1 where also the pressure of N neutral noninteracting pairs is indicated.

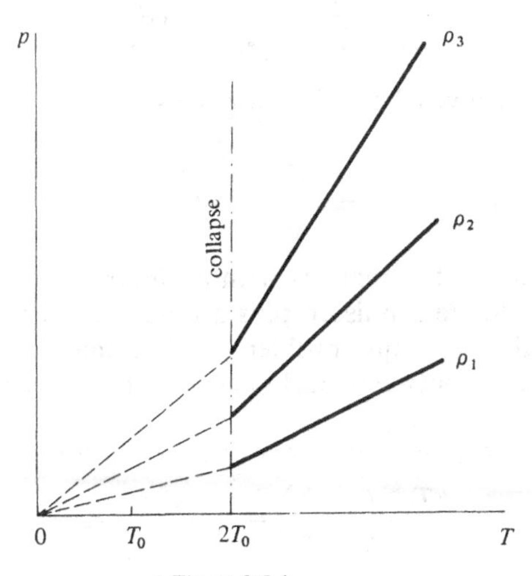

Figure 9.5.1.

9.6 As a simple model of a gas with hard core interaction consider the following **lattice gas**:

N particles move in a volume v divided into cells, each of volume b. The potential energy is assumed to be $+\infty$ if two or more particles are in the same cell, zero otherwise.

Determine the equation of state in the thermodynamic limit.

Calculate the nth virial coefficient.

Solution

In this case the configuration integral Q_N is simply determined by the number of different ways of distributing the N particles over the $\nu = v/b$ cells without multiple occupancy, each allowed configuration contributing a factor $b^N/N!$ to Q_N.

The number of allowed configurations is $\nu(\nu-1)\ldots(\nu-N+1)$, since the first particle has ν cells available, the second particle $(\nu-1)$, etc. Hence,

$$Q_N = \nu(\nu-1)\ldots(\nu-N+1)\,b^N = \frac{\nu!\,b^N}{(\nu-N)!}\,.$$

Using Stirling's formula $N! \sim N^N e^{-N}$ for the factorials we obtain for large N and v

$$\frac{Q_N}{N!} = \left(1-\frac{Nb}{v}\right)^{-v/b}\left(\frac{v}{N}-b\right)^N\,.$$

This yields the following equation of state:

$$\frac{p}{kT} = \frac{\partial \ln Q_N}{\partial v} = -\frac{1}{b}\ln\left(1-\frac{Nb}{v}\right).$$

Expansion of the logarithm (with $N/v = \rho$),

$$\frac{p}{kT} = \sum_{n=1}^{\infty} b^{n-1}\frac{\rho^n}{n}\,,$$

shows that the nth virial coefficient equals

$$B_n = \frac{b^{n-1}}{n}\,.$$

9.7 As a model of a real gas with an intermolecular interaction consisting of a hard-core repulsion plus a long-range attraction consider the lattice gas of the preceding problem, and assume an additional attractive interaction $-2a/v$ between each pair of the N particles. Here a is a constant.

(a) Show that this gas model obeys the equation of state

$$p = -\frac{kT}{b}\ln\left(1-\frac{b}{\bar{v}}\right)-\frac{a}{\bar{v}^2}\,,\qquad \bar{v} \equiv \frac{v}{N} = \text{finite}$$

in the thermodynamic limit.

(b) Determine also the critical point and the critical ratio $\kappa_c = p_c\bar{v}_c/kT_c$ for this gas model.

[Note the analogies with the van der Waals gas, Problem 1.11, and with the hole theory of liquids, Problem 6.5.]

Solution

For this gas model the attractive part of the total potential energy is independent of the configuration of the particles and is equal to $(2a/v)$ times $\frac{1}{2}N(N-1)$, the number of pairs of particles. This yields a factor

$$\exp\left[\frac{2aN(N-1)}{v}\frac{1}{2}\frac{1}{kT}\right]$$

in the configuration integral Q_N. Hence, when both the repulsion and the attraction are taken into account (see the preceding problem),

$$\frac{Q_N}{N!} = b^N\binom{v/b}{N}\exp\left[\frac{aN(N-1)}{vkT}\right]$$

$$\approx \left(1-\frac{Nb}{v}\right)^{-v/b}\left(\frac{v}{N}-b\right)^N \exp\left(\frac{aN^2}{vkT}\right).$$

In the last expression both N and v are assumed to be very large.

The equation of state

$$p = kT\frac{\partial}{\partial v}\ln Q_N = -\frac{kT}{b}\ln\left(1-\frac{Nb}{v}\right)-\frac{aN^2}{v^2},$$

or (with $\bar{v} = v/N$)

$$p = -\frac{kT}{b}\ln\left(1-\frac{b}{\bar{v}}\right)-\frac{a}{\bar{v}^2}$$

is of a similar structure as van der Waals' equation.

(b) The critical point is determined by

$$\left(\frac{\partial p}{\partial \bar{v}}\right)_T = -\frac{kT}{\bar{v}(\bar{v}-b)}+\frac{2a}{\bar{v}^3} = 0$$

together with

$$\left(\frac{\partial^2 p}{\partial \bar{v}^2}\right)_T = \frac{kT(2\bar{v}-b)}{\bar{v}^2(\bar{v}-b)^2}-\frac{6a}{\bar{v}^4} = 0.$$

Elimination of T yields

$$\bar{v}_c = 2b.$$

The critical temperature

$$T_c = \frac{2a(\bar{v}_c-b)}{k\bar{v}_c^2} = \frac{a}{2kb}.$$

The equation of state furnishes the critical pressure

$$p_c = \frac{kT_c\ln 2}{b}-\frac{a}{4b^2} = \frac{a}{b^2}\left(\frac{\ln 2}{2}-\frac{1}{4}\right).$$

The critical ratio is a dimensionless number, in this case

$$\kappa_c = \frac{p_c \tilde{v}_c}{kT_c} = \ln 4 - 1 = 0\cdot 386 \ldots .$$

This is close to the critical ratio for a van der Waals gas, which is $\kappa_c = \frac{3}{8} = 0\cdot 375$ (see Problem 1.11).

THE VIRIAL EXPANSION

9.8 (a) The expression for the pressure in the grand canonical ensemble given in the comment on Problem 3.5 is

$$\frac{pv}{kT} = \ln \sum_{N=0}^{\infty} Q_N \frac{z^N}{N!} \qquad (Q_0 \equiv 1) .$$

It may be rearranged to yield a power series expansion in the **fugacity** z,

$$\frac{p}{kT} = \sum_{l=1}^{\infty} b_l z^l . \tag{9.8.1}$$

Establish that the three simplest **cluster integrals** b_l are given by

$$b_1 = \frac{Q_1}{v} ;$$

$$b_2 = \frac{Q_2 - Q_1^2}{2v} ;$$

$$b_3 = \frac{Q_3 - 3Q_1Q_2 + 2Q_1^3}{6v} .$$

(b) For an intermolecular pair potential of finite range the b_l's stay finite in the thermodynamic limit $v \to \infty$. Verify this for the three b_l's above.

(c) Show that the grand canonical average number density $\rho = \langle N \rangle / v$ is given by

$$\rho = \sum_{l=1}^{\infty} l b_l z^l . \tag{9.8.2}$$

(d) Inverting this series expansion, to obtain an expansion $z = z(\rho)$, and inserting into Equation (9.8.1) one obtains the virial expansion (or density expansion) of the pressure

$$\frac{p}{kT} = \sum_{n=1}^{\infty} B_n \rho^n .$$

(Here the limit $v \to \infty$ is assumed.)

Show that the first **virial coefficients** B_n are given by

$$B_1 = 1 ;$$

$$B_2 = -\tfrac{1}{2} \int f(r) d\mathbf{r} ;$$

$$B_3 = -\tfrac{1}{3} \int f(r_{12})\, f(r_{23})\, f(r_{31}) d\mathbf{r}_2\, d\mathbf{r}_3 ;$$

where we have introduced the **Mayer function**

$$f(r) = \exp[-\varphi(r)/kT] - 1 .$$

Solution

(a) An easy algebraic result if one makes use of the expansion

$$\ln(1+x) = -\sum_1^\infty \frac{(-x)^n}{n} ,$$

and keeps terms up to order z^3.

(b) Recall the definition of the configuration integrals,

$$Q_N = \int_v \exp\left[-\sum_{1 \leqslant i < j \leqslant N} \frac{\varphi(r_{ij})}{kT}\right] d\mathbf{r}_1 \ldots d\mathbf{r}_N .$$

In particular $Q_1 = v$, i.e. $b_1 = 1$.

Inserting for the configuration integrals we may write the second and third cluster function

$$b_2 = \frac{1}{2v} \int (e_{12} - 1) d\mathbf{r}_1\, d\mathbf{r}_2$$

$$b_3 = \frac{1}{6v} \int (e_{12}e_{23}e_{31} - e_{12} - e_{23} - e_{31} + 2) d\mathbf{r}_1\, d\mathbf{r}_2\, d\mathbf{r}_3$$

with

$$e_{ik} = \exp\left[-\frac{\varphi(r_{ik})}{kT} - 1\right] .$$

Note that the first integrand vanishes if $r_{12} > R$, and that the second integrand vanishes if two relative distances exceed R. Integration over the variables $\mathbf{r}_2$ and $\mathbf{r}_3$ yields a result independent of $\mathbf{r}_1$ and the subsequent integration of $\mathbf{r}_1$ cancels the factor v^{-1}. The argument fails when particle 1 is within a distance R ($2R$ in the last case) from the wall, but this possibility can be neglected when $v \to \infty$.

Hence,

$$b_2 \to \tfrac{1}{2} \int (e_{12} - 1) d\mathbf{r}_2$$

$$b_3 \to \tfrac{1}{6} \int (e_{12}e_{23}e_{31} - 3e_{12} + 2) d\mathbf{r}_2\, d\mathbf{r}_3 ,$$

in the thermodynamic limit.

(c) Use the probability of finding precisely N particles in the grand canonical ensemble

$$p_N = \frac{z^N Q_N/N!}{\sum_N (z^N Q_N/N!)}$$

to obtain the average number of particles

$$\langle N \rangle = \sum_N N p_N = z\frac{\partial}{\partial z}\left[\ln \sum_N (z^N Q_N/N!)\right]$$
$$= z\frac{\partial}{\partial z}\left[\frac{p(z)v}{kT}\right] .$$

Hence

$$\rho = \frac{\langle N \rangle}{v} = z\frac{\partial}{\partial z}\frac{p(z)}{kT} = \sum_{l=1}^{\infty} l b_l z^l .$$

(d) Write $z = \rho + c_2\rho^2 + c_3\rho^3$, and determine the unknown coefficients c_2 and c_3 by inserting into $\rho = z + 2b_2z^2 + 3b_3z^3$. This gives $c_2 = -2b_2$, $c_3 = 8b_2^2 - 3b_3$. If now $z = \rho - 2b_2\rho^2 + (8b_2^2 - 3b_3)\rho^3$ is inserted into the expression for the pressure, Equation (9.8.1), we get to order ρ^3

$$\frac{p}{kT} = \rho - b_2\rho^2 + (4b_2^2 - 2b_3)\rho^3 .$$

Using the results for b_2 and b_3 obtained in part (b), we obtain (in the thermodynamic limit)

$$B_2 = -b_2 = -\tfrac{1}{2}\int \mathrm{f}(r)\,d\mathbf{r} ,$$

and $B_3 = -2b_3 + 4b_2^2 = -\frac{1}{3}\int (e_{12}e_{23}e_{31} + 3e_{12} - 1 - 3e_{12}e_{31})\,d\mathbf{r}_2\,d\mathbf{r}_3$. This is easily shown to be identical to

$$B_3 = -\tfrac{1}{3}\int \mathrm{f}(r_{12})\,\mathrm{f}(r_{23})\,\mathrm{f}(r_{31})\,d\mathbf{r}_2\,d\mathbf{r}_3 ,$$

using $\mathrm{f}(r_{ik}) = e_{ik} - 1$.

9.9 Using the results of Problem 9.8(d), calculate the second and third virial coefficient for a gas of hard spheres of diameter d.

Solution

The expressions for B_2 and B_3 are given in the preceding problem. For hard spheres the Mayer function is simply

$$\mathrm{f}(r) = \exp\left[-\frac{\varphi(r)}{kT}\right] - 1 = \begin{cases} -1 & \text{if } r < d \\ 0 & \text{otherwise} . \end{cases}$$

Hence,

$$B_2 = -\tfrac{1}{2}\int_0^d (-1)4\pi r^2\,dr = \tfrac{2}{3}\pi d^3$$

In the expression for B_3,

$$B_3 = -\tfrac{1}{3}\int f(r_{12})\,f(r_{23})\,f(r_{31})\,d\mathbf{r}_2\,d\mathbf{r}_3\,,$$

integrate over $\mathbf{r}_3$ first. Because of the step function character of the Mayer function, this integral is precisely the intersection volume $I(r)$ of two spheres each with radius d centred a distance $r = |\mathbf{r}_1 - \mathbf{r}_2|$ apart. It is an easy exercise to show that for $r \leqslant d$

$$I(r) = \tfrac{1}{3}\pi(4d^3 - 3d^2r + \tfrac{1}{4}r^3)\,.$$

The final integration over r_2 yields

$$B_3 = -\tfrac{1}{3}(-1)^3\int_0^d I(r)4\pi r^2\,dr = \frac{5\pi^2 d^6}{18} = \frac{5B_2^2}{8}\,.$$

The equation of state for hard spheres thus has the virial expansion

$$p = kT\rho(1 + B_2\rho + \tfrac{5}{8}B_2^2\rho^2 + \ldots)\,,$$

where B_2 is four times the volume of one sphere.

9.10 (a) Prove that the second virial coefficient $B_2(T)$ as a function of temperature can have at most one extremum.

(b) Prove that $B_2(T)$ is monotonically decreasing with temperature if the pair potential $\varphi(r)$ is non-negative.

(c) Find the necessary and sufficient condition for $B_2(T)$ to exhibit a maximum.

Solution

(a) It is convenient to use $\beta = 1/kT$ as variable rather than the temperature:

$$B_2(\beta) = \tfrac{1}{2}\int\{1 - \exp[-\beta\varphi(r)]\}d\mathbf{r}\,.$$

If the interaction potential contains a hard core of diameter d we write

$$B_2(\beta) = \frac{2\pi d^3}{3} + \tfrac{1}{2}\int_{r>d}\{1 - \exp[-\beta\varphi(r)]\}d\mathbf{r}\,.$$

The second derivative with respect to β equals

$$B_2''(\beta) = -\tfrac{1}{2}\int\varphi^2\exp(-\beta\varphi)\,d\mathbf{r} < 0\,,$$

and the derivative can therefore change sign at most once.

(b) The assertion follows from the fact that the derivative,

$$B_2'(\beta) = \tfrac{1}{2} \int_{r>d} \varphi(r) \exp(-\beta\varphi)\, \mathrm{d}r\,,$$

is always positive if $\varphi(r)$ is non-negative.

(c) Since it follows from part (b) that a maximum is only possible when the potential is negative somewhere we have for $T \to 0$ that

$$B_2'(\infty) = -\infty\,.$$

In part (a) we showed that $B_2'(\beta)$ is a monotonically decreasing function of β, and consequently it has, by continuity, a zero if and only if

$$B_2'(0) = \tfrac{1}{2} \int_{r>d} \varphi\, \mathrm{d}r > 0\,.$$

The two necessary and sufficient conditions for a maximum in $B_2(T)$ are thus

$$\varphi < 0 \text{ for some distances, but } \int_{r>d} \varphi\, \mathrm{d}r > 0\,.$$ [(2)]

9.11 Calculate the second virial coefficient for two gases, one with the intermolecular potential

$$\varphi_1(r) = \begin{cases} +\infty & \text{for } r < d \\ c(r^3 - R^3) & \text{for } d \leqslant r \leqslant R \\ 0 & \text{for } R < r\,, \end{cases}$$

the other with

$$\varphi_2(r) = \begin{cases} +\infty & \text{for } r < d \\ c(d^3 - r^3) & \text{for } d \leqslant r \leqslant R \\ 0 & \text{for } R < r\,. \end{cases}$$

[The identity of the results demonstrates the important fact that the function $B_2(T)$ does *not* determine the intermolecular potential uniquely.]

Solution

Introducing $r^3 = s$ as new integration variable we find *in both cases*

$$B_2(T) = \frac{2\pi R^3}{3} + \frac{2\pi kT}{3c}\left[1 - \exp\frac{c(R^3 - d^3)}{kT}\right].$$

(2) H. L. Frisch and E. Helfand, *J. Chem. Phys.*, **32**, 269 (1959).

Note that

$$B_2 = \begin{cases} \frac{2}{3}\pi R^3 & \text{for } c = -\infty, \\ \frac{2}{3}\pi d^3 & \text{for } c = 0\,, \end{cases}$$

as it must be since both these limiting cases correspond to a hard sphere gas.

9.12 Calculate the second virial coefficient for the following intermolecular potentials:

(a) $$\varphi(r) = \frac{c}{r^6}$$

(b) $$\varphi(r) = \begin{cases} \infty & \text{for } r \leqslant d \\ -\epsilon & \text{for } d < r \leqslant \lambda d \\ 0 & \text{for } \lambda d < r\,. \end{cases}$$

(c) $$\varphi(r) = \begin{cases} \infty & \text{for } r \leqslant d \\ -U(2-r/d) & \text{for } d < r \leqslant 2d \\ 0 & \text{for } 2d < r\,. \end{cases}$$

Solution

(a) $$4\pi\left(\frac{c\pi}{kT}\right)^{1/2}$$

(b) $$\frac{2\pi d^3}{3}[\lambda^3 + (1-\lambda^3)\exp(\epsilon/kT)]$$

(c) $$\frac{2\pi d^3}{\alpha^3}[\tfrac{8}{3}\alpha^3 - 2 - 4\alpha - 4\alpha^2 - (2+2\alpha+\alpha^2)\mathrm{e}^{\alpha}]\,,\ \text{with } \alpha = U/kT\,.$$

9.13 (a) Show that the difference $F - F_\mathrm{i}$ between the free energy for a real gas and for an ideal classical gas of N particles can be written

$$F - F_\mathrm{i} = -kT\ln\left\{1 + v^{-N}\int[\exp(-U/kT) - 1]\,\mathrm{d}\mathbf{r}_1 \ldots \mathrm{d}\mathbf{r}_N\right\}.$$

Here $U = \sum_{i<k} \varphi(r_{ik})$ is the interaction energy.

(b) Assume now that the real gas is so dilute that only one pair of particles are close together (i.e. within the range of the potential φ) at the same time. More precisely assume that in the expression for $F - F_\mathrm{i}$ the integrand (which vanishes if no pair of particles are close together) may be replaced by the sum over all pairs,

$$\exp(-U/kT) - 1 \approx \sum_{i<k} \{\exp[-\varphi(r_{ik})] - 1\}\,.$$

Assume further that there are so few particles in the volume v that the first term in the expansion of the logarithm above suffices.

Show that these approximations lead to

$$\frac{F-F_{\mathrm{i}}}{N} \approx \frac{NkT}{2v}\int\{1-\exp[-\varphi(r)/kT]\}d\mathbf{r}\,.$$

(c) Show that this expression for the free energy yields a first-order correction to the pressure of an ideal gas in agreement with the virial expansion of Problem 9.8.

(d) Calculate the corresponding lowest order correction term to the ideal gas results for each of the following quantities for a real gas: Gibbs free energy G, the entropy S, the internal energy U, the enthalpy H, and the heat capacity at constant volume C_v.

Solution

(a) The free energy F is given by the partition function

$$\exp(-F/kT) = Z_N = \frac{1}{N!h^{3N}}\int\exp(-E/kT)d\mathbf{p}_1 \ldots d\mathbf{p}_N\, d\mathbf{r}_1 \ldots d\mathbf{r}_N$$

$$= \frac{1}{N!h^{3N}}\int\exp(-K/kT)d\mathbf{p}_1 \ldots d\mathbf{p}_N\int\exp(-U/kT)d\mathbf{r}_1 \ldots d\mathbf{r}_N\,.$$

We have stated that the energy is the sum of the kinetic energy $K(p)$ and the interaction energy $U(r)$. For an ideal gas $U = 0$. Hence

$$\exp\left(-\frac{F-F_{\mathrm{i}}}{kT}\right) = \frac{\int\exp(-U/kT)d\mathbf{r}_1 \ldots d\mathbf{r}_N}{\int d\mathbf{r}_1 \ldots d\mathbf{r}_N} = v^{-N}\int\exp(-U/kT)d\mathbf{r}_1 \ldots d\mathbf{r}_N\,.$$

Putting $\exp(-U/kT) = 1+[\exp(-U/kT)-1]$, and taking the logarithm, we have

$$F-F_{\mathrm{i}} = -kT\ln\left\{1+v^{-N}\int[\exp(-U/kT)-1]d\mathbf{r}_1 \ldots d\mathbf{r}_N\right\}.$$

(b) Replacement of $\exp(-U/kT)-1$ by $\sum_{i<k}\{\exp[-\varphi(r_{ik})/kT]-1\}$ yields $\frac{1}{2}N(N-1)$ equal contributions. In each term $N-2$ of the integrations are trivial, leading to

$$F-F_{\mathrm{i}} = -kT\ln\left[\!\left[1+\frac{N(N-1)}{2v^2}\int\{\exp[-\varphi(r_{12})/kT]-1\}d\mathbf{r}_1 d\mathbf{r}_2\right]\!\right].$$

Here we may replace $N-1$ by N. For a large volume the integration over $\mathbf{r}_1$ may be taken to be independent of $\mathbf{r}_2$ and equal to

$$\int\{\exp[-\varphi(r)/kT]-1\}d\mathbf{r} = -2B_2(T)\,,$$

in the notation of Problem 9.8. The subsequent integration over $\mathbf{r}_2$

yields a factor v, leading to

$$F - F_\mathrm{i} \approx -kT \ln\left[1 - \frac{N^2}{v} B_2(T)\right] \approx \frac{N^2 kT}{v} B_2(T) .$$

We have used the fact that $\ln(1+x) \approx x$ for $x \ll 1$. The free energy per particle,

$$\frac{F - F_\mathrm{i}}{N} \approx \frac{NkT}{v} B_2(T) ,$$

is an intensive quantity, as it should be.

(c) The pressure is given by $p = -(\partial F/\partial v)_T$. In this case the excess pressure (compared with the ideal gas) is

$$p - p_\mathrm{i} = \frac{N^2 kT}{v^2} B_2(T) ,$$

or

$$\frac{p}{kT} - \frac{1}{\tilde{v}} = \frac{B_2(T)}{\tilde{v}^2} .$$

This is the equation of state of the gas in the approximation we are considering. It is identical with the first two terms of the virial expansion of Problem 9.8.

(d) Using the results obtained above,

$$F - F_\mathrm{i} = \frac{NkTB_2(T)}{\tilde{v}} , \quad p - p_\mathrm{i} = \frac{kTB_2(T)}{\tilde{v}^2} ,$$

we find easily

$$G = F + pv = G_\mathrm{i} + \frac{2NkTB_2}{\tilde{v}} ,$$

and

$$S = -\left(\frac{\partial F}{\partial T}\right)_v = S_\mathrm{i} - \frac{Nk[B_2(T) + TB_2'(T)]}{\tilde{v}} ,$$

$$U = F + TS = U_\mathrm{i} - \frac{NkT^2 B_2'(T)}{\tilde{v}} ,$$

$$H = U + pv = H_\mathrm{i} + \frac{NkT[B_2 - TB_2'(T)]}{\tilde{v}} ,$$

$$C_v = \left(\frac{\partial U}{\partial T}\right)_v = C_{v,\mathrm{i}} - \frac{NkT[2B_2'(T) + TB_2''(T)]}{\tilde{v}} .$$

PAIR DISTRIBUTION FUNCTION. VIRIAL THEOREM

9.14 (a) Show that, if the configuration of a gas of N molecules is observed, the probability of finding one particle within the volume element $d\mathbf{r}_1$ at $\mathbf{r}_1$ and another particle in the volume element $d\mathbf{r}_2$ at $\mathbf{r}_2$ is given by

$$n_2(\mathbf{r}_1, \mathbf{r}_2)d\mathbf{r}_1 d\mathbf{r}_2 = \frac{N(N-1)}{Q_N} d\mathbf{r}_1 d\mathbf{r}_2 \int_v \dots \int_v d\mathbf{r}_3 \dots d\mathbf{r}_N \exp(-U_N/kT).$$

Here U_N is the interaction energy of the N particles and Q_N is the configuration integral. No external fields are present.

In the thermodynamic limit $\mathcal{L}$ this distribution function will depend upon the relative distance between $\mathbf{r}_1$ and $\mathbf{r}_2$:

$$\mathcal{L}n_2(\mathbf{r}_1, \mathbf{r}_2) = n_2(r_{12}).$$

$n_2(r)$ is called the **pair distribution function** and the function

$$g(r) = \bar{v}^2 n_2(r) - 1 \quad (\bar{v} \equiv v/N)$$

is often called the **correlation function.**

Show that $g(r)$ vanishes for an ideal gas. Explain why $g(r)$ should vanish when $r \to \infty$ in a homogeneous system (one phase).

(b) Consider a system consisting of a gas and a liquid phase, both of macroscopic extent. Give a probabilistic argument showing that one expects the following form for the pair distribution function in the two-phase region:

$$n_2(r; \bar{v}) = \bar{v}^{-1}[x_l \bar{v}_l n_2(r; \bar{v}_l) + x_g \bar{v}_g n_2(r; \bar{v}_g)].$$

Here $\bar{v}_l$ and $\bar{v}_g$ are the specific volumes of the coexisting liquid and gas phases (see Figure 9.14.1). x_l and x_g are the corresponding mole fractions so that $\bar{v} = x_l \bar{v}_l + x_g \bar{v}_g$. The pair correlation function $n_2(r)$ depends clearly upon the state variables ($\bar{v}$, T), and the notation $n_2(r; \bar{v})$ is used above to indicate explicitly the value of $\bar{v}$.

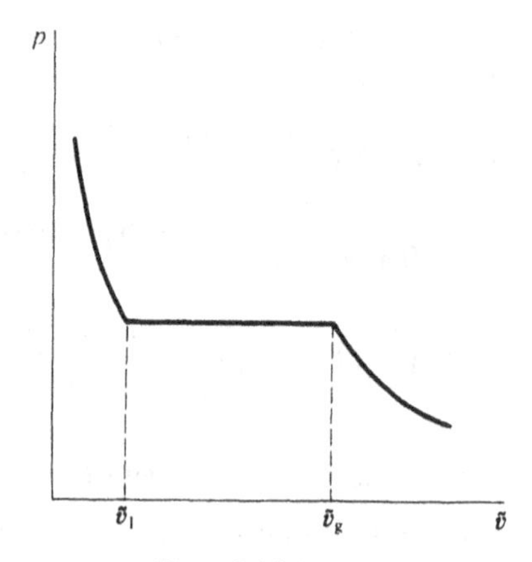

Figure 9.14.1.

Solution

(a) In the canonical ensemble the probability distribution of all positions and momenta is given by

$$\rho(p_1 \ldots r_N) = \frac{\exp[-E(p,r)/kT]}{\int \exp[-E(p,r)/kT]\,\mathrm{d}\mathbf{p}_1 \ldots \mathrm{d}\mathbf{r}_N} .$$

Integration over all momenta and over $\mathbf{r}_3, \mathbf{r}_4, \ldots \mathbf{r}_N$ yields the probability $P(\mathbf{r}_1, \mathbf{r}_2)\mathrm{d}\mathbf{r}_1\,\mathrm{d}\mathbf{r}_2$ of finding particle 1 in a volume element $\mathrm{d}\mathbf{r}_1$ at $\mathbf{r}_1$ and particle 2 in $\mathrm{d}\mathbf{r}_2$ at $\mathbf{r}_2$. We obtain

$$P(\mathbf{r}_1, \mathbf{r}_2) = \frac{\int \mathrm{d}\mathbf{r}_3 \ldots \mathrm{d}\mathbf{r}_N \exp(-U/kT)}{\int \mathrm{d}\mathbf{r}_1\,\mathrm{d}\mathbf{r}_2 \ldots \mathrm{d}\mathbf{r}_N \exp(-U/kT)} .$$

$$= \frac{1}{Q_N} \mathrm{d}\mathbf{r}_3 \ldots \mathrm{d}\mathbf{r}_N \exp(-U/kT) .$$

The probability of finding one particle (not necessarily particle 1) in $\mathrm{d}\mathbf{r}_1$ and another in $\mathrm{d}\mathbf{r}_2$ is simply $N(N-1)\,P(\vec{\mathbf{r}}_1\vec{\mathbf{r}}_2)\mathrm{d}\mathbf{r}_1\,\mathrm{d}\mathbf{r}_2$, since there are N choices for the particle in $\mathrm{d}\mathbf{r}_1$, and then $(N-1)$ possibilities for the second particle. Hence

$$n_2(\mathbf{r}_1\mathbf{r}_2) = \frac{N(N-1)}{Q_N} \int \mathrm{d}\mathbf{r}_3 \ldots \mathrm{d}\mathbf{r}_N \exp(-U/kT) .$$

For the ideal gas this reduces to

$$n_2(\mathbf{r}_1, \mathbf{r}_2) = \frac{N(N-1)}{v^2}$$

since $U = 0$ in this case. Taking the thermodynamic limit we obtain

$$n_2(r) = \frac{1}{\bar{v}^2} \; ; \quad g(r) = \bar{v}^2 n_2 - 1 = 0 .$$

One expects two particles in a gas or liquid to become completely independent as the distance between them increases. Hence the joint probability of finding one particle in $\mathrm{d}\mathbf{r}_1$ and another particle in $\mathrm{d}\mathbf{r}_2$ approaches the product of the probabilities of each event, namely

$$\left(\frac{N}{v}\mathrm{d}\mathbf{r}_1\right)\left(\frac{N}{v}\mathrm{d}\mathbf{r}_2\right) = \frac{1}{\bar{v}^2}\mathrm{d}\mathbf{r}_1\,\mathrm{d}\mathbf{r}_2 .$$

(b) The probability of finding a particle in $\mathrm{d}\mathbf{r}_1$ is still $(1/\bar{v})\mathrm{d}\mathbf{r}_1$, since the positions of the two phases are random when there is no outside field of force. Now x_g and x_l are the *a priori* probabilities that the first particle belongs to the gas and liquid phase. If the first particle belongs to the gas phase then the other particle at a microscopic distance r will also be in the gas phase (of macroscopic size). Hence the conditional probability of finding this second particle in $d\mathbf{r}_2$, given that the first particle is at $\mathbf{r}_1$, is $\tilde{v}_g n_2(r;\tilde{v}_g)d\mathbf{r}_2$. The same argument applies to the liquid phase. We obtain thus for the joint probability distribution function in the two-phase system the following expression:

$$n_2(r;\tilde{v}) = \frac{x_g \tilde{v}_g n_2(r;\tilde{v}_g) + x_1 \tilde{v}_1 n_2(r;\tilde{v}_1)}{\tilde{v}} .$$

9.15 Prove that

$$p = \frac{kT}{\tilde{v}} - \tfrac{1}{6}\int r\varphi'(r) n_2(r)\, d\mathbf{r}$$

by differentiating the configuration integral with respect to the volume. Here $n_2(r)$ is the pair distribution function of Problem 9.14 and $\varphi(r)$ is the intermolecular pair potential.

[Hint: To facilitate the differentiation procedure introduce new variables $\hat{\mathbf{r}}_i = \mathbf{r}_i v^{-1/3}$ in the configuration integral. The above result is often called the **virial theorem of statistical mechanics.**]

Solution

Introduce the new variables $\hat{\mathbf{r}}_i = \mathbf{r}_i/L$, with $L^3 = v$, in the configuration integral:

$$Q_N = L^{3N}\int d\hat{\mathbf{r}}_1 \ldots d\hat{\mathbf{r}}_N \exp\left[-\sum_{i<k}\varphi(L\hat{r}_{ik})/kT\right] .$$

Note that the integration limits are now independent of v. Hence

$$\frac{p}{kT} = \frac{1}{Q_N}\frac{\partial Q_N}{\partial v} = \frac{1}{3vQ_N}L\frac{\partial Q_N}{\partial L}$$

$$= \frac{1}{Q_N}\int d\hat{\mathbf{r}}_1 \ldots d\hat{\mathbf{r}}_N\left[\frac{N}{v}L^{3N} - \frac{L^{3N+1}}{3kTv}\frac{\partial\varphi(L\hat{r}_{ik})}{\partial L}\right]\exp\left[-\sum_{i<k}\frac{\varphi(L\hat{r}_{ik})}{kT}\right] .$$

In the second term use

$$\frac{\partial\varphi(L\hat{r}_{ik})}{\partial L} = \hat{r}_{ik}\frac{\partial\varphi(L\hat{r}_{ik})}{\partial(L\hat{r}_{ik})} ,$$

and transform back to the original coordinates:

$$\frac{p}{kT} = \frac{1}{Q_N}\int d\mathbf{r}_1 \ldots d\mathbf{r}_N\left[\frac{N}{v} - \frac{1}{3kTv}\sum_{i<k} r_{ik}\frac{\partial\varphi(r_{ik})}{\partial r_{ik}}\right]\exp\left[-\sum_{i<k}\frac{\varphi(r_{ik})}{kT}\right] .$$

The first term is simply $N/v = \rho$, and by symmetry the second term consists of $\frac{1}{2}N(N-1)$ equal integrals. Hence,

$$p = kT\rho - \frac{N(N-1)}{6vQ_N}\int d\mathbf{r}_1 \ldots d\mathbf{r}_N r_{12}\frac{\partial\varphi(r_{12})}{\partial r_{12}}\exp\left[-\sum_{i<k}\frac{\varphi(r_{ik})}{kT}\right]$$

$$= kT\rho - \frac{1}{6v}\int d\mathbf{r}_1\, d\mathbf{r}_2\, r_{12}\frac{\partial\varphi(r_{12})}{\partial r_{12}} n_2(\mathbf{r}_1, \mathbf{r}_2) ,$$

using the definition of the pair distribution function (see Problem 9.14).

In the thermodynamic limit the pair distribution function is a function of $r_{12} = r$. The **virial theorem of statistical mechanics,**

$$p = kT\rho - \tfrac{1}{6}\int \mathrm{d}\mathbf{r}\, r\varphi'(r) n_2(r)\,,$$

also called the **thermal equation of state,** follows.

9.16 (a) Prove Clausius' **virial theorem** in classical mechanics (for bounded motion)

$$\bar{E}_{\mathrm{kin}} = -\tfrac{1}{2}\sum_{n=1}^{N} \overline{\mathbf{r}_n \mathbf{F}_n}\,,$$

where $\mathbf{F}_n$ is the force acting upon particle n. The bar denotes an average over time.

(b) Apply Clausius' theorem to an imperfect gas in equilibrium, assuming that the total force can be derived from the wall potential and the intermolecular potential $\varphi(r_{ik})$ acting between each pair i, k of particles. Assuming that the time averages involved may be evaluated as phase averages in the canonical ensemble, show that the result of Problem 9.15 follows.

[For related considerations see Problems 3.7 and 3.8.]

Solution

(a) Use

$$\mathbf{r}_n \mathbf{F}_n = \mathbf{r}_n \frac{\mathrm{d}\mathbf{p}_n}{\mathrm{d}t} = \frac{\mathrm{d}}{\mathrm{d}t}(\mathbf{r}_n \mathbf{p}_n) - \mathbf{p}_n \frac{\mathrm{d}\mathbf{r}_n}{\mathrm{d}t} = \frac{\mathrm{d}}{\mathrm{d}t}(\mathbf{r}_n \mathbf{p}_n) - \frac{\mathbf{p}_n^2}{m}\,.$$

The last term is twice the kinetic energy of particle n. The time average of the first term on the right hand side vanishes for a motion where $\mathbf{r}_n$ and $\mathbf{p}_n$ are bounded since

$$\frac{1}{T}\int_0^T \frac{\mathrm{d}}{\mathrm{d}t}(\mathbf{r}_n \mathbf{p}_n)\mathrm{d}t = \frac{1}{T}[\mathbf{r}_n(T)\mathbf{p}_n(T) - \mathbf{r}_n(0)\mathbf{p}_n(0)] \to 0$$

when $T \to \infty$. Summation over all particles yields

$$\sum_n \overline{\mathbf{r}_n \mathbf{F}_n} = -2\bar{E}_{\mathrm{kin}}\,.$$

(b) The kinetic energy consists of $3N$ quadratic terms and the equipartition theorem yields the ensemble average

$$\bar{E}_{\mathrm{kin}} = \tfrac{3}{2}NkT\,.$$

The force exerted by a wall element $\mathrm{d}S$ equals $-p\,\mathrm{d}S\mathbf{n}$. Here $\mathbf{n}$ is a unit vector normal to $\mathrm{d}S$ pointing outward. The contribution of the wall force to $\sum_n \overline{\mathbf{r}_n \mathbf{F}_n}$ equals

$$-p\int \mathbf{r}\mathbf{n}\,\mathrm{d}S = -p\int_v \nabla\mathbf{r}\,\mathrm{d}\mathbf{r} = -3pv\,.$$

Here we have used Gauss' theorem and $\nabla \mathbf{r} = 3$.

The contribution to $\overline{\sum \mathbf{r}_n \mathbf{F}_n}$ from the intermolecular forces is the sum of $\frac{1}{2}N(N-1)$ equal terms, one from each pair of particles,

$$-\tfrac{1}{2}N(N-1)\,\overline{(\mathbf{r}_1\nabla_1+\mathbf{r}_2\nabla_2)\,\varphi(r_{12})}\,.$$

Here ∇_k operates on the coordinates $\mathbf{r}_k$. Using

$$\nabla_1\varphi(r_{12}) = (\nabla_1 r_{12})\frac{\partial\varphi}{\partial r_{12}} = \frac{\mathbf{r}_1-\mathbf{r}_2}{r_{12}}\varphi'(r_{12})\,,$$

and $\mathbf{r}_1(\mathbf{r}_1-\mathbf{r}_2)-\mathbf{r}_2(\mathbf{r}_2-\mathbf{r}_1) = r_{12}^2$, we may write for this contribution

$$-\tfrac{1}{2}N(N-1)\,\overline{r_{12}\varphi'(r_{12})}\,.$$

Collecting the three contributions we obtain

$$pv = NkT-\tfrac{1}{6}N(N-1)\,\overline{r_{12}\varphi'(r_{12})}\,.$$

The ensemble average of $r_{12}\varphi'(r_{12})$ is given by

$$\overline{r_{12}\varphi'(r_{12})} = \frac{1}{Q_N}\int d\mathbf{r}_1 \ldots d\mathbf{r}_N r_{12}\varphi'(r_{12})\exp(-U_N/kT)$$

$$= \int d\mathbf{r}_1\, d\mathbf{r}_2\, r_{12}\varphi'(r_{12})\frac{1}{N(N-1)}n_2(\mathbf{r}_1,\mathbf{r}_2)\,,$$

from the definition of the pair distribution function (Problem 9.14). Thus,

$$pv = NkT-\tfrac{1}{6}\int d\mathbf{r}_1\, d\mathbf{r}_2 r_{12}\varphi'(r_{12})n_2(\mathbf{r}_1\mathbf{r}_2)\,,$$

which is the same form of the virial theorem as was obtained in the solution of the preceding problem.

9.17 Assume that for a very dilute gas the pair distribution function $n_2(r)$ is well approximated by the Boltzmann factor

$$\bar{v}^{-2}\exp\left[-\frac{\varphi(r)}{kT}\right].$$

Here $\varphi(r)$ is the interaction energy of the two particles. The constant in front of the exponential is chosen so as to yield the correct limiting behaviour $n_2 \to (N/v)^2 = \bar{v}^{-2}$ as $r \to \infty$ (see Problem 9.14).

Use this in conjunction with the virial theorem of Problem 9.15 to obtain the following first order correction to the ideal gas equation of state:

$$\frac{p}{kT} = \frac{1}{\bar{v}}+\frac{2\pi}{\bar{v}^2}\int_0^\infty\left\{1-\exp\left[-\frac{\varphi(r)}{kT}\right]\right\}r^2\,dr\,.$$

[Note that this agrees with the virial expansion of Problem 9.8(d) to order ρ^2.]

Solution

Introduce in the virial theorem of Problem 9.15 the low density form of the pair distribution function,

$$n_2 = \frac{\exp[-\varphi(r)/kT]}{\bar{v}^2} .$$

We obtain

$$\frac{p}{kT} = \frac{1}{\bar{v}} + \frac{1}{6kT\bar{v}^2}\int_0^\infty r\varphi'(r)\exp(-\varphi/kT)4\pi r^2\,\mathrm{d}r .$$

An integration by parts yields

$$\frac{p}{kT} = \frac{1}{\bar{v}} + \frac{2\pi}{\bar{v}^2}\int_0^\infty [1-\exp(-\varphi/kT)]r^2\,\mathrm{d}r .$$

This agrees with the usual expression for the second virial coefficient given in Problem 9.8(d).

9.18 Show that the virial theorem of Problem 9.15 may be written

$$\frac{p}{kT} = \frac{1}{\bar{v}} + \frac{2}{3}\pi d^3 n_2(d_+) - \frac{2\pi}{3kT}\int_{d+}^\infty \mathrm{d}r\, r^3\varphi'(r)n_2(r)$$

when the interaction potential φ contains a hard core of diameter d. Here $n_2(d_+)$ means the limiting value when $r \to d$ from above.

[Hint: Rewrite the virial theorem of Problem 9.15 as follows:

$$\frac{p}{kT} = \frac{1}{\bar{v}} + \frac{2}{3}\pi\int_0^\infty \mathrm{d}r\, r^3 n_2(r)\exp\left[\frac{\varphi(r)}{kT}\right]\frac{\mathrm{d}}{\mathrm{d}r}\exp\left[-\frac{\varphi(r)}{kT}\right] ,$$

and show that the first bracket is finite everywhere, while the second bracket vanishes for $r < d$ and has a δ-function contribution at $r = d$.]

Solution

The function $\exp(-\varphi/kT)$ vanishes for $r < d$ and jumps at $r = d$ from zero to the value $\exp[-\varphi(d_+)/kT]$, so that its derivative equals

$$\frac{\mathrm{d}}{\mathrm{d}r}\exp(-\varphi/kT) = \begin{cases} \exp[-\varphi(d_+)/kT]\,\delta(r-d) & \text{for } r \leqslant d \\ \exp[-\varphi(r)/kT] & \text{for } r > d , \end{cases}$$

where $\delta(r-d)$ is Dirac's delta function.

Recall also the definition of the pair distribution function (Problem 9.14):

$$n_2(\mathbf{r}_1, \mathbf{r}_2) = \frac{N(N-1)}{Q_N}\int \mathrm{d}\mathbf{r}_3 \ldots \mathrm{d}\mathbf{r}_N \exp\left[-\sum_{i<k}\varphi(r_{ik})/kT\right] .$$

Here a factor $\exp[-\varphi(r_{12})/kT]$ can be taken outside the integral, which shows that

$$\exp[\varphi(r)/kT]n_2(r)$$

is for all r finite (and even continuous).

Splitting the range of integration in the virial theorem at $r = d_+$ we obtain

$$\frac{p}{kT} = \frac{1}{\bar{v}} + \tfrac{2}{3}\pi \int_0^{d+} dr r^3 n_2(r) \exp[\varphi(r)/kT] \exp[-\varphi(d_+)/kT] \delta(r-d)$$

$$-\tfrac{2}{3}\pi \int_{d+}^{\infty} dr r^3 \varphi'(r) n_2(r)$$

$$= \frac{1}{\bar{v}} + \tfrac{2}{3}\pi d^3 n_2(d_+) - \tfrac{2}{3}\pi \int_{d+}^{\infty} dr r^3 \varphi'(r) n_2(r) \,.$$

In particular, the equation of state of a gas of hard spheres is determined completely by the pair distribution function *at contact*,

$$\frac{p}{kT} = \frac{1}{\bar{v}} + \tfrac{2}{3}\pi d^3 n_2(d_+) \ .$$

9.19 Consider an interaction potential with a hard-core repulsion plus a weak exponential attraction of long range (see Figure 9.19.1)

$$\varphi = \begin{cases} \infty & \text{for } r < d \\ -\dfrac{a}{4\pi}\gamma^3 e^{-\gamma r} & \text{for } r > d \end{cases}$$

where a and γ are constants. Assume only one phase to be present so that the pair distribution function $n_2 \to \bar{v}^{-2}$ when $r \to \infty$ (Problem 9.14).

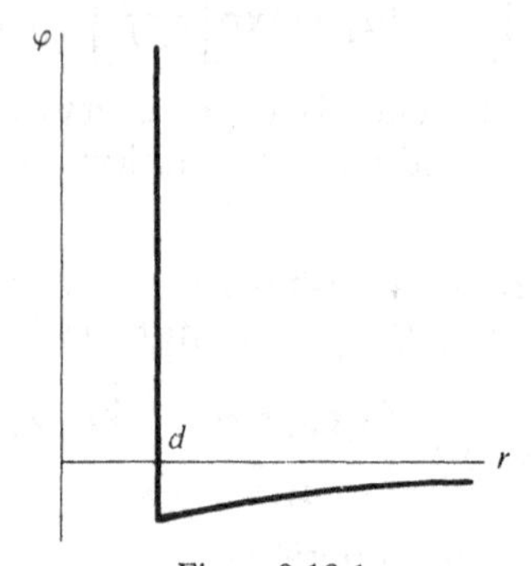

Figure 9.19.1.

Use the virial theorem of the preceding problem to obtain the following equation of state in the limit $\gamma \to 0$ (very weak attraction of very long range):

$$p = p_s - \frac{a}{\bar{v}^2} \ .$$

Here p_s is the pressure of a gas of hard spheres at the same temperature and density. Assume that because of the weakness of the attraction $n_2(d_+)$ is determined by the hard core alone.

Solution

The virial theorem yields

$$p = \frac{kT}{\bar{v}} + \tfrac{2}{3}\pi d^3 n_2(d_+) - \tfrac{1}{6}\gamma^4 \int_d^\infty r^3 e^{-\gamma r} n_2(r)\,dr\,.$$

In the integral introduce the new variable $s = \gamma r$, and take the limit $\gamma \to 0$:

$$\gamma^4 \int_d^\infty r^3 e^{-\gamma r} n_2(r)\,dr = \int_\gamma^\infty s^3 e^{-s} n_2(s\gamma^{-1})\,ds$$

$$\to \int_0^\infty s^3 e^{-s} n_2(\infty)\,ds = 6\bar{v}^{-2}\,,$$

using $n_2(\infty) = 1/\bar{v}^2$. With the assumption that when $\gamma \to 0$, $n_2(d_+) \to n_{2,s}(d_+)$, the pair distribution function of a pure hard sphere gas, we obtain in the limit

$$p = \frac{kT}{\bar{v}} + \tfrac{2}{3}\pi d^3 n_{2,s}(d_+) - \frac{a}{\bar{v}^2}\,.$$

Note that the first two terms are the pressure p_s of a hard sphere gas. Hence,

$$p = p_s - \frac{a}{\bar{v}^2}\,.$$

Note that the last term which embodies the effect of the attraction is precisely the same term that occurs in van der Waals' equation[3].

[3] For a rigorous proof of this result from the partition function see J. L. Lebowitz and O. Penrose, *J. Math. Phys.*, 7, 98 (1966). For a complete discussion of the one-dimensional version of this model see M. Kac, G. E. Uhlenbeck, and P. C. Hemmer, *J. Math. Phys.*, **4**, 216 (1963).

10
The imperfect quantum gas[1]

D.ter HAAR
(*University of Oxford, Oxford*)

THE EQUATION OF STATE

10.1 In Problem 1.11 the virial expansion $pv = E_1 + E_2 p + E_3 p^2 + \cdots$ has been considered. An alternative virial expansion is

$$pv = NkT\left(A + \frac{B}{v} + \frac{C}{v^2} + \cdots\right)$$

where the A, B, ... are volume-independent virial coefficients.

Find expressions for A, B, C in terms of the first three E's.

Solution

Straightforward series expansions give

$$A = \frac{E_1}{NkT}, \qquad B = \frac{E_1 E_2}{NkT}, \qquad C = \frac{E_2^2 E_1 + E_3 E_1^2}{NkT}.$$

10.2 Consider a system of identical particles governed by the following Hamiltonian:

$$H = H_0 + H_1, \tag{10.2.1}$$

where H_0 is given by the equation

$$H_0 = \sum_i \left(-\frac{\hbar^2}{2m}\nabla_i^2\right), \tag{10.2.2}$$

and H_1 by the equation

$$H_1 = \tfrac{1}{2}\sum_{i \neq j} U(r_{ij}), \tag{10.2.3}$$

with $r_{ij} = |\mathbf{r}_i - \mathbf{r}_j|$.

Here m is the mass of the particles, $\mathbf{r}_i$ their position (we assume them to be point particles which may have spin), and we have assumed that the interactions between the particles are binary in nature so that H_1 gives the complete interaction Hamiltonian: we thus neglect any three-body forces.

(1) For a further discussion, see D.ter Haar, *Elements of Statistical Mechanics* (Holt, Rinehart, and Winston, New York), 1954.

To obtain the equation of state, we must evaluate the grand partition function

$$\Xi = \mathrm{Tr}\exp(\alpha N_{op} - \beta H), \tag{10.2.4}$$

where β is again $1/kT$, α/β the partial thermal potential μ, and N_{op} the number operator. In fact, it is more convenient to work with the so-called q potential

$$q = \ln\Xi, \tag{10.2.5}$$

which satisfies the equations (see Problems 2.4 and 2.6c)

$$q = \beta pv, \tag{10.2.6}$$

and

$$N = \left(\frac{\partial q}{\partial \alpha}\right)_{v,T}. \tag{10.2.7}$$

Let φ_n be a complete orthonormal set of functions for a system of N identical fermions—which means that they are completely antisymmetric in the coordinates of the various particles. We now introduce a set of functions $W_N(\mathbf{r}_1, ..., \mathbf{r}_N)$ defined by the equations

$$W_N(\mathbf{r}_1, ..., \mathbf{r}_N) = N!\sum_n \varphi_n(\mathbf{r}_i)\exp(-\beta H_N)\varphi_n(\mathbf{r}_i), \tag{10.2.8}$$

where we have written H_N to emphasise that there are N particles in the system. [The sum on the right-hand side of Equation (10.2.8) is called a **Slater sum.**]

We shall introduce the quantities

$$W(\mathbf{r}_i; \mathbf{r}_i') = \sum_n \psi_n^*(\mathbf{r}_i)\exp(-\beta H_N)\psi_n(\mathbf{r}_i') \tag{10.2.9}$$

$$\Delta(\mathbf{r}_i; \mathbf{r}_i') = \sum_n \psi_n^*(\mathbf{r}_i)\psi_n(\mathbf{r}_i'), \tag{10.2.10}$$

and

$$\Delta_{qu}(\mathbf{r}_i; \mathbf{r}_i') = \frac{1}{N!}\sum_P \epsilon_P \sum_n \psi_n^*(\mathbf{r}_{Pi})\psi_n(\mathbf{r}_i), \tag{10.2.11}$$

where the summation is over all $N!$ permutations Pi of the N values of i and where $\epsilon_P = +1$ or -1 for even or odd permutations, and where in Equations (10.2.9) to (10.2.11) the ψ_n form some complete orthonormal set which does not satisfy any symmetry conditions.

Prove that

(i)
$$W_N = N!\,W(\mathbf{r}_i; \mathbf{r}_i); \tag{10.2.12}$$

(ii)
$$W(\mathbf{r}_i; \mathbf{r}_i') = \exp(-\beta H')\Delta(\mathbf{r}_i; \mathbf{r}_i'), \tag{10.2.13}$$

where the prime on the H' indicates that it operates on the $\mathbf{r}_i'$ but not on the $\mathbf{r}_i$,

(iii) if $H_1 = 0$, that is, for the case of a perfect fermion gas, then

$$W_N^{(0)}(\mathbf{r}_i) = \frac{1}{N!v_0^N}\sum_P \epsilon_P \exp\left[-\frac{\sum_i(\mathbf{r}_i - \mathbf{r}_{Pi})^2}{\lambda^2}\right], \tag{10.2.14}$$

where

$$\lambda = \left(\frac{2\beta\hbar^2}{m}\right)^{1/2} \tag{10.2.15}$$

is essentially the **thermal de Broglie wavelength,** and

$$v_0 = \pi^{3/2}\lambda^3 . \tag{10.2.16}$$

Solution

(iii) We note that

$$\Delta(\mathbf{r}_i;\mathbf{r}_i') = \Delta_{qu}(\mathbf{r}_i;\mathbf{r}_i') ,$$

when the proper symmetry requirements have been taken into consideration. In Equation (10.2.11) the summation over n is over a complete set irrespective of symmetry requirements. In that case, we can use the completeness relation for the ψ_n and write instead of Equation (10.2.11)

$$\Delta_{qu}(\mathbf{r}_i;\mathbf{r}_i') = \frac{1}{N!}\sum_P \epsilon_P \prod_i \delta(\mathbf{r}_{Pi}-\mathbf{r}_i') .$$

In the case of a perfect gas H is given by Equation (10.2.2), and, if we write the three-dimensional Dirac delta function in the form

$$\delta(\mathbf{r}-\mathbf{r}') = \frac{1}{(2\pi)^3}\int \exp[i(\mathbf{k}\cdot\mathbf{r}-\mathbf{r}')]\,d^3\mathbf{k} ,$$

we have

$$\nabla'^2\delta(\mathbf{r}-\mathbf{r}') = \int -\frac{k^2}{(2\pi)^3}\exp[i(\mathbf{k}\cdot\mathbf{r}-\mathbf{r}')]\,d^3\mathbf{k} .$$

and hence

$$\exp(-\beta H_0)\,\delta(\mathbf{r}-\mathbf{r}') = \frac{1}{v_0}\exp\left[-\frac{(\mathbf{r}-\mathbf{r}')^2}{\lambda^2}\right] ,$$

where v_0 and λ are given by Equations (10.2.16) and (10.2.15).

Combining the various results, we now get the result (10.2.14).

10.3 Show that the above results remain valid for a system of N bosons, when the φ_n are symmetric in the particles, and $\epsilon_P = 1$.

Solution

The proof proceeds as in Problem 10.2.

10.4 N identical particles are said to form separate clusters of N_1 and N_2 particles ($N_1+N_2 = N$) if the particles of the first are separated from each of the particles of the second by at least a distance D, where D is such that $U(r) = 0$ when $r > D$, and also $D \gg \lambda$. Prove from Problem 10.2 that

$$W_N = W_{N_1} W_{N_2} . \tag{10.4.1}$$

Solution

The proof consists of two parts. First of all, we note that, if the N particles form two clusters, all terms where the kth particle and the Pkth particle are in different clusters will have a factor $(\mathbf{r}_k - \mathbf{r}_{Pk})^2/\lambda^2$ in the exponent and this will mean that those terms can be neglected, as $|\mathbf{r}-\mathbf{r}_{Pk}| > D \gg \lambda$. We have thus

$$W_N^{(0)} = \frac{1}{v_0^{N_1}} \sum_{P_1} \epsilon_{P_1} \exp\left[-\sum_{i_1} \frac{(\mathbf{r}_{i_1} - \mathbf{r}_{P_1 i_1})^2}{\lambda^2}\right] \times \frac{1}{v_0^{N_2}} \sum_{P_2} \epsilon_{P_2} \exp\left[-\sum_{i_2} \frac{(\mathbf{r}_{i_2} - \mathbf{r}_{P_2 i_2})^2}{\lambda^2}\right]$$

$$= W_{N_1}^{(0)} \cdot W_{N_2}^{(0)} ,$$

where the sums over P_1 (and i_1) and P_2 (and i_2) extend only over the first and second cluster, respectively, and where we have used the fact that $\epsilon_P = \epsilon_{P_1} \epsilon_{P_2}$.

The above argument suffices for the case of a perfect gas. If there are interactions, we note that, if the N particles form two clusters, we have

$$H_N = H_{N_1} + H_{N_2} \quad \text{or} \quad \exp(-\beta H_N) = \exp(-\beta H_{N_1}) \exp(-\beta H_{N_2}) ,$$

and because of this and the product property of the W_N for the case of a perfect gas, Equation (10.4.1) follows.

10.5 Show that the W satisfy the **Bloch equation**

$$\frac{\partial W(\mathbf{r}_i ; \mathbf{r}_i')}{\partial \beta} = -H' W(\mathbf{r}_i ; \mathbf{r}_i') , \tag{10.5.1}$$

where H' operates on the $\mathbf{r}_i'$ and not on the $\mathbf{r}_i$.

Solution

From Equation (10.2.13) we see that

$$W(\mathbf{r}_i ; \mathbf{r}_i') |_{\beta=0} = \Delta(\mathbf{r}_i ; \mathbf{r}_i') ,$$

and hence that

$$W(\mathbf{r}_i ; \mathbf{r}_i' ; \beta) = \exp(-\beta H') W(\mathbf{r}_i ; \mathbf{r}_i' ; 0) .$$

Equation (10.5.1) follows immediately.

10.6 Prove that q and W_N are related through the equation

$$e^q = \sum_n \frac{e^{n\alpha}}{n!} \int W_N(\mathbf{r}_i) d^3\mathbf{r}_1 \dots d^3\mathbf{r}_N . \tag{10.6.1}$$

Solution

Equation (10.6.1) follows immediately by using Equation (10.2.8).

10.7 Use the result of Problems 10.5 and 10.6 to prove that in the classical limit ($\hbar \to 0$ or $\lambda \to 0$) the W_N go over into the W_N of classical theory. From Equation (10.6.1) and the expressions in Chapters 2 and 9 we see that

$$W_N^{\text{cl}} = v_0^{-N} \exp(-\beta H_1) . \tag{10.7.1}$$

In the proof it is convenient to write the $W_N(\mathbf{r}_i)$ in the form

$$W_N(\mathbf{r}_i) = \sum_P \epsilon_P \exp g(\mathbf{r}_i; \mathbf{r}_{Pi}) , \tag{10.7.2}$$

where

$$g(\mathbf{r}_i; \mathbf{r}_i') = \ln\left[\exp(-\beta H')\prod_i \delta(\mathbf{r}_i - \mathbf{r}_i')\right] . \tag{10.7.3}$$

Solution

The function g satisfies the equation

$$\frac{\partial g}{\partial \beta} = \sum_i \frac{\hbar^2}{2m}\nabla_i'^2 g + \sum_i \frac{\hbar^2}{2m}(\nabla_i' g \cdot \nabla_i' g) - H_1 g .$$

This equation can be solved as follows:

$$g = -\frac{m}{2\beta\hbar^2}\sum_i (\mathbf{r}_i - \mathbf{r}_i')^2 - N\ln v_0 - \beta H_1 + \text{power series in } \frac{\beta\hbar^2}{2m} .$$

The first part is the solution when H_1 = constant and for details of obtaining the power series, we refer to the solution to Problem 10.9. In the limit as $\lambda \to 0$ (or $v_0 \to 0$) we need retain in the sum over P in the expression for $W_N(\mathbf{r}_i)$ only the term with the identical permutation. Moreover, in the same limit, we can drop the power series in $\beta\hbar^2/2m$. Hence we find for W_N

$$W_N \to \exp g(\mathbf{r}_i, \mathbf{r}_i) \to N\ln v_0 - \beta H_1 .$$

10.8 As the equation of state follows from Equations (10.2.6) and (10.2.7) and as Equation (10.6.1) is formally the same as in the classical case with the W_N of Equation (10.2.8) replacing the W_N of Equation (10.7.1), we get formally the same expressions for the virial coefficients.

Find the expression for the second virial coefficient B for a quantum gas.

Solution

From the cluster expansion for the equation of state we have

$$B = \frac{Nv_0^2}{2v}\int [W_2(\mathbf{r}_1, \mathbf{r}_2) - W_1(\mathbf{r}_1)W_1(\mathbf{r}_2)]\, d^3\mathbf{r}_1 d^3\mathbf{r}_2 ,$$

which can be written in the form

$$B = -\frac{Nv_0^2}{2v}\int [2\exp(-\beta H_2')\Delta(\mathbf{r}_1, \mathbf{r}_2; \mathbf{r}_1', \mathbf{r}_2')$$
$$- \exp(-\beta H_2^{(0)\prime})\Delta(\mathbf{r}_1; \mathbf{r}_1')\Delta(\mathbf{r}_2; \mathbf{r}_2')]_{\mathbf{r}_i' = \mathbf{r}_i} d^3\mathbf{r}_1 d^3\mathbf{r}_2 .$$

We note that quantum effects and interaction effects are completely mixed up.

10.9 Expand the expression for B found in the preceding problem, for high temperatures, in a power series in $\hbar$, up to the term in $\hbar^2$.

Solution

The second term in the square brackets can be found easily using the result of Problem 10.2 and we can thus rewrite the expression for B as follows:

$$B = \frac{N}{2v}\int [1 - 2v_0^2\{\exp(-\beta H_2')\Delta(\mathbf{r}_1, \mathbf{r}_2; \mathbf{r}_1', \mathbf{r}_2')\}]_{\mathbf{r}_i' = \mathbf{r}_i} \mathrm{d}^3\mathbf{r}_1 \mathrm{d}^3\mathbf{r}_2 \, .$$

Introducing centre of mass and relative coordinates, we can integrate over the former which gives a factor v, and we are left with

$$B = \tfrac{1}{2}N\int [1 - \mathrm{f}(\mathbf{r}_{12}; \beta)]\, \mathrm{d}^3\mathbf{r}_{12} \, ,$$

where

$$\mathrm{f}(\mathbf{r};\beta) = 2^{3/2} v_0 \left[\exp\left(\beta\frac{\hbar^2}{m}\nabla'^2 - \beta U\right)\{\delta(\mathbf{r}-\mathbf{r}') \pm \delta(\mathbf{r}+\mathbf{r}')\}\right]_{\mathbf{r}'=\mathbf{r}} ,$$

where the upper (lower) sign refers to the case of bosons (fermions). We stress that it will pay the reader to derive the above results in detail.

We note that the above equation for $\mathrm{f}(\mathbf{r};\beta)$ can be written in the form

$$\mathrm{f}(\mathbf{r};\beta) = \exp\left(\beta\frac{\hbar^2}{m}\nabla^2 - \beta U\right) \mathrm{f}(\mathbf{r};0)$$

and that $\mathrm{f}(\mathbf{r};\beta)$ thus satisfies a Bloch-type equation

$$\frac{\partial \mathrm{f}}{\partial \beta} = \frac{\hbar^2}{m}\nabla^2 \mathrm{f} - U\mathrm{f} \, .$$

If we introduce a function $\mathrm{g}(\mathbf{r};\mathbf{r}')$ by the equation

$$\mathrm{g}(\mathbf{r};\mathbf{r}') = \ln\left[\exp\left(\beta\frac{\hbar^2}{m}\nabla'^2 - \beta U\right)\delta(\mathbf{r}-\mathbf{r}')\right] ,$$

we have

$$\mathrm{f}(\mathbf{r};\beta) = 2^{3/2} v_0 \{\exp[\mathrm{g}(\mathbf{r};\mathbf{r})] \pm \exp[\mathrm{g}(\mathbf{r};-\mathbf{r})]\} ,$$

while g satisfies the equation

$$\frac{\partial \mathrm{g}}{\partial \beta} = \frac{\hbar^2}{m}\nabla^2 \mathrm{g} + \frac{\hbar^2}{m}(\nabla \mathrm{g} \cdot \nabla \mathrm{g}) - U \, .$$

We look for a solution in the form

$$\mathrm{g}(\mathbf{r};\mathbf{r}') = \frac{m(\mathbf{r}-\mathbf{r}')^2}{4\beta\hbar^2} - \tfrac{3}{2}\ln\frac{4\pi\beta\hbar^2}{m} - \sum_{n=1}^{\infty} a_n \beta^n \, .$$

Substituting this expression into the equation for g we can solve it by successive approximations, and we find for $g(\mathbf{r};\mathbf{r})$

$$g(\mathbf{r};\mathbf{r})+\tfrac{3}{2}\ln\frac{4\pi\beta\hbar^2}{m} = -\beta U+\frac{\beta^2\hbar^2}{m}\left[-\tfrac{1}{6}\nabla^2 U+\frac{\beta}{12}(\nabla U\cdot\nabla U)\right]+\cdots .$$

We do not need the expression for $g(\mathbf{r};-\mathbf{r})$ as we are looking for a high-temperature expression in which case $\exp[g(\mathbf{r};-\mathbf{r})]$ will contain a factor $\exp(-mr^2/\beta\hbar^2)$ which will lead to an extra factor v_0 which is negligible.

We finally find for $f(\mathbf{r};\beta)$

$$f(\mathbf{r};\beta) = \exp[-\beta U(r)]\left\{1+\frac{\hbar^2}{m}\left[-\frac{\beta^2}{6}\nabla^2 U+\frac{\beta^3}{12}(\nabla U\cdot\nabla U)\right]+\cdots\right\}.$$

Substituting this expression into the equation for B we find

$$B = B_{\text{cl}}+\hbar^2 B_1+\cdots ,$$

where B_{cl} is the classical result [see Problem 9.8(d)].

10.10 Split the expression for B found in Problem 10.8 into two parts: B_{perf} corresponding to a perfect quantum gas, and B_{imp} containing both quantum and interaction effects. Prove that the second part can be written in the form

$$B_{\text{imp}} = -\frac{2^{3/2}}{\pi}Nv_0\sum_l (2l+1)\delta_l\int_0^\infty \frac{d\eta}{dk}\exp\left(-\frac{\beta\hbar^2 k^2}{m}\right)dk , \quad (10.10.1)$$

where $\delta_l = 1(0)$ when l is even and $0(1)$ when l is odd for the case of bosons (fermions), and where η is the phase of the asymptotic solution of the radial Schrödinger equation,

$$\frac{d^2R}{dr^2}+\left[k^2-\frac{m}{\hbar^2}U(r)-\frac{l(l+1)}{r^2}\right]R = 0 \quad (10.10.2)$$

and

$$R \sim \sin[kr+\eta(k,l)] \ \text{as}\ r\to\infty . \quad (10.10.3)$$

Solution

We can rewrite the second expression as follows:

$$B = B_{\text{perf}}+B_{\text{imp}} ,$$

with

$$B_{\text{perf}} = -\frac{Nv_0^2}{2v}\int[\exp(-\beta H_2^{(0)\prime})\{2\Delta(\mathbf{r}_1,\mathbf{r}_2;\mathbf{r}_1',\mathbf{r}_2') - \Delta(\mathbf{r}_1;\mathbf{r}_1')\Delta(\mathbf{r}_2;\mathbf{r}_2')\}]_{\mathbf{r}_i'=\mathbf{r}_i}d^3\mathbf{r}_1 d^3\mathbf{r}_2 ,$$

$$B_{\text{imp}} = -\frac{Nv_0^2}{v}\int[\{\exp(-\beta H_2')-\exp(-\beta H_2^{(0)\prime})\}\Delta(\mathbf{r}_1,\mathbf{r}_2;\mathbf{r}_1',\mathbf{r}_2')]_{\mathbf{r}_i'=\mathbf{r}_i}d^3\mathbf{r}_1 d^3\mathbf{r}_2 .$$

It is a straightforward exercise to evaluate B_{perf} with the result

$$B_{\text{perf}} = \mp \frac{Nv_0}{2^{5/2}} ,$$

with the upper (lower) sign again referring to bosons (fermions).

In fact, it is more convenient to write B_{imp} in the form

$$B_{\text{imp}} = -\frac{Nv_0^2}{v}[\text{Tr}\exp(-\beta H_2) - \text{Tr}\exp(-\beta H_2^{(0)})] ,$$

or

$$B_{\text{imp}} = -\frac{Nv_0^2}{v}\left[\sum_k \exp(-\beta \epsilon_k) - \sum_k \exp(-\beta \epsilon_k^{(0)})\right] ,$$

where the $\epsilon_k(\epsilon_k^{(0)})$ are the eigenvalues of $H_2(H_2^{(0)})$.

Introducing centre of mass $[\mathbf{R} = \frac{1}{2}(\mathbf{r}_1 + \mathbf{r}_2)]$ and relative $(\mathbf{r}_{12})$ coordinates, we get the eigenfunctions of H_2 and $H_2^{(0)}$ in the form

$$\varphi_k(\mathbf{r}_1, \mathbf{r}_2) = \exp\left[\frac{i}{\hbar}(\mathbf{P}\cdot\mathbf{R})\right]\frac{R_{nl}(r_{12})}{r_{12}}\delta_l Y_l^m(\omega) ,$$

$$\epsilon_k = \frac{p^2}{4m} + E_{n,l} ,$$

where $\omega = \mathbf{r}_{12}/r_{12}$, Y_l^m is a spherical harmonic, and $\delta_l = 1(0)$, l = even; $\delta_l = 0(1)$, l = odd for bosons (fermions).

The $R_{n,l}$ satisfy the equations

$$\frac{d^2R}{dr^2} + \left\{\frac{m}{\hbar^2}[E_{n,l} - U(r)] - \frac{l(l+1)}{r^2}\right\}R = 0 ,$$

and the same equation with $U = 0$ for $R^{(0)}$.

The contribution from the centre of mass gives a factor $2^{3/2}V/v_0$, and we have

$$B_{\text{imp}} = -2^{3/2}Nv_0\sum_l (2l+1)\delta_l \int_0^\infty \exp(-\beta E)\rho(E,l)\,dE ,$$

where we have assumed (i) that there are no discrete energy levels, (ii) that we may replace the summation over n by an integration over E, and (iii) that the numbers of energy levels E_{nl} and $E_{nl}^{(0)}$ between E and $E+dE$ is given by $\rho_1(E;l)\,dE$ and $\rho_0(E;l)\,dE$ with $\rho(E;l) = \rho_1(E;l) - \rho_0(E;l)$.

Introducing the wave number k by

$$E = \frac{\hbar^2k^2}{m} ,$$

we have

$$B_{\text{imp}} = -2^{3/2}Nv_0\sum_l (2l+1)\delta_l \int_0^\infty \exp\left[-\frac{\beta\hbar^2k^2}{m}\right]g(k,l)\,dk ,$$

and the equations for the $R_{n,l}$ are now

$$\frac{d^2R}{dr^2}+\left[k^2-\frac{m}{\hbar^2}U(r)-\frac{l(l+1)}{r^2}\right]R = 0 .$$

For large values of r, $U(r)$ will vanish, and the asymptotic solution for R is

$$R = \sin[kr+\eta(k,l)] ,$$

with the boundary condition (corresponding to the vanishing of the wave-function at the boundary of the volume) $R(r_0) = R^{(0)}(r_0) = 0$, where r_0 is large. This boundary condition leads to

$$kr_0+\eta(k,l) = n\pi ,$$

$$kr_0+\eta^{(0)}(k,l) = n\pi . \qquad (10.10.4)$$

If Δk and $\Delta k^{(0)}$ are the changes in k when n changes by unity, we have clearly

$$\rho(k,l) = \frac{1}{\Delta k}-\frac{1}{\Delta k^{(0)}} ,$$

and from Equation (10.10.4) we have

$$\left(r_0+\frac{d\eta}{dk}\right)\Delta k = \pi ,$$

$$\left(r_0+\frac{d\eta^{(0)}}{dk}\right)\Delta k^{(0)} = \pi ,$$

and hence Equation (10.10.1) follows.

SECOND QUANTISATION FORMALISM[2]

10.11 For many purposes it is convenient to use a formalism which is independent of the number of particles in the system. Let φ_n be now a complete orthonormal set (c.o.s.) of single-particle functions (including the spin dependence where necessary). For a system of N identical systems one can then use as a c.o.s. of basis functions the set

$$|i_1, i_2 ..., i_N\rangle = \frac{1}{N!}\sum_P \epsilon_P \varphi_{i_1}(P1)\varphi_{i_2}(P2) ... \varphi_{i_N}(PN) . \qquad (10.11.1)$$

This is a properly symmetrised set. The bra set corresponding to the ket set (10.11.1) is

$$\langle i_1, i_2, ..., i_N| = \frac{1}{N!}\sum_P \epsilon_P \varphi^*_{i_1}(P1)\varphi^*_{i_2}(P2) ... \varphi^*_{i_N}(PN) . \qquad (10.11.2)$$

[2] For a general reference, see J.de Boer, *Progress in Low Temperature Physics,* Vol.III, (North-Holland, Amsterdam), 1965, p.215.

(a) Prove that the sets (10.11.1) and (10.11.2) satisfy the orthonormality relation

$$\langle i'_1, i'_2, ..., i'_N \mid i_1, i_2, ..., i_N \rangle = \frac{1}{N!}\sum_P \epsilon_P \delta(i_1 - i'_{P1}) \dots \delta(i_N - i'_{PN}), \quad (10.11.3)$$

and a completeness or closure relation. We assumed here that the index i is a continuous parameter. If it is a discrete one, the Dirac δ functions must be replaced by Kronecker ones.

(b) If $|\Psi\rangle$ is a properly symmetrised wave function of the N-particle system, express $|\Psi\rangle$ in terms of the set (10.11.1). Also give an expression for an operator Ω operating on functions in the Hilbert space spanned by the set (10.11.1).

Solution

(b)
$$|\Psi\rangle = \int \dots \int \mathrm{d}i_1 \dots \mathrm{d}i_N |i_1, ..., i_N\rangle\langle i_1, ..., i_N|\Psi\rangle,$$

where $\int \mathrm{d}i$ indicates integrating over the coordinates of the ith particle (and if necessary summing over its spin variables).

$$\Omega = \int \dots \int \mathrm{d}i_1 \dots \mathrm{d}i_N \mathrm{d}i'_1 \dots \mathrm{d}i'_N |i_1, .., i_N\rangle\langle i_1, ..., i_N|\Omega|i'_1, ..., i'_N\rangle\langle i'_1, ..., i'_N| .$$

10.12 (a) Consider systems with an arbitrary number of particles, that is, consider the Hilbert space which is the (direct) product space of the Hilbert spaces corresponding to 0, 1, ..., N, ... particles. The c.o.s. spanning this Hilbert space will be the set

$$|0\rangle, |i\rangle, |i_1, i_2\rangle, ..., |i_1, i_2, ..., i_N\rangle, ... , \quad (10.12.1)$$

where $|0\rangle$ is the vacuum state. We now introduce **creation** (or construction) **operators** which will produce an eigenvector corresponding to $N+1$ particles from one corresponding to N particles, as follows:

$$\mathrm{a}^{+}(i)|i_1, ..., i_N\rangle = (N+1)^{1/2}|i, i_1, ..., i_N\rangle . \quad (10.12.2)$$

Express the $|i_1, ..., i_N\rangle$ in terms of the $\mathrm{a}^{+}(i)$ and the vacuum state.

(b) If $[\mathrm{A}, \mathrm{B}]_- = \mathrm{AB} - \mathrm{BA}$ is the commutator, and $[\mathrm{A}, \mathrm{B}]_+ = \mathrm{AB} + \mathrm{BA}$ the anticommutator, of the two operators A and B, prove that

$$[\mathrm{a}^{+}(i), \mathrm{a}^{+}(j)]_- = 0 \quad \text{for bosons,}$$

and (10.12.3)

$$[\mathrm{a}^{+}(i), \mathrm{a}^{+}(j)]_+ = 0 \quad \text{for fermions.}$$

(c) We now introduce an operator $\mathrm{a}(i)$ by the equation

$$\langle i_1, ..., i_N|\mathrm{a}(i) = (N+1)^{1/2}\langle i, i_1, ..., i_N| . \quad (10.12.4)$$

Prove that

$$[a(i), a(j)]_- = 0 \quad \text{for bosons,}$$

and (10.12.5)

$$[a(i), a(j)]_+ = 0 \quad \text{for fermions.}$$

(d) Find an expression for $a(i)|i_1, ..., i_n\rangle$.
(e) Prove that

$$[a(i), a^+(j)]_- = \delta(i-j) \text{ for bosons,}$$

and (10.12.6)

$$[a(i), a^+(j)]_+ = \delta(i-j) \text{ for fermions.}$$

(f) If Ω is an operator of the form

$$\Omega = \sum_i \Omega_i^{(1)} + \tfrac{1}{2} \sum_{i,j} \Omega_{ij}^{(2)}, \tag{10.12.7}$$

where the $\Omega_i^{(1)}$ and $\Omega_{ij}^{(2)}$ are, respectively, single-particle and two-particle operators which differ only in the particles on which they operate, express Ω in terms of the $a^+(i)$ and $a(i)$.

Solution

(a)

$$|i_1, ..., i_N\rangle = \frac{1}{N!} a^+(i_1) a^+(i_2) \ldots a^+(i_N)|0\rangle$$

(b) This follows from Equations (10.12.1) and (10.11.1).
(c) This follows from Equations (10.12.4) and (10.11.1).
(d) By considering the matrix element $\langle i_1', ..., i_{N-1}' | a(i) | i_1', ..., i_N\rangle$ and bearing in mind that the resulting expression is valid for any $\langle i_1', ..., i_{N-1}' |$ we find

$$a(i)|i_1, ..., i_N\rangle = N^{-1/2}\{\delta(i-i_1)|i_2, ..., i_N\rangle + \ldots + (\pm 1)^{N-1}\delta(i-i_N)|i_1, ..., i_{N-1}\rangle\},$$

where the upper (lower) sign again refers to the case of bosons (fermions). From this equation it follows that the $a(i)$ are **annihilation operators.**

(e) This follows from Equations (10.12.1), (10.12.4), and the equation given under (d).

(f)

$$\Omega = \int di\, di' \langle i|\Omega^{(1)}|i'\rangle a^+(i) a(i')$$

$$+ \tfrac{1}{2} \int di\, dj\, di'\, dj' \langle ij|\Omega^{(2)}|i'j'\rangle a^+(i) a^+(j) a(j') a(i') . \tag{10.12.8}$$

10.13 Consider a system contained in a finite volume v so that we can take for the original c.o.s. a set of plane waves,

$$\varphi_i \to v^{-1/2} \exp[i(\mathbf{k} \cdot \mathbf{r})] . \tag{10.13.1}$$

In that case, one usually writes $a_{\mathbf{k}}^{+}$ and $a_{\mathbf{k}}$ rather than $a^{+}(\mathbf{k})$ and $a(\mathbf{k})$.

If Ω_i and Ω_{ij} are of the form

$$\Omega_i = -\frac{\hbar^2}{2m}\nabla_i^2 + U(r), \quad \Omega_{ij} = V(r_{ij}), \tag{10.13.2}$$

find an expression for Ω in terms of the $a_{\mathbf{k}}^{+}$ and $a_{\mathbf{k}}$.

Solution

$$\Omega = \sum_{\mathbf{k}} \frac{\hbar^2 k^2}{2m} a_{\mathbf{k}}^{+} a_{\mathbf{k}}^{+} + \frac{1}{v}\sum_{\mathbf{k},\mathbf{q}} U(\mathbf{q}) a_{\mathbf{k}}^{+} a_{\mathbf{k}-\mathbf{q}} + \frac{1}{2v}\sum_{\mathbf{k},\mathbf{k}',\mathbf{q}} V(\mathbf{q}) a_{\mathbf{k}}^{+} a_{\mathbf{k}'}^{+} a_{\mathbf{k}'+\mathbf{q}} a_{\mathbf{k}-\mathbf{q}},$$

where

$$U(\mathbf{q}) = \int d^3\mathbf{r} \exp[-i(\mathbf{q}\cdot\mathbf{r})] U(r), \quad V(\mathbf{q}) = \int d^3\mathbf{r} \exp[-i(\mathbf{q}\cdot\mathbf{r})] V(r).$$

10.14 Give an expression for

$$a^{+}(i)a(i)\,|\,i_1, \ldots, i_N\rangle,$$

and discuss the physical meaning of the operator

$$n(i) = a^{+}(i)a(i). \tag{10.14.1}$$

Solution

$$a^{+}(i)a(i)\,|\,i_1, \ldots, i_N\rangle = N(i)\,|\,i_1, \ldots, i_N\rangle,$$

where $N(i)$ is an integer which tells us how often i occurs among the numbers $i_1, \ldots, i_N$. The operator $n(i)$ is thus an occupation number operator.

11

Phase transitions

D.ter HAAR
(*University of Oxford, Oxford*)

11.0 Let us remind the reader that for a grand canonical ensemble we have (Problem 10.2)

$$q = \ln \Xi , \tag{10.2.5}$$

$$q = \beta p v , \tag{10.2.6}$$

and

$$n = \frac{1}{v}\left(\frac{\partial q}{\partial \alpha}\right)_{v,T} , \tag{10.2.7'}$$

where $n = N/v = 1/v_1$, with $v_1 = v/N$ the specific volume.

EINSTEIN CONDENSATION OF A PERFECT BOSON GAS[1]

11.1 For a perfect boson gas with single-particle quantum states j of energy E_j we have (see Problem 3.12a)

$$q = -\sum_j \ln(1-t_j) , \tag{11.1.1}$$

where

$$t_j = \exp(\alpha - \beta E_j) . \tag{11.1.2}$$

If we consider the case of spin-zero point particles in a force-free volume, the number of energy levels dZ lying between E and $E+\mathrm{d}E$ is given by the expression (compare Problem 3.13c)

$$\mathrm{d}Z = 2\pi\left(\frac{2m}{h^2}\right)^{3/2} v E^{1/2}\, \mathrm{d}E . \tag{11.1.3}$$

From Equation (11.1.2) it is clear that we must always have

$$\alpha < \beta E_j \tag{11.1.4}$$

so that t_j is always less than unity.

Put

$$y = \mathrm{e}^{\alpha} \tag{11.1.5}$$

[1] For a general discussion see, e.g. D.ter Haar, *Elements of Thermostatistics* (Holt, Rinehart, and Winston, New York), 1966.

and show that the equation of state follows from the equations

$$\beta p v_0 = \mathrm{f}(y)\,, \tag{11.1.6}$$

$$\frac{v_0}{v_1} = y\mathrm{f}'(y), \tag{11.1.7}$$

where

$$\mathrm{f}(y) = \sum_{n=1}^{\infty} \frac{y^n}{n^{5/2}}\,, \tag{11.1.8}$$

and where v_0 is given by Equation (10.2.16).

Obtain a virial expansion in the form of a series expansion of $\beta p v_1$ in a power series of v_1^{-1}.

Solution

Combining Equations (11.1.1), (11.1.2), (11.1.3), and (10.2.6) and going over to an integral from the sum in Equations(11.1.1), we find

$$q = -2\pi\left(\frac{2m}{h^2}\right)^{3/2} v \int_0^\infty \sqrt{E}\ln(1-\mathrm{e}^{\alpha-\beta E})\,\mathrm{d}E\,. \tag{11.1.9}$$

Expanding the logarithm, and integrating, we have

$$q = \left(\frac{2\pi m}{\beta h^2}\right)^{3/2} v \sum_{n=1}^{\infty} \frac{\mathrm{e}^{n\alpha}}{n^{5/2}}\,. \tag{11.1.10}$$

From Equation (10.2.7′) it then follows that

$$N = \left(\frac{2\pi m}{\beta h^2}\right)^{3/2} v \sum_{n=1}^{\infty} \frac{\mathrm{e}^{n\alpha}}{n^{3/2}}\,, \tag{11.1.11}$$

and Equations (11.1.6) and (11.1.7) follow immediately.

To obtain the virial expansion, we must express y as a power series in v_1^{-1} from Equation (11.1.11) and substitute this into Equation (11.1.10). This can be done either by tedious sorting out, or more elegantly as follows. If we write $x = v_0/v_1$, we have

$$x = y\mathrm{f}'(y)\,, \tag{11.1.12}$$

and we want to find the expansion

$$\frac{\mathrm{f}(y)}{y\mathrm{f}'(y)} = \frac{\mathrm{f}}{x} = \sum a_n x^n\,. \tag{11.1.13}$$

Let y_0 be the value of y satisfying Equation (11.1.12). From the theory of functions of a complex variable it follows that

$$\frac{1}{2\pi\mathrm{i}}\oint \mathrm{f}(y)\,\mathrm{d}\ln[y\mathrm{f}'(y)-x] = \mathrm{f}(y_0)\,, \tag{11.1.14}$$

where the contour lies in the complex y-plane and is going around the origin and around y_0.

If we write

$$yf'(y)-x = yf'(y)\left[1-\frac{x}{yf'(y)}\right] , \tag{11.1.15}$$

we can expand the logarithm and we find

$$f(y_0) = \frac{1}{2\pi i}\oint f(y)d\left\{\ln yf'(y) - \sum_{n=0}^{\infty}\frac{1}{n+1}\left[\frac{x}{yf'(y)}\right]^{n+1}\right\}$$

$$= \frac{1}{2\pi i}\sum_{n=0}^{\infty}\frac{x^{n+1}}{n+1}\oint\frac{[f'(y)]^{-n}dy}{y^{n+1}} , \tag{11.1.16}$$

from which it follows that $(n+1)a_n$ is the coefficient of y^n in $[f'(y)]^{-n}$. In this way we find

$$\beta pv_1 = N\left[1-\frac{x}{2^{5/2}}+\cdots\right] . \tag{11.1.17}$$

11.2 It has been shown[2] that when $y < 1$ the functions $f(y)$ and $f'(y)$ can be written in the form

$$f(y) = 2{\cdot}36\,(-\ln y)^{3/2}+1{\cdot}34+2{\cdot}61\ln y-0{\cdot}73\,(\ln y)^2+\cdots , \tag{11.2.1}$$

$$f'(y) = -3{\cdot}54\,(-\ln y)^{1/2}+2{\cdot}61-1{\cdot}46\ln y-0{\cdot}10\,(\ln y)^2+\cdots . \tag{11.2.2}$$

We see that these functions have a branch point at $y = 1$ and are not defined for $y > 1$ (the power series no longer converge). As long as the specific volume v_1 is larger than a critical value v_c given by the equation

$$\frac{1}{v_c} = \frac{1}{v_0}f'(1) = \frac{2{\cdot}61}{v_0} , \tag{11.2.3}$$

the parametric form of the equation of state of Problem 11.1, Equations (11.1.6) and (11.1.7), can be used.

We must now discuss what happens when $v_1 < v_c$. The difficulty arises because, when v_1 is decreased from very large values ($v_1 \gg v_c$), the solution y of Equation (11.1.7) will increase and will approach 1 as v_1 approaches v_c. This means that α approaches zero starting from large negative values for $v_1 \gg v_c$. This in turn means that for the lowest energy level E_0, which we had put equal to zero when deriving Equation (11.1.6), $t_j = t_0$ will approach unity, and $\ln(1-t_0)$ will approach $-\infty$. This entails that the corresponding term in the sum on the right-hand side of Equation (11.1.1) dominates. This term represents the number of particles in the level E_0 (see Problem 3.12).

Assuming N to be large but finite, find the value of y for the case $v_1 < v_c$, by splitting off from the sum in Equation (11.1.2) the term with $j = 0$. Assume the state $j = 0$ to be non-degenerate. Hence find an expression for that part of the isotherm for which $v_1 < v_c$.

[2] W.Opechowski, *Physica*, 4, 722 (1937).

Solution

From Equations (11.1.1), (11.1.2), and (10.2.7′) we have

$$N = \sum_j \frac{1}{\exp(-\alpha+\beta E_j)-1}$$
$$= \frac{1}{\exp(-\alpha+\beta E_0)-1} + \sum_{j\neq 0} \frac{1}{\exp(-\alpha+\beta E_j)-1} . \qquad (11.2.4)$$

If we choose $E_0 = 0$ [corresponding to the lower limit on the integral in Equation (11.1.9)], we can write Equation (11.2.4) in the form

$$N = \frac{1}{e^{-\alpha}-1} + \sum_{j\neq 0} \frac{1}{\exp(-\alpha+\beta E_j)-1} . \qquad (11.2.5)$$

Using the fact that the first term is the number N_0 of bosons in the lowest energy state, and replacing the sum in Equation (11.2.5) by an integral, we get

$$N = N_0 + \frac{v}{v_0} y f'(y) , \qquad (11.2.6)$$

and using Equation (11.2.3) we find

$$1 = \frac{N_0}{N} + \frac{v_1}{v_c}\frac{y f'(y)}{f'(1)} . \qquad (11.2.7)$$

If $v_1 > v_c$, we expect y to be about equal to 1. Let us put $y = 1$ in the last term on the right-hand side of Equation (11.2.7) and afterwards verify that the errors made are negligible. For N_0 we write

$$N_0 = \frac{1}{e^{-\alpha}-1} = \frac{y}{1-y} \approx \frac{1}{1-y} . \qquad (11.2.8)$$

We then have from Equation (11.2.7)

$$1 = \frac{1}{N(1-y)} + \frac{v_1}{v_c} , \quad (v_1 < v_c) \qquad (11.2.9)$$

or

$$y = 1 - \frac{1}{N}\frac{v_c}{v_c - v_1} , \qquad (11.2.10)$$

which, indeed, as $N \to \infty$ leads to $y \to 1$. From the expansions (11.2.1) and (11.2.2) we can easily estimate the magnitude of the terms which we have neglected.

Substituting Equation (11.2.10) into Equation (11.1.6) we find that for $v_1 < v_c$ the isotherm is given by the equation

$$\beta p = \frac{1}{v_0} f(1) , \qquad (11.2.11)$$

that is, the isotherm is horizontal.

We also note that now

$$\frac{N_0}{N} = 1 - \frac{v_1}{v_c}, \tag{11.2.12}$$

that is, a finite fraction of the system is in the lowest energy state. This is called the **Einstein condensation.**

11.3 Using the expansions (11.2.1) and (11.2.2) show that on the isotherm at $v = v_c$, $\partial^n p/\partial v^n$ is of the order of $N^{(n-2)/3}$ for large N.[3]

Solution

If $v_1 \approx v_c$, we must take, instead of only the leading term, $2{\cdot}61 \equiv f'(1)$, on the right-hand side of Equation (11.2.2), the first two terms, and we get instead of Equation (11.2.9) the equation

$$N = \frac{1}{-\alpha} + \frac{Nv_1}{v_0}\left[\frac{v_0}{v_c} - 3{\cdot}54(-\alpha)^{1/2}\right], \tag{11.3.1}$$

so that for $v_1 = v_c$ we have

$$-\alpha = \left(\frac{3{\cdot}54Nv_c}{v_0}\right)^{-2/3}.$$

The terms neglected in Equation (11.3.1) can readily be shown to be small compared with those retained. From Equation (11.1.6) we can now find the partial derivatives of p with respect to α and we find that $\partial p/\partial\alpha$ is finite and $\partial^m p/\partial\alpha^m$ $(m > 1)$ is of the order $N^{(2n-1)/3}$. Similarly, from Equation (11.1.7) it follows that $\partial^n v_1/\partial\alpha^n$ is of the order $N^{(2n-1)/3}$. Combining these results we find by straightforward calculation that $\partial^n p/\partial v_1^n$ is of the order $N^{(n-2)/3}$. This means that the isotherm has a horizontal tangent at $v = v_c$.

11.4 Estimate at what temperature a perfect boson gas of molecular weight 4 and density $0{\cdot}15$ g cm^{-3} (the density of liquid helium) will, at constant volume, show the Einstein condensation phenomenon.

Solution

The transition temperature follows from Equations (10.2.16) and (11.2.3) and we find

$$T_c = \frac{h^2}{2\pi mk}\left(\frac{\rho}{2{\cdot}61m}\right)^{2/3}, \tag{11.4.1}$$

where ρ is the density, or

$$T_c = 115\rho^{2/3}M^{-5/3}, \tag{11.4.2}$$

where ρ is the density in g cm^{-3}, M the molecular weight, and where T_c is in °K.

[3] See D.ter Haar, *Proc. Roy. Soc.*, **A212**, 552 (1952).

If we substitute $\rho = 0 \cdot 15$ g cm^{-3}, $M = 4$, we find

$$T_c = 3 \cdot 2°\text{K}\,, \tag{11.4.3}$$

a value sufficiently close to the λ-temperature, where ^{4}He I goes over into ^{4}He II, to suggest to many people that the λ-transition of liquid ^{4}He is, indeed, something in the nature of an Einstein condensation. Among other things, experiment diverges from this simple theory by the fact that for a perfect boson gas the specific heat stays finite at T_c (see Problem 11.6), while at the λ-point the specific heat of liquid helium becomes infinite.

11.5 Find the locus in the (p, v_1) diagram of all condensation points and prove that it is an isentrope.

Solution

The locus is obtained from Equations (11.1.6) and (11.1.7) by putting $y = 1$ and eliminating β. The result is

$$pv_1^{5/3} = \frac{h^2}{2\pi m}\frac{\mathrm{f}(1)}{[\mathrm{f}'(1)]^{5/3}}\,. \tag{11.5.1}$$

The q-potential, $q = \beta p v$, is related to $\alpha (= \beta\mu)$, $\beta(= 1/kT)$, and the entropy S by the equation (compare Problem 1.20)

$$q = \frac{S}{k} + \alpha N - \beta U\,, \tag{11.5.2}$$

so that we have for the entropy

$$S = kq - k\alpha N + k\beta U\,. \tag{11.5.3}$$

For a classical perfect gas we have (see Problem 3.12)

$$U = \tfrac{3}{2}pv\,, \tag{11.5.4}$$

and using Equation (11.1.1) we finally obtain

$$\frac{S}{kN} = \tfrac{5}{2}\beta p v_1 - \alpha\,. \tag{11.5.5}$$

Using Equations (11.1.6) and (11.1.7) we can write Equation (11.5.5) in the form

$$\frac{S}{kN} = \tfrac{5}{2}\frac{\mathrm{f}(y)}{y\mathrm{f}'(y)} - \ln y\,, \tag{11.5.6}$$

so that the isentropes are the curves y = constant. As the locus (11.5.1) is one of those curves, it must be an isentrope.

11.6 Calculate the specific heat of a perfect boson gas.

Solution

As long as the temperature is above T_c, given by Equation (11.4.1), we have for the pressure

$$p = \frac{kT}{v_0} f(y) = CT^{5/2} f(y) , \tag{11.6.1}$$

with

$$C = \left(\frac{2\pi mk}{h^2}\right)^{3/2} k , \tag{11.6.2}$$

and for the energy, from Equation (11.5.4),

$$U = \tfrac{3}{2} C v T^{5/2} f(y) , \tag{11.6.3}$$

while the specific heat per particle is given by

$$C_V = \frac{1}{N}\frac{dU}{dT} = \tfrac{3}{2} v_1 \frac{dp}{dT} , \tag{11.6.4}$$

which is a monotonically decreasing function of T for $T > T_c$.

Below T_c we have $y = 1$ and hence

$$p = CT^{5/2} f(1) , \tag{11.6.5}$$

and hence

$$C_V = \tfrac{15}{4} CT^{3/2} v_1 f(1) . \tag{11.6.6}$$

We note that in the perfect boson gas C_V behaves as $T^{3/2}$ at low temperatures, while the specific heat of liquid helium behaves as T^3, another reason to hesitate before identifying the λ-transition with the Einstein condensation.

At $T = T_c$ we find from Equations (11.6.6) and (11.4.1):

$$C_V = \tfrac{15}{4} k \frac{f(1)}{f'(1)} = 1{\cdot}9k . \tag{11.6.7}$$

From Equation (11.6.4) and the behaviour of p (compare Problem 11.3) it follows that C_V is a continuous function of T, while $\partial C_V/\partial T$ is discontinuous.

11.7 Calculate the fraction of particles in a perfect boson gas, which is in the lowest energy state, as function of temperature below the transition temperature.

Solution

From Equations (11.2.2) and (11.4.1) written in the form

$$T_c^{3/2} = \left(\frac{h^2}{2\pi m}\right)^{3/2} \frac{1}{v f'(1)} , \tag{11.7.1}$$

Equation (11.1.7) with $y = 1$, and Equation (10.2.16) we find

$$N_0 = N\left[1 - \left(\frac{T}{T_c}\right)^{3/2}\right], \quad T < T_c . \tag{11.7.2}$$

11.8 Show that a two-dimensional perfect boson gas does not show an Einstein condensation.

Solution

In a two-dimensional system, the energy levels satisfy the equation

$$E_j = \frac{h^2}{8mL^2}(n_1^2 + n_2^2) , \tag{11.8.1}$$

where the n_i are positive integers.

Instead of Equation (11.1.3) we now have (now $v = L^2$, the two-dimensional volume)

$$dZ = \frac{2\pi m}{h^2} v \, dE . \tag{11.8.2}$$

Equation (11.1.9) becomes

$$q = -\frac{2\pi m}{h^2} v \int_0^\infty \ln(1 - e^{\alpha - \beta E}) dE , \tag{11.8.3}$$

leading to

$$q = \frac{2\pi m}{\beta h^2} v \sum_{n=1}^\infty \frac{e^{n\alpha}}{n^2} , \tag{11.8.4}$$

and from Equation (10.2.7′) we now get

$$\frac{N}{v} = \frac{2\pi m}{\beta h^2} \sum_{n=1}^\infty \frac{e^{n\alpha}}{n} . \tag{11.8.5}$$

As $\sum_{n=1}^\infty n^{-1}$ diverges, we can accommodate any number of particles in a given volume, and there will be no condensation.

[The occurrence or non-recurrence of condensation in perfect boson systems depends on the density of states function[4].]

VAPOUR CONDENSATION

11.9 Consider a model[5] of a vapour consisting of non-interacting drops where each drop takes up a volume W_l in which l atoms move independently in a smoothed-out negative potential $-\chi_l$ ($l = 1, 2, ...$) and where χ_l and W_l vary with l as follows (χ, w_1 are constants):

$$\chi_l = \frac{l-1}{l} \chi , \tag{11.9.1}$$

$$W_l = W_1 l^{(l-2)/(l-1)} . \tag{11.9.2}$$

(4) P.T.Landsberg, *Thermodynamics* (Interscience, New York), 1961, p.313 and Appendix D.
(5) See H.Wergeland, *Avhandl. Norske Videnskaps-Akad. Oslo, Mat.-Naturv. Kl.*, No.11 (1943).

Prove that the free energy f_l of a drop is given by the expression

$$\exp(-\beta f_l) = \frac{v}{W_1}\frac{\exp(-\beta\chi)}{l!}\left[\frac{W_1}{v_0}\exp(\beta\chi)\right]^l l^{l-2}\,. \qquad (11.9.3)$$

Solution

From the theory of canonical ensembles it follows that f_l is given by the equation

$$\exp(-\beta f_l) = \frac{1}{l!}\int \exp(-\beta\epsilon)\frac{d\Omega}{h^{3l}}\,, \qquad (11.9.4)$$

where $d\Omega$ is an element of the $6l$-dimensional phase space and where ϵ is the energy of the drop. The integration over the momenta gives a factor $(2\pi m/\beta)^{3l/2}$ as shown in Problem 3.5. Together with the factor h^{3l} this leads to v_0^{-l} with v_0 given by Equation (10.2.16). To evaluate the integral over the coordinates we note that the drop itself has the whole volume v at its disposal, but the atoms in the drop only the volume W_l. Moreover, each atom in the drop contributes a factor $\exp(\beta\chi_l)$.

The final result is then given by Equation (11.9.3), if we use Equations (11.9.1) and (11.9.2).

11.10 (a) If α_l/β is the chemical potential of a drop of l atoms in Problem 11.9, justify the following relation for the q-potential of the vapour [as the number of drops can change, we must use a grand canonical ensemble to discuss the system]:

$$q = \sum_{l=1}^{\infty} \exp(\alpha_l - \beta f_l)\,, \qquad (11.10.1)$$

with

$$\alpha_l = l\alpha\,. \qquad (11.10.2)$$

(b) Show that the equation of state follows from the equations

$$p = \frac{e^{-\beta\chi}}{\beta v_1}\sum_l \frac{l^{l-2}}{l!}z^l\,. \qquad (11.10.3)$$

$$N = \frac{V}{v_1}e^{-\beta\chi}\sum_l \frac{l^{l-1}}{l!}z^l\,, \qquad (11.10.4)$$

where

$$z = \frac{W_1}{v_0}e^{\alpha+\beta\chi}\,. \qquad (11.10.5)$$

(c) Using the relation

$$\frac{x^x}{x!} = \frac{1}{2\pi i}\oint \frac{du}{u^{x+1}}e^{xu}\,, \qquad (11.10.6)$$

eliminate z.

Discuss briefly the shape of the isotherms.

Solution

(a) Equation (11.10.1) follows, if we bear in mind that from the general thermodynamics of reactions it follows that the chemical potential of a drop of l atoms must be l times the chemical potential of an atom, α, expressed by Equation (11.10.2).

(b) Combining Equations (11.9.3), (11.10.1), and (11.10.2) we obtain

$$q = \frac{V}{v_1} e^{-\beta\chi} \sum_l \frac{l^{l-2}}{l!} z^l \,. \tag{11.10.7}$$

The equation of state (11.10.3) and (11.10.4) then follows from Equations (10.2.6) and (10.2.7′).

(c) To eliminate z we define quantities ξ, η, ζ by the equations

$$\xi = \sum_{l=1}^{\infty} \frac{l^{l-2}}{l!} z^l \,, \tag{11.10.8}$$

$$\eta = \sum_{l=1}^{\infty} \frac{l^{l-1}}{l!} z^l \,, \tag{11.10.9}$$

$$\zeta = \sum_{l=1}^{\infty} \frac{l^{l}}{l!} z^l \,. \tag{11.10.10}$$

We note that

$$\xi = \int_0^z \eta(z) \frac{dz}{z} \,, \quad \eta = \int_0^z \zeta(z) \frac{dz}{z} \,. \tag{11.10.11}$$

Using Equation (11.10.6) we find

$$1+\zeta = \frac{1}{2\pi i} \oint \frac{du}{u} \sum_{l=0}^{\infty} \left(\frac{ze^u}{u}\right)^l \,, \tag{11.10.12}$$

or

$$1+\zeta = \frac{1}{2\pi i} \oint \frac{du}{u - ze^u} \,. \tag{11.10.13}$$

The pole u_0 thus satisfies the equation

$$u_0 = ze^{u_0} \,, \tag{11.10.14}$$

and we get from Equations (11.10.13) and (11.10.14)

$$\zeta = \frac{u_0}{1-u_0} \,. \tag{11.10.15}$$

From Equation (11.10.11) we now get

$$\eta = \int_0^z \frac{u_0}{1-u_0} \frac{dz}{z} = \int_0^{u_0} du_0 = u_0 \,, \tag{11.10.16}$$

and

$$\xi = \int_0^z \frac{u_0}{z} dz = \int_0^{u_0} (1-u_0)\, du_0 = u_0 - \tfrac{1}{2}u_0^2 \,. \tag{11.10.17}$$

From Equations (11.10.7), (11.10.4), (11.10.8), (11.10.9), (11.10.16), and (11.10.17) we now get by eliminating u_0 the (exact) equation of state ($v_1 = v/N$)

$$\beta p v_1 = 1 - \tfrac{1}{2}\frac{W_1}{v_1}e^{\beta\chi} . \qquad (11.10.18)$$

The derivation of Equation (11.10.18) is valid as long as the series for p and N converge. This is the case as long as $z < 1/e$. If we write Equation (11.10.4) in the form

$$\frac{1}{v_1} = \frac{1}{W_1}e^{-\beta\chi}\sum_l \frac{l^{l-1}}{l!} z^l , \qquad (11.10.19)$$

we note its resemblance to Equation (11.1.7). Indeed, using the Stirling approximation (see Problem 2.11)

$$l! \approx l^{l+\frac{1}{2}} e^{-l} , \qquad (11.10.20)$$

we see that the general term in the sum on the right-hand side of Equation (11.10.19) is of the form $(z/e)^l l^{-\frac{3}{2}}$, as compared to $y^n n^{-\frac{3}{2}}$ in $f'(y)$ occurring in Equation (11.1.7). One would expect that the analysis of Problem 11.2 can be repeated with minor alterations. The isotherm will be given by Equation (11.10.18) as long as v_1 is larger than a critical volume v_c given by the equation

$$\frac{1}{v_c} = \frac{e^{-\beta\chi}}{W_1} . \qquad (11.10.21)$$

This relation follows from the fact that for $z = 1/e$, $u_0 = 1$ from Equation (11.10.14), and thus from Equation (11.10.16), $\eta = 1$ so that the relationship (11.10.21) follows from Equations (11.10.4) and (11.10.9). Similarly, we find that for $v_1 < v_c$ the isotherm will be horizontal with a pressure $p = p_c$ given by the equation

$$p_c = \frac{e^{-\beta\chi}}{2\beta v_1} . \qquad (11.10.22)$$

11.11 Using the results of the preceding problem show that the mean number $\langle m_l \rangle$ of drops of l atoms at temperature T is given by the equation

$$\langle m_l \rangle = N\frac{l^{l-2}}{l!}\eta^{l-1}e^{-l\eta} \qquad (11.11.1)$$

with η the solution of the equation

$$\eta e^{-\eta} = z \qquad (11.11.2)$$

and z given by Equation (11.10.5).

Solution

From the general theory of grand ensembles it follows that the average value of m_l satisfies the equation

$$\langle m_l \rangle = \frac{\partial q}{\partial \alpha_l}, \qquad (11.11.3)$$

and if we write Equation (11.10.7) in the form

$$q = \frac{V}{v_1} \mathrm{e}^{-\beta \chi} \sum_l \frac{l^{l-2}}{l!} \left(\frac{v_1}{v_0} \mathrm{e}^{-\beta \chi} \right)^l \mathrm{e}^{\alpha_l}, \qquad (11.11.4)$$

we have

$$\langle m_l \rangle = \frac{V}{v_1} \mathrm{e}^{-\beta \chi} \frac{l^{l-2}}{l!} z^l. \qquad (11.11.5)$$

We note in passing that combining Equations (11.11.5) and (11.10.3) we obtain

$$\beta p v = \sum_l \langle m_l \rangle, \qquad (11.11.6)$$

which is, of course, a consequence of our assumption of non-interacting drops.

Using Equations (11.10.14), (11.10.16), (11.10.7), and (11.10.9), we can write Equation (11.11.5) in the form (11.11.1).

We saw earlier that at the condensation point $\eta = 1$, while $\eta \to 0$ as $v_1 \to \infty$ [compare Equation (11.10.19)] so that η measures the degree of saturation. We note that, as $v_1 \to \infty$, we find

$$\langle m_1 \rangle = N; \ \langle m_l \rangle = 0, \quad l > 1, \qquad (11.11.7)$$

that is, every atom is single!

Even at the condensation point where it follows from Equation (11.10.20) that $\langle m_l \rangle$ is proportional to $l^{-5/2}$, only the smallest drops are present in appreciable amounts.

HARD SPHERE GAS[6]

11.12 Consider a gas of N small hard spheres interacting through two-particle interactions which are long-range and smoothly varying. Moreover, assume the interaction potential, ϕ, to be everywhere negative or zero. Consider the canonical partition function for such a gas. If we want to derive the equation of state from it, we must find its volume dependence. We have seen on many occasions that the integration over momenta which occurs in the expression yields only a multiplying factor v_0^{-N} (compare Problems 3.5 and 11.9), which we can leave out as it is unimportant for our discussion. To evaluate the configurational partition function, divide the volume v into cells of volume Δ so small as to make ϕ practically constant inside Δ, but large enough for each cell to contain a large number of particles. Let $\mathbf{r}_i$ be the position of the ith cell and N_i the number of particles in it. If δ measures the volume of the hard

[6] See N.G.van Kampen, *Phys.Rev.*, **135**, A362 (1964).

spheres, and if $\omega(N_i)$ is the amount of phase space of N_i hard spheres in a volume Δ, we have for $\omega_i(N_i)$ in one dimension (see Problem 9.3)

$$\omega(N_i) = (\Delta - N_i\delta)^{N_i}. \tag{11.12.1}$$

Assume that this expression can still be used for the three-dimensional case. Van Kampen has shown that this assumption is sufficient for the discussion of the condensation of the system considered.

Write down an expression for the canonical partition function, Q_N, omitting the kinetic energy contribution in the form of a summation over configurations $\{N_i\}$, that is sets of numbers N_i,

$$Q_N = \sum \exp \Phi\{N_i\}. \tag{11.12.2}$$

Give an expression for $\Phi\{N_i\}$.

Use the standard procedure of statistical mechanics (see Problem 2.11) to evaluate the sum over configurations by finding its largest term. Find the condition to be satisfied by the configuration corresponding to this maximum of Φ, say Φ_s.

Assuming that the density is homogeneous, show that Φ_s satisfies the equation

$$\Phi_s = N \ln \frac{v - N\delta}{N} + N - \tfrac{1}{2}\beta\phi_0 \frac{N^2}{v}, \tag{11.12.3}$$

with

$$\phi_0 = \sum_j \phi_{ij}\Delta = \int \phi(\mathbf{r})\, d^3\mathbf{r}, \tag{11.12.4}$$

where we have assumed that the interaction forces are central forces,

$$\phi(\mathbf{r}) = \phi(r), \tag{11.12.5}$$

and where

$$\phi_{ij} = \phi(\mathbf{r}_i, \mathbf{r}_j) = \phi(|\mathbf{r}_i - \mathbf{r}_j|). \tag{11.12.6}$$

Hence derive the van der Waals equation

$$\beta p = \frac{N}{v - N\delta} + \tfrac{1}{2}\beta\phi_0 \frac{N^2}{v^2}, \tag{11.12.7}$$

and discuss the result, comparing Equation (11.12.7) with the usual form of the van der Waals equation.

Solution

We first of all note that the N_i must satisfy the condition

$$\sum_i N_i = N, \tag{11.12.8}$$

and that the energy of a given $\{N_i\}$ configuration is

$$\tfrac{1}{2}\sum_{ij} \phi(\mathbf{r}_i, \mathbf{r}_j) N_i N_j = \tfrac{1}{2}\sum_{ij} \phi_{ij} N_i N_j. \tag{11.12.9}$$

The partition function will thus be

$$Q_N = \frac{1}{N!}\sum \frac{N!}{\prod_i N_i!}\left[\prod_i \omega(N_i)\right] \exp\left\{-\tfrac{1}{2}\beta \sum_{ij} \phi_{ij} N_i N_j\right\}, \quad (11.12.10)$$

which is of the form (11.12.2) and where the sum is over all configurations satisfying condition (11.12.8). The function $\Phi\{N_i\}$ in Equation (11.12.2) is then given by

$$\Phi\{N_i\} = \sum_i [N_i \ln(\Delta - N_i\delta) - N_i \ln N_i + N_i] - \tfrac{1}{2}\beta \sum_{ij} \phi_{ij} N_i N_j, \quad (11.12.11)$$

where we have used Equation (11.12.1) and Stirling's formula for the factorial.

The maximum term in Equation (11.12.10) with the N_i satisfying condition (11.12.8) is found in the usual way by varying the N_i and using a Lagrangian multiplier γ, to be determined from the condition (11.12.8), to take the subsidiary condition into account. The result is

$$\ln\frac{\Delta - N_i\delta}{N_i} - \frac{N_i\delta}{\Delta - N_i\delta} - \beta\sum_j \phi_{ij} N_j = \gamma. \quad (11.12.12)$$

If the density is homogeneous we put

$$N_i = \frac{N\Delta}{v}, \quad (11.12.13)$$

and if we determine γ from the equation

$$\ln\frac{v - N\delta}{N} - \frac{N\delta}{v - N\delta} - \beta\phi_0\frac{N}{v} = \gamma, \quad (11.12.14)$$

with ϕ_0 given by Equation (11.12.4), all Equations (11.12.12) are satisfied. This means that relation (11.12.13) is a possible solution of Equation (11.12.12), and that the uniform-density states are thermodynamic states as they make Q_N stationary. It remains to be determined whether these states are stable, metastable, or unstable. Substituting Equation (11.12.13) into Equation (11.12.11) we obtain the result (11.12.3).

If Φ_s, given by Equation (11.12.3), is an absolute maximum, we have $Q_N = \exp\Phi_s$ and the free energy F is given by $-\beta F = \Phi_s$, so that one finds p by taking the derivative with respect to v from which the relation (11.12.7) follows. We can write this equation in the form

$$\beta\left(p - \tfrac{1}{2}\phi_0\frac{N^2}{v^2}\right)(v - N\delta) = N, \quad (11.12.15)$$

which is the usual van der Waals form. We see clearly the excluded volume term and the term due to attractive forces (ϕ_0 is negative!): $-\tfrac{1}{2}\phi_0$ is the work that a particle has to do against the attraction of the other particles to reach the boundary.

11.13 Consider the function $\mathrm{f}(n)$

$$\mathrm{f}(n) = n \ln\left(\frac{1-n}{n}\right) - \tfrac{1}{2}\beta\phi_0 n^2 , \tag{11.13.1}$$

which is related to Φ_s of Equation (11.12.3), when we put $\delta = 1$ and write $n = N/v$.

(a) Prove that, if A is a real symmetrical matrix of the form

$$A_{ij} = a_i\delta_{ij} - b_{ij}, \quad b_{ij} = b_{ji} > 0 , \tag{11.13.2}$$

it is positive definite if

$$A_i > \sum_j b_{ij} \tag{11.13.3}$$

for all i. Prove also that A is not positive definite, if

$$a_i < \sum_j b_{ij}$$

for all i.

(b) Use the results of (a) to show that, when $\mathrm{f}''(n)$ is negative, the solution leading to condition (11.12.3) corresponds to a (relative) maximum of Φ, but, if $\mathrm{f}''(n)$ is positive, this is not the case.

(c) Show that the condition on β that $\mathrm{f}''(n)$ is always negative is

$$-\beta\phi_0 < \tfrac{27}{4} , \tag{11.13.4}$$

and show that the maximum is an absolute maximum. Discuss condition (11.13.4) on β.

Solution

If we write $n = N/v$ and use units in which $\delta = 1$ so that $0 < n < 1$, we have $\Phi_s = v\mathrm{f}(n)$ with f given by Equation (11.13.1). Equation (11.12.14) for γ then has the form

$$\mathrm{f}'(n) = \gamma , \tag{11.13.5}$$

with

$$\mathrm{f}'(n) = \ln\frac{1-n}{n} - \frac{n}{1-n} - \beta\phi_0 n . \tag{11.13.6}$$

In the following we also need $\mathrm{f}''(n)$ which is given by the equation

$$\mathrm{f}''(n) = -\frac{1}{n(1-n)^2} - \beta\phi_0 . \tag{11.13.7}$$

(a) To prove this let λ be an eigenvalue and $\{x_i\}$ the corresponding eigenvector of A so that we have

$$\lambda x_i = \sum_j A_{ij}x_j = a_i x_i - \sum_j b_{ij}x_j . \tag{11.13.8}$$

Let x_1 be the component of $\{x_i\}$ with the largest absolute magnitude and choose the arbitrary multiplying factor of all x_i such that x_1 is real and positive. From Equation (11.13.3) it then follows that

$$(a_1 - \lambda)x_1 = \sum_j b_{ij}x_{ij} < x_1 \sum b_{ij} < x_1 a_1 . \tag{11.13.9}$$

Hence each eigenvalue is positive, and A is positive definite.

On the other hand, if one considers the vector $\{1, 1, ..., 1\} \equiv \{y_i\}$ we see that $\sum_{ij} y_i A_{ij} y_j < 0$ if $a_i < \sum_j b_{ij}$ for all i.

(b) To find out whether a solution gives a maximum of Φ we construct the matrix of the second functional derivatives

$$\frac{\partial^2 \Phi}{\partial N_i \partial N_j} = -\left[\delta_{ij}\frac{\Delta^2}{N_i(\Delta - N_i)^2} + \beta\phi_{ij}\right] , \tag{11.13.10}$$

or, if we consider the uniform-density case

$$\frac{\partial^2 \Phi}{\partial N_i \partial N_j} = \frac{-\delta_{ij}}{n(1-n)^2\Delta} + \beta\phi_{ij} . \tag{11.13.11}$$

This matrix will be negative definite so that the solution leads to a (relative) maximum—and hence to a stable or at least metastable thermodynamic state—if

$$\frac{1}{n(1-n)^2\Delta} > -\beta \sum_j \phi_{ij} = -\frac{\beta\phi_0}{\Delta} , \tag{11.13.12}$$

and we see that this condition is satisfied if $f''(n)$ is negative. On the other hand, if $f''(n)$ is positive, the condition for the second part of the lemma holds, and the solution does not correspond to a maximum: the corresponding thermodynamic state is unstable.

(c) From Equation (11.13.7) it follows easily that $f''(n)$ has a maximum for $n = \frac{1}{3}$, which is equal to $-\frac{27}{4} - \beta\phi_0$, from which condition (11.13.4) follows. From a similar calculation it also follows that

$$\frac{\Delta^2}{N_i(\Delta - N_i)^2} \geqslant \frac{27}{4\Delta} , \tag{11.13.13}$$

so that the matrix (11.13.10) is negative definite for any configuration $\{N_i\}$. The function $\Phi\{N_i\}$ is thus convex and can have only one maximum. Provided condition (11.13.4) holds, the stable state is thus the uniform-density state.

In the previous problem we noted the relation between Equation (11.12.7) and the van der Waals equation of state. If we express van der Waals' constants a and b in terms of our parameters, we have

$$a = -\tfrac{1}{2}\phi_0 N^2, \quad b = N\delta , \tag{11.13.14}$$

and condition (11.13.4) becomes

$$\beta < \beta_c \quad \text{or} \quad T > T_c \tag{11.13.15}$$

with the critical temperature T_c satisfying the van der Waals relation

$$NkT_c = \frac{8a}{27b} \,. \tag{11.13.16}$$

This means that the uniform-density state is thus the stable state above the critical temperature. Temperatures below T_c are considered in the next two problems.

11.14 Consider temperatures for which condition (11.13.4) is not satisfied. Introduce now a position-dependent density $n(\mathbf{r})$, so that Φ becomes a functional of $n(\mathbf{r})$. Assume that $n(\mathbf{r})$ varies so slowly that $n(\mathbf{r})-n(\mathbf{r}')$ can be expanded in powers of $\mathbf{r}-\mathbf{r}'$. Show that Φ becomes a maximum, if $n(\mathbf{r})$ satisfies the equation

$$\mathrm{f}'(n)-\tfrac{1}{2}\beta\phi_2\nabla^2 n = \gamma \,, \tag{11.14.1}$$

where γ is a Lagrangian multiplier and

$$\phi_2 = \tfrac{1}{3}\int r^2\phi(r)\,\mathrm{d}^3\mathbf{r} \,. \tag{11.14.2}$$

Assume now that $n(\mathbf{r})$ depends on only one coordinate, x say, so that Equation (11.14.1) becomes

$$-\tfrac{1}{2}\beta\phi_2\frac{\mathrm{d}^2 n}{\mathrm{d}x^2} = -\mathrm{f}'(n)+\gamma \,. \tag{11.14.3}$$

Note that this equation can be interpreted as the equation of motion of a classical point mass with coordinate n, mass $-\frac{1}{2}\beta\phi_2(\phi_2 < 0!)$, time x, moving in the potential

$$\psi(n) = \mathrm{f}(n)-\gamma n \,. \tag{11.14.4}$$

Prove that $\mathrm{f}'(n)$ has one minimum and one maximum in the range $0 < n < 1$. Consider a value of γ lying between those two extrema and find

(i) the two solutions of Equation (11.14.3) with $n(x)$ = constant and
(ii) the solution with $n = n_{\mathrm{I}}$ as $x \to -\infty$ and $n = n_{\mathrm{II}} \neq n_{\mathrm{I}}$ as $x \to +\infty$.

Prove that this last solution is possible only, when one applies Maxwell's equal-area rule.

Solution

If we assume that the density depends on position, so that we can write

$$N_i = n(\mathbf{r})\Delta \,, \tag{11.14.5}$$

we get instead of Equation (11.12.11)

$$\Phi\{n(\mathbf{r})\} = \int n(\mathbf{r})\left\{\ln\frac{1-n(\mathbf{r})}{n(\mathbf{r})}+1\right\}\mathrm{d}^3\mathbf{r}-\tfrac{1}{2}\beta\iint\phi(\mathbf{r},\mathbf{r}')n(\mathbf{r})n(\mathbf{r}')\,\mathrm{d}^3\mathbf{r}\,\mathrm{d}^3\mathbf{r}' \,, \tag{11.14.6}$$

while the condition (11.12.8) becomes

$$\int n(\mathbf{r})\,\mathrm{d}^3\mathbf{r} = N\,. \tag{11.14.7}$$

Equation (11.14.6) can be written in the form

$$\Phi\{n(\mathbf{r})\} = \int \mathrm{f}(n)\,\mathrm{d}^3\mathbf{r} + \tfrac{1}{4}\beta \iint \phi(\mathbf{r},\mathbf{r}')[n(\mathbf{r})-n(\mathbf{r}')]^2\,\mathrm{d}^3\mathbf{r}\,\mathrm{d}^3\mathbf{r}'\,, \tag{11.14.8}$$

where $\mathrm{f}(n)$ is given by Equation (11.13.1).

To find the extremum we take functional derivatives and take condition (11.14.7) into account through a Lagrangian multiplier, and the result is

$$\mathrm{f}'(n) + \beta \int \phi(\mathbf{r},\mathbf{r}')[n(\mathbf{r})-n(\mathbf{r}')]\,\mathrm{d}^3\mathbf{r}' = \gamma\,. \tag{11.14.9}$$

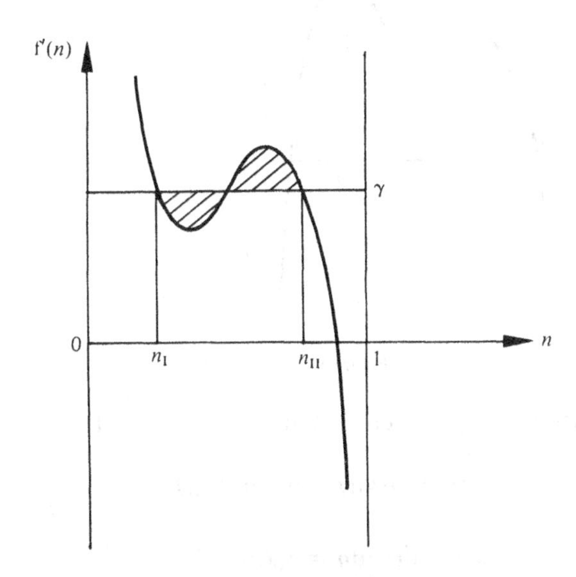

Figure 11.14.1.

If $n(\mathbf{r})$ is varying sufficiently slowly so that we can expand $n(\mathbf{r})-n(\mathbf{r}')$, Equation (11.14.9) leads to the result (11.14.1). If we then assume that $n(\mathbf{r})$ depends on x only, we get the solution (11.14.3).

In the preceding problem we considered some of the properties of $\mathrm{f}(n)$ and we saw there that $\mathrm{f}''(n)$ had one extremum at $n = \frac{1}{3}$ and, if condition (11.13.4) is not satisfied, f'' will be positive for some range of n-values. As $\mathrm{f}''(n)$ is negative for $n = 0$ and $n = 1$, we have proved that $\mathrm{f}'(n)$ behaves as shown in Figure 11.14.1 and hence $\psi(n)$ given by Equation (11.14.4) behaves as shown in Figure 11.14.2.

It follows that if $n = n_{\mathrm{I}} =$ constant or $n = n_{\mathrm{II}} =$ constant, $\mathrm{d}^2n/\mathrm{d}x^2 = 0$ as they correspond to maxima of $\psi(n)$, and these two values of n thus correspond to uniform-density states.

To find a solution with $n = n_{\mathrm{I}}$ as $x \to -\infty$ and $n = n_{\mathrm{II}}$ as $x \to +\infty$, we consider the problem posed by Equation (11.14.3) and see that such a solution is possible only if the 'particle', initially at rest on top of the first 'potential peak' ($n = n_{\mathrm{I}}$), 'moves' to the other 'peak' and comes to rest there. From the analogy used here it follows that this is possible only if the two peaks have the same 'height', or

$$\psi(n_{\mathrm{I}}) = \psi(n_{\mathrm{II}}) \,. \tag{11.14.10}$$

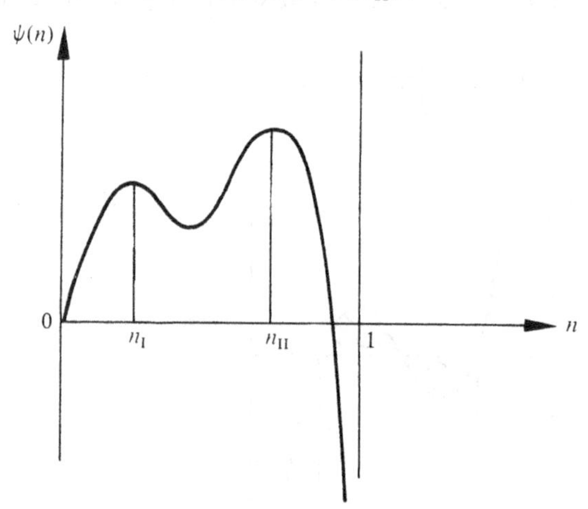

Figure 11.14.2.

This is an equation for γ. There is thus only one solution $\gamma = \gamma_0$ such that

$$\mathrm{f}(n_{\mathrm{I}}) - \mathrm{f}(n_{\mathrm{II}}) = \gamma_0(n_{\mathrm{I}} - n_{\mathrm{II}}) \,, \tag{11.14.11}$$

or

$$\int_{n_{\mathrm{I}}}^{n_{\mathrm{II}}} \mathrm{f}'(n)\mathrm{d}n = \gamma_0(n_{\mathrm{II}} - n_{\mathrm{I}}) \,, \tag{11.14.12}$$

indicating that γ_0 is found by the equal-area construction applied to $\mathrm{f}'(n)$: the two shaded areas in Figure 11.14.1 must be equal. This condition implies that p is the same for n_{I} and n_{II}, as in the uniform-density case we have $\Phi_{\mathrm{s}} = v\mathrm{f}(n)$, and thus

$$\beta p = \frac{\partial \Phi_{\mathrm{s}}}{\partial v} = \mathrm{f}(n) - n\mathrm{f}'(n) = \mathrm{f}(n) - \gamma n = \psi(n) \,, \tag{11.14.13}$$

where we have used the fact that for the uniform-density case $\mathrm{f}'(n) = \gamma$. The equality of $\psi(n_{\mathrm{I}})$ and $\psi(n_{\mathrm{II}})$ thus implies that $p_{\mathrm{I}} = p_{\mathrm{II}}$, cf Equation (7.3.1) of Problem 7.3.

11.15 For given values of the total volume v and total number of particles N find the fractions of space occupied by the densities n_{I} and n_{II}, respectively, of the previous problem, neglecting the transition region. Give an expression for the free energy of this two-density state, neglecting the free energy of the transition region.

Prove that the pressure is the same for all two-density states. Also prove that the two-density state has a lower free energy than the uniform-density state for the same N and β[7].

Solution

Let v_{I} (v_{II}) be the volume occupied by density n_{I} (n_{II}). Neglecting the transition region we have

$$v_{\mathrm{I}} > 0, \quad v_{\mathrm{II}} > 0, \quad v_{\mathrm{I}}+v_{\mathrm{II}} = v\,. \tag{11.15.1}$$

For any value of v_{I} we have a solution of condition (11.15.1) corresponding to the same $\gamma = \gamma_0$. The number of particles satisfies the equation

$$N = n_{\mathrm{I}}v_{\mathrm{I}}+n_{\mathrm{II}}v_{\mathrm{II}}\,, \tag{11.15.2}$$

so that for all values of N/v in the range

$$n_{\mathrm{I}} < N/v < n_{\mathrm{II}} \tag{11.15.3}$$

there is a two-density state. The densities n_{I} and n_{II} are determined, as we have seen, by the conditions

$$\mathrm{f}'(n_{\mathrm{I}}) = \mathrm{f}'(n_{\mathrm{II}}) \tag{11.15.4}$$

and

$$\mathrm{f}(n_{\mathrm{I}})-n_{\mathrm{I}}\mathrm{f}'(n_{\mathrm{I}}) = \mathrm{f}(n_{\mathrm{II}})-n_{\mathrm{II}}\mathrm{f}'(n_{\mathrm{II}})\,. \tag{11.15.5}$$

From Equation (11.14.8) it follows that, neglecting the transition region, we have for the free energy

$$-\beta F = \Phi\{n\} = v_{\mathrm{I}}\mathrm{f}(n_{\mathrm{I}})+v_{\mathrm{II}}\mathrm{f}(n_{\mathrm{II}})\,, \tag{11.15.6}$$

or

$$\mathrm{F} = \mathrm{F}_{\mathrm{I}}+\mathrm{F}_{\mathrm{II}}\,, \tag{11.15.7}$$

and for the pressure we find

$$\beta p = \frac{\partial\Phi}{\partial v} = \frac{\partial v_{\mathrm{I}}}{\partial v}p_{\mathrm{I}}+\frac{\partial v_{\mathrm{II}}}{\partial v}p_{\mathrm{II}} = \beta p_{\mathrm{I}} \quad (= \beta p_{\mathrm{II}})\,, \tag{11.15.8}$$

which proves that the pressure is the same for all the two-density states.

To prove that for all values of N/v satisfying condition (11.15.3) the free energy of the two-density state is higher than that for the uniform-density state for the same N and β, we must prove that

$$v_{\mathrm{I}}\mathrm{f}(n_{\mathrm{I}})+v_{\mathrm{II}}\mathrm{f}(n_{\mathrm{II}}) > v\mathrm{f}\left(\frac{v_{\mathrm{I}}n_{\mathrm{I}}+v_{\mathrm{II}}n_{\mathrm{II}}}{v}\right), \tag{11.15.9}$$

[7] Other characteristics of the unstable uniform-density state are examined in Problem 7.2(e).

or, that in an $f(n)$-n diagram (see Figure 11.15.1) the double tangent lies above the curve $f(n)$ which is obviously the case.

In conclusion, we note that Problems 11.12 to 11.15 give an (approximate) proof of the van der Waals isotherm, *including the horizontal part*.

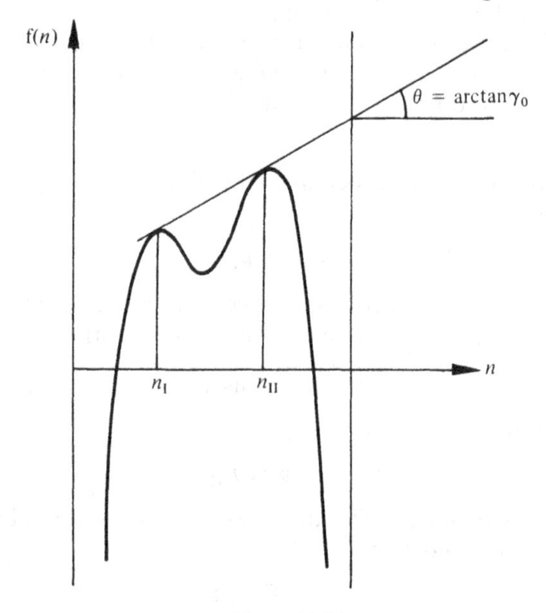

Figure 11.15.1.

The treatment is approximate but can be made rigorous[8]. However, it is restricted to long-range attractive forces while actual interatomic forces are usually short-range.

[8] N.G.van Kampen, *Phys.Rev.*, 135, A362 (1964).

12

Cooperative phenomena[1]

D.ter HAAR
(University of Oxford, Oxford)

12.1 For our purpose we define a ferromagnet as a lattice on which spins are situated. In this chapter we shall be extensively concerned with two models of ferromagnets: the **Ising model** and the **Heisenberg model.** In general we can write for the Hamiltonian of the interaction between the spins

$$H = -2\sum_{\mathbf{f},\mathbf{g}} [I_x(\mathbf{f}-\mathbf{g})S_\mathbf{f}^x S_\mathbf{g}^x + I_y(\mathbf{f}-\mathbf{g})S_\mathbf{f}^y S_\mathbf{g}^y + I_z(\mathbf{f}-\mathbf{g})S_\mathbf{f}^z S_\mathbf{g}^z], \quad (12.1.1)$$

where the f and g are the position vectors of the lattice sites, the $S_\mathbf{f}^x$, $S_\mathbf{f}^y$, $S_\mathbf{f}^z$ the components of $\mathbf{S_f}$, the spin vector corresponding to lattice site f (we express all spins in units ħ) and the $I_i(\mathbf{f}-\mathbf{g})$ are functions of $\mathbf{f}-\mathbf{g}$ only.

The **Ising model** in its simplest form is obtained by putting

$$I_x(\mathbf{f}-\mathbf{g}) = I_y(\mathbf{f}-\mathbf{g}) = 0 . \quad (12.1.2)$$

$$I_z(\mathbf{f}-\mathbf{g})\begin{cases} = I, \text{ if f and g are nearest neighbours}, \\ = 0, \text{ otherwise}, \end{cases} \quad (12.1.3)$$

and assuming that the particles have spin $\frac{1}{2}$ so that $S_\mathbf{f}^z$ only takes on the values $+\frac{1}{2}$ and $-\frac{1}{2}$.

The **Heisenberg model** in its simplest, isotropic form is obtained by putting

$$I_x = I_y = I_z = I(\mathbf{f}-\mathbf{g}) \quad (12.1.4)$$

so that the Hamiltonian (12.1.1) becomes

$$H = -2\sum_{\mathbf{f},\mathbf{g}} I(\mathbf{f}-\mathbf{g})(\mathbf{S_f}\cdot\mathbf{S_g}) . \quad (12.1.5)$$

If there is an external magnetic field (induction B), we must write instead of the expression (12.1.5)

$$H = -g\mu_\mathrm{B}\sum_\mathbf{f}(\mathbf{B}\cdot\mathbf{S_f}) - 2\sum_{\mathbf{f},\mathbf{g}} I(\mathbf{f}-\mathbf{g})(\mathbf{S_f}\cdot\mathbf{S_g}) , \quad (12.1.6)$$

[1] See F.C.Nix and W.Shockley, *Rev.Mod.Phys.*, **10**, 1 (1938); D.ter Haar, *Elements of Statistical Mechanics* (Holt, Rinehart, and Winston, New York), 1954, Ch.12; H.J.Goldsmid (Ed.), *Problems in Solid State Physics* (Pion, London), 1968.

where μ_B is the **Bohr magneton** and g the **Landé g-factor** and where we have assumed the field to be uniform.

The simplest way of progressing further is through the **molecular field approximation.** In this approximation one writes the Hamiltonian (12.1.6) first of all in the form

$$H = -g\mu_B \sum_{\mathbf{f}} \left[\mathbf{S}_{\mathbf{f}} \cdot \left(\mathbf{B} + \frac{2}{g\mu_B} \sum_{\mathbf{g}} I(\mathbf{f}-\mathbf{g})\mathbf{S}_{\mathbf{g}} \right) \right] . \tag{12.1.7}$$

One now replaces the $\mathbf{S}_{\mathbf{g}}$ in the sum over $\mathbf{g}$ by their average value $\langle \mathbf{S} \rangle$ which must be independent of $\mathbf{g}$, as the system is invariant under translations. We see from expression (12.1.7) that the external field is replaced by an effective field

$$\mathbf{B}_{\text{eff}} = \mathbf{B} + \mathbf{B}' , \tag{12.1.8}$$

where $\mathbf{B}'$ is the so-called **molecular field.**

To find the behaviour of the system, we must first study the behaviour of a magnetic dipole of moment μ in an external field $\mathbf{B}$, when its energy is $-(\boldsymbol{\mu} \cdot \mathbf{B})$. Show that its average value at a given temperature is

$$\langle \mu \rangle = \mu \left(\coth z - \frac{1}{z} \right), \tag{12.1.9}$$

with

$$z = \beta \mu B . \tag{12.1.10}$$

Solution

The energy of the dipole is

$$H = -(\boldsymbol{\mu} \cdot \mathbf{B}) , \tag{12.1.11}$$

and the average value of μ in the direction of $\mathbf{B}$ is given by the expression

$$\langle \mu \rangle = \frac{\int e^{-\beta H} \mu \cos\theta \, d\cos\theta}{\int e^{-\beta H} d\cos\theta} , \tag{12.1.12}$$

where θ is the angle between μ and $\mathbf{B}$. This yields expression (12.1.9).

At high temperatures the **Langevin function,** $\coth z - 1/z$, behaves as $\frac{1}{3}z$ and we have therefore

$$\langle \mu \rangle \sim \frac{\mu^2 B}{3kT} \qquad (z \ll 1) , \tag{12.1.13}$$

while at low temperatures we get

$$\langle \mu \rangle \sim \mu \qquad (z \gg 1) . \tag{12.1.14}$$

12.2 A spin $-S$ particle in a field B has spin energy levels

$$E_M = -g\mu_B BM\ ,\ \ M = -S, ..., +S\ . \qquad (12.2.1)$$

Neglecting all other effects, show that its partition function Z is given by the equation

$$Z = \frac{\sinh(2S+1)z}{\sinh z}\ , \qquad (12.2.2)$$

where now

$$z = \tfrac{1}{2}\beta g\mu_B B\ . \qquad (12.2.3)$$

Use Equation (12.2.2) to find an expression for its average magnetic moment.

Solution

The partition function (12.2.2) follows directly from the equation

$$Z = \sum_M \exp(\beta g\mu_B BM)\ . \qquad (12.2.4)$$

The average magnetic moment is given by the equation

$$\langle\mu\rangle = \frac{1}{\beta}\frac{\partial \ln Z}{\partial B} = g\mu_B S B_S(z)\ , \qquad (12.2.5)$$

where the **Brillouin function** B_S is given by the equation

$$B_S(z) = \frac{2S+1}{2S}\coth[(2S+1)z] - \frac{1}{2S}\coth z\ . \qquad (12.2.6)$$

At high temperatures, $z \ll 1$, we find

$$B_S(z) \sim \tfrac{2}{3}(S+1)z\ , \qquad (12.2.7)$$

and thus

$$\langle\mu\rangle \sim \frac{S(S+1)g^2\mu_B^2}{3kT}B\ , \qquad (12.2.8)$$

which is the equivalent of expression (12.1.13), if we bear in mind that $S(S+1)g^2\mu_B^2$ is the quantum-mechanical equivalent of μ^2 in the classical case.

12.3 Use the molecular field approximation to find an expression for the average magnetisation $\langle M\rangle$ of a spin-$\frac{1}{2}$ ferromagnet.

(i) Show that in this approximation there exists a **transition point** or **Curie point** T_C such that in the limit as $B \to 0$, $\langle M\rangle = 0$ is the only value of $\langle M\rangle$ for $T > T_C$, while $\langle M\rangle$ can be non-zero for $T < T_C$. If $\langle M\rangle \neq 0$ in the limit as $B \to 0$ the material is said to exhibit **spontaneous magnetisation.** Find an expression for T_C [2].

[2] The terms transition temperature and Curie temperature are used synonymously in this section.

(ii) Determine the behaviour of the average spontaneous magnetisation near T_C.

(iii) Determine the behaviour of the average spontaneous magnetisation in the low-temperature region.

(iv) Determine the susceptibility $\langle M \rangle / B$ at very high temperatures.

Solution

Instead of the Hamiltonian (12.1.1) we now have

$$H = -(\mathbf{M} \cdot \mathbf{B}_{\text{eff}}) , \tag{12.3.1}$$

where the magnetic moment $\mathbf{M}$ of the whole system is

$$\mathbf{M} = g\mu_B \sum_{\mathbf{f}} \mathbf{S}_{\mathbf{f}} , \tag{12.3.2}$$

and where

$$\mathbf{B}_{\text{eff}} = \mathbf{B} + \frac{2}{g\mu_B} \langle \mathbf{S} \rangle \sum_{\mathbf{g}} I(\mathbf{g}) . \tag{12.3.3}$$

From Equations (12.2.5) and (12.2.6) we get a system of N spin-$\frac{1}{2}$ particles

$$\langle M \rangle = N g\mu_B \langle S \rangle \tag{12.3.4}$$

or

$$\langle M \rangle = \tfrac{1}{2} N g\mu_B \tanh(\tfrac{1}{2} B g\mu_B B_{\text{eff}}) . \tag{12.3.5}$$

Hence from Equations (12.3.4) and (12.3.3) we get the following implicit equation for $\langle M \rangle$

$$\langle M \rangle = \tfrac{1}{2} N g\mu_B \tanh[\tfrac{1}{2}\beta g\mu_B (B + q\langle M \rangle)] , \tag{12.3.6}$$

where

$$q = \frac{2\sum_{\mathbf{g}} I(\mathbf{g})}{N(g\mu_B)^2} . \tag{12.3.7}$$

(i) To find the Curie temperature we must solve Equation (12.3.6) for the case where $B = 0$, that is, we must solve the equation

$$\frac{\langle M \rangle}{M_0} = \tanh\left[\frac{\sum_{\mathbf{g}} I(\mathbf{g})}{2kT} \frac{\langle M \rangle}{M_0}\right] , \tag{12.3.8}$$

where $M_0 = \frac{1}{2} N g\mu_B$ in the magnetisation at $T = 0$. We note first of all that $\langle M \rangle = 0$ is always a solution. Also, we see that as the curve

$$y = \frac{\langle M \rangle}{M_0}$$

has unit slope in a y *versus* $(\langle M \rangle / M_0)$ diagram, while the curve

$$y = \tanh\left[\frac{\sum_{\mathbf{g}} I(\mathbf{g})}{2kT} \frac{\langle M \rangle}{M_0}\right]$$

has a slope equal to $\sum_{\mathbf{g}} I(\mathbf{g})/2kT$ at the origin in the same diagram, and as the latter curve is convex, the solution $\langle M \rangle = 0$ will be the only one as long as

$$kT > \tfrac{1}{2} \sum_{\mathbf{g}} I(\mathbf{g}) . \tag{12.3.9}$$

However, for temperatures below T_C given by the equation

$$T_C = \sum_{\mathbf{g}} I(\mathbf{g})/2k , \tag{12.3.10}$$

there will also be a non-vanishing solution to Equation (12.3.8). One can show by evaluating the free energy (compare Problem 12.4) that this solution is the equilibrium solution, so that the system shows spontaneous magnetisation below the Curie temperature T_C.

(ii) Near, but below, T_C, $\langle M \rangle$ will be small so that we can expand the tanh in Equation (12.3.8):

$$\frac{\langle M \rangle}{M_0} = \frac{T_C}{T}\frac{\langle M \rangle}{M_0} - \frac{T_C^3}{T^3}\frac{\langle M \rangle^3}{M_0^3} + \cdots , \tag{12.3.11}$$

or, using the fact that $T \approx T_C$ and $\langle M \rangle \approx 0$,

$$\langle M \rangle = M_0 \left(\frac{T_C - T}{T_C} \right)^{1/2} . \tag{12.3.12}$$

(iii) As $T \to 0$, $\langle M \rangle \to M_0$. Expanding the tanh for large values of its argument, we find

$$\langle M \rangle = M_0 \left[1 - 2 \exp\left(-\frac{T_C}{T} \right) \right] , \tag{12.3.13}$$

while experimentally it is found that $M_0 - \langle M \rangle$ behaves as $T^{3/2}$.

(iv) At temperatures above T_C there is no spontaneous magnetisation. When $T \gg T_C$, we can expand the tanh in Equation (12.3.6) and we have

$$\frac{\langle M \rangle}{M_0} = \frac{g\mu_B B}{2kT} + \frac{T_C}{T}\frac{\langle M \rangle}{M_0} + \cdots , \tag{12.3.14}$$

or

$$\frac{\langle M \rangle}{M_0} = \frac{g\mu_B B}{2k(T - T_C)} , \tag{12.3.15}$$

whence we get for the susceptibility per spin $\chi = \langle M \rangle / BN$:

$$\chi = \frac{(\frac{1}{2} g\mu_B)^2}{k(T - T_C)} , \tag{12.3.16}$$

the so-called **Curie-Weiss law.**

12.4 An Ising ferromagnet contains N spin-$\frac{1}{2}$ particles. Let N_+ (N_-) be the number of spins with z-components $+\frac{1}{2}$ ($-\frac{1}{2}$) and let the up- and down-spins be distributed randomly over the lattice. Let

$$R = \frac{N_+ - N_-}{N} \tag{12.4.1}$$

be an order parameter. Show that the entropy and internal energy are given by

$$S = -kN[\tfrac{1}{2}(1+R)\ln\tfrac{1}{2}(1+R)+\tfrac{1}{2}(1-R)\ln\tfrac{1}{2}(1-R)], \tag{12.4.2}$$

$$E = -\tfrac{1}{4}zINR^2, \tag{12.4.3}$$

where z is the coordination number, that is, the number of nearest neighbours per spin.

Minimise the free energy with respect to R to find the equilibrium value of R.

Solution

If the distribution of + and − spins is assumed to be completely random, the probability W for N_+ + spins and N_- − spins is

$$W = \frac{N!}{N_+!N_-!}, \tag{12.4.4}$$

and the entropy S is given by the expression

$$S = k\ln W = k(\ln N! - \ln N_+! - \ln N_-!). \tag{12.4.5}$$

Using the Stirling approximation (see Problem 2.11) in the form

$$\ln N! = N\ln N - N, \tag{12.4.6}$$

we have

$$S = k(N\ln N - N_+\ln N_+ - N_-\ln N_-), \tag{12.4.7}$$

whence introducing R from Equation (12.4.1), and using the fact that $N_+ + N_- = N$, we get expression (12.4.2).

To find the free energy, we must evaluate the internal energy. We have

$$H = -\tfrac{1}{2}I\sum_{\mathbf{f},\mathbf{g}} \mu_{\mathbf{f}}\mu_{\mathbf{g}}, \tag{12.4.8}$$

where the summation is only over nearest neighbour pairs and where we have introduced new variables $\mu_{\mathbf{f}}$ which are +1 (−1) if the spin on the f site is in the positive (negative) z direction. From Equation (12.4.8) we can write the energy in the form

$$E = -\tfrac{1}{2}I(Q_{++} + Q_{--} - Q_{+-}), \tag{12.4.9}$$

where Q_{++} (Q_{--}) is the number of nearest neighbour pairs where both spins are in the positive (negative) z direction while Q_{+-} is the number of pairs with one spin in the positive and the other spin in the negative

z direction. On the assumption that the + and − spins are randomly distributed, we find for Q_{++} (z is the coordination number, that is, the number of nearest neighbours per spin):

$$Q_{++} = \tfrac{1}{2}zN_+p_+ = \tfrac{1}{2}z\frac{N_+^2}{N}, \tag{12.4.10}$$

where p_+ is the probability that a spin is in the positive z direction. Similarly we have

$$Q_{--} = \tfrac{1}{2}zN_-p_- = \tfrac{1}{2}z\frac{N_-^2}{N} \tag{12.4.11}$$

and

$$Q_{+} = \tfrac{1}{2}zN_+p_- + \tfrac{1}{2}zN_-p_+ = z\frac{N_+N_-}{N}. \tag{12.4.12}$$

Combining Equations (12.4.9), (12.4.12), and (12.4.1), we obtain expression (12.4.3). The free energy is thus given by the equation

$$F = E - TS$$
$$= -\tfrac{1}{4}zINR^2 + NkT[\tfrac{1}{2}(1+R)\ln\tfrac{1}{2}(1+R) + \tfrac{1}{2}(1-R)\ln\tfrac{1}{2}(1-R)]. \tag{12.4.13}$$

From the condition $\partial F/\partial R = 0$ we then get

$$\ln\frac{1+R}{1-R} = \frac{zI}{kT}R, \tag{12.4.14}$$

or

$$R = \tanh\frac{zI}{2kT}R. \tag{12.4.15}$$

Bearing in mind (i) that for the Ising model $\Sigma I(g) = zI$, and (ii) that R from expression (12.4.1) is directly proportional to $\langle M\rangle$, we see that we have rederived the molecular field equation (12.3.8).

12.5 Use the results of Problem 12.4 to find a parametric expression for the specific heat of an Ising ferromagnet in the molecular field approximation.

Find the jump in the specific heat at T_C, and the behaviour of the specific heat as $T \to 0$.

Solution

From Equation (12.4.13) we find for the specific heat c_v:

$$c_v = -\tfrac{1}{2}zINR\frac{dR}{dT}, \tag{12.5.1}$$

while R satisfies Equation (12.4.15) which for our present purpose we shall write in the form

$$R = \tanh\frac{RT_C}{T}, \tag{12.5.2}$$

whence

$$\frac{dR}{dT} = -\frac{RT_C}{T^2}\Big/\left(\cosh^2\frac{RT_C}{T} - \frac{T_C}{T}\right), \tag{12.5.3}$$

so that the specific heat is determined by the parametric equations

$$c_v = \frac{Nkr^2}{\cosh^2 r - r\coth r}, \tag{12.5.4}$$

$$r\coth r = \frac{T_C}{T}, \tag{12.5.5}$$

where r is related to R by the equation $r = RT_C/T$.

To find the behaviour near T_C we note that as $T \to T_C$, $r \to 0$, so that we can expand the various hyperbolic functions occurring in Equation (12.5.4); the final result is

$$c_v \to \tfrac{3}{2}Nk \text{ as } T \to T_C\,. \tag{12.5.6}$$

As $c_v = 0$ for $T > T_C$ in the molecular field approximation [$E = 0$; see Equation (12.4.3)] the jump in c_v is $\frac{3}{2}Nk$ at T_C.

When $T \ll T_C$, we see from Equation (12.5.5) that $r \approx T_C/T$ and we thus obtain from Equation (12.5.4)

$$c_v \approx 4Nk\left(\frac{T_C}{T}\right)^2 \exp\left(-\frac{2T_C}{T}\right). \tag{12.5.7}$$

12.6 In a **ferromagnetic** substance $I(\mathbf{f}-\mathbf{g})$ is positive when $\mathbf{f}$ and $\mathbf{g}$ are nearest neighbours. If this quantity is negative, we are dealing with an **antiferromagnetic**. To consider an antiferromagnetic we assume that the crystal can be divided into two sublattices such that all the nearest neighbours of a spin on one sublattice are on the other sublattice. We shall now have two averages which will be different for the two sublattices, as a negative value of $I(\mathbf{f}-\mathbf{g})$ implies a preference for an antiparallel alignment of nearest neighbours. Use the molecular field approximation to find the susceptibility of an antiferromagnetic at high temperatures and to find the **Néel temperature**, that is, the temperature below which both sublattices show spontaneous magnetisation. Consider again the spin-$\frac{1}{2}$ case.

Solution

Taking into account that there are two sublattices, we write the Hamiltonian in the form

$$\begin{aligned} H = &-g\mu_B\sum_{\mathbf{f}_1}\left\{\mathbf{S}_{\mathbf{f}_1}\cdot\left[\mathbf{B}+\frac{2}{g\mu_B}\sum_{\mathbf{g}_1}I(\mathbf{f}_1-\mathbf{g}_1)\mathbf{S}_{\mathbf{g}_1}+\frac{2}{g\mu_B}\sum_{\mathbf{g}_2}I(\mathbf{f}_1-\mathbf{g}_2)\mathbf{S}_{\mathbf{g}_2}\right]\right\} \\ &-g\mu_B\sum_{\mathbf{f}_2}\left\{\mathbf{S}_{\mathbf{f}_2}\cdot\left[\mathbf{B}+\frac{2}{g\mu_B}\sum_{\mathbf{g}_1}I(\mathbf{f}_2-\mathbf{g}_1)\mathbf{S}_{\mathbf{g}_1}+\frac{2}{g\mu_B}\sum_{\mathbf{g}_2}I(\mathbf{f}_2-\mathbf{g}_2)\mathbf{S}_{\mathbf{g}_2}\right]\right\}, \end{aligned} \tag{12.6.1}$$

where the $\mathbf{f}_1$, $\mathbf{g}_1$ ($\mathbf{f}_2$, $\mathbf{g}_2$) are lattice sites on the first (second) sublattice. In the molecular field approximation we replace $\mathbf{S}_{\mathbf{g}_1}$ and $\mathbf{S}_{\mathbf{g}_2}$ by their averages $\langle \mathbf{S}_1 \rangle$ and $\langle \mathbf{S}_2 \rangle$, which will now be different, and we rewrite the Hamiltonian (12.6.1) in the form

$$H = -(\mathbf{M}_1 \cdot \mathbf{B}_{\text{eff}}^{(1)}) - (\mathbf{M}_2 \cdot \mathbf{B}_{\text{eff}}^{(2)}) , \tag{12.6.2}$$

where

$$\mathbf{M}_i = g\mu_{\text{B}} \sum_{\mathbf{f}_i} \mathbf{S}_{\mathbf{f}_i} , \tag{12.6.3}$$

and

$$\left.\begin{aligned} \mathbf{B}_{\text{eff}}^{(1)} &= \mathbf{B} - q_1 \langle \mathbf{M}_1 \rangle - q_2 \langle \mathbf{M}_2 \rangle , \\ \mathbf{B}_{\text{eff}}^{(2)} &= \mathbf{B} - q_2 \langle \mathbf{M}_1 \rangle - q_1 \langle \mathbf{M}_2 \rangle , \end{aligned}\right\} \tag{12.6.4}$$

with

$$\langle \mathbf{M}_i \rangle = \tfrac{1}{2} N g\mu_{\text{B}} \langle \mathbf{S}_i \rangle , \tag{12.6.5}$$

the extra factor $\frac{1}{2}$ deriving from the fact that the N spins are divided evenly over the two sublattices, and

$$\left.\begin{aligned} q_1 &= -2\sum_{\mathbf{g}_1} \frac{I(\mathbf{f}_1 - \mathbf{g}_1)}{N(g\mu_{\text{B}})^2} = -2\sum_{\mathbf{g}_2} \frac{I(\mathbf{f}_2 - \mathbf{g}_2)}{N(g\mu_{\text{B}})^2} , \\ q_2 &= -2\sum_{\mathbf{g}_1} \frac{I(\mathbf{f}_2 - \mathbf{g}_1)}{N(g\mu_{\text{B}})^2} = -2\sum_{\mathbf{g}_2} \frac{I(\mathbf{f}_1 - \mathbf{g}_2)}{N(g\mu_{\text{B}})^2} , \end{aligned}\right\} \tag{12.6.6}$$

where we have included an extra minus sign to take into account that the dominant interaction leads to antiferromagnetism.

As the two sublattices are now decoupled in Equation (12.6.2) the results of Problem 12.2 can be used for each of the sublattices, and we get

$$\left.\begin{aligned} \langle M_1 \rangle &= \tfrac{1}{4} N g\mu_{\text{B}} \tanh[\tfrac{1}{2}\beta g\mu_{\text{B}}(B - q_1\langle M_1 \rangle - q_2 \langle M_2 \rangle)] , \\ \langle M_2 \rangle &= \tfrac{1}{4} N g\mu_{\text{B}} \tanh[\tfrac{1}{2}\beta g\mu_{\text{B}}(B - q_2\langle M_1 \rangle - q_1 \langle M_2 \rangle)] . \end{aligned}\right\} \tag{12.6.7}$$

At high temperatures we can write $\tanh x \approx x$, or

$$\left.\begin{aligned} \langle M_1 \rangle &\approx \tfrac{1}{8}\beta N(g\mu_{\text{B}})^2 (B - q_1 \langle M_1 \rangle - q_2 \langle M_2 \rangle) , \\ \langle M_2 \rangle &\approx \tfrac{1}{8}\beta M(g\mu_{\text{B}})^2 (B - q_2 \langle M_1 \rangle - q_1 \langle M_2 \rangle) , \end{aligned}\right\} \tag{12.6.8}$$

so that we get for the total magnetisation

$$\langle M \rangle = \langle M_1 \rangle + \langle M_2 \rangle \approx \frac{2B\Theta}{q(T+\Theta)} , \tag{12.6.9}$$

with

$$q = q_1 + q_2 = -\frac{2\sum I(\mathbf{g})}{N(g\mu_{\text{B}})^2} , \tag{12.6.10}$$

which is the same as expression (12.3.7) apart from the sign, and

$$\Theta = \frac{qN(g\mu_{\text{B}})^2}{8k} . \tag{12.6.11}$$

To find the Néel temperature, T_N, we put $B = 0$ and expand the tanh, as just below T_N the sublattice magnetisations will be small [compare the solution of Problem 12.3 (ii)]. We then have

$$\left.\begin{aligned} \langle M_1\rangle &\approx -\frac{\Theta}{qT}(q_1\langle M_1\rangle + q_2\langle M_2\rangle) + \text{term of third order in } \langle M_i\rangle\,, \\ \langle M_2\rangle &\approx -\frac{\Theta}{qT}(q_2\langle M_2\rangle + q_1\langle M_1\rangle) + \cdots\,. \end{aligned}\right\} \tag{12.6.12}$$

For an antiferromagnetic we always have a vanishing total magnetisation when there is no external field so that we can put in Equations (12.6.12) $\langle M_1\rangle = -\langle M_2\rangle$ and we then see that the sublattice magnetisations vanish at T_N given by the equation

$$T_N = \frac{q_2 - q_1}{q}\Theta\,. \tag{12.6.13}$$

We note that if we had assumed that there were only nearest neighbour interactions q_1 would be zero and $q_2 = q$ so that T_N and Θ would be the same. We also note that in the unlikely case where $q_1 > q_2$ we would have no transition; this is not surprising, as $q_1 > q_2$ would mean that there would be a strong tendency for spins in the **same** sublattice to align antiparallel. In that case, the model used would clearly be a very poor one.

12.7 Find for an antiferromagnetic of the type considered in the preceding problem the perpendicular and parallel susceptibility $\chi_\perp$ and $\chi_\parallel$ (for the cases where the applied field is perpendicular or parallel to the sublattice magnetisations) at temperatures below the Néel temperature.

Solution

Let

$$\langle \mathbf{M}_1\rangle_0 = -\langle \mathbf{M}_2\rangle_0 = \mathbf{m}\,, \tag{12.7.1}$$

where the index 0 indicates the absence of an external field. Consider first the case

$$\mathbf{B} \perp \mathbf{m}\,. \tag{12.7.2}$$

The effective fields are again given by Equations (12.6.4).

We are interested in the susceptibility, that is, in the value of the magnetisation when there is a small external field present, and as $\mathbf{B} \perp \mathbf{m}$, we expect that $\langle \mathbf{M}_1\rangle$ and $\langle \mathbf{M}_2\rangle$ will be no longer strictly antiparallel. Let ϵ be the angle between $\langle \mathbf{M}_1\rangle$ and $-\langle \mathbf{M}_2\rangle$. We expect $\epsilon \ll 1$. The effective fields $\mathbf{B}_{\text{eff}}^{(i)}$ will be parallel to the $\langle \mathbf{M}_i\rangle$ so that from Equation (12.6.4) it follows that $\mathbf{B} - q_2\langle \mathbf{M}_2\rangle$ must be parallel to $\langle \mathbf{M}_1\rangle$ and $\mathbf{B} - q_2\langle \mathbf{M}_1\rangle$

parallel to $\langle \mathbf{M}_2 \rangle$ (see Figure 12.7.1). The total magnetisation $\langle \mathbf{M}_1 \rangle + \langle \mathbf{M}_2 \rangle$ will be parallel to $\mathbf{B}$, and its magnitude, divided by B, will give us $\chi_\perp$. From Figure 12.7.1 we then see that

$$\chi_\perp = \frac{1}{q_2} . \tag{12.7.3}$$

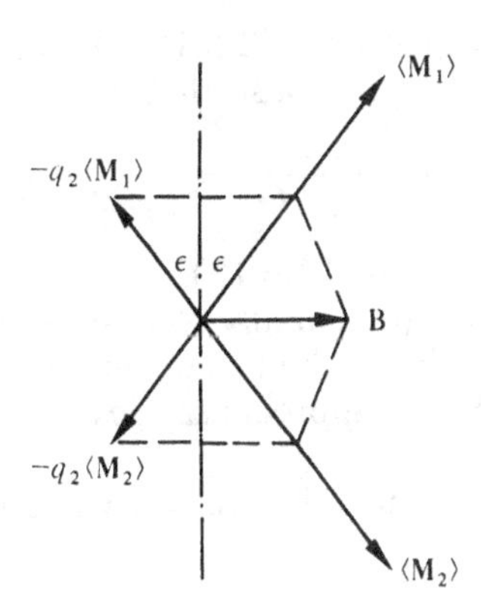

Figure 12.7.1.

Consider now

$$\mathbf{B} \parallel \mathbf{m} . \tag{12.7.4}$$

We then expect all vectors to be parallel (or antiparallel) and, moreover, for small B we expect $\langle \mathbf{M}_1 \rangle$ and $-\langle \mathbf{M}_2 \rangle$ to be nearly equal to $\mathbf{m}$ so that we can write

$$\langle \mathbf{M}_1 \rangle = \mathbf{m} + \delta \mathbf{m}_1 , \quad \langle \mathbf{M}_2 \rangle = -\mathbf{m} + \delta \mathbf{m}_2 . \tag{12.7.5}$$

We use again Equation (12.6.7) and write for the argument of the first tanh

$$\tfrac{1}{2}\beta g \mu_B (B - q_1 \langle M_1 \rangle - q_2 \langle M_2 \rangle) = \tfrac{1}{2}\beta g \mu_B m(q_2 - q_1) + \tfrac{1}{2}\beta g \mu_B (B - q_1 \delta m_1 - q_2 \delta m_2) , \tag{12.7.6}$$

and similarly for the argument of the second tanh. Expanding in terms of the small quantity which contains B, δm_1, and δm_2, we obtain from Equation (12.6.7)

$$\left.\begin{aligned} \delta m_1 &= \tfrac{1}{8}\beta N (g\mu_B)^2 (B - q_1 \delta m_1 - q_2 \delta m_2) \cosh^{-2}[\tfrac{1}{2}\beta g \mu_B m(q_2 - q_1)] , \\ \delta m_2 &= \tfrac{1}{8}\beta N (g\mu_B)^2 (B - q_1 \delta m_2 - q_1 \delta m_1) \cosh^{-2}[\tfrac{1}{2}\beta g \mu_B m(q_2 - q_1)] . \end{aligned}\right\} \tag{12.7.7}$$

Adding these equations, we find the total magnetisation

$$\langle M_1 \rangle + \langle M_2 \rangle = \delta m_1 + \delta m_2 ,$$

and dividing by B we find $\chi_{\|}$. The result is

$$\chi_{\|} = \frac{2\Lambda}{q(T+\Lambda)}, \tag{12.7.8}$$

where

$$\Lambda = \Theta \cosh^{-2}[\tfrac{1}{2}\beta g\mu_B m(q_2-q_1)], \tag{12.7.9}$$

with Θ and q given by Equations (12.6.10) and (12.6.11).

We note that as $T \to T_N$, $m \to 0$, so that $\Lambda \to \Theta$, and

$$\chi_{\|} \to \frac{2\Theta}{q(T_N+\Theta)} = \frac{1}{q_2} \quad (= \chi_{\perp}), \tag{12.7.10}$$

where we have used Equation (12.6.13).

As $T \to 0$, Λ vanishes exponentially so that $\chi_{\|} \to 0$.

Finally, we note from Equation (12.6.9) that as T approaches T_N from above, the susceptibility also approaches $1/q_2$.

12.8 Consider the Ising model. We can write the Hamiltonian in the form

$$H = -\tfrac{1}{2}g\mu_B \sum_{\mathbf{f}} \mu_{\mathbf{f}} - \tfrac{1}{2}\sum_{(\mathbf{f},\mathbf{g})} I\mu_{\mathbf{f}}\mu_{\mathbf{g}}, \tag{12.8.1}$$

where now the sum over **f** and **g** is such that only terms where **f** and **g** are nearest neighbours occur, and where we have introduced new variables $\mu_{\mathbf{f}}$ which can take on the values $+1$ and -1.

The molecular field approximation consists in replacing the Hamiltonian (12.8.1) by

$$H = -\tfrac{1}{2}g\mu_B B \sum_{\mathbf{f}} \mu_{\mathbf{f}} - \tfrac{1}{2}z\sum_{\mathbf{f}} I\mu_{\mathbf{f}}\bar{\mu}, \tag{12.8.2}$$

where $\bar{\mu}$ is the average value of μ and z is the **coordination number**, that is, the number of nearest neighbours per spin. This approximation essentially reduces the problem to that of a single spin problem, or, as we saw in Problem 12.4 to assuming that the $+1$ and -1 values are randomly distributed over the lattice.

The next approximation takes into account that because of the interaction there will be a preference for $++$ pairs or $--$ pairs, rather than $+-$ pairs.

Let Q be the total number of pairs in the lattice, so that

$$Q = \tfrac{1}{2}zN. \tag{12.8.3}$$

Let Q_{++}, Q_{+-}, Q_{--} be, respectively, the number of $++$, $+-$, and $--$ pairs. Prove the following relations for Q_{++}, Q_{+-}, and Q_{--}:

$$\left.\begin{aligned} 2Q_{++}+Q_{+-} &= zN_+, \\ 2Q_{--}+Q_{+-} &= zN_-, \end{aligned}\right\} \tag{12.8.4}$$

where N_+ and N_- were defined in Problem 12.4. We have, of course, also the relation

$$Q_{++}+Q_{--}+Q_{+-} = Q\,, \qquad (12.8.5)$$

which is a consequence of Equations (12.8.4). We now introduce a **short-range** parameter σ by the equation

$$Q_{++}+Q_{--}-Q_{+-} = \sigma Q\,. \qquad (12.8.6)$$

A given state of the system is now characterised by the **long-range** parameter R and the short-range parameter σ.

To proceed further we consider instead of a one-spin system the system of $z+1$ spins where we take into account explicitly the interactions between a central spin and its z neighbours, but take the influence of the other spins in the lattice into account through a mean field B'. That is, we consider the Hamiltonian.

$$H_{z+1} = -\tfrac{1}{2}g\mu_B B\left(\mu_0+\sum_{j=1}^{z}\mu_j\right)-\tfrac{1}{2}g\mu_B B'\sum_{j=1}^{z}\mu_j-\tfrac{1}{2}I\mu_0\sum_{j=1}^{z}\mu_j\,, \qquad (12.8.7)$$

where 0 indicates the central spin and $j = 1, ..., z$ its nearest neighbours.

(i) Prove that in this approximation

$$\frac{\langle Q_{+-}\rangle^2}{\langle Q_{++}\rangle\langle Q_{--}\rangle} = \tfrac{1}{4}x^2\,, \qquad (12.8.8)$$

where

$$x = e^{-\beta I}\,. \qquad (12.8.9)$$

(ii) Find σ as a function of x and R.

(iii) Use the fact that for self-consistency $\langle\mu_0\rangle$ must equal $\langle\mu_j\rangle$, which determines the mean field B', to find the Curie temperature in this approximation.

(iv) Show that in the limit as $z\to\infty$ the critical values for x of the present approximation and of the molecular field approximation become the same and that also in that limit $\sigma = R^2$ so that the expressions for the energy in the present and in the molecular field approximation become the same. Note that as $z\to\infty$ one must assume that $I\to 0$ while zI remains finite, as zI determines the transition temperature which we want to keep finite.

Solution

Equation (12.8.4) follows immediately by seeing that each spin has z nearest neighbours and that each +-spin must lead to either a ++ or a +− pair, and that, if we are not careful, we shall count certain pairs twice.

(i) To prove the Equation (12.8.8) we write down the partition function Z:

$$Z = \sum_{\mu_0=\pm1}\sum_{\mu_1=\pm1}\cdots\sum_{\mu_z=\pm1}\exp(-\beta H_{z+1})\,. \qquad (12.8.10)$$

If we write

$$K = \tfrac{1}{2}\beta g\mu_B B\,, \quad K' = \tfrac{1}{2}\beta g\mu_B(B+B')\,, \quad J = \tfrac{1}{2}\beta I\,, \qquad (12.8.11)$$

we have

$$Z = \sum_{\mu_0=\pm 1}\sum_{\mu_1=\pm 1}\cdots\sum_{\mu_z=\pm 1}\exp\left(K\mu_0+K'\sum_j\mu_j+J\sum\mu_0\mu_j\right) \qquad (12.8.12)$$

$$= \sum_{\mu_0=\pm 1}\sum_{\mu_1=\pm 1}\exp(K\mu_0+K'\mu_1+J\mu_0\mu_1)\prod_{j=2}^{z}\exp[(K'+J\mu_0)\mu_j] \qquad (12.8.13)$$

$$= \sum_{\mu_0=\pm 1}\sum_{\mu_1=\pm 1}\exp(K\mu_0+K'\mu_1+J\mu_0\mu_1)[2\cosh(K'+J\mu_0)]^{z-1}\,. \qquad (12.8.14)$$

Let us denote the four terms corresponding, respectively, to the μ_0, μ_j values: $+1, +1$; $+1, -1$; $-1, +1$; $-1, -1$ by Z_{++}, Z_{+-}, Z_{-+}, Z_{--}. We have

$$\left.\begin{aligned} Z_{++} &= \exp(K+K'+J)Z_{z-1}^{(+)}\,,\\ Z_{+-} &= \exp(K-K'-J)Z_{z-1}^{(+)}\,,\\ Z_{-+} &= \exp(-K+K'-J)Z_{z-1}^{(-)}\,,\\ Z_{--} &= \exp(-K-K'+J)Z_{z-1}^{(-)}\,, \end{aligned}\right\} \qquad (12.8.15)$$

where

$$Z_{z-1}^{(\pm)} = [2\cosh(K' \pm J)]^{z-1}\,. \qquad (12.8.16)$$

From the general theory of partition functions (compare Chapters 4 and 6) it follows that

$$Z_{++}:Z_{+-}:Z_{-+}:Z_{--} = \langle Q_{++}\rangle:\tfrac{1}{2}\langle Q_{+-}\rangle:\tfrac{1}{2}\langle Q_{-+}\rangle:\langle Q_{--}\rangle\,. \qquad (12.8.17)$$

Equation (12.8.8) follows immediately.

In this derivation we have used a symmetry argument to argue that Z_{+-} must equal Z_{-+}. One could say that this is the argument which determines K' in terms of K and J, and we shall see presently, that indeed, a consistency requirement leads to K' such that $Z_{+-} = Z_{-+}$ (see part (iii)].

(ii) From Equations (12.8.3), (12.8.4), and (12.4.1) we find easily

$$\left.\begin{aligned} Q_{++} &= \tfrac{1}{8}zN(1+\sigma+2R)\,,\\ Q_{--} &= \tfrac{1}{8}zN(1+\sigma-2R)\,,\\ Q_{+-} &= \tfrac{1}{4}zN(1-\sigma)\,, \end{aligned}\right\} \qquad (12.8.18)$$

and hence from Equation (12.8.8) we find

$$\frac{(1+\sigma+2R)(1+\sigma-2R)}{(1-\sigma)^2} = x^{-2}\,, \qquad (12.8.19)$$

whence

$$\sigma = \frac{1+x^2-2x[1-R^2+R^2x^2]^{1/2}}{1-x^2}\,. \qquad (12.8.20)$$

We note that when $T > T_C$, $R = 0$ so that then

$$\sigma = \frac{1-x}{1+x} \, . \tag{12.8.21}$$

(iii) To find the transition temperature, one carries out first the μ_1-summation in Equation (12.8.14) to find

$$Z = \sum_{\mu_0 = \pm 1} \exp(K\mu_0)[2\cosh(K+K'+J\mu_0)]^z \tag{12.8.22}$$

$$= Z_+ + Z_- \, , \tag{12.8.23}$$

with

$$Z_\pm = e^{\pm K} Z_z^{(\pm)} \, . \tag{12.8.24}$$

From Equation (12.8.12) it follows, on the one hand, that

$$\langle \mu_0 \rangle = \frac{\partial \ln Z}{\partial K} = \frac{Z_+ - Z_-}{Z} \, . \tag{12.8.25}$$

On the other hand, we have

$$\sum_j \langle \mu_j \rangle = \frac{\partial \ln Z}{\partial K'} = z \frac{Z_+ \tanh(K'+J) + Z_- \tanh(K'-J)}{Z} \, . \tag{12.8.26}$$

From the consistency condition

$$\langle \mu_0 \rangle = \frac{1}{z} \sum_j \langle \mu_j \rangle \, , \tag{12.8.27}$$

we then find

$$\frac{Z_+}{Z_-} = \frac{1 + \tanh(K'-J)}{1 - \tanh(K'+J)} \, , \tag{12.8.28}$$

or, using Equations (12.8.24) and (12.8.16),

$$\frac{\cosh(K'+J)}{\cosh(K'-J)} = \exp\frac{2(K'-K)}{z-1} \, , \tag{12.8.29}$$

from which the equality of Z_{+-} and Z_{-+} also follows.

To find the transition temperature we put $B = 0$ in Equation (12.8.29) so that now

$$K' = \tfrac{1}{2}\beta g \mu_B B' \, , \tag{12.8.30}$$

while the equation for determining K' becomes

$$\frac{\cosh(K'+J)}{\cosh(K'-J)} = \exp\frac{2K'}{z-1} \, . \tag{12.8.31}$$

We note that $K' = 0$ is always a solution. The transition temperature is that temperature for which a non-vanishing solution for K' becomes possible. Expanding both sides of Equation (12.8.31) for small values of K' up to K'^3 we obtain, after taking the logarithm,

$$\ln\frac{\cosh(J+K')}{\cosh(J-K')} = \frac{2K'}{z-1}, \tag{12.8.32}$$

whence

$$\ln(1+2K'\tanh J+2K'^2\tanh^2 J-\tfrac{2}{3}K'^3\tanh J+\cdots) \tag{12.8.33}$$

or

$$2K'\tanh J-\tfrac{2}{3}K'^3\tanh J(1+2\tanh^2 J)+\cdots = \frac{2K'}{z-1}. \tag{12.8.34}$$

As $K' \to 0$, it follows that

$$\tanh J_C = \frac{1}{z-1}, \tag{12.8.35}$$

or

$$\frac{1-x_C}{1+x_C} = \frac{1}{z-1}, \tag{12.8.36}$$

whence

$$x_C = 1-\frac{2}{z}. \tag{12.8.37}$$

At temperatures just below T_C, we find from Equation (12.8.34) that K'^2 satisfies the equation

$$K'^2 = \frac{3(\tanh J-\tanh J_C)}{\tanh J+2\tanh^3 J}. \tag{12.8.38}$$

(iv) From Equation (12.3.10), the fact that for the Ising model $\sum_{\mathbf{g}} I(\mathbf{g})$ becomes zI, and the definition (12.8.9) of x it follows that in the molecular field approximation

$$x_C = \exp(-\beta_C I) = \exp(-2/z), \tag{12.8.39}$$

which is the same as expression (12.8.37) in the limit as $z \to \infty$.

To find σ in that limit we note that it follows from Equations (12.8.9) and (12.8.11) that

$$x = \exp(-2J) \approx 1-2J+\cdots \tag{12.8.40}$$

in the limit as $I \to 0$ (and also $J \to 0$). We then get from Equation (12.8.20) after some expansions

$$\sigma \approx R^2. \tag{12.8.41}$$

We note that we can write Equation (12.4.9) in the form

$$E = -\tfrac{1}{2}IQ\sigma, \tag{12.8.42}$$

which is now the same as Equation (12.4.3), if (12.8.41) holds.

In fact, one can prove generally that all results in the present approximation become those of the molecular field approximation in the limit as $z \to \infty$.

13
Green function methods[1]

D.ter HAAR
(*University of Oxford, Oxford*)

MATHEMATICAL PRELIMINARIES

13.1 Show that, if $[A, B]_- \equiv AB - BA = K$, where A and B are operators and K is a *c*-number, and if λ is a *c*-number,

$$e^{\lambda(A+B)} = e^{\lambda A} e^{\lambda B} e^{-\frac{1}{2}\lambda^2 K} . \tag{13.1.1}$$

Solution

Let $\varphi(\lambda) \equiv e^{\lambda A} e^{\lambda B}$. We then have

$$\frac{\partial \varphi(\lambda)}{\partial \lambda} = A e^{\lambda A} e^{\lambda B} + e^{\lambda A} B e^{\lambda B} = (A+B)\varphi(\lambda) + [e^{\lambda A}, B]_- e^{\lambda B} . \tag{13.1.2}$$

We further have

$$[e^{\lambda A}, B]_- = \sum_{n=0}^{\infty} \frac{\lambda^n}{n!}[A^n, B]_- = \sum_{n=0}^{\infty} \frac{\lambda^n}{(n-1)!} K A^{n-1} = \lambda K e^{\lambda A} . \tag{13.1.3}$$

Hence

$$\frac{\partial \varphi}{\partial \lambda} = [A + B + \lambda K]\varphi , \tag{13.1.4}$$

or

$$\varphi = e^{\lambda(A+B) + \frac{1}{2}\lambda^2 K} C , \tag{13.1.5}$$

where C is an arbitrary constant operator. By letting $\lambda \to 0$, we find that C is the unit operator, which concludes the proof.

If $[A, B]_- = D$, where D is an operator, expression (13.1.1) becomes more complicated.

We note that we can only commute e^A and e^B, if A and B themselves commute.

13.2 Show that if $\theta(t)$ is the **unit step function,**

$$\theta(t) = 1, \quad t > 0; \quad \theta(t) = 0, \quad t < 0 , \tag{13.2.1}$$

we have

$$\dot{\theta}(t - t') \equiv \frac{d}{dt}[\theta(t - t')] = -\dot{\theta}(t' - t) = \delta(t - t') . \tag{13.2.2}$$

[1] For a general discussion see: V.L.Bonch-Bruevich and S.V.Tyablikov, *Green Function Methods in Statistical Mechanics* (North-Holland, Amsterdam), 1962, and D.ter Haar in *Fluctuation, Relaxation and Resonance in Magnetic Systems* (Ed.D.ter Haar) (Oliver and Boyd, Edinburgh), 1961.

13.3 Make it plausible that

$$y \equiv \lim_{\epsilon\to 0} \frac{1}{x \pm i\epsilon} = \mathscr{P}\left(\frac{1}{x}\right) \mp i\pi\delta(x)\,, \tag{13.3.1}$$

where x and ϵ are real and $\mathscr{P}$ indicates that in an x-integration the principal value of the integral must be taken.

Solution

To prove the equality (13.3.1) we consider the integral

$$I = \lim_{\epsilon\to 0} \int_{-\infty}^{+\infty} \frac{f(x)}{x - i\epsilon} dx\,, \tag{13.3.2}$$

where we assume that $f(x)$ has no singularities on the real axis.

We write this in the form

$$I = \lim_{\epsilon\to 0} \lim_{\delta\to 0} \left(\int_{-\infty}^{-\delta} + \int_{+\delta}^{+\infty}\right) \frac{f(x)}{x - i\epsilon} dx + \lim_{\epsilon\to 0} \lim_{\delta\to 0} \int_{-\delta}^{+\delta} \frac{f(x)}{x - i\epsilon} dx\,. \tag{13.3.3}$$

The first term gives the principal value of

$$\int_{-\infty}^{+\infty} \frac{f(x)\,dx}{x}$$

and the second term gives

$$\lim_{\epsilon\to 0} \lim_{\delta\to 0} \int_{-1}^{+1} \frac{f(\delta y)\,dy}{y - i(\epsilon/\delta)} = f(0) \lim_{\epsilon\to 0} \ln \frac{y - i\epsilon}{y + i\epsilon} = \pi i f(0)\,.$$

GENERAL FORMALISM

13.4 In statistical mechanics one is often interested in averages taken over a grand canonical ensemble. Sometimes these are averages of products of operators

$$\langle AB\rangle = \frac{\mathrm{Tr}\,AB\exp(-\beta H + \alpha N_{\mathrm{op}})}{\Xi}\,, \tag{13.4.1}$$

where the symbols have the same meaning as in Equation (10.2.4). It is usually difficult to evaluate $\langle AB\rangle$ exactly and approximation methods must be used. Apart from $\langle AB\rangle$ one is also often interested in **correlation functions** such as $\langle A(t)B(t')\rangle$ or $\langle B(t')A(t)\rangle$, which are not the same. Although one can write down an equation of motion for $\langle A(t)B(t')\rangle$, it turns out that it is more convenient to use the so-called **retarded and advanced Green functions** defined by the equations

$$\langle\langle A(t);\ B(t')\rangle\rangle_{\substack{r\\a}} = \mp\frac{i}{\hbar}\theta(\pm t \mp t')\langle A(t)B(t')\rangle \pm \frac{i\eta}{\hbar}\theta(\pm t \mp t')\langle B(t')A(t)\rangle\,, \tag{13.4.2}$$

where $\theta(t)$ is the step-function of Equation (13.2.1), η is at the moment a disposable parameter which we shall choose to be either $+1$ or -1

according to our convenience, and the $A(t)$ and $B(t')$ are **time-dependent (Heisenberg) operators**, defined by the equation

$$A(t) = \exp\left[\frac{i(H-\mu N_{op})t}{\hbar}\right] A \exp\left[-\frac{i(H-\mu N_{op})t}{\hbar}\right], \qquad (13.4.3)$$

with $\mu = \alpha/\beta$.

The generalisation from the usual Heisenberg operators is necessary whenever H and N_{op} do not commute, as should be clear from the solution of the present problem and the result obtained in Problem 13.1.

Prove that $\langle\langle A(t);\ B(t')\rangle\rangle_{\substack{r\\a}}$ as well as the correlation functions

$$F_{BA} = \langle B(t')A(t)\rangle \quad \text{and} \quad F_{AB} = \langle A(t)B(t')\rangle \qquad (13.4.4)$$

are functions of $t-t'$ only.

Solution

This follows by writing out explicitly the grand canonical averages, using the facts that (i) inside a trace operators can be cyclically permuted, and (ii) the various factors commute.

13.5 Prove that the $\langle\langle A(t);\ B(t')\rangle\rangle_{\substack{r\\a}}$ satisfy the following equation of motion:

$$i\hbar\langle\langle A(t);\ B(t')\rangle\rangle = \delta(t-t')\langle AB-\eta BA\rangle + \langle\langle [A, H-\mu N_{op}]_-(t);\ B(t')\rangle\rangle\ . \qquad (13.5.1)$$

Solution

Equation (13.5.1) follows from the result of Problem 13.2 and the relation

$$i\hbar\dot{A} = [A, H-\mu N_{op}]_-\ . \qquad (13.5.2)$$

13.6 Introducing the Fourier transform $\langle\langle A;\ B\rangle\rangle_{\substack{r\\a}E}$ of $\langle\langle A(t);\ B(t')\rangle\rangle_{\substack{r\\a}}$ by the equation

$$\langle\langle A;\ B\rangle\rangle_E = \frac{1}{2\pi}\int_{-\infty}^{+\infty} \langle\langle A(t);\ B(t')\rangle\rangle \exp\frac{iE(t-t')}{\hbar}\, d(t-t')\ , \qquad (13.6.1)$$

so that

$$\langle\langle A(t);\ B(t')\rangle\rangle = \int_{-\infty}^{+\infty} \langle\langle A;\ B\rangle\rangle_E \exp\left[-\frac{iE(t-t')}{\hbar}\right] d\left(\frac{E}{\hbar}\right), \qquad (13.6.2)$$

find the equation for motion for $\langle\langle A;\ B\rangle\rangle_E$.

Solution

Straightforward Fourier transform of Equation (13.5.1) leads to

$$E\langle\langle A;\ B\rangle\rangle_E = \frac{1}{2\pi}\langle AB-\eta BA\rangle + \langle\langle [A, H-\mu N_{op}]_-;\ B\rangle\rangle_E\ . \qquad (13.6.3)$$

13.7 Introducing the **spectral representation** (Fourier transform) $J(\omega)$ of F_{BA} by the equation

$$F_{BA}(t,t') = \int_{-\infty}^{+\infty} J(\omega)\mathrm{e}^{-\mathrm{i}\omega(t-t')}\mathrm{d}\omega\,, \tag{13.7.1}$$

and similarly introducing the Fourier transform $J'(\omega)$ of F_{AB}, prove that

$$J'(\omega) = J(\omega)\exp(\beta\hbar\omega)\,. \tag{13.7.2}$$

Solution

If the $|\mu\rangle$ are the eigenfunctions of $H-\mu N_{\text{op}}$ and if

$$\langle\nu|H-\mu N_{\text{op}}|\mu\rangle = \delta_{\nu\mu}E_\nu\,,$$

we find

$$J(\omega) = \frac{\hbar}{\Xi}\sum_{\nu,\mu}\langle\nu|B|\mu\rangle\langle\mu|A|\nu\rangle\exp(-\beta E_\nu)\delta(\omega-E_\nu+E_\mu)\,.$$

Repeating the calculation for F_{AB}, we find expression (13.7.2).

13.8 If $G(E)$ is an analytic function of E which is equal to $\langle\langle A;B\rangle\rangle_r$ [2] in the upper half-plane and equal to $\langle\langle A;B\rangle\rangle_a$ in the lower half-plane, use the equation

$$\theta(t) = \mathrm{e}^{-\epsilon t} \quad (\epsilon\to 0+),\quad t>0;\quad \theta(t)=0,\quad t<0\,, \tag{13.8.1}$$

and (13.3.1) to express $J(\omega)$ in terms of $G(E)$.

Solution

From Equations (13.6.1), (13.4.2), (13.7.1) and the similar equation for F_{ab}, the relation between $J(\omega)$ and $J'(\omega)$, and Equation (13.8.1), we find

$$\langle\langle A;\ B\rangle\rangle_{\substack{r\\a}} = \frac{1}{2\pi}\int_{-\infty}^{+\infty}[\exp(\beta\hbar\omega)-\eta]J(\omega)\frac{\mathrm{d}\omega}{E-\hbar\omega\pm\mathrm{i}\epsilon} \quad (\epsilon\to 0+)\,.$$

The function $G(E)$ defined in the problem is thus given by

$$G(E) = \frac{1}{2\pi}\int_{-\infty}^{+\infty}[\exp(\beta\hbar\omega)-\eta]J(\omega)\frac{\mathrm{d}\omega}{E-\hbar\omega}\,, \tag{13.8.2}$$

and using Equation (13.3.1) one finds

$$J(\omega) = \frac{\mathrm{i}}{\exp(\beta\hbar\omega)-\eta}\lim_{\epsilon\to 0+}[G(\omega+\mathrm{i}\epsilon)-G(\omega-\mathrm{i}\epsilon)]\,. \tag{13.8.3}$$

13.9 (a) If we use for the φ_i in Equation (10.11.1) the eigenfunctions of $\Omega^{(1)}$, and if we are dealing with a system of non-interacting particles so that $\Omega^{(2)} = 0$, we have instead of Equation (10.12.8)

$$\Omega = \int \mathrm{d}i\Omega_i a^\dagger(i)a(i)\,, \tag{13.9.1}$$

[2] The suffix E, introduced in Equation (13.6.1), is now dropped.

where the Ω_i are the eigenvalues of $\Omega^{(1)}$. Apply this to the operator $H-\mu N_{\text{op}}$ for the case of a perfect gas, and use the properties of the operator $n(i)$, to write this operator in the form

$$H-\mu N_{\text{op}} = \sum_n (\epsilon_n - \mu) a_n^+ a_n \ , \tag{13.9.2}$$

where the ϵ_n are the single-particle energies and where we have assumed that the ϵ_n form a discrete set.

(b) Prove that

$$\langle\langle a_k ;\, a_k^+ \rangle\rangle_E = \frac{1}{2\pi(E-\epsilon_k+\mu)} \ , \tag{13.9.3}$$

where the quantum number k includes spin dependence, if necessary.

Hint: Use $\eta = +1$ for bosons and $\eta = -1$ for fermions.

(c) Use the result of (b) to prove

$$\langle n_k \rangle = \langle a_k^+ a_k \rangle = \frac{1}{\exp[\beta(\epsilon_k - \mu)] - \eta} \ . \tag{13.9.4}$$

Solution

(b) Using the relations (10.12.3), (10.12.4), and (10.12.6) as well as the equation of motion (13.6.3), one finds the expression (13.9.3).

(c) From the expression (13.8.3) for $J(\omega)$ and Equations (13.3.1) and (13.7.1) we get

$$\langle a_k^+(t') a_k(t) \rangle = \frac{\exp[-\mathrm{i}(\epsilon_k - \mu)(t'-t)]}{[\exp\beta(\epsilon_k - \mu) - \eta]} \ ,$$

and hence, by putting $t = t'$ for $\langle n_k \rangle$, the usual boson and fermion distributions (13.9.4).

THE KUBO FORMULA

13.10 (a) Apart from their application in studying equilibrium properties, Green functions are also useful for deriving **kinetic coefficients.** To see how this can be done consider a system with a Hamiltonian $H^{(0)}$ perturbed by a periodic term $U\mathrm{e}^{\mathrm{i}\omega t+\epsilon t}$ ($\epsilon \to 0+$). Let at $t = -\infty$ the density matrix $\rho(-\infty)$ be the equilibrium density matrix $\rho^{(0)}$ corresponding to $H^{(0)}$,

$$\rho(-\infty) = \rho^{(0)} = Z^{(0)-1} \exp(-\beta H^{(0)}), \quad Z^{(0)} = \mathrm{Tr}\exp(-\beta H^{(0)}) \ . \tag{13.10.1}$$

Using the equation of motion for $\rho(t)$,

$$\mathrm{i}\hbar\dot{\rho} = [H, \rho]_- \ , \tag{13.10.2}$$

writing

$$\rho(t) = \rho^{(0)} + \Delta\rho \ , \tag{13.10.3}$$

and neglecting second order terms in U and $\Delta\rho$, write down the equation of motion for $\Delta\rho$. Solve that equation, using as an intermediate quantity $\Delta\rho'$ given by the equation

$$\Delta\rho' = \exp\left(\frac{\mathrm{i}H^{(0)}t}{\hbar}\right)\Delta\rho\exp\left(-\frac{\mathrm{i}H^{(0)}t}{\hbar}\right). \tag{13.10.4}$$

(b) Using the solution for $\Delta\rho$ found under (a) prove the so-called **Kubo formula**[3].

$$\langle G(t)\rangle = \langle G\rangle^{(0)} - 2\pi \mathrm{e}^{\mathrm{i}\omega t+\epsilon t}\langle\langle G;U\rangle\rangle_{-\omega}, \tag{13.10.5}$$

for the average value of a physical quantity G, where $\langle\ldots\rangle^{(0)}$ indicates the average taken with the equilibrium density matrix $\rho^{(0)}$ and where the Green function is the retarded Green function with $\eta = +1$.

Solution

(a) From Equations (13.10.2) and (13.10.3) we get, neglecting a term involving $[U, \Delta\rho]_-$,

$$\mathrm{i}\hbar\Delta\dot{\rho} = [U, \rho^{(0)}]_- \mathrm{e}^{\mathrm{i}\omega t+\epsilon t} + [H^{(0)}, \Delta\rho]_- .$$

Substituting expression (13.10.4) into this equation, integrating the resulting equation and again using expression (13.10.4), we find

$$\Delta\rho = -\frac{\mathrm{i}}{\hbar}\int_{-\infty}^{t}\exp\left[\frac{\mathrm{i}H^{(0)}(\tau-t)}{\hbar}\right][U,\rho^{(0)}]_-\exp\left[-\frac{\mathrm{i}H^{(0)}(\tau-t)}{\hbar}\right]\mathrm{e}^{\mathrm{i}\omega\tau+\epsilon\tau}\,\mathrm{d}\tau .$$

(b) For $\langle G(t)\rangle$ we find

$$\begin{aligned}\langle G(t)\rangle &= \mathrm{Tr}\,\rho G = \mathrm{Tr}\,\rho^{(0)}G + \mathrm{Tr}\,\Delta\rho G\\ &= \langle G\rangle^{(0)} - \frac{\mathrm{i}}{\hbar}\int_{-\infty}^{t}\langle[G'(t), U'(\tau)]_-\rangle^{(0)}\mathrm{e}^{\mathrm{i}\omega\tau+\epsilon\tau}\,\mathrm{d}\tau ,\end{aligned}$$

from which expression (13.10.5) follows. In this equation we have

$$G'(t) = \exp\left(\frac{\mathrm{i}H^{(0)}t}{\hbar}\right)G\exp\left(-\frac{\mathrm{i}H^{(0)}t}{\hbar}\right),$$

$$U'(\tau) = \exp\left(\frac{\mathrm{i}H^{(0)}\tau}{\hbar}\right)U\exp\left(-\frac{\mathrm{i}H^{(0)}\tau}{\hbar}\right),$$

and in deriving expression (13.10.5) one uses the relation

$$\int_{-\infty}^{t}\mathrm{f}(\tau)\,\mathrm{d}\tau = \int_{-\infty}^{+\infty}\theta(t-\tau)\mathrm{f}(\tau)\,\mathrm{d}\tau .$$

THE HEISENBERG FERROMAGNET

13.11 We shall now apply the Green function formalism to the case of the isotropic Heisenberg ferromagnet which is described by the Hamiltonian (compare Chapter 12):

$$H = -\frac{g\mu_{\mathrm{B}}B}{\hbar}\sum_{\mathbf{f}}S_{\mathbf{f}}^{z} - \frac{2}{\hbar^2}\sum_{\mathbf{f},\mathbf{g}}I(\mathbf{f}-\mathbf{g})(\mathbf{S}_{\mathbf{f}}\cdot\mathbf{S}_{\mathbf{g}}), \tag{13.11.1}$$

[3] A simple introduction to this result is given in Problem 24.9.

where g is the Landé g factor, μ_B the Bohr magneton, and B the strength of a uniform magnetic induction. We shall take the z axis along the magnetic induction and we shall restrict our discussion to the spin $\frac{1}{2}$ case. In that case $\mathbf{S}_f$ is the spin-operator vector for the spin on lattice site f which has the components

$$S_f^x = \tfrac{1}{2}\hbar\begin{pmatrix} 0 & 1 \\ 1 & 0 \end{pmatrix}, \quad S_f^y = \tfrac{1}{2}\hbar\begin{pmatrix} 0 & -i \\ i & 0 \end{pmatrix}, \quad S_f^z = \tfrac{1}{2}\hbar\begin{pmatrix} 1 & 0 \\ 0 & -1 \end{pmatrix}, \tag{13.11.2}$$

where we have omitted the unit two-by-two matrices, which are factors of the components of $\mathbf{S}_f$ and which refer to all lattice sites bar f. The $I(\mathbf{f}-\mathbf{g})$ are exchange integrals which depend only on $\mathbf{f}-\mathbf{g}$; assuming that the system has inversion symmetry—as we shall assume to be the case—we have

$$I(\mathbf{f}-\mathbf{g}) = I(\mathbf{g}-\mathbf{f}), \tag{13.11.3}$$

while we can also put

$$I(0) = 0. \tag{13.11.4}$$

(a) Introducing the operators

$$b_f = \begin{pmatrix} 0 & 1 \\ 0 & 0 \end{pmatrix}, \quad b_f^+ = \begin{pmatrix} 0 & 0 \\ 1 & 0 \end{pmatrix}, \tag{13.11.5}$$

express the components of $\mathbf{S}_f$ in terms of b_f and b_f^+, and discuss the physical meaning of b_f, b_f^+ and the operator n_f given by

$$n_f = b_f^+ b_f. \tag{13.11.6}$$

(b) Find expressions for

$$[b_f, b_g]_-, \quad [b_f^+, b_g^+]_-, \quad [b_f, b_g^+]_-,$$
$$[b_f, b_g]_+, \quad [b_f^+, b_g^+]_+, \quad [b_f, b_g^+]_+.$$

(c) Express the Hamiltonian (13.4.1) in terms of the b_f and b_f^+, putting $g = 2$ for the sake of simplicity.

Solution

(a) $S_f^x = \frac{1}{2}\hbar(b_f + b_f^+)$, $S_f^y = \frac{1}{2}\hbar(b_f^+ - b_f)$, $S_f^z = \frac{1}{2}\hbar(1 - 2n_f)$.

If we write

$$b_f = \frac{1}{\hbar}S_f^+ = \frac{1}{\hbar}(S_f^x + iS_f^y), \quad b_f^+ = \frac{1}{\hbar}S_f^- = \frac{1}{\hbar}(S_f^x - iS_f^y)$$

we see that the b_f and b_f^+ are the usual raising and lowering operators. The physical meaning of n_f follows from its connection with S_f^z.

(b) $[b_f, b_g]_- = [b_f^+, b_g^+]_- = 0 \qquad [b_f, b_g]_+ = 2b_f b_g(1 - \delta_{fg})$,

$[b_f, b_g^+]_- = [1 - 2n_f]\delta_{fg}, \qquad [b_f^+, b_g^+]_+ = 2b_f^+ b_g^+(1 - \delta_{fg})$,

$$[b_f, b_g^+]_+ = 2b_f b_g^+(1 - \delta_{fg}) + \delta_{fg}.$$

We note that if **f** and **g** are different lattice sites, the $b_{\mathbf{f}}$ and $b_{\mathbf{f}}^{+}$ behave as boson operators, but if **f** and **g** are the same site, they behave as fermion operators. They are sometimes called **Pauli operators.**

(c)

$$H = -\mu_{\mathrm{B}} BN - 2\mu_{\mathrm{B}} B \sum_{\mathbf{f}} n_{\mathbf{f}} - 2\sum_{\mathbf{f},\mathbf{g}} I(\mathbf{f}-\mathbf{g}) b_{\mathbf{f}}^{+} b_{\mathbf{g}} - \tfrac{1}{2} N \sum_{\mathbf{f}} I(\mathbf{f})$$

$$+2\left[\sum_{\mathbf{f}} I(\mathbf{f})\right]\left[\sum_{\mathbf{g}} n_{\mathbf{g}}\right] - 2\sum_{\mathbf{f},\mathbf{g}} I(\mathbf{f}-\mathbf{g}) n_{\mathbf{f}} n_{\mathbf{g}} \, ,$$

where N is the total number of lattice sites in the system.

13.12 (a) Find the equation of motion for the Green function $\langle\langle b_{\mathbf{g}};\, b_{\mathbf{f}}^{+}\rangle\rangle$, putting $\eta = +1$.

(b) In order to solve the equations found under (a) we must make approximations. The so-called **random-phase approximation** (RPA) first introduced by Bogolyubov and Tyablikov consists in writing

$$\langle\langle n_{\mathbf{f}_1} b_{\mathbf{f}_2};\, b_{\mathbf{f}_3}^{+}\rangle\rangle = \langle n_{\mathbf{f}_1}\rangle \langle\langle b_{\mathbf{f}_2};\, b_{\mathbf{f}_3}^{+}\rangle\rangle \, . \qquad (13.12.1)$$

Using the fact that H is translationally invariant, we can put

$$\langle n_{\mathbf{f}}\rangle = \bar{n} \quad \text{(independent of } \mathbf{f}) \, , \qquad (13.12.2)$$

and we can Fourier transform with respect to the lattice sites. Using the fact that

$$\delta_{\mathbf{fg}} = \frac{1}{N} \sum_{\mathbf{q}} e^{i(\mathbf{q}\cdot\mathbf{g}-\mathbf{f})} \, , \qquad (13.12.3)$$

where N is the total number of spins in the system, and writing

$$\langle\langle b_{\mathbf{g}};\, b_{\mathbf{f}}^{+}\rangle\rangle = \frac{1}{N} \sum_{\mathbf{q}} e^{i(\mathbf{q}\cdot\mathbf{g}-\mathbf{f})} G_{\mathbf{q}} \qquad (13.12.4)$$

find the equation for $G_{\mathbf{q}}$. In Equations (13.12.3) and (13.12.4) the summation over **q** is over the first Brillouin zone.

(c) From the equation for $G_{\mathbf{q}}$ find an implicit equation for $\bar{n}$ and hence for the quantity $2\langle S^z\rangle/\hbar = \sigma$.

Solution

(a)

$$E\langle\langle b_{\mathbf{g}};\, b_{\mathbf{f}}^{+}\rangle\rangle = \frac{1}{2\pi} \delta_{\mathbf{fg}} [1 - 2\langle n_{\mathbf{f}}\rangle] + \left[2\mu_{\mathrm{B}} B + 2\sum_{\mathbf{f}} I(\mathbf{f})\right] \langle\langle b_{\mathbf{g}};\, b_{\mathbf{f}}^{+}\rangle\rangle$$

$$-2\sum_{\mathbf{p}} I(\mathbf{p}-\mathbf{g}) \langle\langle b_{\mathbf{p}};\, b_{\mathbf{f}}^{+}\rangle\rangle + 4\sum_{\mathbf{p}} I(\mathbf{p}-\mathbf{g}) [\langle\langle n_{\mathbf{g}} b_{\mathbf{p}};\, b_{\mathbf{f}}^{+}\rangle\rangle - \langle\langle n_{\mathbf{p}} b_{\mathbf{g}};\, b_{\mathbf{f}}^{+}\rangle\rangle] \, .$$

(b) From Equations (13.12.1) and (13.12.2) and the last equation we get

$$E\langle\langle b_{\mathbf{g}};\, b_{\mathbf{f}}^{+}\rangle\rangle = \frac{1-2\bar{n}}{2\pi} \delta_{\mathbf{fg}} + \left[2\mu_{\mathrm{B}} B + 2(1-2\bar{n}) \sum_{\mathbf{f}} I(\mathbf{f})\right] \langle\langle b_{\mathbf{g}};\, b_{\mathbf{f}}^{+}\rangle\rangle$$

$$-2(1-2\bar{n}) \sum_{\mathbf{p}} I(\mathbf{p}-\mathbf{g}) \langle\langle b_{\mathbf{p}};\, b_{\mathbf{f}}^{+}\rangle\rangle \, .$$

Using now Equations (13.12.3) and (13.12.4) we get

$$G_{\mathbf{q}}\{E-2\mu_{\mathrm{B}}B-2(1-2\bar{n})[K(0)-K(\mathbf{q})]\} = \frac{1-2\bar{n}}{2\pi},$$

$$K(\mathbf{q}) = \sum_{\mathbf{f}} I(\mathbf{f})\mathrm{e}^{\mathrm{i}(\mathbf{f}\cdot\mathbf{q})},$$

$$G_{\mathbf{q}} = \frac{1-2\bar{n}}{2\pi(E-E_{\mathbf{q}})},$$

with

$$E_{\mathbf{q}} = 2\mu_{\mathrm{B}}B+2(1-2\bar{n})[K(0)-K(\mathbf{q})].$$

(c) From Equation (13.12.4), the equation for $G_{\mathbf{q}}$, and Equations (13.8.3) and (13.3.1) we find

$$\langle b_{\mathbf{f}}^{+}(t')b_{\mathbf{g}}(t)\rangle = \frac{1-2\bar{n}}{N}\sum_{\mathbf{q}}\frac{\exp[\mathrm{i}(\mathbf{q}\cdot\mathbf{g}-\mathbf{f})-\mathrm{i}E_{\mathbf{q}}(t-t')/\hbar]}{\exp(\beta E_{\mathbf{q}})-1}.$$

Putting $t = t'$, $\mathbf{f} = \mathbf{g}$, and changing from a summation over $\mathbf{q}$ to an integration

$$\frac{1}{N}\sum_{\mathbf{q}}\ldots \rightarrow \frac{v'}{(2\pi)^3}\int \mathrm{d}^3\mathbf{q}\ldots,$$

where $v' = v/N$ is the volume per spin, we get the equation for $\bar{n}$

$$\frac{\bar{n}}{1-2\bar{n}} = \frac{v'}{(2\pi)^3}\int\frac{\mathrm{d}^3\mathbf{q}}{\exp(\beta E_{\mathbf{q}})-1}. \tag{13.12.5}$$

Using the fact that the number of reciprocal lattice sites in the first Brillouin zone is N, we have

$$1 = \frac{v'}{(2\pi)^3}\int \mathrm{d}^3\mathbf{q},$$

and we find then the following implicit equation for σ:

$$\frac{1}{\sigma} = \frac{v'}{(2\pi)^3}\int \coth(\tfrac{1}{2}\beta E_{\mathbf{q}})\,\mathrm{d}^3\mathbf{q}.$$

13.13 (a) Use the equation for σ obtained in the last problem to evaluate the transition temperature T_{C}, that is, the temperature at which the magnetisation vanishes in zero magnetic field.

(b) Find the first term in the expansion of σ in terms of $T_{\mathrm{C}}-T$ for temperatures satisfying the inequality $(T_{\mathrm{C}}-T)/T_{\mathrm{C}} \ll 1$.

(c) Find the first two terms in a series expansion of σ in terms of T for $T \ll T_{\mathrm{C}}$, that is, the first deviation from the saturation value.

(d) Find the leading term in a series expansion for the susceptibility χ in terms of T^{-1} for $T \gg T_{\mathrm{C}}$.

Solution

(a) Near, but below the Curie (or transition) temperature T_C, corresponding to β_C, σ will be small and in zero magnetic field, which means that $E_{\mathbf{q}}$ will be small. We can thus expand the hyperbolic tangent in a power series in σ. The result is

$$\frac{1}{\sigma} = \frac{v'}{(2\pi)^3}\int d^3\mathbf{q}\left[\frac{2}{\sigma\beta K(0)\eta(\mathbf{q})}+\tfrac{1}{6}\beta\sigma K(0)\eta(\mathbf{q})+\cdots\right], \quad (13.13.1)$$

where

$$\eta(\mathbf{q}) = 1-\frac{K(\mathbf{q})}{K(0)}.$$

The functions $F(n)$, defined by the equations

$$F(n) = \frac{v'}{(2\pi)^3}\int \eta^n(\mathbf{q})\, d^3\mathbf{q},$$

will depend on the crystal structure, and have been computed for several lattices and values of n. From Equation (13.13.1) we find

$$\frac{1}{\sigma} = \frac{2F(-1)}{\beta\sigma K(0)}+\tfrac{1}{6}\beta\sigma K(0)F(1)+\cdots, \quad (13.13.2)$$

and hence, letting σ tend to zero, for β_C

$$\beta_C = \frac{2F(-1)}{K(0)}.$$

(b) From Equation (13.13.2) we find for $\beta > \beta_C$, but $(\beta-\beta_C)/\beta_C \ll 1$

$$\sigma \approx \frac{3}{F(1)F(-1)}\sqrt{\frac{\beta-\beta_C}{\beta_C}}+\cdots.$$

(c) At low temperatures we expect σ to be about equal to unity and we expect only $\mathbf{q}$-values close to the origin to contribute to the integral in Equation (13.12.5) so that we can without loss of accuracy extend the integration over the whole of $\mathbf{q}$-space. Using polar angles θ and φ to characterise $\mathbf{q}$ we have

$$\frac{\bar{n}}{1-2\bar{n}} = \frac{v'}{(2\pi)^3}\int_0^{2\pi} d\varphi \int_0^{\pi} \sin\theta\, d\theta \int_0^{\infty} q^2 dq \sum_{r=1}^{\infty} \exp[-2r\beta\sigma K(0)\eta(\mathbf{q})].$$

Expanding $\eta(\mathbf{q})$ in powers of $\mathbf{q}$, retaining only the first term (which is proportional to q^2), and integrating, we get a power series in β^{-1}. From the definition of $K(\mathbf{q})$ we get, for instance, for the case of a simple cubic lattice

$$K(\mathbf{q}) = \tfrac{1}{3}K(0)[\cos(q_x a)+\cos(q_y a)+\cos(q_z a)],$$

where a is the nearest-neighbour distance. (One can easily write down the analogous expressions for the cases of face-centred and body-centred lattices. We leave this as an exercise to the reader.) Hence we get in this case for $\eta(\mathbf{q})$

$$\eta(\mathbf{q}) = \tfrac{1}{2}a^2q^2+\cdots,$$

and hence (note that for a simple cubic lattice $v' = a^3$)

$$\frac{\bar{n}}{1-2\bar{n}} \approx \frac{v'}{(2\pi)^3}4\pi \sum_{r=1}^{\infty} \int_0^{\infty} \mathrm{e}^{-r\alpha q^2} q^2 \,\mathrm{d}q, \quad \alpha = \beta\sigma K(0)a^2$$

$$\approx \frac{v'}{(2\pi)^3}4\pi[\beta\sigma K(0)a^2]^{-3/2} \sum_{r=1}^{\infty} r^{-3/2}$$

or

$$\sigma = 1 - AT^{3/2} - \cdots,$$

where

$$A = \left[\frac{k_B}{4\pi K(0)}\right]^{3/2} \zeta(\tfrac{3}{2});$$

$\zeta(n)$ is the Riemann zeta function, and k_B the Boltzmann constant.

(d) In this case we must have $B \neq 0$ as otherwise σ would vanish. We write

$$\coth \tfrac{1}{2}\beta E_{\mathbf{q}} = \frac{1+t_0 t_1}{t_0+t_1},$$

where

$$t_0 = \tanh \beta\mu_B B, \quad t_1 = \tanh[\beta\sigma K(0)\eta(\mathbf{q})].$$

Expanding the hyperbolic cotangent in powers of t_1 and then t_1 in powers of β, we find from Equation (13.12.5), again for the simple cubic case,

$$\frac{\bar{n}}{1-2\bar{n}} = \frac{1}{2t_0}[1 - \frac{\beta K(0)}{t_0} + \cdots],$$

and finally for the susceptibility χ per spin

$$\chi = \beta\mu_B^2[1 + \tfrac{1}{2}K(0)\beta + \cdots],$$

which can be written (approximately) in the Curie-Weiss form

$$\chi = \frac{\mu_B^2}{k_B(T-\Theta)}, \quad \Theta = \frac{K(0)}{2k_B}.$$

13.14 When one studies ferromagnetic resonance, one is dealing with a sample which has a finite size. To take boundary effects into account most simply, even though only approximately, we add to the Hamiltonian a term involving the demagnetisation factors, N_x, N_y, and N_z, which depend on the shape of the sample assumed to be ellipsoidal with the principal axes along the x, y, and z axes. We must thus add to the Hamiltonian (13.11.1) a term

$$\tfrac{1}{2}(N_x M_x^2 + N_y M_y^2 + N_z M_z^2), \tag{13.14.1}$$

where $\mathbf{M}$ is the magnetisation vector,

$$\mathbf{M} = \frac{2\mu_B}{\hbar} \sum_{\mathbf{f}} \mathbf{S}_{\mathbf{f}}. \tag{13.14.2}$$

(a) Express the term (13.14.1) in the b_f and b_f^+.

(b) This is a situation envisaged in Problem 13.10. Apart from a steady field B along the z axis there will be an r.f. field $\mathbf{B}_1$ in the xy plane. If that field has components $b\cos\omega t$, $b\sin\omega t$, 0, the perturbing Hamiltonian H' will be of the form

$$H' = -\frac{2\mu_B}{\hbar}\sum_f(\mathbf{S}_f\cdot\mathbf{B}_1) = U e^{i\omega t} + U^* e^{-i\omega t}, \qquad (13.14.3)$$

with

$$U = -\frac{\mu_B b}{\hbar}\sum_f S_f^- \quad (S_f^{\pm} = S_f^x \pm iS_f^y). \qquad (13.14.4)$$

The r.f. field will produce an additional magnetisation in the xy plane:

$$\delta M_{\pm} = \delta(M_x \pm iM_y) = N\chi_{\pm} b e^{\pm i\omega t}. \qquad (13.14.5)$$

Express $\chi_{\pm}$ in terms of a Green function.

(c) Writing

$$\chi_{\pm} = \chi' \pm i\chi'', \qquad (13.14.6)$$

express the energy absorbed per unit time, W, in terms of χ' and χ''.

(d) Determine $\chi_{\pm}$, using the equation of motion for the relevant Green function and using not only Equation (13.12.1) but also the approximate equation

$$\langle\langle n_{f_1} b_{f_2}^+;\, b_{f_3}\rangle\rangle = \langle n_{f_1}\rangle\langle\langle b_{f_2}^+;\, b_{f_3}\rangle\rangle. \qquad (13.14.7)$$

Solution

(a)

$$\tfrac{1}{2}(N_x M_x^2 + N_y M_y^2 + N_z M_z^2) = \tfrac{1}{2}N^2 N_z \mu_B^2 - 2NN_z\mu_B^2\sum_f n_f$$
$$+ (N_x + N_y)\mu_B^2 \sum_{f,g} b_f^+ b_g + N_z\mu_B^2\sum_{f,g} n_f n_g$$
$$+ \left[\tfrac{1}{2}\mu_B^2(N_x - N_y)\sum_{f,g}(b_f b_g + b_f^+ b_g^+)\right].$$

(b)

$$\chi_+ = \chi_-^* = -\frac{4\pi\mu_B^2}{\hbar^2}\sum_f\langle\langle S_g^+;\, S_f^-\rangle\rangle - 4\pi\mu_B^2\sum_f\langle\langle b_f;\, b_g^+\rangle\rangle.$$

(c)

$$W = (\mathbf{B}_1\cdot\delta\dot{\mathbf{M}}) = N\frac{b^2\omega}{2i}(\chi_+ - \chi_-) = Nb^2\omega\chi''.$$

(d) If we use the decoupling expressions (13.12.1) and (13.14.7) we find that the equation of motion for $\langle\langle b_g;\, b_f^+\rangle\rangle$ contains the Green functions $\langle\langle b_p^+;\, b_f^+\rangle\rangle$. However, if we write down the equations of motion for both the $\langle\langle b_g;\, b_f^+\rangle\rangle$ and the $\langle\langle b_p^+;\, b_f^+\rangle\rangle$, we get a closed set, provided we use Equations (13.12.1) and (13.14.7):

$$E\langle\langle b_{\mathbf{g}};\ b_{\mathbf{f}}^{+}\rangle\rangle = \frac{\sigma}{2\pi}\delta_{\mathbf{fg}} + [2\mu_{\mathrm{B}}B - 2\sigma N N_z \mu_{\mathrm{B}}^2 + 2\sigma K(0)]\langle\langle b_{\mathbf{g}};\ b_{\mathbf{f}}^{+}\rangle\rangle$$

$$+\sum_{\mathbf{p}}[-2\sigma I(\mathbf{g}-\mathbf{p}) + \sigma\mu_{\mathrm{B}}^2(N_x + N_y)]\langle\langle b_{\mathbf{p}};\ b_{\mathbf{f}}^{+}\rangle\rangle$$

$$+\sigma\mu_{\mathrm{B}}^2(N_x - N_y)\sum_{\mathbf{p}}\langle\langle b_{\mathbf{p}}^{+};\ b_{\mathbf{f}}^{+}\rangle\rangle\ ,$$

$$E\langle\langle b_{\mathbf{g}}^{+};\ b_{\mathbf{f}}^{+}\rangle\rangle = -[2\mu_{\mathrm{B}}B - 2\sigma N N_z \mu_{\mathrm{B}}^2 + 2\sigma K(0)]\langle\langle b_{\mathbf{g}}^{+};\ b_{\mathbf{f}}^{+}\rangle\rangle$$

$$-\sum_{\mathbf{p}}[2\sigma I(\mathbf{g}-\mathbf{p}) + \sigma\mu_{\mathrm{B}}^2(N_x + N_y)]\langle\langle b_{\mathbf{p}}^{+};\ b_{\mathbf{f}}^{+}\rangle\rangle$$

$$-\sigma\mu_{\mathrm{B}}^2(N_x - N_y)\sum_{\mathbf{p}}\langle\langle b_{\mathbf{p}};\ b_{\mathbf{f}}^{+}\rangle\rangle\ .$$

Using again Equations (13.12.3) and (13.12.4) and the equations

$$\langle\langle b_{\mathbf{g}}^{+};\ b_{\mathbf{f}}^{+}\rangle\rangle\frac{1}{N}\sum_{\mathbf{q}} \mathrm{e}^{\mathrm{i}(\mathbf{q}\cdot\mathbf{f}-\mathbf{q})}\Gamma_{\mathbf{q}}\ ,$$

$$\delta_{\mathbf{q}0} = \frac{1}{N}\sum_{\mathbf{f}} \mathrm{e}^{\mathrm{i}(\mathbf{f}\cdot\mathbf{q})}\ ,$$

we find

$$(E - \bar{E}_{\mathbf{q}})G_{\mathbf{q}} = \frac{\sigma}{2\pi} + \sigma\mu_{\mathrm{B}}^2(N_x - N_y)N\delta_{\mathbf{q}0}\Gamma_{\mathbf{q}}\ ,$$

$$(E + \bar{E}_{\mathbf{q}})\Gamma_{\mathbf{q}} = -\sigma\mu_{\mathrm{B}}^2(N_x - N_y)N\delta_{\mathbf{q}0}G_{\mathbf{q}}\ ,$$

where

$$\bar{E}_{\mathbf{q}} = E_{\mathbf{q}} + N\sigma\mu_{\mathrm{B}}^2[(N_x + N_y)\delta_{\mathbf{q}0} - 2N_z]\ .$$

We note that for $\mathbf{q} \neq 0$, $\bar{E}_{\mathbf{q}}$ differs from $E_{\mathbf{q}}$ in that B is replaced by $B - N_z(N\sigma\mu_{\mathrm{B}})$, that is by the field including the demagnetisation.

Solving the equations for $\Gamma_{\mathbf{q}}$ and $G_{\mathbf{q}}$, we find

$\mathbf{q} \neq 0$:

$$\Gamma_{\mathbf{q}} = 0,\quad G_{\mathbf{q}} = \frac{\sigma}{2\pi(E - \bar{E}_{\mathbf{q}})} \quad \text{(much as in Problem 13.12)}$$

$\mathbf{q} = 0$:

$$G_0 = \frac{\sigma}{2\pi}\frac{E + \bar{E}_0}{E^2 - E_r^2}\ ,$$

where

$$E_r^2 = 4\mu_{\mathrm{B}}^2[B + N\sigma\mu_{\mathrm{B}}(N_x - N_z)][B + N\sigma\mu_{\mathrm{B}}(N_y - N_z)]\ .$$

For χ_+ we get now

$$\chi_+ = -4\pi\mu_B^2 \sum_f \langle\langle b_g;\ b_f^+ \rangle\rangle = -\frac{4\pi\mu_B^2}{N} \sum_f \sum_q G_q \, e^{i(f-g\cdot g)}$$

$$= \lim_{\epsilon\to 0+} \left[-\frac{4\pi\mu_B^2}{N} \sum_{q\neq 0} \frac{\sigma}{2\pi} \frac{1}{\hbar(\omega+i\epsilon)-\bar{E}_q} \sum_f e^{i(q\cdot f-g)} \right.$$

$$\left. -\frac{4\pi\mu_B^2}{N} \sum_f \frac{\sigma}{2\pi} \frac{\hbar(\omega+i\epsilon)-\bar{E}_0}{\hbar^2(\omega+i\epsilon)^2-E_r^2} \right]$$

$$= \lim_{\epsilon\to 0+} \left\{ -\frac{\sigma\mu_B^2}{E_r} [\hbar(\omega+i\epsilon)-\bar{E}_0] \left[\frac{1}{\hbar(\omega+i\epsilon)-E_r} - \frac{1}{\hbar(\omega+i\epsilon)+E_r} \right] \right\},$$

and, as both ω and E_r are positive so that $\delta(\hbar\omega+E_r) = 0$, we get for χ''

$$\chi'' = \frac{2\pi\mu_B^2(\hbar\omega+\bar{E}_0)}{E_r} \delta(\hbar\omega-E_r),$$

showing that absorption takes place only at $\omega = E_r/\hbar$, with zero line-width in the present approximation.

14

The plasma[1]

D.ter HAAR
(*University of Oxford, Oxford*)

14.1 A **plasma** is a fluid consisting of positively and negatively charged particles. We shall consider the following model of a plasma: a gas of negatively charged particles, charge $-e$, is moving in a neutralising background with uniform charge density $n_0 e$. Let the average number density of the negatively charged particles be n_0. Consider now an additional infinitesimal point charge q which we shall—for the sake of simplicity—assume to be at the origin. Let this charge give rise to a small spherically symmetric change $\phi(r)$ in the electrostatic potential. Show that

$$\phi(r) = \frac{q}{r} e^{-\kappa r} , \tag{14.1.1}$$

where

$$\kappa^2 = 4\pi n_0 \beta e^2 , \quad \beta = \frac{1}{kT} . \tag{14.1.2}$$

Discuss under what conditions the argument used is valid, and show that one of the conditions is that

$$r_D \gg d , \tag{14.1.3}$$

where the Debye radius r_D is given by the equation

$$r_D = \kappa^{-1} , \tag{14.1.4}$$

while

$$d \sim n^{-1/3} , \tag{14.1.5}$$

where n is the particle density, that is, d is a length of the order of the interparticle density. A plasma for which the condition (14.1.3) is satisfied is called a **hot dilute plasma**. In the following we shall assume that we are dealing with hot dilute plasmas. This kind of theory is called the **Debye theory** or **Debye approximation**.

From expression (14.1.1) we see that ϕ changes appreciably in the so-called **Debye sphere**, which is the sphere of radius r_D within which the potential $\phi(r)$ changes appreciably.

[1] For general references see, e.g. N.G.van Kampen and B.U.Felderhof, *Theoretical Methods in Plasma Physics* (North-Holland, Amsterdam), 1967.

Solution

The electrostatic potential $\phi(\mathbf{r})$ must satisfy Poisson's equation

$$\nabla^2\phi = 4\pi e n(\mathbf{r}) - 4\pi e n_0 - 4\pi q\delta(\mathbf{r}) . \tag{14.1.6}$$

At equilibrium, $n(\mathbf{r})$ is related to $\phi(\mathbf{r})$ through Boltzmann's formula

$$n(r) = A\, e^{\beta e\phi} , \tag{14.1.7}$$

with A to be determined from the condition

$$\int n(\mathbf{r})\mathrm{d}^3\mathbf{r} = n_0 v. \tag{14.1.8}$$

In the thermodynamic limit as the volume of the system $v \to \infty$, ϕ will tend at infinity to a constant value $\phi(\infty)$, which we can take to be zero, so that we have $A = n_0$.

Substituting Boltzmann's formula (14.1.7) into Poisson's equation (14.1.6), and using the assumption that $\beta e\phi \ll 1$ to expand the exponential and to retain only the term linear in ϕ, we get

$$\nabla^2\phi - \kappa^2\phi = -4\pi q\delta(\mathbf{r}) , \tag{14.1.9}$$

where κ is given by Equation (14.1.2).

The solution of Equation (14.1.9) which has the required spherical symmetry is the expression (14.1.1). We note that this is a shielded potential with range r_D. The density $n(\mathbf{r})$ is the average number of negatively charged particles in a volume element divided by the volume of the element. This only has a physical meaning, if the element can be chosen sufficiently large so that it contains many particles. This in turn means that, if d is the average interparticle distance, so that the relationship (14.1.5) holds, we have as a first condition for the applicability of the present considerations:

$$\mathrm{d}|\nabla n| \ll n \tag{14.1.10}$$

or

$$|\nabla n| \ll n^{4/3} . \tag{14.1.11}$$

As we have to consider $n(\mathbf{r})$ only as an average over a sufficiently large volume, it is a coarse-grained quantity. The same is true for $\phi(\mathbf{r})$, and we must also require for ϕ that it varies slowly over a distance d. This means that r_D must be large compared to d, so that expression (14.1.3) is the second condition. Using expressions (14.1.4) and (14.1.2), we get from it the condition

$$n_0\beta e^2 d^2 \ll 1 , \tag{14.1.12}$$

or from expression (14.1.5)

$$kT \gg \frac{e^2}{d}$$

or

$$kT \gg e^2 n_0^{1/3} , \tag{14.1.13}$$

which shows that the average kinetic energy of a particle must be large compared to its average potential energy. Systems obeying this condition are called **hot dilute plasmas**.

14.2 Prove that the number density of negative charges satisfies the equation

$$n(\mathbf{r}) = n_0 + \frac{q\kappa^2}{4\pi e}\phi(r)\,, \tag{14.2.1}$$

when the additional infinitesimal point charge is present.

Solution

Equation (14.2.1) follows from Boltzmann's formula (14.1.7) under the assumption that $\beta e\phi \ll 1$.

14.3 Find in the Debye theory the total number n_{exc} of excess negatively charged particles in the Debye sphere.

Solution

$$n_{\text{exc}} = \int [n(\mathbf{r}) - n_0]\mathrm{d}^3\mathbf{r} = \frac{q\kappa^2}{e}\int_0^\infty \mathrm{e}^{-\kappa r} r\,\mathrm{d}r = \frac{q}{e}\,. \tag{14.3.1}$$

14.4 Find in the Debye theory the energy E_{int} of the interaction between q and the charge in the Debye sphere.

Solution

$$E_{\text{int}} = \int \frac{qe[n(\mathbf{r}) - n_0]}{r}\mathrm{d}^3\mathbf{r} = q^2\kappa\,. \tag{14.4.1}$$

14.5 Find in the Debye theory the energy E_{D} of the charge density in the Debye sphere.

Solution

$$E_{\text{D}} = \tfrac{1}{2}\int \mathrm{d}^3\mathbf{r}\int \mathrm{d}^3\mathbf{r}'\frac{e^2[n(\mathbf{r}) - n_0][n(\mathbf{r}') - n_0]}{|\mathbf{r} - \mathbf{r}'|}\,. \tag{14.5.1}$$

Using the expansion in Legendre functions $P_n(\mu)$, where μ is the cosine of the angle between $\mathbf{r}$ and $\mathbf{r}'$,

$$\begin{aligned}\frac{1}{|\mathbf{r} - \mathbf{r}'|} &= \frac{1}{r}\sum_{n=0}^{\infty}\left(\frac{r'}{r}\right)^n P_n(\mu), \quad r' < r\,, \\ &= \frac{1}{r'}\sum_{n=0}^{\infty}\left(\frac{r}{r'}\right)^n P_n(\mu), \quad r' > r\,,\end{aligned} \tag{14.5.2}$$

we get

$$\begin{aligned}E_{\text{D}} &= \tfrac{1}{2}q^2\kappa^4\left[\int_0^\infty \mathrm{d}r\,\mathrm{e}^{-\kappa r}\int_0^r r'\,\mathrm{d}r'\,\mathrm{e}^{-\kappa r'} + \int_0^\infty r\,\mathrm{d}r\,\mathrm{e}^{-\kappa r}\int_r^\infty \mathrm{d}r'\,\mathrm{e}^{-\kappa r'}\right] \\ &= \tfrac{1}{4}q^2\kappa\,.\end{aligned} \tag{14.5.3}$$

14.6 Estimate for the Debye theory the average number n_D of negative particles in the Debye sphere.

Solution

$$n_D \approx \tfrac{4}{3}\pi n_0 r_D^3 = \frac{1}{3\sqrt{4\pi}}\left(\frac{kT}{e^2 n^{1/3}}\right)^{3/2},$$

which, as follows from Equation (14.2.6), is large compared to unity for all cases where the Debye theory is applicable.

14.7 Estimate for the Debye theory the ratio of the energy found in the preceding problem to the fluctuations in energy in the Debye sphere.

Solution

From Problems 14.4 and 14.5 it follows that the energy attributable to each particle is of the order of $e^2\kappa$ which is of the same order as E_D. The total electrostatic energy in the Debye sphere will thus be of the order $n_D e^2\kappa$ and the fluctuations of the order $\sqrt{n_D}e^2\kappa$, so that the required ratio is of order $\sqrt{n_D}$ which, as we saw in Problem 14.6, is a large number.

14.8 Find for a hot dilute plasma an expression for the Debye length in a mixture of ionised gases.

Solution

Using arguments similar to those which led to expression (14.1.2) we find

$$\kappa^2 = \frac{4\pi}{kT}\sum_j e_j^2 n_j, \tag{14.8.1}$$

where the summation is over all kinds of ions while e_j and n_j are their respective charges and densities.

14.9 In Problem 9.16 it is shown that one can use the virial theorem to write the equation of state in the form

$$pv = NkT + \tfrac{1}{3}\langle W\rangle_{t\,\mathrm{av}}, \tag{14.9.1}$$

where W is the virial deriving from intermolecular forces, that is,

$$W = \sum_i (\mathbf{F}_i \cdot \mathbf{r}_i), \tag{14.9.2}$$

where the summation is over all particles in the system, $\mathbf{r}_i$ is the position of the ith particle, and $\mathbf{F}_i$ the force on the ith particle due to intermolecular forces. Finally $\langle ...\rangle_{t\,\mathrm{av}}$ indicates an average both over the time and over all particles in the system.

Use Equation (14.9.1) to derive the following plasma equation of state:

$$p = n_0 kT \left[1 - \frac{1}{18 n_D}\right], \tag{14.9.3}$$

with n_D the quantity defined in Problem 14.6.

Solution

For a plasma the evaluation of $\langle W \rangle_{t\,av}$ is simplified by the fact that all forces are Coulomb forces. For W we have

$$W = \sum_i (\mathbf{r}_i \cdot \mathbf{F}_i^C) = -\sum_i (\mathbf{r}_i \cdot \nabla_i U), \tag{14.9.4}$$

where the superscript C indicates that where we are dealing with Coulomb forces and where U is the total electrostatic potential energy. As U is a homogeneous function of degree -1, we thus have from Equation (14.9.4)

$$W = U. \tag{14.9.5}$$

To get the time average of U, we write it in the form

$$U = \tfrac{1}{2} \sum_{i \neq j} \frac{e_i e_j}{r_{ij}} = \tfrac{1}{2} \sum_i e_i \varphi_i', \tag{14.9.6}$$

where e_i is the charge of the ith particle, $r_{ij} = |\mathbf{r}_i - \mathbf{r}_j|$, and φ_i' is the potential at $\mathbf{r}_i$ due to all other ions, so that we can write,

$$\varphi_i'(\mathbf{r}_i) = \left[\frac{\varphi_i(\mathbf{r}) - e_i}{|\mathbf{r} - \mathbf{r}_i|}\right]_{r = r_i}, \tag{14.9.7}$$

which is the total potential acting at r_i less the potential due to the ith ion itself. For $\varphi_i(r)$ we can take the Debye potential (14.1.1) and taking the average [compare expression (14.4.1)] we get

$$\langle W \rangle_{t\,av} = \langle U \rangle_{t\,av} = \tfrac{1}{2} \sum_i e_i \langle \varphi_i' \rangle_{t\,av} = \tfrac{1}{2} N \langle \varphi' \rangle_{av} = -\tfrac{1}{2} N e^2 \kappa. \tag{14.9.8}$$

From Equations (14.9.1) and (14.1.2) we thus get for the equation of state

$$P = n_0 kT \left(1 - \frac{1}{18 n_D}\right), \tag{14.9.9}$$

where n_D is given in the solution to Problem 14.6. We note that the deviations from the perfect gas law are small whenever the Debye theory is applicable.

14.10 Write the pressure p of a plasma in the following form:

$$p = p_{e=0} + p^e, \tag{14.10.1}$$

where the first term is the perfect gas term (we assume that the only interactions are Coulomb interactions) while the second term derives from the electrostatic interactions. Prove that

$$p^{\mathrm{e}} = \frac{U^{\mathrm{e}}}{3v} , \qquad (14.10.2)$$

where U^{e} is the total electrostatic potential energy, by considering the configurational partition function, introducing new variables $\mathbf{r}_i' = \mathbf{r}_i/L$, where L^3 = the volume occupied by the system, and finally using the fact that U^{e} is a homogeneous function.

Solution

We have the following relations:

$$Q = \int \mathrm{d}^3\mathbf{r}_1 \ldots \mathrm{d}^3\mathbf{r}_N \exp(-\beta U^{\mathrm{e}}) , \qquad (14.10.3)$$

where the integration is over the volume v of the system,

$$U^{\mathrm{e}} = \tfrac{1}{2} \sum_{i \neq j} \frac{e_i e_j}{r_{ij}} , \qquad (14.10.4)$$

and

$$\beta p = \frac{\partial \ln Q}{\partial v} . \qquad (14.10.5)$$

Putting $v = L^3$, $\mathbf{r}_i' = \mathbf{r}_i/L$, and using the fact that

$$U^{\mathrm{e}}(\mathbf{r}_i) = U^{\mathrm{e}}(\mathbf{r}_i' L) = -\frac{1}{L} U^{\mathrm{e}}(\mathbf{r}_i') , \qquad (14.10.6)$$

we have

$$\beta p = \frac{1}{Q} \frac{1}{3L^2} \frac{\partial}{\partial L} L^{3N} \int' \mathrm{d}^3\mathbf{r}_1' \ldots \mathrm{d}^3\mathbf{r}_N' \exp\left[-\frac{\beta U^{\mathrm{e}}(\mathbf{r}_i')}{L}\right] , \qquad (14.10.7)$$

where the integration is now over a unit volume. The only L-dependence behind the $\partial/\partial L$ sign lies now in the factor L^{3N} and in the L in the index of the exponential, and there is nowhere an implicit dependence. Hence we get, after taking the derivative with respect to L and changing back to the old coordinates,

$$\beta p = \frac{1}{Q} \int \mathrm{d}^3\mathbf{r}_1 \ldots \mathrm{d}^3\mathbf{r}_N \frac{1}{v} [N + \tfrac{1}{3}\beta U^{\mathrm{e}}] \exp(-\beta U^{\mathrm{e}}) , \qquad (14.10.8)$$

or

$$pv = NkT + \tfrac{1}{3}\langle U^{\mathrm{e}} \rangle . \qquad (14.10.9)$$

14.11 From Equation (14.10.2) and the thermodynamic equation of state (see Problem 1.9a)

$$\left(\frac{\partial U^{\mathrm{e}}}{\partial v}\right)_T = T\left(\frac{\partial p^{\mathrm{e}}}{\partial T}\right)_v - p^{\mathrm{e}} , \qquad (14.11.1)$$

prove that for a hot dilute plasma

$$vT^3 = \text{adiabatic invariant} . \qquad (14.11.2)$$

Solution

If we use expression (14.10.2) and write

$$u^e = \frac{U^e}{v} , \tag{14.11.3}$$

we can write Equation (14.11.1) in the form

$$u^e = \tfrac{1}{3}T\frac{du^e}{dT} - \tfrac{1}{3}u^e , \tag{14.11.4}$$

with the solution

$$U^e = u^e v = aT^4 v . \tag{14.11.5}$$

Using the Maxwell relation (see Problem 1.7a)

$$\left(\frac{\partial S^e}{\partial v}\right)_T = \left(\frac{\partial P^e}{\partial T}\right)_v , \tag{14.11.6}$$

and Equations (14.10.2) and (14.11.5) we get

$$S^e = \frac{4U^e}{3T} = \tfrac{4}{3}aT^3 v , \tag{14.11.7}$$

which proves the expression (14.11.2).

14.12 Considering the charges in a plasma to be quantities which can be changed adiabatically (scale transformation), and considering the total electrostatic potential energy U^e as a function of the entropy S, the volume, and the magnitude of the charges e, find a general expression for U in the form

$$U^e = \frac{Ne^2}{v_1^{1/3}} f(x) \tag{14.12.1}$$

where $v_1 = v/N$, and x is a dimensionless adiabatic invariant.

Solution

From Equation (14.10.4) it follows that

$$\left(\frac{\partial U^e}{\partial e}\right)_{S,v} = \frac{2U^e}{e} , \quad \left(\frac{\partial U^e}{\partial v}\right)_{S,e} = -\tfrac{1}{3}\frac{U^e}{v} . \tag{14.12.2}$$

Applying a general procedure, explained for instance in Problem 1.20, we have

$$dU^e = \frac{2U^e}{e}de - \frac{U^e}{3v}dv + TdS^e , \tag{14.12.3}$$

or

$$\frac{U^e v^{1/3}}{e^2} = \text{adiabatic invariant} . \tag{14.12.4}$$

Relation (14.12.1) now follows immediately.

14.13 Use the result of Problem 14.11 and Equation (14.9.8) to find first an expression for x and then the form of f(x) in the Debye approximation. [Use dimensional analysis.]

Solution

From Equation (14.11.2), and the fact that the only natural constants which can occur in x are e and k, we find by dimensional analysis

$$x = \frac{\beta e^2}{v_1^{1/3}} \ . \tag{14.13.1}$$

It is of some interest to write this adiabatic invariant in a slightly different form by using Equations (14.1.2) and (14.1.4):

$$x = \kappa^2 v_1^{2/3} = \frac{v_1^{2/3}}{r_{\mathrm{D}}^2} \ . \tag{14.13.2}$$

The adiabatic invariance of v/r_{D}^3 is plausible, as one would expect that a simultaneous slow change in the volume and in the Debye cloud would preserve the degree of order and hence the entropy.

From Equations (14.13.1) and (14.9.8) it now follows at once that in the Debye approximation

$$\mathrm{f}(x) = -\sqrt{\pi x} \ . \tag{14.13.3}$$

15

Negative temperatures and population inversion

U.M.TITULAER
(*Rijksuniversiteit, Utrecht*)

15.0 The concept of negative temperature is introduced by means of a statistical model of a spin system. The thermodynamic peculiarities are discussed and the possible use of negative temperature systems and systems with non-thermal level occupation as amplifiers of radiation is indicated. Ramsey's classical paper on the subject[(1)] may serve as a general reference, especially for subjects treated in the first few problems in this chapter.

15.1 Consider a system of n identical weakly interacting spins. Each of the spins occupies one out of $2s+1$ equally spaced non-degenerate energy levels with energy mW ($m = -s, -s+1, \ldots, +s$). The average energy of each spin is u_0. Find the maximum entropy distribution and the relation between u_0 and the temperature parameter β. Calculate the partition function of the system and the specific heat as functions of β for positive as well as negative values of β. [Use the relation $\beta = 1/kT$ to define temperature for negative β.] Find the simplified expressions for the case $s = \frac{1}{2}$.

Solution

As for an ideal gas the partition function is simply the product of n identical factors, one for each spin. A straightforward application of the procedure used in Problem 2.2 gives for the probability p_m of occupying level m of any single spin

$$p_m = \frac{1}{Z(\beta)} e^{-\beta m W}$$

with

$$Z(\beta) = \sum_{m=-s}^{+s} e^{-\beta m W} = \frac{\sinh(s+\frac{1}{2})\beta W}{\sinh\frac{1}{2}\beta W} .$$

β will be determined from the relation

$$u_0 = -\frac{\partial \ln Z(\beta)}{\partial \beta} = -W[(s+\tfrac{1}{2})\coth(s+\tfrac{1}{2})\beta W - \tfrac{1}{2}\coth\tfrac{1}{2}\beta W] .$$

(1) N.F.Ramsey, *Phys. Rev.*, **103**, 20 (1956).

As a result of the concave character of the function $\coth x$ the sign of u_0 is negative when β is positive and conversely. For the specific heat we obtain by differentiating once more

$$C = -nk\beta^2 \frac{\partial u_0}{\partial \beta} = nk\beta^2 W^2 [\tfrac{1}{4}\,\mathrm{cosech}^2\,\tfrac{1}{2}\beta W - (s+\tfrac{1}{2})^2 \mathrm{cosech}^2 (s+\tfrac{1}{2})\beta W] \,.$$

For $s = \frac{1}{2}$ we obtain the simplified expressions

$$Z(\beta) = 2\cosh\tfrac{1}{2}\beta W \,,$$
$$u_0(\beta) = -W\tanh\tfrac{1}{2}\beta W \,,$$
$$C = \tfrac{1}{2}nk\beta^2 W^2 \mathrm{sech}^2\,\tfrac{1}{2}\beta W \,.$$

The entropy for a system with $s = \frac{1}{2}$ is given by

$$S = n(k\beta u_0 + k\ln Z)$$
$$= nk[\beta W \tanh\tfrac{1}{2}\beta W + \ln(2\cosh\tfrac{1}{2}\beta W)] \,.$$

For the system with $s = \frac{1}{2}$ the expressions are simpler and their general features (signs of u_0 and β, zero of C at $\beta = 0$) are more easily found. On the other hand any system which can be described by means of a set of occupation probabilities $p_{-1/2}$ and $p_{+1/2}$ can also be described by means of some inverse temperature $k\beta$; the additional information that $\ln p_m$ is proportional to the energy of level m, which follows from entropy maximisation, does not imply any additional constraint in the case $s = \frac{1}{2}$. This was our reason for not taking $s = \frac{1}{2}$ from the beginning.

15.2 (a) Show from general formulae for the canonical ensemble that negative temperature distributions are possible if and only if the density of possible states of the system decreases fast enough as a function of energy; specify 'fast enough' quantitatively.

(b) Show that this condition is violated if the energy of the system includes the kinetic energy of a particle or the energy of a mode of the radiation field. [The results of Problems 3.13(c) and 3.14(c) may be used.] For a spin system this implies that negative temperatures are meaningful only if the coupling of the spins to the kinetic degrees of freedom in the 'lattice' and to the radiation field can be neglected. In practical terms this means that the time in which the spins reach equilibrium among themselves should be much shorter than the time in which equilibrium with the lattice or the radiation field is established.]

(c) Consider a system for which the density of possible states increases exponentially: $\rho(E) \propto \mathrm{e}^{\alpha E}$, $\alpha > 0$; show that only temperatures $T < 1/k\alpha$ are permissible.

[An exponentially rising density of states was postulated[2] in a statistical model of strong interactions at high energies. The possible consequences of the resulting maximal temperature were also discussed.]

[2] R.Hagedorn, *Nuovo Cimento Suppl.* III, 147 (1965).

Solution

(a) Let $\rho(E)\mathrm{d}E$ denote the number of states with energy between E and $E+\mathrm{d}E$. The average value of the energy is given (in a canonical ensemble) by

$$U_0 = \int \mathrm{e}^{-\beta E} E\rho(E)\mathrm{d}E \bigg/ \int \mathrm{e}^{-\beta E}\rho(E)\mathrm{d}E \ .$$

Unless $\rho(E)$ decreases at least exponentially with E this expression diverges for any negative value of β. Even the total probability that the system occupies any state with energy lower than any given value E_0 would be vanishingly small. Such distributions are physically unacceptable.

(b) For the kinetic energy of a particle we found in Problem 3.1 that the number of states with a velocity between V and $V+\mathrm{d}V$ is proportional to $V^2\mathrm{d}V$. This corresponds to $E^{1/2}\mathrm{d}E$, which is not exponentially decreasing. For the quantum mechanical case a similar result was derived in Problem 3.13(c). For a single mode of the radiation field $\rho(E)$ is a constant; for the photon field as a whole $\rho(E) \propto E^2$ [see Problem 3.14(c)]. Finally, if a system consists of two parts such that $E = E^{(1)}+E^{(2)}$, then $\rho(E) = \int_0^E \mathrm{d}E'\rho^{(1)}(E')\rho^{(2)}(E-E')$. In this formula $\rho^{(1)}(E)$ and $\rho^{(2)}(E)$ are the density of states for the subsystems. It is obvious that $\rho(E)$ will only decrease exponentially if both $\rho^{(1)}$ and $\rho^{(2)}$ decrease at least exponentially.

(c) If $\rho(E) \propto \mathrm{e}^{\alpha E}$ a comparison with (a) shows that values of β such that $\beta \leqslant \alpha$ are unacceptable. Temperatures higher than $T_0 = 1/k\alpha$ are therefore impossible in this model.

15.3 Suppose that the level spacing constant W in a system of n spins $\frac{1}{2}$ is proportional to an external magnetic field: $W = \mu H$. Calculate the amount of heat and work which has to be supplied to the system if we want to keep it at a constant temperature T_1 while the field is changed from H_1 to H_2. Find also the amount of work needed to increase the field adiabatically and the relation between temperature and field during such a process. What is the net effect of a Carnot cycle operating between two negative temperatures?

[The heat absorbed is given by the relation $đQ = T\mathrm{d}S$, for both positive and negative temperatures.]

Solution

From the formulae derived in Problem 15.1 we find for the increase in U and S during isothermal magnetisation

$$(\Delta U)_{T_1} = n\mu[H_2 \tanh x_2 - H_1 \tanh x_1] \ ,$$

$$(\Delta S)_{T_1} = k\beta_1(\Delta U)_{T_1} + nk[\ln Z(x_2) - \ln Z(x_1)] \ ,$$

$$x_{1,2} = \tfrac{1}{2}\mu\beta_1 H_{1,2} \ .$$

The amount of heat supplied to the system equals $T_1(\Delta S)_{T_1}$. The amount of work exerted on the system is given by

$$(\Delta A)_{T_1} = -nkT_1[\ln Z(x_2) - \ln Z(x_1)] .$$

The relation between T and H along an adiabatic line follows from the observation that S depends only on the combination $x = \frac{1}{2}\beta\mu H$. Thus x must be constant along an adiabatic and H and T are proportional to one another. The amount of work absorbed during an adiabatic increase of H is

$$(\Delta A)_{S_1} = n\mu(H_2 - H_1)\tanh x = n\mu H_1 \frac{T_2 - T_1}{T_1}\tanh x .$$

We now consider the Carnot cycle

$$\begin{array}{ccccc}
(H_2, T_1) & \xrightarrow{\text{adiabatic}} & \left(H_3 = \dfrac{T_2}{T_1}H_2, T_2\right) & & x_2 = \frac{1}{2}\beta_1\mu H_2 \\
\uparrow \text{isothermal} & & \downarrow \text{isothermal} & & \\
(H_1, T_1) & \xleftarrow{\text{adiabatic}} & \left(H_4 = \dfrac{T_2}{T_1}H_1, T_2\right) & & x_1 = \frac{1}{2}\beta_1\mu H_1
\end{array}$$

By adding the amounts of work done on the system in each of the four stages one finally obtains

$$(\Delta A)_{\text{tot}} = \frac{T_2 - T_1}{T_1}(\Delta Q)_{T_1} = \frac{T_1 - T_2}{T_2}(\Delta Q)_{T_2} .$$

Both for positive and for negative values of T_1 and T_2 there are two possible results:

(1) an amount of heat ΔQ is absorbed from a bath with temperature T_1; a fraction $(1 - T_2/T_1)$ is converted into work, the rest is discarded into a bath with $|T_2| < |T_1|$;

(2) work is exerted on the system while heat is transported from a bath with low $|T|$ to one with higher $|T|$.

The meaning of these results and their relation to various formulations of the second law will be discussed in the next problem.

Notice that we do not consider Carnot cycles between positive and negative values of T. In our model such processes would be possible if the adiabatics would cross at $H = 0$, $T = 0$. However in this region our model is no longer thermodynamically correct. In real spin systems the interaction between the spins would no longer be negligible and some sort of magnetic ordering would occur in the immediate neighbourhood of $H = 0$, $T = 0$. A similar phenomenon was observed in the case of the classical ideal gas (Problem 1.26).

15.4 (a) Two systems of n_1 and n_2 spins $\frac{1}{2}$ respectively, are initially at different temperatures and are then allowed to exchange energy. Determine the final equilibrium temperature and the direction of heat flow. [The level spacing may be taken the same in both systems.]

(b) Combine this result with that of the preceding problem and see whether Kelvin's and Clausius' formulation of the second law can be maintained if negative absolute temperatures are admitted. If not, propose a modification. It is reasonable to call a temperature T_1 hotter than T_2 if the energy content of the system at T_1 is higher than at T_2. With this convention, negative temperatures are hotter than positive ones and -0 indicates the hottest conceivable temperature. Like $+0$ it cannot be reached in a finite number of steps.

Solution

(a) The total energy content of the system is equal to $n_1 u(\beta_1)+n_2 u(\beta_2)$. At equilibrium each spin is at the same temperature; this final temperature corresponds to

$$u(\beta_f) = \frac{[n_1 u(\beta_1)+n_2 u(\beta_2)]}{(n_1+n_2)}$$

$$\beta_f = -\frac{2}{W}\operatorname{argtanh}\left(-\frac{n_1 \tanh\frac{1}{2}\beta_1 W + n_2 \tanh\frac{1}{2}\beta_2 W}{n_1+n_2}\right).$$

Since $u(\beta)$ is a monotonically decreasing function, equilibrium can be reached only if energy is transported from lower to higher values of β. If we adopt the definition of hotter and colder proposed in the problem, then the natural direction of energy flow is from hotter to colder temperatures. If we recall that $đQ = T\mathrm{d}S$ and notice that $(\mathrm{d}U/\mathrm{d}S)_W < 0$ for $T < 0$ we see that the same is true for the natural direction of heat flow. The sign conventions used for $đQ$ and the ordering of temperatures according to 'hotness' are reasonable but not unavoidable[3].

(b) With our definition of 'hotter' the possible results of Carnot cycle between negative temperatures as calculated in Problem 15.3 can be viewed differently:

(1) when heat is converted into work there is a heat flow from a cool to a hot reservoir;

(2) when work is converted into heat some heat has to flow from a hot to a cool reservoir.

By combining this with the possibility of natural heat flow from a hot to a cool reservoir we see that Kelvin's principle can be violated. It must be changed into the modified statement:

It is impossible to devise an engine which, working in a cycle, would produce no effect other than:

(1) the extraction of heat from a positive temperature reservoir and the production of an equal amount of mechanical work,

[3] The consequences of alternative choices are discussed in Section 13 of P.T.Landsberg, *Thermodynamics with Quantum Statistical Illustrations* (Interscience, New York), 1961.

(2) the rejection of heat into a negative temperature reservoir while an equal amount of work is done on the engine.

Clausius' formulation of the second law remains unaltered.

DYNAMIC POLARIZATION

The most direct way of bringing a spin system from a positive to a negative temperature is by means of a sudden reversal of the magnetic field. The spins cannot follow the reversal and are left with a polarization opposite to the field. Since this method is essentially not quasistatic and not completely reversible it is not suitable for obtaining very hot negative temperatures. The alternative method of dynamic polarization, sketched in this problem, is analogous to the techniques of optical pumping.

15.5 Consider an impurity in a magnetic crystal with three relevant energy levels. The transitions between the levels may occur via different mechanisms. Those inducing transitions between levels 1 and 3 and between levels 2 and 3 are represented by heat baths with temperatures T_1 and T_2. The transition between levels 1 and 2 can occur by means of energy exchange with the spin system. Find the temperature of the spin system at which there is no longer any net heat flow between the reservoirs, and between the reservoirs and the spin system.

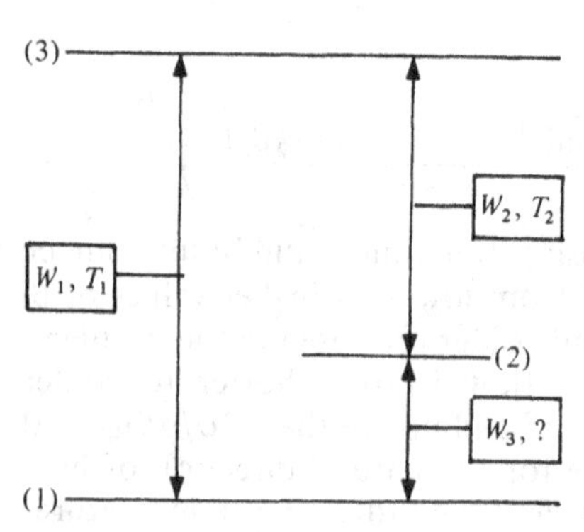

Solution

The transition between levels 1 and 3 is in equilibrium with a heat bath at temperature T_1 if the occupation probabilities of the two levels satisfy the relation

$$p_3 = p_1 \exp(-\beta_1 W_1) .$$

In the same manner we find

$$p_2 = p_3 \exp(+\beta_2 W_2) = p_1 \exp(-\beta_1 W_1 + \beta_2 W_2) .$$

Equilibrium is possible only if the temperature of the spin system satisfies the relation

$$\beta_3(W_1 - W_2) = \beta_1 W_1 - \beta_2 W_2 .$$

This equilibrium temperature will be negative if β_1 is small enough compared to β_2; in terms of the temperatures T_1 and T_2 we must have

$$T_1 > T_2 \frac{W_1}{W_2} .$$

The model discussed in this problem was introduced with reference to thermodynamical aspects of laser action [4]. The relation with laser amplification will be discussed in the next problems.

15.6 A beam of N photons is directed at a sample containing n two-level systems at temperature T. The frequency ν of the photons is resonant with the level splitting ($h\nu = W$); the probability that a single photon is absorbed by a system in its lower state and excites it to its upper state is equal to A. Using Problem 3.20, calculate the attenuation or amplification of the photon beam. Neglect spontaneous emission by the system and effects of the size of the sample.

[The attenuation or amplification of resonant radiation is a sensitive indicator of the temperature of a spin system. The presence of negative spin temperatures in a sample and the relaxation to positive temperatures were first demonstrated in this way by Purcell and Pound [5].]

Solution

According to Problem 3.20 the probability of absorption by a system in the ground state is equal to the probability of stimulated emission by the same system in the excited state. The net increase (or decrease) in the number of photons is therefore equal to

$$\Delta N = NnA(-p_{-1/2} + p_{+1/2}) .$$

In this formula $p_{-1/2}$ and $p_{+1/2}$ are the probabilities that a two-level system is in its upper or lower state. A comparison with the result of Problem 15.1 gives

$$\frac{\Delta N}{N} = -nA \tanh \tfrac{1}{2}\beta W .$$

If the temperature of the two-level systems is negative the beam of photons is amplified by it. This is the principle of maser or laser amplification. Neglect of spontaneous emission is justified if the number of photons in the relevant modes of the radiation field is large compared to unity. For radio and microwave frequencies this is usually the case; in the optical region it is often necessary to include effects of spontaneous emission.

15.7 In Problem 15.6 we saw that the energy in a two-level system at negative temperature can be extracted by means of stimulated emission. In this way it may be used for the build-up of a coherent electromagnetic oscillation. The system discussed in Problem 15.5 could be operated in such a way that heat from the reservoir at T_1 is partially converted into coherent oscillation. Show that the efficiency of this energy conversion cannot exceed that of a Carnot process operating between T_1 and T_2.

[4] P.Aigrain in *Quantum Optics and Electronics*, Les Houches 1964, Eds. C.DeWitt, A.Blandin, and C.Cohen-Tannoudji (Gordon and Breach, New York), 1965, p.527.
[5] E.M.Purcell and R.V.Pound, *Phys. Rev.*, **81**, 279 (1961).

Solution

From the total energy furnished by the heat bath at T_1 a fraction W_2/W_1 has to go into the reservoir at T_2. Stimulated emission can only exceed absorption if at least the equilibrium temperature of the spin system is negative. According to Problem 15.5 this implies that $W_2/W_1 > T_2/T_1$. Consequently the conversion efficiency η has to obey the relation[6]

$$\eta \leqslant \frac{W_3}{W_1} < 1 - \frac{T_2}{T_1} .$$

A MODEL OF LASER ACTION

15.8 The model of Problem 15.5 can be extended to provide a simple model of a laser. For this purpose we suppose that the sample of n three-level systems is enclosed in a resonant cavity. Losses of this cavity are taken into account by giving a finite lifetime τ_c to the photons in the resonant mode of the cavity.

(a) Find the rate of inversion $(p_2 - p_1)$ at which the increase of the number of photons from stimulated emission compensates the cavity losses. Notice that the expression contains the stimulated emission rate B_{21}, but not the number of photons N.

(b) Next we consider the equations for the three-level systems. Suppose the temperature T_1 is so large that the upward and downward transition rates between levels 1 and 3 caused by this pumping system are equal: $A_{13} = A_{31} = P$. Suppose further that the reservoir at T_2 and the systems causing transitions between levels 1 and 2 (other than the radiation field) are so cold that the upward transition rates can be neglected. Give equations for the change in occupation probability of the three levels for given values of P, A_{32}, A_{21}, and B_{21} and a given number N of photons in the resonant cavity mode. By putting the rates of change equal to zero the degree of inversion in the stationary state can be derived. Simplify this expression by assuming that A_{32} is much larger than the other transition rates.

(c) By comparing the results of parts (a) and (b) the number of photons in the cavity N can be found. For which values of the pumping rate P is laser action possible? Express the laser output in terms of $P - P_{cr}$.

Solution

(a) By adding a term for the losses to the result derived in Problem 15.6 we find for the change in the number of photons

$$\frac{dN}{dt} = -\frac{N}{\tau_c} + nB_{21}(p_2 - p_1)N .$$

[6] H.E.D.Scovil and E.O.Schulz-DuBois, *Phys. Rev. Letters*, **2**, 262 (1959).

For a stationary state we must have

$$\Delta n_{cr} = n(p_2 - p_1)_{cr} = \frac{1}{B_{21}\tau_c}.$$

(b) The equations for the change in the occupation probabilities are

$$\frac{dp_1}{dt} = A_{21}p_2 + NB_{21}(p_2 - p_1) - P(p_1 - p_3),$$

$$\frac{dp_2}{dt} = -A_{21}p_2 + NB_{21}(p_1 - p_2) + A_{32}p_3,$$

$$\frac{dp_3}{dt} = P(p_1 - p_3) - A_{32}p_3.$$

Putting the left hand sides equal to zero and using $p_1 + p_2 + p_3 = 1$ we obtain for the stationary value of the inversion

$$(p_2 - p_1)_{st} = \frac{P(A_{32} - A_{21}) - A_{21}A_{32}}{(P + A_{32})(2NB_{21} + A_{21}) + P(2NB_{21} + A_{32} + A_{21})}.$$

If we assume that A_{32} is much larger than all other transition rates, this expression becomes

$$(p_2 - p_1)_{st} = \frac{P - A_{21}}{2NB_{21} + A_{21} + P}.$$

(c) A stationary regime is only possible if this inversion rate is equal to the critical inversion rate calculated in part (a):

$$\frac{1}{nB_{21}\tau_c} = \frac{P - A_{21}}{2NB_{21} + A_{21} + P}$$

or

$$N = \frac{n\tau_c}{2}(P - A_{21}) - \frac{1}{2B_{21}}(P + A_{21}).$$

The critical pumping rate is equal to (put $N = 0$):

$$P_{cr} = A_{21}\frac{1 + 1/nB_{21}\tau_c}{1 - 1/nB_{21}\tau_c}.$$

The power output of the laser equals N/τ_c. It can be rewritten in the form

$$N/\tau_c = \tfrac{1}{2}(n - \Delta n_{cr})(P - P_{cr}).$$

Notice that $\frac{1}{2}(n - \Delta n_{cr})$ is exactly the number of systems in level 1 when the critical inversion is reached. The quantity np_1P is the number of transitions to level 3 induced by the pumping heat bath. A fixed number of these excited systems is evidently needed to overcome losses due to spontaneous transitions; the remaining ones give rise to photons leaking out of the cavity. Modifications and improvements of this laser model can be found in any textbook on the subject[7].

[7] e.g. A. Yariv, *Quantum Electronics* (John Wiley, New York), 1967, Chapter 15.

16

Recombination rate theory in semiconductors

J.S.BLAKEMORE
(*Florida Atlantic University, Boca Raton, Florida*)

16.1 (a) For non-interacting or 'one-electron' states at energy E, write down an expression for the probability that any such state is occupied. You may employ the knowledge of Fermi–Dirac statistics gained from Problem 3.12. In the present chapter, we shall want to express occupancy probabilities in terms of an electrochemical potential or Fermi energy E_F for equilibrium situations.

(b) Given that a band of permitted electron states extends upwards from energy E_c with a density of states $g(E) = A(E-E_c)^{1/2}$, express in integral form a condition for E_F at temperature T when n_0 electrons occupy the band at equilibrium. How can this be simplified for the case of n_0 quite small?

(c) The kind of band we have considered is usually referred to as the **conduction band** of a semiconductor, containing a few electrons in the lowest states, and many empty states. When thermal equilibrium is perturbed, the conduction electron density n may well be different from n_0. Use a **quasi-Fermi level** F_n as a normalizing parameter for the electron conduction density under such circumstances. At an energy well below E_c in a semiconductor we expect to find a band of allowed electron states which is almost completely full, extending downwards from energy E_v. Describe the total number of 'holes' in this **valence band** in terms of the Fermi energy E_F for equilibrium (p_0 for holes) and of a hole quasi-Fermi level F_p for a non-equilibrium situation (p holes). Show the sense of departure of F_n and F_p from E_F when the free electron and hole populations are enlarged.

(d) Show that at equilibrium the product n_0p_0 depends on temperature but not on n_0 or p_0 provided that the Fermi level lies within the **intrinsic energy gap**, i.e. between the top of the valence band and the bottom of the conduction band. Why is the quantity $n_i = (n_0p_0)^{1/2}$ called the **intrinsic** pair density? Relate the energy difference $F_n - F_p$ for a non-equilibrium situation to the product np (**mass action law**).

Solution

(a) According to the Pauli principle, no two electrons in a system can be in the same quantum state. (Two electrons of opposing spin can enjoy the same translational wavefunction; we regard these as distinguishable quantum states.) Fermi–Dirac statistics are appropriate for describing the occupancy of 'one-electron' states in accordance with the Pauli principle. An electron which occupies a 'one-electron' state acts dynamically as an independent particle, not interacting with the other electrons in 'one-electron' states, although the properties (e.g. energies) of the 'one-electron' states are affected by the average behaviour of all of the electrons in the system.

As discussed in Chapter 3, the time-averaged probability that a state of energy E be occupied by an electron is of the form

$$f(E) = \frac{1}{1+\alpha\exp(E/\beta)} \tag{16.1.1}$$

which has the necessary maximum occupancy of one electron per state for states of low energy. Conformity with Boltzmann statistics for the high-energy limit of small occupancy enables us to identify the parameter β with kT. The parameter α in Equation (16.1.1) performs a normalization function, for if $g(E)\,\mathrm{d}E$ distinguishable one-electron states are found in an energy range $\mathrm{d}E$ at energy E then in equilibrium at temperature T we can always find one (and only one) value for α such that

$$N = \int_{-\infty}^{+\infty} g(E)f(E)\,\mathrm{d}E \tag{16.1.2}$$

is equal to the total number of electrons in the system.

The normalization concept can in practice be expressed much more conveniently in terms of a parameter with the dimensions of energy. This is the **electrochemical potential**, or **Fermi energy**

$$E_{\mathrm{F}} = -kT\ln(\alpha)\,. \tag{16.1.3}$$

This can replace α in Equation (16.1.1) in writing the equilibrium probability of electron occupancy as

$$f(E) = \frac{1}{1+\exp[(E-E_{\mathrm{F}})/kT]}\,. \tag{16.1.4}$$

We may note from this that the occupation probability is 50% for a state at energy E_{F} itself. States well above E_{F} are sparsely occupied with electrons. States well below E_{F} are almost all filled; they contain few 'holes' among the ranks of filled states.

(b) Given that a band of states starts from minimum energy E_{c} and that $g(E) = A(E-E_{\mathrm{c}})^{1/2}$ for higher energies, we know that at thermal equilibrium the distribution of the total electron supply n_0 over states of various energies must conform with

$$n_0 = \int_{E_{\mathrm{c}}}^{\infty} \frac{A(E-E_{\mathrm{c}})^{1/2}}{1+\exp[(E-E_{\mathrm{F}})/kT]}\,\mathrm{d}E\,. \tag{16.1.5}$$

This inserts the form of Equation (16.1.4) into Equation (16.1.2). Equation (16.1.5) serves as a definition of the resulting Fermi energy for any combination of n_0 and temperature.

Given further that n_0 is very small, we can see that Equation (16.1.5) must be satisfied with E_F much lower in energy than E_c. In this asymptotic case,

$$\begin{aligned} n_0 &= A(kT)^{3/2}\exp[(E_F - E_c)/kT] \int_0^\infty Y^{1/2}\exp(Y)\mathrm{d}Y \\ &= A(kT)^{3/2}(\tfrac{1}{4}\pi)^{1/2}\exp[(E_F - E_c)/kT] \\ &= N_c \exp[(E_F - E_c)/kT]\,. \end{aligned} \qquad (16.1.6)$$

It may be noted that n_0 of Equation (16.1.6) is the same as though the band were replaced by N_c states, all at energy E_c. The conditions of Equation (16.1.6) are met provided that $n_0 \ll N_c$; this is a temperature dependent inequality which requires that E_F be several kT lower than E_c.

(c) We now consider the possibility of a departure from equilibrium, to provide for $n \neq n_0$ electrons in this conduction band. It is apparent that a normalizing parameter F_n with the dimensions of energy can be used to define the relationship of n and the crystal temperature by

$$n = \int_{E_c}^{\infty} \frac{g(E)}{1+\exp[(E-F_n)/kT]}\mathrm{d}E \qquad (16.1.7)$$

as an adaptation of Equation (16.1.5). Equation (16.1.7) places no apparent restriction on the value of n. The quantity F_n serves as a quasi-Fermi level for non-equilibrium situations, and becomes coincident with E_F for equilibrium itself.

Provided that the total non-equilibrium population is rather small (so that F_n is lower in energy than E_c), the procedure of Equation (16.1.6) can be applied to Equation (16.1.7) in writing

$$n = N_c \exp[(F_n - E_c)/kT] \qquad (n \ll N_c)\,. \qquad (16.1.8)$$

It is worth observing that the electrons distributed over band states for a non-equilibrium situation may have a velocity distribution which is incompatible with a thermal one for any real temperature. However, Equations (16.1.7) and (16.1.8) are concerned only with characterizing the *total* conduction electron density for a given *crystal* temperature T and a given density of states. In practice, we hope that excess electrons thrown into a conduction band will thermalize their speeds within a time of 10^{-11} seconds or less, while in many semiconductors the time taken for excess conduction electrons to return to states in lower bands is very much longer.

When equilibrium is disturbed in a semiconductor, it is usually in the sense of making $n > n_0$. From Equations (16.1.7) or (16.1.8), this makes F_n higher in energy than the equilibrium E_F, as shown in Figure 16.1.1.

The same figure shows a valence band of states lying below the equilibrium Fermi energy. Such a band will be almost filled with electrons at equilibrium, i.e. will have a relatively small density p_0 of 'free holes'.

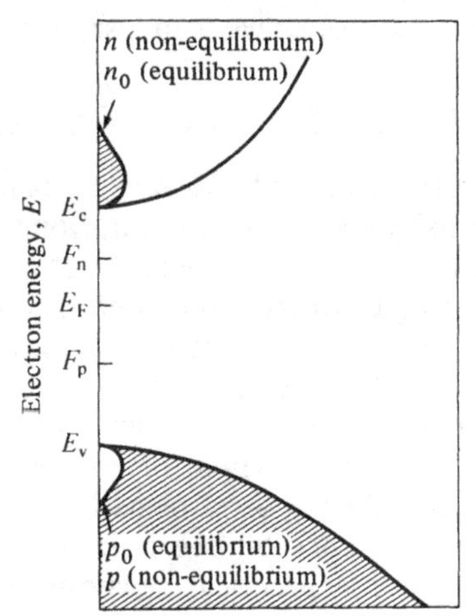

Figure 16.1.1. An idealized valence band (almost full) and conduction band (almost empty) for a semiconductor with an intrinsic gap extending from energy E_v to energy E_c. The equilibrium Fermi energy E_F is compatible with the densities n_0, p_0 for free electrons and holes at temperature T. For any non-equilibrium densities n, p, which represent an enlargement of the free carrier densities, the quasi-Fermi levels separate from E_F in the sense $F_n > E_F > F_p$.

From the preceding arguments, it will be clear that p_0, E_F, and T are related by a condition

$$p_0 = \int_{-\infty}^{E_v} \frac{g(E)}{1+\exp[(E_F - E)/kT]} \mathrm{d}E \,, \qquad (16.1.9)$$

since the factor of $g(E)\mathrm{d}E$ is clearly the probability of a state *not* being occupied by an electron. When the Fermi energy is considerably higher than E_v (so that even the uppermost states in the band have rather few free holes), then we can expect to apply the procedure used in connection with Equation (16.1.6) for electrons, and write

$$p_0 = N_v \exp[(E_v - E_F)/kT] \qquad (p_0 \ll N_v)\,. \qquad (16.1.10)$$

In this limiting situation, the free hole density is the same as though the valence band were replaced by N_v states all at energy E_v.

Evidently, if the valence band contains $p \neq p_0$ free holes for a non-equilibrium situation, the density p can be described in terms of a hole quasi-Fermi level F_p by

$$\left.\begin{aligned} p &= \int_{-\infty}^{E_v} \frac{g(E)}{1+\exp[(F_p - E)/kT]} \mathrm{d}E && \text{(arbitrary } p) \\ p &= N_v \exp[(E_v - F_p)/kT] && \text{(small } p) \end{aligned}\right\} \qquad (16.1.11)$$

as adaptations of Equations (16.1.9) and (16.1.10). The quantity F_p will coincide with E_F at equilibrium, but will be lower than E_F (as shown in Figure 16.1.1) when $p > p_0$.

(d) For equilibrium conditions, we can use Equations (16.1.6) and (16.1.10) to write

$$n_0 p_0 = N_c N_v \exp[(E_v - E_c)/kT] \qquad (n_0 \ll N_c,\ p_0 \ll N_v) \quad (16.1.12)$$

which depends only on the width of the valence band/conduction band intrinsic gap, on the density of states configurations near the extrema of these bands, and on the temperature. The product $n_0 p_0$ does not depend explicitly on the value of n_0 or of p_0 if the cited inequalities are met, and these equalities require that the Fermi energy should be appreciably higher than E_v yet appreciably lower than E_c.

An **intrinsic semiconductor** is one for which the densities of free electrons and free holes are equal, which of course is the case if excitation of electrons from the valence band into the conduction band is the dominant reason for the existence of free holes and free electrons. An intrinsic semiconductor is for practical purposes independent of the existence of any extrinsic features (such as the presence of foreign impurities or lattice defects which may in less perfect materials provide localized states for the provision of *either* free holes *or* free electrons). Thus in an intrinsic semiconductor, the densities of free holes and free electrons are each the quantity

$$n_i = (n_0 p_0)^{1/2} = (N_c N_v)^{1/2} \exp[(E_v - E_c)/2kT] . \quad (16.1.13)$$

For a semiconductor (which may or may not be intrinsic) containing excess electrons and holes as a result of some departure from equilibrium, the product np can be written as

$$\begin{aligned} np &= N_c N_v \exp[(F_n - E_c)/kT] \exp[(E_v - F_p)/kT] \\ &= n_i^2 \exp[(F_n - F_p)/kT] \end{aligned} \quad (16.1.14)$$

by employing Equations (16.1.8), (16.1.11), and (16.1.13). Instead of n_i^2 on the right side of Equation (16.1.14) we could equally well write $n_0 p_0$, remembering that n and p must be written in the integral forms of Equations (16.1.7) and (16.1.11) if the free carrier densities are large enough to bring a quasi-Fermi level close to the edge of a band. At any rate, when F_n and F_p still lie within the intrinsic gap,

$$F_n - F_p = kT \ln(np/n_0 p_0) . \quad (16.1.15)$$

16.2 Consider a semiconducting material for which the natural processes of energy transformation at a certain temperature are sufficient to maintain a population of n_0 free electrons per unit volume in a conduction band which can accommodate many more electrons. The density n_0 results from a balance at thermal equilibrium of two opposing processes: (i) the generation (at rate = g) of conduction electrons by excitation from filled electron states at lower energies, and (ii) the recombination (at rate = r) of free electrons by their de-excitation to any empty states at lower energies.

Now suppose that thermal equilibrium is perturbed, and the conduction electron density changed to $n = n_0 + n_e$. We refer to n_e as the excess electron density. Whenever n_e is non-zero, then $g \neq r$, and usually both g and r are then modified from their equilibrium values. Such a perturbation may result from externally induced excess generation at a rate g_e, or by a flow of excess electrons from elsewhere. The net effect of the various influences can be expressed in an electron **continuity equation**

$$\frac{\partial n}{\partial t} = \frac{\partial n_e}{\partial t} = g_e + g - r + \frac{1}{e}\nabla \cdot \mathbf{I}_n ;$$

where $\mathbf{I}_n$ denotes the current density due to electron flow. In discussions of recombination in a semiconductor, it is convenient to define the electron lifetime[1] τ as the excess electron density per unit rate of $(r-g)$. In terms of τ, the electron continuity equation is

$$\frac{\partial n_e}{\partial t} = g_e - \frac{n_e}{\tau} + \frac{1}{e}\nabla \cdot \mathbf{I}_n .$$

(a) Comment on the conditions under which this becomes an ordinary and linear differential equation. What is then the general form of solution for any period of time-invariant excess generation?

(b) Describe the build-up of n_e when g_e starts abruptly at time $t = T_1$ and continues at a constant rate for a long time thereafter. What happens when g_e ceases equally abruptly at time $t = T_2$?

(c) From these results, show how n_e varies when g_e changes abruptly from g_{e1} to g_{e2} at time $t = T_3$. Sketch the time dependence for g_{e2} larger than or smaller than g_{e1}.

(d) When in practice the excess generation rate g_e is an arbitrary function of time, express n_e as an integral with respect to prior generation. Use this to show how n_e responds to a pulse of generation which is an isolated half sine wave as a function of time.

(e) In the same way, show how n_e declines when g_e is an exponentially decreasing function of time.

Solution

(a) The continuity equation for electrons becomes an ordinary differential equation if the excess electron density n_e is a function of time but not of location. This can be true if we consider a region well within a large, homogeneous single crystal, a region far from any surfaces, p–n junctions, or contacts. It must also be presumed that the mechanism

[1] The various meanings of 'excess carrier lifetime' are explored in detail with respect to the various processes of generation and recombination in the books by Ryvkin and by Blakemore cited in the bibliography at the end of the section.

responsible for the excess generation rate g_e is spatially uniform; thus if g_e results from the absorption of photons of suitable energy directed from without, the photons must be ones which are absorbed rather weakly by the solid.

Since we assume spatial uniformity and remoteness from surfaces and contacts, the electron current will be solenoidal (non-divergent) and the continuity equation reduces to

$$\frac{dn_e}{dt} = g_e - \frac{n_e}{\tau} \, . \qquad (16.2.1)$$

This equation is also *linear* if the lifetime τ is a constant, i.e. if the difference between the natural rates of recombination and generation is directly proportional to the departure from an equilibrium free electron density. (In practice it is found that τ has *some* dependence on n_e for any dominant recombination mechanism, though for certain recombination regimes τ behaves as a constant over a fairly wide range of n_e.)

When conditions of spatial uniformity and constant τ are imposed, Equation (16.2.1) can be integrated immediately for any period of time in which g_e is maintained at a constant value. The general solution is of the form

$$n_e = g_e\tau + C\exp(-t/\tau) \qquad (16.2.2)$$

where the quantity C is determined by the initial conditions.

(b) If there is no excess generation prior to time $t = T_1$, and the constant rate g_e thereafter, then C in Equation (16.2.2) must have such a value that $n_e = 0$ at $t = T_1$. This condition is satisfied if

$$C = -g_e\tau\exp(T_1/\tau)$$

so that

$$n_e = g_e\tau\{1 - \exp[(T_1 - t)/\tau]\} \qquad (t > T_1)\, . \qquad (16.2.3)$$

This has reached the value

$$n_e(T_2) = g_e\tau\{1 - \exp[(T_1 - T_2)/\tau]\}$$

at the time T_2 when excess generation abruptly ends. From Equation (16.2.2) it is clear that n_e will subsequently decay in accordance with

$$\begin{aligned} n_e &= C\exp(-t/\tau) \\ &= n_e(T_2)\exp[(T_2 - t)\tau] \\ &= g_e\tau[\exp(T_2/\tau) - \exp(T_1/\tau)]\exp(-t/\tau) \qquad (t > T_2)\, . \quad (16.2.4) \end{aligned}$$

(c) In this problem, we assume that the lifetime is independent of n_e, so that the continuity equation (16.2.1) is linear. Thus if g_e operates at a value g_{e1} during the period $T_1 \leqslant t \leqslant T_3$, and at a different value g_{e2} for $t > T_3$, then n_e at times later than T_3 can be described as a simple sum of contributions in the forms of Equations (16.2.4) and (16.2.3). The former of these will describe that component of n_e which derives from acts of generation prior to T_3, and the latter to the consequences of the more recent generation. Thus

$$n_e = g_{e1}\tau[\exp(T_3/\tau)-\exp(T_1/\tau)]\exp(-t/\tau)+g_{e2}\tau\{1-\exp[(T_3-t)/\tau] \quad (t > T_3)\,. \quad (16.2.5)$$

This can be separated into the sum of a constant component and a transient component in the form

$$n_e = g_{e2}\tau-[(g_{e2}-g_{e1})\tau\exp(T_3/\tau)+g_{e1}\tau\exp(T_1/\tau)]\exp(-t/\tau) \quad (t > T_3)\,. \quad (16.2.6)$$

The consequences of this equation are illustrated by the curves of Figure 16.2.1 for situations in which g_e increases, decreases, or remains unchanged at the time T_3.

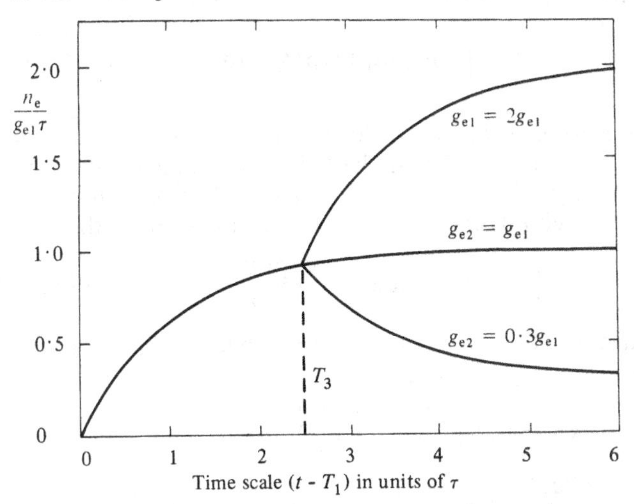

Figure 16.2.1. The transient behaviour of the excess pair density when excess generation starts at a time T_1 and changes to a different rate at time T_3. The curves here follow Equations (16.2.5) and (16.2.6) after time T_3 for situations in which $g_{e2}/g_{e1} = 2$, 1, and 0·3.

(d) We must now consider the solution of Equation (16.2.1) when g_e is an arbitrary function of time. This solution can be envisaged by considering g_e to be a sequence of instantaneous generation events, each of which can be described by a delta function. Thus suppose that

$$\int_{-\infty}^{\infty} g_e \, dt = N\delta(t-t_0)\,.$$

For this generation alone, Equation (16.2.1) has the solution

$$n_e(t) = N\exp[(t_0-t)/\tau]\,.$$

Now since the lifetime t is assumed to be a constant in this problem, Equation (16.2.1) is linear, and the solution for $n_e(t)$ for many delta-function acts of generation at different times is simply the sum of the values for $n_e(t)$ derived from each generative act separately. Thus when g_e is any arbitrary (not necessarily continuous) function of time, then the integral representation for $n_e(t)$ is

$$n_e(t) = \int_{-\infty}^{t} g_e(t_0)\exp[(t_0-t)/\tau]\,dt_0 \tag{16.2.7}$$

which is a very simple example of a Green's function form of solution.

We can use Equation (16.2.7) to determine how n_e responds to sinusoidally varied creation. Suppose that we have a single half sine wave of generation:

$$g_e(t_0) = G\sin(\omega t_0) \qquad (0 < t_0 < \pi/\omega)\,,$$

and that there is no excess generation at other times. Then, from Equation (16.2.7), we have

$$n_e(t) = G\exp(-t/\tau)\int_0^t \sin(\omega t_0)\exp(t_0/\tau)\,dt_0 \qquad (0 < t < \pi/\omega) \tag{16.2.8}$$

as a description of n_e while the generation process is still going on. In order to describe n_e after the half sine wave of generation has ceased, we must write Equation (16.2.8) with π/ω as the upper limit of integration.

Now through integration by parts we can establish that

$$\int \exp(ax)\sin x\,dx = \frac{\exp(ax)}{1+a^2}(a\sin x - \cos x)\,,$$

and Equation (16.2.8) is of this form when we change to the dimensionless variable $x = \omega t_0$. Thus

$$\begin{aligned} n_e(t) &= \frac{G}{\omega}\exp(-t/\tau)\int_0^{\omega t} \exp(x/\omega\tau)\sin x\,dx \\ &= \frac{G\omega\tau^2}{1+\omega^2\tau^2}\left[\frac{\sin(\omega t)}{\omega\tau} - \cos(\omega t) + \exp(-t/\tau)\right] \qquad (0 < t < \pi/\omega)\,, \end{aligned} \tag{16.2.9}$$

while the generation is in progress, and

$$\begin{aligned} n_e(t) &= \frac{G}{\omega}\exp(-t/\tau)\int_0^{\pi} \exp(x/\omega\tau)\sin x\,dx \\ &= \frac{G\omega\tau^2}{1+\omega^2\tau^2}[1+\exp(\pi/\omega\tau)]\exp(-t/\tau) \qquad (t > \pi/\omega) \end{aligned} \tag{16.2.10}$$

for the subsequent monotonic decay. The form of the response while the generation is still in progress can be seen more clearly by defining a phase angle $\theta = \tan^{-1}(\omega\tau)$, and rewriting Equation (16.2.9) as

$$n_e(t) = G\tau\cos\theta[\sin(\omega t-\theta)+\exp(-t/\tau)\sin\theta] \quad (0 < t < \pi/\omega). \tag{16.2.11}$$

Figure 16.2.2 illustrates the rise and fall of n_e in accordance with Equations (16.2.9) through (16.2.11) when $\omega\tau = 1$, to make the phase angle $\theta = \tan^{-1}(\omega\tau) = \frac{1}{4}\pi$.

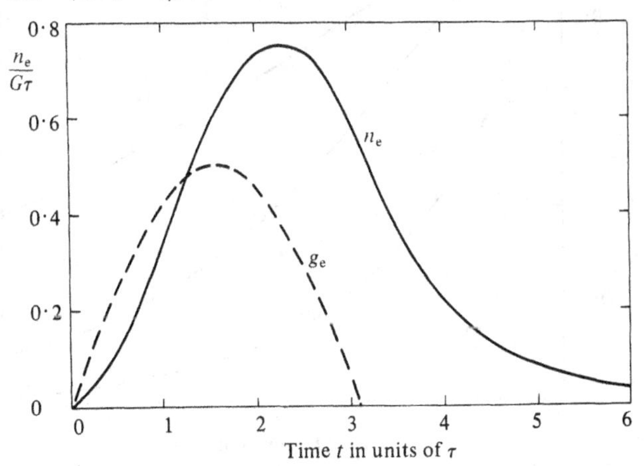

Figure 16.2.2. The response of the excess carrier pair density to a half sine wave of generation, in accordance with Equations (16.2.9) through (16.2.11). The dashed curve illustrates g_e for a situation of $\omega\tau = 1$, and the solid curve is the result for n_e.

As a corollary of the preceding discussion, it can readily be established that generation in the form of a continuous sine wave results in an excess carrier density comprising the sum of a constant term and a sinusoidal term. The sinusoidal component lags behind the phase of the generating sine wave by the angle $\theta = \tan^{-1}(\omega\tau)$.

(e) We should now like to know how n_e decays when g_e is itself an exponentially decreasing function of time. Suppose that g_e has been maintained at the constant value G for all negative time (so that $n_e = G\tau$ at time $t = 0$), and that $g_e = G\exp(-t/T)$ for all positive t. Then from Equation (16.2.7) we have

$$n_e(t) = (G\tau)\exp(-t/\tau) + \int_0^t G\exp(-t_0/T)\exp[(t_0 - t)/\tau]\,dt_0 \quad (t > 0)\,.$$

Provided that T and τ are not exactly the same, the integration yields

$$n_e(t) = G\tau\frac{T\exp(-t/T) - \tau\exp(-t/\tau)}{T - \tau} \qquad (t > 0)\,, \quad (16.2.12)$$

while for the special case of $T = \tau$ the result is

$$n_e(t) = G(t + \tau)\exp(-t/\tau) \qquad (t > 0)\,. \quad (16.2.13)$$

Figure 16.2.3 displays the decay of n_e according to Equation (16.2.13) for the particular case of $T = \tau$, compared with the conventional exponential decay for $T = 0$ and the greatly slowed decay when $T = 3\tau$. Only if T is smaller than, or equal to, τ is it possible to determine τ from the decay rate at large values of t.

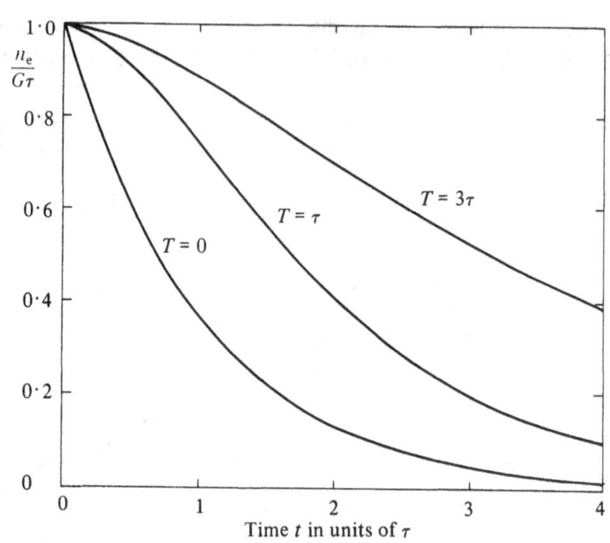

Figure 16.2.3. The decay of the excess carrier density upon the cessation of or attenuation of the excess generation rate. The lowest curve is the exponential decay for generation which stops abruptly. The middle curve follows Equation (16.2.13) for generation which declines exponentially with time constant $T = \tau$. The upper curve fits Equation (16.2.12) for $T = 3\tau$.

16.3 In this problem we consider the direct radiative transitions in a semiconductor which result in the creation of, or annihilation of, a hole–electron pair. The term **direct radiative** means that a photon can create a hole and an electron whose difference in momenta is just the (negligibly small) momentum of the photon itself. An electron state and hole state which are connected by a direct transition probability have the same wavevector, as for example the states of energy E_u and E_l in Figure 16.3.1. The semiconductor drawn in this figure is a **direct gap** semiconductor in that the allowed states forming the upper and lower borders of the intrinsic gap E_i are of the same wavevector.

(a) Use considerations of mass action and detailed balance in discussing the rates at which transitions between E_l and E_u will occur by stimulated and spontaneous radiative processes. Show that the ratio of stimulated to spontaneous rates of radiative recombination between these energies is just the number $\overline{N}$ of photons per mode for photon energy

$$h\nu = E_u - E_l\,,$$

(b) From this result, show that stimulated transitions have an effect on the radiative recombination rate between E_u and E_l which can be described by multiplying the spontaneous rate by a factor of

$$1 - \overline{N}\left[\exp\left(\frac{h\nu + F_p - F_n}{kT}\right) - 1\right] .$$

What does this tell us about a threshold condition for supremacy of stimulated recombination? (See also Problems 3.20, 15.6, and 15.7 for related considerations.)

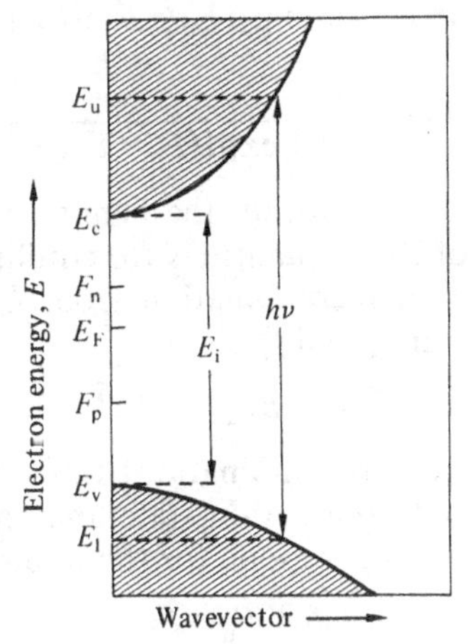

Figure 16.3.1. The model of a direct gap semiconductor assumed in Problem 16.3. The electrochemical potential or Fermi level is E_F for thermodynamic equilibrium, and the gross populations of the two bands for relatively steady-state non-equilibrium are characterized by F_n and F_p.

[Of course, operation of a semiconductor laser is also contingent on the elimination of competing (non-radiative) recombination mechanisms, and on an optical configuration which takes proper advantage of recombination radiation.]

Solution

(a) The radiative transitions we must consider between E_l and E_u lie in three categories:

dr_{sp} = rate of spontaneous events of electron–hole recombination;
dr_{st} = rate of recombination events stimulated by the recombination radiation field;
dg_{st} = rate of generative events stimulated by the recombination radiation field.

For the spontaneous recombination process, we may use the principle of mass action in writing the rate in a form

$$dr_{sp} = Af_u(1-f_l). \tag{16.3.1}$$

The parameter A contains information about the densities of states at energies E_u and E_l, and about the possibilities for electron–hole annihilation when an electron of kinetic energy $E_u - E_c$ encounters a hole of kinetic energy $E_v - E_l$. In Equation (16.3.1), f_u denotes the probability that a state at E_u contains an electron, and $1-f_l$ is the probability that a state at E_l contains a hole. If it can be assumed that a disturbance of thermodynamic equilibrium changes the magnitudes of the electron and hole populations but not the character of the velocity distributions within the co-existing electron and hole gases, then

$$f_u = \frac{1}{1+\exp[(E_u - F_n)/kT]}$$

and

$$1 - f_l = \frac{1}{1+\exp[(F_p - E_l)/kT]}$$

for any values of E_u and E_l within the bands. We shall denote by f_{u0} and $1-f_{l0}$ the values of these quantities for equilibrium.

The rate of stimulated recombination from E_u to E_l has the same functional dependence on f_u and f_l:

$$dr_{st} = Bf_u(1-f_l)\overline{N}, \tag{16.3.3}$$

since (as with spontaneous recombination) this rate depends on the co-existence of an occupied conduction state and an empty valence state. $\overline{N}$ is the number of photons per mode for the photon energy

$$h\nu = E_u - E_l,$$

and we know that in thermodynamic equilibrium $\overline{N}$ must reduce to the Planck result

$$\overline{N}_0 = \frac{1}{\exp(h\nu/kT)-1}. \tag{16.3.4}$$

We are interested in the relationship of the parameter B in Equation (16.3.3) to A in Equation (16.3.1). The quantity B must be used again in describing the rate of radiative transitions stimulated *upwards* by the presence of a recombination radiation field. Since this latter rate depends upon a full valence state and an empty conduction state, we have

$$dg_{st} = Bf_l(1-f_u)\overline{N} \tag{16.3.5}$$

from E_l to E_u.

The principle of detailed balance—which is tantamount to a statement of the second law of thermodynamics—requires that $r_{sp}+r_{st}-g_{st}$ vanish in thermodynamic equilibrium, not only *in toto*, but also between any selected groups of initial and final states. From Equations (16.3.1), (16.3.3), and (16.3.5) we can write that

$$[Af_{u0}(1-f_{l0})+Bf_{u0}(1-f_{l0})\overline{N}_0 - Bf_{l0}(1-f_{u0})\overline{N}_0] = 0$$

at equilibrium. Thus the ratio of stimulated to spontaneous coefficients is

$$\frac{B}{A} = \frac{f_{u0}(1-f_{l0})}{\overline{N}_0(f_{l0}-f_{u0})}.$$

An insertion of f_{u0}, f_{l0}, and $\overline{N}_0$ from Equations (16.3.2) and (16.3.4) and a brief manipulation of these factors yields the very simple result that

$$\frac{B}{A} = 1. \tag{16.3.6}$$

Thus at equilibrium, or away from it, the ratio of stimulated to spontaneous recombination rates between E_u and $E_u - h\nu$ is just

$$\frac{dr_{st}}{dr_{sp}} = \overline{N}_{h\nu} \tag{16.3.7}$$

from the ratio of the right sides of Equations (16.3.3) and (16.3.1).

(b) When the conditions are of a steady-state non-equilibrium, then the *net* recombination rate from E_u to E_l is

$$dr_{net} = dr_{sp} + dr_{st} - dg_{st} = dr_{sp}\left(1 - \frac{dg_{st} - dr_{st}}{dr_{sp}}\right)$$
$$= dr_{sp}\left[1 - \frac{\overline{N}(f_l - f_u)}{f_u(1 - f_l)}\right],$$

since we have demonstrated from detailed balance arguments at equilibrium that A and B must be the same. Substitutions of f_u and f_l from Equation (16.3.2) then immediately yields that

$$dr_{net} = dr_{sp}\left\{1 - \overline{N}\left[\exp\left(\frac{h\nu + F_p - F_n}{kT}\right) - 1\right]\right\}. \tag{16.3.8}$$

A modest departure from thermal equilibrium will make $F_n - F_p$ a positive quantity, but a quantity smaller than $h\nu = E_u - E_l$. Under these circumstances, the net recombination rate dr_{net} is smaller than dr_{sp} alone. A more severe violation of equilibrium which makes $F_n - F_p$ just equal to $h\nu$ (that is, a violation which makes f_u the same as f_l) causes stimulated transitions to occur in the upwards and downwards directions at exactly cancelling rates. For this threshold value of $F_n - F_p$, dr_{net} is identical to dr_{sp}.

When the excess densities of free electrons and holes are made still larger, f_u is actually larger than f_l. This is the case when $F_n - F_p > h\nu$. Obviously this condition can be met most easily for the states of energy E_v and $E_c = E_v + E_i$, and with progressively more difficulty for states separated by larger photon energies. The creation of **lasing conditions** in a semiconductor requires among other things[2] that there be a **population inversion** (upper energy states more heavily occupied with electrons than lower states, or $f_u > f_l$) between the highest valence band states and the lowest conduction band states. When a population inversion is in existence, then the net radiative transition rate dr_{net} is appreciably larger than dr_{sp} and is dominated by stimulated downward transitions.

[2] In order that the recombination radiation produced by downward transitions should be amplified by stimulating further downward transitions, a population inversion is a necessary but not sufficient condition. The semiconductor must also be one for which non-radiative recombination processes are rather feeble and present little competition to the radiative reactions. Mirror surfaces must be arranged to give an 'optical gain' which outweighs the inevitable optical losses for some direction of propagation of the photon stream.

16.4 In this problem, as in Problem 16.3, we concern ourselves with direct band-to-band generation of hole–electron pairs and their direct recombination. For convenience, we shall again assume that radiative transitions provide the dominant mechanism for energy transformation.

Suppose a large homogeneous semiconducting crystal with no non-uniform currents, so that the continuity equation for excess electrons and p_0 holes is not a function of position. This medium contains n_0 electrons and p_0 holes per cm^3 at equilibrium, and n_0+n_e electrons and p_0+n_e holes in a non-equilibrium situation. The quantities n_0, p_0, and n_e are all small enough to ensure that E_F, F_n, and F_p all lie several kT within the intrinsic gap (as was true for the situation illustrated in Figure 16.3.1). Show that the radiative lifetime τ_R of excess carrier pairs is proportional to

$$\frac{n_0 p_0}{n_0+p_0+n_e} \; .$$

Solve the equation for transient decay when externally provoked generation is terminated (using the symbol τ_{R0} for the small-modulation lifetime when n_e is sufficiently small). Show that an arbitrarily large disturbance of equilibrium must decay to a small-modulation situation within the time τ_{R0}.

Solution

Since we are given that the equilibrium free carrier densities and the excess pair density n_e are small enough to keep $E_c - F_n$ and $F_p - E_v$ both larger than kT, it is apparent that the intrinsic gap E_i must be considerably larger than kT. Accordingly, the photon occupancy number per mode under equilibrium conditions must be small compared with unity,

$$\overline{N}_0 \approx \exp(-h\nu/kT) \ll 1 \tag{16.4.1}$$

for any photon energy relevant to band-to-band transitions. Moreover, since $F_n - F_p$ is required to be appreciably smaller than E_i, it seems probable from the conclusions of Problem 16.3 that $\overline{N}$ in a non-equilibrium situation is unlikely to be substantially different from $\overline{N}_0$ of equilibrium itself. Under these circumstances, stimulated recombination dr_{st} from an upper state E_u to a lower state E_l can be ignored in comparison with the spontaneous recombination rate dr_{sp} between the same states. Thus the *net* radiative recombination rate is

$$\begin{aligned} dr_{net} &= dr_{sp} - dg_{st} = Af_u(1-f_l) - Af_l(1-f_u)\overline{N} \\ &\approx Af_u(1-f_l) - Af_l(1-f_u)\overline{N}_0 \, , \end{aligned} \tag{16.4.2}$$

where we have demonstrated in Problem 16.3 that the same coefficient A should be used for the spontaneous and stimulated terms.

If the electron and hole populations have quasi-Maxwellian speed distributions, even when excess free carriers are present, then

$$f_u = \frac{1}{1+\exp[(E_u - F_n)/kT]} \approx \exp[(F_n - E_u)/kT] \ll 1 \quad (16.4.3)$$

for any conduction band states, and

$$1-f_l = \frac{1}{1+\exp[(F_p - E_l)/kT]} \approx \exp[(E_l - F_p)/kT] \ll 1 \quad (16.4.4)$$

for any valence band states. When Equations (16.4.3) and (16.4.4) are inserted into Equation (16.4.2), the net recombination rate can be written as

$$\begin{aligned} \mathrm{d}r_{\text{net}} &= A\exp(-h\nu/kT)\{\exp[(F_n - F_p)/kT]-1\} \\ &= A\exp(-h\nu/kT)\{\exp[(F_n - E_F)/kT]\exp[(E_F - F_p)/kT]-1\} \\ &= A\exp(-h\nu/kT)\left[\left(\frac{n_0+n_e}{n_0}\right)\left(\frac{p_0+n_e}{p_0}\right)-1\right]. \end{aligned} \quad (16.4.5)$$

Note that the factor inside the square brackets depends only on the total carrier densities, not on the energy separation of the participating states. Thus this factor is preserved in common when a summation is made over all transition energies. Accordingly,

$$r_{\text{net}} = \sum \mathrm{d}r_{\text{net}} = \frac{n_e}{\tau_R} = C\left[\frac{(n_0+n_e)(p_0+n_e)}{n_0 p_0} - 1\right]. \quad (16.4.6)$$

The quantity C in the above equation is determined by the interband matrix element, the densities of states, etc. Thus for given densities of free carriers, the value of C will determine whether the radiative lifetime τ_R is large or small. We can arrange Equation (16.4.6) to write the radiative lifetime in the form

$$\tau_R = \frac{n_0 p_0}{C(n_0+p_0+n_e)}, \quad (16.4.7)$$

which is functionally controlled by the factor cited in the question.

Since we are permitted to assume that the continuity equations for free electrons and free holes are spatially homogeneous in this problem, the continuity equation for each free carrier species is

$$\frac{\mathrm{d}n_e}{\mathrm{d}t} = g_e - \frac{n_e}{\tau_R} \quad (16.4.8)$$

if non-radiative processes can be ignored. In Equation (16.4.8), g_e is the rate of pair generation caused by some external provocation, as discussed in Problem 16.2. Solutions of Equation (16.4.8) are aided by writing the lifetime in the form

$$\tau_R = \frac{\tau_{R0}}{1+n_e/(n_0+p_0)}, \quad (16.4.9)$$

where

$$\tau_{\mathrm{R0}} = \frac{n_0 p_0}{C(n_0+p_0)} \qquad (16.4.10)$$

is the **small-modulation lifetime** which is operative whenever n_e is small compared with either one of n_0 and p_0. Insertion of Equation (16.4.9) into Equation (16.4.8) gives us the **continuity equation**

$$\frac{\mathrm{d}n_e}{\mathrm{d}t} = g_e - \frac{n_e}{\tau_{\mathrm{R0}}}\left(1+\frac{n_e}{n_0+p_0}\right) \qquad (16.4.11)$$

and this can be solved simply by separation of variables following the end of any period of excess generation. [Solution during a period of time-varying generation is often possible, but can become involved!]

If g_e is set as zero for all positive time, and it is assumed that $n_e = N$ at time $t = 0$, then separation of variables gives

$$\int_0^t \frac{\mathrm{d}t}{\tau_0} = \frac{t}{\tau_0} = \int_{n_e}^{N} \frac{\mathrm{d}n_e}{n_e[1+n_e/(n_0+p_0)]} = \ln\frac{N(n_e+n_0+p_0)}{n_e(N+n_0+p_0)} \cdot \qquad (16.4.12)$$

Accordingly, n_e is given explicitly by

$$n_e = \frac{(n_0+p_0)N}{(N+n_0+p_0)\exp(t/\tau_0)-N} \cdot \qquad (16.4.13)$$

No matter how large N is made, n_e is forced to become smaller than n_0+p_0 in a time of $0{\cdot}7\tau_0$ or less, and the subsequent decay is essentially exponential. The initial decay of n_e is hyperbolic if $N \gg n_0+p_0$.

16.5 Consider the semiconductor situation of Figure 16.5.1. We say that this semiconductor is **extrinsic** in that the free electron density is derived from impurities (flaws) rather than by excitation from the conduction band, and that it is **n-type extrinsic** in that negatively charged free electrons dominate. Assume that localized flaw states other than those shown in the figure can be neglected, and that the valence band has a negligible number of free holes. Assume further that the free electron population is **non-degenerate**[3] both at equilibrium and in the presence of an excess electron density n_e (cf also Problem 3.13).

(a) Obtain a relation between the density of free electrons and the quasi-Fermi level for the conduction band, using the symbol N_c to denote the quantity $2(2\pi m_c kT/h^2)^{3/2}$ as a density of states which could be imagined at E_c as a replacement for the distribution of allowed states in the band. (This is a repetition of part of Problem 16.1.) Also obtain

[3] A degenerate electron gas has properties (such as specific heat) which are degenerated from classical predictions because n_0 is very large, and the Fermi energy lies above the bottom of the band. Thus a non-degenerate situation is one for which the electron density is small enough to keep the Fermi energy below the bottom of the band; the Fermi occupancy factor is then essentially a Boltzmann factor for any energy corresponding with band states.

an expression for the density of neutral donor impurities in terms of the Fermi energy (at equilibrium) or a quasi-Fermi level (away from equilibrium). Note from the figure that any donor can accommodate one electron at energy $E_l = -E_d$ for neutrality, and that there are β_l choices for the wavefunction of this localized electron. We ignore the possibility that any donor may be neutral but with its electron in an excited state.

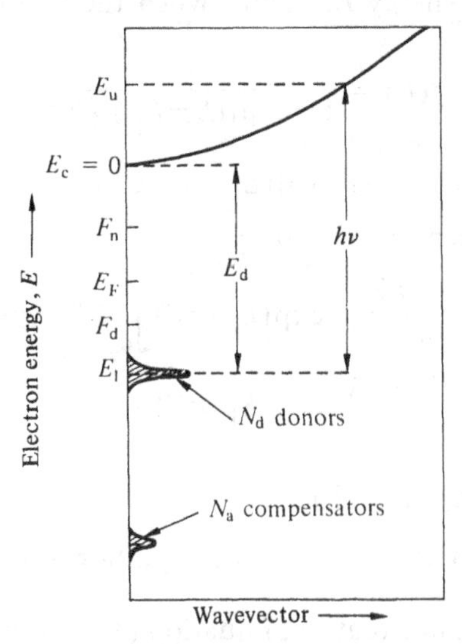

Figure 16.5.1. The model of a simple extrinsic semiconductor assumed for Problem 16.5. The solid contains N_d monovalent donor impurities per unit volume, and **compensation** equivalent to N_a monovalent compensating acceptors ($N_a < N_d$). Electrons can be excited from donor bound states into the conduction band, and we arbitrarily set zero energy to be the base of the conduction band, $E_c = 0$. The band itself is isotropic, characterized by a scalar effective mass m_c. The donor ground state energy is $E_l = -E_d$, and a donor can accommodate an electron at this energy in any one of β_l ways; excited states of donors are to be ignored in this problem. The quasi-Fermi levels F_n, F_d, for conduction and donor states respectively, converge on the Fermi level E_F for thermodynamic equilibrium.

(b) Show accordingly that the free electron density n_0 for equilibrium satisfies

$$\frac{n_0(n_0+N_a)}{N_d-N_a-n_0} = \frac{N_c}{\beta_l}\exp(-E_d/kT) = n_1$$

as a conservation law for electrons.

Solution

(a) The conduction band assumed in this problem is a simple one, characterized by a scalar effective mass m_c. This mass parameter sets the density of states per unit energy interval to be $g(E) = 4\pi(2m_c/h^2)^{3/2}E^{1/2}$ from the known density of quantum states in reciprocal space[4]. The

[4] Quantum mechanics tells us that a maximum of two electrons (of opposing spin) can be associated with a volume h^3 of momentum space in a crystal of unit volume. The quoted result for $g(E)$ follows by transforming to the energy variable $E = p^2/2m_c$.

total number of conduction band electrons for equilibrium at temperature T is

$$n_0 = \int_0^\infty f(E)g(E)\mathrm{d}E\,, \tag{16.5.1}$$

where $f(E)$ denotes the Fermi-Dirac probability of occupancy for non-interacting states of energy E. Thus, when the free electron gas is non-degenerate,

$$\begin{aligned} f(E) &= \frac{1}{1+\exp[(E-E_F)/kT]} \\ &\approx \exp[(E_F-E)/kT]\,. \end{aligned} \tag{16.5.2}$$

Equation (16.5.1) then reduces to

$$\begin{aligned} n_0 &= 4\pi\left(\frac{2m_c kT}{h^2}\right)^{3/2} \exp(E_F/kT)\int_0^\infty Y^{1/2}\exp(-Y)\mathrm{d}Y \\ &= 2\left(\frac{2\pi m_c kT}{h^2}\right)^{3/2} \exp(E_F/kT) \\ &= N_c\exp(E_F/kT)\,, \end{aligned} \tag{16.5.3}$$

since the integral on the first line of Equation (16.5.3) has a value of $(\frac{1}{4}\pi)^{1/2}$.

In a non-equilibrium situation, Equation (16.5.3) must be modified to take into account a total of $n = n_0+n_e$ free electrons, and then defines the electron quasi-Fermi energy F_n through

$$\begin{aligned} n = n_0+n_e &= N_c\exp(F_n/kT) \\ &= n_0\exp[(F_n-E_F)/kT]. \end{aligned} \tag{16.5.4}$$

However, we remember that F_n can properly be substituted back into Equation (16.5.2) to give the occupation probability for a particular energy only if the recombination time (lifetime) is very long compared with the time taken for a disturbed free electron distribution to thermalize itself.

We now consider the occupancy of the bound donor states in terms of a Fermi level or quasi-Fermi level. If N_{di} donors per unit volume are ionized, then $N_{dn} \equiv N_d - N_{di}$ are neutral; since excited states of the donors are to be ignored, then each of the neutral donors must be in one of its ground states. We note that a previously ionized donor can be made neutral by placement of an electron in *any one* of the β_1 states at the energy $E_1 = -E_d$. Thus we may say that at energy E_1 we have N_{dn} occupied states, compared with $\beta_1 N_{di}$ unoccupied and available states[5].

[5] Only one of the β_1 states of a given donor can be occupied at any one time. However, for an ionized donor all β_1 states are available for the initiation of recombinative transitions.

For thermal equilibrium at temperature T, the ratio of occupied to available states is set by the energy separation of E_1 and the electrochemical potential:

$$N_{dn} : \beta_1 N_{di} = \exp[(E_F - E_1)/kT] : 1 .$$

This ratio requires that the total density of neutral donors be

$$N_{dn} \equiv N_d - N_{di} = \frac{N_d}{1 + \frac{1}{\beta_1}\exp[(E_1 - E_F)/kT]}$$

$$= \frac{N_d}{1 + \frac{1}{\beta_1}\exp[(-E_d - E_F)/kT]} . \qquad (16.5.5)$$

For a non-equilibrium situation we can define a 'donor quasi-Fermi level' F_d as the energy which must be substituted for E_F in Equation (16.5.5) in order to reproduce correctly the density of neutral donors.

For the extrinsic semiconductor of this problem, an electron removed from a donor impurity must be either on one of the N_a compensator centers or be one of the n_0 electrons in the conduction band. Thus we have

$$\left.\begin{aligned} N_{dn} &= N_d - N_a - n_0 \\ N_{di} &= N_a + n_0 \end{aligned}\right\} \qquad (16.5.6)$$

in equilibrium. Away from equilibrium, the expressions of Equation (16.5.6) must be written with $n = n_0 + n_e$ replacing n_0.

(b) We can develop a simple expression for n_0 which does not explicitly involve E_F for our non-degenerate n-type semiconductor at equilibrium, by using Equations (16.5.3), (16.5.5), and (16.5.6). The first of these permits us to write n_0/N_c for $\exp(E_F/kT)$ in the expression of Equation (16.5.5) for the density of neutral donors. Since this density is also $N_d - N_a - n_0$, then

$$N_d - N_a - n_0 = \frac{N_d}{1 + \frac{N_c}{n_0 \beta_1}\exp(-E_d/kT)} \qquad (16.5.7)$$

which can be simply rearranged to the form

$$\frac{n_0(n_0 + N_a)}{N_d - N_a - n_0} = \frac{N_c}{\beta_1}\exp(-E_d/kT) \qquad (16.5.8)$$

as required. The symbol n_1 is used in the question to denote the quantity $(N_c/\beta_1)\exp(-E_d/kT)$. n_1 is often referred to as the **mass action constant** for the interaction between the donors and the band, and we shall find n_1 a useful symbol in Problem 16.6.

16.6 This problem continues to study free and localized electron densities associated with the extrinsic semiconductor situation of Figure 16.5.1.

(a) Show that the electron conservation equation in a non-degenerate situation (16.5.7) is a consequence of the law of mass action, from the balance of natural generation and recombination in equilibrium. Show further that the total thermal generation rate has the form

$$g = Cn_1(N_d - N_a - n)$$

whether n is the equilibrium value or not, provided that stimulated radiative recombination can be ignored, and that recombination events involving more than one conduction band state can similarly be ignored. Identify the quantity C in microscopic terms.

(b) Show that, for these conditions, the lifetime characterizing an excess electron density n_e is inversely proportional to $N_a + 2n_0 + n_1 + n_e$. Imagine a steady-state spatially homogeneous situation for which light of intensity I_0 results in a quasi-uniform excess generation rate g_e; show that n_e is proportional to I_0 for weak illumination and to $I_0^{1/2}$ for stronger illumination, and that n_e must eventually saturate no matter how intense the light.

Solution

The conservation law for electrons in a non-degenerate extrinsic semiconductor controlled by a single species of impurity has been shown to be

$$\frac{n_0(n_0 + N_a)}{N_d - N_a - n_0} = \frac{N_c}{\beta_1}\exp(-E_d/kT) \equiv n_1 \tag{16.6.1}$$

in the terminology of Problem 16.5. The above result was derived in that problem from the requirement that the same electrochemical potential should dictate the occupancies of free and bound states at equilibrium. However, Equation (16.6.1) is of the form to be expected from mass action considerations.

For we know that the rates of thermal generation and of recombination must balance in detail and *in toto* at equilibrium. Since we can assume for this problem that the free electron gas is non-degenerate (i.e. states near the bottom of the band are more likely to be empty than full), then the recombination rate should be proportional to the total free electron density[(6)] n and to the number $\beta_1(n + N_a)$ of available states on ionized donors. We can write this rate as

$$r = \bar{v}\bar{\sigma}\beta_1(n + N_a)n \tag{16.6.2}$$

where $\bar{v}$ denotes the thermal speed of a free electron (averaged with respect to a Boltzmann distribution) and $\bar{\sigma}$ is the (similarly averaged)

[(6)] If an electron could suffer capture at an impurity site by donating the recombination energy to one or more other electrons (in a manner which conserves energy and momentum) then the recombination would have terms dependent on n^2, n^3, etc. The question specifically excludes such multi-electron processes from our consideration.

capture cross-section for a free electron presented by an unoccupied donor state. Expression of r in the form of Equation (16.6.2) would not be possible in such a simple way if stimulated radiative recombination were important, but this complication is explicitly excluded.

The opposing rate of generation is determined by N_{dn}, the number of electrons capable of excitation, and the thermal environment for such excitation. However, it will not depend on the occupancy of states within the band, provided that this occupancy is small for all band states. Thus

$$g = A(N_d - N_a - n) \tag{16.6.3}$$

where the quantity A should depend on the temperature, on the minimum excitation energy E_d, and on the density of conduction states, but *not* on N_d, N_a, or n.

Since $g = r$ at equilibrium, Equations (16.6.2) and (16.6.3) can be equated for this condition to yield

$$\frac{n_0(n_0 + N_a)}{N_d - N_a - n_0} = \frac{A}{\bar{v}\bar{\sigma}\beta_1}, \tag{16.6.4}$$

which is of the form of Equation (16.6.1). Evidently we should identify $A/\bar{v}\bar{\sigma}\beta_1$ with the mass action constant n_1, so that the rate of generation is

$$g = \bar{v}\bar{\sigma}\beta_1 n_1(N_d - N_a - n) \tag{16.6.5}$$

whether or not the conditions are of equilibrium.

It may be noted that Equation (16.6.1) would be recreated from mass action considerations even if generation and recombination processes involving two or more electrons were considered; but then Equations (16.6.2) and (16.6.3) would involve higher powers of n.

(b) The difference between the rates of natural recombination and generation can be used in the definition of an excess electron lifetime τ in the continuity equation

$$\frac{dn_e}{dt} = g_e + (g - r) = g_e - \frac{n_e}{\tau} \tag{16.6.6}$$

for excess electrons in a situation of spatial uniformity. Here g_e is once again an excess generation rate caused by an external influence. Provided that r and g are given by Equations (16.6.2) and (16.6.5), then

$$\tau = \frac{n_e}{r - g} = \frac{n_e}{\bar{v}\bar{\sigma}\beta_1[(n_0 + n_e)(n_0 + n_e + N_a) - n_1(N_d - N_a - n_0 - n_e)]}.$$

There is a major cancellation among the terms in the denominator of the right side, since $n_0(n_0 + N_a)$ is equal to $n_1(N_d - N_a - n_0)$. Thus

$$\tau = \frac{1}{\bar{v}\bar{\sigma}\beta_1(N_a + 2n_0 + n_1 + n_e)} \tag{16.6.7}$$

in the continuity equation (16.6.6).

This continuity equation requires in steady state that

$$n_e(N_a + 2n_0 + n_1 + n_e) = \frac{g_e}{\bar{v}\bar{\sigma}\beta_1}$$

as a response to spatially uniform generation.

Now suppose that excess electrons are created at the rate g_e (per unit volume and time) as a result of an incident stream of I_0 photons (per unit area and time), each with an energy of at least E_d. Excitation can take place only from neutral donors, thus we may expect a quasi-uniform[7] excess generation rate

$$g_e = \sigma_{ph} N_{dn} I_0 ,$$

where σ_{ph} is the photo-ionization cross section of a neutral donor. If not all incident photons have the same energy, then σ_{ph} is a suitable average with respect to the excitation probabilities to the various upper states $E_u = h\nu - E_d$. Since $N_{dn} = N_d - N_a - n$, then

$$\frac{n_e(N_a + 2n_0 + n_1 + n_e)}{N_d - N_a - n_0 - n_e} = I_0 \frac{\sigma_{ph}}{v\sigma\beta_1} \qquad (16.6.8)$$

controls the relationship between I_0 and n_e. This relationship is shown graphically in the log–log plot of Figure 16.6.1. Part (a) of the curve shows the linear dependence of n_e on light intensity for weak illumination. The curve changes character at point (b), which corresponds with

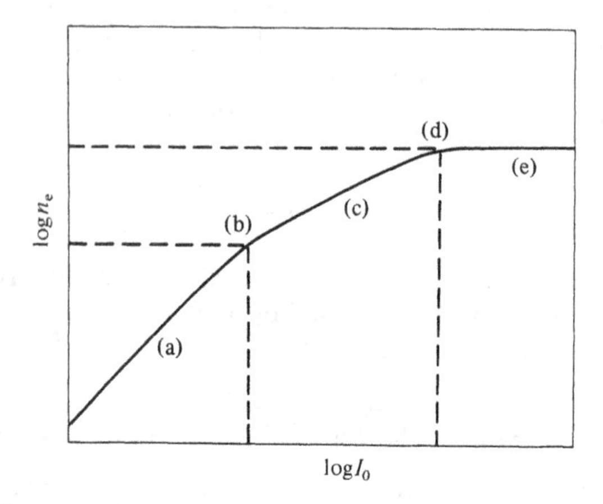

Figure 16.6.1. Variation of n_e with incident photon flux I_0 as required by Equation (16.6.8).

[7] In a more rigorous version of this problem, we should have to concern ourselves with the attenuation of the photon flux with increasing distance through the crystal, to provide a depth-dependent g_e.

the condition

$$\left.\begin{aligned} n_e &\approx N_a + 2n_0 + n_1 \\ I_0 &\approx \frac{2\bar{v}\bar{\sigma}\beta_1(N_a+2n_0+n_1)^2}{\sigma_{ph}(N_d-2N_a-3n_0-n_1)} \end{aligned}\right\} . \quad (16.6.9)$$

For a semiconductor crystal with relatively weak compensation, in which $N_a \ll N_d$, and for a temperature low enough to make the mass action constant n_1 small compared with N_d, it is possible for n_e to be substantially larger than $N_a + 2n_0 + n_1$ yet substantially smaller than N_d. Such a combination of requirements is necessary in order that the region (c) in the curve of Figure 16.6.1 be prominent. Within that range (when it exists), the left side of Equation (16.6.8) is approximately n_e^2/N_d; thus under these conditions n_e varies essentially as $I_0^{1/2}$. The upper limit of the range occurs at the point (d) on the curve, when

$$\left.\begin{aligned} I_0 &\approx N_d v\beta_1\frac{\bar{\sigma}}{\sigma_{ph}} \\ n_e &\to N_d - N_a - n_0 \end{aligned}\right\} \quad (16.6.10)$$

and as section (e) of the curve indicates, a further increase of photon flux cannot make $n_0 + n_e$ any larger than the available electron supply of $N_d - N_a$.

The behaviour of the system when $n_0 + n_e$ becomes comparable with $N_d - N_a$ would be more complicated if stimulated recombination played a significant role.

16.7 Consider steady state conditions in a semiconductor crystal which has free carrier densities n_0 and p_0 at equilibrium. We shall assume again that the semiconducting medium is spatially homogeneous, and shall assume further that the free carrier populations are non-degenerate (so that $n_0p_0 = n_i^2$). Electron–hole recombination in this material is assumed to be dominated by the activity of N_r monovalent recombination centres. Each centre has just two states of charge: 'empty', or 'filled' with a single electron. We assume that the centre can be filled in only one way, so that the filling probability is a Fermi–Dirac occupancy factor. The energy of the localized state is such that at equilibrium the ratio of filled centres to empty centres is p_1/p_0. Along with the definition of the quantity p_1, we can define the companion quantity $n_1 = n_i^2/p_1$, so that the ratio of filled to empty centres is also n_0/n_1. It will be obvious that the free carrier densities are just n_1 and p_1 when the Fermi energy coincides with the energy of the localized state.

Suppose that an empty center has a cross-section σ_n for the capture of a free electron of speed v_n. Then we say that the centre has a **capture coefficient** $v_n\sigma_n$ for a free electron of this speed. Averaged over the thermal distribution of electron speeds, the capture coefficient of an empty centre may usefully be written $c_n \equiv \langle v_n \sigma_n \rangle$. In the same manner, we may speak of $c_p \equiv \langle v_p \sigma_p \rangle$ as the capture coefficient of a full centre for a free hole, averaged over the speeds of all free holes.

(a) When an externally provoked process of excess generation causes the free carrier densities to change to $n = n_0 + n_e$ and to $p = p_0 + p_e$, then show that the fraction of filled recombination centres is

$$\frac{n_0}{n_1 + n_0} + \frac{p_e - n_e}{N_r} .$$

(b) Demonstrate that a steady excess generation rate g_e produces values for n and p such that

$$g_e = \frac{N_r c_n c_p (np - n_i^2)}{c_n(n+n_1) + c_p(p+p_1)} .$$

(c) Provided that N_r is small enough so that n_e and p_e will never be appreciably different, show that the common lifetime for excess electrons and holes is of the form

$$\tau = \frac{\tau_0(n_0+p_0) + \tau_\infty n_e}{n_0+p_0+n_e} ,$$

where τ_0 is the lifetime for very small n_e and τ_∞ is the limiting lifetime for very large n_e.

Solution

(a) The energy of the localized state associated with a recombination centre in this problem is defined in terms of the numbers n_1 and p_1. From the manner of the definition, it is evident that $n_1 p_1 = n_0 p_0 = n_i^2$. Since the ratio of filled centres to empty centres at equilibrium is n_0/n_1, the fraction of all centres filled at equilibrium is

$$f_{r0} = \frac{n_0/n_1}{1+n_0/n_1} ,$$

or

$$f_{r0} = \frac{n_0}{n_0+n_1} .$$

This fraction will remain unchanged when equilibrium is disturbed if conditions are such that $n_e = p_e$. However, if $n_e \neq p_e$, the recombination centres must assume $p_e - n_e$ additional electronic charges (assuming that the occupancy changes of other species of localized states can be ignored). Thus the non-equilibrium situation involves a *fractional* occupancy change for recombination centers of $(p_e - n_e)/N_r$. The resulting fraction of 'full' recombination centres is

$$f_r = \frac{n_0}{n_0+n_1} + \frac{p_e - n_e}{N_r} . \quad (16.7.1)$$

(b) Assuming spatial homogeneity for this problem, then continuity equations for excess electrons and excess holes can be expressed as

$$\left.\begin{aligned} \frac{dn_e}{dt} &= g_e+(g-r) = g_e-\frac{n_e}{\tau_n} \\ \frac{dp_e}{dt} &= g_e+(g'-r') = g_e-\frac{p_e}{\tau_p} \end{aligned}\right\} \tag{16.7.2}$$

In these equations, g_e is an externally provoked generation rate, and τ_n and τ_p are the lifetimes associated with excess populations of electrons and holes. Since electron–hole recombination is dominated by the given set of recombination centres, then

r = capture rate of free electrons by empty recombination centres,
g = thermal generation rate of free electrons from occupied centres,
r' = capture rate of free holes by electron-occupied centres,
g' = thermal generation rate of free holes from 'empty' centres.

It is clear from Equation (16.7.2) that $g = r$ and $g' = r'$ at thermal equilibrium. Moreover, $g-r$ must equal $g'-r'$ for a steady-state non-equilibrium situation.

We concentrate first on the generation-recombination traffic between the centres and the conduction band. In terms of the capture coefficient c_n, we can write an electron capture rate

$$r = nN_r(1-f_r)c_n$$

into the $N_r(1-f_r)$ available centers. The opposing rate of thermal generation must be proportional to the density of occupied centres:

$$g = AN_rf_r\ .$$

The thermal activation parameter A can be expressed in terms of n_1 and c_n, since it is necessary that $g = r$ at equilibrium when f_r reduces to $f_{r0} = n_0/(n_0+n_1)$, and n reduces to n_0. Thus

$$\frac{AN_rn_0}{n_0+n_1} = \frac{n_0N_rn_1c_n}{n_0+n_1}$$

or

$$A = n_1c_n.$$

In view of this, we can summarize the electron traffic as

$$\left.\begin{aligned} g &= n_1N_rf_rc_n \\ r &= nN_r(1-f_r)c_n \\ r-g &= N_rc_n[n-f_r(n+n_1)] \end{aligned}\right\}. \tag{16.7.3}$$

in terms of an occupancy factor f_r which remains to be determined.

A similar appeal to the balance of r' and g' at equilibrium permits us to describe the rate of hole generation in terms of p_1. The generation–

recombination traffic of holes can then be summarized by

$$\left.\begin{aligned} g' &= p_1 N_r (1-f_r) c_p \\ r' &= p N_r f_r c_p \\ r'-g' &= N_r c_p [f_r(p+p_1)-p_1] \end{aligned}\right\} . \qquad (16.7.4)$$

The occupancy factor f_r for a steady-state non-equilibrium situation is then constrained by the requirement that $r-g = r'-g'$, or

$$N_r c_n [n - f_r(n+n_1)] = N_r c_p [(p+p_1) f_r - p_1]$$

which yields

$$f_r = \frac{n c_n + p_1 c_p}{c_n(n+n_1) + c_p(p+p_1)} . \qquad (16.7.5)$$

Equation (16.7.5) can then be reinserted into Equation (16.7.3) to obtain $(r-g)$ in terms of n and p:

$$r-g = N_r c_n \left[n - \frac{(n+n_1)(n c_n + p_1 c_p)}{c_n(n+n_1)+c_p(p+p_1)} \right]$$

$$= \frac{N_r c_n c_p (np - n_1 p_1)}{c_n(n+n_1)+c_p(p+p_1)} . \qquad (16.7.6)$$

Since the situation under discussion is a steady state one, it is apparent from Equation (16.7.2) that the result of Equation (16.7.6) must be identified with the excess generation rate g_e, and with n_e/τ_n and p_e/τ_p. If we choose to write $n_1 p_1$ as the square of the intrinsic carrier density, we obtain

$$g_e = \frac{n_e}{\tau_n} = \frac{p_e}{\tau_p} = \frac{N_r c_n c_p (np - n_i^2)}{c_n(n+n_1)+c_p(p+p_1)} . \qquad (16.7.7)$$

The result contained in Equation (16.7.7) was described first by Hall[8] and by Shockley and Read[9].

(c) As it stands, Equation (16.7.7) specifies τ_n or τ_p in terms of both n_e and p_e. By a further use of Equations (16.7.1) and (16.7.5) it is possible to secure a relationship between τ_n and n_e which does not involve p_e, or a relationship between τ_p and p_e which does not involve n_e, but such expressions are far from simple[10].

However, if N_r is small, then n_e and p_e are substantially the same, whether the excess density is large or small compared with the thermal densities of free carriers. The fraction f_r of occupied centers is still a function of the departure from equilibrium, but if $N_r(f_r - f_{r0})$ is small compared with n_e then electrons and holes enjoy a common steady-state

[8] R.N.Hall, *Phys. Rev.*, 87, 387 (1952).

[9] W.Shockley and W.T.Read, *Phys.Rev.*, 87, 835 (1952).

[10] See, for example, J.S.Blakemore, *Semiconductor Statistics* (Pergamon Press, Oxford), 1962, pp.277-281.

lifetime. Making $n_e = p_e$ in Equation (16.7.7), we have that

$$\tau_n = \tau_p = \frac{c_n(n_0+n_1+n_e)+c_p(p_0+p_1+n_e)}{N_r c_n c_p(n_0+p_0+n_e)} . \tag{16.7.8}$$

For a very small departure from equilibrium, the low-level lifetime is

$$\tau_0 = \frac{c_n(n_0+n_1)+c_p(p_0+p_1)}{N_r c_n c_p(n_0+p_0)} , \tag{16.7.9}$$

while for a departure very large compared with the thermal free carrier densities the lifetime asymptotically approaches

$$\tau_\infty = \frac{1}{N_r c_p}+\frac{1}{N_r c_n} . \tag{16.7.10}$$

Then, for any magnitude of modulation, the quantities defined as τ_0 and τ_∞ can be used in a description of the carrier pair lifetime as

$$\tau = \frac{\tau_0(n_0+p_0)+\tau_\infty n_e}{n_0+p_0+n_e} \tag{16.7.11}$$

as suggested in the question.

GENERAL REFERENCES

S. M. Ryvkin, *Photoelectric Effects in Semiconductors* (Consultants Bureau, New York), 1964.

J. S. Blakemore, *Semiconductor Statistics* (Pergamon Press, Oxford), 1962.

P. T. Landsberg, "Problems in recombination statistics", in *Festkörperprobleme,* Vol.6 (Ed. O. Madelung) (Academic Press, New York), 1967.

A. Rose, *Concepts in Photoconductivity and Allied Problems* (Wiley-Interscience, New York), 1963.

A. Many and R. Bray, "Lifetime of excess carriers in semiconductors", in *Progress in Semiconductors*, Vol.3 (Ed. A. Gibson) (John Wiley, New York), 1958.

17
Transport in gases

D.J.GRIFFITHS
(University of Exeter, Exeter)

17.1 (a) A foreign molecule is initially at the point $r = 0$ in an isotropic, stationary gas. After making N collisions with other molecules it is at the point $\mathbf{r}_N = (x_N, y_N, z_N)$. If $\Delta x_N \equiv (x_N - x_{N-1})$, assume for the averages over many possible paths that

(i) $\langle \Delta x_N^2 \rangle = \langle \Delta y_N^2 \rangle = \langle \Delta z_N^2 \rangle$ (i.e. the gas is isotropic),

(ii) $\langle \Delta x_N^2 \rangle$ ($\equiv \frac{1}{3}\lambda^2$, say) is independent of N (i.e. the gas is homogeneous and stationary),

(iii) $\langle \Delta x_i \Delta x_j \rangle = 0$ for $i \neq j$ (i.e. successive free paths are uncorrelated).

Show that the mean square displacement $\langle x_N^2 \rangle$ in the x direction is $\frac{1}{3}N\lambda^2$. Hence show that

$$\langle x^2(t) \rangle = \frac{\lambda^2 t}{3\tau}$$

if $1/\tau$ is the mean collision frequency. Express in words the meaning of λ.

(b) Let $f(x, t)$ be the probability distribution of the foreign molecule at time t [i.e. $f(x, t)\,dx$ is the probability that the molecule lies between the planes x and $x + dx$ at time t, irrespective of its y- and z-coordinates].

Calculate the mean and the variance of $f(x, t)$.

(c) According to the **central limit theorem** of statistics, $f(x, t)$ is, asymptotically for large t, a normal (Gaussian) distribution with zero mean, as introduced in Problem 2.9. Show that in this case it satisfies the equation

$$\frac{\partial f}{\partial t} = D\frac{\partial^2 f}{\partial x^2}$$

where $D = \lambda^2/6\tau$.

[The foreign molecule is said to perform a **random walk**[1].]

Solution

(a) Since

$$x_N = x_{N-1} + \Delta x_N\ ,$$

then

$$\langle x_N^2 \rangle = \langle x_{N-1}^2 \rangle + 2\langle x_{N-1}\Delta x_N \rangle + \langle \Delta x_N^2 \rangle\ .$$

[1] F.Reif, *Fundamentals of Statistical and Thermal Physics,* Ch.12 (McGraw-Hill, New York), 1965.

Now

$$x_{N-1} = \sum_{i=1}^{N-1} \Delta x_i.$$

Therefore

$$\langle x_{N-1}\Delta x_N\rangle = \sum_{i=1}^{N-1} \langle \Delta x_i \Delta x_N\rangle = 0$$

since successive displacements are uncorrelated. Therefore

$$\langle x_N^2\rangle = \langle x_{N-1}^2\rangle + \tfrac{1}{3}\lambda^2 .$$

By induction, since $x_0 = 0$,

$$\langle x_N^2\rangle = \tfrac{1}{3}N\lambda^2.$$

In time t, the foreign molecule makes $N = t/\tau$ collisions on average. Therefore

$$\langle x^2(t)\rangle = \frac{\lambda^2 t}{3\tau} ;$$

$$\lambda^2 = \langle \Delta x_N^2\rangle + \langle \Delta y_N^2\rangle + \langle \Delta z_N^2\rangle = \langle \Delta x_N^2 + \Delta y_N^2 + \Delta z_N^2\rangle .$$

Thus λ^2 is the mean square displacement between collisions, i.e. λ is the root mean square free path.

(b) Since the gas is isotropic and stationary, the mean displacement is zero, by symmetry. The variance is $\langle x^2(t)\rangle = \lambda^2 t/3\tau$, by definition.

(c) The normal distribution for a variable x with mean zero and variance σ^2 is

$$f(x) = (2\pi\sigma^2)^{-1/2} \exp\left(-\frac{x^2}{2\sigma^2}\right) \qquad \text{(cf Problem 2.9)} .$$

Taking logarithms, differentiating, and using the result [from part (b)] that $\sigma^2 = \lambda^2 t/3\tau$, the desired result follows at once, with

$$D = \frac{\sigma^2}{2t} = \frac{\lambda^2}{6\tau} .$$

17.2 (a) A uniform, isotropic, stationary gas contains a small proportion of foreign molecules, whose number density $n(x, t)$ is independent of y and z. Assuming that the foreign molecules perform independent random walks of the type discussed in the preceding problem, construct an integral equation for $n(x, t)$ (for large t) in terms of $n(x, 0)$.

[Hint: $f(x, 0) = \delta(x)$.]

(b) Using the result of Problem 17.1(c) transform this integral equation to a partial differential equation for $\partial n/\partial t$.

(c) Hence show that the mean flux of foreign molecules in the x direction is proportional to the concentration gradient $\partial n/\partial x$, and identify the diffusion coefficient. [This is **Fick's law of diffusion.**]

(d) What is the equilibrium distribution of foreign molecules?

Solution

(a) For $t = 0$ the distribution function $f(x, t)$ becomes a δ function, describing a molecule located with certainty at $x = 0$, i.e.

$$f(x, 0) = \delta(x) ,$$

Now

$$n(x, 0) = \int_{-\infty}^{\infty} n(\xi, 0)\, \delta(x - \xi)\, d\xi$$

by a standard property of the δ-function. This expresses $n(x, 0)$ as a weighted sum of δ-functions, each representing one molecule on the plane $\xi = x$. Thus

$$n(x, 0) = \int_{-\infty}^{\infty} n(\xi, 0)\, f(x - \xi, 0)\, d\xi .$$

Since the proportion of foreign molecules is small, each behaves independently, and each $f(x - \xi, 0)$ evolves independently in time as discussed in Problem 17.1, i.e.

$$n(x, t) = \int_{-\infty}^{\infty} n(\xi, 0)\, f(x - \xi, t)\, d\xi ,$$

where for large $t \gg \tau$, $f(x, t)$ is the normal distribution with zero mean and variance $\lambda^2 t/3\tau$. This is the required integral equation.

(b) From part (a), we have

$$\frac{\partial n(x, t)}{\partial t} = \int_{-\infty}^{\infty} n(\xi, 0) \frac{\partial f(x - \xi, t)}{\partial t}\, d\xi ,$$

and from Problem 17.1(c)

$$\frac{\partial f(x - \xi, t)}{\partial t} = D \frac{\partial^2 f(x - \xi, t)}{\partial (x - \xi)^2} = D \frac{\partial^2 f(x - \xi, t)}{\partial x^2} .$$

Therefore

$$\frac{\partial n(x, t)}{\partial t} = D \frac{\partial^2}{\partial x^2} \int_{-\infty}^{\infty} n(\xi, 0)\, f(x - \xi, t)\, d\xi = D \frac{\partial^2 n(x, t)}{\partial x^2} .$$

This is the required differential equation.

(c) Let the flux of foreign molecules in the x direction be Φ_x per unit area. By symmetry arguments $\Phi_y = \Phi_z = 0$.

$$\frac{\partial n(x, t)}{\partial t} = -\frac{\partial \Phi_x}{\partial x} ,$$

since the number of molecules is conserved. Hence, from part (b),

$$-\frac{\partial \Phi_x}{\partial x} = D \frac{\partial^2 n}{\partial x^2} .$$

Integrating, and remembering that, by symmetry, Φ_x vanishes when $\partial n/\partial x$ is zero,

$$\Phi_x = -D\frac{\partial n}{\partial x} .$$

The diffusion coefficient is D, where, from Problem 17.1,

$$D = \frac{\lambda^2}{6\tau} .$$

(d) In equilibrium, $\Phi_x = 0$ and so

$$\frac{\partial n}{\partial x} = 0 .$$

Similarly

$$\frac{\partial n}{\partial y} = \frac{\partial n}{\partial z} = 0 .$$

Therefore the equilibrium density distribution is uniform.

17.3 (a) Assuming that all molecules behave as rigid elastic spheres, show that the mean free path of a foreign molecule of radius r_1 in a gas of molecules of number density n_2 and radius r_2 is approximately

$$\frac{1}{\pi(r_1+r_2)^2 n_2} .$$

(b) Hence show that the coefficient of mutual diffusion D for a small proportion of foreign molecules of mass m_1 and radius r_1 in a gas of molecules of number density n_2 and radius r_2 at temperature T can be written as

$$D = \frac{\alpha}{6\pi n_2(r_1+r_2)^2}\left(\frac{3kT}{m_1}\right)^{1/2} ,$$

where α is a dimensionless constant of order unity and k is the Boltzmann constant. [Use the result of Problem 17.2(c) and the equipartition theorem of Problem 3.6(c).]

(c) Carbon monoxide (CO), ethylene (C_2H_4), and nitrogen (N_2) each have molecular weight 28. The coefficient of mutual diffusion for a small proportion of CO in C_2H_4 gas at NTP is

$$D(\mathrm{CO};\ \mathrm{C_2H_4}) = 0{\cdot}129\ \mathrm{cm^2\ s^{-1}} .$$

Similarly, under the same conditions

$$D(\mathrm{CO};\ \mathrm{N_2}) = 0{\cdot}176\ \mathrm{cm^2\ s^{-1}}$$

and

$$D(\mathrm{C_2H_4};\ \mathrm{N_2}) = 0{\cdot}129\ \mathrm{cm^2\ s^{-1}} .$$

Show that the coefficient of self-diffusion for nitrogen gas at NTP is

$$D(\mathrm{N_2};\ \mathrm{N_2}) = 0{\cdot}176\ \mathrm{cm^2\ s^{-1}} .$$

(d) Can one speak of a coefficient of self-diffusion for a gas of molecules which are quantum-mechanically indistinguishable?

Solution

(a) The foreign molecule collides with all molecules whose centres would otherwise come within (r_1+r_2) of its centre, that is, with all molecules whose centres lie within a circular cylinder of which the cross-section is $\pi(r_1+r_2)^2$, and of which the axis coincides with the path of the centre of the foreign molecule. The mean free path is, to a good approximation, equal to the mean separation between molecules whose centres lie within this cylinder, which is

$$\frac{1}{\pi(r_1+r_2)^2 n_2} .$$

Because of differences in averaging, the root mean square free path λ may differ from this value by a factor of order unity.

(b) From Problem 17.2(c),

$$D = \frac{\lambda^2}{6\tau} = \frac{\lambda}{6} \cdot \frac{\lambda}{\tau} .$$

But λ/τ is of the order of the mean molecular speed, i.e.

$$\frac{\lambda}{\tau} = \alpha(\overline{v^2})^{1/2} ,$$

where α is a dimensionless constant of order unity. From the equipartition theorem [cf Problem 3.6(c)],

$$\tfrac{1}{2}m_1\overline{v^2} = \tfrac{3}{2}kT .$$

Therefore

$$\frac{\lambda}{\tau} = \alpha\left(\frac{3kT}{m_1}\right)^{1/2} .$$

From this result and that of part (a) the expression for D follows.

(c) All three gases have the same molecular mass and temperature. The major constituent has the same pressure in each case and therefore the same number density—to the approximation that the ideal gas equation is obeyed. From part (c) the diffusion coefficients differ only through the effective molecular radii. Since

$$D(\mathrm{CO};\ \mathrm{C_2H_4}) = D(\mathrm{C_2H_4};\ \mathrm{N_2}) ,$$

we have

$$r(\mathrm{CO})+r(\mathrm{C_2H_4}) = r(\mathrm{C_2H_4})+r(\mathrm{N_2}) ,$$

i.e.

$$r(\mathrm{CO}) = r(\mathrm{N_2}) .$$

In the corresponding expression to that in part (c) for the self-diffusion coefficient of nitrogen $D(\mathrm{N_2};\ \mathrm{N_2})$,

$$r_1+r_2 = 2r(\mathrm{N_2}) = r(\mathrm{N_2})+r(\mathrm{CO}) ,$$

whence

$$D(N_2, N_2) = D(CO; N_2) = 0{\cdot}176 \text{ cm}^2 \text{ s}^{-1} .$$

(d) Experimentally, all diffusion is mutual diffusion. Self-diffusion is a convenient term for the mutual diffusion of two distinguishable groups of molecules, where there is no significant difference between the groups in the properties which determine the diffusion coefficient (i.e. the molecular mass and the interaction with other molecules of either group). Thus at an experimental level the question does not arise. At a deeper level, diffusion is an irreversible process associated with an increase in entropy. The mixing (self-diffusion) of identical molecules does not lead to an increase in entropy. Hence one cannot, strictly, speak of a coefficient of self-diffusion.

[This is the origin of the Gibbs paradox[(2)].]

17.4 (a) To what extent can thermal conduction be treated as a random-walk process?

(b) A pure gas has coefficient of thermal conductivity K, mass density ρ, and specific heat at constant volume c_v per unit mass. By exploiting the analogy between Fick's law for diffusion [Problem 17.2(c)] and the heat conduction equation, extend the arguments of Problems 17.1 and 17.2 to show that

$$\frac{K}{\rho c_v} = \frac{\lambda_u^2}{6\tau} ,$$

where λ_u is the effective root mean square free path for the transport of thermal energy.

(c) To what extent can the phenomenon of viscosity be treated as a random-walk process?

(d) If a pure gas has coefficient of viscosity η, show by an argument similar to that in part (b), that

$$\frac{\eta}{\rho} = \frac{\lambda_p^2}{6\tau} ,$$

where λ_p is the effective root mean square free path for momentum transport.

(e) Hence calculate the ratio K/η for a gas and compare it critically with the data shown in Table 17.4.1, for three gases at NTP. C_v is the molar specific heat at constant volume and R is the gas constant.

Table 17.4.1.

Gas	Molecular Weight	K (J cm^{-1} s^{-1} degK^{-1})	η (g cm^{-1} s^{-1})	C_v/R
Ne	20·2	$4{\cdot}54 \times 10^{-4}$	$2{\cdot}98 \times 10^{-4}$	1·5
N_2	28·0	$2{\cdot}43 \times 10^{-4}$	$1{\cdot}67 \times 10^{-4}$	2·5
Kr	82·9	$0{\cdot}89 \times 10^{-4}$	$2{\cdot}33 \times 10^{-4}$	1·5

(2) See, for example, K.Huang, *Statistical Mechanics* (Wiley, New York), 1963, p.153.

Solution

(a) Thermal conduction is the result of the transfer of thermal energy from molecule to molecule in collisions. Like the number of foreign molecules in the case of mutual diffusion, thermal energy is conserved in a collision. Unlike foreign molecules, it is necessarily present in a gas in equilibrium at temperature T. Thus it is any local excess of thermal energy which spreads by a random-walk process. Further, a given excess of thermal energy can (classically) be indefinitely subdivided amongst the molecules. Hence the units which perform the random walk must be taken to be of infinitesimal size. These differences from the diffusion case should leave unaltered the asymptotic behaviour of the probability distributions, and therefore of the differential equations governing the transport, provided one substitutes 'excess thermal energy' for 'number of foreign molecules', and λ_u for λ, where λ_u is the effective root mean square free path for the transfer of thermal energy. One expects that λ_u/λ is of order unity, the exact value depending on the efficiency of thermal energy transfer in collisions.

(b) The analogue of Fick's law of diffusion is

$$Q_x = -D_u \frac{\partial u}{\partial x} ,$$

where Q_x is the excess thermal energy flux per unit area in the x direction, and u is the excess thermal energy density. (It is assumed for simplicity that $\partial u/\partial y = \partial u/\partial z = 0$.) The diffusion coefficient for thermal energy, D_u, is given by

$$D_u = \frac{\lambda_u^2}{6\tau} .$$

Now

$$\mathrm{d}u = \rho c_v \,\mathrm{d}T ,$$

whence

$$Q_x = -D_u \rho c_v \frac{\partial T}{\partial x} .$$

But

$$Q_x = -K \frac{\partial T}{\partial x}$$

by definition of K. Therefore

$$\frac{K}{\rho c_v} = D_u = \frac{\lambda_u^2}{6\tau} .$$

(c) Like thermal energy, a given momentum component is conserved in a collision. Again, classically, it is indefinitely subdivisible amongst the molecules. The transport of a given momentum component normal to its own direction only can be treated as a straightforward random-walk process. Parallel transport can occur only when $\mathrm{div}\,\mathbf{v} \neq 0$: thus density variations appear and the behaviour is dominated by the propagation of sound waves.

(d) Consider the transport of y momentum in the x direction, where x and y axes are normal Cartesian axes. The analogue of Fick's law is

$$P_{yx} = -D_p \frac{\partial P_y}{\partial x} ,$$

where P_{yx} is the flux of y momentum in the x direction per unit area normal to the x axis, and p_y is the density of y momentum. The diffusion coefficient for transverse momentum, D_p, is given by

$$D_p = \frac{\lambda_p^2}{6\tau} ,$$

where λ_p is the effective root mean square free path for the process. λ_p/λ should be of order unity.

Since

$$P_y = \rho v_y ,$$

where v_y is the mean molecular velocity in the y direction, we have

$$P_{yx} = -D_p \rho \frac{\partial v_y}{\partial x} .$$

But

$$P_{yx} = -\eta \frac{\partial v_y}{\partial x}$$

by definition of η, whence

$$\frac{\eta}{\rho} = D_p = \frac{\lambda_p^2}{6\tau} .$$

(e) From parts (a) and (c)

$$\frac{K}{\eta} = \frac{c_v \lambda_u^2}{\lambda_p} = \frac{C_v \lambda_u^2}{M \lambda_p^2} ,$$

where M is the molecular weight of the gas. λ_u^2/λ_p^2 should be a number of order unity, the same for all gases with similar collision transfer processes for energy and momentum. From the figures given,

$$\frac{KM}{\eta C_v} = 2{\cdot}47 \text{ for Ne} ,$$
$$= 1{\cdot}96 \text{ for } N_2 ,$$
$$= 2{\cdot}54 \text{ for Kr} ,$$

i.e. for both the monatomic gases λ_u^2/λ_p^2 is approximately 2·5. For the diatomic gas the lower value reflects the difference between the transfer of the thermal energy associated with the internal (rotational) degrees of freedom, and that associated with the translational degrees of freedom.

17.5 (a) Use the results of Problems 17.3 and 17.4(b) to show that the coefficient of viscosity, at temperature T, of a gas of atoms which behave as rigid elastic spheres of radius r and mass m is approximately

$$\frac{(3mkT)^{1/2}}{24\pi r^2} .$$

(b) How should the coefficient of viscosity depend on the pressure of the gas?

(c) The variation of the coefficient of viscosity of ^{4}He gas with temperature is given in Table 17.5.1.

Explain qualitatively this temperature variation.

(d) Assuming that the force between two helium atoms is predominantly repulsive, and varies inversely as the nth power of their separation, use a dimensional argument to determine n from the data.

Table 17.5.1. Coefficient of viscosity of ^{4}He gas as a function of temperature.

T (°K)	15·0	75·5	171	291	457	665	1090
η (g cm^{-1} s^{-1}) x 10^6	29·5	81·8	139	197	268	339	470

Solution

(a) From Problem 17.4(b),

$$\frac{\eta}{\rho} = \frac{\lambda_p^2}{6\tau}$$

$$\approx \frac{\lambda^2}{6\tau} = D ,$$

where D is the coefficient of self-diffusion of the gas.

From Problem 17.3(c) and (e),

$$D = \frac{\alpha}{6\pi n(2r)^2}\left(\frac{3kT}{m}\right)^{1/2}$$

Hence

$$\eta \approx \frac{\rho}{24\pi n r^2}\left(\frac{3kT}{m}\right)^{1/2} = \frac{(3mkT)^{1/2}}{24\pi r^2} .$$

(b) η is independent of density, and hence of pressure, at a given temperature.

(c) For a gas of rigid spheres $\eta \propto T^{1/2}$. From a plot of $\lg\eta$ versus $\lg T$ for ^{4}He gas, we have

$$\lg\eta \approx s \lg T + \text{constant} ,$$

where

$$s = 0 \cdot 66 \pm 0 \cdot 02 > 0 \cdot 5 .$$

^{4}He atoms interact through a repulsive force which varies rapidly with their separation. As the temperature is raised the atoms have greater mean kinetic energy and therefore approach more closely at collisions.

Hence the effective rigid-sphere radius r decreases, and η varies more rapidly with temperature than $T^{1/2}$, as observed. (The effect of the longer-range attractive interaction between ^{4}He atoms is negligible at temperatures much above the critical temperature of about 5°K.)

(d) The mean atomic kinetic energy at temperature T is approximately kT. The force between two atoms with separation x is $-C/x^n$, where C is constant. Therefore their potential energy at separation x is given by $C/(n-1)x^{n-1}$ $(n > 1)$. The effective rigid-sphere radius r is determined by these two energy terms. The only dimensionless combination of r, C, and kT is a function of $C/(kTr^{n-1})$. Hence

$$r \propto \left(\frac{1}{T}\right)^{1/(n-1)},$$

and

$$\eta \propto \frac{T^{1/2}}{r^2} \propto T^{1/2 + 2/(n-1)}.$$

From the data,

$$\frac{1}{2} + \frac{2}{n-1} = 0 \cdot 66 \pm 0 \cdot 02$$

whence

$$n = 13 \cdot 7 \pm 1 \cdot 6.$$

[The value $n = 13$ is used for the repulsive part of the familiar Lennard-Jones potential. The dimensional argument was first given by Lord Rayleigh (3).]

17.6 (a) Deuterium and helium, both of molecular weight 4·0, have coefficients of viscosity at 0°C of $1 \cdot 2 \times 10^{-4}$ and $1 \cdot 9 \times 10^{-4}$ g cm^{-1} s^{-1} respectively. Estimate the ratio of their second virial coefficients B_2 at this temperature, assuming that their molecular interactions may be treated as those of rigid spheres.

[Note from Problem 9.9 that B_2 is proportional to the molecular volume.]

(b) An ideal gas of given molecular weight at a given temperature and pressure may be considered as the limit of a series of hypothetical imperfect gases with the same molecular weight, temperature and pressure. What is the theoretical value of the coefficient of viscosity of the ideal gas, considered as this limit?

(c) The coefficient of viscosity of a gas may be determined from its rate of flow through a capillary tube of diameter d under a pressure gradient. Outline qualitatively the result of a series of such determinations on the hypothetical series of gases.

(d) In what range of (mean) pressure would ^{4}He behave like an ideal gas in this experiment, if $d = 10^{-4}$ cm?

(3) Lord Rayleigh, *Proc.Roy.Soc.*, **66**, 68 (1900).

Solution

(a) From Problem 17.5(a)

$$\left[\frac{r(\text{He})}{r(\text{D}_2)}\right]^2 = \frac{1\cdot 2 \times 10^{-4}}{1\cdot 9 \times 10^{-4}} \cdot$$

The second virial coefficient B_2 is proportional to the molecular volume for a rigid-sphere interaction (cf Problem 9.9). Hence the estimated ratio is

$$\frac{B_2(\text{He})}{B_2(\text{D}_2)} = \left(\frac{1\cdot 2}{1\cdot 9}\right)^{3/2} \approx 0\cdot 50\,.$$

(b) In the ideal-gas limit, B_2 (and higher coefficients) vanish. Therefore $r \to 0$. Hence the theoretical coefficient of viscosity is given by [cf Problem 17.5(a)]

$$\eta = \lim_{r\to 0} \frac{(3mkT)^{1/2}}{24\pi r^2}\,.$$

i.e. $\eta \to \infty$ as $r \to 0$.

(c) $\eta \to \infty$ because $\lambda \to \infty$ as $r \to 0$ [cf Problem 17.3(a)]. In a tube of diameter d, the coefficient of viscosity is a valid concept only when $\lambda \ll d$. In this case, $\eta \propto 1/r^2$. For $\lambda \gg d$ the flow rate is determined by the collisions of molecules with the tube walls, and is independent of r. Thus the *apparent* viscosity (as judged from the flow rate) would reach a finite limiting value as $r \to 0$.

(d) ^{4}He would behave like an ideal gas in the pressure range for which intermolecular collisions were unimportant, i.e. for which $\lambda \gg d$. From Problem 17.3(a),

$$\lambda \approx \frac{1}{4\pi r^2 n}\,.$$

From the ideal gas equation of state

$$p = \frac{\rho RT}{M} = nkT$$

where the symbols have been defined in previous problems. Hence the required pressure range is

$$p \ll \frac{kT}{4\pi r^2 d} \approx \frac{6\eta}{d}\left(\frac{kT}{3m}\right)^{1/2},$$

since

$$\eta \approx \frac{(3mkT)^{1/2}}{24\pi r^2}\,.$$

With the given figures for ^{4}He this gives the pressure range

$$p \ll 5 \times 10^5 \text{ dyn cm}^{-2} \approx \tfrac{1}{2}\text{atm}\,.$$

17.7 (a) Two atoms of equal mass, with initial velocities $\mathbf{v}_1$ and $\mathbf{v}'_1$, respectively, collide elastically. What velocities are possible after the collision?

(b) If the angle of scattering in the centre-of-mass reference frame is θ, what is the change $\underset{\sim}{\Delta}$ in the velocity of one of the atoms?

(c) A spatially uniform monatomic gas of number density n has an isotropic velocity distribution $\mathrm{f}(\mathbf{v})$ with the following properties:

$$\int \mathrm{f}(\mathbf{v})\,d\mathbf{v} = n\,,$$

$$\frac{1}{n}\int \mathbf{v}\mathrm{f}(\mathbf{v})\,d\mathbf{v} = \overline{\mathbf{v}} = 0\,,$$

$$\frac{1}{n}\int v^2\mathrm{f}(\mathbf{v})\,d\mathbf{v} = \overline{v^2} = V^2\,.$$

In a collision between two atoms, the distribution of θ is independent of their relative velocity, and the mean value of θ is ϕ.

Calculate the average value $\langle\Delta\rangle$ of $|\Delta|$ for the collisions of an atom with initial velocity $\mathbf{u} = 0$, to first order in ϕ. [$\langle\rangle$ denotes an average over collisions.]

(d) Hence show that a group of atoms initially having zero velocity will begin to 'diffuse' outwards in velocity space, with a 'diffusion coefficient' D_v equal to $A^2V^2\phi^2/\tau$, where $1/\tau$ is the mean collision frequency for an atom with zero velocity, and A is a dimensionless constant. [Use the results of Problem 17.1.]

Solution

(a) In the centre-of-mass reference frame the atoms have initial velocities $\pm\mathbf{w}_1 = \pm\frac{1}{2}(\mathbf{v}'_1 - \mathbf{v}_1)$. From conservation of total momentum, the final velocities also are equal and opposite. From conservation of total kinetic energy the magnitudes of the initial and final velocities are equal. Hence the initial velocities are equal and opposite, and so are the final velocities, and all four lie on a sphere in velocity space of radius w_1, centred at the centre-of-mass velocity.

(b)
$$\underset{\sim}{\Delta} = \mathbf{w}_2 - \mathbf{w}_1\,,$$

where $2\mathbf{w}_2$ is the relative velocity after the collision, i.e.

$$\mathbf{w}_2 = \mathbf{w}_1 = \tfrac{1}{2}|\mathbf{v}'_1 - \mathbf{v}_1|$$

and

$$\mathbf{w}_1 \cdot \mathbf{w}_2 = \mathbf{w}_1\mathbf{w}_2\cos\theta\,.$$

Resolving $\underset{\sim}{\Delta}$ parallel and perpendicular to $\mathbf{w}_1$ we obtain

$$\Delta_\parallel = -w_1(1-\cos\theta)\,,$$

$$\Delta_\perp = w_1(\sin\theta)\,.$$

(c) To first order in θ,

$$\Delta_\perp = w_1\theta\,, \quad \Delta_\parallel = 0.$$

Therefore

$$|\Delta| = w_1\theta\,,$$

and after averaging over θ

$$|\Delta| = w_1\phi\,.$$

The probability per unit time that an atom of velocity zero is struck by one of velocity **v** is proportional to **v** f(**v**). Thus the mean value of $|\Delta|$ is

$$\langle\Delta\rangle = \frac{\int \Delta v\,\mathrm{f}(\mathbf{v})\,\mathrm{d}\mathbf{v}}{\int v\,\mathrm{f}(\mathbf{v})\,\mathrm{d}\mathbf{v}} \qquad \text{(averaged over } \theta) \,.$$

Since $w_1 = v$

$$\langle\Delta\rangle = \frac{1}{2}\phi\frac{\int v^2\mathrm{f}(\mathbf{v})\,\mathrm{d}\mathbf{v}}{\int v\,\mathrm{f}(\mathbf{v})\,\mathrm{d}\mathbf{v}} = \frac{1}{2}\phi\frac{V^2}{\overline{v}}\,.$$

Now

$$\overline{v} = \overline{(v^2)^{1/2}} = \frac{V}{A'}\,,$$

where A' is a dimensionless constant. Hence

$$\langle\Delta\rangle = \tfrac{1}{2}A'\phi V\,.$$

(d) Because the gas is isotropic, $\underset{\sim}{\Delta}$ is random in direction. Hence an atom with zero velocity begins to perform a random walk in velocity space with steps of average length $\langle\Delta\rangle$, occurring with mean frequency $1/\tau$. In consequence its velocity departs from zero, and the step length may change as the random walk progresses. The initial diffusion coefficient is, from Problem 17.1,

$$D_v \approx \frac{\langle\Delta\rangle^2}{6\tau}\,,$$

whence

$$D_v = \frac{A^2\phi^2V^2}{\tau}\,,$$

where A is a new dimensionless constant.

[Some processes in metals where such a random-walk treatment of collisions in velocity space is particularly appropriate have been discussed recently[4].]

[4] A.B.Pippard, *Proc.Roy.Soc.*, **A305**, 291 (1968).

17.8 (a) For the situation of Problem 17.7(c), calculate the average value of $\underset{\sim}{\Delta}$ for the collisions of an atom with low velocity **u**, to second order in ϕ.

(b) Hence show that a group of atoms with low velocity **u** will acquire through collisions an average acceleration

$$\langle \dot{\mathbf{u}} \rangle = -\frac{B^2\phi^2}{\tau}\mathbf{u},$$

where B is a dimensionless constant.

(c) Assuming that the expressions obtained for D_v and $\langle \dot{\mathbf{u}} \rangle$ are correct for arbitrary **u**, show that in general there is a flux of atoms (per unit 'area' of velocity space) given by

$$\left(A^2 V^2 \frac{\partial \mathrm{f}}{\partial \mathbf{v}} + B^2 \mathbf{v}\mathrm{f}\right)$$

per unit volume of real space per unit time.

(d) Hence show that the velocity distribution function f(**v**) in the equilibrium state of the gas is the Maxwellian distribution of Problem 3.1.

(e) From your results identify f_0 and estimate t^* in the following common approximation for the effect of collisions on f(**v**):

$$\frac{\partial \mathrm{f}}{\partial t} = -\frac{\mathrm{f}-\mathrm{f}_0}{t^*}.$$

[This is the **relaxation-time approximation**[5].

(f) Estimate the order of magnitude of t^* in a gas of classical rigid spheres, for which all scattering angles are equally probable in the centre-of-mass reference frame.

Solution

(a) Consider a collision of two atoms with initial velocities **u** and **v**. Let $\mathbf{w} = \frac{1}{2}(\mathbf{u}-\mathbf{v})$. Resolve the change $\underset{\sim}{\Delta}$ in velocity of the first atom into vectors parallel and perpendicular to **w**. Since $\underset{\sim}{\Delta}_\perp$ may lie in any direction in the plane $\perp\mathbf{w}$ with equal probability,

$$\langle \underset{\sim}{\Delta}_\perp \rangle = 0;$$

from Problem 17.7(b), $\Delta_\parallel = -\mathbf{w}(1-\cos\theta) = \frac{1}{2}\mathbf{w}\theta^2$ to second order in θ. Since the gas is isotropic, $\langle \underset{\sim}{\Delta} \rangle = \langle \underset{\sim}{\Delta}_\parallel \rangle$ must be parallel to **u**: let this be the x-direction. Then

$$\langle \Delta \rangle = \langle \underset{\sim}{\Delta}_\parallel \cdot \mathbf{u}/u \rangle = -\tfrac{1}{4}\langle \theta^2 \rangle\langle u - v_x \rangle = -\tfrac{1}{4}\langle \theta^2 \rangle(u - \langle v_x \rangle).$$

[5] It is discussed further in F.Reif, *loco cit.*, Ch.13.

From here the result follows by a dimensional argument, or from the following more detailed analysis:

$$\langle \mathrm{v}_x \rangle = \frac{\int v_x |\mathbf{v}-\mathbf{u}| \mathrm{f}(\mathbf{v})\, d\mathbf{v}}{\int |\mathbf{v}-\mathbf{u}| \mathrm{f}(\mathbf{v})\, d\mathbf{v}} .$$

For small u,

$$|\mathbf{v}-\mathbf{u}| = v - u\frac{\partial v}{\partial x} + \mathrm{O}(u^2) = v - u\frac{v_x}{v} + \mathrm{O}(u^2) .$$

Hence

$$\int |\mathbf{v}-\mathbf{u}| \mathrm{f}(\mathbf{v})\, d\mathbf{v} = n[\bar{v} + \mathrm{O}(u^2)] , \qquad (17.8.1)$$

since, by symmetry, $\overline{v_x/v} = 0$. Also

$$\int v_x |\mathbf{v}-\mathbf{u}| \mathrm{f}(\mathbf{v})\, d\mathbf{v} = \int \left[v_x v - u\frac{v_x^2}{v} + \mathrm{O}(u^2) \right] \mathrm{f}(\mathbf{v})\, d\mathbf{v} = n\left[-u\overline{\left(\frac{v_x^2}{v}\right)} + \mathrm{O}(u^2 \right] .$$

Since $\mathrm{f}(\mathbf{v})$ is isotropic,

$$\overline{\left(\frac{v_x^2}{v}\right)} = \overline{\left(\frac{v_y^2}{v}\right)} = \overline{\left(\frac{v_z^2}{v}\right)} = \tfrac{1}{3}\overline{\left(\frac{v^2}{v}\right)} = \tfrac{1}{3}\bar{v} .$$

Therefore

$$\langle v_x \rangle = -\frac{nu\bar{v}}{3n\bar{v}} + \mathrm{O}(u^2) .$$

Hence

$$\langle \underset{\sim}{\Delta} \rangle = -\tfrac{1}{4}\langle \theta^2 \rangle \cdot \tfrac{4}{3}\mathbf{u} + \mathrm{O}(u^2) = -B^2\phi^2\mathbf{u} + \mathrm{O}(u^2) ,$$

where B^2 is a dimensionless constant.

(b)

$$\langle \dot{\mathbf{u}} \rangle = \langle \underset{\sim}{\Delta} \rangle / \tau(u) ,$$

where $1/\tau(u)$ is the mean collision frequency for an atom with velocity $\mathbf{u}$. $1/\tau(u)$ is proportional to the integrated molecular flux given by Equation (17.8.1). Hence

$$\frac{1}{\tau(u)} = \frac{1}{\tau} + \mathrm{O}(u^2) ,$$

where $\tau = \tau(0)$. Hence

$$\langle \dot{\mathbf{u}} \rangle = -\frac{B^2\phi^2}{\tau}\mathbf{u} + \mathrm{O}(u^2) .$$

(c) The diffusive flux of atoms through velocity space is given by the analogue of Fick's law (in three dimensions) as $-D_v \partial \mathrm{f}/\partial \mathbf{v}$ atoms per unit 'area' of velocity space per unit volume of real space per unit time, where D_v has been calculated in part (a).

The net drift calculated in part (b) causes a flux through velocity space given by

$$f(\mathbf{v})\langle\dot{\mathbf{v}}\rangle = -\frac{B^2\phi^2}{\tau}\mathbf{v}f(\mathbf{v}) ,$$

in the same units. Hence the result for the total net flux follows.

(d) In equilibrium there is zero net flux for all **v**. From the result of part (c), $\partial f/\partial \mathbf{v} \parallel \mathbf{v}$. This is consistent with the assumed isotropy of f(**v**), i.e. f(**v**) is a function of $|v|$ only. Thus

$$A^2V^2\frac{\partial f}{\partial v}+B^2vf = 0 .$$

i.e.

$$\ln f = -\frac{B^2v^2}{2A^2V^2}+\text{constant}$$

or

$$f(\mathbf{v}) = \text{constant} \times \exp\left(-\frac{B^2v^2}{2A^2V^2}\right) .$$

It is easily shown that, with this f(**v**),

$$\overline{v^2} = \frac{3A^2V^2}{B^2} .$$

But $\overline{v^2} = V^2$ by definition. Therefore $A^2/B^2 = \frac{1}{3}$. Using the theorem of the equipartition of energy to relate V^2 to the temperature T,

$$f(\mathbf{v}) = \text{constant} \times \exp\left(-\frac{mv^2}{2kT}\right) .$$

The constant can be determined, from the normalisation condition

$$\int f(\mathbf{v})\,d\mathbf{v} = n ,$$

as

$$n\left(\frac{m}{2\pi kT}\right)^{3/2} .$$

Thus the equilibrium distribution is Maxwellian.

[The result is correct despite our approximations.]

(e) $f_0(\mathbf{v})$ is the equilibrium form towards which f(**v**) relaxes through collisions, i.e. the Maxwellian distribution from part (d). (For a moving gas, modification of (d) is required.)

The relaxation time t^* is the time necessary for local departures from $f_0(\mathbf{v})$ to be communicated, by a diffusive process, to the whole velocity distribution, which has linear dimensions of order V in velocity space. From Problem 17.1(b) and (c), an initially localised irregularity is distributed after time t, over a region of mean square width of order $2D_v t$.

Hence

$$2D_v t^* \approx V^2$$

i.e.

$$t^* \approx \frac{\tau}{A^2\phi^2} \approx \frac{\tau}{\phi^2} \; .$$

(f) For rigid spheres, $\phi \approx 1$ radian. Hence the condition $\phi \ll 1$ is not satisfied. To order of magnitude, however,

$$t^* \approx \tau \; .$$

[This is the value of t^* usually assumed in the relaxation-time approximation.]

17.9 (a) Atoms of a gas with velocity distribution function $f(\mathbf{v})$ are acted upon by a uniform force $\mathbf{F}$. What is the resulting flux (i.e. that due to $\mathbf{F}$ alone) of atoms per unit 'area' of velocity space, per unit volume of real space per unit time?

(b) Using the result of Problem 17.8(d), write down the equilibrium distribution function for a gas at temperature T with number density n, moving with uniform mean velocity $\overline{\mathbf{v}}\,(\ll c)$.

(c) A gas of atoms of mass m contains a few ions of the same mass, each carrying charge e. A small uniform electric field $\mathbf{E}$ is switched on, and the system comes to a steady state where the neutral gas is at rest at temperature T and the ions are drifting through it. What is the distribution function of ionic velocities?

[Use the result of Problem 17.8(c).]

(d) Calculate the mean ionic velocity $\overline{\mathbf{u}}$.

(e) Hence calculate the ionic mobility $\mu = \overline{u}/E$.

Solution

(a) Under a force $\mathbf{F}$ each atom acquires an acceleration $\dot{\mathbf{v}} = \mathbf{F}/m$. Hence the required flux is

$$\dot{\mathbf{v}}f(\mathbf{v}) = \frac{f(\mathbf{v})\mathbf{F}}{m} \; ,$$

in the given units.

(b) The required distribution function is obtained by shifting the origin of velocity coordinates from zero to $\overline{\mathbf{v}}$, i.e.

$$f(\mathbf{v}) = \text{constant} \times \exp\left[-\frac{m}{2kT}(\mathbf{v}-\overline{\mathbf{v}})^2\right] ,$$

where the value of the constant is unchanged, since

$$\int f(\mathbf{v})\,d\mathbf{v} = \int f(\mathbf{v})\,d(\mathbf{v}-\overline{\mathbf{v}}) = n \; .$$

This transformation is correct for $v \ll c$. For large v relativistic effects are important.

(c) In an electric field the ionic velocity distribution is distorted from the equilibrium Maxwellian form given in Problem 17.8(d), because of the flux calculated in part (a). Collisions of ions with atoms of the stationary gas tend to restore this equilibrium distribution through the flux calculated in Problem 17.8(c). Hence the net flux is (remembering that $B^2 = 3A^2$)

$$-\frac{A^2\phi^2}{\tau}\left(V^2\frac{\partial \mathrm{f}}{\partial \mathbf{v}}+3\mathbf{v}\mathrm{f}\right)+\frac{e\mathbf{E}\mathrm{f}}{m}\ .$$

In the steady state, tne net flux is zero. Hence

$$V^2\frac{\partial \mathrm{f}}{\partial \mathbf{v}} = \left(\frac{e\mathbf{E}\tau}{mA^2\phi^2}-3\mathbf{v}\right)\mathrm{f}\ .$$

For the ionic velocity components normal to **E** this equation is identical with that in Problem 17.8(d) and these components vanish. For the ionic velocity component u parallel to **E** we have

$$\frac{V^2}{3}\frac{\partial \mathrm{f}}{\partial u} = \left(\frac{eE\tau}{3mA^2\phi^2}-u\right)\mathrm{f}(u)\ .$$

If the electric field is small, $\frac{1}{3}V^2$ is approximately equal to its equilibrium value of kT/m (theorem of equipartition of energy, Problem 3.6). Hence

$$\ln \mathrm{f} = \frac{mu}{kT}\left(\frac{eE\tau}{3mA^2\phi^2}-\frac{u}{2}\right)\ .$$

i.e.

$$\mathrm{f} = \text{constant} \times \exp\left(-\frac{mu^2}{2kT}+\frac{eE\tau u}{3kTA^2\phi^2}\right)\ .$$

(d) This expression may be rewritten as

$$\mathrm{f} = \text{constant} \times \exp\left[-\frac{m(u-\bar{u})^2}{2kT}\right],$$

where

$$\bar{u} = \frac{eE\tau}{3mA^2\phi^2}\ .$$

Hence the mean ionic drift velocity is

$$\bar{\mathbf{u}} = \frac{e\mathbf{E}\tau}{3mA^2\phi^2}\ .$$

(e)

$$\mu = \frac{e\tau}{3mA^2\phi^2}\ .$$

17.10 (a) A gas of atoms of mass m, subject to no externally applied force, has a velocity distribution function which is a function of position $\mathbf{r}$, i.e. its mean temperature, density and velocity vary with position.

If the velocity distribution function at time t_0 is $f(\mathbf{v}, \mathbf{r}, t_0)$, what is the distribution function at time t_1, if the effect of collisions is neglected?

(b) What differential equation governs the rate of change of the distribution function, if it is unaffected by collisions?

(c) In the steady state, what is the distribution function $f(\mathbf{v}, \mathbf{r})$, if it is assumed that collisions establish local equilibrium at each point?

(d) Show that this distribution function is not, in general, an exact steady solution of the differential equation set up in part (b). When is it an approximate steady solution?

(e) Use the relaxation-time approximation of Problem 17.8(e) and the differential equation of part (b) to set up the following better approximation for the rate of change of $f(\mathbf{v}, \mathbf{r}, t)$:

$$\frac{\partial f}{\partial t} = -\mathbf{v} \cdot \frac{\partial f}{\partial \mathbf{r}} - \frac{f - f_0}{t^*} \ .$$

In what circumstances is this approximation a good one?

[The equation under (e) is one approximate form of the **Boltzmann transport equation**. The equation under (b) is a form of the 'collisionless' Boltzmann equation[6].

Solution

(a) At time t_1 an atom which was at $\mathbf{r}$ at t_0 with velocity $\mathbf{v}$ will be at $\mathbf{r}' = \mathbf{r} + \mathbf{v}(t_1 - t_0)$, still with velocity $\mathbf{v}$, if collisions are neglected. Since the number of atoms remains constant,

$$f(\mathbf{v}, \mathbf{r}', t_1) d\mathbf{r}' d\mathbf{v} = f(\mathbf{v}, \mathbf{r}, t_0) d\mathbf{r} \, d\mathbf{v} \ ,$$

where $d\mathbf{r}'$ is the volume element at time t_1 corresponding to $d\mathbf{r}$ at time t_0. Since $\mathbf{v}$ is constant, $d\mathbf{r}' = d\mathbf{r}$. Therefore

$$f(\mathbf{v}, \mathbf{r} + \mathbf{v}(t_1 - t_0), t_1) - f(\mathbf{v}, \mathbf{r}, t_0) = 0 \ .$$

(b) Let $t_1 \to t_0$. Then the equation becomes

$$\mathbf{v} \cdot \frac{\partial f}{\partial \mathbf{r}} + \frac{\partial f}{\partial t} = 0 \ ,$$

i.e. $\partial f/\partial t$ is equal to $-\mathbf{v} \cdot \partial f/\partial \mathbf{r}$, a term due entirely to the unimpeded motion of the atoms with the given distribution function.

(c) If there is local equilibrium at each point, the distribution is everywhere Maxwellian with the appropriate local values of temperature, density and mean velocity. We have therefore

$$f(\mathbf{v}, \mathbf{r}) = f_0(\mathbf{v}, \mathbf{r}) = n(\mathbf{r}) \left[\frac{m}{2\pi k T(\mathbf{r})} \right]^{3/2} \exp \left\{ -\frac{m[\mathbf{v} - \bar{\mathbf{v}}(\mathbf{r})]^2}{2kT(\mathbf{r})} \right\} .$$

[6] Further discussion of the equation and its approximations is given in K.Huang, *Statistical Mechanics* (John Wiley, New York), 1963, Ch.5 and 6, and in F.Reif, *loco cit.*, Ch.13.

(d) In the steady state $\partial \mathrm{f}/\partial t = 0$. Therefore $\mathbf{v} \cdot \partial \mathrm{f}/\partial \mathbf{r} = 0$, for all $\mathbf{v}$, whence $\partial \mathrm{f}/\partial \mathbf{r} = 0$.

The function written down in part (c) does not satisfy this condition. Therefore it is not a steady solution of the differential equation, but it becomes a better approximation as $\partial \mathrm{f}/\partial \mathbf{r} \to 0$, i.e. as the spatial non-uniformity decreases.

(e) When the effect of collisions is considered, the equation set up in part (b) becomes

$$\frac{\partial \mathrm{f}}{\partial t} = -\mathbf{v} \cdot \frac{\partial \mathrm{f}}{\partial \mathbf{r}} + \left(\frac{\partial \mathrm{f}}{\partial t}\right)_{\text{collisions}} .$$

The next approximation after (b) is to assume that the second term on the right hand side depends only on *local* departures from the Maxwellian distribution $\mathrm{f}_0(\mathbf{v}, \mathbf{r})$. Thus, in the relaxation-time approximation,

$$\frac{\partial \mathrm{f}}{\partial t} = -\mathbf{v} \cdot \frac{\partial \mathrm{f}}{\partial \mathbf{r}} - \frac{\mathrm{f} - \mathrm{f}_0}{t^*} .$$

Since collisions tend to establish local equilibrium quickly over distances of order λ (the mean free path) such a local approximation is good when $\mathrm{f}(\mathbf{v}, \mathbf{r}, t)$ varies slowly over such distances, i.e. as $\partial \mathrm{f}/\partial \mathbf{r} \to 0$.

17.11 (a) A stationary gas of atoms of mass m at uniform temperature T is bounded by a wall in the plane $x = 0$, from which foreign atoms of the same mass are released at a steady rate. Hence a steady concentration gradient $\partial n/\partial x$ exists, where $n(x)$, the number density of foreign atoms at distance x from the wall, is everywhere small in comparison with total number density of the gas.

Write down the steady velocity distribution function $\mathrm{f}_0(\mathbf{v}, x)$ for the foreign atoms, assuming that local equilibrium exists.

(b) Calculate the first-order correction to this distribution. using the method of Problem 17.10(e).

(c) Hence calculate the mean velocity of the foreign atoms, to first order.

(d) Hence calculate the coefficient of mutual diffusion D for the foreign atoms in the gas. Compare the result with that of Problem 17.1(c). Why is the present result more valuable?

Solution

(a)

$$\mathrm{f}_0(\mathbf{v}, x) = n(x)\left(\frac{m}{2\pi kT}\right)^{3/2} \exp\left(-\frac{mv^2}{2kT}\right) .$$

[Note that $\bar{\mathbf{v}} = 0$ in this approximation.]

(b) Let the first-order corrected velocity distribution function be $\mathbf{f}_1$. Then, from Problem 17.10(e),

$$\mathbf{v} \cdot \frac{\partial \mathbf{f}_1}{\partial \mathbf{r}} = -\frac{\mathbf{f}_1 - \mathbf{f}_0}{t^*},$$

since $\partial \mathbf{f}/\partial t = 0$ in the steady state. To first order, $\partial \mathbf{f}_1/\partial \mathbf{r}$ can be replaced by $\partial \mathbf{f}_0/\partial \mathbf{r}$ on the left hand side. Now

$$\mathbf{v} \cdot \frac{\partial \mathbf{f}_0}{\partial \mathbf{r}} = v_x \frac{\partial n}{\partial x} \left(\frac{m}{2\pi kT} \right)^{3/2} \exp\left(-\frac{mv^2}{2kT} \right).$$

Therefore, to first order,

$$\mathbf{f}_1 - \mathbf{f}_0 = -\frac{t^* v_x}{n} \frac{\partial n}{\partial x} \mathbf{f}_0,$$

i.e.

$$\mathbf{f}_1(\mathbf{v}, x) = \mathbf{f}_0(\mathbf{v}, x) \left(1 - \frac{t^* v_x}{n} \frac{\partial n}{\partial x} \right).$$

(c)

$$\overline{v_x} = \frac{1}{n} \int v_x \mathbf{f}_1(\mathbf{v}, x) \, d\mathbf{v}.$$

Now

$$\int v_x \mathbf{f}_0 \, d\mathbf{v} = 0.$$

Therefore

$$\overline{v_x} = -\frac{t^*}{n} \frac{\partial n}{\partial x} \frac{1}{n} \int v_x^2 \mathbf{f}_0 \, d\mathbf{v} = -\frac{t^*}{n} \frac{\partial n}{\partial x} \overline{v_x^2}$$
$$= -\frac{t^* kT}{mn} \frac{\partial n}{\partial x},$$

using the theorem of the equipartition of energy (Problem 3.6). It is easily verified that $\overline{v_y} = \overline{v_z} = 0$.

(d) The flux of molecules in the x direction per unit area per unit time is

$$\Phi_x = n\overline{v_x} = -\frac{t^* kT}{m} \frac{\partial n}{\partial x} = -D \frac{\partial n}{\partial x}$$

by definition of D. Therefore

$$D = \frac{t^* kT}{m}.$$

Alternatively, since $\overline{v_x^2} = \frac{1}{3}\overline{v^2}$,

$$D = \tfrac{1}{3} t^* \overline{v^2}.$$

Since $t^* \approx \tau$, for atoms scattering through large angles at collisions, and $\overline{v^2} \approx (\lambda/\tau)^2$, this expression is essentially the same as the previous one. It is more valuable, however, because the atomic velocity distribution has been taken into account in its derivation, and because t^* is generally a well-defined quantity, whereas λ can be defined precisely only for atoms which interact like rigid bodies.

17.12 (a) A gas of atoms of mass m at temperature T contains a small proportion of ions with the same mass, carrying charge e. The mobility of the ions in the gas is μ, their coefficient of mutual diffusion is D, and their number density at the point $\mathbf{r}$ is $n(\mathbf{r})$. If an electric field $\mathbf{E}(\mathbf{r})$ exists in the gas, what is the condition that the mean ionic velocity is everywhere zero?

(b) Assuming that the ions obey classical statistics, show that

$$\mu = \frac{eD}{kT},$$

where k is the Boltzmann constant.

[This is the **Nernst-Einstein relation**[7].]

(c) By applying the Nernst-Einstein relation to the results of Problems 17.9(e) and 17.11(d) find a more exact relation between t^* and τ than the estimate obtained in Problem 17.8(e).

Solution

(a) The mean ionic flux due to the electric field is $n\mu\mathbf{E}$. That due to diffusion is $-D\partial n/\partial\mathbf{r}$. Hence, in the stationary state,

$$n\mu\mathbf{E} - D\frac{\partial n}{\partial \mathbf{r}} = 0\,.$$

(b)

$$\mathbf{E} = -\frac{\partial V}{\partial \mathbf{r}},$$

where $V(\mathbf{r})$ is the electrostatic potential at point $\mathbf{r}$. Therefore

$$\ln n = -\frac{\mu V}{D} + \text{constant},$$

and

$$-n\mu\frac{\partial V}{\partial \mathbf{r}} - D\frac{\partial n}{\partial \mathbf{r}} = 0\,.$$

i.e.

$$n = n_0 \exp\left(-\frac{\mu V}{D}\right),$$

where n_0 is a constant. However, the stationary state is one of equilibrium in the given electric field. Therefore the ionic number density is given by the classical Boltzmann (canonical) distribution

$$n = n_0 \exp\left(-\frac{eV}{kT}\right),$$

where $eV(\mathbf{r})$ is the potential energy of an ion at $\mathbf{r}$. By comparison, the Nernst-Einstein relation follows.

[7] This is discussed further in F.Reif, *loco cit.*, Ch.15, and in P.T.Landsberg, *Proc.Roy.Soc.*, A213, 226 (1952).

(c) From Problem 17.9(e),

$$\mu = \frac{e\tau}{3mA^2\phi^2} .$$

From Problem 17.11(d)

$$D = \frac{t^* kT}{m} ,$$

whence

$$\frac{\mu}{D} = \frac{e}{kT} \frac{\tau}{3A^2\phi^2 t^*} .$$

This gives

$$t^* = \frac{\tau}{3A^2\phi^2} .$$

[The previous estimate was $t^* \approx \tau/A^2\phi^2$.]

GENERAL REFERENCES

A comprehensive account of elementary (mean free path) kinetic theory is given in:

J. Jeans, *Kinetic Theory of Gases* (Cambridge University Press, Cambridge), 1940.

The following textbooks deal with the Boltzmann equation:

E. A. Desloge, *Statistical Physics,* Part III (Holt, Rinehart, and Winston, New York), 1966.

F. Reif, *Fundamentals of Statistical and Thermal Physics* (McGraw-Hill, New York), 1965, Ch.12-14.

K. Huang, *Statistical Mechanics* (John Wiley, New York), 1963, Ch.3-6.

The classical treatise on the subject is perhaps

S. Chapman and T. G. Cowling, *Mathematical Theory of Non-uniform Gases* (Cambridge University Press, Cambridge), 1939.

For recent advances (e.g. the treatment of dense gases) consult:

I. Prigogine (Ed.), *Proc. International Symposium on Transport Processes in Statistical Mechanics* (Interscience, New York), 1958.

W. E. Brittin (Ed.), *Lectures in Theoretical Physics,* Vol.IX C (*Kinetic Theory*) (Gordon and Breach, New York), 1967.

18
Transport in metals[1]

J.M.HONIG
(*Purdue University, Lafayette, Indiana*)

INTRODUCTORY COMMENTS

18.0 In Chapters 18 and 19 we consider problems which deal with the response of mobile electrons in solids to externally applied electric fields, magnetic fields, and temperature gradients.

In the standard approximation utilized here the electrons are considered to be independent particles subject to Fermi-Dirac statistics. In zero order approximation the solid is viewed as a 'box' or container, within which the electrons move as a 'gas'; this is the so-called **Sommerfeld model**. The effect of the solid is introduced more realistically in first order approximation by regarding the periodic potential of the lattice as a perturbation on the nearly free electrons. Alternatively, one may proceed from the opposite assumption: the electrons are considered rather tightly bound to the atomic cores in the solid but able to move through the lattice by virtue of some overlap among orbitals associated with adjacent atoms. In either case the following conclusions are reached: there is an alternation between closely spaced energy levels (energy bands) and forbidden energy ranges (energy gaps), corresponding to ranges where the Schrödinger wave equation does or does not admit of solutions. The demarcation line between allowed and forbidden levels is termed a **band edge**. The ψ functions can always be represented as free electron wave functions modulated by a function which has the lattice periodicity.

Of cardinal importance is the specification of the energy $\mathcal{E}$ of the electrons in the solid and of the dependence of $\mathcal{E}$ on independent variables and on parameters. As in the case of free electrons the energy depends on the wave number vector **k**. In what follows we shall always treat a very special case, namely **bands of standard form** for which

$$\mathcal{E} = \mathcal{E}_c + \frac{\hbar^2 k^2}{2m} = \mathcal{E}_c + \epsilon \tag{18.0.1}$$

where $\mathcal{E}_c$ is the lower band edge and the second term is a kinetic energy formally identical to the expression obtained for free particles, in which m is the mass of the particle, h Planck's constant, and $\hbar \equiv h/2\pi$.

[1] In Chapters 18 and 19 Boltzmann's constant is denoted by $\mathscr{k}$ to avoid confusion with the modulus of the wave vector $k = |\mathbf{k}|$.

However, in the present context, m is not the free electron mass but, rather, an **effective mass** which depends on the band structure of the solid. By this method it is possible to dispense with the explicit consideration of interactions of charge carriers with the lattice.

The dynamics of the particle is introduced as follows: it may be shown from very general considerations that the velocity of an electron in the crystal is specified by

$$\mathbf{v} = \frac{\nabla_{\mathbf{k}} \mathcal{E}(k)}{\hbar} \tag{18.0.2}$$

where $\nabla_{\mathbf{k}}$ is the gradient operator with respect to the independent variable $\mathbf{k}$. The time derivative $\dot{\mathbf{k}}$ of the wave number vector is related to the externally applied force $\mathbf{F}$ through the equation $\dot{\mathbf{k}} = \mathbf{F}/\hbar$.

The interaction of an electron to an externally applied field is handled in the present context by setting up and solving a **Boltzmann transport equation**, in which the electric, magnetic, and temperature fields appear explicitly as parameters. In a perfect periodic lattice the electron encounters no resistance to its motion; however, impurities, lattice vibrations, and other types of imperfections introduce scattering mechanisms which must also be handled through the Boltzmann equation. A standard procedure here consists in introducing a relaxation time τ related to the mean free path l by $\tau = l/|\mathbf{v}|$. This approach can be shown to apply under certain rather restrictive conditions, and the results are equivalent to the linear theory of irreversible thermodynamics. Adjustment of τ along the lines shown in later problems simulates different scattering mechanisms.

As will be shown in the last problem of this section, a set of band states that is very nearly completely occupied may equally well be described in terms of the remaining unoccupied band states which may be associated with fictitious particles, termed **holes**. These may be regarded as charge carriers that have a positive charge and a kinetic energy, and Fermi level which is referred to the upper band edge.

Thus, for holes,

$$\mathcal{E} = \mathcal{E}_v - \frac{\hbar^2 k^2}{2m} \equiv \mathcal{E}_v - \epsilon \tag{18.0.3}$$

where $\mathcal{E}_v$ is the upper band edge. The reader should carefully distinguish in subsequent sections between the total energy $\mathcal{E}$ of a charge carrier and its 'kinetic energy' $\hbar^2 k^2/2m$ which is designated as ϵ.

Finally, a specimen is said to be n type or p type according as electrons or holes are responsible for conduction .

18.1 This problem is designed to acquaint the reader (a) with the concept of electrochemical potential which is extensively used in Chapters 18 and 19, and (b) with methods commonly used to describe on a macroscopic thermodynamic scale the response of charge carriers in conductors to externally applied forces.

(a) The **electrochemical potential** ζ for electrons is defined by

$$\zeta \equiv \mu_n - e\varphi_s \tag{18.1.1}$$

where μ_n is the **chemical potential,** $-e$ the charge on the electron, and φ_s the electrostatic potential. Express the chemical potential in terms of the **activity** of the electrons in the system, which is related to the concentration of electrons, c_n, by $a_n = \gamma_n c_n$, where γ_n is the activity coefficient. Show that for a uniform material at constant temperature the gradient of the electrochemical potential per unit electronic charge coincides with the electrostatic field.

(b) Let l_λ be the length of the sample specimen in the direction $\lambda = x, y, z$; let $\mathbf{V} \equiv \nabla(\zeta/e)$, $\mathbf{U} \equiv \nabla\varphi_s$; let J_λ and C_λ be the current density and heat flux along the direction λ; and let T be the temperature. Referring to Figure 18.3.1 (page 414), note the orientation of the rectangular parallelepiped relative to the Cartesian axes; let L be the distance between the voltage probes at C and D, and let H_z denote the magnetic field, aligned along the z direction. The following definitions will now be introduced for uniform isotropic materials under the open-circuit conditions $J_y = J_z = 0$ and for the isothermal conditions $\nabla_y T = \nabla_z T = 0$:

1. The **electrical resistivity**

$$\rho = \frac{\nabla_x(\zeta/e)}{J_x} \quad \text{with } \nabla_x T = 0\,. \tag{18.1.2}$$

Show that this reduces to the common formulation of ρ in the terms of the sample resistance R_s.

2. The **Hall coefficient**

$$R \equiv \frac{\nabla_y(\zeta/e)}{J_x H_z} \quad \text{with } \nabla_x T = 0\,. \tag{18.1.3}$$

Show how this quantity is related to the potential difference across y, and to the current along x. Demonstrate why it is desirable to make the samples as thin as possible along the magnetic field direction. Interpret the results.

3. The **Seebeck coefficient**

$$\alpha \equiv \frac{\nabla_x(\zeta/e)}{\nabla_x T} \quad \text{with } J_x \equiv 0\,. \tag{18.1.4}$$

State whether α may be re-expressed in terms of electrostatic potential differences; discuss the significance of Equation (18.1.4).

4. The **Nernst coefficient**

$$N \equiv \frac{\nabla_y(\zeta/e)}{H_z \nabla_x T} \quad \text{with } J_x \equiv 0\,. \tag{18.1.5}$$

State whether N may be re-expressed in terms of electrostatic potential differences; discuss the significance of Equation (18.1.5).

5: The **thermal conductivity** is defined by

$$\kappa \equiv \frac{-C_x}{\nabla_x T} \text{ with } J_x \equiv 0\,. \tag{18.1.6}$$

Interpret this relation.

Solution

(a) From standard thermodynamics we find that

$$\mu_{\rm n} = \mu_{\rm n}^0 + kT \ln a_{\rm n} \tag{18.1.7}$$

where $a_{\rm n}$ is the activity of the electrons in the system under study, and $\mu_{\rm n}^0$ is the standard chemical potential of the electrons when $a_{\rm n} = 1$.

Taking the gradient of the potential (18.1.1) at constant temperature we find

$$\nabla(\zeta/e) = \frac{kT}{e}\nabla(\ln a_{\rm n}) - \nabla\varphi_{\rm s} \tag{18.1.8}$$

where $\nabla\varphi_{\rm s}$ is the gradient of the electrostatic potential.

Since T is constant and the material is uniform, $\nabla \ln a_{\rm n} \equiv 0$; then

$$\mathbf{V} \equiv \nabla(\zeta/e) = -\nabla\varphi_{\rm s} = -\frac{\mathbf{U}}{L} = \mathbf{E} \tag{18.1.9}$$

where $\mathbf{U}$ is the electrostatic potential difference, L the distance over which this quantity is measured, and $\mathbf{E}$ is the electrostatic field.

(b) 1. Let I_x be the current along the x direction and utilize the defining symbols introduced earlier. Then, with $J_x = I_x/l_y l_z$ and $l_y l_z \equiv A_x$ we may rewrite the definition (18.1.2) as

$$\rho = \frac{V_x l_y l_z}{I_x} = -\frac{U_x l_y l_z}{I_x L} = -\frac{U_x A_x}{I_x L}\,. \tag{18.1.10}$$

On introducing Ohm's law, $-U_x/I_x = R_{\rm s}$, we find

$$\rho = \frac{R_{\rm s} A_x}{L} \tag{18.1.11}$$

which is the required formulation connecting ρ and $R_{\rm s}$.

2. In terms of the symbols introduced earlier, we have from definition (18.1.3) and Figure 18.1.1

$$R = \frac{V_y l_y l_z}{I_x H_z} = -\frac{U_y l_z}{I_x H_z} \tag{18.1.12}$$

where l_z is the thickness of the sample. Note that according to Equation (18.1.12), for a given material with fixed R, the potential difference is proportional to $1/l_z$; hence, the thinner the sample, the larger $|U_y|$, and the easier it is to measure U_y. Difficulties arise, however, when l_z becomes comparable to the mean free path of the electrons. According to the definition (18.1.3) based on a phenomenological approach to irreversible thermodynamics, the application of a current along the positive x direction, and of a magnetic field along the positive z direction, leads to the establishment of a gradient in electrochemical potential (or of a difference in electrostatic potential in the case of uniform materials) along the y axis. The magnitude of this gradient (or difference in potential) is given by the Hall coefficient under the assumed boundary conditions.

3. Even for uniform materials we may no longer set $\nabla_x(\zeta/e) = -U_x/l_x$ because in the presence of a temperature gradient it is no longer appropriate to drop the first term on the right in Equation (18.1.8).

According to the definition (18.1.4) based on the phenomenological approach to irreversible thermodynamics, the establishment, within a sample, of a temperature gradient, subject to the conditions detailed earlier, leads to the existence of a gradient of electrochemical potential (of Fermi level) within the sample, whose magnitude is specified by α. In view of Equation (18.1.8) this results in the concomitant establishment of a gradient in activity (or concentration) and of an electric field for electrons within the sample.

4. For reasons discussed above, it is not permissible to replace $\nabla_y(\zeta/e)$ with $-U/l_y$ here.

According to Equation (18.1.5) based on the phenomenological approach to irreversible thermodynamics, the establishment of a temperature gradient along x and of a magnetic field along z leads to the presence of a gradient in electrochemical potential along the y axis. For reasons given in Part 3, this produces both an electric field and a non-uniform distribution of activity (or charge density) along y. The magnitude of this effect in isotropic materials is specified by N.

5. According to irreversible thermodynamics the thermal conductivity measures the heat flux in a material in response to an imposed temperature gradient. Note that the heat flow is in a direction opposite to the temperature gradient.

18.2 In Problem 18.1 we considered the phenomenological description of the response of charge carriers to applied forces. Here we attack the problem from the microscopic viewpoint.

It is assumed that the model described in the introduction holds and that the Boltzmann transport equation also alluded to there has been solved [(2)] for the distribution function

$$f_{\mathbf{k}} = f_0 - \mathbf{v} \cdot \Psi\left(\frac{\partial f_0}{\partial \epsilon}\right) \tag{18.2.1}$$

[(2)] T.C.Harman and J.M.Honig, *Thermoelectric and Thermomagnetic Effects and Applications* (McGraw-Hill, New York), 1967, pp.178-183.

with

$$f_0 \equiv \frac{1}{1+\exp[(\epsilon-\mu)/kT]} \tag{18.2.2}$$

and

$$\Psi \equiv \frac{\tau[\mathbf{F}+(Ze\tau/mc)\mathbf{F}\times\mathbf{H}+(e\tau/mc)^2\mathbf{H}(\mathbf{F}\cdot\mathbf{H})]}{1+(e\tau H/mc)^2} \tag{18.2.3}$$

and

$$\mathbf{F} \equiv Ze\nabla_{\mathbf{r}}(\zeta/e)-\frac{(\epsilon-\mu)\nabla_{\mathbf{r}}T}{T} . \tag{18.2.4}$$

The quantity $f_{\mathbf{k}}$ represents the probability that an electron with wave number vector **k** is actually encountered in the crystal. As is seen from Equation (18.2.1), this quantity is given in terms of the equilibrium Fermi-Dirac distribution function (18.2.2) and further involves a term which represents in first order the departure from equilibrium. Here **v** is the carrier velocity, ϵ the energy, and μ the Fermi energy relative to the lower band edge if one deals with electrons ($Z = -1$) or relative to the upper band edge if one deals with holes ($Z = +1$); k is Boltzmann's constant, and T is the temperature. The correction term in Equation (18.2.1) also involves a function Ψ given by Equation (18.2.3), in which the electric charge $|e|$, carrier mass m, velocity of light c, applied magnetic field **H**, and relaxation time τ appear explicitly. Ψ also depends on a quantity **F** which in turn involves the spatial gradient of the electrochemical potential per unit charge $\nabla_{\mathbf{r}}(\zeta/e)$ and of the temperature, $\nabla_{\mathbf{r}}T$. It is through the quantity **F** that the externally applied forces are introduced into the problem. As stated in the introduction, the resistive effect of the medium enters through the relaxation time τ. A detailed theoretical analysis shows that where this concept is applicable the relaxation time is specified by $\tau = \tau_0\epsilon^{r-1/2}$, where τ_0 is a collection of constants and r is a **scattering parameter** which has the values 0, 1, or 2 according as scattering through acoustic vibrational modes, optical vibrational modes, or ionized impurities predominates.

In proceeding with the problems cited in this section the reader should note how the function (18.2.1) is used to construct an expression for the current density and heat flux in (a), and how the definitions of Problem 18.1 are used to formulate the transport coefficients in terms of a certain set of integrals, in (b). These integrals are then evaluated under the special set of conditions referred to in (c) and (d). On the basis of the above:

(a) Derive phenomenological equations which specify current and heat flux in a crystal subjected to magnetic fields and to gradients in electrochemical potential and in temperature. Utilize the distribution function specified above.

(b) From (a) identify in terms of appropriate transport integrals the following transport coefficients:

1. the resistivity for $H = 0$;
2. the resistivity for $H \neq 0$;
3. the Hall coefficient;
4. the Seebeck coefficient for $H = 0$;
5. the Seebeck coefficient for $H \neq 0$;
6. the Nernst coefficient;
7. the thermal conductivity for $H = 0$;
8. the charge carrier density.

(c) Specialize part (b) as follows: introduce the mobility defined by $u \equiv e\tau/m$. Further, specify the relaxation time through the assumed relation $\tau = \tau_0 \epsilon^{r-\frac{1}{2}}$ referred to in the introduction. Finally, take the limit $H_z \to 0$ and apply the limiting case of either classical $(-\eta \gg 1)$ or highly degenerate $(\eta \gg 1)$ statistics in Equation (18.2.2). Tabulate the resulting transport coefficients in terms of atomic parameters. Note that for highly degenerate statistics and for $r = 0$ the results turn out to be identical with those based on the Sommerfeld model.

(d) Repeat (c) for the limit $H_z \to \infty$.

Solution

(a) The rate of transport of charge past unit cross-section is given by

$$\mathbf{J} = \sum_{\mathbf{k}} Ze\mathbf{v}_{\mathbf{k}} f_{\mathbf{k}} = \frac{Ze}{4\pi^3} \int \mathbf{v}_{\mathbf{k}} f_{\mathbf{k}} \, d^3\mathbf{k} \, . \qquad (18.2.5)$$

On the right hand side the summation over discrete $\mathbf{k}$ is replaced by the integration $\frac{1}{4\pi^3} \int d^3\mathbf{k}$; the numerical factor arises from the counting of states with discrete pseudomomentum $\hbar\mathbf{k}$ [(3)].

Now write out the dot product in Equation (18.2.2) and switch from Cartesian to spherical coordinates in $\mathbf{k}$ space; the latter step is permissible if the material under study is isotropic. Thus, we replace v_x, v_y, v_z with $V_r \equiv V, V_\theta, V_\phi$; we then introduce expression (18.2.2) in Equation (18.2.1) and use $k_r \equiv k, \theta_k, \phi_k$ as variables of integration. On integrating over ϕ_k and θ_k and simplifying one obtains

$$J_\lambda = -\frac{4Ze}{3h^2} \int_0^\infty \Psi_\lambda \left(\frac{d\epsilon}{dk}\right)^2 \left(\frac{\partial f_0}{\partial \epsilon}\right) k^2 dk \, . \qquad (18.2.6)$$

We now specialize to the 'transverse case' where the 'forces' $\mathbf{F}$, subject to experimental control, are restricted to lie in a plane normal to the magnetic field which is aligned with the z axis.

(3) C.Kittel, *Introduction to Solid State Physics* (Wiley, New York), 1966, 3rd Edition, p.207; J.S.Blakemore, *Solid State Physics* (Saunders, Philadelphia), 1969, p.157; T.C.Harman and J.M.Honig, *loco cit.*, p.157.

Specializing Equations (18.2.3) and (18.2.4) in this manner, substituting (18.2.4) in (18.2.3) and the resultant for Ψ_λ in Equation (18.2.6), we then obtain by straightforward algebraic manipulations:

$$J_x = e^2K_1\nabla_x(\zeta/e)+Ze^3H_zG_1\nabla_y(\zeta/e)+\frac{Ze}{T}(K_1\mu_B-K_2)\nabla_xT + \frac{e^2H_z}{T}(G_1\mu_B-G_2)\nabla_yT \tag{18.2.7a}$$

$$J_y = -Ze^2H_zG_1\nabla_x(\zeta/e)+e^2K_1\nabla_y(\zeta/e)-\frac{e^2H_z}{T}(G_1\mu_B-G_2)\nabla_xT + \frac{Ze}{T}(K_1\mu_B-K_2)\nabla_yT \tag{18.2.7b}$$

where ζ may be identified with the electrochemical potential (18.1.1), and where we have introduced the general **transport integrals**

$$K_i \equiv -\frac{4}{3h^2}\int_0^\infty \frac{\epsilon^{i-1}\tau k^2}{1+\omega^2\tau^2}\frac{\partial f_0}{\partial\epsilon}\frac{d\epsilon}{dk}d\epsilon \tag{18.2.8a}$$

$$G_j \equiv -\frac{4}{3h^2mc}\int_0^\infty \frac{\epsilon^{j-1}\tau^2k^2}{1+\omega^2\tau^2}\frac{\partial f_0}{\partial\epsilon}\frac{d\epsilon}{dk}d\epsilon \tag{18.2.8b}$$

and

$$\omega \equiv \frac{ZeH_z}{mc} . \tag{18.2.8c}$$

It should be noted that this particular formulation can be used only for materials characterized by parabolic band shapes in which the effective carrier mass m is a constant.

A parallel analysis may be carried out for the heat flux C due to the motion of charge carriers:

$$\mathbf{C} = \sum_{\mathbf{k}} \epsilon_{\mathbf{k}}v_{\mathbf{k}}f_{\mathbf{k}} = \frac{1}{4h^3}\int \epsilon_{\mathbf{k}}v_{\mathbf{k}}f_{\mathbf{k}}\, d^3\mathbf{k} . \tag{18.2.9}$$

This leads to the relations

$$C_x = ZeK_2\nabla_x(\zeta/e)+e^2H_zG_2\nabla_y(\zeta/e)+\frac{K_2\mu_B-K_3}{T}\nabla_xT + \frac{ZeH_z}{T}(G_2\mu_B-G_3)\nabla_yT \tag{18.2.10a}$$

$$C_y = -e^2H_zG_2\nabla_x(\zeta/e)+ZeK_2\nabla_y(\zeta/e)-\frac{ZeH_z}{T}(G_2\mu_B-G_3)\nabla_xT + \frac{K_2\mu_B-K_3}{T}\nabla_yT . \tag{18.2.10b}$$

Observe that Equations (18.2.7) and (18.2.10) specify the transport of charge and of heat in response to externally applied gradients of electrochemical potential and of temperature, and thus represent phenomenological relations.

(b) Using the boundary conditions of Problem 18.1 and the definitions listed there, we can now determine the various transport coefficients in terms of the integrals K_i and G_j:

1. The conductivity is given by $1/\rho \equiv \sigma = J_x/\nabla_x(\zeta/e) = J_y/\nabla_y(\zeta/e)$ with $\nabla_x T = \nabla_y T = 0$. With $H_z \equiv 0$, Equations (18.2.7) show that $J_\lambda = e^2 K_1 \nabla_\lambda(\zeta/e)$ ($\lambda = x$ or y), whence

$$\frac{1}{\rho} \equiv \sigma = e^2 K_1 . \tag{18.2.11}$$

2,3. The extension of this procedure to non-vanishing fields is straightforward. We set $\nabla_x T = \nabla_y T = 0$ in Equations (18.2.7a,b) and $J_y = 0$ in (18.2.7b). This permits us to eliminate $\nabla_y(\zeta/e)$ in favour of $\nabla_x(\zeta/e)$ or vice versa in (18.2.7a,b). We then find

$$\sigma(H_z) = \frac{(e^2K_1)^2 + (e^3G_1H_z)^2}{e^2K_1} \tag{18.2.12a}$$

and

$$R(H_z) = \frac{\nabla_y(\zeta/e)}{H_z J_x} = \frac{Ze^3G_1}{(e^2K_1)^2 + (e^3G_1H_z)^2} . \tag{18.2.12b}$$

4. The Seebeck coefficient is given by $\alpha = \nabla_x(\zeta/e)/\nabla_x T = \nabla_y(\zeta/e)/\nabla_y T$ with $J_x = J_y = 0$. Since we also set $H_z = 0$, we see that either Equation (18.2.7a) or (18.2.7b) yields

$$\alpha = \frac{Z}{eT}\left[\frac{K_2}{K_1} - \mu_B\right] . \tag{18.2.13}$$

5,6. The extension of the above procedure to non-vanishing fields is straightforward. Set $J_x = J_y = \nabla_y T = 0$ in Equation (18.2.7) and eliminate either $\nabla_y(\zeta/e)$ or $\nabla_x(\zeta/e)$ from the resultant pair of equations. One can readily determine

$$\alpha(H_z) = \frac{\nabla_x(\zeta/e)}{\nabla_x T} = -\frac{Ze^3(K_1\mu_B - K_2)K_1 + Ze^5H_z^2(G_1\mu_B - G_2)G_1}{T[(e^2K_1)^2 + (e^3G_1H_z)^2]} \tag{18.2.14}$$

$$N(H_z) = \frac{\nabla_y(\zeta/e)}{\nabla_x T} = \frac{e^4(K_2G_1 - K_1G_2)}{T[(e^2K_1)^2 + (e^3G_1H_z)^2]} . \tag{18.2.15}$$

7. To determine the thermal conductivity in zero field, we shall set $J_x = J_y = \nabla_y T = H_z = 0$ and define $\kappa_e = -C_x/\nabla_x T$. On applying the boundary conditions to Equations (18.2.7a) and (18.2.10a) one can eliminate $\nabla_x(\zeta/e)$ between them and thus express C_x solely in terms of $\nabla_x T$.

This yields (κ_1 is the lattice contribution)

$$\kappa = \kappa_l + \kappa_e = \kappa_l + \frac{K_3 K_1 - K_2^2}{TK_1} . \qquad (18.2.16)$$

8. The charge carrier density is given by

$$n = \frac{2}{8\pi^3}\int_0^\infty f_0\, d^3k = \frac{1}{\pi^2}\int_0^\infty f_0 k^2 dk = \frac{1}{3\pi^2}\int_0^\infty \left(-\frac{\partial f_0}{\partial \epsilon}\right) k^3 d\epsilon \ ; \qquad (18.2.17)$$

the term on the right is found by integrating $f_0 k^2$ by parts; the term $f_0 k^3/3\pi^2$ vanishes at both limits of the integral.

(c) To evaluate the quantities listed in (b), we introduce the mobility $u = e\tau/m$ and the relation $(d\epsilon/dk) = \hbar^2 k/m$. We further utilize the assumed relation for the relaxation time $\tau = \tau_0 \epsilon^{r-½} = \tau_0 (kT)^{r-½} x^{r-½}$, where $x \equiv \epsilon/kT$. Next, we define a new transport integral

$$L(g) \equiv \int_0^\infty g(u, H, \epsilon) k^3(\epsilon) \left(-\frac{\partial f_0}{\partial \epsilon}\right) d\epsilon . \qquad (18.2.18)$$

This permits us to write

$$K_i = \frac{1}{3\pi^2 e} L\left[\frac{\epsilon^{i-1} u}{1 + u^2 H^2/c^2}\right] ; \qquad G_j = \frac{1}{3\pi^2 e^2 c} L\left[\frac{\epsilon^{j-1} u^2}{1 + u^2 H^2/c^2}\right] . \qquad (18.2.19)$$

Then in the limit of vanishing magnetic field we obtain the entries in Table 18.2.1, which are found by applying relations (18.2.19) in this limit to Equations (18.2.11)–(18.2.17). The designation $L(xu)$ is to imply that we have set $g = \epsilon u/kT$ in Equation (18.2.18) before carrying out the integration over ϵ.

Table 18.2.1. $\left(H_z \to 0, \quad x \equiv \frac{\epsilon}{kT}, \quad \eta \equiv \frac{\mu}{kT}, \quad \bar{u} \equiv \frac{L(u)}{L(1)}\right)$.

Transport coefficient	Expressions in terms of L
n	$\frac{1}{3\pi^2} L(1)$
$\sigma(0)$	$\left(\frac{e}{3\pi^2}\right) L(u) = ne\frac{L(u)}{L(1)} = ne\bar{u}$
$R(0)$	$\frac{1}{Zenc}\frac{L(u^2)L(1)}{L^2(u)}$
$\alpha(0)$	$\frac{k}{Ze}\left[\frac{L(xu)}{L(u)} - \eta\right]$
$N(0)$	$\frac{k\bar{u}}{ec} L(1) \frac{L(xu)L(u^2) - L(u)L(xu^2)}{L^3(u)}$
$\kappa(0)$, electronic contribution	$\frac{k^2 T}{e^2}\sigma(0)\left[\frac{L(x^2 u)}{L(u)} - \frac{L^2(xu)}{L^2(u)}\right]$

For classical statistics where $-\eta \gg 1$, we have $-\partial f_0/\partial\epsilon \approx (kT)^{-1}e^{\eta}e^{-x}$; with $u = (e\tau_0/m)(kTx)^{r-\frac{1}{2}}$, the transport integrals $L(u^n x^l)$ become

$$L(u^n x^l) = N_n^c \Gamma(l+n(r-\tfrac{1}{2})+\tfrac{5}{2}) \tag{18.2.20a}$$

where $\Gamma(q)$ is the gamma function $\int_0^\infty e^x x^{q-1}\,dx$ and

$$N_n^c \equiv \frac{e^{\eta}}{\hbar^3}(2mkT)^{3/2}\left(\frac{e\tau_0}{m}\right)^n (kT)^{n(r-\frac{1}{2})}. \tag{18.2.20b}$$

With this result and the identities $\Gamma(n+1) = n\Gamma(n)$ the entries in Table 18.2.1 specialize to those listed in Table 18.2.2.

Table 18.2.2. $\left[H_z \to 0, \text{ classical statistics}, \bar{u} \equiv \dfrac{L(u)}{L(1)}\right]$.

Transport coefficient	Expression
n	$2\left(\dfrac{2\pi mkT}{h^2}\right)^{3/2} e^{\eta}$
$\sigma(0)$	$ne\bar{u} \equiv 16\pi e^2\tau_0(2m)^{1/2}(kT)^{r+1}\dfrac{e^{\eta}\Gamma(r+2)}{3h^3}$
$R(0)$	$\dfrac{1}{Zenc} \times \dfrac{3\sqrt{\pi}\,\Gamma(2r+\frac{3}{2})}{4\ \Gamma^2(r+2)}$
$\alpha(0)$	$\dfrac{k}{Ze}(r+2-\eta)$
$N(0)$	$\dfrac{k\bar{u}}{ec}(\tfrac{1}{2}-r) \times \dfrac{3\sqrt{\pi}\,\Gamma(2r+\frac{3}{2})}{4\ \Gamma^2(r+2)}$
$\kappa(0)$, electronic contribution	$\dfrac{k^2}{e^2}T\sigma(0)(r+2)$

For highly degenerate materials, $\eta \gg 1$, and it then is appropriate to employ the so-called Bethe-Sommerfeld approximation to represent L as

$$L(u^n x^l) = N_n^d \eta^q [1+\tfrac{1}{6}\pi^2 q(q-1)\eta^{-2}+\cdots] \tag{18.2.21a}$$

with

$$N_n^d \equiv \frac{(2mkT)^{3/2}}{\hbar^3}\left(\frac{e\tau_0}{m}\right)^n (kT)^{n(r-\frac{1}{2})} \tag{18.2.21b}$$

$$q \equiv l+n(r-\tfrac{1}{2})+\tfrac{3}{2}. \tag{18.2.21c}$$

The scattering index r appears because we have again used the relation $u = (e\tau_0/m)(kTx)^{r-\frac{1}{2}}$. Applying Equation (18.2.21) to the entries in

Table 18.2.1, we obtain the entries in Table 18.2.3.

Table 18.2.3. $\left[H_z \to 0, \text{degenerate statistics}, \bar{u} \equiv \dfrac{L(u)}{L(1)} \right]$.

Transport coefficient	Expression
n	$\dfrac{8\pi}{3h^3}(2mkT)^{3/2}\eta^{3/2}\left(1+\dfrac{\pi^2}{8\eta^2}\right)$
$\sigma(0)$	$\dfrac{16\pi e^2}{3h^3}(2m)^{1/2}(kT)^{r+1}\tau_0\eta^{r+1}\left[1+\dfrac{\pi^2 r(r+1)}{6\eta^2}\right]$
$R(0)$	$\dfrac{1}{Zenc}\left[1+\dfrac{\pi^2}{3\eta^2}(r-\frac{1}{2})^2\right]$
$\alpha(0)$	$\dfrac{k}{Ze}\left[\dfrac{\pi^2(1+r)}{3\eta}\right]$
$N(0)$	$\dfrac{k\bar{u}}{ec}\left[\dfrac{\pi^2(\frac{1}{2}-r)}{3\eta}\right]$
$\kappa(0)$, electronic contribution	$\dfrac{k^2 T\sigma(0)}{e^2}\times\dfrac{\pi^2}{3}$

(d) In the limit of very large magnetic fields we obtain the entries shown in Table 18.2.4 by applying Equation (18.2.19) in this limit to Equations (18.2.11)-(18.2.17).

Table 18.2.4. $(H_z \to \infty)$.

Transport coefficient	Expression in terms of L
$\sigma(\infty)$	$\dfrac{e}{3\pi^2}\dfrac{L^2(1)}{L(u^{-1})} = \sigma(0)\dfrac{L^2(1)}{L(u)L(u^{-1})}$
$R(\infty)$	$\dfrac{3\pi^2}{ZecL(1)} = \dfrac{1}{Zenc}$
$\alpha(\infty)$	$\dfrac{k}{Ze}\left[\dfrac{L(x)}{L(1)}-\eta\right]$
$N(\infty)$	0
$\kappa(\infty)$, electronic contribution	0 [4]

[4] T.C.Harman and J.M.Honig, *loco cit.*, p.228.

Using the procedure described in (c), the above results specialize in the limit of classical statistics as shown in Table 18.2.5, and for highly degenerate statistics as shown in Table 18.2.6.

Table 18.2.5. ($H_z \to \infty$, classical statistics).

Transport coefficient	Expression
$\sigma(\infty)$	$\sigma(0)\dfrac{\frac{9}{16}\pi}{\Gamma(3-r)\Gamma(2+r)}$
$R(\infty)$	$\dfrac{1}{Zenc}$
$\alpha(\infty)$	$\dfrac{k}{Ze}(\frac{5}{2}-\eta)$
$N(\infty)$	0
$\kappa(\infty)$, electronic contribution	0

Table 18.2.6. ($H_z \to \infty$, degenerate statistics).

Transport coefficient	Expression
$\sigma(\infty)$	$\sigma(0)\left[1-\dfrac{\pi^2(r-\frac{1}{2})^2}{3\eta^2}\right]$
$R(\infty)$	$\dfrac{1}{Zenc}$
$\alpha(\infty)$	$\dfrac{k}{Ze}\dfrac{\pi^2}{2\eta}$
$N(\infty)$	0
$\kappa(\infty)$, electronic contribution	0

18.3 Problems 18.3 and 18.4 are designed to provide readers with practice in evaluating certain of the transport coefficients, introduced in Problems 18.1 and 18.2, in terms of experimental measurement as carried out under typical conditions.

(a) A homogeneous isotropic sample in the form of a rectangular parallelepiped of dimensions $l_x = 3$ cm, $l_y = 1$ cm, $l_z = 0{\cdot}5$ cm is maintained at a uniform temperature and connected to a power supply. A current of 10 mA is passed through the sample along the x direction.

Voltage leads, 1·5 cm apart, are connected to a potentiometer, which registers a potential difference of 73 mV. Calculate the sample resistivity and conductivity.

(b) The sample is now placed in a magnetic field of 25 kG, oriented along the positive z axis. With a 10 mA current passing along the positive x direction the potential difference along the positive y direction is $-85\ \mu$V. Calculate the magnitude of the Hall coefficient, the Hall mobility, and the corresponding density of charge carriers.

(c) Is the above sample n type or p type? Explain your answer.

(d) Describe sources of experimental errors and how they may be minimized.

Solution

(a) The resistivity ρ is given by the relation [see Equation (18.1.10)]

$$\rho = \frac{R_s A}{L} = -\frac{U_x}{I_x}\frac{l_y l_z}{L} \tag{18.3.1}$$

where $A = l_y l_z$ is the cross-sectional area perpendicular to the current flow direction, $-U_x$ is the potential drop across the voltage leads along the x direction, L is the separation distance between the voltage leads, I_x is the current, and R_s the total measured resistance of that portion of the sample whose dimensions are L, l_y, l_z. Note that the quantity l_x does not enter this problem. The various dimensions are depicted in Figure 18.3.1. On inserting the values of $L = 1{\cdot}5$ cm, $l_y = 1$ cm, $l_z = 0{\cdot}5$ cm, $I_x = 0{\cdot}010$ A and $U_x = 0{\cdot}073$ V one obtains

$$\rho = 2{\cdot}4 \text{ ohm cm}; \quad \sigma = 1/\rho = 0{\cdot}42 \text{ ohm}^{-1}\text{ cm}^{-1}. \tag{18.3.2}$$

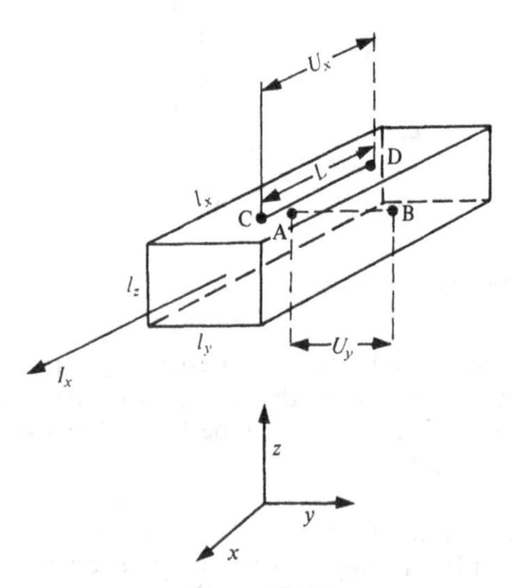

Figure 18.3.1.

(b) The Hall coefficient R is defined by [see Equation (18.1.3)]

$$\nabla_y(\zeta/e) \equiv V_y = H_z R J_x \tag{18.3.3}$$

where $\nabla_y(\zeta/e) \equiv V_y$ is the gradient of the electrochemical potential (ζ) per unit electronic charge which develops under steady state and open circuit conditions along the y direction when a current density J_x is maintained along the x axis and a magnetic field H_z is impressed along the z axis. Now write $V_y = -U_y/l_y$, where $-U_y$ is the potential drop across the y axis (see Figure 18.3.1) and set $J_x = I_x/l_y l_z$. On substitution in Equation (18.3.3) and rearrangement we obtain

$$R = \frac{l_z(-U_y)}{I_x H_z} \ . \tag{18.3.4}$$

For consistency we use the c.g.s. e.m.u. system of units. Here, $l_z = 0{\cdot}5$ cm, $U_y = -0{\cdot}000085 \times 10^8$ abV, $I_x = 0{\cdot}010 \times 10^{-1}$ abA, and $H_z = 25\,000$ G. Insertion into the above formula yields for the magnitude of the Hall coefficient

$$|R| = 1{\cdot}7 \times 10^2 \text{ cm}^3 \text{ abC}^{-1} = 17 \text{ cm}^3 \text{ C}^{-1} \ . \tag{18.3.5}$$

Using this result in the relation $|R| = 1/ne$, where n is the density of carriers, and writing $e = 1{\cdot}60 \times 10^{-19}$ coulombs we find

$$n = 1/|R|e = 3{\cdot}7 \times 10^{17} \text{ cm}^{-3} \ . \tag{18.3.6}$$

The mobility $\bar{u}$ is computed according to the relations $\sigma = ne\bar{u}$, $|R| = 1/ne$, whence

$$\bar{u} = R\sigma = 17 \text{ cm}^3 \text{ C}^{-1} \times 0{\cdot}42 \text{ ohm}^{-1} \text{ cm}^{-1} = 7{\cdot}1 \text{ cm}^2 \text{ V}^{-1} \text{ s}^{-1} \ . \tag{18.3.7}$$

(c) The fact that, with I_x and H_z positive, U_y is found to be negative implies that the potential at A in Figure 18.3.1 is higher than the potential at B, and that the electric field within the sample points along the positive y axis. Therefore under steady state conditions excess positive charge accumulates at A and excess negative charge accumulates at B. But according to the Lorentz relation, the force on charge carriers, regardless of their sign, is given by $\mathbf{F} = \mathbf{I} \times \mathbf{H}/c = I_x H_z/c = -F_y$. If the sample were n type, electrons would be pushed in the direction of face A of Figure 18.3.1; if the sample were p type, holes would be pushed in this direction. Since the latter alternative is consistent with the observation that excess positive charge accumulates on face A, it is concluded that the sample is p type.

We show that this is consistent with the definition of R via Equation (18.3.3). For a homogeneous sample at constant temperature we have $\nabla_y(\zeta/e) = -\nabla_y \varphi_s \equiv -U_y/l_y$, where φ_s is the electrostatic potential. Since $U_y < 0$ under the experimental conditions, $\nabla_y(\zeta/e) > 0$; moreover, both H_z and J_x are positive by assumption. Equation (18.3.3) now shows that $R > 0$, as is consistent with our earlier conclusion.

(d) Aside from the implication that the material is homogeneous and isotropic, as stated in the problem, the following tacit assumptions were introduced:

1. The open circuit voltage U_y is assumed to be due solely to the Hall voltage and no errors arise from probe misalignments (failure of the leads to lie on an equipotential line) or from thermal e.m.f. To minimize these sources of error one conventionally takes U_y readings with current flow and magnetic fields in the positive and negative directions. Suitable averaging of the results will largely eliminate these errors.

2. The formula $n = 1/|R|e$ was assumed to be applicable; this is the case only if the solid can be characterized by carriers in a single band contributing to conduction. For the more general case of carriers in several bands participating in such processes the reader is referred to Problem 19.6. Moreover, the above formula applies to one-band models only in the limit of high magnetic fields; a more rigorous analysis shows that $n = A/|R|e$, where the coefficient A may deviate from unity by as much as 20% in extreme cases. For calculations requiring high precision this factor must be taken into consideration.

3. Steady state conditions are assumed.

18.4 (a) A homogeneous, isotropic sample is cut in the form of a rectangular parallelepiped of length $l = 0\cdot 8$ cm, width $w = 0\cdot 2$ cm, and thickness $t = 0\cdot 1$ cm. Copper-Constantan thermocouples attached to the sample indicate a temperature of $273\cdot 1$°K at one end and a temperature of $278\cdot 4$°K at the other end. When a potentiometer is connected across the copper leads of the thermocouples, a potential difference of $0\cdot 104$ mV is registered on the instrument. To achieve balance, the negative end of the potentiometer had to be connected to the cold end of the sample. What is the Seebeck coefficient for the sample? Is the material n or p type?

(b) Discuss three experimental errors incurred in this type of measurement and how each can be circumvented. What other assumptions are made as regards these measurements?

(c) The above sample is placed in a transverse magnetic field of 30000 G which is oriented parallel to the t dimension. The thermocouples register a temperature difference of 4 deg along the positive l dimension. A potentiometer hooked to leads attached across the positive w dimension registers a potential difference of $-0\cdot 036$ mV. What is the magnitude of the Nernst coefficient for the sample?

(d) Describe briefly several sources of experimental error in the above measurement.

Solution

(a) For homogeneous, isotropic materials the Seebeck coefficient is defined through the relation [see Equation (18.1.4)]

$$\nabla_x(\zeta/e) \equiv V_x = \alpha \nabla_x T \tag{18.4.1}$$

where in the notation of Problem 18.1, V_x is the gradient of electrochemical potential per unit charge, and $\nabla_x T$ the gradient of temperature along the length of the sample. We now approximate $\nabla_x T$ with $\Delta_x T/l$ and V_x with $-U_x/l$ where $\Delta_x T$ is the temperature difference between the ends of the sample and U_x is the corresponding potential difference. We can then rewrite Equation (18.4.1) as

$$\alpha = -\frac{U_x}{\Delta_x T} \,. \tag{18.4.2}$$

Inserting the values $\Delta_x T = 278{\cdot}4 - 273{\cdot}1 = 5{\cdot}3$ deg and $U_x = 104\ \mu\mathrm{V}$ we find

$$\alpha = -104/5{\cdot}3 = -20\ \mu\mathrm{V\ deg^{-1}} \,. \tag{18.4.3}$$

Since the cold end is connected to the negative end of the potentiometer, $\alpha < 0$ and the material is n type.

(b) One should imbed the thermocouples away from the ends of the sample to avoid spurious temperature readings due to end effects. Also, instead of relying on a single U_x measurement at one $\Delta_x T$, one should take a set of U_x values corresponding to different $\Delta_x T$, and compute α as the slope from a plot of these data. This procedure circumvents problems which arise if, due to experimental difficulties, U_x does not vanish when $\Delta_x T = 0$. Finally, the value of α obtained through the above procedure includes a contribution due to the lead wires; this matter is discussed in great detail in several sources in the literature[5]. If α_x and α_L are the Seebeck coefficients of the test specimen and of the lead wire at the average temperature of $275{\cdot}7^\circ$K, then $\alpha = \alpha_x - \alpha_L$, whence $\alpha_x = \alpha + \alpha_L$. For copper lead wires $\alpha_L \approx 2\ \mu\mathrm{V\ deg^{-1}}$, showing that the correction is by no means negligible.

Other assumptions are implied in the statement of the problem. The material is assumed to be isotropic and homogeneous; this permits us to use Equation (18.4.1) and the approximation implied in (18.4.2). Further, 'isothermal' conditions are assumed to hold along the two directions perpendicular to the temperature gradient.

(c) The Nernst coefficient may be defined by [see Equation (18.1.5)]

$$N = \frac{\nabla_y(\zeta/e)}{H_z \nabla_x T} = -\frac{U_y}{w} \Big/ H_z \frac{\Delta_x T}{l} \,. \tag{18.4.4}$$

On inserting the values $U_y = -36\ \mu\mathrm{V} = -36 \times 10^2$ abV, $\Delta_x T = 4$ deg, $w = 0{\cdot}2$ cm, $l = 0{\cdot}8$ cm, $H_z = 30000$ G, we obtain for the magnitude of the Nernst coefficient

$$N = 0{\cdot}12\ \mathrm{abV\ G^{-1}\ deg^{-1}} = 0{\cdot}12 \times 10^{-8}\ \mathrm{V\ G^{-1}\ deg^{-1}} \,. \tag{18.4.5}$$

[5] C.A.Domenicali, *Rev. Mod. Phys.*, **26**, 237 (1954); T.C.Harman and J.M.Honig, *loco cit.*, Section 1.18, pp.41-47.

(d) The effects discussed in (b) must be modified to the extent that 'isothermal' conditions are assumed only parallel to the applied magnetic field. A further source of experimental error arises from spurious voltages due to misalignment of the voltage probes across the middle of the sample. On reversing both $\nabla_x T$ and/or H_z and taking suitable averages this particular error may be cancelled out.

18.5 This problem is designed to acquaint the reader with the relation between the electronic contribution to the thermal conductivity and the electrical conductivity of a metal. According to the last entry of Table 18.2.1, $\kappa(0) = (k^2T/e^2)\mathcal{L}\sigma(0)$; the proportionality between κ and σ is referred to as the **Wiedemann-Franz law**, and the proportionality constant is called the **Lorenz number**. [The symbol $\mathcal{L}$ introduced here is not to be confused with the symbol L introduced in Problem 18.2 for the transport integral (18.2.18).] In the last line of Table 18.2.1 $\mathcal{L}$ is specified by the quantity in brackets.

(a) Determine the Lorenz number $\mathcal{L}$ in zero order approximation for a highly degenerate metal (Sommerfeld model).

(b) Calculate the electronic contribution at 300°K to the total thermal conductivity of a metal (Sommerfeld model) whose resistivity at that temperature is $5{\cdot}01 \times 10^{-6}$ ohm cm.

(c) Show that $\mathcal{L}$ as determined in part (b) has as appropriate units the dimensions of W deg^{-1} cm^{-1}.

Solution

(a) For a highly degenerate material the Lorenz number is a universal constant given by (see Table 18.2.3, last entry)

$$\mathcal{L} = \frac{k^2\pi^2}{3e^2} . \qquad (18.5.1)$$

Upon substituting the various natural constants, one obtains

$$\mathcal{L} = 2{\cdot}45 \times 10^6 \text{ erg}^2 \text{ deg}^{-2} \text{ C}^{-2}. \qquad (18.5.2)$$

(b) Since

$$\kappa_e = \mathcal{L}T\sigma = \frac{\mathcal{L}T}{\rho} , \qquad (18.5.3)$$

insertion of Equation (18.5.2), $T = 300$°K, and $\rho = 5{\cdot}01 \times 10^{-6}$ ohm cm yields

$$\kappa_e = 1{\cdot}46 \times 10^{14} \text{ erg}^2 \text{ deg}^{-1} \text{ ohm}^{-1} \text{ cm}^{-1} \text{ C}^{-2} \qquad (18.5.4a)$$

$$= 1{\cdot}46 \text{ W deg}^{-1} \text{ cm}^{-1}. \qquad (18.5.4b)$$

(c) In expression (18.5.4a) we adopted the units obtained by straightforward substitution of the usual dimensions for k, e, T, and σ. We now divide both numerator and denominator by s^2 to obtain

$$(\text{erg s}^{-1})^2 (\text{C s}^{-1})^{-2} \text{ ohm}^{-1} \text{ deg}^{-1} \text{ cm}^{-1} .$$

Since

$$1 \text{ erg s}^{-1} = 10^{-7} (\text{J s}^{-1}) = 10^{-7} \text{ W}$$

and since

$$1 \text{ C}^2 \text{ s}^{-2} \text{ ohm}^{-1} = 1 \text{ A}^2 \text{ ohm} = 1 \text{ W},$$

we see that

$$1 \text{ erg}^2 \text{ C}^{-2} \text{ ohm}^{-1} = 10^{-14} \text{ W},$$

which leads us directly to expression (18.5.4b).

18.6 This problem is designed to familiarize the reader further with information available from the determination of the Seebeck and Nernst transport coefficients (see Problem 18.1). For simplicity it is to be assumed that all of the conditions leading up to the entries of Table 18.2.3 apply.

(a) The Seebeck coefficient of a highly degenerate metal is found to be $-7\cdot1$ μV deg^{-1} at $77\cdot8$°K under conditions where acoustic mode scattering predominates. Determine the Fermi level for this material.

(b) Clarify the nature of the Fermi level calculated according to part (a). Briefly describe the basic assumptions made in the numerical evaluation of the Fermi level.

(c) The metal is now placed in a very strong magnetic field. Neglecting quantization effects and any variation of Fermi energy with magnetic field, redetermine the Seebeck coefficient; observe the magnitude of the change.

(d) Calculate the isothermal transverse voltage developed across the width of the sample (y direction) placed in a magnetic field of 10000 G which is oriented along the z axis. The two ends of the sample along the x axis are maintained at 302 and 298°K respectively, and the drift mobility is 12000 cm^2 V^{-1} s^{-1}; the sample dimensions are l_x = 8 mm, l_y = 1 mm, l_z = $1\cdot5$ mm.

Solution

(a) According to the Sommerfeld model introduced in Problem 18.2 the Seebeck coefficient is related to the Fermi level, μ, by[6] [see Table 18.2.3, fourth entry, with $r = 0$]

$$\alpha = \pm\frac{\pi^2 k^2 T}{3e\mu}. \tag{18.6.1}$$

Solving for μ, inserting $k/e = 86\cdot4$ μV deg^{-1}, $k = 8\cdot62 \times 10^{-5}$ eV deg, $T = 77\cdot8$°K, $\alpha = -7\cdot1$ μV deg^{-1} we obtain

$$\mu = 0\cdot27 \text{ eV}. \tag{18.6.2}$$

[6] N.Cusack, *The Electrical and Magnetic Properties of Solids* (Longmans, Green & Co., London), 1958, Section 5.9, pp.111-117; T.C.Harman and J.M.Honig, *loco cit.*, Section 3.14, pp.142-147.

(b) The quantity μ specified in Equation (18.6.1) is the Fermi level relative to the appropriate band edge, $\mathcal{E}_B$. For n type materials, $\mathcal{E}_B$ is the conduction band edge; for p type materials, it is the valence band edge.

Equation (18.6.1) applies only to a highly degenerate electron (or hole) gas. Thus the above relation holds only if the Fermi level lies well inside a single band of standard form whose associated charge carriers are responsible for conduction phenomena. Furthermore, acoustic phonon scattering is presumed to be the dominant scattering mechanism, which means that $r = 0$ in Problem 18.2(c).

(c) In the limit of very high magnetic fields and for the model discussed in part (b), Equation (18.6.1) must be altered to read [see Table 18.2.6, third entry]

$$\alpha = \pm \frac{\pi^2 k^2 T}{2e\mu} . \qquad (18.6.3)$$

Hence, on substituting the values cited in part (a) and $\mu = 0 \cdot 27$ eV we obtain

$$\alpha = -11 \ \mu\text{V deg}^{-1} . \qquad (18.6.4)$$

This represents an increase by a factor $\frac{3}{2}$ over the value observed in zero magnetic field.

(d) The transverse voltage arises from the Nernst effect according to the relation [see Equation (18.1.5)]

$$\nabla_y(\zeta/e) = NH_z \nabla_x T . \qquad (18.6.5)$$

On the assumption that the material is homogeneous and isotropic, the above may be reduced to

$$-U_y = NH_z \Delta_x T \frac{l_y}{l_x} . \qquad (18.6.6)$$

Assuming that one is operating in the regime of low magnetic fields, that the material is highly degenerate and characterized by a single band of standard form, and that acoustic phonon scattering predominates, the Nernst coefficient may be written as [see Table 18.2.3, fifth entry with $r = 0$]

$$N = \tfrac{1}{2}\left(\frac{\pi^2}{3}\right)\left(\frac{k}{e}\right)\left(\frac{kT}{\mu}\right)\frac{\bar{u}}{c} \qquad (18.6.7)$$

where $\bar{u}$ is the drift mobility; and μ the Fermi level relative to the appropriate band edge.

For consistency, we employ the c.g.s. e.m.u. system of units. Values we shall insert are:

$$k/e = 86 \cdot 4 \ \mu\text{V deg}^{-1} = 8 \cdot 64 \times 10^3 \text{ abV deg}^{-1} ,$$
$$\bar{u}/c = 12 \times 10^3 \text{ cm}^2 \text{ V}^{-1} \text{ s}^{-1} = 1 \cdot 2 \times 10^{-4} \text{ cm}^2 \text{ abV}^{-1} \text{ s}^{-1} ,$$
$$kT = 0 \cdot 0259 \text{ eV for } T = 300\text{°K} ,$$
$$\mu = 0 \cdot 27 \text{ eV (assuming that there is a negligible shift in Fermi level from 78 to 300°K)} .$$

This yields

$$N = 0 \cdot 16 \text{ cm}^2 \text{ deg}^{-1} \text{ s}^{-1} . \tag{18.6.8}$$

Using this value in Equation (18.6.6), along with $\Delta_x T = 4$ deg, $l_y/l_x = \frac{1}{8}$, we obtain

$$-U_y = 8 \cdot 1 \times 10^2 \text{ abV} = 8 \cdot 1 \times 10^{-6} \text{ V} = 8 \cdot 1\ \mu\text{V} . \tag{18.6.9}$$

One should observe that Equation (18.6.7) is only marginally applicable since the 'low field regime' holds only as long as $\bar{u}H/c < 1$. In our case $\bar{u}H/c = 1 \cdot 2$, but it may be presumed that the above calculation is a reasonable first order approximation to the correct answer.

18.7 This problem is designed to introduce the reader to the general linear theory of irreversible thermodynamics as applied to the flow of current and of heat through a solid.

One may show from rather general considerations[(7)] that when a solid is subjected to a temperature gradient ∇T and a gradient in electrochemical potential per unit charge $\nabla(\zeta/e) \equiv \mathbf{V}$, there occurs a flux of entropy $\mathbf{J}_S$ and of electric charge (current density) $\mathbf{J}$. In the above $T\mathbf{J}_S = \mathbf{C} + \alpha\mathbf{J}$; as earlier, $\mathbf{C}$ denotes the heat flux and α represents the Seebeck coefficient of the specimen. It should be noted that in the absence of current, $T\mathbf{J}_S$ and $\mathbf{C}$ coincide.

It is generally assumed that $T\mathbf{J}_S$ and $\mathbf{J}$ depend linearly on $\mathbf{V}$ and ∇T; thus, relations of the form $T\mathbf{J}_S = A\nabla T + B\mathbf{V}$; $\mathbf{J} = C\nabla T + D\mathbf{V}$ may be expected to apply, where A, B, C, D are constants. However, these interrelations may be simplified by the so-called **Onsager reciprocity theorem**: as applied to the present case, the theorem states that if one replaces ∇T with $\nabla(1/T)$ and $\mathbf{V}$ with $\mathbf{V}/T$ then the equations specifying the response to these particular 'forces' read

$$T\mathbf{J}_S = L_{11}\nabla(1/T) + \frac{L_{12}}{T}\mathbf{V} \tag{18.7.1}$$

$$\mathbf{J} = L_{12}\nabla(1/T) + \frac{L_{22}}{T}\mathbf{V} \tag{18.7.2}$$

where the off diagonal coefficients are identical. The above equations form the basis of our subsequent development.

On the basis of the above:

(a) Show how the various L_{ij} may be identified in terms of transport coefficients that are experimentally measurable.

(b) Rewrite the phenomenological equations with the various L_{ij} replaced by transport coefficients.

(7) See Chapters 25 and 28. See also D.D.Fitts, *Non-equilibrium Thermodynamics* (McGraw-Hill, New York), 1962, Ch.3; T.C.Harman and J.M.Honig, *loco cit.*, Section 1.10, pp.20–22.

Solution

(a) At constant temperature and for homogeneous materials we have $\mathbf{J} = (L_{22}/T)\mathbf{V} = (L_{22}/T)\mathbf{E}$ as was demonstrated in Equation (18.1.9). This relation between $\mathbf{J}$ and $\mathbf{E}$ represents a formulation of Ohm's law, whence

$$\frac{L_{22}}{T} = \sigma \quad \text{or} \quad L_{22} = T\sigma \tag{18.7.3}$$

where σ is the electrical conductivity in isotropic media.

In the absence of net current flow Equation (18.7.2) reads ($L_{21} = L_{12}$)

$$\mathbf{V} = -\frac{TL_{21}}{L_{22}}\nabla(1/T) = \frac{L_{21}}{L_{22}T}\nabla T\,. \tag{18.7.4}$$

The Seebeck coefficient α is defined by $\mathbf{V} = \alpha\nabla T$, for $\mathbf{J} = 0$ [see Equation (18.1.4)]; comparison with Equations (18.7.4) and (18.7.3) shows that

$$L_{21} = L_{12} = T^2\alpha\sigma\,. \tag{18.7.5}$$

The thermal conductivity κ is defined by the relation [see Equation (18.1.6)] $T\mathbf{J}_S = -\kappa\nabla T = \kappa T^2\nabla(1/T)$, with the subsidiary condition $\mathbf{J} = 0$. The latter condition allows us to substitute for $\mathbf{V}$ in Equation (18.7.1) from (18.7.4), and thereby to obtain

$$T\mathbf{J}_S = \left(L_{11} - \frac{L_{12}^2}{L_{22}}\right)\nabla(1/T) = \kappa T^2\nabla(1/T)\,. \tag{18.7.6}$$

On introducing the results (18.7.3) and (18.7.5) and solving for L_{11} we find

$$L_{11} = T^2(\kappa + T\alpha^2\sigma)\,. \tag{18.7.7}$$

(b) On substituting the expressions (18.7.3), (18.7.5), and (18.7.7) in Equations (18.7.1) and (18.7.2) one obtains

$$T\mathbf{J}_S = T^2(\kappa + T\alpha^2\sigma)\nabla(1/T) + T^2\alpha\sigma(\mathbf{V}/T) \tag{18.7.8}$$

$$\mathbf{J} = T^2\alpha\sigma\nabla(1/T) + T\sigma(\mathbf{V}/T) \tag{18.7.9}$$

or

$$\mathbf{J}_S = -\left(\frac{\kappa}{T} + \alpha^2\sigma\right)\nabla T + \alpha\sigma\mathbf{V} \tag{18.7.10}$$

$$\mathbf{J} = -\alpha\sigma\nabla T + \sigma\mathbf{V}\,. \tag{18.7.11}$$

18.8 In this problem we consider the transport phenomena in solids where electrons in a conduction band and holes in a valence band simultaneously participate in conduction processes. In what follows let the subscript or superscript $i = 1$ or $i = 2$ refer to holes or electrons respectively, and assume steady state conditions so that $\mathbf{V} = \nabla(\zeta/e)$ and ∇T is the same for charge carriers in each band. All of the simplifications leading up to Table 18.2.3 (the Sommerfeld model) are assumed to apply.

(a) Express the total σ, α, κ in terms of densities, effective masses, and mobilities of carriers in each of the two bands, assuming sufficient carriers in each band to render the statistics highly degenerate, and assuming acoustic mode scattering to predominate for both sets of carriers. Utilize the results shown in Table 18.2.3.

(b) Specialize part (a) to a metal with 'mirror image' bands.

Solution

(a) Equations (18.7.10) and (18.7.11) apply to each band separately. Let these be denoted by subscripts or superscripts 1 and 2 respectively. Then in the notation of Problem 18.7,

$$\mathbf{J}_S^{(i)} = -\left(\frac{\kappa_i}{T} + \alpha_i^2\sigma_i\right)\nabla T + \alpha_i\sigma_i\mathbf{V}\,, \tag{18.8.1}$$

$$(i = 1, 2)$$

$$\mathbf{J}^{(i)} = -\alpha_i\sigma_i\nabla T + \sigma_i\mathbf{V}\,. \tag{18.8.2}$$

Steady-state conditions are assumed which allow us to assign to the gradients of T and of ζ/e the same value in each band. Since the fluxes are additive we find that

$$\mathbf{J}_S = \mathbf{J}_S^{(1)} + \mathbf{J}_S^{(2)} = -\frac{\kappa_1 + \kappa_2}{T}\nabla T - (\alpha_1^2\sigma_1 + \alpha_2^2\sigma_2)\nabla T + (\alpha_1\sigma_1 + \alpha_2\sigma_2)\mathbf{V} \tag{18.8.3}$$

$$\mathbf{J} = \mathbf{J}^{(1)} + \mathbf{J}^{(2)} = -(\alpha_1\sigma_1 + \alpha_2\sigma_2)\nabla T + (\sigma_1 + \sigma_2)\mathbf{V}\,. \tag{18.8.4}$$

For $T = 0$ and for uniform materials we have (see Problem 18.7)

$$\mathbf{J} = (\sigma_1 + \sigma_2)\mathbf{V} = (\sigma_1 + \sigma_2)\mathbf{E}\,. \tag{18.8.5}$$

Since this is a formulation of Ohm's Law,

$$\sigma = \sigma_1 + \sigma_2\,. \tag{18.8.6}$$

For $\mathbf{J} = 0$, Equation (18.8.4) becomes

$$\mathbf{V} = \frac{\alpha_1\sigma_1 + \alpha_2\sigma_2}{\sigma_1 + \sigma_2}\nabla T\,. \tag{18.8.7}$$

With the definition $\mathbf{V} = \alpha\nabla T$ and in view of Equation (18.8.6) it is evident that

$$\alpha = \frac{\alpha_1\sigma_1 + \alpha_2\sigma_2}{\sigma}\,. \tag{18.8.8}$$

The thermal conductivity is found from the relation $T\mathbf{J}_S = -\kappa\nabla T$ with $\mathbf{J} = 0$. In view of this latter requirement we may use Equation (18.8.7) to eliminate $\mathbf{V}$ from Equation (18.8.3). This yields

$$T\mathbf{J}_S = -\left[(\kappa_1+\kappa_2)+T(\alpha_1^2\sigma_1+\alpha_2^2\sigma_2)-T\frac{(\alpha_1\sigma_1+\alpha_2\sigma_2)^2}{\sigma}\right]\nabla T . \quad (18.8.9)$$

In conjunction with Equation (18.8.8) this leads to

$$\kappa = \kappa_1+\kappa_2+T(\alpha_1^2\sigma_1+\alpha_2^2\sigma_2-\alpha^2\sigma) . \quad (18.8.10)$$

On the basis of Equation (18.8.6) this may be rewritten as

$$\kappa = \kappa_1+\kappa_2+T\frac{\sigma_1\sigma_2}{\sigma}(\alpha_1-\alpha_2)^2 . \quad (18.8.11)$$

Actually, the above relation is strictly correct only for the electronic contributions to the thermal conductivity, since it is not possible to partition the contribution of the lattice κ in the same manner as for the electrons. Thus, for the total thermal conductivity we must adjoin the lattice contribution to the value given by expression (18.8.11).

For the two bands in question we now introduce the following specific relations

$$\sigma_i = n_i e u_i \quad (i = 1, 2) \quad (18.8.12)$$

where u_i is the drift mobility of the carriers and n_i their density in the ith band. The Seebeck coefficient for a degenerate electron gas in a band of standard form is given by (see Table 18.2.3)

$$\alpha_i = \frac{Z_i\pi^2 k^2 T}{3e\mu_i} \quad (18.8.13)$$

where μ_i is the position of the Fermi level relative to the valence band edge ($i = 1$) or conduction band edge ($i = 2$), $Z_1 = 1$ and $Z_2 = -1$, and where we have set $r = 0$.

For μ_i we substitute the relation appropriate for a highly degenerate electron gas in a band of standard form (see Table 18.2.3, neglecting the term in η^{-2}):

$$\mu_i = \frac{h^2}{2m_i}\left(\frac{3n_i}{8\pi}\right)^{2/3} \quad (18.8.14)$$

where m_i is the effective mass of carriers in the ith band. Then

$$\alpha_i = \frac{2Z_i k^2 T m_i}{3e\hbar^2}\left(\frac{\pi}{3n_i}\right)^{2/3} \quad (18.8.15)$$

where $\hbar \equiv h/2\pi$. Finally, for the electronic contribution to the thermal conductivity we use the Wiedemann-Franz law (see Problem 18.5):

$$\kappa_i = \frac{k^2\pi^2}{3e^2}T\sigma_i = \frac{k^2\pi^2 T n_i u_i}{3e} . \quad (18.8.16)$$

On introducing relations (18.8.12,15,16) into Equations (18.8.6,8,11) and simplifying one obtains the following results:

$$\sigma = e(n_1 u_1 + n_2 u_2) \tag{18.8.17}$$

$$\alpha = \frac{2}{3}\frac{k^2 T}{\hbar^2 e}\left(\frac{\pi}{3}\right)^{2/3}\left[\frac{n_1^{1/3} m_1 u_1 - n_2^{1/3} m_2 u_2}{n_1 u_1 + n_2 u_2}\right] \tag{18.8.18}$$

$$\kappa = \frac{k^2 T \pi^2}{e\ 3}(n_1 u_1 + n_2 u_2) + \frac{4}{9}\frac{k^4 T^3}{e\hbar^4}\left(\frac{\pi}{3}\right)^{4/3}\frac{(u_1 u_2)(m_1 n_2^{2/3} + m_2 n_1^{2/3})^2}{(n_1 n_2)^{1/3}(n_1 u_1 + n_2 u_2)} . \tag{18.8.19}$$

(b) We now specialize to the case of 'mirror image bands', where $n_1 = n_2 \equiv n'$, $m_1 = m_2 \equiv m'$, $u_1 = u_2 \equiv u'$. Equations (18.8.17,18,19) then simplify to

$$\sigma = 2n'eu' \tag{18.8.20}$$

$$\alpha = 0 \tag{18.8.21}$$

$$\kappa = \frac{k^2 T}{e}\ \frac{2\pi^2}{3}\ n'u' + \frac{8}{9}\frac{k^4 T^3}{e\hbar^4}\left(\frac{\pi}{3}\right)^{4/3}\frac{(m')^2 u'}{(n')^{1/3}} . \tag{18.8.22}$$

18.9 So far we have utilized special approximations in transport theory that permitted us to obtain final results in closed, analytical form. The remaining problems, 18.9 to 18.11, are designed to provide the reader with a less specialized approach to the subject matter. The basic model discussed in Problem 18.2 is utilized below. The principal change relative to Problem 18.2 is that we do not evaluate the transport integrals for various limiting cases.

(a) Obtain an expression for the electrical conductivity of a material characterized by a single band of standard form, in terms of integrals pertaining to Fermi-Dirac statistics. Assume that the relaxation time formalism is applicable, and that no magnetic field is present.

(b) Obtain an expression for the Seebeck coefficient of a material in the same terms as under part (a). Show that this measurement specifies the location of the Fermi level relative to the appropriate band edge.

(c) Discuss briefly in further detail how the present treatment differs from that of Problem 18.2.

Solution

(a) Standard analysis[8] as well as the exposition leading to Equation (18.2.11) shows that the electrical conductivity is given by

$$\sigma = e^2 K_1^0 \tag{18.9.1}$$

[8] A.H.Wilson, *The Theory of Metals* (Cambridge University Press, Cambridge), 1954, 2nd Edition, Section 3.3, pp.70-73; E.H.Putley, *The Hall Effect and Related Phenomena* (Butterworths, London), 1960, Section 8.2, pp.196-198; T.C.Harman and J.M.Honig, *loco cit.*, Sections 4.11, 4.14, pp.183-186, 193-195.

where K_1^0 is the transport integral specified according to Equation (18.2.8a) with $H_z = \omega = 0$. Thus,

$$K_i^0 \equiv -\frac{4}{3h^2}\int \epsilon^{i-1}\tau k^2 \frac{\partial f_0}{\partial \epsilon}\frac{d\epsilon}{dk}d\epsilon \,. \tag{18.9.2}$$

As in Problem 18.2, we introduce the relation

$$\tau = \tau_0 \epsilon^{r-\frac{1}{2}} \tag{18.9.3}$$

as well as expression (18.2.2) for f_0. Then

$$\frac{\partial f_0}{\partial \epsilon} = -\frac{\exp[(\epsilon-\mu)/kT]}{\{1+\exp[(\epsilon-\mu)/kT]\}^2 kT} \,. \tag{18.9.4}$$

It is also necessary to specify the dependence of ϵ on k; for bands of standard form the functional relation is

$$\epsilon = \frac{\hbar^2 k^2}{2m} \tag{18.9.5}$$

where m is the effective mass of carriers in band b; in the case under study this quantity is a constant.

Inserting expressions (18.9.3,4,5) into (18.9.1,2) yields

$$\sigma = \frac{16\pi e^2\sqrt{2m}}{3h^3}(kT)^{r+1}\tau_0 \int_0^\infty \frac{x^{r+1}e^{x-\eta}}{(1+e^{x-\eta})^2}dx \tag{18.9.6}$$

where we have also introduced the changes in variable $x \equiv \epsilon/kT$, $\eta \equiv \mu/kT$. The integral on the right may be determined by parts:

$$\int_0^\infty \frac{x^{r+1}e^{x-\eta}}{(1+e^{x-\eta})^2}dx = (r+1)\int_0^\infty \frac{x^r}{1+e^{x-\eta}}dx \equiv (r+1)F_r(\eta) \tag{18.9.7}$$

where the integral defined by F_r is termed the **Fermi-Dirac integral.** Extensive numerical evaluations of F_r are available in the literature. In view of expression (18.9.7) we may write

$$\sigma = \frac{16\pi e^2\sqrt{2m}}{3h^3}\tau_0(kT)^{r+1}(r+1)F_r(\eta) \,. \tag{18.9.8}$$

(b) The Seebeck coefficient may be expressed in terms of transport integrals as follows [9] [see Equation (18.2.13)]:

$$\alpha = \pm\frac{k}{e}\left(\frac{K_2}{kTK_1} - \eta\right) . \tag{18.9.9}$$

On substituting expressions (18.9.3-7) one obtains

$$\alpha = \pm\frac{k}{e}\left[\frac{(r+2)F_{r+1}(\eta)}{(r+1)F_r(\eta)} - \eta\right] . \tag{18.9.10}$$

[9] A.H.Wilson, *loco cit.*, Section 3.4, pp.73-77; T.C.Harman and J.M.Honig, *loco cit.*, Sections 4.13, 4.14, pp.190-197.

Equation (18.9.10) shows explicitly that α depends functionally on η alone; once r is specified, a measurement of α fixes η experimentally.

(c) Aside from the fact that $H = 0$ here, the transport integral K_i^0, Equation (18.9.2), has not been reformulated in terms of the mobility, as was done in Problem 18.2(c), where the related integrals $L(x^l u^n)$ were introduced. As a result, it is possible to specify σ and α (as well as κ) solely in terms of the integrals F_r defined in Equation (18.9.7), which do not involve $u(\epsilon)$. Tabulations of $F_r(\eta)$ are available in the literature, so that numerical calculations based on the use of $F_r(\eta)$ are more readily done than those based on the integrals L.

18.10 (a) On the basis of transport integrals which are valid for isotropic materials characterized by a band of standard form, show that the conductivity in the presence of a small magnetic field H varies as H^2, and establish the condition under which this approximation holds.

(b) Specialize the above to the case of a highly degenerate electron gas. Obtain an expression for the conductivity in the absence of a magnetic field and relate it to the more general treatment of problem.

(c) Determine the magnetoresistance in the approximation of part (b).

Solution

(a) In an extension[10] of the theory leading to Equation (18.9.6) [see also Equation (18.2.8a)], the conductivity of an isotropic material in a magnetic field is given by

$$\sigma = \frac{16\pi e^2\sqrt{2m}}{3h^3}(kT)^{r+1}\tau_0\int_0^\infty \frac{x^{r+1}}{1+\omega^2\tau^2}\left(-\frac{\partial f_0}{\partial x}\right)\mathrm{d}x \qquad (18.10.1)$$

where the nomenclature of Problem 18.9 is retained. Here $\omega \equiv \pm eH/mc$, where H is the externally applied field and c is the velocity of light. Furthermore, $f_0 = 1/[1+\exp(x-\eta_B)]$ is the Fermi-Dirac function, and $x \equiv \epsilon/kT$ is the reduced energy. It is assumed that the material is characterized by a band of standard form. With $\tau = \tau_0\epsilon^{r-1/2}$ and for $(\omega\tau)^2 \ll 1$, we can approximate expression (18.10.1) by setting

$$\frac{1}{1+\omega^2\tau^2} \approx 1-\omega^2\tau_0^2(kT)^{2r-1}x^{2r-1}$$

and by writing for later convenience

$$\sigma = C\int_0^\infty h(x)\left(-\frac{\partial f_0}{\partial x}\right)\mathrm{d}x \qquad (18.10.2)$$

where

$$C \equiv \frac{16\pi e^2\sqrt{2m}(kT)^{r+1}\tau_0}{3h^3} \qquad (18.10.3)$$

[10] E.H.Putley, *loco cit.*, Section 3.5, pp.71–77; T.C.Harman and J.M.Honig, *loco cit.*, Section 4.14, pp.193–198.

and

$$h(x) \equiv x^{r+1}[1-\omega^2\tau_0^2(kT)^{2r-1}x^{2r-1}]. \tag{18.10.4}$$

Insertion of expression (18.10.4) into (18.10.2) shows that we may write

$$\sigma = C\left[\int_0^\infty x^{r+1}\left(-\frac{\partial f_0}{\partial x}\right)\mathrm{d}x - \left(\frac{eH\tau_0}{mc}\right)^2 (kT)^{2r-1}\int_0^\infty x^{3r}\left(-\frac{\partial f_0}{\partial x}\right)\mathrm{d}x\right] \tag{18.10.5}$$

where the first term represents the conductivity in zero magnetic field and the second represents the first order correction which is valid for $\omega^2\tau^2 \ll 1$. Note that σ varies as H^2.

(b) For a highly degenerate electron gas $(-\partial f_0/\partial x)$ approaches the Dirac delta function. Accordingly, we substitute for Equation (18.10.2) the Bethe-Sommerfeld approximation

$$\sigma = Ch(\eta_B) + \frac{C\pi^2}{6}\frac{\partial^2 h(\eta_B)}{\partial x^2} \tag{18.10.6}$$

where η_B is the Fermi level relative to the appropriate band edge, divided by kT.

On substituting into expression (18.10.6) from (18.10.4) we obtain

$$\begin{aligned}\sigma(H) &= C\eta_B^{r+1}[1-\omega^2\tau_0^2(kT)^{2r-1}\eta_B^{2r-1}] \\ &\quad + \tfrac{1}{6}C\pi^2[r(r+1)\eta_B^{r-1} - \omega^2\tau_0^2(kT)^{2r-1}3r(3r-1)\eta_B^{3r-2}] \\ &= C\eta_B^{r+1}\{[1+(\tfrac{1}{6}\pi^2)r(r+1)\eta_B^{-2}] \\ &\quad - \omega^2\tau_0^2(kT)^{2r-1}\eta_B^{2r-1}[1+(\tfrac{1}{6}\pi^2)3r(3r-1)\eta_B^{-2}]\}.\end{aligned} \tag{18.10.7}$$

We note that the only factor depending on H is ω. Equation (18.10.7) is the specialization of Equation (18.10.5) to a highly degenerate electron gas.

When $\omega = H \equiv 0$, Equation (18.10.7) reduces to

$$\sigma(0) = C\eta_B^{r+1}[1+(\tfrac{1}{6}\pi^2)r(r+1)\eta_B^{-2}] \tag{18.10.8}$$

which represents the specialization of Equation (18.9.8) to a highly degenerate electron gas.

(c) The magnetoresistance is defined by

$$\frac{\Delta\rho}{\rho_0} = \frac{\rho(H)-\rho(0)}{\rho(0)} = \frac{\sigma(0)-\sigma(H)}{\sigma(H)}.$$

On introducing expressions (18.10.7,8) into this definition we obtain

$$\frac{\Delta\rho}{\rho_0} = \frac{C\eta_B^{3r}\omega^2\tau_0^2(kT)^{2r-1}[1+(\tfrac{1}{6}\pi^2)3r(3r-1)\eta_B^{-2}]}{C\eta_B^{r+1}\{[1+(\tfrac{1}{6}\pi^2)r(r+1)\eta_B^{-2}] - \omega^2\tau_0^2(kT)^{2r-1}\eta_B^{2r-1}[1+(\tfrac{1}{6}\pi^2)3r(3r-1)\eta_B^{-2}]\}} \tag{18.10.9}$$

Under the original assumption the second term in the denominator is small compared to the first. Since we intend to keep only terms of order ω^2, we may drop the second term in the denominator and simplify Equation (18.10.9):

$$\frac{\Delta\rho}{\rho_0} = \omega^2\tau_0^2(kT)^{2r-1}\eta_B^{2r-1}\frac{1+(\frac{1}{6}\pi^2)3r(3r-1)\eta_B^{-2}}{1+(\frac{1}{6}\pi^2)r(r+1)\eta_B^{-2}} \;. \quad (18.10.10)$$

Equation (18.10.10) shows that the magnetoresistance increases parabolically with ω or H, so long as the fundamental assumption $\omega^2\tau^2 \ll 1$ holds.

18.11 By examination of the energy flux establish that if $\mathcal{E}(\mathbf{k}, \mathbf{r})$ is the energy associated with an electron of wave number vector $\mathbf{k}$ at position $\mathbf{r}$, then the energy which must be associated with the corresponding hole is $-\mathcal{E}(-\mathbf{k}, \mathbf{r})$.

Solution

This question may be settled by determining the rate of transport of energy through a group of electrons associated with wave number vectors in the range $\mathbf{k}$ to $\mathbf{k}+d^3\mathbf{k}$

$$\mathbf{J}_\epsilon(\mathbf{k}) = \mathcal{E}(\mathbf{k}, \mathbf{r})\mathbf{v}_n(\mathbf{k})\frac{f_n(\mathbf{k}, \mathbf{r})}{4\pi^3}\, d^3\mathbf{k} \;. \quad (18.11.1)$$

Here $f_n(\mathbf{k}, \mathbf{r})d^3\mathbf{k}/(4\pi^3)$ is the density of electrons at $\mathbf{r}$ whose wave number vectors lie in the range $d^3\mathbf{k}$ about $\mathbf{k}$. The total energy $\mathcal{E}(\mathbf{k}, \mathbf{r})$ of the electron may be rewritten as

$$\mathcal{E}(\mathbf{k}, \mathbf{r}) = \mathcal{E}_c(\mathbf{r}) + \epsilon_n(\mathbf{k}) \quad (18.11.2)$$

where $\mathcal{E}_c$ is the lower (conduction) band edge energy and ϵ_n is the energy of the electron relative to $\mathcal{E}_c$. The velocity of these electrons is given by the standard expression[11]

$$\mathbf{v}_n(\mathbf{k}) = \hbar^{-1}\nabla_\mathbf{k}\, \epsilon_n(\mathbf{k}) \;. \quad (18.11.3)$$

It should be noted that $\mathcal{E}(\mathbf{k}, \mathbf{r}) = \mathcal{E}(-\mathbf{k}, \mathbf{r})$, $\epsilon_n(\mathbf{k}) = \epsilon_n(-\mathbf{k})$ are even functions in the wave number vector, while $\mathbf{v}_n(\mathbf{k}) = -\mathbf{v}_n(-\mathbf{k})$ is an odd function in $\mathbf{k}$.

Consider now the total energy flux obtained from Equation (18.11.1) as

$$\mathbf{J}_E = \frac{1}{4\pi^3\hbar}\int_b \mathcal{E}(\mathbf{k}, \mathbf{r})\nabla_\mathbf{k}[\epsilon_n(\mathbf{k})]f_n(\mathbf{k}, \mathbf{r})d^3\mathbf{k} \quad (18.11.4)$$

and subtract from this the quantity

$$0 = \frac{1}{4\pi^3\hbar}\int_b \mathcal{E}(\mathbf{k}, \mathbf{r})\nabla_\mathbf{k}[\epsilon_n(\mathbf{k})]d^3\mathbf{k} \quad (18.11.5)$$

[11] A.H.Wilson, *loco cit.*, Section 2.8, pp.43-45; T.C.Harman and J.M.Honig, *loco cit.*, Section 4.4, pp.164-165.

which vanishes since $\mathcal{E}$ is even and its derivative is odd in **k**. We now have

$$\mathbf{J}_E = \frac{1}{4\pi^3\hbar}\int_b -\mathcal{E}(\mathbf{k},\mathbf{r})\nabla_\mathbf{k}[\epsilon_\mathrm{n}(\mathbf{k})][1-f_\mathrm{n}(\mathbf{k},\mathbf{r})]d^3\mathbf{k}\,. \quad (18.11.6)$$

At this stage it is important to relate the properties of various functions at **k** to their properties at −**k**. Towards this end we rewrite expression (18.11.2) as

$$\mathcal{E}(\mathbf{k},\mathbf{r}) = \mathcal{E}(-\mathbf{k},\mathbf{r}) = \mathcal{E}_v(\mathbf{r}) - \epsilon_\mathrm{p}(-\mathbf{k}) \quad (18.11.7)$$

where $\mathcal{E}_v$ is the upper (valence) band edge of the same band and ϵ_p is the energy of the electrons relative to $\mathcal{E}_v$. Comparison with expression (18.11.2) establishes that

$$\epsilon_\mathrm{p}(\mathbf{k}) = \epsilon_\mathrm{p}(-\mathbf{k}) = \mathcal{E}_v(\mathbf{r}) - \mathcal{E}_c(\mathbf{r}) - \epsilon_\mathrm{n}(\mathbf{k})\,. \quad (18.11.8)$$

It now follows that

$$\nabla_\mathbf{k}\epsilon_\mathrm{p}(k) = -\nabla_\mathbf{k}\epsilon_\mathrm{n}(\mathbf{k}) = -\nabla_{-\mathbf{k}}\epsilon_\mathrm{p}(-\mathbf{k}) \quad (18.11.9)$$

where the term on the right is obtained from that on the left by the substitution of $\epsilon_\mathrm{p}(-k)$ for $\epsilon_\mathrm{p}(k)$ and of $-\nabla_{-\mathbf{k}}$ for $\nabla_\mathbf{k}$. Finally, we study a new distribution function introduced by

$$f_\mathrm{p}(-\mathbf{k},\mathbf{r}) \equiv 1 - f_\mathrm{n}(\mathbf{k},\mathbf{r})\,. \quad (18.11.10)$$

On inserting expressions (18.11.7,9,10) into (18.11.6) we find that the energy flux is given by

$$\mathbf{J}_E = \frac{1}{4\pi^3\hbar}\int_\mathrm{b} -\mathcal{E}(-\mathbf{k},\mathbf{r})[-\nabla_{-\mathbf{k}}\epsilon_\mathrm{p}(-\mathbf{k})]f_\mathrm{p}(-\mathbf{k},\mathbf{r})d^3(-\mathbf{k})\,. \quad (18.11.11)$$

In carrying out the final step we have replaced $\int\int\int_{-\infty}^{+\infty} dk_x\, dk_y\, dk_z$ by $-\int\int\int_{+\infty}^{-\infty} d(-k_x)d(-k_y)d(-k_z)$, and then reversed the integration limits to obtain $\int\int\int_{-\infty}^{\infty} d(-k_x)d(-k_y)d(-k_z)$. With the understanding that the integration is always carried out in the direction of increasing values for the variable of integration, this procedure allowed us to replace $\int_\mathrm{b} d^3\mathbf{k}$ by $\int_\mathrm{b} d^3(-\mathbf{k})$.

Examination shows that the original formulation for $\mathbf{J}_E$, Equation (18.11.4), has been rewritten in Equation (18.11.11) in such a way that the integration over positive **k** is now replaced by one over negative **k**. Further, the distribution function $f_\mathrm{n}(\mathbf{k},\mathbf{r})$ over occupied states of different **k** has been replaced with one, $f_\mathrm{p}(-\mathbf{k},\mathbf{r})$, over unoccupied states [see Equation (18.11.10)] of different −**k**. The velocity function for electrons has been replaced with one involving entities associated with −**k**.

Finally, as is seen from Equation (18.11.8), the 'kinetic' energy $\epsilon_n(\mathbf{k})$ is replaced with a negative 'kinetic' energy $-\epsilon_p(-\mathbf{k})$.

The above leads to the important conclusion that a summation of contributions of all electrons to the transport of energy $\mathcal{E}$ in a given band can be replaced by a summation over hole states in the same band. However, comparison of Equation (18.11.11) with (18.11.4) shows that in so doing $\mathcal{E}(\mathbf{k}, \mathbf{r})$ must be replaced with $-\mathcal{E}(-\mathbf{k}, \mathbf{r})$. In other words the total energy associated with a hole must be the negative of that associated with the electron. Moreover, since $\mathbf{k}$ in expression (18.11.4) is converted to $-\mathbf{k}$ in expression (18.11.11), the momentum of free holes, $-\hbar\mathbf{k}$, is the negative of the momentum of free electrons, $+\hbar\mathbf{k}$.

19

Transport in semiconductors

J.M.HONIG
(*Purdue University, Lafayette, Indiana*)

INTRODUCTORY COMMENTS

19.0 The reader is referred to Problems 16.1, 16.5, Section 18.0, and Problems 18.1 and 18.2 for a review of basic concepts required in later problems. The following summary and additional commentary is offered for subsequent use.

Both metals and semiconductors are characterized by partially filled energy bands. However, for metals the degree of filling is so extensive that quantum (**degenerate**, or **Fermi-Dirac**) statistics must be invoked in handling problems in statistical thermodynamics or in transport theory. The Fermi level falls within one or more bands so that even at the absolute zero of temperature the metal is a conductor (in many instances superconductivity also sets in at low temperatures). By contrast, the degree of filling of energy bands of semiconductors can be so small that classical statistics may be an excellent first approximation in problems pertaining to statistical thermodynamics (see Problem 16.5) or transport theory. The Fermi level now lies within a gap, so that at the absolute zero of temperature all bands are either completely filled or completely empty; the material is thus an insulator at 0°K.

We distinguish between an *extrinsic* (Problem 16.5) and an *intrinsic* (Problem 16.1) semiconductor at temperature T according to whether the band gap separating the nearly filled from the nearly empty bands is considerably greater than, or nearly comparable to, kT. In the former case the material would normally still be an insulator, were it not for the fact that impurities are always present which act as a source of electrons or holes. The mechanism by which this occurs is explored in Problem 19.1, together with an elementary sample calculation showing that charge carriers are freed from such impurity centres at very low temperatures. The extrinsic semiconductor is said to be n or p type according to whether the impurities create electrons in the lowest lying empty band, termed the **conduction band**, or create holes in the highest lying filled band, termed the **valence band**. A semiconductor is said to be **compensated** if both types of impurities are present in nearly equal concentration.

When kT becomes comparable to the band gap it is possible to promote electrons thermally from the top of the valence band to the bottom of the conduction band. As already discussed at length in Problem 18.11 (see also Problem 19.4), the voids left in the valence band may be considered as holes. The intrinsic semiconductor must thus be treated as a material in which two sets of charge carriers are simultaneously present. A number of the above concepts have already been explained in connection with various problems in Chapter 16.

In what follows we shall always consider *bands of standard form*; the energy ϵ, taken relative to the band edge energy varies quadratically with wave number vector **k**. Specifically, we set

$$\epsilon = \frac{\hbar^2 k^2}{2m} \tag{19.0.1}$$

where m is the effective mass of the charge carrier in the lattice (see Section 18.0). Further, we assume that the relaxation time formalism described in Section 18.0 and Problem 18.2 may be invoked.

For a definition of the various transport coefficients the reader is referred to Problem 18.1.

Problem 19.7 deals with a different type of semiconductor. Here orbital overlap is sufficiently small, relative to the internuclear distances in the lattice, that charge carriers can no longer move freely through the lattice in states characterized by energy bands. Rather, in first approximation, the carriers remain in residence at certain lattice sites and require an activation energy ϵ_a to overcome the energy barrier that separates this site from equivalent ones in the immediate neighbourhood. Materials in which conduction occurs through a diffusion-type transfer are said to be characterized by a **hopping mechanism**. Clearly, both excess carriers and equivalent empty lattice sites must be present for this mechanism to be operative.

19.1 This problem deals with the processes whereby charge carriers are promoted from impurity levels to appropriate bands.

(a) Give qualitative arguments to show that the presence of arsenic impurities in germanium converts this material into an n type extrinsic semiconductor. Show how the hydrogen atom model may be invoked to estimate the energy required to generate freely mobile charge carriers.

(b) Estimate the temperature range in which complete ionization can be expected for donor impurities in germanium. Within the framework of the hydrogenic model what factors alter this range in other host lattices such as Si, InSb, or ZnTe?

(c) Give qualitative arguments to show that the presence of gallium impurities converts germanium into a p type extrinsic semiconductor. Show that the hydrogenic model may be invoked to estimate the energy required to free holes in the valence band.

Solution

(a) Arsenic possesses one more valence electron than germanium; hence, for each arsenic atom incorporated substitutionally in the germanium lattice there remains one extra electron that does not fit the bonding pattern of the host lattice. At low temperatures the extra electrons remain localized at their respective impurity centres. At sufficiently high temperatures the electrons become detached and move freely through the lattice in energy states falling within the conduction band. The detachment can be represented symbolically by the relation $As \rightarrow As^{+} + e^{-}$, which is akin to the ionization of a hydrogen atom embedded in the germanium host lattice. Impurities that generate free electrons in a conduction band are termed **donor impurities.**

In this very simple model, the energy required to free the electron from the impurity site is given by the Bohr formula

$$\epsilon_{\mathrm{i}} = \frac{m_{\mathrm{n}} e^4}{2\hbar^2 \kappa^2} \tag{19.1.1}$$

where e is the electronic charge, $\hbar \equiv h/2\pi$, h is Planck's constant, m_{n} is the effective free electron mass in germanium ($m_{\mathrm{n}} \approx 0{\cdot}25\, m_0$), m_0 is the rest mass of the electron in free space), and $\kappa \approx 16$ is the dielectric constant of the lattice. Thus, we may write

$$\epsilon_{\mathrm{i}} = \epsilon_{\mathrm{H}} \frac{m_{\mathrm{n}}}{m_0} \frac{1}{\kappa^2}\,, \tag{19.1.2}$$

where ϵ_{H} ($\approx 13{\cdot}6$ eV) is the ionization energy of the hydrogen atom. It is evident that $\epsilon_{\mathrm{i}}/\epsilon_{\mathrm{H}} \approx \frac{1}{4} \times \frac{1}{256} = \frac{1}{1024}$, whence

$$\epsilon_{\mathrm{i}} \approx 0{\cdot}013 \text{ eV} \tag{19.1.3}$$

for the ionization energy of As in Ge.

(b) In the crude model used here, no distinction is made between various impurity ions in the germanium lattice. Accordingly, it is predicted that they would all be characterized by ionization energies in the range of 0·01-0·02 eV, as is indeed verified experimentally for at least a large class of impurities. These energies correspond to temperatures in the range 100-200°K, in which range one can expect most of the impurity centres to have been ionized. As one passes to other types of host lattices the quantities m_{n} and κ are altered, and the temperature range where complete ionization is encountered is altered accordingly. The temperature region where ionization of all impurities is essentially complete before the onset of the intrinsic régime is termed the **exhaustion range.**

(c) Gallium has one less valence electron than germanium; hence, when it is incorporated substitutionally in germanium, there occurs one lacuna per Ga in the bonding pattern of the host lattice. At low temperatures each void remains in the vicinity of its respective impurity, but at elevated temperatures the void may be filled by electrons from neighbouring bonds. In this process the vacant site is shifted to the point of origin of the electron that took its place. Other electrons now are free to move to the location of the displaced void, which thereby is further displaced in the opposite direction. As has been argued in Problem 18.11, the absence of an electron in a 'sea' of surrounding electrons may be treated as a positively charged particle, termed a hole. Hence, the process just described may be represented symbolically by the equation $Ga \rightarrow Ga^- + e^+$, where e^+ represents the hole and Ga^-, the impurity site with one more electron than its normal complement of three. It is intuitively evident that the process is akin to the ionization of an atom of 'antihydrogen', embedded in germanium, into an antiproton and a hole. Hence, Equations (19.1.1) and (19.1.3) should again be applicable, with m_n replaced by m_p. Impurities that create holes in valence bands are termed **acceptor impurities.**

19.2 (a) Specialize the standard transport integral K_i as defined by Equation (18.2.8a) to the case of zero magnetic field, bands of standard form, and classical statistics. Obtain an expression in terms of the mean free path parameters and the scattering index r introduced in Problem 18.2(c).

(b) With the transport integral as found in part (a), obtain an expression for the conductivity σ of the solid in terms of (i) the Fermi level and (ii) the charge carrier density.

(c) Employing the expressions for the transport integrals K_i obtained in part (a), obtain an expression for the Seebeck coefficient α.

Solution

(a) Under the provisions of the problem,

$$\omega \equiv 0, \quad \tau = \frac{l}{v} = l\hbar \Big/ \frac{d\epsilon}{dk}, \quad k^2 = \frac{2m\epsilon}{\hbar^2} .$$

Here l is the mean free path, $v \equiv (1/\hbar)(d\epsilon/dk)$ is the velocity of the carrier whose wave number vector is $\mathbf{k}$, m is the effective mass, ϵ is the energy, and $\hbar \equiv h/2\pi$, where h is Planck's constant. Equation (18.2.8a) then reduces to

$$K_i = -\frac{16\pi m}{3h^3}\int_0^\infty \epsilon^i l \frac{\partial f_0}{\partial \epsilon} d\epsilon . \qquad (19.2.1)$$

Now write $l = l_0\epsilon^r$ [this corresponds to the formulation $\tau = \tau_0\epsilon^{r-1/2}$ of Problem 18.2(c)] and $f_0 \approx e^{(\mu-\epsilon)/kT}$; ϵ and μ both refer to the appropriate band edge, k is Boltzmann's constant, and T is the temperature. Then

$$K_i = \frac{16\pi m l_0}{3h^3 kT} e^{\mu/kT} \int_0^\infty \epsilon^{i+r} e^{-\epsilon/kT} d\epsilon . \qquad (19.2.2)$$

The integral in this equation converges to the value $(kT)^{i+r+1}\Gamma(i+r+1)$ where Γ is the standard gamma function. Accordingly,

$$K_i = 16\pi m l_0 (kT)^{r+1}\Gamma(r+1+i)\frac{e^{\mu/kT}}{3h^3}\ . \qquad (19.2.3)$$

(b) i. In zero magnetic field one finds, according to Equation (18.2.12a), that $\sigma = e^2K_1$. Utilizing Equation (19.2.3) one obtains

$$\sigma = 16\pi m e^2 l_0 (kT)^{r+1}\Gamma(r+2)\frac{e^{\mu/kT}}{3h^3}\ . \qquad (19.2.4)$$

ii. One may now eliminate the exponential term from Equation (19.2.4) by use of Entry 1 of Table 18.2.2 or Equation (16.5.3); this yields

$$\sigma = 4nl_0 e^2 (kT)^r \frac{\Gamma(r+2)}{3(2\pi m kT)^{1/2}}\ . \qquad (19.2.5)$$

(c) According to Equation (18.2.13), $\alpha = (Z/eT)(K_2/K_1 - \mu)$, where $Z = \pm 1$. On inserting Equation (19.2.3) one obtains

$$\alpha = \frac{Z}{eT}\left[kT\frac{\Gamma(r+3)}{\Gamma(r+2)} - \mu\right] = Z\frac{k}{e}\left(r+2-\frac{\mu}{kT}\right)\ . \qquad (19.2.6)$$

19.3 (a) On the basis of the discussion in Problem 16.5 obtain an expression for the density of electrons n_n promoted from donor impurities into the conduction band of an extrinsic semiconductor at temperature T. Assume that classical statistics is applicable, that bands are of standard form, and write the final results in terms of densities of donors N_d, densities of acceptors N_a, the effective mass m_n of the promoted electrons, and the activation energy ϵ_d, required for the promotion. For simplicity, set $r = 0$.

(b) Using the results of part (a) find an expression for the conductivity σ of an extrinsic n type semiconductor in terms of the above mentioned parameters and in terms of τ_0 introduced in Table 18.2.2.

Solution

(a) According to the standard statistical theory[1] pertaining to impurity levels in semiconductors, the probability of finding a given level occupied by a charge carrier is given by

$$f = \frac{1}{1+g_n \exp[(\epsilon_d - \mu_c)/kT]} \equiv \frac{\Delta_{occ}}{\Delta}\ . \qquad (19.3.1)$$

In the above $\mu_c \equiv \zeta - \epsilon_c$ is the Fermi level and ϵ_d is the ionization energy taken relative to the conduction band edge (see Problem 19.1), and g_n is a statistical weight factor with a value of $\frac{1}{2}$ for electrons ($g_p = 2$ for

[1] P.T.Landsberg in *Semiconductors and Phosphors*, M.Schön and H.Welker, Eds. (Vieweg, Braunschweig and Interscience, New York), 1958, p.45. E.Spenke, *Electronic Semiconductors* (McGraw-Hill, New York), 1958, pp.393-394.

holes). On the right, $\Delta \equiv N_d - N_a$ is the net number of donor centres per unit volume occupied by electrons when $T = 0°K$, and Δ_{occ} is the same quantity for $T > 0$, N_d and N_a being the densities of donor and acceptor impurities. Let $\Delta_+ = \Delta - \Delta_{occ}$ represent the net density of ionized donor centres. On this basis, consider the ratio $\Delta_+ n_n/\Delta_{occ}$ when n_n is the density of charge carriers in the conduction band. Rewrite this quantity as $(\Delta_+/\Delta)n_n/(\Delta_{occ}/\Delta)$ and substitute for n_n from Equation (16.5.3) and utilize Equation (19.3.1) in the limit where $-\mu_c/kT \gg 1$. Since $n_n = \Delta_+$, we find

$$\frac{\Delta_+ n_n}{\Delta_{occ}} = \frac{n_n^2}{\Delta - n_n} = 2g_n\left(\frac{2\pi m_n kT}{h^2}\right)^{3/2} \exp(-\epsilon_d/kT)\,. \qquad (19.3.2)$$

When $kT \ll \epsilon_d$, it follows that $n_n \ll \Delta$; we then obtain

$$n_n = [2g_n(N_d - N_a)]^{1/2}\left(\frac{2\pi m_n kT}{h^2}\right)^{3/4} \exp(-\epsilon_d/2kT)\,. \qquad (19.3.3)$$

(b) Starting with entries 1 and 2 of Table 18.2.2 we may eliminate e^{η} in favour of n to find the following expression for the conductivity

$$\sigma(0) \equiv \sigma = \frac{4e^2\tau_0 n_n}{3m_n(\pi kT)^{1/2}} \quad (r = 0)\,. \qquad (19.3.4)$$

On substituting now for n_n from Equation (19.3.3) and simplifying, we find that

$$\sigma = \tfrac{4}{3}\left(\frac{2}{m_n}\right)^{1/2}\frac{e^2\tau_0}{h^{3/2}}[2g_n(N_d - N_a)]^{1/2}(2\pi mkT)^{1/4}\exp(-\epsilon_d/2kT) \qquad (19.3.5)$$

which is the desired relation.

(c) If the variation of the product $\tau_0 T^{1/4}$ with T may be neglected relative to that of the exponential term, then a plot of $\ln\sigma$ versus $1/T$ should yield a straight line with slope $-\epsilon_d/2k$; ϵ_d may then be determined from the optimal fit of the data to Equation (19.3.5).

19.4 This problem is designed to acquaint the reader with the electrical conductivity of intrinsic semiconductors.

(a) Derive an expression for the conductivity σ of an intrinsic semiconductor in terms of the effective masses (m_n, m_p), mobilities (u_n, u_p), and energy gap (ϵ_g), under conditions where extrinsic contributions from impurity centres may be neglected. Assume classical statistics to hold and that the bands are of standard form.

(b) Show from the results derived in part (a) how the energy gap of the intrinsic semiconductor may be determined from the observed variation of σ with temperature.

(c) Assuming that the material is electrically neutral, derive a general expression for the conductivity of a semiconductor in which electrons and holes in comparable numbers participate in the conduction process, writing the result in terms of donor and acceptor densities. Assume a temperature range such that most of the impurities are ionized. Under what conditions does this expression reduce to that obtained in part (a)?

(d) Specialize part (c) to the temperature range before the intrinsic characteristics become significant, i.e. to the exhaustion range. Comment briefly on the temperature dependence of conductivity for this class of materials.

Solution

According to the discussion of Problem 16.5, the density of electrons in the conduction band multiplied by the density of holes in the valence band is a quantity given by Equation (16.1.12). In the slightly different notation employed here[2], this expression reads

$$n_{\mathrm{i}} \equiv (n_{\mathrm{n}} n_{\mathrm{p}})^{1/2} = N_{\mathrm{c}} N_{\mathrm{v}} \exp(-\epsilon_{\mathrm{g}}/2kT) \tag{19.4.1}$$

where $\epsilon_{\mathrm{g}} \equiv \epsilon_{\mathrm{c}} - \epsilon_{\mathrm{v}}$ is the energy gap. For the problem under discussion, Equation (16.5.3) or Entry 1 of Table 18.2.2 is applicable, according to which

$$N_{\mathrm{c}} \equiv 2\left(\frac{2\pi m_{\mathrm{n}} kT}{h^2}\right)^{3/2} \tag{19.4.2a}$$

$$N_{\mathrm{v}} \equiv 2\left(\frac{2\pi m_{\mathrm{p}} kT}{h^2}\right)^{3/2} . \tag{19.4.2b}$$

The symbols in the above expression have been introduced in Problems 19.2 and 19.3.

Then

$$n_{\mathrm{i}} = (n_{\mathrm{n}} n_{\mathrm{p}})^{1/2} = 2\left(\frac{2\pi kT}{h^2}\right)^{3/2} (m_{\mathrm{n}} m_{\mathrm{p}})^{3/4} \exp(-\epsilon_{\mathrm{g}}/2kT) . \tag{19.4.3}$$

(a) For an intrinsic material $n_{\mathrm{n}} = n_{\mathrm{p}} \equiv n_{\mathrm{i}}$; hence the conductivity of the sample due to the holes in the valence band and due to the electrons in the conduction band is given by

$$\begin{aligned} \sigma &= n_{\mathrm{n}} e u_{\mathrm{n}} + n_{\mathrm{p}} e u_{\mathrm{p}} = n_{\mathrm{i}} e (u_{\mathrm{n}} + u_{\mathrm{p}}) \\ &= 2e\left(\frac{2\pi kT}{h^2}\right)^{3/2} (m_{\mathrm{n}} m_{\mathrm{p}})^{3/4} (u_{\mathrm{n}} + u_{\mathrm{p}}) \exp(-\epsilon_{\mathrm{g}}/2kT) \end{aligned} \tag{19.4.4}$$

where u is the mobility of the carrier.

(b) If the variation of $T^{3/2}(u_{\mathrm{n}} + u_{\mathrm{p}})$ with temperature may be neglected relative to that of the exponential term, then a plot of $\ln\sigma$ versus $1/T$ should yield a straight line with the slope $-\epsilon_{\mathrm{g}}/2k$.

(c) We base the derivation on the law of electroneutrality

$$n_{\mathrm{p}} - n_{\mathrm{n}} + N_{\mathrm{d}} - N_{\mathrm{a}} + n_{\mathrm{p}}' - n_{\mathrm{n}}' = 0 . \tag{19.4.5}$$

[2] Comparison of symbols

Problem 16.5	n_0	p_0	E	E_{v}	E_{c}	m_{c}	m_{v}	N_{c}	N_{v}	E_{F}
Chapter 19	n_{n}	n_{p}	e	ϵ_{v}	ϵ_{c}	m_{n}	m_{p}	N_{c}	N_{v}	$\mu_{\mathrm{c}}, \mu_{\mathrm{v}}$

Here, n'_p and n'_n are densities of holes and electrons bound to their respective impurities, and N_d and N_a are the densities of donor and acceptor levels. Under the conditions specified in the problem statement, n'_p and n'_n may be neglected relative to N_a and N_d respectively. On replacing n_p with n_i^2/n_n in the simplified electroneutrality equation and solving for n_n one obtains

$$n_n = \tfrac{1}{2}(N_d - N_a) + [\tfrac{1}{4}(N_d - N_a)^2 + n_i^2]^{1/2} \,. \tag{19.4.6}$$

By similar techniques it may be shown that

$$n_p = \tfrac{1}{2}(N_a - N_d) + [\tfrac{1}{4}(N_a - N_d)^2 + n_i^2]^{1/2} \tag{19.4.7}$$

where n_i is specified by Equation (19.4.3).

The conductivity due to electrons in the conduction band and to holes in the valence band is then given by

$$\sigma = n_n e u_n + n_p e u_p = e u_p (b n_n + n_p) \tag{19.4.8}$$

where u_n and u_p are the mobilities, and $b \equiv u_n/u_p$. On substituting from Equations (19.4.6) and (19.4.7) one finds

$$\sigma = e u_p \{\tfrac{1}{2}(b-1)(N_d - N_a) + (b+1)[\tfrac{1}{4}(N_d - N_a)^2 + n_i^2]^{1/2}\} \,. \tag{19.4.9}$$

When $N_d = N_a$ the above reduces to Equation (19.4.4).

(d) In the 'exhaustion range', $n_i \ll |N_c - N_a|$ and Equation (19.4.9) reduces to

$$\sigma \approx (N_d - N_a) b e u_p = (N_d - N_a) e u_n \,. \tag{19.4.10}$$

In this approximation, σ varies with temperature to the same degree as u_n does; the charge carrier density is effectively given by the constant quantity $(N_d - N_a)$. This is to be contrasted with the cases dealt with in the earlier parts, where the charge carrier density varies sensitively with temperature.

19.5 This problem deals with thermoelectric phenomena in extrinsic and intrinsic semiconductors.

(a) Derive an expression for the Seebeck coefficient α_n of an n type extrinsic semiconductor in terms of the charge carrier density n. Derive a similar expression in terms of the density of donor and acceptor impurity centres, N_d, N_a. Do the same for a p type extrinsic semiconductor. Assume classical statistics to hold and that the bands are of standard form.

(b) Show how the ionization energy ϵ_d or ϵ_a for creation of free holes or electrons may be determined from a knowledge of the variation of α_n or α_p with temperature.

(c) Obtain an expression for the Seebeck coefficient α_n of an extrinsic n type semiconductor in the exhaustion range. How does this coefficient vary with temperature in this range?

(d) Derive expressions for the Seebeck coefficient α of a nearly intrinsic semiconductor in terms of appropriate band parameters, effective masses, densities, and mobilities for carriers in two bands. Specialize to the case of intrinsic semiconductors.

(e) How may the band-gap be determined from the variation of α with temperature?

Solution

(a) Under the stated conditions, Entry 4 of Table 18.2.2 or Equation (19.2.6) is relevant. When we substitute from entry 1 of Table 18.2.2 to eliminate $\mu_c/kT \equiv \eta$ we obtain

$$\alpha_n = -\frac{k}{e}\left[r+2-\ln\frac{n_n h^3}{2(2\pi m_n kT)^{3/2}}\right], \qquad (19.5.1)$$

where all symbols have been defined in Problems 19.2 and 19.3. On applying Equation (19.3.3) this may be rewritten as

$$\alpha_n = -\frac{k}{e}\left\{r+2+\frac{\epsilon_d}{2kT}+\frac{3}{4}\ln T-\ln\frac{[h^3 g_n(N_d-N_a)/2]^{1/2}}{(2\pi m_n k)^{3/4}}\right\}. \qquad (19.5.2)$$

For holes we set $\eta \equiv (\epsilon_v - \zeta)kT \equiv \mu_v/kT$ and proceed similarly. This yields

$$\alpha_p = \frac{k}{e}\left[r+2-\ln\frac{n_p h^3}{2(2\pi m_p kT)^{3/2}}\right] \qquad (19.5.3)$$

and

$$\alpha_p = \frac{k}{e}\left\{r+2+\frac{\epsilon_a}{2kT}+\frac{3}{4}\ln T-\ln\frac{[h^3 g_p(N_a-N_d)/2]^{1/2}}{(2\pi m_p k)^{3/4}}\right\}. \qquad (19.5.4)$$

(b) In ordinary circumstances the term $\epsilon_c/2kT$ or $\epsilon_a/2kT$ outweighs the term $\frac{3}{4}\ln T$; hence, in first approximation, a plot of $-\alpha_n$ versus $1/T$, or of α_p versus $1/T$, should yield a straight line with slope $\epsilon_d/2k$ or $\epsilon_a/2k$ respectively.

(c) In the exhaustion range $n_n = N_d - N_a$, as is intuitively evident; the same result is also found from Equation (19.4.6) on neglecting n_i^2 relative to $\frac{1}{4}(N_d - N_a)^2$. Inserting this result into Equation (19.5.1) yields

$$\alpha_n = -\frac{k}{e}\left[r+2-\ln\frac{(N_d-N_a)h^3}{2(2\pi m_n k)^{3/2}}+\frac{3}{2}\ln T\right] \qquad (19.5.5)$$

which shows that in the exhaustion range $-\alpha_n$ varies with temperature as $\frac{3}{2}(k/e)\ln T$.

(d) For intrinsic semiconductors one must take into account the participation of both electrons and holes in conduction processes. According to Equation (18.8.8), the overall Seebeck coefficient is given by

$$\alpha = \frac{\alpha_n \sigma_n + \alpha_p \sigma_p}{\sigma_n + \sigma_p}, \qquad (19.5.6)$$

where the transport coefficients with subscripts refer to the one-band contributions. On introducing Equation (19.2.6) and its counterpart for holes into the above, one obtains ($\sigma \equiv \sigma_n + \sigma_p$)

$$\alpha = \frac{k}{e\sigma}\left[-\sigma_n\left(r+2-\frac{\mu_c}{kT}\right)+\sigma_p\left(r+2-\frac{\mu_v}{kT}\right)\right]. \tag{19.5.7}$$

But with $\zeta = \mu_c + \epsilon_c = \epsilon_v - \mu_v$ it follows that $\mu_v = -\mu_c - \epsilon_g$, whence

$$\alpha = \frac{k}{e\sigma}\left[(r+2)(\sigma_p - \sigma_n) + \sigma\frac{\mu_c}{kT} + \sigma_p\frac{\epsilon_g}{kT}\right]. \tag{19.5.8}$$

We may simplify further by dividing expression (16.1.8) by (16.1.11); this yields

$$\frac{n_n}{n_p} = \frac{N_c}{N_v}\exp\frac{\mu_c - \mu_v}{kT}, \tag{19.5.9}$$

which may be solved for

$$\frac{\mu_c - \mu_v}{kT} \equiv \frac{2\mu_c + \epsilon_g}{kT} = -\frac{3}{2}\ln\frac{m_n}{m_p} + \left[\ln\frac{n_n}{n_p}\right], \tag{19.5.10}$$

where the definitions (19.4.2a) and (19.4.2b) were utilized. Now solve Equation (19.5.10) for μ_c and substitute this result in Equation (19.5.8) or (19.5.7) to find

$$\alpha = -\frac{k}{e}\left[\frac{\sigma_n - \sigma_p}{\sigma_n + \sigma_p}\left(r+2+\frac{\epsilon_g}{2kT}\right) + \frac{3}{4}\ln\frac{m_n}{m_p} - \frac{1}{2}\ln\frac{n_n}{n_p}\right]. \tag{19.5.11}$$

We now consider the quantity ($b \equiv u_n/u_p$)

$$\frac{\sigma_n - \sigma_p}{\sigma_n + \sigma_p} = \frac{n_n u_n - n_p u_p}{n_n u_n + n_p u_p} = \frac{bn_n/n_p - 1}{bn_n/n_p + 1}, \tag{19.5.12}$$

and on substitution in Equation (19.5.11) we obtain

$$\alpha = -\frac{k}{e}\left[\frac{bn_n/n_p - 1}{bn_n/n_p + 1}\left(r+2+\frac{\epsilon_g}{2kT}\right) + \frac{3}{4}\ln\frac{m_n}{m_p} - \frac{1}{2}\ln\frac{n_n}{n_p}\right]. \tag{19.5.13}$$

Finally, for intrinsic semiconductors where $n_n = n_p$ the above reduces to

$$\alpha = -\frac{k}{e}\left[\frac{b-1}{b+1}\left(r+2+\frac{\epsilon_g}{2kT}\right) + \frac{3}{4}\ln\frac{m_n}{m_p}\right]. \tag{19.5.14}$$

(e) If the dependence of b on T is neglected, Equation (19.5.14) shows that a plot of $-\alpha$ versus $1/T$ should yield a straight line with slope

$$\frac{\epsilon_g}{2k}\frac{b-1}{b+1}\frac{k}{e}.$$

19.6 (a) Show that the Wiedemann-Franz law cited in Problem 18.5 applies to extrinsic semiconductors in the limit of classical statistics and derive an expression for the Lorenz number.

(b) Contrast the numerical values of the Lorenz number when classical or when highly degenerate statistics is applicable to materials where carriers in a single band participate in conduction. Discuss the relative contribution of charge carriers and of the lattice to the overall thermal conductivity of extrinsic semiconductors and of metals.

(c) Derive a general expression for the electronic contribution due to holes and electrons in an intrinsic semiconductor in terms of the one-band thermal conductivities, one-band electrical conductivities, and band gap. Assume classical statistics to apply.

(d) Is the total electronic contribution to the thermal conductivity of a solid invariably greater than the sum of the one-band contributions? Under what conditions may one expect this total to be very much greater than the one-band contributions?

Solution

(a) Under the specified conditions, the last Entry of Table 18.2.2 may be utilized:

$$\kappa_e = \left(\frac{k}{e}\right)^2 T(r+2)\sigma\,, \qquad (19.6.1)$$

from which it immediately follows that the Wiedemann-Franz law applies and that the Lorenz number is given by

$$\mathcal{L} = \left(\frac{k}{e}\right)^2 (r+2)\,. \qquad (19.6.2)$$

(b) On setting $r = 0$, and inserting standard numerical values the Lorenz numbers for the two extreme cases specified by Equations (18.5.1) and (19.6.2), we obtain:

$$\mathcal{L} = 1{\cdot}49 \times 10^{-8}\ \mathrm{V}^2\ \mathrm{deg}^{-2}, \text{ classical statistics,} \qquad (19.6.3a)$$

$$\mathcal{L} = 2{\cdot}45 \times 10^{-8}\ \mathrm{V}^2\ \mathrm{deg}^{-2}, \text{ degenerate statistics,} \qquad (19.6.3b)$$

where we have used the value $k/e = 86{\cdot}4\ \mu\mathrm{V}\ \mathrm{deg}^{-1}$.

In both cases the one-band thermal conductivities arising from the circulation of charge carriers are proportional to the corresponding electrical conductivities; the latter differ by many orders of magnitude as one passes from semiconductors to metals. By contrast, the lattice thermal conductivities of semiconductors do not differ enormously from those of metals. Hence, as a rough rule of thumb, one may state that the electronic contribution to the total thermal conductivity exceeds that of the lattice contribution in the case of metals, but that the opposite situation obtains in extrinsic semiconductors.

(c) Here we apply Equation (18.8.11), into which we substitute from Equation (19.2.6) and its counterpart for holes; this yields

$$\kappa_e = \kappa_n + \kappa_p + \left(\frac{k}{e}\right)^2 \frac{T\sigma_n\sigma_p}{\sigma_n + \sigma_p}\left[2(r+2) + \left(\frac{\epsilon_g}{kT}\right)\right]^2\,, \qquad (19.6.4)$$

where $\epsilon_g = \epsilon_c - \epsilon_v$.

(d) Since the last term on the right-hand side of Equation (19.6.4) is positive, the 'ambipolar' thermal conductivity κ_e always exceeds the sum of the one-band contributions $\kappa_n + \kappa_p$. When (i) σ_p and σ_n are both large, (ii) the temperature is high, and (iii) ϵ_g/kT is large, the ambipolar contribution may make κ_e much larger than the sum $\kappa_n + \kappa_p$.

19.7 This problem relates to the physical characteristics of solids in which electrons are transferred in the 'hopping-type' process discussed in Section 19.0.

(a) Starting with the generalized version of Ohm's law, Problem 18.1(b), and with Equation (18.0.1) for the electrochemical potential, derive the Einstein formula for the diffusion constant of a particle in terms of its mobility, for a nondegenerate material.

(b) Obtain an expression for the diffusion constant of particles essentially localized about lattice sites, but being able to jump to equivalent neighbouring sites along a certain specified direction, in terms of the lattice spacing b and attempt frequency ν for hopping.

(c) Derive an expression for the electrical conductivity of a solid in which excess charge carriers can move between equivalent lattice sites by an activated diffusion-type process. Show explicitly how the temperature enters this expression.

(d) Discuss briefly the variation of the electrical conductivity derived in the last part: (i) with temperature when the number of excess carriers is fixed, and (ii) with excess carrier density at a fixed temperature. Interpret these results physically.

Solution

(a) As discussed in Problem 18.1(b), Ohm's law in generalized form reads: $\mathbf{J} = \sigma\nabla(\zeta/e)$ where $\mathbf{J}$ is the current density, σ is the conductivity, ∇ the gradient operator, e the electronic charge, and ζ, as specified by Equation (18.1.1), is the electrochemical potential. Now, in the relation $\zeta = \mu_n - e\phi_s$, substitute for the chemical potential μ_n the standard expression of thermodynamics[3]

$$\mu_n = \mu_0 + kT\ln a_n\,, \tag{19.7.1}$$

where k is Boltzmann's constant, T is the temperature, a_n is the activity of the electron, and μ_0 is the chemical potential of the standard state. Then Ohm's law may be reformulated as follows:

$$\mathbf{J} = \sigma\nabla(\zeta/e) = \sigma\left(\frac{kT}{ea_n}\nabla a_n - \nabla\phi_s\right) = -e\mathbf{J}_n \tag{19.7.2}$$

where on the right we have introduced the particle flux $\mathbf{J}_n \equiv \mathbf{J}/(-e)$. In

[3] F.A.MacDougall, *Thermodynamics and Chemistry* (Wiley, New York), 1939, Chapter 13. G.N.Lewis and M.Randall (revised by K.S.Pitzer and L.Brewer), *Thermodynamics* (McGraw-Hill, New York), 1961, 2nd Ed., Chapter 20. S.Glasstone, *Thermodynamics for Chemists* (Van Nostrand, New York), 1947, Chapter 15. E.A.Guggenheim, *Thermodynamics* (North Holland, Amsterdam), 1967, Chapter 5.

the absence of an externally applied electric field Equation (19.7.2) reduces to

$$\mathbf{J}_\mathrm{n} = -\frac{\sigma \mathit{k} T}{e^2 a_\mathrm{n}} \nabla a_\mathrm{n} \ . \tag{19.7.3}$$

The above relation may be compared with Fick's law of diffusion in the formulation $\mathbf{J} = -D\nabla a_\mathrm{n}$, where D is the diffusion coefficient. Comparison with Equation (19.7.3) shows that

$$D = \frac{\sigma \mathit{k} T}{e^2 a_\mathrm{n}} \ . \tag{19.7.4}$$

Writing the conductivity σ as $a_\mathrm{n} e u_\mathrm{n}$, where u_n is the mobility, we obtain Einstein's formulation for the diffusion coefficient as

$$D = \frac{\mathit{k} T u_\mathrm{n}}{e} \ . \tag{19.7.5}$$

A generalization of the above result to cover the case of degenerate materials has been discussed by Landsberg(4).

(b) Consider three parallel adjacent planes passing through rows of atoms in a lattice. Let θ be the fraction of atomic sites surrounded by excess charge carriers, let ν be the attempt frequency for a jump to adjacent sites, and δ_x be the probability that a successful jump occurs in the two directions perpendicular to the above-mentioned planes and along the direction of the applied field. For a transfer to be successful the prospective site in either direction must be empty. Accordingly, the probability of successful jumps in unit time along the specified direction is $\nu\delta_x(1-\theta)$.

Let $n(x)$ be the density of carriers on the plane x at time t, and let $n(x \pm b)$ be the corresponding density on the two adjacent planes separated from the central one by the distance b. Then, expanding $n(x \pm b)$ in Taylor's series we find

$$n(x \pm b) = n(x) \pm bn'(x) + \tfrac{1}{2} b^2 n''(x) + \cdots \ . \tag{19.7.6}$$

The net change in carrier density on plane x in time $\mathrm{d}t$ is proportional to the difference in net inflow and net outflow, which in turn is governed by the probability factor derived earlier. Thus,

$$\mathrm{d}n(x) = \mathrm{d}t[n(x+b) + n(x-b) - 2n(x)]\delta_x(1-\theta)\nu \ . \tag{19.7.7}$$

On substituting from (19.7.6) and simplifying we find

$$\frac{\partial n}{\partial t} = \delta_x(1-\theta)\nu b^2 \frac{\partial^2 n}{\partial x^2} \tag{19.7.8}$$

The above differential equation is of the form $\partial n/\partial t = D(\partial^2 n/\partial x^2)$, encountered in diffusion theory. Hence we may set

$$D = \delta_x(1-\theta)\nu b^2 \ . \tag{19.7.9}$$

(4) P.T.Landsberg, *Proc. Roy. Soc. (London)*, **A213**, 226, 1952.

(c) Assuming that the density of carriers remains constant and that b is sensibly independent of temperature, D varies with temperature as ν does. According to the standard theory of reaction rates, we specify this dependence by use of the relation

$$\nu = \nu_0 \exp(-\epsilon_a/kT)\,, \tag{19.7.10}$$

where ϵ_a is the activation energy needed to effect the transition of the electron from one site to an adjacent one. On substituting expression (19.7.10) into (19.7.9) and the resultant into expression (19.7.4), one obtains

$$\sigma = a_n \frac{(1-\theta)\nu_0 e^2 \delta_x b^2}{kT} \exp(-\epsilon_a/kT)\,. \tag{19.7.11}$$

Let there be C lattice sites which can accommodate the electrons in a crystal of total volume V, and let γ_n be the activity coefficient of the charge carriers in the lattice; then,

$$n_n \equiv \frac{a_n}{\gamma_n} = \frac{C\theta}{V} = \frac{c\theta}{V_0} \tag{19.7.12}$$

where c is the number of centres available in a unit cell whose volume is V_0. Finally, define a quantity $\delta_3 \equiv V_0/b^3$; then

$$a_n = \frac{c\theta\gamma_n}{b^3\delta_3}\,. \tag{19.7.13}$$

On substituting expression (19.7.13) in (19.7.11) we find

$$\sigma = \frac{c\nu_0\gamma_n\delta_x}{\delta_3}\left(\frac{\theta(1-\theta)e^2}{bkT}\right)\exp(-\epsilon_a/kT)\,. \tag{19.7.14}$$

(d) If one ignores the dependence of $\gamma_n\delta_x/\delta_3$ on T, the above analysis predicts that σ should vary with temperature as $T^{-1}\exp(-\epsilon_a/kT)$, i.e. essentially exponentially. This is consistent with a diffusion-controlled activated mobility process. At constant temperature, σ varies with the fraction θ of lattice sites that have associated excess carriers, according to the relation $\theta(1-\theta)$; thus, $\sigma \to 0$ for $\theta \to 0$ and $\theta \to 1$, and σ is maximal for $\theta = \frac{1}{2}$. This may be understood on the basis that the excess carriers must have empty sites to which they can move. If there are either no excess carriers or no available empty sites there can be no movement of carriers; the case $\theta = \frac{1}{2}$ represents the optimal balance between carriers and empty sites into which they can move.

19.8 This problem is designed to acquaint readers with the thermoelectric properties of materials in which conduction occurs by a 'hopping' mechanism.

(a) Show, from the thermodynamic interpretation concerning the Fermi level, how the Seebeck coefficient α of a charge carrier is related to its partial molal entropy $\overline{S}$.

(b) Derive an expression for the configurational entropy corresponding to the random distributions of excess charge carriers among lattice sites. From this result and the expression obtained in part (a), derive an expression for the Seebeck coefficient in terms of the thermal contribution to the partial molal entropy of the carriers, and in terms of the fractional occupation of available lattice sites by such carriers.

(c) Discuss the variation of the Seebeck coefficient in materials characterized by a 'hopping' conduction mechanism: (i) as a function of the fraction of cations in valence states, (n) and ($n+1$), at constant temperature; (ii) as a function of temperature at constant ratio of cations in two valence states, (n) and ($n+1$).

(d) Provide a physical interpretation of the result discussed in part (c, i).

Solution

(a) According to standard thermodynamic theory[5], the expression for the differential of the electrochemical potential is given by

$$d\zeta = -\overline{S}\,dT + \overline{V}\,dp + q\,d\phi_s\,, \tag{19.8.1}$$

where $\overline{S}$, $\overline{V}$, and q represent the partial molal entropy, partial molal volume, and charge respectively. Thus, the derivative $(\partial\zeta/\partial T)_{p,\,\phi_s}$ is given by $-\overline{S}$. To the extent to which it is permissible to replace $\alpha \equiv \nabla(\zeta/e)/\nabla T$ by $\alpha = \partial(\zeta/e)/\partial T$ we may write

$$\alpha = -\frac{\overline{S}}{e}\,. \tag{19.8.2}$$

(b) In the diffusion-type transport model we may in first approximation regard the n carriers as nearly localized about the N fixed lattice sites. Using the standard Boltzmann combinatorial formula we then find for this system

$$S = S_T + S_c = S_T + k\ln W \tag{19.8.3}$$

where S_T represents the thermal and S_c the configurational contribution to the entropy of the charge carriers on the lattice sites. According to the usual combinatorial statistics, the total configurational entropy reads

$$S_c = k\ln W = k\ln\frac{N!}{(N-n)!\,n!}\,. \tag{19.8.4}$$

On introducing Stirling's approximation, $\ln a! \approx a\ln a - a$, valid for large a, we obtain

$$S_c = N\ln N - (N-n)\ln(N-n) - n\ln n\,. \tag{19.8.5}$$

Accordingly,

$$\overline{S}_c \equiv \left(\frac{\partial S_c}{\partial n}\right)_N = -\ln\frac{n}{N-n}\,. \tag{19.8.6}$$

[5] See footnote (3).

On inserting Equation (19.8.6) into (19.8.3) and employing expression (19.8.2) we find

$$\alpha = -\frac{\bar{S}_T}{e} + \frac{k}{e}\ln\frac{n}{N-n} = -\frac{\bar{S}_T}{e} + \frac{k}{e}\ln\frac{\theta}{1-\theta}, \tag{19.8.7}$$

where

$$\theta \equiv \frac{n}{N}.$$

(c) According to Equation (19.8.7),

$$\alpha_c \equiv \alpha + \frac{\bar{S}_T}{e} = \frac{k}{e}\ln\frac{\theta}{1-\theta} \tag{19.8.8}$$

is a large positive quantity for $\theta \to 1$, passes through zero at $\theta = \frac{1}{2}$, and becomes a large negative quantity for $\theta \to 0$.

Let us take as a simple model a material, such as an oxide in which the cation M can be in one of two valence states, for example $M^{(n+1)}$ and $M^{(n)}$. Here the superscripts $(n+1)$ or (n) refer to the formal valence states of the cation. Since the higher-valent cation contains one less electron, $\theta = [M^{(n)}]/\{[M^{(n)}]+[M^{(n+1)}]\}$, where the square brackets denote concentrations. This predicts that α_c as specified by Equation (19.8.8) is negative for $[M^{(n)}] < \{[M^{(n)}]+[M^{(n+1)}]\}$ and positive otherwise; $\alpha_c \to -\infty$ as $[M^{(n)}] \to 0$ and $\alpha_c \to \infty$ as $[M^{(n+1)}] \to 0$.

Inasmuch as, at fixed $[M^{(n)}]$ and $[M^{(n+1)}]$, α_c is independent of T, α varies with T only to the extent that $\bar{S}_T$ does. Generally, this is a weak dependence, so that in zero order approximation α is nearly independent of T.

(d) The variation of α with θ may be understood on the basis that $\theta \to 0$ corresponds to the presence of cations primarily in the valence state $(n+1)$, with a sprinkling of cations in the valence state (n), containing an additional electron each. These excess carriers, being able to move to corresponding empty sites on adjacent cations in the $(n+1)$ valence state, are clearly n type carriers. Since the sign of α reflects the sign of the dominant carrier species, one would expect α to be negative for $\theta < \frac{1}{2}$. When $\theta \to 1$, nearly all cations are in the valence state (n), with a sprinkling of cations in the $(n+1)$ state. The latter may be characterized by the absence of an electron, in a background of a preponderant majority of cations, all containing an additional electron. The carriers here may be considered to be holes; hence one expects α to be positive for $\theta < \frac{1}{2}$. The precise θ value where α shifts from the positive to the negative range depends on the value of $-\bar{S}_T/e$ relative to that of α_c.

20
Fluctuations of energy and number of particles

C.W.McCOMBIE
(*University of Reading, Reading*)

20.1 A system, which may be macroscopic or microscopic, is in contact with a heat bath at temperature T, so that the probability p_r of the system being in its rth quantum state, energy E_r, is given by the canonical expression

$$p_r = \frac{\exp(-E_r/kT)}{\sum_s \exp(-E_s/kT)} .$$

Write down expressions for the mean energy $\overline{E}$, the mean square energy $\overline{E^2}$ and the mean square fluctuation in energy $\overline{\Delta E^2}$ (where $\Delta E = E - \overline{E}$) and establish the relation

$$\overline{\Delta E^2} = kT^2 \frac{\mathrm{d}\overline{E}}{\mathrm{d}T} .$$

Discuss the applicability of this result, with $\overline{E}$ interpreted as the internal energy, to a macroscopic system: (a) at fixed volume, (b) at zero pressure, and (c) under a fixed non-zero pressure.

Solution

Recall first the very simple but important relation between mean value, mean square value, and mean square fluctuation from the mean which is obtained as follows:

$$\overline{\Delta E^2} = \overline{(E-\overline{E})^2} = \overline{E^2 - 2E\overline{E} + \overline{E}^2} = \overline{E^2} - 2\overline{E}^2 + \overline{E}^2 = \overline{E^2} - \overline{E}^2 .$$

Substituting the expression given for p_r in $\overline{E} = \sum_r E_r p_r$, we get

$$\overline{E} \sum_s \exp(-E_s/kT) = \sum_s E_s \exp(-E_s/kT) .$$

Differentiating each side with respect to T and dividing the result throughout by $\sum_s \exp(-E_s/kT)$ gives

$$\frac{1}{kT^2}\overline{E}^2 + \frac{\mathrm{d}\overline{E}}{\mathrm{d}T} = \frac{1}{kT^2}\overline{E^2} ,$$

i.e.

$$kT^2\frac{d\overline{E}}{dT} = \overline{E^2} - \overline{E}^2 = \overline{\Delta E^2}\ ,$$

which is the required result.

The above considerations clearly apply to a system at fixed volume or at zero pressure, since in both these cases the Hamiltonian operator is such that the eigenvalue will be just the internal energy in the state. In the case of constant non-zero pressure, however, the Hamiltonian of the system will include the potential associated with the forces producing the pressure. The corresponding eigenvalues cannot be interpreted as the internal energy and one must proceed differently. [See Problem 22.1 for further discussion.]

20.2 Apply the preceding result to obtain the mean square fluctuation in energy, and the mean square fractional fluctuation in energy of: (a) a harmonic oscillator (it is convenient to take the zero of energy to coincide with the ground-state energy so that the zero-point energy does not appear explicitly), (b) a collection of N identical harmonic oscillators, and (c) a non-degenerate perfect gas of N particles in a container of fixed volume, all supposed in contact with a heat bath at temperature T. In cases (a) and (b) consider the form of the results in the high-temperature limit.

Solution

(a) For the oscillator, natural angular frequency ω_0, one has $E_r = r\hbar\omega_0$ $(r = 0, 1, 2 \ldots)$ and one finds from the canonical expression for p_r

$$\overline{E} = \sum_{r=0}^{\infty} E_r p_r = \frac{\hbar\omega_0}{\exp(\hbar\omega_0/kT) - 1}$$

so that

$$\overline{\Delta E^2} = kT^2\frac{d\overline{E}}{dT} = (\hbar\omega_0)^2\frac{\exp(\hbar\omega_0/kT)}{[\exp(\hbar\omega_0/kT) - 1]^2}\ .$$

The fractional fluctuation $\Delta E/\overline{E}$ has mean square value

$$\frac{\overline{\Delta E^2}}{\overline{E}^2} = \exp(\hbar\omega_0/kT)\ .$$

Expressed in terms of $\overline{E}$ this becomes

$$\frac{\overline{\Delta E^2}}{\overline{E}^2} = 1 + \frac{\hbar\omega_0}{\overline{E}}\ .$$

At high temperaturés ($kT \gg \hbar\omega_0$) these results become

$$\overline{E} = kT\ ,$$

$$\overline{\Delta E^2} = (kT)^2\ ,$$

$$\frac{\overline{\Delta E^2}}{\overline{E}^2} = 1\ .$$

Note that the fractional fluctuations tend to unity, not to zero, as the mean energy becomes large.

(b) Writing E_N for the energy of N oscillators, and E_1 for the energy of one of the oscillators, so that the results for E_1 are given by the expressions already obtained, we have,

$$\bar{E}_N = N\bar{E}_1 ,$$

$$\overline{\Delta E_N^2} = kT^2 \frac{d\bar{E}_N}{dT} = NkT^2 \frac{d\bar{E}_1}{dT} = N\overline{\Delta E_1^2} ,$$

and

$$\frac{\overline{\Delta E_N^2}}{\bar{E}_N^2} = \frac{1}{N} \frac{\overline{\Delta E_1^2}}{\bar{E}_1^2} = \frac{1}{N}\left(1 + \frac{\hbar\omega_0}{\bar{E}}\right) .$$

Note that if N is large the fractional fluctuations are small. This is consistent with the requirement that the fractional fluctuations in energy of a macroscopic system in contact with a heat bath should be predicted to be small, in agreement with experience.

(c) The equipartition result (Problem 3.6) gives at once for the non-degenerate gas

$$\bar{E} = \tfrac{3}{2}NkT .$$

It follows that

$$\overline{\Delta E^2} = kT^2 \frac{d\bar{E}}{dT} = \tfrac{3}{2}Nk^2T^2$$

and

$$\frac{\overline{\Delta E^2}}{\bar{E}^2} = \frac{2}{3} \times \frac{1}{N} .$$

Note again the decrease in fractional fluctuations with increasing N.

20.3 Use the Debye model to find how the fractional fluctuations in vibrational energy of a crystal of fixed volume depend on temperature at low temperatures. Determine the fraction of the Debye temperature at which the root-mean-square fractional fluctuations become 1% for 10^{-4} gram-molecule of a monatomic crystal.

Solution

For s gram-molecules the Debye model specific heat at constant volume is given by

$$C_v = s(\tfrac{12}{5}\pi^4 R)\left(\frac{T}{\theta}\right)^3 ,$$

where θ is the Debye temperature and R is the gas constant. The corresponding energy is

$$\bar{E} = \int_0^T C_v \, dT = s(\tfrac{12}{5}\pi^4 R)\frac{T^4}{4\theta^3} .$$

So we have

$$\frac{\overline{\Delta E^2}}{\overline{E}^2} = \frac{kT^2 C_v}{\overline{E}^2} = \frac{16}{s(\frac{12}{5}\pi^4 R)} k \left(\frac{\theta}{T}\right)^3 = \frac{16}{s(\frac{12}{5}\pi^4 N)} \left(\frac{\theta}{T}\right)^3 ,$$

where N is Avogadro's number.

Putting $\overline{\Delta E^2}/\overline{E}^2 = 10^{-4}$, $s = 10^{-4}$ and using $N = 6 \times 10^{23}$, we find

$$\frac{T}{\theta} \approx 2 \times 10^{-6} .$$

[The expression used for C_v can be linked up with the result

$$E_{\text{th}} = \frac{\frac{4}{15}\pi^5 v k^4 T^4}{h^3} \sum_{s=1}^{3} \frac{1}{a_s^3}$$

of Problem 3.11(d) if

$$\frac{1}{3} \sum_{s=1}^{3} \frac{1}{a_s^3} \equiv \frac{1}{a_{\text{D}}^3} .$$

Then

$$E_{\text{th}} = \frac{\frac{4}{5}\pi^5 v k^4 T^4}{h^3 a_{\text{D}}^3} = \frac{3}{5}\pi^4 R s \frac{T^4}{\theta^3} ,$$

provided

$$k\theta = \left(\frac{3}{4\pi} \frac{Rs}{vk}\right)^{1/3} h a_{\text{D}} .$$

Rs/k can be interpreted as the number of atoms, N, in the crystal. This formula gives an expression for the Debye temperature θ.]

20.4 Derive the canonical-ensemble result,

$$\overline{\Delta E^2} = kT^2 \frac{\mathrm{d}\overline{E}}{\mathrm{d}T} ,$$

starting from the general expression for the mean value of an observable A (corresponding operator also denoted by A)

$$\overline{A} = \mathrm{Tr}(\rho A)$$

with

$$\rho = \frac{\exp(-H/kT)}{\mathrm{Tr}[\exp(-H/kT)]} ,$$

where H is the Hamiltonian operator.

Solution

We have

$$\overline{E} = \overline{H} = \frac{\mathrm{Tr}[\exp(-H/kT)H]}{\mathrm{Tr}[\exp(-H/kT)]} .$$

Since all operators which appear are either H or functions of H, they all commute and so the ordinary manipulations can be carried through. Thus

$$\frac{\mathrm{d}\overline{E}}{\mathrm{d}T} = \frac{1}{kT^2} \frac{\mathrm{Tr}[\exp(-H/kT)H^2]}{\mathrm{Tr}[\exp(-H/kT)]} - \frac{1}{kT^2} \frac{\{\mathrm{Tr}[\exp(-H/kT)H]\}^2}{\{\mathrm{Tr}[\exp(-H/kT)]\}^2} .$$

If we write down the expression for $\overline{E^2}$,

$$\overline{E^2} = \frac{\mathrm{Tr}[\exp(-H/kT)H^2]}{\mathrm{Tr}[\exp(-H/kT)]} ,$$

and that for $\overline{E}$ given above, then comparing $\overline{\Delta E^2} = \overline{E^2} - \overline{E}^2$ and the expression for $\mathrm{d}\overline{E}/\mathrm{d}T$ gives the required result.

20.5 A system with fixed volume (or at zero pressure) is in contact with a heat bath at temperature T and a particle reservoir with which it can exchange one kind of particle, the chemical potential of such particles in the reservoir being μ. The probability p_r of the rth quantum state of the system (number of particles n_r and energy E_r) is given by the **grand canonical ensemble** expression (cf Problem 2.4)

$$p_r = \frac{\exp[-(E_r - \mu n_r)/kT]}{\sum_s \exp[-(E_s - \mu n_s)/kT]} .$$

Deduce the following results for the fluctuations in the number of particles n, and the energy E [$\overline{n}$ and $\overline{E}$ being written n and E in the derivative[(1)]]:

(a) $$\overline{\Delta n^2} = kT\left(\frac{\partial n}{\partial \mu}\right)_T ,$$

(b) $$\overline{\Delta E \Delta n} = kT\left(\frac{\partial E}{\partial \mu}\right)_T ,$$

(c) $$\overline{\Delta E^2} = kT^2\left(\frac{\partial E}{\partial T}\right)_\mu + kT\mu\left(\frac{\partial E}{\partial \mu}\right)_T .$$

If, corresponding to the fluctuations in energy and number of particles, we define fluctuations ΔT in temperature of the system by using the thermodynamic expression relating the temperature to energy and number of particles, show further that

(d) $$\overline{\Delta E \Delta T} = kT^2 ,$$

and

(e) $$\overline{\Delta n \Delta T} = 0 .$$

Solution

(a) Differentiating the expression for $\overline{n}$,

$$\overline{n} = \frac{\sum_r n_r \exp[-(E_r - \mu n_r)/kT]}{\sum_s \exp[-(E_s - \mu n_s)/kT]} ,$$

(1) For a macroscopic system this can be understood as identifying $\overline{n}$ and $\overline{E}$ with the thermodynamic equilibrium values n and E. In applying the results to a microscopic system n and E must be replaced by $\overline{n}$ and $\overline{E}$.

with respect to μ at fixed T gives (writing n for $\bar{n}$ in the derivative)

$$\left(\frac{\partial n}{\partial \mu}\right)_T = \frac{1}{kT}\frac{\sum_r n_r^2 \exp[-(E_r-\mu n_r)/kT]}{\sum_s \exp[-(E_s-\mu n_s)/kT]}$$

$$-\frac{1}{kT}\left[\frac{\sum_r n_r \exp\{-(E_r-\mu n_r)/kT\}}{\sum_s \exp\{-(E_s-\mu n_s)/kT\}}\right]^2 = \frac{1}{kT}(\overline{n^2}-\bar{n}^2) = \frac{1}{kT}\overline{\Delta n^2}\ .$$

(b) Differentiating the expression for $\bar{E}$,

$$\bar{E} = \frac{\sum_r E_r \exp[-(E_r-\mu n_r)/kT]}{\sum_s \exp[-(E_s-\mu n_s)/kT]}\ ,$$

with respect to μ at fixed T gives similarly

$$\left(\frac{\partial E}{\partial \mu}\right)_T = \frac{1}{kT}\frac{\sum_r E_r n_r \exp[-(E_r-\mu n_r)/kT]}{\sum_s \exp[-(E_s-\mu n_s)/kT]}$$

$$-\frac{1}{kT}\frac{\sum_r E_r \exp[-(E_r-\mu n_r)/kT]}{\sum_s \exp[-(E_s-\mu n_s)/kT]}\,\frac{\sum_s n_s \exp[-(E_s-\mu n_s)/kT]}{\sum_s \exp[-(E_s-\mu n_s)/kT]}$$

$$= \frac{1}{kT}(\overline{En}-\bar{E}\bar{n}) = \frac{1}{kT}\overline{(E-\bar{E})(n-\bar{n})} = \frac{1}{kT}\overline{\Delta E \Delta n}\ .$$

(c) Differentiating the above expression for E with respect to T at fixed μ gives, in a very similar way,

$$\left(\frac{\partial E}{\partial T}\right)_\mu = \frac{1}{kT^2}[\overline{E(E-\mu n)}-\bar{E}(\overline{E-\mu n})] = \frac{1}{kT^2}[\overline{E^2}-\bar{E}^2-\mu(\overline{En}-\bar{E}\bar{n})]$$

$$= \frac{1}{kT^2}[\overline{\Delta E^2}-\mu\overline{\Delta E \Delta n}]\ .$$

Combining this with the expression for $\overline{\Delta E \Delta n}$ derived above gives the required result.

(d) We get a reasonably convenient form for the coefficients in the expression for ΔT in terms of ΔE and Δn as follows:

$$\Delta E = T\Delta S + \mu \Delta n = T\left(\frac{\partial S}{\partial T}\right)_n \Delta T + \left[T\left(\frac{\partial S}{\partial n}\right)_T + \mu\right]\Delta n$$

$$= C_n \Delta T + \left[T\left(\frac{\partial S}{\partial n}\right)_T + \mu\right]\Delta n\ ,$$

where C_n is the heat capacity at constant n. This gives

$$\Delta T = \frac{1}{C_n}\Delta E - \frac{1}{C_n}\left[T\left(\frac{\partial S}{\partial n}\right)_T + \mu\right]\Delta n\,.$$

Hence

$$\begin{aligned}\overline{\Delta E \Delta T} &= \frac{1}{C_n}\left\{\overline{\Delta E^2} - \left[T\left(\frac{\partial S}{\partial n}\right)_T + \mu\right]\overline{\Delta E \Delta n}\right\}\\ &= \frac{1}{C_n}\left\{kT^2\left(\frac{\partial E}{\partial T}\right)_\mu + kT\mu\left(\frac{\partial E}{\partial \mu}\right)_T - kT\left[T\left(\frac{\partial S}{\partial n}\right)_T + \mu\right]\left(\frac{\partial E}{\partial \mu}\right)_T\right\}\\ &= \frac{kT^2}{C_n}\left[\left(\frac{\partial E}{\partial T}\right)_\mu - \left(\frac{\partial S}{\partial n}\right)_T\left(\frac{\partial E}{\partial \mu}\right)_T\right].\end{aligned}$$

One can show that the last expression in square brackets is C_n, which is $T(\partial S/\partial T)_n$, as follows:

$$\begin{aligned}T\mathrm{d}S = \mathrm{d}E - \mu\,\mathrm{d}n &= \left(\frac{\partial E}{\partial T}\right)_\mu \mathrm{d}T + \left(\frac{\partial E}{\partial \mu}\right)_T \mathrm{d}\mu - \mu\,\mathrm{d}n\\ &= \left(\frac{\partial E}{\partial T}\right)_\mu \mathrm{d}T + \left(\frac{\partial E}{\partial \mu}\right)_T\left(\frac{\partial \mu}{\partial T}\right)_n \mathrm{d}T + \left(\frac{\partial E}{\partial \mu}\right)_T\left(\frac{\partial \mu}{\partial n}\right)_T \mathrm{d}n - \mu\,\mathrm{d}n\,.\end{aligned}$$

The required expression for C_n is now easily obtained:

$$\begin{aligned}C_n = T\left(\frac{\partial S}{\partial T}\right)_n &= \left(\frac{\partial E}{\partial T}\right)_\mu + \left(\frac{\partial E}{\partial \mu}\right)_T\left(\frac{\partial \mu}{\partial T}\right)_n\\ &= \left(\frac{\partial E}{\partial T}\right)_\mu - \left(\frac{\partial E}{\partial \mu}\right)_T\left(\frac{\partial S}{\partial n}\right)_T,\end{aligned}$$

where we have used the relation $(\partial S/\partial n)_T = -(\partial\mu/\partial T)_n$ which is derivable from $\mathrm{d}(E-TS) = \mu\,\mathrm{d}n - S\mathrm{d}T$.

(e) Using the above expression for ΔT we obtain

$$\begin{aligned}\overline{\Delta n \Delta T} &= \frac{1}{C_n}\left\{\overline{\Delta n \Delta E} - \left[T\left(\frac{\partial S}{\partial n}\right)_T + \mu\right]\overline{\Delta n^2}\right\}\\ &= \frac{kT}{C_n}\left\{\left(\frac{\partial E}{\partial \mu}\right)_T - \left[T\left(\frac{\partial S}{\partial n}\right)_T + \mu\right]\left(\frac{\partial n}{\partial \mu}\right)_T\right\}\\ &= \frac{kT}{C_n}\left\{\left(\frac{\partial E}{\partial \mu}\right)_T - T\left(\frac{\partial S}{\partial \mu}\right)_T - \mu\left(\frac{\partial n}{\partial \mu}\right)_T\right\}.\end{aligned}$$

This is zero, since $\mathrm{d}E - T\mathrm{d}S - \mu\,\mathrm{d}n = 0$ gives

$$\mathrm{d}E = \left[T\left(\frac{\partial S}{\partial \mu}\right)_T + \mu\left(\frac{\partial n}{\partial \mu}\right)_T\right]\mathrm{d}\mu + \left[T\left(\frac{\partial S}{\partial T}\right)_\mu + \mu\left(\frac{\partial n}{\partial T}\right)_\mu\right]\mathrm{d}T$$

so that

$$\left(\frac{\partial E}{\partial \mu}\right)_T = T\left(\frac{\partial S}{\partial \mu}\right)_T + \mu\left(\frac{\partial n}{\partial \mu}\right)_T.$$

20.6 Apply the preceding result to obtain the fluctuations in the occupation number for a single particle state when the particles are Fermi-Dirac and when they are Einstein-Bose.

Solution

We take the system of the previous problem to be the single-particle state concerned and suppose this state has energy ϵ. In the Fermi-Dirac case

$$\bar{n} = \frac{1}{\exp[(\epsilon - \mu)/kT] + 1}$$

and so, since the system is microscopic, n must be replaced by $\bar{n}$ in the partial derivative (cf footnote to Problem 20.5),

$$\overline{\Delta n^2} = kT\left(\frac{\partial \bar{n}}{\partial \mu}\right)_T = \frac{\exp[(\epsilon - \mu)/kT]}{\{\exp[(\epsilon - \mu)/kT] + 1\}^2} = \bar{n} - \bar{n}^2 .$$

In the Einstein-Bose case

$$\bar{n} = \frac{1}{\exp[(\epsilon - \mu)/kT] - 1}$$

and so

$$\overline{\Delta n^2} = kT\left(\frac{\partial \bar{n}}{\partial \mu}\right)_T = \frac{\exp[(\epsilon - \mu)/kT]}{\{\exp[(\epsilon - \mu)/kT] - 1\}^2} = \bar{n} + \bar{n}^2 .$$

20.7 If, in the system considered, the particles which can be exchanged with the surroundings are non-interacting and the temperature is high enough for corrected classical counting of states [cf Problem 3.4(c)] to be used, show that the fluctuations in the number of particles obey a Poisson distribution and verify the expression for $\bar{N}$ which follows from comparison with the standard form of the Poisson distribution.

Solution

Let the single particle partition function be z_1, so that

$$z_1 = \sum_s \exp(-\epsilon_s/kT) ,$$

where s numbers the single-particle states and ϵ_s is the energy of the sth single-particle state.

If r numbers the states of the whole system, the energy and number of particles in the rth state being E_r and N_r respectively, the probability $P(N)$ that there will be N particles in the system is

$$P(N) = C \sum_{(N_r = N)} \exp[-(E_r - \mu N_r)/kT] ,$$

where the notation implies that the summation is over all states (i.e. all r) for which $N_r = N$. The normalisation constant C is just the inverse of the sum over all states (all r).

This gives, on writing Z_N for the canonical partition function of the system when it is constrained to have precisely N particles in it,

$$P(N) = C\exp(\mu N/kT)Z_N = C\exp(\mu N/kT)\frac{z_1^N}{N!}\ .$$

This is of the form

$$P(N) = C\frac{\overline{N}^N}{N!}\ ,$$

with

$$\overline{N} = \exp(\mu/kT)z_1$$

and normalisation gives

$$C = \exp(-\overline{N})$$

so that the distribution is the Poisson distribution

$$P(N) = \exp(-\overline{N})\frac{\overline{N}^N}{N!}\cdot$$

We can calculate $\overline{N}$ directly by writing $\overline{N} = \sum_s \overline{n}_s$ where $\overline{n}_s$ is the occupation number of the sth single particle state. The Fermi-Dirac and Einstein-Bose expressions for the mean occupation number both take the form $\overline{n}_s = \exp[-(\epsilon_s - \mu)/kT]$ in the circumstances under which classical counting is appropriate. Thus

$$\overline{N} = \sum_s \overline{n}_s = \exp(\mu/kT)\sum_s \exp(-\epsilon_s/kT) = \exp(\mu/kT)z_1$$

in agreement with the expression already obtained.

20.8 Compare the fluctuation in occupation number of a single particle state for non-interacting Einstein-Bose particles with the fluctuations in quantum number for a harmonic oscillator. Identical probability distributions imply the same mean square deviations from the mean, i.e. the same mean square fluctuations.

Solution

For Einstein-Bose particles the probability distribution for the occupation number n of a single particle state of energy ϵ is given, as follows from the grand canonical result, by

$$p_n = C\exp[-(n\epsilon - \mu n)/kT] = C\{\exp[-(\epsilon - \mu)/kT]\}^n\ .$$

The canonical distribution result gives for the probability p_r that an oscillator of angular frequency ω will be in its rth quantum state

$$p_r = C'\exp[-(r+\tfrac{1}{2})\hbar\omega/kT] = C''[\exp(-\hbar\omega/kT)]^r\ .$$

These distributions agree if we take $\epsilon = \hbar\omega$ and $\mu = 0$. The constants C and C'', since they are determined by normalisation, must clearly be the same when ϵ and μ are chosen as above.

21

Fluctuations of general classical mechanical variables

C.W.McCOMBIE
(*University of Reading, Reading*)

21.1 Show that if X is a coordinate of an entirely classical system in contact with a heat bath at temperature T and $\Delta X = X - \overline{X}$ then

$$\overline{\Delta X^2} = kT\left(\frac{\partial \overline{X}}{\partial F}\right)_T,$$

where F is the generalised force on X due to external forces.
[Hint: Consider the canonical distribution functions for the undisturbed system, and for the system with the force F applied, so that there is an extra term $-FX$ in the Hamiltonian.]

Solution

In the presence of the force F we have

$$\overline{X} = \frac{\int X(\mathbf{p}, \mathbf{q}) \exp[-(H_0 - FX)/kT] \mathrm{d}\mathbf{p} \mathrm{d}\mathbf{q}}{\int \exp[-(H_0 - FX)/kT] \mathrm{d}\mathbf{p} \mathrm{d}\mathbf{q}},$$

where we have introduced canonical coordinates and momenta and have expressed X as a function of them. Differentiation gives at once

$$\left(\frac{\partial \overline{X}}{\partial F}\right)_T = \frac{1}{kT} \frac{\int X^2(\mathbf{p}, \mathbf{q}) \exp[-(H_0 - FX)/kT] \mathrm{d}\mathbf{p} \mathrm{d}\mathbf{q}}{\int \exp[-(H_0 - FX)/kT] \mathrm{d}\mathbf{p} \mathrm{d}\mathbf{q}}$$

$$- \frac{1}{kT} \left\{ \frac{\int X(\mathbf{p}, \mathbf{q}) \exp[-(H_0 - FX)/kT] \mathrm{d}\mathbf{p} \mathrm{d}\mathbf{q}}{\int \exp[-(H_0 - FX)/kT] \mathrm{d}\mathbf{p} \mathrm{d}\mathbf{q}} \right\}^2$$

$$= \frac{1}{kT}(\overline{X^2} - \overline{X}^2) = \frac{1}{kT} \overline{\Delta X^2}.$$

We shall frequently wish to apply this result to determine the magnitude of the fluctuations at $F = 0$, in which case the derivative with respect to F will be taken at $F = 0$, but it is useful to note that the result holds more generally. The more general result will be applied later (see Problem 21.3) to determining the fluctuations in volume of a system under a fixed non-zero pressure.

21.2 We consider now the derivation of the result of Problem 21.1 in the case in which the system has to be described quantum mechanically but the variable X behaves classically. This means that all components of frequency ω in the time variation of X will be negligible unless $\hbar\omega \ll kT$[1]. Contributions of frequency ω in the time dependence of X arise from matrix elements of X between stationary states which differ in energy by $\hbar\omega$, so take the classical behaviour of X to imply it has negligible matrix elements between states which differ in energy by more than a very small fraction of kT. Use this assumption to derive the the result of Problem 21.1.

Solution

We again write the Hamiltonian of the system as $H_0 - FX$ and we let r number the eigenstates $|r\rangle$, eigenvalues E_r, of H_0. The mean value of X when the force is F can then be written

$$\bar{X} = \frac{\mathrm{Tr}\{X\exp[-(H_0 - FX)/kT]\}}{\mathrm{Tr}\{\exp[-(H_0 - FX)/kT]\}}$$

$$= \frac{\sum_r \langle r|X\exp[-(H_0 - FX)/kT]|r\rangle}{\sum_r \langle r|\exp[-(H_0 - FX)/kT]|r\rangle}$$

$$= \frac{\sum_{r,s} \langle r|X|s\rangle\langle s|\exp[-(H_0 - FX)/kT]|r\rangle}{\sum_r \langle r|\exp[-(H_0 - FX)/kT]|r\rangle} .$$

It is not a straightforward matter to differentiate this with respect to F, because X and H_0 need not commute. The classical behaviour which has been postulated for X allows us, however, to throw the expression into an equivalent form which can be differentiated without difficulty. If we suppose the exponential operator expanded in powers of its exponent, a typical term in $\langle s|\exp[-(H_0 - FX)/kT]|r\rangle$ will be

$$\frac{(-1)^n}{n!(kT)^n} \sum_{p_1 \ldots p_{n-1}} \langle s|H_0 - FX|p_1\rangle\langle p_1|H_0 - FX|p_2\rangle \ldots \langle p_{n-1}|H_0 - FX|r\rangle .$$

X and therefore $H_0 - FX$ (since H_0 has only diagonal matrix elements) will have appreciable matrix elements only between states which differ in energy by much less than kT. It follows that all the states which appear in terms contributing appreciably to the above sum will have eigenvalues

[1] L.D.Landau and E.M.Lifshitz, *Statistical Physics* (Pergamon Press, London), 1968, p.345.

of H_0 which differ negligibly from E_r. We can, therefore, replace H_0 by E_r in all such terms and so in the exponential. Noting that $\exp(-E_r/kT)$ is then simply a numerical multiplying factor, we have

$$\overline{X} = \frac{\sum_{r,s} \exp(-E_r/kT)\langle r|X|s\rangle\langle s|\exp(FX/kT)|r\rangle}{\sum_r \exp(-E_r/kT)\langle r|\exp(FX/kT)|r\rangle}$$

$$= \frac{\sum_r \exp(-E_r/kT)\langle r|X\exp(FX/kT)|r\rangle}{\sum_r \exp(-E_r/kT)\langle r|\exp(FX/kT)|r\rangle} .$$

Since there is now no difficulty about commuting operators we can differentiate with respect to F in a straightforward way and find

$$\left(\frac{\partial \overline{X}}{\partial F}\right)_T = \frac{1}{kT}\frac{\sum_r \exp(-E_r/kT)\langle r|X^2\exp(FX/kT)|r\rangle}{\sum_r \exp(-E_r/kT)\langle r|\exp(FX/kT)|r\rangle}$$

$$-\frac{1}{kT}\left[\frac{\sum_r \exp(-E_r/kT)\langle r|X\exp(FX/kT)|r\rangle}{\sum_r \exp(-E_r/kT)\langle r|\exp(FX/kT)|r\rangle}\right]^2 .$$

But it is clear that the argument applied above to throw $\overline{X}$ into an alternative form will, when applied to $\overline{X^2}$, yield just the first term in the preceding equation. Consequently

$$kT\left(\frac{\partial \overline{X}}{\partial F}\right)_T = \overline{X^2} - \overline{X}^2 = \overline{\Delta X^2} .$$

It may be convenient to include part of the term FX in H_0: this will not affect the above argument. If, for example, X is the volume so that F is the negative of the pressure one may wish to avoid considering a system under zero pressure since such a system (e.g. a perfect gas) may not have finite volume.

21.3 Use the result of Problem 21.2 to find (a) the fluctuations in volume of a system, at a fixed external pressure which need not be zero, (b) the mean square fluctuation in deflection of a galvanometer suspension, (c) the mean square fluctuation in charge on a condenser of capacity C with a resistor connected across its plates, and (d) the mean square fluctuation in the distance r between adjacent atoms in a one-dimensional chain of atoms of one kind interacting by nearest neighbour harmonic forces. All the systems are supposed in contact with a heat bath at temperature T.

Solution

(a) The generalised force associated with volume v is the negative pressure, $-p$. Thus the general result gives

$$\overline{\Delta v^2} = -kT\left(\frac{\partial v}{\partial p}\right)_T .$$

This relates the volume fluctuations to the isothermal compressibility.

(b) The generalised force associated with deflection θ is the couple M. The general result therefore gives

$$\overline{\Delta\theta^2} = kT\left(\frac{\partial\theta}{\partial M}\right)_T .$$

If c is the (isothermal) torsion constant this gives

$$\overline{\Delta\theta^2} = \frac{kT}{c} .$$

This agrees, of course, with the equipartition result

$$\tfrac{1}{2}c\overline{\Delta\theta^2} = \tfrac{1}{2}kT$$

but the derivation from the general formula avoids doubts about the validity of treating the macroscopic suspended system as having just one degree of freedom.

(c) The generalised force corresponding to charge q on the condenser plates may be taken to be the potential difference V across a battery introduced in series with the condenser, the positive terminal of the battery being connected to the condenser plate on which the charge is q (the work done by the battery is then $V\delta q$). The general result therefore gives

$$\overline{\Delta q^2} = kT\left(\frac{\partial q}{\partial V}\right)_T = kTC .$$

This agrees with the equipartition result which is

$$\tfrac{1}{2}\frac{\overline{\Delta q^2}}{C} = \tfrac{1}{2}kT .$$

(d) The generalised force associated with the distance r between the neighbours is F if a pair of forces of magnitude F act outward on the two atoms. If the increase in energy of the bond between the neighbouring atoms is $\tfrac{1}{2}\kappa(\delta r)^2$ then the result

$$\overline{\Delta r^2} = kT\left(\frac{\partial r}{\partial F}\right)_T$$

becomes $\overline{\Delta r^2} = \kappa kT$, it being assumed that the chain has free ends or is very long with fixed ends.

21.4 Show that the result established in Problems 21.1 and 21.2 can break down if X does not behave as a classical variable by supposing that X is the x component of the dipole moment of a large number N of negligibly interacting charged particles of mass m and charge e bound to centres such that their frequency of oscillation is ω_0.

Solution

If we denote the x component of displacement of the rth particle by x_r and the x component of the total dipole moment by P_x, we have

$$P_x = \sum_r e x_r \,.$$

Since x_r and x_s $(r \neq s)$ are independent and, in the absence of an external field, have mean value zero

$$\overline{x_r x_s} = 0\,.$$

Thus

$$\overline{P_x^2} = e^2 \sum_{r,s} \overline{x_r x_s} = e^2 \sum_r \overline{x_r^2} = N e^2 \overline{x^2}\,.$$

The mean potential energy of an oscillator is half the mean energy, so

$$\tfrac{1}{2} m \omega_0^2 \overline{x^2} = \frac{1}{2}\left[\tfrac{1}{2}\hbar\omega_0 + \frac{\hbar\omega_0}{\exp(\hbar\omega_0/kT) - 1}\right].$$

This gives

$$\overline{P_x^2} = \frac{Ne^2}{m\omega_0^2}\left[\tfrac{1}{2}\hbar\omega_0 + \frac{\hbar\omega_0}{\exp(\hbar\omega_0/kT) - 1}\right].$$

Since $\overline{P}_x$ is zero in the absence of a field, this is also $\overline{\Delta P_x^2}$. Turning now to an application of the result derived for a classical variable, we note that the generalised force associated with P_x is E_x, the x component of the electric field. In an electric field

$$\overline{P}_x = Ne\overline{x} = \frac{Ne^2 E_x}{m\omega_0^2}\,,$$

so that

$$\frac{\partial \overline{P}_x}{\partial E_x} = \frac{Ne^2}{m\omega_0^2}\,.$$

We see that the result

$$\overline{\Delta P_x^2} = kT\left(\frac{\partial P_x}{\partial E_x}\right)_T$$

holds only for $kT \gg \hbar\omega_0$. Notice that this means that the states of the system between which P_x has non-zero matrix elements, namely states in which one oscillator has changed its quantum number by ± 1, will then differ in energy by much less than kT. This agrees with the form of the criterion for classical behaviour which we have used in preceding problems.

21.5 Suppose that X can be expressed as a linear superposition of normal coordinates q_r of the system (kinetic energy $\sum_r \frac{1}{2}\dot{q}_r^2$, potential energy $\sum_r \frac{1}{2}\omega_r^2 q_r^2$):

$$X = \sum_r \alpha_r q_r$$

(X might be, for example, the transverse displacement of the middle point of a stretched string). The mean square fluctuation in X may be calculated either by applying the result of Problem 21.2 or by using the familiar results for $\overline{q_r^2}$. Show that if X is classical (and what this implies for the α's should be considered), then the two methods will lead to the same result.

Solution

Consider first the requirement that X is classical. It will have matrix elements proportional to $\alpha_r/\omega_r^{1/2}$ between states differing in energy by $\hbar\omega_r$. These matrix elements will be negligible for states differing in energy by more than a small fraction of kT, provided $\alpha_r/\omega_r^{1/2}$ is negligible, except when $\hbar\omega_r \ll kT$. This means that all oscillators which play an appreciable part in the problem can be treated classically.

To apply the result of Problem 21.2 we note that a generalised force F_x applied to X will result in a generalised force $F_x\alpha_r$ on the coordinate q_r and so a displacement $F_x\alpha_r/\omega_r^2$ of the mean value of this coordinate. It follows that the resulting displacement of $\overline{X}$ (from zero) will be given by

$$\overline{X} = F_x \sum_r \frac{\alpha_r^2}{\omega_r^2} .$$

This implies

$$\overline{X^2} = kT\frac{\partial \overline{X}}{\partial F_x} = kT\sum_r \frac{\alpha_r^2}{\omega_r^2} .$$

The second method (since $\overline{q_r q_s} = 0, r \neq s$) gives

$$\overline{X^2} = \sum_r \alpha_r^2 \overline{q_r^2} .$$

Now $\frac{1}{2}\omega_r^2\overline{q_r^2} = \frac{1}{2}kT$, since, as already discussed, all oscillators which contribute appreciably can be treated classically. Substituting for $\overline{q_r^2}$ gives the same final result as the first method.

21.6 Show that if X_1 and X_2 are two coordinates of an entirely classical system and the corresponding generalised forces are F_1 and F_2 then the correlation function for the fluctuations of the two coordinates when the system is in contact with a heat bath at temperature T is given by

$$\overline{\Delta X_1 \Delta X_2} = \left(\frac{\partial \overline{X}_1}{\partial F_2}\right)_T = \left(\frac{\partial \overline{X}_2}{\partial F_1}\right)_T .$$

[Note: There is no difficulty in establishing the same result for a quantum mechanical system, X_1 and X_2 being assumed to behave classically. It is necessary only to modify slightly the argument used in Problem 21.2.]

Solution

We take the Hamiltonian to be

$$H_0 - F_1 X_1 - F_2 X_2 ,$$

where H_0, X_1, and X_2 are all functions of the generalised coordinates and conjugate momenta $\mathbf{q}, \mathbf{p}$. Then

$$\overline{X}_1 = \frac{\int X_1 \exp[-(H_0 - F_1 X_1 - F_2 X_2)/kT]\,\mathrm{d}\mathbf{q}\,\mathrm{d}\mathbf{p}}{\int \exp[-(H_0 - F_1 X_1 - F_2 X_2)/kT]\,\mathrm{d}\mathbf{q}\,\mathrm{d}\mathbf{p}} .$$

This gives

$$\left(\frac{\partial X_1}{\partial F_2}\right)_T = \frac{1}{kT}\frac{\int X_1 X_2 \exp[-(H_0 - F_1 X_1 - F_2 X_2)/kT]\,\mathrm{d}\mathbf{q}\,\mathrm{d}\mathbf{p}}{\int \exp[-(H_0 - F_1 X_1 - F_2 X_2)/kT]\,\mathrm{d}\mathbf{q}\,\mathrm{d}\mathbf{p}}$$

$$- \frac{1}{kT}\frac{\int X_1 \exp[-(H_0 - F_1 X_1 - F_2 X_2)/kT]\,\mathrm{d}\mathbf{q}\,\mathrm{d}\mathbf{p}}{\int \exp[-(H_0 - F_1 X_1 - F_2 X_2)/kT]\,\mathrm{d}\mathbf{q}\,\mathrm{d}\mathbf{p}}$$

$$\times \frac{\int X_2 \exp[-(H_0 - F_1 X_1 - F_2 X_2)/kT]\,\mathrm{d}\mathbf{q}\,\mathrm{d}\mathbf{p}}{\int \exp[-(H_0 - F_1 X_1 - F_2 X_2)/kT]\,\mathrm{d}\mathbf{q}\,\mathrm{d}\mathbf{p}}$$

$$= \frac{1}{kT}(\overline{X_1 X_2} - \overline{X}_1 \overline{X}_2) = \frac{1}{kT}\overline{\Delta X_1\, \Delta X_2} .$$

Clearly the same calculation applied to $(\partial X_2/\partial F_1)_T$ leads to the same expression.

21.7 A conductor is connected by a thin wire to a very large conductor, the two conductors and the connecting wire all being formed from the same metal. The large conductor acts as a heat bath (temperature T) and electron reservoir (chemical potential μ) with which the smaller conductor is in contact via the wire. The smaller conductor, in the presence of the large one, has electrical capacity $\mathscr{C}$.

The fluctuations Δn in the number of electrons on the smaller conductor may be determined either by applying the result for the fluctuations in charge on a capacitor (Problem 21.3) or by using the expression for the fluctuations in n derived from the grand canonical ensemble (Problem 20.5). Verify that the two approaches yield the same result.

Solution

The mean square value of the fluctuation $e\Delta n$ in the charge is given by

$$\frac{1}{2}\frac{e^2\overline{\Delta n^2}}{\mathscr{C}} = \tfrac{1}{2}kT$$

or

$$\overline{\Delta n^2} = \frac{\mathscr{C}kT}{e^2}\ .$$

On the other hand the result derived from the grand canonical ensemble gives

$$\overline{\Delta n^2} = kT\left(\frac{\partial n}{\partial \mu}\right)_T\ .$$

The chemical potential which appears here is of course the chemical potential of the electrons in the particle reservoir, which will also be the chemical potential of the electrons in the smaller conductor (fluctuations being irrelevant to evaluating the right-hand side). The chemical potential can be raised, at constant temperature, by adding electrons to the system. These additional electrons will reside on the surface of the metal system and will raise the chemical potential merely by adding electrostatic energy to the electrons of the system. If δn is the increase in the number of electrons on the smaller conductor, the electrostatic potential will increase from zero to $e\delta n/\mathscr{C}$, and the additional energy of the electrons will be $e^2\delta n/\mathscr{C}$. This is just the increase $\delta\mu$ in the chemical potential, so that

$$\delta\mu = \frac{e^2\delta n}{\mathscr{C}}\ .$$

Thus

$$\left(\frac{\partial n}{\partial \mu}\right)_T = \frac{\mathscr{C}}{e^2}$$

and the two expressions derived above are consistent.

22

Fluctuations of thermodynamic variables: constant pressure systems, isolated systems

C.W.McCOMBIE
(*University of Reading, Reading*)

22.1 In Problem 20.1 we considered the fluctuations in energy of a system at zero external pressure in contact with a heat bath. We turn now to considering the fluctuations in thermodynamic quantities for a system at arbitrary fixed external pressure in contact with a heat bath. To avoid the difficulties associated with the introduction of a pressure ensemble (cf Problem 3.18) we shall regard the fixed external pressure p as introducing an extra term pv into the Hamiltonian, where v is the operator associated with the volume. This enables us to use the canonical ensemble.

(a) The observable associated with the Hamiltonian of the system $H_0 + pv$ is $E + pv$ where E is the internal energy. Fluctuations in this quantity, at fixed p, can be investigated in exactly the same way as were energy fluctuations in the zero pressure case (Problem 20.1). Show that this leads to the result for entropy fluctuations

$$\overline{\Delta S^2} = kC_p \ .$$

(b) Show also that

$$\overline{\Delta S \Delta v} = kT\left(\frac{\partial v}{\partial T}\right)_p \ .$$

[Recall the result (Problem 21.3): $\overline{\Delta v^2} = -kT(\partial v/\partial p)_T$.]

Solution

(a) The argument which gave for the zero pressure case (Problem 20.1)

$$\overline{\Delta E^2} = kT^2\left(\frac{\partial E}{\partial T}\right)$$

now gives, since p is fixed,

$$\overline{(\Delta E + p\Delta v)^2} = kT^2\left[\frac{\partial(E+pv)}{\partial T}\right]_p \ .$$

But the first law of thermodynamics gives

$$T\Delta S = \Delta E + p\Delta v \ ,$$

so that the left-hand side becomes $T^2\overline{\Delta S^2}$ while the right-hand side is $kT^2 C_p$. We thus have

$$\overline{\Delta S^2} = kC_p \ .$$

Notice that we have made use of the following way of attaching a meaning to the fluctuation of a thermodynamic quantity such as entropy. We determine from ordinary thermodynamic considerations a linear relation between variations in the thermodynamic quantity and variations in energy and volume. We then use this linear relation to associate fluctuations in the thermodynamic quantity with the unambiguous fluctuations in the purely mechanical quantities energy and volume.

(b) We have, denoting the eigenstates of H_0+pv by $|r\rangle$,

$$\bar{v} = \frac{\mathrm{Tr}\{v\exp[-(H_0+pv)/kT]\}}{\mathrm{Tr}\{\exp[-(H_0+pv)/kT]\}} = \frac{\sum_{r,s}\langle r|v|s\rangle\langle s|\exp[-(H_0+pv)/kT]|r\rangle}{\sum_r\langle r|\exp[-(H_0+pv)/kT]|r\rangle} \ .$$

Differentiating with respect to T at constant p, and noting that (H_0+pv) and $\exp[-(H_0+pv)/kT]$ have only diagonal elements, we have

$$\left(\frac{\partial\bar{v}}{\partial T}\right)_p = \frac{1}{kT^2}\left\{\frac{\sum_r\langle r|v|r\rangle\langle r|H_0+pv|r\rangle\langle r|\exp[-(H_0+pv)/kT]|r\rangle}{\sum_r\langle r|\exp[-(H_0+pv)/kT]|r\rangle}\right.$$

$$-\frac{\sum_r\langle r|v|r\rangle\langle r|\exp[-(H_0+pv)/kT]|r\rangle}{\sum_r\langle r|\exp[-(H_0+pv)/kT]|r\rangle}$$

$$\left.\times\frac{\sum_r\langle r|H_0+pv|r\rangle\langle r|\exp[-(H_0+pv)/kT]|r\rangle}{\sum_r\langle r|\exp[-(H_0+pv)/kT]|r\rangle}\right\}$$

$$= \frac{1}{kT^2}[\overline{v(E+pv)} - \bar{v}\overline{(E+pv)}] = \frac{1}{kT^2}\overline{\Delta v(\Delta E+p\Delta v)}$$

$$= \frac{1}{kT^2}T\overline{\Delta v\Delta S} \ .$$

Thus

$$\overline{\Delta v\Delta S} = kT\left(\frac{\partial v}{\partial T}\right)_p \ .$$

22.2 Deduce from the results of the previous question the values of $\overline{\Delta E^2}$ and $\overline{\Delta E\Delta v}$ for a system at fixed external pressure in contact with a heat bath.

Solution

We have

$$\Delta E = T\Delta S - p\Delta v\,.$$

It follows that

$$\begin{aligned}\overline{\Delta E^2} &= T^2\overline{\Delta S^2} - 2pT\overline{\Delta S\Delta v} + p^2\overline{\Delta v^2}\\ &= kT^2C_p - 2pkT^2\left(\frac{\partial v}{\partial T}\right)_p - p^2kT\left(\frac{\partial v}{\partial p}\right)_T\,.\end{aligned}$$

Also

$$\overline{\Delta E\Delta v} = T\overline{\Delta S\Delta v} - p\overline{\Delta v^2} = kT^2\left(\frac{\partial v}{\partial T}\right)_p + pkT\left(\frac{\partial v}{\partial p}\right)_T\,.$$

These results can, of course, be put in a variety of different forms.

Note that if p is zero we reproduce the result kT^2C_p for $\overline{\Delta E^2}$, which we derived in Problem 20.1.

22.3 Deduce from the general result of the previous question that the fluctuations in internal energy and volume of a perfect classical gas at fixed external pressure and in contact with a heat bath at temperature T satisfy

$$\overline{\Delta E^2} = kT^2C_v$$

and

$$\overline{\Delta E\Delta v} = 0\,.$$

Solution

For a perfect gas of N particles

$$pv = NkT$$

and so

$$\left(\frac{\partial v}{\partial T}\right)_p = \frac{Nk}{p}$$

$$\left(\frac{\partial v}{\partial p}\right)_T = -\frac{NkT}{p^2}\,.$$

Substituting these expressions and

$$C_p = C_v + Nk$$

in the general expressions for $\overline{\Delta E^2}$ and $\overline{\Delta E\Delta v}$ leads at once to the required results.

These results may be obtained directly from the fact that the classical canonical distribution is such that the total internal energy of a perfect gas (translational kinetic energy of the molecules plus the internal energy of the molecules) has a probability distribution independent of any requirement placed on the positions of the molecules.

22.4 Starting from the result that, if X is a classical variable associated with an isolated system, then the probability $p(X)\mathrm{d}X$ that the variable has a value between X and $X+\mathrm{d}X$ is given by

$$p(X) \propto \exp[S(X)/k]\,,$$

where $S(X)$ is the entropy of the system constrained to have the value X for the variable, show that for an isolated system

$$\overline{\Delta X^2} = kT\left(\frac{\partial X}{\partial F}\right)_S,$$

where F is a generalised force associated with the variable, this force being supposed applied externally (cf Problems 21.1 and 22.6).

[Hint: Expand the entropy as far as quadratic terms in ΔE and ΔX about the energy E_0 and the equilibrium value X_0 of the variable appropriate to the energy E_0. Show that the coefficients of ΔX, ΔE, and $-\Delta X^2$ are zero, $1/T$, and $\frac{1}{2}k/\overline{\Delta X^2}$ respectively. To determine the relation between F and ΔX suppose that F is applied reversibly. This means that the entropy stays constant and that (for small F) the increase in energy ΔE will be $\frac{1}{2}F\Delta X$. The required relation follows at once.]

Solution

Since, when the force F is applied, the internal energy E can vary from its initial value E_0 as the force does work on the system, we must consider the entropy to be a function of E as well as X, say $S(E, X)$. Expanding this about E_0 and X_0 (the equilibrium value of X when F is zero) we get, retaining only terms up to the quadratic,

$$S(E,X) = S(E_0, X_0) + A\Delta E + B\Delta X - \tfrac{1}{2}(C\Delta E^2 + 2D\Delta E\Delta X + G\Delta X^2)\,.$$

Since, when $E = E_0$ ($\Delta E = 0$) the entropy has a maximum at $X = X_0$ ($\Delta X = 0$), we have

$$B = 0.$$

Also

$$\frac{1}{T} = \left(\frac{\partial S}{\partial E}\right)_X = A\,,$$

where we have used in analogy with $đQ = \mathrm{d}U + p\,\mathrm{d}v$ (Problem 1.5)

$$T\mathrm{d}S = \mathrm{d}E - F\mathrm{d}X\,.$$

Again, when F is zero, so that ΔE is zero, the stated relation between probability and entropy gives that the probability distribution $p(X)$ for X will be proportional to $\exp[-\frac{1}{2}(G/k)\Delta X^2]$. Comparing this with the standard form of the normal distribution, $\exp(-\Delta X^2/2\overline{\Delta X^2})$, we get

$$G = \frac{k}{\overline{\Delta X^2}}\,.$$

When F is non-zero, and is supposed to have been applied reversibly,

$$\Delta E = \tfrac{1}{2}F\Delta X\,,$$

and the entropy is given, to second order in the small quantities F and ΔX, by

$$S(E_0+\tfrac{1}{2}F\Delta X, \Delta X) = S(E_0, X_0)+\tfrac{1}{2}\frac{F}{T}\Delta X-\tfrac{1}{2}G\Delta X^2\,.$$

Since the process is reversible, the entropy must remain equal to its initial value $S(E_0, X_0)$, and this implies

$$\Delta X = \frac{F}{GT} = F\frac{\overline{\Delta X^2}}{kT}\,.$$

Thus, at $F = 0$,

$$\left(\frac{\partial X}{\partial F}\right)_S = \frac{\overline{\Delta X^2}}{kT}\,,$$

which is the required result.

22.5 Suppose that the entropy S is a function of a number of classical variables $X_1, X_2, ..., X_n$, which we shall take to be zero at equilibrium (i.e. at maximum entropy). Thus

$$S = S_0-\tfrac{1}{2}\sum_{r,s}a_{rs}X_rX_s$$

where only quadratic terms have been retained and the quadratic form $\sum_{r,s}a_{rs}X_rX_s$ is positive definite.

Corresponding to each X_r define a force F_r by

$$F_r = -\frac{\partial S}{\partial X_r} = \sum_s a_{rs}X_s$$

so that each F_r is a linear combination of the X_s.

Starting from the $\exp(S/k)$ form for the probability distribution, prove that

$$\overline{F_uX_v} = k\delta_{u,v}\,,$$

where $\delta_{u,v}$ is unity if $u = v$, and zero otherwise.

[This result plays an important part in establishing the Onsager relations (Problem 25.7).]

Solution

We have

$$\overline{F_uX_v} = \int F_uX_vC\exp(S/k)\prod_s \mathrm{d}X_s$$

where C is a normalising constant such that

$$\int C\exp(S/k)\prod_s \mathrm{d}X_s = 1 .$$

Since

$$F_u \exp(S/k) = -\frac{\partial S}{\partial X_u}\exp(S/k) = -k\frac{\partial}{\partial X_u}\exp(S/k) ,$$

we can integrate the expression for $\overline{F_u X_v}$ by parts with respect to X_u. The integrated part vanishes because $\exp(S/k)$ goes to zero at the limits, and we get

$$\overline{F_u X_v} = k\int \frac{\partial X_v}{\partial X_u} C\exp(S/k)\prod_s \mathrm{d}X_s = k\delta_{u,v} .$$

22.6 We can consider the fluctuations in volume of a body under zero pressure: (a) if it is isolated with an energy E such that its temperature is T, and (b) when it is in contact with a heat bath at temperature T. Write down the mean square fluctuations in volume in the two cases in terms of appropriate compressibilities.

It seems reasonable to suppose that the fluctuations in the case of contact with a heat bath will be greater than in the case of isolation, because there will be fluctuations in energy of the body in the first case. These can be interpreted as fluctuations in temperature which, by producing thermal expansion, will give rise to fluctuations in volume. Verify by direct calculation that the mean square fluctuation in volume, due to energy fluctuations and resulting thermal expansion, is equal to the excess of isothermal mean square fluctuations (case (b)) over adiabatic fluctuations (case (a)).

Solution

If we recall that the generalised force associated with volume v is $-p$, we can use the general results of Problems 22.4 and 21.1 to write down the mean square fluctuations in the two cases:

(a) $$(\overline{\Delta v^2})_S = -kT\left(\frac{\partial v}{\partial p}\right)_S ,$$

(b) $$(\overline{\Delta v^2})_T = -kT\left(\frac{\partial v}{\partial p}\right)_T ,$$

so that the fluctuations in the two cases depend on the adiabatic and isothermal compressibilities, respectively.

The volume fluctuations due to energy fluctuations may be determined as follows

$$\overline{\Delta v^2} = \left(\frac{\partial v}{\partial T}\right)_p^2 \frac{\overline{\Delta E^2}}{C_p^2} = \left(\frac{\partial v}{\partial T}\right)_p^2 \frac{kT^2}{C_p} .$$

We have therefore

$$(\overline{\Delta v^2})_T = (\overline{\Delta v^2})_S + \overline{\Delta v^2} ,$$

provided

$$\left(\frac{\partial v}{\partial p}\right)_S = \left(\frac{\partial v}{\partial p}\right)_T + \frac{T}{C_p}\left(\frac{\partial v}{\partial T}\right)_p^2 ,$$

and this is a standard thermodynamic result.

[Since the adiabatic compressibility $K_a \equiv K_S$ is $(1/v)(\partial v/\partial p)_a$ (see Problem 1.4), this thermodynamic relation is

$$K_T - K_S = Tv\frac{\alpha_p^2}{C_p} ,$$

where α_p is the coefficient of volume expansion. This relation is established in Problem 1.8(b).]

23

Time dependence of fluctuations: correlation functions, power spectra, Wiener-Khintchine relations

C.W.McCOMBIE
(*University of Reading, Reading*)

23.1 We consider a fluctuating quantity $y(t)$ (deflection of suspended mirror executing **Brownian fluctuations**, component of velocity of a diffusing particle, or the like) with statistical properties independent of time. We shall assume that time averages and ensemble averages are equal. Putting

$$\Delta y(t) = y(t) - \overline{y}$$

we define the **correlation function** $\psi_y(\tau)$ by

$$\psi_y(\tau) = \overline{\Delta y(t)\Delta y(t+\tau)}\,,$$

from which it follows at once that $\psi_y(0) = \overline{\Delta y^2}$ and $\psi_y(\tau) = \psi_y(-\tau)$. It will be assumed that there is a time τ_c such that $\psi_y(\tau)$ is negligibly small for $|\tau| > \tau_c$.

Assuming $S \gg \tau_c$, show that the mean square fluctuation from the mean of $\int_0^S y(t)\,dt$ is $2S\int_0^\infty \psi_y(\tau)\,d\tau$.

Solution

The fluctuation from the mean, $\int_0^S y(t)\,dt - \overline{\int_0^S y(t)\,dt}$, is clearly $\int_0^S \Delta y(t)\,dt$. The required mean square value is therefore

$$\overline{\left(\int_0^S \Delta y(t)\,dt\right)^2} = \overline{\int_0^S \int_0^S \Delta y(t_1)\Delta y(t_2)\,dt_1\,dt_2}$$

$$= \int_0^S \int_0^S \overline{\Delta y(t_1)\Delta y(t_2)}\,dt_1\,dt_2\,.$$

On changing variables to $t = t_1$ and $\tau = t_2 - t_1$, this becomes

$$\int_0^S dt \int_{-t}^{S-t} \psi_y(\tau)\,d\tau\,.$$

Provided both t and $S-t$ are greater than τ_c (and, since $S \gg \tau_c$ this holds for all except negligibly small portions of the range of t values between 0 and S) one can extend the range of integration of the integral over τ to $-\infty$ and $+\infty$ without changing the value of the whole expression. Thus

$$\overline{\left(\int_0^S \Delta y(t)\,\mathrm{d}t\right)^2} = \int_0^S \mathrm{d}t \int_{-\infty}^{\infty} \psi_y(\tau)\,\mathrm{d}\tau = S\int_{-\infty}^{\infty} \psi_y(\tau)\,\mathrm{d}\tau = 2S\int_0^{\infty} \psi_y(\tau)\,\mathrm{d}\tau\,.$$

23.2 Assuming in each case[1] that the correlation function for positive τ is proportional to the way in which the quantity decays macroscopically (i.e. with neglect of fluctuations) from an initial non-zero value to zero (the initial rate of change of the quantity, if necessary for specifying the macroscopic decay, being taken to be zero), determine the correlation functions at temperature T for

(a) the x component of the velocity $\mathbf{v}$ of a particle of mass m moving in a medium such that the force resisting its motion is $-\kappa\mathbf{v}$,

(b) the deflection θ of a suspended system of moment of inertia I subject to a damping couple $-\gamma\dot{\theta}$ and a restoring couple $c\theta$. Assume damping less than critical.

Solution

(a) The macroscopic equation of motion is

$$m\dot{v}_x + \kappa v_x = 0.$$

This has the solution

$$v_x = v_x(0)\exp(-\kappa t/m)\,.$$

Thus for positive τ the correlation function is proportional to $\exp(-\kappa\tau/m)$ and, since $\psi_{v_x}(\tau) = \psi_{v_x}(-\tau)$, this implies

$$\psi_{v_x}(\tau) = A\exp(-\kappa|\tau|/m)\,.$$

But we have $A = \psi_{v_x}(0)$ and by a previous result $\psi_{v_x}(0) = \overline{v_x^2}$.

From the equipartition result we have $\frac{1}{2}m\overline{v_x^2} = \frac{1}{2}kT$, so $\overline{v_x^2} = kT/m$. Thus

$$\psi_{v_x}(\tau) = \frac{kT}{m}\exp(-\kappa|\tau|/m)\,.$$

(b) The macroscopic equation of motion is now

$$I\ddot{\theta} + \gamma\dot{\theta} + c\theta = 0\,.$$

[1] It can be shown by general arguments or—as we shall see later (Problem 24.1)—in special cases by detailed calculation, that the macroscopic-decay correlation-function assumption made here is equivalent to assuming that the fluctuations can be regarded as due to the action of a fluctuating force with correlation time τ_c which is negligible on the time scale of the macroscopic decay.

The solution of this with $\dot{\theta}(0) = 0$ is easily found to be

$$\theta(t) = \theta(0)\exp(-\beta t)\left[\cos\omega' t + \frac{\beta}{\omega'}\sin\omega' t\right],$$

where $\beta = \gamma/2I$ and $\omega' = (4Ic - \gamma^2)^{1/2}/2I$.

It follows that the correlation function is given by

$$\psi_\theta(\tau) = \frac{kT}{c}\exp(-\beta|\tau|)\left[\cos\omega'|\tau| + \frac{\beta}{\omega'}\sin\omega'|\tau|\right].$$

23.3 Use the results of Problems 23.1 and 23.2(a) to determine the mean square x displacement in time S of a particle moving at temperature T in a medium such that its mobility is μ $(= F_x/v_x)$.

Combine this with the solution

$$\rho(r, t) = n(4\pi Dt)^{-3/2}\exp(-r^2/4Dt)$$

of the diffusion equation for n particles concentrated at the origin at $t = 0$ to relate mobility μ and diffusion constant D.

Solution

Since the x component of the displacement in time S is related to the x component of velocity by

$$x_S = \int_0^S v_x(t)\,dt,$$

we have

$$\overline{x_S^2} = \overline{\left(\int_0^S v_x(t)\,dt\right)^2} = 2S\int_0^\infty \psi(\tau)\,d\tau,$$

where $\psi(\tau)$ is the correlation function for v_x.

Since $F_x = (1/\mu)v_x$ we see that $1/\mu$ plays the part of the damping constant κ, so that the result of Problem 23.2(a) gives here

$$\psi(\tau) = \frac{kT}{m}\exp(-|\tau|/\mu m).$$

The integral of this from 0 to ∞ is μkT and the above expression for $\overline{x_S^2}$ becomes

$$\overline{x_S^2} = 2\mu kTS.$$

Turning now to the given solution of the diffusion equation and substituting $r^2 = x^2 + y^2 + z^2$ we note that integrating over the y and z coordinates gives that the probability distribution at time t of the x coordinate of particles at the origin at $t = 0$ is proportional to

$$\exp(-x^2/4Dt).$$

Comparing this with the standard normal distribution $\exp(-x^2/2\overline{\Delta x^2})$ gives $2Dt$ for the mean square displacement. Thus

$$\overline{x_S^2} = 2DS ,$$

and comparing with the previous expression for $\overline{x_S^2}$ gives

$$\frac{D}{\mu} = kT .$$

This is, of course, the **Einstein relation,** which can be obtained in other ways, see for example Problem 17.12.

23.4 Assuming the relation between correlation function and macroscopic decay for the current in a circuit of inductance L and resistance R, determine the mean square fluctuation in the charge passing round the circuit in time S.

Solution

The macroscopic equation for the current I,

$$L\frac{\mathrm{d}I}{\mathrm{d}t} + RI = 0 ,$$

gives

$$I = I_0 \exp[-(R/L)t] .$$

Combining this with $\frac{1}{2}L\overline{I^2} = \frac{1}{2}kT$ we get

$$\psi_I(\tau) = \frac{kT}{L}\exp[-(R/L)|\tau|] .$$

It follows, using the result of Problem 23.1, that

$$\overline{Q^2} \equiv \overline{\left(\int_0^S I(t)\,\mathrm{d}t\right)^2} = 2S\int_0^\infty \psi_I(\tau)\,\mathrm{d}\tau = \frac{2kTS}{R} .$$

23.5 Our previous discussion of the fluctuations in charge passing round a circuit of resistance R was based on the relation between correlation function and macroscopic decay function. We now consider a kinetic theory calculation which verifies the result of Problem 23.4 for a very particular model of the circuit and the resistive element in it.

Suppose that the electrons in the circuit can be treated classically and that the resistance arises entirely from the presence of a symmetric potential barrier of height V_0. It will be supposed that the velocity distribution of particles in planes on either side of the barrier can be treated as equilibrium distributions for the purpose of calculating the rates at which particles move into the barrier region. Collisions in the barrier region will be supposed negligible.

Show that the number of particles capable of surmounting a barrier of height V which pass into the barrier from one side per unit time is equal to $C\exp(-V/kT)$ where C is independent of V. Determine, in terms of

V_0 and C, the net current across the barrier when an electrostatic potential $2\delta E$ is applied across it and hence determine the resistance of the barrier. Determine also the fluctuations in net current across the barrier integrated over the time S and so verify that the previously derived relation is satisfied.

Solution

The number of particles per unit volume with momenta in the range dp_x, dp_y, dp_z at p_x, p_y, p_z is

$$A\exp[-(p_x^2+p_y^2+p_z^2)/2mkT]\,dp_x\,dp_y\,dp_z\ .$$

The number with p_z in dp_z at p_z (z will be supposed normal to the plane of the barrier) will therefore be

$$B\exp(-p_z^2/2mkT)\,dp_z\ .$$

The number per unit time with z component of momentum in this range which pass into the barrier (supposed of area σ) will be the number having z component of momentum in the required range which lie in a volume $(p_z/m)\sigma$. Define p_z' by $p_z'^2/2m = V$. The number of electrons with momentum greater than p_z' which pass into the barrier per unit time will be

$$B\int_{p_z'}^{\infty}\exp(-p_z^2/2mkT)\frac{p_z}{m}\sigma\,dp_z = B\sigma\int_V^{\infty}\exp(-x/kT)\,dx$$

$$= B\sigma kT\exp(-V/kT) = C\exp(-V/kT)\,.$$

If an electrostatic potential $2\delta E$ is applied across the symmetric barrier, the heights of the barrier viewed from the two sides will be $V_0-e\delta E$ and $V_0+e\delta E$. The mean net current across will therefore be

$$Ce\exp[-(V_0-e\delta E)/kT]-Ce\exp[-(V_0+e\delta E)/kT]\,.$$

For small δE this becomes

$$2Ce^2\exp(-V_0/kT)\frac{\delta E}{kT}\ .$$

The resistance R of the barrier is therefore given by

$$\frac{1}{R} = \frac{Ce^2}{kT}\exp(-V_0/kT)\,.$$

Because of the independence of the classical particles we can suppose that the number of particles crossing the barrier from either side will have a Poisson distribution.

For each direction the mean number crossing in time S will be

$$CS\exp(-V_0/kT)$$

and this will also be the mean square fluctuation in the number. If the numbers are denoted by n_1 and n_2, and the charge passed by Q, then

$$Q = en_1 - en_2 ,$$

$$\overline{\Delta Q^2} = e^2\overline{\Delta n_1^2} + e^2\overline{\Delta n_2^2} = 2e^2 C\exp(-V_0/kT)S .$$

Comparison with the expression for $1/R$ gives

$$\overline{\Delta Q^2} = \frac{2kT}{R}S .$$

This agrees with the previous result.

23.6 A fluctuating quantity $y(t)$ is made up from a series of identical pulses of form $\mathrm{f}(t)$ randomly distributed in time.

If we suppose the time scale divided up by a series of closely and equally spaced instants t_r, such that $t_{r+1} - t_r = \delta$, and that n_r denotes the number of pulses 'centred on' times between t_r and t_{r+1}, we may write [if $\mathrm{f}(t)$ is 'centred on' $t = 0$]:

$$y(t) = \sum_r n_r \mathrm{f}(t - t_r) .$$

Since the quantities n_r are independent and each has a Poisson distribution we have

$$\overline{\Delta n_r^2} = \overline{n_r} = \lambda\delta$$

and

$$\overline{\Delta n_r \Delta n_s} = 0 \quad r \neq s ,$$

where λ is the mean number of pulses per second. Deduce that

$$\overline{y(t)} = \lambda \int_{-\infty}^{+\infty} \mathrm{f}(t')\mathrm{d}t'$$

and

$$\psi_y(\tau) = \lambda \int_{-\infty}^{+\infty} \mathrm{f}(t)\mathrm{f}(t+\tau)\mathrm{d}t .$$

Solution

The first result is easily obtained:

$$\overline{y(t)} = \sum_r \overline{n_r}\mathrm{f}(t - t_r) = \lambda\delta \sum_r \mathrm{f}(t - t_r) = \lambda \int_{-\infty}^{+\infty} \mathrm{f}(t')\mathrm{d}t' ,$$

where we have let δ tend to zero to turn the sum into an integral. The second is derived as follows. We have

$$\Delta y(t) = y(t) - \overline{y(t)} = \sum_r n_r \mathrm{f}(t - t_r) - \sum_r \overline{n_r}\mathrm{f}(t - t_r) = \sum_r \Delta n_r \mathrm{f}(t - t_r)$$

and so

$$\overline{\Delta y(t)\Delta y(t+\tau)} = \overline{\left[\sum_r \Delta n_r \mathrm{f}(t-t_r)\right]\left[\sum_s \Delta n_s \mathrm{f}(t+\tau-t_s)\right]}$$

$$= \sum_{r,s} \overline{\Delta n_r \Delta n_s} \mathrm{f}(t-t_r)\mathrm{f}(t+\tau-t_s)$$

$$= \lambda \sum_r \delta \mathrm{f}(t-t_r)\mathrm{f}(t+\tau-t_r)$$

$$= \lambda \int_{-\infty}^{+\infty} \mathrm{f}(t)\mathrm{f}(t+\tau)\,\mathrm{d}t\,,$$

where we have again let δ tend to zero to make the last step.

23.7 A current $I(t)$ consists of a random series of pulses occurring at a mean rate λ per second, each consisting of a constant current I_0 lasting for a time t_0. Find the correlation function.

Solution

From the previous result we have

$$\psi_I(\tau) = \begin{cases} \lambda I_0^2(t_0 - |\tau|) & |\tau| < t_0 \\ 0 & \text{otherwise}\,. \end{cases}$$

since the integrand has the value I_0^2 throughout the region for which two pulses, displaced with respect to one another by an amount τ, overlap, and is zero elsewhere. The length of the overlap region will be $t_0 - |\tau|$ if $|\tau| < t_0$ and it will be zero otherwise.

23.8 The **power spectrum** $G_y(\omega)$ of a real fluctuating quantity $y(t)$, for which $\bar{y}$ has been made zero by suitable choice of origin, is defined as follows. We first define a truncated quantity $y_S(t)$ which is equal to $y(t)$ in a range of t of length S and zero elsewhere. The Fourier transform of $y_S(t)$ is denoted by $Y_S(\omega)$, so that

$$y_S(t) = \frac{1}{\sqrt{2\pi}} \int_{-\infty}^{+\infty} Y_S(\omega) \exp(i\omega t)\,\mathrm{d}\omega\,.$$

Then $G_y(\omega)$ is defined by

$$G_y(\omega) = \lim_{S\to\infty} \frac{2}{S} \overline{|Y_S(\omega)|^2}\,,$$

where the bar denotes an ensemble average. Show that

$$\overline{y^2} = \int_0^{\infty} G_y(\omega)\,\mathrm{d}\omega\,.$$

Evaluate $G_y(\omega)$ for the case in which $y(t)$ is formed from a series of identical pulses occurring at random times, a pulse being assumed for simplicity to have zero time integral, so that $\overline{y(t)}$ will be zero and readjustment of the zero is not necessary. Show from this example that it is necessary to take both the limit of large S and the ensemble average in order that the definition of $G(\omega)$ should be satisfactory.

Solution

We have, from a standard Fourier transform result,

$$\int_{-\infty}^{+\infty} |y_S(t)|^2 \mathrm{d}t = \int_{-\infty}^{+\infty} |Y_S(\omega)|^2 \mathrm{d}\omega\,.$$

In the limit of large S the left-hand side, which is the integral of y^2 over a time S, becomes $S\overline{y^2}$. Since $y_S(t)$ is real $Y_S(\omega) = Y_S^*(-\omega)$ so that $|Y_S(\omega)|^2 = |Y_S(-\omega)|^2$. Thus the right-hand side can be expressed as an integral from zero to infinity and we have

$$\overline{y^2} = \lim_{S\to\infty} \frac{2}{S} \int_0^{\infty} |Y_s(\omega)|^2 \mathrm{d}\omega\,.$$

Clearly, taking an ensemble average of the right-hand side will not invalidate this equation, so we get

$$\overline{y^2} = \int_0^{\infty} G_y(\omega) \mathrm{d}\omega\,.$$

Suppose now that

$$y(t) = \sum_{r=-\infty}^{+\infty} \mathrm{f}(t-t_r)$$

where the t_r occur at random, there being on average λ values of t_r per unit time.

If S is very long compared with the duration of a pulse we may write (neglecting only very small end effects)

$$y_S(t) = \sum_r \mathrm{f}(t-t_r)\,,$$

the summation being over all r such that t_r lies in the interval of length S used in defining $y_S(t)$. This gives

$$\begin{aligned} Y_S(\omega) &= \sum_r \frac{1}{\sqrt{2\pi}} \int_{-\infty}^{+\infty} \mathrm{f}(t-t_r)\exp(-\mathrm{i}\omega t)\mathrm{d}t \\ &= \sum_r \frac{1}{\sqrt{2\pi}} \int_{-\infty}^{+\infty} \mathrm{f}(t-t_r)\exp[-\mathrm{i}\omega(t-t_r)]\mathrm{d}(t-t_r)\exp(-\mathrm{i}\omega t_r) \\ &= \mathrm{F}(\omega)\sum_r \exp(-\mathrm{i}\omega t_r)\,, \end{aligned}$$

where $\mathrm{F}(\omega)$ is the Fourier transform of a pulse ‘centred on’ the origin,

i.e.

$$F(\omega) = \frac{1}{\sqrt{2\pi}}\int_{-\infty}^{+\infty} f(t)\exp(-i\omega t)dt\,.$$

This gives

$$|Y_S(\omega)|^2 = |F(\omega)|^2 \left|\sum_r \exp(-i\omega t_r)\right|^2 .$$

If S is very long compared with $1/\omega$, the summation on the right-hand side will be over complex numbers of modulus unity with random phases, there being on average λS terms in the sum. The square of the modulus of this sum will be the square of the distance from the origin to the final point of a two-dimensional random walk consisting of on average λS steps of unit length. No matter how large S, these distances squared will fluctuate wildly. It is only if we average over the directions of the steps by averaging over the possible values of the t_r (i.e. taking an ensemble average) that we get a definite value, namely the mean square displacement in a random walk of λS unit steps, which is just λS. Thus for sufficiently large S

$$\overline{|Y_S(\omega)|^2} = \lambda S|F(\omega)|^2$$

and so

$$G_y(\omega) = \frac{2}{S}\overline{|Y_S(\omega)|^2} = 2\lambda|F(\omega)|^2 .$$

It is clear that it has been necessary both to take S large and to take an ensemble average. Thus a definition of $G_y(\omega)$ which did not include both the limit of large S and an ensemble average would not lead to a sensible result in the particular example considered, and so would not be a satisfactory definition.

23.9 A linear system has a **response function** $A(\omega)$, i.e. an input $x(t) = \exp(i\omega t)$ gives an output $y(t) = A(\omega)\exp(i\omega t)$.

Determine the power spectrum $G_O(\omega)$ of the output, if a fluctuating input with power spectrum $G_I(\omega)$ is applied. Hence express the mean square value of the output in terms of the input power spectrum.

Solution

If the Fourier transform of the truncated input (defined as in Problem 23.8) is denoted by $X_S(\omega)$, we have

$$G_I(\omega) = \lim_{S\to\infty}\frac{2}{S}\overline{|X_S(\omega)|^2} .$$

The Fourier transform of the output produced by this truncated input will be

$$A(\omega)X_S(\omega) .$$

For sufficiently long S the output for the truncated input will differ from the truncated output $y_S(t)$ only by a negligible end effect. We can, therefore, write

$$Y_S(\omega) = A(\omega)X_S(\omega)$$

and the output power spectrum $G_O(\omega)$ is then given by

$$G_O(\omega) = \lim_{S\to\infty} \frac{2}{S}\overline{|Y_S(\omega)|^2} = |A(\omega)|^2 \lim_{S\to\infty} \frac{2}{S}\overline{|X_S(\omega)|^2} = |A(\omega)|^2 G_I(\omega)\,.$$

The importance of the power spectrum concept stems in large measure from the simplicity of this relation between output and input power spectra. The relation may of course be regarded as obvious since one may think of $G(\omega)\delta\omega$ as being the mean square value of the narrow band fluctuations obtained by eliminating all Fourier components of the fluctuations except those with frequency in the range $\delta\omega$ at ω. (From the point of view adopted here, this follows from the expression which has been derived for the mean square value in terms of an integral over the power spectrum; one may, alternatively, start the discussion from this less formal definition of the power spectrum.) Since a narrow band fluctuation is sinusoidal with slowly varying amplitude and phase, $|A(\omega)|$ will give the ratio of the output amplitude to the input amplitude for the frequency range considered and the result follows at once.

The mean square value of the output is given by

$$\overline{y^2} = \int_0^\infty G_O(\omega)\mathrm{d}\omega = \int_0^\infty |A(\omega)|^2 G_I(\omega)\mathrm{d}\omega\,.$$

23.10 Calculate the response function $A(\omega)$ where the system is a capacitance and resistance in series, the input being the voltage across them and the output the current through them. Calculate also the response function for a damped suspended system, the input being the applied couple and the output the deflection.

Solution

In the case of the capacitance C and resistance R in series, the voltage $V = V_0 e^{i\omega t}$ and the current $I = I_0 e^{i\omega t}$ are related by

$$V_0 = Z(\omega)I_0\,,$$

where $Z(\omega)$, the impedance, is given by elementary a.c. theory,

$$Z(\omega) = R + \frac{1}{i\omega C}\,.$$

But, by the definition of $A(\omega)$,

$$I_0 = A(\omega)V_0$$

so that

$$A(\omega) = \frac{1}{Z(\omega)} = \frac{1}{R + 1/i\omega C}\,.$$

For the damped suspended system the deflection θ is related to the applied couple P by an equation of the form

$$I\ddot{\theta} + \kappa\dot{\theta} + c\theta = P(t) .$$

If we put $P(t) = \mathrm{e}^{i\omega t}$, then θ will be given, after the decay of transients, by

$$\theta = \frac{\mathrm{e}^{i\omega t}}{c - I\omega^2 + i\kappa\omega},$$

so that

$$A(\omega) = \frac{1}{c - I\omega^2 + i\kappa\omega} .$$

23.11 Calculate $A(\omega)$ for the case in which the input $x(t)$ produces an output $y(t) = x(t+\tau) - x(t)$. Hence obtain the mean square value of the output if the power spectrum of the input is $G_x(\omega)$, and so obtain the correlation function for x, $\psi_x(\tau)$, in terms of the power spectrum of x, $G_x(\omega)$. This gives a somewhat unorthodox derivation of the **Wiener-Khintchine relation** between correlation function and power spectrum. Show that the relation obtained can be inverted to give the power spectrum in terms of the correlation function.

Note: No physical system could have the response assumed here (τ being assumed positive), since it implies a response to, say, an input δ function which occurs before the input is applied. This is, however, unimportant for the purely mathematical discussion involved in the question.

Solution

An input $x(t) = \mathrm{e}^{i\omega t}$ will produce an output

$$y(t) = \mathrm{e}^{i\omega(t+\tau)} - \mathrm{e}^{i\omega t} = (\mathrm{e}^{i\omega\tau} - 1)\mathrm{e}^{i\omega t},$$

so that

$$A(\omega) = \mathrm{e}^{i\omega\tau} - 1 .$$

Expressing the mean square value of the output in terms of its power spectrum, $|A(\omega)|^2 G_x(\omega)$, gives

$$\overline{[x(t+\tau) - x(t)]^2} = \int_0^\infty |\mathrm{e}^{i\omega\tau} - 1|^2 G_x(\omega)\mathrm{d}\omega$$

$$= 2\int_0^\infty (1 - \cos\omega\tau) G_x(\omega)\mathrm{d}\omega .$$

But the left-hand side is

$$\overline{x(t+\tau)^2} + \overline{x(t)^2} - 2\overline{x(t)x(t+\tau)} = 2\overline{x^2} - 2\psi_x(\tau)$$

$$= 2\int_0^\infty G_x(\omega)\mathrm{d}\omega - 2\psi_x(\tau) .$$

It follows that

$$\psi_x(\tau) = \int_0^\infty G_x(\omega)\cos\omega\tau\,\mathrm{d}\omega\,.$$

To invert this, starting from the more familiar exponential form of the Fourier transform relations, we may proceed as follows. We define $G_x(\omega)$ for negative ω by $G_x(-\omega) = G_x(\omega)$ and rewrite the above relation as

$$\psi_x(\tau) = \frac{1}{\sqrt{2\pi}}\int_{-\infty}^{+\infty}\frac{\sqrt{2\pi}}{2}G_x(\omega)\mathrm{e}^{\mathrm{i}\omega\tau}\,\mathrm{d}\omega\,.$$

The inverse of this relation then gives

$$\frac{\sqrt{2\pi}}{2}G_x(\omega) = \frac{1}{\sqrt{2\pi}}\int_{-\infty}^{+\infty}\psi_x(\tau)\mathrm{e}^{-\mathrm{i}\omega\tau}\,\mathrm{d}\tau$$

or

$$G_x(\omega) = \frac{2}{\pi}\int_0^\infty \psi_x(\tau)\cos\omega\tau\,\mathrm{d}\tau\,.$$

23.12 Show that if the correlation function has dropped practically to zero for $\tau > \tau_c$, then the power spectrum will be constant up to frequencies of the order $1/\tau_c$. τ_c is called the **correlation time.**

Solution

We have the Wiener-Khintchine relation derived in Problem 23.11

$$G(\omega) = \frac{2}{\pi}\int_0^\infty \psi(\tau)\cos\omega\tau\,\mathrm{d}\tau\,.$$

Because $\psi(\tau)$ is negligible for $\tau > \tau_c$, we can restrict the range of integration over τ to that between zero and τ_c. If $\omega \ll 1/\tau_c$, $\omega\tau$ will be much less than unity for all τ in the above range, so that $\cos\omega\tau$ may be replaced by unity in the integrand.

Thus to a good approximation

$$G(\omega) = \frac{2}{\pi}\int_0^\infty \psi(\tau)\,\mathrm{d}\tau$$

for values of ω up to a value less than (but of the order of) $1/\tau_c$. This expression is independent of ω.

23.13 Evaluate the power spectrum of the fluctuations associated with a random series of identical pulses by applying the Wiener-Khintchine relation (Problem 23.11) to the expression for the correlation function obtained in Problem 23.6. Check that the result agrees with that obtained by direct application of the definition of the power spectrum in Problem 23.8: notice that in this alternative derivation we do not make the assumption (introduced in Problem 23.8 for simplicity) that the time integral of a pulse is zero.

Solution

We have from Problem 23.6

$$\psi_y(\tau) = \lambda \int_{-\infty}^{+\infty} \mathrm{f}(t)\mathrm{f}(t+\tau)\,\mathrm{d}t\,.$$

The Wiener-Khintchine relation gives

$$\begin{aligned} G_y(\omega) &= \frac{2}{\pi}\int_0^{\infty} \psi_y(\tau)\cos\omega\tau\,\mathrm{d}\tau = \frac{1}{\pi}\int_{-\infty}^{+\infty} \psi_y(\tau)\mathrm{e}^{\mathrm{i}\omega\tau}\,\mathrm{d}\tau \\ &= \frac{\lambda}{\pi}\int_{-\infty}^{+\infty}\int_{-\infty}^{+\infty} \mathrm{f}(t)\mathrm{f}(t+\tau)\mathrm{e}^{\mathrm{i}\omega\tau}\,\mathrm{d}t\,\mathrm{d}\tau \\ &= \frac{\lambda}{\pi}\int_{-\infty}^{+\infty}\int_{-\infty}^{+\infty} \mathrm{f}(t_1)\mathrm{f}(t_2)\mathrm{e}^{\mathrm{i}\omega(t_2-t_1)}\,\mathrm{d}t_1\,\mathrm{d}t_2 \\ &= \frac{\lambda}{\pi}\left[\int_{-\infty}^{+\infty} \mathrm{f}(t)\mathrm{e}^{\mathrm{i}\omega t}\mathrm{d}t\right]\left[\int_{-\infty}^{+\infty} \mathrm{f}(t)\mathrm{e}^{-\mathrm{i}\omega t}\,\mathrm{d}t\right] \\ &= 2\lambda\,|\mathrm{F}(\omega)|^2\,, \end{aligned}$$

where $\mathrm{F}(\omega)$ denotes the Fourier transform of $\mathrm{f}(t)$, and we used in the last step the fact that $\mathrm{f}(t)$ is real. This agrees with the result obtained previously.

23.14 The remaining problems of this section lead up to an important result concerning the response functions of physical systems. This result will be made use of in the following section. The response of a linear system may be specified by giving the output $\mathrm{f}(t)$ produced by a delta function input $\delta(t)$. By taking the Fourier transforms of the output and the input, when the input is $\delta(t)$, show that the response function $A(\omega)$ for the system is the Fourier transform of $\mathrm{f}(t)$ multiplied by $\sqrt{2\pi}$.

The Fourier transform $\mathrm{F}(\omega)$ of $\mathrm{f}(t)$ is taken to be given by

$$\mathrm{F}(\omega) = \frac{1}{\sqrt{2\pi}}\int_{-\infty}^{+\infty} \mathrm{f}(t)\mathrm{e}^{-\mathrm{i}\omega t}\,\mathrm{d}t\,.$$

Solution

The Fourier transform of the input $\delta(t)$ is

$$\frac{1}{\sqrt{2\pi}}\int_{-\infty}^{+\infty} \delta(t)\mathrm{e}^{-\mathrm{i}\omega t}\,\mathrm{d}t = \frac{1}{\sqrt{2\pi}}\,.$$

The Fourier transform of the output $\mathrm{f}(t)$ will be

$$\mathrm{F}(\omega) = \frac{1}{\sqrt{2\pi}}\int_{-\infty}^{+\infty} \mathrm{f}(t)\,\mathrm{e}^{-\mathrm{i}\omega t}\,\mathrm{d}t\,.$$

But the ratio of the Fourier component of the output at frequency ω to that of the input at the same frequency must be $A(\omega)$, and it follows that

$$A(\omega) = \sqrt{2\pi}\mathrm{F}(\omega)$$

which is the required result.

23.15 Consider a system for which the response to an input $\delta(t)$ is a delta function delayed by a time τ, $\delta(t-\tau)$. Determine the response function $A(\omega)$ for the system and show that if it is written in the form

$$A(\omega) = A'(\omega) - \mathrm{i}A''(\omega)$$

with $A'(\omega)$ and $A''(\omega)$ real, then $A'(\omega)$ and $A''(\omega)$ satisfy the relations

$$A'(\omega) = \frac{1}{\pi}P\int_{-\infty}^{+\infty}\frac{A''(\omega')\mathrm{d}\omega'}{\omega'-\omega},$$

$$A''(\omega) = -\frac{1}{\pi}P\int_{-\infty}^{+\infty}\frac{A'(\omega')\mathrm{d}\omega'}{\omega'-\omega},$$

if, and only if, the delay time τ is positive. The symbol P indicates that the Cauchy principal value of the integral which follows is to be taken.

[Note: $\int_{-\infty}^{+\infty}\frac{\sin ax}{x}\mathrm{d}x$ has the value π, 0, or $-\pi$ respectively when a is positive, zero, or negative.]

Solution

In this case $\mathrm{f}(t) = \delta(t-\tau)$ and, we have from Problem 23.14

$$A(\omega) = \sqrt{2\pi}\mathrm{F}(\omega) = \sqrt{2\pi}\frac{1}{\sqrt{2\pi}}\int_{-\infty}^{+\infty}\delta(t-\tau)\mathrm{e}^{-\mathrm{i}\omega t}\mathrm{d}t$$

$$= \mathrm{e}^{-\mathrm{i}\omega\tau} = \cos\omega\tau - \mathrm{i}\sin\omega\tau\,.$$

Thus

$$A'(\omega) = \cos\omega\tau\,, \qquad A''(\omega) = \sin\omega\tau\,.$$

It follows that

$$P\int_{-\infty}^{+\infty}\frac{A''(\omega')\mathrm{d}\omega'}{\omega'-\omega} = P\int_{-\infty}^{+\infty}\frac{\sin\omega'\tau\,\mathrm{d}\omega'}{\omega'-\omega}$$

$$= P\int_{-\infty}^{+\infty}\frac{\sin[(\omega'-\omega)\tau+\omega\tau]\mathrm{d}(\omega'-\omega)}{\omega'-\omega}$$

$$= \cos\omega\tau\int_{-\infty}^{+\infty}\frac{\sin\tau x}{x}\mathrm{d}x + \sin\omega\tau P\int_{-\infty}^{+\infty}\frac{\cos\tau x}{x}\mathrm{d}x$$

$$= \begin{cases} \pi\cos\omega\tau & \tau > 0 \\ 0 & \tau = 0 \\ -\pi\cos\omega\tau & \tau < 0 \end{cases}$$

and since $A'(\omega)$ is $\cos\omega\tau$, we see that the first relation is satisfied if, and only if, τ is positive. We have used the fact that $P\int_{-\infty}^{+\infty}[(\cos\tau x)/x]\,dx$ is zero because $(\cos\tau x)/x$ is odd, and that, since $\int_{-\infty}^{+\infty}[(\sin\tau x)/x]\,dx$ has no singularity, the operation of taking the principal value does not affect it. The other relation can be verified similarly.

23.16 If $f(t)$ describes the response of a physical system to an input in the form of a delta function at zero time, it is clear that $f(t)$ must be zero for negative t as the system cannot respond before the signal is applied. But if $f(t)$ is zero for negative t it can be represented as a superposition of delta functions $\delta(t-\tau)$ delayed by times τ, with all τ positive or zero. Assuming that the instantaneous response of the system may be neglected, so that only $\tau > 0$ need be considered, use the result of the previous problem to show that the real and imaginary parts of the response function for the physical system will satisfy the **Kramers-Krönig relations**

$$A'(\omega) = \frac{1}{\pi}P\int_{-\infty}^{+\infty}\frac{A''(\omega')\,d\omega'}{\omega'-\omega},$$

$$A''(\omega) = -\frac{1}{\pi}P\int_{-\infty}^{+\infty}\frac{A'(\omega')\,d\omega'}{\omega'-\omega}.$$

Show that, if one wishes to get relations which take account of the possibility that the function $f(t)$ contains an instantaneous response part $C\delta(t)$, one has to replace $A'(\omega)$ by $A'(\omega)-A'(\infty)$ in the relations written above.

Solution

As $A(\omega)$ is $\sqrt{2\pi}$ times the Fourier transform of $f(t)$, the $A(\omega)$ corresponding to a superposition (i.e. linear combination) of delta functions will be the same linear combination of the associated response functions. Since the delta functions all have delay $\tau > 0$, the associated response functions will all satisfy the given relations and, as the relations are linear in $A'(\omega)$ and $A''(\omega)$, a superposition of response functions which satisfy these will also satisfy them. This establishes the result, on the assumption of no instantaneous response.

It will be useful to note here that, if there is no instantaneous response, $A(\infty)$ will be zero: an individual delta function with delay τ will contribute $\cos\omega\tau - i\sin\omega\tau$ to $A(\omega)$ and this oscillates as ω tends to infinity; the total $A(\omega)$ will, however, be obtained by integrating over τ the product of this into some reasonably smooth function of τ [in fact the multiplying function is $f(\tau)$]: the rapid oscillations with τ for large ω will cause the integral to tend to zero as ω tends to infinity.

On the other hand a part $C\delta(t)$ in $f(t)$ will contribute a part C to $A(\omega)$. This does lead to a contribution at infinite frequency so we can write $A(\infty) = C$. Thus the contribution of $C\delta(t)$ to $A(\omega)$ may be written $A'(\infty)$. To get the part of $A'(\omega)$ which satisfies the above relations we must subtract out the instantaneous response part, i.e. replace $A(\omega)$ by $A(\omega) - A'(\infty)$.

23.17 Verify by direct calculation that the response function relating input voltage to output current for a capacitance and resistance in series (Problem 23.10) satisfies the Kramers-Krönig relations (Problem 23.16).

Solution

We have (Problem 23.10)

$$A(\omega) = \frac{1}{R + 1/\mathrm{i}\omega C} = \mathrm{i}\omega C \frac{1}{1 + \mathrm{i}\omega RC} = \frac{\omega^2 RC^2 + \mathrm{i}\omega C}{1 + \omega^2 R^2 C^2} \,.$$

Thus

$$A'(\omega) = \frac{\omega^2 RC^2}{1 + \omega^2 R^2 C^2} \,,$$

$$A''(\omega) = -\frac{\omega C}{1 + \omega^2 R^2 C^2} \,.$$

Note that

$$A'(\infty) = \frac{1}{R}$$

so that

$$A'(\omega) - A'(\infty) = -\frac{1}{R(1 + \omega^2 R^2 C^2)} \,.$$

We verify first that

$$A'(\omega) - A'(\infty) = \frac{1}{\pi} P \int_{-\infty}^{+\infty} \frac{A''(\omega')\,\mathrm{d}\omega'}{\omega' - \omega} \,.$$

The right-hand side is

$$-\frac{C}{\pi} P \int_{-\infty}^{+\infty} \frac{\omega'\,\mathrm{d}\omega'}{(\omega' - \omega)(1 + \omega'^2 R^2 C^2)} \,.$$

The integral can be evaluated either by contour integration or by elementary methods. We outline both procedures, contour integration first. The integrand has a pole at $\omega' = \omega$ with residue $\omega(1 + \omega^2 R^2 C^2)^{-1}$ and poles at $\omega' = \pm \mathrm{i}/RC$ with residues $-\frac{1}{2}(\omega + \mathrm{i}/RC)(1 + R^2 C^2 \omega^2)^{-1}$ and $-\frac{1}{2}(\omega - \mathrm{i}/RC)(1 + R^2 C^2 \omega^2)^{-1}$ respectively. We consider a contour which traverses the real axis from $-L$ to $+L$ except that it passes above the pole on the real axis in a small semicircle centred on the pole. The contour is completed by a large semicircle of radius L centred on the origin and lying in the upper half plane. The contribution of the large semicircle part of the contour integral will go to zero as L increases and we therefore have by the residue theorem

$$P\int_{-\infty}^{+\infty}\frac{\omega'\,d\omega'}{(\omega'-\omega)(1+\omega'^2R^2C^2)}-\frac{\pi i\omega}{1+\omega^2R^2C^2}=-\frac{\pi i(\omega+i/RC)}{1+\omega^2R^2C^2},$$

i.e.

$$P\int_{-\infty}^{+\infty}\frac{\omega'\,d\omega'}{(\omega'-\omega)(1+\omega'^2RC^2)}=\frac{\pi}{RC}\frac{1}{1+\omega^2R^2C^2}\,.$$

Consequently the right-hand side in the Kramers-Krönig relation being considered is $-(1/R)(1+\omega^2R^2C^2)^{-1}$, which is also the value of the left-hand side.

The other relation can be verified similarly.

An elementary evaluation can be based on expressing the integrand in partial fractions by means of the identity

$$\frac{\omega'}{(\omega'-\omega)(1+R^2C^2\omega'^2)}=\frac{1}{1+R^2C^2\omega^2}\left(\frac{\omega}{\omega'-\omega}+\frac{1-R^2C^2\omega\omega'}{1+R^2C^2\omega'^2}\right).$$

Since

$$P\int_{-\infty}^{+\infty}\frac{d\omega'}{\omega'-\omega}=P\int_{-\infty}^{+\infty}\frac{d\omega'}{\omega'}=\lim_{\epsilon\to 0}\left(\int_{-\infty}^{-\epsilon}\frac{d\omega'}{\omega'}+\int_{\epsilon}^{\infty}\frac{d\omega'}{\omega'}\right)=0$$

and

$$\int_{-\infty}^{+\infty}\frac{\omega'\,d\omega'}{1+R^2C^2\omega'^2}=0\,,$$

the required integral is equal to

$$\frac{1}{1+R^2C^2\omega^2}\int_{-\infty}^{+\infty}\frac{d\omega'}{1+R^2C^2\omega'^2}=\frac{\pi}{RC(1+\omega^2R^2C^2)}$$

in agreement with what we had previously.

24
Nyquist's theorem and its generalisations

C.W.McCOMBIE
(*University of Reading, Reading*)

24.1 The electrons in a resistor may be supposed to receive impulses from the atoms of the lattice. These impulses are equivalent to a fluctuating voltage with correlation time (i.e. the τ_c of Problem 23.12) which will be of the order of the duration of an impulse. Since this duration will be very short we may suppose the power spectrum of the equivalent voltage to be constant up to very high frequencies (constant in fact for all frequencies for which the resistor has a resistance independent of frequency).

By considering a circuit consisting of a resistance and capacitance in series, and requiring that the power spectrum of the voltage fluctuations associated with the resistor should lead to the equipartition result for the fluctuations in charge on the capacitance, determine the constant value of the power spectrum (i.e. obtain Nyquist's theorem).

Show that this implies a correlation function for the fluctuations in charge which agrees with the decay function assumption made earlier (Problem 23.2).

Solution

As we saw in Problem 23.10 the response function relating input voltage to output current for a resistance R and capacity C in series is

$$\frac{1}{R+1/\mathrm{i}\omega C}\,.$$

The response function relating (input) current to (output) charge is $1/\mathrm{i}\omega$ since charge is the integral of the current. Thus the required response function $A(\omega)$, relating voltage to charge, is given by

$$A(\omega)=\left(\frac{1}{\mathrm{i}\omega}\right)\frac{1}{R+1/\mathrm{i}\omega C}\,,$$

so that

$$|A(\omega)|^2=\frac{C^2}{1+R^2C^2\omega^2}\,.$$

If the constant power spectrum of the fluctuating voltage is G_V it follows that the power spectrum $G_q(\omega)$ of the fluctuating charge will be given by

$$G_q(\omega)=|A(\omega)|^2G_V=\frac{C^2G_V}{1+R^2C^2\omega^2}\,.$$

This gives

$$\overline{q^2} = \int_0^\infty G_q(\omega)\,\mathrm{d}\omega = \int_0^\infty \frac{C^2 G_V \mathrm{d}\omega}{1+R^2C^2\omega^2} = \int_0^\infty \frac{(1/R^2)G_V \mathrm{d}\omega}{\omega^2+(1/RC)^2} = \tfrac{1}{2}\pi\frac{C}{R}G_V\,.$$

But the charge fluctuations on the capacitor satisfy

$$\overline{q^2} = CkT\,.$$

Equating the two expressions for $\overline{q^2}$ gives

$$G_V = \frac{2}{\pi}RkT\,.$$

This is **Nyquist's theorem.**

The correlation function for q, $\psi_q(\tau)$, is obtained from $G_q(\omega)$ by the Wiener-Khintchine relation

$$\psi_q(\tau) = \int_0^\infty G_q(\omega)\cos\omega\tau\,\mathrm{d}\omega = C^2 G_V \int_0^\infty \frac{\cos\omega\tau}{1+R^2C^2\omega^2}\mathrm{d}\omega\,.$$

The integral is easily evaluated by contour integration

$$\int_0^\infty \frac{\cos\omega\tau}{1+R^2C^2\omega^2}\mathrm{d}\omega = \tfrac{1}{2}\int_{-\infty}^{+\infty} \frac{\exp(\mathrm{i}\omega\tau)\,\mathrm{d}\omega}{1+R^2C^2\omega^2} = \tfrac{1}{2}\int_{\mathrm{C}} \frac{\exp(\mathrm{i}z\tau)}{1+R^2C^2z^2}\mathrm{d}z$$

where, for positive τ, the contour C is the real axis and the semicircle at infinity in the upper half plane: the integral round this semicircle is zero. The only pole inside the contour C is at $z = \mathrm{i}/RC$ and the residue there is $\exp(-\tau/RC)/2\mathrm{i}RC$. The value of the required integral is, therefore, $\frac{1}{2}\pi\exp(-\tau/RC)/RC$ and so, substituting also for G_V,

$$\psi_q(\tau) = CkT\exp(-\tau/RC) = \overline{q^2}\exp(-\tau/RC) \qquad (\tau > 0)\,.$$

The result for negative τ is obtained similarly and we find

$$\psi_q(\tau) = \overline{q^2}\exp(-|\tau|/RC)\,.$$

Now q satisfies the equation

$$\frac{\mathrm{d}q}{\mathrm{d}t} + \frac{q}{RC} = 0$$

which has the solution

$$q = q_0\exp(-t/RC)\,.$$

Thus the correlation function and the decay function are related in the way postulated earlier (Problem 23.2).

24.2 The fluctuations of a galvanometer mirror may be regarded as arising from both bombardment by air molecules and current fluctuations in the circuit containing its coil. Either of these acting alone would presumably maintain the fluctuations at equipartition level. The two acting together must also result in equipartition fluctuations. Explain.

Solution

In systems in thermodynamic equilibrium each source of fluctuating force will have a damping associated with it. Thus, so far as maintaining the mean square value of the fluctuations of the system is concerned, the tendency of an added fluctuating force to increase the fluctuations is just compensated by the tendency of the associated damping to reduce them. The ratio of the power spectrum of the total fluctuating force to the total damping force remains fixed.

It is instructive to examine the constancy of the ratio of power spectrum to damping more explicitly in the case of the couple acting on a galvanometer coil. If the coil has effective area A in a magnetic field H, the field direction being in the plane of the coil when the deflection is zero, then a current I will give rise to a couple AHI. An angular velocity $\dot{\theta}$ will give rise, if the deflection is small, to an induced e.m.f. of magnitude $AH\dot{\theta}$: if the resistance in the circuit of the coil is R this will give rise to a current $AH\dot{\theta}/R$ and so to a couple of magnitude $(A^2H^2/R)\dot{\theta}$. The actual sign can be determined from the requirement that the net result is a damping. If we denote the air damping constant by κ, the fluctuating couple due to bombardment by air molecules by $P(t)$, and the fluctuating voltage associated with the resistance in the circuit by $V(t)$, the fluctuating deflection $\theta(t)$ will be determined by

$$K\ddot{\theta}+\left(\kappa+\frac{A^2H^2}{R}\right)\dot{\theta}+c\theta = P(t)+\frac{AH}{R}V(t)\,.$$

The power spectrum of $P(t)$, G_P, will be $(2/\pi)\kappa kT$ (this can be derived by requiring that the mean square deflection of a system with, for simplicity, zero K should have its equipartition value) while that of $(AH/R)V(t)$ will be

$$\frac{A^2H^2}{R^2}G_V = \frac{A^2H^2}{R^2}\frac{2RkT}{\pi}\,.$$

The power spectra of the independent sources of fluctuating couple will add, so the power spectrum of the fluctuating couple in the right-hand side will be

$$\frac{2}{\pi}kT\left(\kappa+\frac{A^2H^2}{R}\right).$$

This has a ratio to the damping constant $\kappa+A^2H^2/R$ which is independent of the values of κ, R, A, and H.

24.3 A body of heat capacity C loses heat to its surroundings at a rate $\alpha\Delta T$, where ΔT is the excess of its temperature over the temperature of its surroundings. Determine the power spectrum (assumed independent of frequency) of the fluctuations in the net exchange of energy with the surroundings in this situation, which corresponds to **Newton's law of cooling.**

Solution

If the net rate of receipt of energy from the surroundings is denoted by $I(t)$ one has

$$C\frac{\mathrm{d}(\Delta T)}{\mathrm{d}t} + \alpha\Delta T = I(t)\,.$$

An input receipt of energy $\exp(\mathrm{i}\omega t)$ will give the steady state variation in ΔT

$$\Delta T = \frac{\exp(\mathrm{i}\omega t)}{\alpha + \mathrm{i}\omega C}\,,$$

so that the response function $A(\omega)$ relating input $I(t)$ to output $\Delta T(t)$ is

$$A(\omega) = \frac{1}{\alpha + \mathrm{i}\omega C}\,.$$

If the power spectrum of the fluctuating $I(t)$ is denoted by G_I and that of $\Delta T(t)$ by $G_{\Delta T}(\omega)$, one has

$$G_{\Delta T}(\omega) = |A(\omega)|^2 G_I = \frac{G_I}{\alpha^2 + \omega^2 C^2}\,.$$

This gives

$$\overline{\Delta T^2} = G_I \int_0^\infty \frac{1}{\alpha^2 + \omega^2 C^2}\mathrm{d}\omega = \frac{\pi}{2\alpha C}G_I\,.$$

But the result of Problem 20.1 gave

$$\overline{\Delta E^2} = CkT^2\,,$$

and, since $\Delta E = C\Delta T$, this implies

$$\overline{\Delta T^2} = \frac{kT^2}{C}\,.$$

Equating the two expressions for $\overline{\Delta T^2}$ gives

$$G_I = \frac{2}{\pi}\alpha k T^2\,.$$

24.4 Consider a radiation detector which operates by measuring the rise in temperature of a black body, of area A, when the radiation is incident upon it. If the black body exchanges energy with its surroundings only by the emission and absorption of radiation, determine the minimum possible mean square error in the measurement of intensity of a steady stream of radiation incident for time S.

[Hint: If the fluctuating net rate of receipt of energy is denoted by $I(t)$, the quantity

$$\frac{1}{S}\int_0^S I(t)\,\mathrm{d}t$$

will be indistinguishable from a part of the intensity of the incident radiation. This quantity therefore simulates part of the intensity of the radiation being measured.]

Solution

The rate of receipt of energy from the surroundings, assumed at temperature T, will be $A\sigma T^4$. The rate of emission of energy will be $A\sigma(T+\Delta T)^4$. To first order in ΔT the net rate of loss will therefore be $4A\sigma T^3\Delta T$ so that α of Problem 24.3 is $4A\sigma T^3$. The power spectrum of $I(t)$, the fluctuating net rate of receipt of energy, will therefore be

$$G_I = \frac{2}{\pi}4A\sigma T^3 \times kT^2 .$$

From Problem 23.1 the mean square value of $\int_0^S I(t)\,\mathrm{d}t$ will be

$$2S\int_0^\infty \psi_I(\tau)\,\mathrm{d}\tau ,$$

so that the mean square value of $\frac{1}{S}\int_0^S I(t)\,\mathrm{d}t$, the simulated part of the rate of incidence of signal energy and hence the error in the intensity measurement, will be

$$\frac{2}{S}\int_0^S \psi_I(\tau)\,\mathrm{d}\tau .$$

This therefore gives the minimum possible mean square error in the intensity measurement. Now the Wiener-Khintchine theorem (Problem 23.11) relating power spectrum to correlation function gives

$$G_I(0) = \frac{2}{\pi}\int_0^\infty \psi_I(\tau)\,\mathrm{d}\tau .$$

Therefore the minimum mean square error may be written

$$\frac{\pi G_I}{S} = \frac{8A\sigma kT^5}{S} .$$

24.5 If a resistor R is connected across the terminals of an impedance $Z = X + \mathrm{i}Y$, both being at temperature T, the rate of transfer of power from R to Z must equal the rate of transfer of power from Z to R. Assuming that this equality must hold in each frequency range (one may think of connecting the two elements by way of a filter) show that, if the fluctuating forces associated with the impedance are represented by a generator of fluctuating voltage in series with Z, the power spectrum of this voltage will be

$$G_{V_Z}(\omega) = \frac{2}{\pi}XkT .$$

Solution

The current due to a voltage $V\mathrm{e}^{\mathrm{i}\omega t}$ in the circuit will be $V\mathrm{e}^{\mathrm{i}\omega t}/(R+Z)$ and the resulting power dissipated in R will be $\frac{1}{2}R|V|^2/|R+Z|^2$, while that dissipated in Z will be $\frac{1}{2}X|V|^2/|R+Z|^2$. This means that the power in frequency range $d\omega$ generated in Z by the fluctuating voltage $G_{V_R}(\omega)$ associated with R will be $\frac{1}{2}XG_{V_R}\, d\omega/|R+Z|^2$, while that generated in R by the fluctuating voltage G_{V_Z} associated with Z will be $\frac{1}{2}RG_{V_Z}\, d\omega/|R+Z|^2$. Equating these expressions gives

$$G_{V_Z} = \frac{X}{R}G_{V_R} = \frac{2}{\pi}XkT\,.$$

24.6 Verify the relation between the power spectrum of the fluctuating force associated with an impedance and the real part of the impedance, for an impedance formed from a resistor and capacitor connected in parallel. Start from the fluctuating voltage associated with the resistor.

Solution

If the impedance is open circuited, a voltage $Ve^{i\omega t}$ in series with the resistor will result in a voltage

$$\left[\frac{V}{i\omega C}\Big/\left(R+\frac{1}{i\omega C}\right)\right]e^{i\omega t}$$

appearing across the terminals. If the impedance is short-circuited the same voltage will result in a current $(V/R)e^{i\omega t}$ through the short circuit. It follows from Thevenin's theorem that the voltage $Ve^{i\omega t}$ in series with the resistor is equivalent to a voltage

$$\left[\frac{V}{i\omega C}\Big/\left(R+\frac{1}{i\omega C}\right)\right]e^{i\omega t}$$

in series with the impedance. Thus a fluctuating voltage with power spectrum $(2/\pi)RkT$ in series with R is equivalent to a fluctuating voltage in series with the impedance with power spectrum

$$\frac{2}{\pi}\frac{RkT}{|i\omega C|^2}\Big/\left|R+\frac{1}{i\omega C}\right|^2 = \frac{2}{\pi}\frac{R}{1+\omega^2R^2C^2}kT = \frac{2}{\pi}XkT\,,$$

where X is the real part of the impedance of the resistor and capacitor in parallel.

24.7 In this example we consider the extension, by analogy of Nyquist's result for an electrical impedance, to a general impedance and express the result in a variety of ways. [Some consideration of a more fundamental justification of the generalised Nyquist result will be the subject of Problem 24.8.]

If F is the generalised force associated with a generalised coordinate x of a system and $F = F_0e^{i\omega t}$ results in $x = x_0e^{i\omega t}$ we write

$$x_0 = [P(\omega) - iQ(\omega)]F_0\,.$$

Taking $\dot{x}$ to be the analogue of current and F to be the analogue of voltage, express the impedance in terms of $P(\omega)$ and $Q(\omega)$. Hence, by using the analogy with Nyquist's result for an electrical impedance, obtain an expression for the power spectrum $G_F(\omega)$ of the fluctuating force F which must be regarded as acting on the system at temperature T. Deduce the power spectrum $G_x(\omega)$ of the fluctuations in x. Finally, writing the rate of absorption of energy by the system from a force $F_0 e^{i\omega t}$ in the form $\frac{1}{2}\alpha(\omega)|F_0|^2$, express the generalised Nyquist relation in the form of a relation between $G_x(\omega)$ and $\alpha(\omega)$: it is this form of the relation which we shall find it convenient to justify theoretically in the following problem.

Solution

Writing $\dot{x} = \dot{x}_0 e^{i\omega t}$, we have $\dot{x}_0 = i\omega x_0$, and so

$$F_0 = \frac{\dot{x}_0}{i\omega(P - iQ)} .$$

Thus the impedance is $1/i\omega(P-iQ)$. Taking the real part of this and applying the analogue of Nyquist's result we have

$$G_F(\omega) = \frac{2}{\pi}\frac{Q(\omega)}{\omega}\frac{1}{P^2(\omega)+Q^2(\omega)}kT .$$

The power spectrum of the resulting fluctuations in x will be

$$G_x(\omega) = (P^2+Q^2)G_F(\omega) = \frac{2}{\pi}\frac{Q(\omega)}{\omega}kT .$$

Elementary considerations of the work done by a generalised force, when it and the corresponding generalised coordinate vary sinusoidally, allow one to deduce from the initial relation between x_0 and F_0 that the rate of absorption of energy is $\frac{1}{2}\omega Q(\omega)|F_0|^2$. Thus $\alpha(\omega) = \omega Q(\omega)$ and we have

$$G_x(\omega) = \frac{2}{\pi}\frac{\alpha(\omega)}{\omega^2}kT .$$

24.8 We now consider a problem which contains the essence of a quantum statistical justification of the **generalised Nyquist relation**. The special assumptions we make about the system concerned merely avoid minor complications.

A system has a non-degenerate ground state of energy E_0 and above it a quasi-continuum of states, there being $\rho(E)\,dE$ states of the continuum with energy between E and $E+dE$. The coordinate x, which is assumed to be a classical variable (cf Problem 21.2), has matrix elements only between the ground state and the states of the continuum, the matrix element of x between the ground state and a state of the continuum with energy E being denoted by $x'(E)$.

Find for temperature T (a) the power spectrum $G_x(\omega)$ of the fluctuations in x, and (b) the rate of absorption of energy by the system when external forces resulting in a generalised force $F_0 e^{i\omega t}$ on x are applied.

Show that these quantities satisfy the form of Nyquist theorem established at the end of the previous problem.

Hint: For any particular quantum state of the system the value of a classical variable as a function of time can be identified with its quantum mechanical mean value calculated as a function of time. Fourier components of frequency ω in the time dependence of such mean values arise from pairs of stationary states (in the expansion of the given state in terms of stationary states) with energy difference $\hbar\omega$ between which the variable has non-zero matrix elements. The contribution to the mean square fluctuations of x from components with frequency in the range ω to $\omega + d\omega$ can therefore be found by putting equal to zero the matrix element of x between all pairs of states for which the energy difference does not lie between $\hbar\omega$ and $\hbar\omega + \hbar d\omega$, and calculating the mean square fluctuations in the ordinary way.

Note: x can be regarded as a classical variable only if components of the fluctuation with frequencies which do not satisfy $\hbar\omega \ll kT$ are negligible. To be consistent it must therefore be assumed that we restrict ω in this way in considering both the power spectrum of the fluctuations and the absorption. It should, perhaps, be emphasised that the definition we have given of a power spectrum breaks down if the variable concerned shows important quantum effects. We cannot then attach a meaning to the value of the quantity as a function of time, since an attempt to observe this would disturb the system. One can, of course, find relations between entities defined in suitable quantum mechanical terms[1] which have essentially the form of the Nyquist relation and which do not assume classical behaviour of the variables concerned. Direct physical significance, however, attaches to expressions, such as scattering cross sections, of which these quantum-mechanically defined entities form part, rather than to the entities themselves. To treat such matters would take us too far afield and we restrict ourselves to the classical form of Nyquist theorem. It is also possible to give a derivation of Nyquist's result for systems described by classical statistical mechanics, but we shall not consider this here. Problems 24.4 and 24.10, however, amount to a demonstration of the result for a very special classical system.

Solution

(a) If H denotes the Hamiltonian and $|r\rangle$ one of its eigenstates with energy E_r, we have

$$\overline{x^2} = \frac{\mathrm{Tr}[x^2 \exp(-H/kT)]}{\mathrm{Tr}\exp(-H/kT)} = \frac{\sum_{r,s} \langle r|x|s\rangle\langle s|x|r\rangle \exp(-E_r/kT)}{\sum_q \exp(-E_q/kT)} .$$

[1] L.D.Landau and E.M.Lifshitz, *Statistical Physics* (Pergamon Press, London), 1968, Ch.XII.

We find $G_x(\omega)\mathrm{d}\omega$ if in evaluating the sum in the numerator we consider only pairs of states with non-vanishing matrix elements for x and energy difference between $\hbar\omega$ and $\hbar\omega+\hbar\mathrm{d}\omega$. There are two possibilities: (1) r is the ground state and s corresponds to one of the $\hbar\rho(E_0+\hbar\omega)\mathrm{d}\omega$ states of the continuum with energy difference from the ground state in the necessary range, and (2) s is the ground state and r represents the same set of states in the continuum as did s in the previous case.

Putting Z for the denominator we find, writing x' for $x'(E_0+\hbar\omega)$,

$$G_x(\omega)\mathrm{d}\omega = \frac{1}{Z}\rho(E_0+\hbar\omega)|x'|^2\{\exp(-E_0/kT)+\exp[-(E_0+\hbar\omega)/kT]\}\hbar\,\mathrm{d}\omega.$$

With the assumption $\hbar\omega \ll kT$, this becomes

$$G_x(\omega) = \frac{2}{Z}\exp(-E_0/kT)\rho(E_0+\hbar\omega)|x'|^2\hbar\,.$$

(b) The applied generalised force which we take to be the real part of $F_0\exp(\mathrm{i}\omega t)$ introduces into the Hamiltonian a term

$$-\tfrac{1}{2}x(F_0\mathrm{e}^{\mathrm{i}\omega t}+F_0\mathrm{e}^{-\mathrm{i}\omega t})\,.$$

Of this, the part in $\mathrm{e}^{-\mathrm{i}\omega t}$ gives rise to absorption of energy by transitions to states of higher energy, while the part in $\mathrm{e}^{\mathrm{i}\omega t}$ gives rise to emission of energy by transitions to lower energy states. Absorptive transitions can take place only from the ground state to the continuum (since x has no non-zero matrix elements between states of the continuum), and the rate of absorption of energy in this way will be

$$\hbar\omega\frac{\exp(-E_0/kT)}{Z}\frac{2\pi}{\hbar}|x'|^2\rho(E_0+\hbar\omega)\frac{|F_0|^2}{4}\,.$$

Here $\hbar\omega$ is the energy absorbed per transition, the next term in brackets is the probability of the system being in its ground state, and the rest is the usual expression for the rate of transition between a discrete state and a continuum resulting from a sinusoidal perturbation.

The calculation of emission of energy due to transitions from continuum states to the ground state will go in exactly the same way, except that the probability of occupation of a continuum state, $\exp[-(E_0+\hbar\omega)/kT]/Z$, will replace the probability of occupation of the ground state. The net rate of absorption will be the difference of the two expressions and this is just the first expression multiplied by $[1-\exp(-\hbar\omega/kT)]$, i.e. by $\hbar\omega/kT$ since $\hbar\omega \ll kT$. Thus the net absorption rate is

$$\frac{\hbar^2\omega^2}{kT}\frac{\exp(-E_0/kT)}{Z}\frac{2\pi}{\hbar}|x'|^2\rho(E_0+\hbar\omega)\frac{|F_0|^2}{4}$$

and comparing this with $\frac{1}{2}\alpha(\omega)|F_0|^2$ gives

$$\alpha(\omega) = \pi\hbar\frac{\omega^2}{kT}\frac{\exp(-E_0/kT)}{Z}|x'|^2\rho(E_0+\hbar\omega)\,.$$

We see that

$$G_x(\omega) = \frac{2}{\pi}\frac{\alpha(\omega)}{\omega^2}kT$$

in agreement with the generalised form of Nyquist's result.

24.9 The generalised Nyquist relation can be used to determine the impedance from a knowledge of the power spectrum (or, what is equivalent, the correlation function) of the associated fluctuations. General formulae for transport coefficients, the **Kubo relations**, embody this idea. This problem illustrates it in a simple case.

The normal modes of a particular imperfect crystal all contribute to the total x component p_x of the electric dipole moment of the crystal so that

$$p_x = \sum_r \alpha_r q_r\,,$$

where the normal coordinates q_r, angular frequency ω_r, are so normalised that the kinetic energy is $\frac{1}{2}\sum_r \dot{q}_r^2$. It is assumed that all the normal modes form a continuum, i.e. there are no localised modes.

Relate the power spectrum of the fluctuations of p_x at a temperature T, high enough for all the modes to behave classically, to a function $a(\omega)$ defined by

$$a(\omega)\,d\omega = \sum_{\omega<\omega_r<\omega+d\omega} \alpha_r^2\,.$$

Use the generalisation of Nyquist's result to obtain the imaginary part $\chi''(\omega)$ of the polarisability of the system. Apply the Kramers-Krönig relations (Problem 23.16) to obtain $\chi'(\omega)$: it may be assumed that $\chi'(\infty)$ is zero since the response of each oscillator will tend to zero as the forcing frequency tends to infinity.

Solution

With the q_r normalised in the way described, the potential energy will be $\frac{1}{2}\sum_r \omega_r^2 q_r^2$. It follows that at the high temperature T,

$$\overline{q_r^2} = \frac{kT}{\omega_r^2}\,.$$

If the power spectrum of the fluctuations of p_x is denoted by $G_{p_x}(\omega)$, then $G_{p_x}(\omega)\,d\omega$ is the contribution to $\overline{p_x^2}$ from modes with frequency in the range ω to $\omega+d\omega$. Thus

$$G_{p_x}(\omega)\,d\omega = \sum_{\omega<\omega_r<\omega+d\omega} \alpha_r^2\overline{q_r^2} = \frac{kT}{\omega^2}\sum_{\omega<\omega_r<\omega+d\omega}\alpha_r^2 = kT\frac{a(\omega)}{\omega^2}\,d\omega\,.$$

In the present case the generalised Nyquist result (Problem 24.7) takes the form

$$G_{p_x}(\omega) = \frac{2}{\pi}\frac{\chi''(\omega)}{\omega}kT\,,$$

and comparison of the two expressions for $G_{p_x}(\omega)$ gives

$$\chi''(\omega) = \frac{\pi}{2}\frac{a(\omega)}{\omega}\,.$$

Since $\chi''(\omega)$ is an odd function of ω, the fact that the above expression defines χ'' for only positive ω presents no difficulties when we come to determine $\chi'(\omega)$ from the Kramers-Krönig relations. We have

$$\begin{aligned}
\chi'(\omega)-\chi'(\infty) &= \frac{1}{\pi}P\int_{-\infty}^{+\infty}\frac{\chi''(\omega')\,\mathrm{d}\omega'}{\omega'-\omega}\\
&= \frac{1}{\pi}P\int_{0}^{\infty}\frac{\chi''(\omega')\,\mathrm{d}\omega'}{\omega'-\omega} + \frac{1}{\pi}P\int_{-\infty}^{0}\frac{\chi''(\omega')\,\mathrm{d}\omega'}{\omega'-\omega}\\
&= \frac{1}{\pi}P\int_{0}^{\infty}\frac{\chi''(\omega')\,\mathrm{d}\omega'}{\omega'-\omega} + \frac{1}{\pi}P\int_{0}^{\infty}\frac{\chi''(\omega')\,\mathrm{d}\omega'}{\omega'+\omega}\\
&= \frac{1}{\pi}P\int_{0}^{\infty}\frac{2\omega'\chi''(\omega')\,\mathrm{d}\omega'}{\omega'^2-\omega^2}\\
&= P\int_{0}^{\infty}\frac{a(\omega')\,\mathrm{d}\omega'}{\omega'^2-\omega^2}\,.
\end{aligned}$$

Since we take $\chi'(\infty)$ to be zero, this is the required result.

24.10 Verify the results of the previous problem by a direct dynamical calculation using the following procedure. Assuming that an oscillating electric field $E = E_0\mathrm{e}^{\mathrm{i}\omega t}$ is applied in the x direction determine the generalised force on each normal coordinate. Introducing a small damping of constant κ (independent of r) on each mode, determine the resulting time dependence after decay of transients of each q_r and hence of p_x. Replace the sum over modes in the resulting expression by an integral, making use of the function $a(\omega)$. Finally determine the real and imaginary parts of the resulting response function in the limit in which κ tends to zero.

Solution

The work done by the electric field E in the x direction when q_r increases by δq_r, q_s remaining fixed for $s \neq r$, is $E\delta p_x = E\alpha_r\delta q_r$. The generalised force on the rth mode is therefore $\alpha_r E_0\mathrm{e}^{\mathrm{i}\omega t}$. The coordinate q_r will satisfy the equation

$$\ddot{q}_r + \kappa\dot{q}_r + \omega_r^2 q_r = \alpha_r E_0\mathrm{e}^{\mathrm{i}\omega t}\,,$$

of which the solution, after decay of transients, is

$$q_r = \frac{\alpha_r}{\omega_r^2 - \omega^2 + i\omega\kappa} E_0 e^{i\omega t}\,.$$

We have, therefore, for the time dependence of the dipole moment

$$p_x = \sum_r \alpha_r q_r = \left(\sum_r \frac{\alpha_r^2}{\omega_r^2 - \omega^2 + i\omega\kappa}\right) E_0 e^{i\omega t}\,.$$

The sum in brackets is the complex polarisability; replacing the sum by an integral, we have

$$\chi'(\omega) - i\chi''(\omega) = \int_0^\infty \frac{a(\omega')\,d\omega'}{\omega'^2 - \omega^2 + i\omega\kappa}\,.$$

The denominator of the integrand has a pole of residue $a(\omega)/2\omega$ just below the point ω on the real axis. We can shift this pole on to the real axis and take the path of integration above it in a small semicircle without altering the value of the integral. The integral along the real axis becomes

$$P\int_0^\infty \frac{a(\omega')\,d\omega'}{\omega'^2 - \omega^2}$$

while that along the small semicircle is

$$-\frac{\pi}{2}\frac{ia(\omega)}{\omega}\,.$$

These two portions of the integral can therefore be seen to give the real and imaginary parts respectively. Consequently

$$\chi'(\omega) = P\int_0^\infty \frac{a(\omega')\,d\omega'}{\omega'^2 - \omega^2}$$

and

$$\chi''(\omega) = \frac{\pi}{2}\frac{a(\omega)}{\omega}$$

in agreement with what we obtained before.

24.11 The x component of dipole moment of a particular system will be denoted by p_x. The corresponding generalised force is then E_x, the x component of the electric field. If the absorption of energy when $E_x = E_0 e^{i\omega t}$ is denoted by $\frac{1}{2}\alpha(\omega)|E_0|^2$, show that the integral of $\alpha(\omega)$ over all positive ω can be related to the static polarisability: the derivation should be based on the relation (Problem 24.7) between the power spectrum of p_x and $\alpha(\omega)$ together with the relation between the mean square fluctuation in p_x and the electrical polarisability derivable from the general result (Problem 21.1). Show that the result can also be derived from the Kramers-Krönig relations, provided $\chi'(\infty)$ is assumed zero: this assumption is equivalent to assuming no instantaneous response of the polarization and it is therefore reasonable (Problem 23.16).

Solution

We had (Problem 24.7)

$$G_{p_x}(\omega) = \frac{2}{\pi}\frac{\alpha(\omega)}{\omega^2}\, kT\,.$$

Consequently

$$\overline{p_x^2} = \int_0^\infty G_{p_x}(\omega)\mathrm{d}\omega = \frac{2}{\pi}kT\int_0^\infty \frac{\alpha(\omega)\mathrm{d}\omega}{\omega^2}\,.$$

But we have from the general result (Problem 21.1)

$$\overline{p_x^2} = kT\frac{\partial \overline{p_x}}{\partial E} = \chi'(0)kT$$

where $\chi'(0)$ is the static polarisability. Comparison of the two expressions for $\overline{p_x^2}$ gives

$$\chi'(0) = \frac{2}{\pi}\int_0^\infty \frac{\alpha(\omega)\mathrm{d}\omega}{\omega^2}\,.$$

Since we have from Problem 24.7 $\alpha(\omega) = \omega Q(\omega)$, which in the present case becomes $\alpha(\omega) = \omega\chi''(\omega)$, we see that the result obtained is a special case of the form of the Kramers-Krönig relation established in Problem 24.9.

25
Onsager relations

C.W.McCOMBIE
(*University of Reading, Reading*)

25.1 This example is intended to bring out in a simple context a point which is important for the derivation of the Onsager relations. We consider a particle of very small mass m moving in a straight line under the action of a restoring force $-cx$ and a damping force $-\kappa\dot{x}$, where x is the coordinate of the particle. It will be assumed that m/κ, which is of the order of the time taken by the particle to attain the terminal velocity appropriate to the force acting on it, is very small compared with κ/c, the relaxation time for return to the undisplaced position.

Use the relation between correlation function and macroscopic decay to write down the correlation function $\psi_x(\tau)$. It will be sufficient to do this in terms of the two roots p_1, p_2 of the equation

$$mp^2 - \kappa p + c = 0$$

and to note that in the circumstances considered one of these, p_2 say, will be approximately κ/m, and the other will go to c/κ in the limit of small m.

Consider now

$$\overline{x(t)[x(t+\tau)-x(t)]/\tau}$$

which is $[\psi_x(\tau)-\psi_x(0)]/\tau$. Show that if τ goes to zero this expression goes to zero, but that if τ, taken to be positive, becomes very small compared with κ/c, while remaining large compared with m/κ, the expression will be given to a very good approximation by $-kT/\kappa$.

We may express this result by saying that $\overline{x\dot{x}}$ is zero, as is required by statistical mechanics, but if we work on a time-scale on which the process of achieving terminal velocity takes negligible time and take the 'derivative' of x forward from the time at which x is evaluated, we will obtain the value $-kT/\kappa$ for $\overline{x\dot{x}}$.

Solution

We obtain from the solution of the equation for x which has finite x but zero $\dot{x}$ at $t = 0$

$$\psi_x(\tau) = \overline{x^2}\left[\frac{p_2}{p_2-p_1}\exp(-p_1|\tau|) - \frac{p_1}{p_2-p_1}\exp(-p_2|\tau|)\right]$$

with $p_2 \sim \kappa/m$, $p_1 \sim c/\kappa$, and $\overline{x^2} = kT/c$.

The limit of $[\psi_x(\tau)-\psi_x(0)]/\tau$ as τ tends to zero is just the zero τ value of $\mathrm{d}\psi_x(\tau)/\mathrm{d}\tau$, and this is easily seen to be zero. If on the other hand $1/p_2 \ll |\tau| \ll 1/p_1$, $\exp(-p_2|\tau|)$ becomes negligible while $\exp(-p_1|\tau|)$ may be replaced by $1-p_1|\tau|$. If we take τ to be positive we have then

$$\psi_x(\tau)-\psi_x(0) = \frac{kT}{c}\left(-\frac{p_1p_2}{p_2-p_1}\tau + \frac{p_1}{p_2-p_1}\right).$$

With p_1/p_2 negligible, p_1 replaced by c/κ and τ such that $p_2\tau \gg 1$, the right-hand side becomes $-kT\tau/\kappa$.

Notice that if τ were taken negative one would get kT/κ for $[\psi_x(\tau)-\psi_x(0)]/\tau$.

25.2 Obtain the expression $-kT/\kappa$ for $\overline{x\dot{x}}$ in the circumstances of the previous question with negligible m, by starting from the macroscopic equation

$$\dot{x} = -\frac{c}{\kappa}x\,.$$

Avoid explicit determination of the correlation function, as this is very easy here but more troublesome in the systems with a number of independent variables which are considered in establishing the Onsager relations.

Solution

Consider a large number of records in all of which x has the value x_0 at time t. Then the average over the records of the values at time $t+\tau$ will be, for small τ,

$$x_0-\frac{c}{\kappa}x_0\tau$$

since the change in x, produced by fluctuating forces associated with the damping, will average to zero. Thus averaging the quantity

$$x(t)\frac{x(t+\tau)-x(t)}{\tau}$$

over these records will give (since $x(t)$ is just the constant x_0 for the records considered)

$$-\frac{c}{\kappa}x_0^2\,.$$

Now doing a further average over x_0 gives the required result.

It should be noted that we get the correct result by the formal procedure of multiplying the given macroscopic equation by x and averaging. It will be useful to make use of this formal procedure in the slightly more complicated cases to be considered later.

25.3 $X(t)$ and $Y(t)$ are two variables associated with a system both of which behave classically. The statistical character of the fluctuations will be unchanged by time reversal [i.e. $X'(t) = X(-t)$ and $Y'(t) = Y(-t)$ will have the same statistical properties as $X(t)$ and $Y(t)$]. Deduce that

$$\overline{\Delta X(t)[\Delta Y(t+\tau)-\Delta Y(t)]} = \overline{\Delta Y(t)[\Delta X(t+\tau)-\Delta X(t)]}\,,$$

so that dividing by τ and taking the limit of small τ gives

$$\overline{\Delta X(t)\Delta \dot{Y}(t)} = \overline{\Delta Y(t)\Delta \dot{X}(t)}\,.$$

Here, and in the examples which follow, it is to be understood that τ becomes small on one time-scale but remains large on another (see Problem 25.2).

Solution

Adding $\overline{\Delta X(t)\Delta Y(t)}$ to both sides of the first form of the result shows that what we have to prove is

$$\overline{\Delta X(t)\Delta Y(t+\tau)} = \overline{\Delta Y(t)\Delta X(t+\tau)}\,.$$

But the invariance of the statistical properties with respect to time means that such average values are independent of time t, so that we can write in particular

$$\overline{\Delta X(t)\Delta Y(t+\tau)} = \overline{\Delta X(t-\tau)\Delta Y(t)}\,.$$

The invariance under time reversal means that we can replace τ by $-\tau$ on the right-hand side, so that this result becomes

$$\overline{\Delta X(t)\Delta Y(t+\tau)} = \overline{\Delta Y(t)\Delta X(t+\tau)}$$

as required.

25.4 Suppose that the rates of change of the variables $X(t)$ and $Y(t)$ are related to their deviations from their mean values by

$$\Delta \dot{X}(t) = a\Delta X + b\Delta Y\,,$$

$$\Delta \dot{Y}(t) = c\Delta X + d\Delta Y\,.$$

Obtain an equation which is satisfied by the coefficients a, b, c, d, the only other quantities involved in the equation being $\overline{\Delta X^2}$, $\overline{\Delta Y^2}$ and $\overline{\Delta X\Delta Y}$.

Solution

Using the formal procedure discussed in Problem 25.2 we multiply the first equation by $\Delta Y(t)$ and average, the second by $\Delta X(t)$ and average. The left-hand sides are then equal by the time reversal result established in the previous problem, and equating the right-hand sides gives

$$a\overline{\Delta X\Delta Y}+b\overline{\Delta Y^2} = c\overline{\Delta X^2}+d\overline{\Delta X\Delta Y}\,.$$

This is a special case of the type of relation derived by Onsager. It is special because the rates of change of ΔX and ΔY are expressed, in the equations under consideration, in terms of ΔX and ΔY themselves. More general equations, to be considered later, relate $\Delta\dot{X}$ and $\Delta\dot{Y}$ to the departures from the equilibrium values of other quantities.

If it happens that $\overline{\Delta X \Delta Y} = 0$ and $\overline{\Delta X^2} = \overline{\Delta Y^2}$ then the above relation reduces to the typical Onsager relation form

$$b = c\ .$$

It must, however, be emphasised that the physical content of the relation between the coefficients is in no way diminished by its not taking this special form. The same is true in the case of the more general equations mentioned above: the fact that it is always possible to choose the quantities in terms of which $\Delta\dot{X}$ and $\Delta\dot{Y}$ are expressed so that the Onsager relation takes its typical form is to some extent incidental.

25.5 A passive electrical network has two pairs of terminals, the d.c. voltage and current at one pair being V_1, I_1 and at the other V_2, I_2. These are related by the equations

$$I_1 = aV_1 + bV_2\ ,$$

$$I_2 = cV_1 + dV_2\ .$$

The network is assumed to be of such a nature that the determinant of the coefficients is non-zero: this means that I_1 and I_2 cannot both be zero unless V_1 and V_2 both vanish.

Obtain the Onsager relation for the coefficients in these equations by the following steps.

(a) Suppose capacitances C_1 and C_2 are connected across the two pairs of terminals. Denote the charges on these capacitances by q_1 and q_2. Show that if the whole system is at temperature T,

$$\overline{q_1^2} = C_1 kT, \quad \overline{q_2^2} = C_2 kT, \quad \text{and } \overline{q_1 q_2} = 0\ .$$

(b) Express the equations giving I_1, I_2 in terms of V_1, V_2 as equations relating $\dot{q}_1$, $\dot{q}_2$ to q_1, q_2.

(c) Use the time reversal result

$$\overline{q_1 \dot{q}_2} = \overline{q_2 \dot{q}_1}$$

to obtain the required relation.

Solution

(a) The generalised forces associated with q_1 and q_2 are voltages, v_1 and v_2 say, introduced in series with the capacitors (the voltages being so connected that the work done by them is $v_1 \delta q_1$ in one case and $v_2 \delta q_2$ in the other). The voltages across the terminals will then be $v_1 - q_1/C_1$ and $v_2 - q_2/C_2$, and when the currents are zero (as they must be in equilibrium) these voltages must be zero. Thus

$$q_1 = C_1 v_1, \qquad q_2 = C_2 v_2\ .$$

It follows that ($\overline{q_1}$ and $\overline{q_2}$ are, of course, zero for zero applied voltages, so that $\Delta q_1 = q_1$, $\Delta q_2 = q_2$)

$$\overline{q_1^2} = kT\left(\frac{\partial q_1}{\partial v_1}\right)_T = kTC_1 ,$$

$$\overline{q_2^2} = kT\left(\frac{\partial q_2}{\partial v_2}\right)_T = kTC_2 ,$$

$$\overline{q_1 q_2} = kT\left(\frac{\partial q_1}{\partial v_2}\right)_T = 0 .$$

(b)

$$I_1 = -\dot{q}_1, \quad I_2 = -\dot{q}_2 ,$$

$$V_1 = \frac{q_1}{C_1}, \quad V_2 = \frac{q_2}{C_2} .$$

The equations relating I_1 and I_2 to V_1 and V_2 can therefore be rewritten

$$\dot{q}_1 = -\frac{a}{C_1}q_1 - \frac{b}{C_2}q_2 ,$$

$$\dot{q}_2 = -\frac{c}{C_1}q_1 - \frac{d}{C_2}q_2 .$$

(c) As in Problem 25.4 multiply the first of these equations by q_2, the second by q_1, and average

$$\overline{q_2\dot{q}_1} = -\frac{a}{C_1}\overline{q_1 q_2} - \frac{b}{C_2}\overline{q_2^2} ,$$

$$\overline{q_1\dot{q}_2} = -\frac{c}{C_1}\overline{q_1^2} - \frac{d}{C_2}\overline{q_1 q_2} .$$

The left-hand sides are equal, and equating the right-hand sides gives, on substituting from (a),

$$-\frac{b}{C_2}C_2 kT = -\frac{c}{C_1}C_1 kT ,$$

that is

$$b = c .$$

In this case, then, we get the standard Onsager reciprocal relation.

25.6 This problem illustrates the use of the Onsager procedure to obtain a relationship between transport coefficients for a material. The coefficients considered will be those relating current density I and thermal current density J to voltage gradient $\mathrm{d}V/\mathrm{d}x$ and temperature gradient $\mathrm{d}T/\mathrm{d}x$ in a metal: the metal is assumed isotropic so that it is not necessary to work with vectors and tensors. The coefficients are those in the equations

$$I = -\sigma\frac{\mathrm{d}V}{\mathrm{d}x} + a\frac{\mathrm{d}T}{\mathrm{d}x} ,$$

$$J = b\frac{\mathrm{d}V}{\mathrm{d}x} + c\frac{\mathrm{d}T}{\mathrm{d}x} ,$$

where σ is the electrical conductivity.

To obtain the required relation consider, as in Problem 21.7, a piece of the same metal connected by a wire of the metal to a very large block, also of the same metal, which acts as a heat bath (temperature T) and electron reservoir (chemical potential μ). Suppose that the wire has cross section s and length l and that the capacity of the first piece of metal in the presence of the large block is C.

Write down, in terms of the coefficients defined above, equations relating the rate of change of the number of electrons on the piece of metal, and of the energy of the metal to the excess number of electrons Δn and excess temperature ΔT for the piece of metal. Multiply the first equation by ΔE (the excess energy of the piece of metal), the second by Δn, and average. Apply the time inversion relationship (Problem 25.3) and some of the mean values determined in Problem 20.5 to obtain the required relationship.

Solution

We have

$$\Delta\dot{n} = -\frac{Is}{e} , \qquad \Delta\dot{E} = -Js ,$$

$$\frac{\mathrm{d}V}{\mathrm{d}x} = \frac{\Delta V}{l} = e\frac{\Delta n}{lC} , \qquad \frac{\mathrm{d}T}{\mathrm{d}x} = \frac{\Delta T}{l} .$$

The equations given for I and J can, therefore, be rewritten as

$$\Delta\dot{n} = +\frac{s\sigma}{lC}\Delta n - \frac{sa}{el}\Delta T ,$$

$$\Delta\dot{E} = -\frac{esb}{lC}\Delta n - \frac{sc}{l}\Delta T .$$

Multiplying the first and second equation by ΔE and Δn respectively, averaging, and using the time inversion relationship $\overline{\Delta E\Delta\dot{n}} = \overline{\Delta n\Delta\dot{E}}$ gives

$$-\frac{\sigma}{C}\overline{\Delta E\Delta n} + \frac{a}{e}\overline{\Delta E\Delta T} = \frac{eb}{C}\overline{\Delta n^2} + c\overline{\Delta n\Delta T} .$$

We saw in Problem 20.5 that $\overline{\Delta E\Delta T} = kT^2$ and $\overline{\Delta T\Delta n} = 0$. We also saw in Problem 21.7 that $\overline{\Delta n^2} = (C/e^2)kT$. This leaves only $\overline{\Delta E\Delta n}$ to be put in appropriate form, and we had in Problem 20.5 that

$$\overline{\Delta E \Delta n} = kT\left(\frac{\partial E}{\partial \mu}\right)_T = kT\left(\frac{\partial E}{\partial n}\right)_T\left(\frac{\partial n}{\partial \mu}\right)_T .$$

But

$$\left(\frac{\partial E}{\partial n}\right)_T = T\left(\frac{\partial S}{\partial n}\right)_T + \mu = -T\left(\frac{\partial \mu}{\partial T}\right)_n + \mu = -T^2\frac{\partial}{\partial T}\left(\frac{\mu}{T}\right)_n ,$$

and at constant temperature

$$\mathrm{d}\mu = e\,\mathrm{d}V = \frac{e^2\,\mathrm{d}n}{C}$$

so that

$$\left(\frac{\partial n}{\partial \mu}\right)_T = \frac{C}{e^2} .$$

Thus

$$\overline{\Delta E \Delta n} = -kT^3\frac{\partial}{\partial T}\left(\frac{\mu}{T}\right)_n\frac{C}{e^2} .$$

The relation between the coefficients is therefore

$$+\frac{\sigma k T^3}{e^2}\frac{\partial}{\partial T}\left(\frac{\mu}{T}\right)_n + \frac{a}{e}kT^2 = \frac{b}{e}kT ,$$

i.e.

$$\frac{\sigma T^2}{e}\frac{\partial}{\partial T}\left(\frac{\mu}{T}\right)_n + aT = b .$$

This relation is not of the standard reciprocal form, but it contains all the physical information derivable from Onsager type considerations.

25.7 Deduce from Problem 22.5 that if we write the rates of change of the coordinates X_r as linear combinations of the associated forces F_s, defined by $F_s = -\partial S/\partial X_s$, then the array of coefficients is symmetric, i.e. if we write

$$\dot{X}_r = \sum_s \beta_{rs} F_s$$

then

$$\beta_{rs} = \beta_{sr} .$$

This is the orthodox form of the **Onsager relations.**

Solution

Multiplying the equation for $\dot{X}_l$ by X_m and averaging gives

$$\overline{X_m \dot{X}_l} = \sum_s \beta_{ls}\overline{X_m F_s} = \sum_s \beta_{ls} k \delta_{ms} = k\beta_{lm} .$$

Similarly

$$\overline{X_l \dot{X}_m} = k\beta_{ml} .$$

So the time reversal relation gives

$$\beta_{lm} = \beta_{ml} .$$

25.8 The aim of this problem is to derive, by an application of the Onsager relations in their conventional form, the relation between the electro-thermal transport coefficients of a metal already derived in a less orthodox way in Problem 25.6.

As in the earlier problem consider a metal body connected by a metal wire to a much larger metal body, all the metals being the same. If the energy and number of electrons on the smaller metal body increase from their equilibrium values by ΔE and Δn respectively, the increase in entropy of the whole system will be, to lowest non-vanishing order, a quadratic function of ΔE and Δn,

$$\Delta S = \tfrac{1}{2}\left[\left(\frac{\partial^2 S}{\partial E^2}\right)_n \Delta E^2 + 2\left(\frac{\partial^2 S}{\partial E \partial n}\right)\Delta E \Delta n + \left(\frac{\partial^2 S}{\partial n^2}\right)_E \Delta n^2\right].$$

Show that the forces associated in the Onsager theory with E and n, F_E and F_n, are given by

$$F_E = -\left(\frac{\partial S}{\partial E}\right)_n = \frac{\Delta T}{T^2},$$

$$F_n = -\left(\frac{\partial S}{\partial n}\right)_E = \Delta\left(\frac{\mu}{T}\right).$$

Re-express the equations relating $\dot{E}$ and $\dot{n}$ to Δn and ΔT, with coefficients as given in Problem 25.6, as equations relating $\dot{E}$ and $\dot{n}$ to F_E and F_n. Show that the Onsager relation, when applied to these equations, yields the same relation between transport coefficients as was obtained before.

Solution

Substituting the expression for ΔS in the expressions for F_E gives

$$F_E = -\left(\frac{\partial^2 S}{\partial E^2}\right)_n \Delta E - \left(\frac{\partial^2 S}{\partial E \partial n}\right)\Delta n = -\frac{\partial}{\partial E}\left(\frac{\partial S}{\partial E}\right)\Delta E - \frac{\partial}{\partial n}\left(\frac{\partial S}{\partial E}\right)\Delta n$$

$$= -\Delta\left(\frac{\partial S}{\partial E}\right) = -\Delta\left(\frac{1}{T}\right) = \frac{\Delta T}{T^2}.$$

Similarly

$$F_n = -\frac{\partial}{\partial E}\left(\frac{\partial S}{\partial n}\right)\Delta E - \frac{\partial}{\partial n}\left(\frac{\partial S}{\partial n}\right)\Delta n = -\Delta\left(\frac{\partial S}{\partial n}\right) = \Delta\left(\frac{\mu}{T}\right).$$

It is necessary to express both F_E and F_n in terms of Δn and ΔT. F_E is already given in this form and F_n is easily expressed similarly if we recall (Problem 21.7) that, for the system considered, $(\partial\mu/\partial n)_T = e^2/C$, where C is the capacity. Thus

$$F_n = \frac{\partial}{\partial T}\left(\frac{\mu}{T}\right)_n \Delta T + \frac{\partial}{\partial n}\left(\frac{\mu}{T}\right)_T \Delta n = \frac{\partial}{\partial T}\left(\frac{\mu}{T}\right)_n \Delta T + \frac{e^2}{TC}\Delta n.$$

Solving for ΔT and Δn in terms of F_E and F_n gives

$$\Delta T = T^2 F_E \ ,$$

$$\Delta n = \frac{TC}{e^2}\left[F_n - T^2 \frac{\partial}{\partial T}\left(\frac{\mu}{T}\right)_n F_E\right] .$$

In the notation of Problem 25.6 we had

$$\Delta \dot{n} = \frac{s\sigma}{lC}\Delta n - \frac{sa}{el}\Delta T ,$$

$$\Delta \dot{E} = -\frac{esb}{lC}\Delta n - \frac{sc}{l}\Delta T .$$

In terms of F_E and F_n these become

$$\Delta \dot{n} = \frac{s}{l}\frac{T}{e^2}\sigma F_n - \frac{s}{l}\left[\frac{T^3}{e^2}\frac{\partial}{\partial T}\left(\frac{\mu}{T}\right)_n \sigma + \frac{T^2}{e}a\right] F_E \ ,$$

$$\Delta \dot{E} = -\frac{s}{l}\frac{T}{e}bF_n + \frac{s}{l}\left[\frac{T^3}{e}\frac{\partial}{\partial T}\left(\frac{\mu}{T}\right)_n b - T^2 c\right] F_E \ .$$

In this case (Problem 25.7) the Onsager relation is just the equality of the off-diagonal coefficients. This gives at once

$$b = Ta + \frac{T^2}{e}\frac{\partial}{\partial T}\left(\frac{\mu}{T}\right)_n \sigma \ ,$$

which is the result obtained before (Problem 25.6).

26

Stochastic methods: master equation and Fokker–Planck equation

I.OPPENHEIM
(*Massachusetts Institute of Technology, Cambridge, Massachusetts*)
K.E.SHULER
(*University of California, San Diego, California*)
G.H.WEISS
(*National Institute of Health, Bethesda, Maryland*)

26.0 In these problems [1] the following notation is used:

$W(n, t|m, s)$ is the conditional probability that a system is in state n at time t, given that it was in state m at time $s < t$. It is normalised so that $\sum_n W(n, t|m, s) = 1$, where the sum is over all possible states n.

$P(n, t)$ is the probability that the system is in state n at time t. For all cases of present interest the condition $\sum_n P(n, t) = 1$ is valid.

$P_2(m, s;\ n, t)$ is the joint probability that the system is in state m at time s, and that it is in state n at time t.

$$A(n, m; t) = \lim_{\Delta \to 0} \frac{1}{\Delta}[W(n, t+\Delta|m, t) - \delta_{nm}] \qquad (26.0.1)$$

is the transition probability per unit time (i.e. the transition rate) for a transition from state m to state n at time t. All of the processes to be considered will be such that $A(n, m;\ t)$ exists.

The probabilities $P(n, t)$ and $P_2(m, s;\ n, t)$ are related by

$$P(n, t) = \sum_m P_2(m, s;\ n, t)\,. \qquad (26.0.2)$$

The conditional probability $W(n, t|m, s)$ is related to P and P_2 by

$$P_2(m, s;\ n, t) = W(n, t|m, s)P(m, s)\,. \qquad (26.0.3)$$

The reader is referred to Problem 2.14 for an introduction to the idea of **master equations.**

26.1 (i) Show that the **master equation** has the form

$$\frac{\partial P(n, t)}{\partial t} = \sum_m A(n, m, t)P(m, t)\,. \qquad (26.1.1)$$

[1] A general reference to this section is I.Oppenheim, K.E.Shuler, and G.H.Weiss, "Stochastic theory of multistage relaxation processes", in *Advances in Molecular Relaxation Processes,* Vol.1, 1967-8, pp.13-68.

(ii) In a radioactivity decay process the probability of emission of a single particle in the time interval $(t, t+\mathrm{d}t)$ is $\lambda\,\mathrm{d}t$, where λ is a constant and the probability of emission of two or more particles in the same interval is zero. The state of the system is described by the number of particles n that have been emitted between time 0 and t.

(a) Calculate the conditional probability $W(n, t+\mathrm{d}t\,|\,m, t)$ for this process.

(b) Calculate the transition rate $A(n, m, t)$ for this process.

(c) Show that $P(n, t)$ satisfies the master equation

$$\begin{aligned} \frac{\partial P(n,t)}{\partial t} &= \lambda[P(n-1,t)-P(n,t)], \quad n = 1,2,3,\ldots \\ \frac{\partial P(0,t)}{\partial t} &= -\lambda P(0,t)\,. \end{aligned} \tag{26.1.2}$$

(d) Calculate the probability $P(n, t)$ from these last equations assuming that $P(0, 0) = 1$.

Solution

(i) By definition

$$P(n,t+\Delta) = \sum_m W(n,t+\Delta\,|\,m,t)P(m,t)\,. \tag{26.1.3}$$

Subtract from this the identity

$$P(n,t) = \sum_m \delta_{nm}P(m,t)\,, \tag{26.1.4}$$

where $\delta_{nm} = 0$ for $n \neq m$ and $\delta_{nm} = 1$ for $n = m$, and divide by Δ. One then obtains

$$\frac{P(n,t+\Delta)-P(n,t)}{\Delta} = \sum_m \frac{1}{\Delta}[W(n,t+\Delta\,|\,m,t)-\delta_{nm}]P(m,t)\,. \tag{26.1.5}$$

Taking the limit $\Delta = 0$ and using the definition of Equation (26.0.1) one finds

$$\frac{\partial P(n,t)}{\partial t} = \sum_m A(n,m;\,t)P(m,t) \tag{26.1.6}$$

as asserted.

(ii) Since the state can only change by 1 during the time interval $(t, t+\mathrm{d}t)$ one has

$$\begin{aligned} &W(n+1,t+\mathrm{d}t\,|\,n,t) = \lambda\,\mathrm{d}t \\ &W(n,t+\mathrm{d}t\,|\,m,t) = 0 \qquad \text{for } m \neq n, n-1 \\ &W(n,t+\mathrm{d}t\,|\,n,t) = 1-W(n+1,t+\mathrm{d}t\,|\,n,t) = 1-\lambda\,\mathrm{d}t\,. \end{aligned} \tag{26.1.7}$$

(b) By combining expressions (26.0.1) with (26.1.7) one obtains

$$A(n+1;n;\,t) = \frac{1}{\Delta}\lambda\Delta = \lambda$$

$$A(n,n;\,t) = -\lambda \tag{26.1.8}$$

$$A(n,m;\,t) = 0 \quad \text{for } m \neq n,\ n-1\ .$$

(c) Substituting the transition rates (26.1.8) into the general master equation (26.1.1) one obtains the specific master equation (26.1.2)

(d) Define a generating function

$$G(z,t) = \sum_{n=0}^{\infty} P(n,t)z^n\ . \tag{26.1.9}$$

If the nth term of the master equation (26.1.2) is multiplied by z^n and the sum taken, it is found that $G(z, t)$ obeys the equation

$$\frac{\partial G}{\partial t} = \lambda(z-1)G \tag{26.1.10}$$

which has the solution

$$G(z,t) = G(z,0)\exp[-(1-z)\lambda t]\ . \tag{26.1.11}$$

For the initial conditions $P(0, 0) = 1$ and $P(n, 0) = 0$ for $n = 1, 2, 3$, one finds $G(z, 0) = 1$ and $G(z, t) = \exp[-(1-z)\lambda t]$. Expansion of this function in a power series in z and comparison with Equation (26.1.9) yields

$$P(n,t) = \exp(-\lambda t)\frac{\lambda t^n}{n!} \tag{26.1.12}$$

for the probability that n particles have been emitted at time t. The distribution function of Equation (26.1.12) is known as the **Poisson distribution.**

26.2 Consider a master equation characterised by a time independent matrix $\mathbf{A} = (A_{nm})$ in which the elements A_{nm} are the rates of transition from states m to states n. Define the vector of state probabilities $\mathbf{P}(t)$ by

$$\mathbf{P}(t) = \begin{pmatrix} P(0,t) \\ P(1,t) \\ P(2,t) \\ \cdot \\ \cdot \\ \cdot \end{pmatrix} \tag{26.2.1}$$

Let the matrix $\mathbf{A}$ have non-degenerate eigenvalues λ_j ($j = 0, 1, 2, \ldots$), right eigenvectors $\mathbf{R}_j$ and left eigenvectors $\mathbf{L}_j$, defined by

$$\mathbf{A}\mathbf{R}_j = \lambda_j \mathbf{R}_j, \quad \mathbf{L}_j \mathbf{A} = \lambda_j \mathbf{L}_j \, . \tag{26.2.2}$$

The master equation can be written in terms of $\mathbf{P}(t)$ and $\mathbf{A}$ as

$$\dot{\mathbf{P}} = \mathbf{A}\mathbf{P} \, . \tag{26.2.3}$$

(i) Show that $\mathbf{P}(t)$ has the formal solution

$$\mathbf{P}(t) = \sum_{j=0}^{\infty} (\mathbf{L}_j \mathbf{P}(0)) \mathbf{R}_j \exp(\lambda_j t) \tag{26.2.4}$$

(ii) Define a **conservative system** as one in which $A(n, m) \geqslant 0$ for $n \neq m$, and for which

$$A(n, n) = -\sum_m{}' A(m, n) \, , \tag{26.2.5}$$

where the prime denotes omission of the $m = n$ term. Show that if

$$\sum_{n=0}^{\infty} P(n, 0) = 1 \tag{26.2.6}$$

it follows that

$$\sum_{n=0}^{\infty} P(n, t) = 1 \tag{26.2.7}$$

for all t [the $A(n, m)$ may depend on time]. Most systems of physical interest are conservative, and all systems to be discussed here are conservative.

(iii) Show that there exists a zero eigenvalue.

(iv) If $\lambda_0 = 0$ and $\lambda_1 \neq 0$, show that

$$\mathbf{P}(\infty) = \mathbf{R}_0 \Big/ \sum_{i=0}^{\infty} (\mathbf{R}_0)_i \, , \tag{26.2.8}$$

where $(\mathbf{R}_0)_i$ is the ith component of $\mathbf{R}_0$. Notice that the equilibrium distribution $\mathbf{P}(\infty)$ is thus independent of the initial condition $\mathbf{P}(0)$.

(v) The condition of **detailed balance** at equilibrium can be expressed as

$$A(m, n) P(n, \infty) = A(n, m) P(m, \infty) \, . \tag{26.2.9}$$

Show that, when Equation (26.2.9) is satisfied, the eigenvalues λ_j are real and negative, and $\lambda_0 = 0$.

[See also Problem 2.14 for additional explanations of detailed balance.]

Solution

(i) Assume a solution for $P(t)$ of the form

$$\mathbf{P}(t) = \sum_{j=0}^{\infty} a_j(t) \mathbf{R}_j \, . \tag{26.2.10}$$

Substitution of this solution into the master equation $\dot{\mathbf{P}} = \mathbf{AP}$ yields

$$\sum_{j=0} [\dot{a}_j(t) - \lambda_j a_j(t)]\mathbf{R}_j = 0 . \tag{26.2.11}$$

Since the $\mathbf{R}_j$ are independent when the λ_j are non-degenerate, one may set each term separately to zero and solve the resulting equations $\dot{a}_j(t) = \lambda_j a_j(t)$ to obtain

$$a_j(t) = a_j(0)\exp(\lambda_j t) . \tag{26.2.12}$$

Thus, the $a_j(0)$ can be calculated from the relation

$$\mathbf{P}(0) = \sum_{j=0}^{\infty} a_j(0)\mathbf{R}_j \tag{26.2.13}$$

in which it is assumed that $\mathbf{P}(0)$ is a known vector. To find the $a_j(0)$ we show that the inner product $\mathbf{L}_n\mathbf{R}_j = 0$ when $n \neq j$. From Equation (26.2.2) it follows that

$$\begin{aligned} \mathbf{L}_n\mathbf{AR}_j &= \lambda_j \mathbf{L}_n\mathbf{R}_j \\ \mathbf{L}_n\mathbf{AR}_j &= \lambda_n \mathbf{L}_n\mathbf{R}_j \end{aligned} \tag{26.2.14}$$

or, subtracting,

$$0 = (\lambda_n - \lambda_j)(\mathbf{L}_n\mathbf{R}_j) . \tag{26.2.15}$$

From the hypothesis of non-degenerate eigenvalues it follows that $(\mathbf{L}_n\mathbf{R}_j) = 0$ when $n \neq j$. In all cases of interest the vectors $\mathbf{L}_n$ and $\mathbf{R}_n$ can be scaled so that $\mathbf{L}_n\mathbf{R}_n = 1$, and we can assume that the eigenvectors are orthonormal. If we now multiply Equation (26.2.13) by $\mathbf{L}_n$ and use the relation of orthonormality we find that

$$a_j(0) = \mathbf{L}_j\mathbf{P}(0) \tag{26.2.16}$$

which leads to the desired formal solution (26.2.4)

$$\mathbf{P}(t) = \sum_{j=0}^{\infty} (\mathbf{L}_j\mathbf{P}(0))\mathbf{R}_j \exp(\lambda_j t) .$$

(ii) Write the master equation out in full as

$$\begin{aligned} \dot{P}(0,t) &= A_{00}P(0,t) + A_{01}P(1,t) + A_{02}P(2,t) + \cdots \\ \dot{P}(1,t) &= A_{10}P(0,t) + A_{11}P(1,t) + A_{12}P(2,t) + \cdots \\ \dot{P}(2,t) &= A_{20}P(0,t) + A_{21}P(1,t) + A_{22}P(2,t) + \cdots \\ &\;\;\vdots \end{aligned} \tag{26.2.17}$$

If these equations are added, taking account of the condition (26.2.5), it is found that

$$\sum_{n=0}^{\infty} \dot{P}(n,t) = 0 . \tag{26.2.18}$$

An integration of this equation yields

$$\sum_{n=0}^{\infty} P(n,t) = \text{constant} = \sum_{n=0}^{\infty} P(n,0) = 1\ . \tag{26.2.19}$$

(iii) Consider the eigenvector

$$\mathbf{L}_0 = \alpha(1,1,1,\ldots)\ . \tag{26.2.20}$$

The condition $\sum_{m=0}^{\infty} A(m,n) = 0$, which is equivalent to Equation (26.2.5), can be written

$$\mathbf{L}_0\mathbf{A} = 0\ ; \tag{26.2.21}$$

hence the eigenvalue λ_0 corresponding to $\mathbf{L}_0$ is 0. The existence of the non-degenerate eigenvalue $\lambda_0 = 0$ insures that the master equation (26.2.3) with the general solution (26.2.4) has a non-zero equilibrium solution $\mathbf{P}(\infty)$.

(iv) Recall the expansion of the solution (26.2.4) which expresses $\mathbf{P}(t)$ in terms of eigenfunctions and eigenvalues. It must be the case that $\text{Re}\,\lambda_j \leqslant 0$ for all j, as otherwise $\mathbf{P}(t)$ would have unbounded components for sufficiently long times. One can therefore separate out the $j = 0$ term to find that

$$\mathbf{P}(t) = [\mathbf{L}_0\mathbf{P}(0)]\mathbf{R}_0 + \sum_{j=1}^{\infty} [\mathbf{L}_j\mathbf{P}(0)]\mathbf{R}_j \exp(\lambda_j t)\ , \tag{26.2.22}$$

where all terms in the second sum tend to zero as $t \to \infty$. Recall that $\mathbf{L}_0$ can be written

$$\mathbf{L}_0 = \alpha(1,1,1,\ldots)\ , \tag{26.2.23}$$

where α is a constant, so that

$$\mathbf{L}_0\mathbf{P}(0) = \alpha\ . \tag{26.2.24}$$

Thus $\mathbf{P}(\infty)$ is proportional to $\mathbf{R}_0$, i.e.

$$\mathbf{P}(\infty) = \alpha\mathbf{R}_0\ . \tag{26.2.25}$$

But, since the system is conservative,

$$\sum_i [\mathbf{P}(\infty)]_i = 1\ ,$$

and

$$\alpha = 1\Big/\sum_i (\mathbf{R}_0)_i\ .$$

(v) In order to prove reality of the eigenvalues λ_j it is sufficient to show that the transition matrix $\mathbf{A}$ is similar to a symmetric matrix $\mathbf{S}$. Since the eigenvalues of a real symmetric matrix are all real, it follows that the eigenvalues of $\mathbf{A}$ are real. Define a matrix $\mathbf{U}$ whose ijth element is $U_{ij} = [P(i,\infty)]^{1/2}\delta_{ij}$, where δ_{ij} is a Kronecker delta, and a matrix $\mathbf{S}$ constructed from $\mathbf{A}$ by $\mathbf{S} = \mathbf{U}^{-1}\mathbf{A}\mathbf{U}$. Since $\mathbf{U}$ is a diagonal matrix, it is easily inverted, and the elements of **S** are found to be

$$S_{nm} = [P(n,\infty)]^{-1/2} A(n,m)[P(m,\infty)]^{1/2} . \qquad (26.2.26)$$

One can now verify from Equation (26.2.9) that $S_{nm} = S_{mn}$, i.e. S is a symmetric matrix.

It is possible to conclude at this point that the λ_j are negative for $j \geqslant 1$ since, if they were not, $P(t)$ as given by expression (26.2.22), could not represent a vector of probabilities for large t. However, the fact that the λ_j must be negative can also be proved by showing that **S** corresponds to a negative semi-definite quadratic form. This can readily be done by expressing the matrix elements $A(n, m)$ in terms of non-negative elements $B(n, m)$ by

$$A(n,m) = (1-\delta_{nm})B(n,m) - \delta_{nm} \sum_{r \neq n} B(r,n) \qquad (26.2.27)$$

with $B(n, m) = A(n, m) \geqslant 0$ for $n \neq m$. If **S** corresponds to a negative definite quadratic form then the function

$$S(\mathrm{y}) = \sum_{n,m} S_{nm} y_n y_m \qquad (26.2.28)$$

must be non-positive. But an explicit calculation shows that $S(\mathrm{y})$ can be expressed as

$$S(\mathrm{y}) = -\sum_{\substack{r,n \\ r \neq n}} B(r,n) y_n^2 + \sum_{\substack{n,m \\ n \neq m}} B(n,m)[P(m,\infty)]^{1/2}[P(n,\infty)]^{-1/2} y_n y_m . \qquad (26.2.29)$$

Using the principle of detailed balance expressed in Equation (26.2.9) one can rewrite Equation (26.2.29) as

$$S(\mathrm{y}) = -\tfrac{1}{2} \sum_{n,m} B(n,m) P(m,\infty) \left\{ \frac{y_n}{[P(n,\infty)]^{1/2}} - \frac{y_m}{[P(m,\infty)]^{1/2}} \right\}^2 \qquad (26.2.30)$$

which is manifestly negative semi-definite.

The condition of detailed balance, which is met with so frequently in physical systems, therefore suffices to insure the real and negative form of the eigenvalues λ_j and so leads to mathematically reasonable forms for the state probabilities.

26.3 The master equation for the relaxation of an ensemble of non-interacting harmonic oscillators in contact with a heat bath at temperature $T(\infty)$ is

$$\frac{\partial P(n,\tau)}{\partial \tau} = \{n e^{-\theta(\infty)} P(n-1,\tau) - [n + (n+1) e^{-\theta(\infty)}] P(n,\tau)$$

$$+ (n+1) P(n+1,\tau)\}, \quad n = 0, 1, 2, \dots \qquad (26.3.1)$$

$$P(-1,\tau) \equiv 0 ,$$

where $P(n, \tau)$ is the probability that the oscillator is in vibrational state n with energy $nh\nu$ at (dimensionless) time τ, and where $\theta(\infty) = h\nu/[kT(\infty)]$. Obtain a solution to Equation (26.3.1) for an initial Boltzmann distribution of oscillators

$$P(n, 0) = (1 - e^{-\theta(0)})e^{-n\theta(0)}, \tag{26.3.2}$$

where $\theta(0) = h\nu/[kT(0)]$ and where $T(0)$ is the vibrational temperature of the ensemble of oscillators at $\tau = 0$. Show that $P(n, \tau)$ is a Boltzmann distribution at all times τ, i.e. that

$$P(n, \tau) = (1 - e^{-\theta(\tau)})e^{-n\theta(\tau)} \tag{26.3.3}$$

and find an explicit expression for $\theta(\tau)$.

Solution

As in Problem 26.1(d), the solution to the difference equation can be obtained through the use of a generating function

$$G(z, t) = \sum_{n=0}^{\infty} z^n P(n, t). \tag{26.3.4}$$

If one multiplies the nth line of Equation (26.3.1) by z^n and sums over all n, one finds that G satisfies a first order partial differential equation

$$\frac{\partial G}{\partial \tau} + (1 - z)(z e^{-\theta(\infty)} - 1)\frac{\partial G}{\partial z} = e^{-\theta(\infty)}(z - 1)G. \tag{26.3.5}$$

This equation can be solved by the method of characteristics, which requires the solution of the system of ordinary differential equations

$$\frac{d\tau}{1} = \frac{dz}{(1 - z)(z e^{-\theta(\infty)} - 1)} = \frac{dG}{e^{-\theta(\infty)}(z - 1)G}. \tag{26.3.6}$$

The second and third of these equations lead to

$$\frac{dG}{G} = \frac{dz}{e^{\theta(\infty)} - z} \tag{26.3.7}$$

with the solution

$$(e^{\theta(\infty)} - z)G(z, \tau) = K_1, \tag{26.3.8}$$

where K_1 is a constant of integration. The first and second equations of (26.3.6) can be solved in a similar way and lead to

$$\frac{e^{\theta(\infty)} - z}{1 - z} \exp[-\tau(1 - e^{-\theta(\infty)})] = K_2 \tag{26.3.9}$$

where K_2 is also a constant of integration. The general solution to Equation (26.3.5) by the method of characteristics can be represented by $K_1 = \psi(K_2)$ where $\psi(x)$ is any arbitrary differentiable function.

Substituting for K_1 and K_2 from Equations (26.3.8) and (26.3.9) one finds

$$G(z,\tau) = \frac{1}{e^{\theta(\infty)}-z}\psi\left\{\frac{e^{\theta(\infty)}-z}{1-z}\exp[-\tau(1-e^{-\theta(\infty)})]\right\}. \quad (26.3.10)$$

The function $\psi(x)$ can be found by setting $\tau = 0$ in this equation. For the initial condition given in Equation (26.3.2), $G(z, 0)$ is obtained from Equation (26.3.4) as

$$G(z,0) = \frac{1-e^{-\theta(0)}}{1-z\,e^{-\theta(0)}} \quad (26.3.11)$$

and after some algebra it is found that $G(z, \tau)$ takes the form

$$G(z,\tau) = \frac{1-A(\tau)}{1-A(\tau)z} = [1-A(\tau)]\sum_{n=0}^{\infty} A^n(\tau)z^n\,, \quad (26.3.12)$$

where

$$A(\tau) = \frac{e^{-\tau}(1-e^{[\theta(\infty)-\theta(0)]})-(1-e^{-\theta(0)})}{e^{-\tau}(1-e^{[\theta(\infty)-\theta(0)]})-e^{\theta(\infty)}(1-e^{-\theta(0)})}\,. \quad (26.3.13)$$

The coefficient of z^n in Equation (26.3.12) is $P(n, \tau)$ which can be written

$$P(n,\tau) = [1-A(\tau)]A^n(\tau) = (1-e^{-\theta(\tau)})e^{-n\theta(\tau)}\,, \quad (26.3.14)$$

where

$$\theta(\tau) = -\log A(\tau)\,. \quad (26.3.15)$$

Thus, $P(n, \tau)$ can be written in the Boltzmann form for all values of the time. The result in Equation (26.3.14) can be interpreted by saying that a vibrational temperature $T(\tau)$ can be defined at every instant of time by

$$T(\tau) = \frac{h\nu}{k\theta(\tau)}\,. \quad (26.3.16)$$

It is easily verified from Equations (26.3.13)-(26.3.16) that

$$\lim_{\tau\to\infty} T(\tau) = T(\infty)$$

as required.

26.4 Derive the second-order partial differential equation to which the set of differential difference equations (26.3.1) reduce in the limit as $\theta(\infty) \to 0$. This differential equation is called the Fokker-Planck equation for this particular system which corresponds to the classical continuum limit of the harmonic oscillator.

Solution

The singlet probability $P(n, t)$ is also a function of the parameter θ so that we rewrite it as $P(n, \theta, t)$ which, in turn, we can write as some function of $n\theta, \theta, t$ as $\hat{P}(n\theta, \theta, t)$. We define a function $p(x, \theta, t)$ of the continuous variable x which has the property that

$$\theta\hat{p}(x,\theta,t) = \hat{P}(n\theta,\theta,t) \tag{26.4.1}$$

when $x = n\theta$.

Equation (26.3.1) can now be rewritten as

$$\frac{\partial\hat{p}(x,\theta,t)}{\partial t} = \frac{1}{\theta}\{xe^{-\theta}\hat{p}(x-\theta,\theta,t)+(x+\theta)\hat{p}(x+\theta,\theta,t)$$

$$-[x+(x+\theta)e^{-\theta}]\hat{p}(x,\theta,t)\}. \tag{26.4.2}$$

Expansion of the right hand side of Equation (26.4.2) in a power series in θ yields

$$\theta\left[x\frac{\partial^2 p(x,t)}{\partial x^2}+(x+1)\frac{\partial p(x,t)}{\partial x}+p(x,t)\right]+O(\theta^2),$$

where $p(x, t) = \hat{p}(x, 0, t)$. We define a new time variable $\tau = \theta t$, and take the limit as $\theta \to 0$ to obtain the **Fokker-Planck equation**

$$\begin{aligned}\frac{\partial p(x,\tau)}{\partial\tau} &= p(x,t)+(x+1)\frac{\partial p(x,\tau)}{\partial x}+x\frac{\partial^2 p(x,\tau)}{\partial x^2}\\ &= \frac{\partial^2}{\partial x^2}[xp(x,\tau)]+\frac{\partial}{\partial x}[(x-1)p(x,\tau)],\end{aligned} \tag{26.4.3}$$

where

$$x = \lim_{\substack{\theta\to 0\\ n\to\infty}}(n\theta) \quad \text{and} \quad \tau = \lim_{\substack{\theta\to 0\\ t\to\infty}}(\theta t) \tag{26.4.4}$$

26.5 (i) Let the probability $P(n, t)$ satisfy a master equation

$$\dot{P}(n,t) = \sum_{m=0}^{\infty} A(n,m)P(m,t), \tag{26.5.1}$$

where the transition rates $A(n, m)$ are independent of time. Give necessary and sufficient conditions for the first moment

$$\mu(t) = \sum_{n=0}^{\infty} nP(n,t) \tag{26.5.2}$$

to have an exponential relaxation of the form

$$\mu(t) = \mu(\infty)+[\mu(0)-\mu(\infty)]e^{-\lambda t} \tag{26.5.3}$$

valid for all values of t.

(ii) Let the probability density $p(x, t)$ be the solution to a Fokker-Planck equation

$$\frac{\partial p(x,t)}{\partial t} = -\frac{\partial}{\partial x}[b_1(x)p(x,t)]+\frac{1}{2}\frac{\partial^2}{\partial x^2}[b_2(x)p(x,t)]. \tag{26.5.4}$$

Give necessary and sufficient conditions for the first moment

$$\mu(t) = \int_{-\infty}^{\infty} xp(x,t)\,\mathrm{d}x \tag{26.5.5}$$

to have the simple relaxation form given in Equation (26.5.3).

Solution

(i) It follows from the master equation (26.5.1) that $\mu(t)$ satisfies the differential equation

$$\dot{\mu}(t) = \sum_{m=0}^{\infty} P(m,t) \sum_{n=0}^{\infty} nA(n,m)\,. \tag{26.5.6}$$

From this equation it is clear that if

$$\sum_{n=0}^{\infty} nA(n,m) = a - bm \tag{26.5.7}$$

for all values of m, where a and b are constants, then

$$\dot{\mu}(t) = a - b\mu(t) \tag{26.5.8}$$

which gives rise to a simple exponential relaxation of the first moment. The identification $b = \lambda$, $a/b = \mu(\infty)$ then leads to the desired form. In addition to being sufficient, the condition (26.5.7) is necessary. Consider the initial condition $P(r,0) = 1$, $P(j,0) = 0$ for $j \neq r$. From Equation (26.5.6) it follows that

$$\dot{\mu}(t)\Big|_{t=0} = \sum_{n=0}^{\infty} nA(n,r)\,. \tag{26.5.9}$$

But if $\mu(t)$ is to be the solution to Equation (26.5.8) it must satisfy

$$\dot{\mu}(t)\Big|_{t=0} = a - br \tag{26.5.10}$$

at $t = 0$. Equating (26.5.9) and (26.5.10) we find that the condition (26.5.7) is necessary as well as sufficient.

The preceding calculations can be generalized to derive necessary and sufficient conditions for the vector of the first k moments, $\boldsymbol{\mu}(t) = \mu_1(t), \mu_2(t), \ldots, \mu_k(t)$ to be the solution to $\dot{\boldsymbol{\mu}}(t) = \mathbf{A} - \mathbf{B}\boldsymbol{\mu}(t)$, where $\mathbf{A}$ and $\mathbf{B}$ are constant matrices.

(ii) Multiplication of the Fokker-Planck equation (26.5.4) by x and integration over all x, followed by a partial integration of the right hand side of the resulting equation, leads to

$$\dot{\mu}(t) = \int_{-\infty}^{\infty} b_1(x)p(x,t)\,\mathrm{d}x + \tfrac{1}{2}\left[xp(x,t)\frac{\partial b_2(x)}{\partial x} + xb_2(x)\frac{\partial p(x,t)}{\partial x} - b_2(x)p(x,t) - 2xb_1(x)p(x,t)\right]\Bigg|_{-\infty}^{\infty}. \tag{26.5.11}$$

In order that the moment relaxation satisfies

$$\dot{\mu}(t) = a - b\mu(t) \tag{26.5.12}$$

we must impose the condition that $p(x, t)$ goes to zero sufficiently rapidly as $x \to \pm\infty$ so that the bracketed terms go to zero at $\pm\infty$. Equation (26.5.11) then reduces to

$$\dot{\mu}(t) = \int_{-\infty}^{\infty} b_1(x)p(x, t)\,\mathrm{d}x\,. \tag{26.5.13}$$

From Problem 26.5(i) it is clear that the form

$$b_1(x) = a - bx \tag{26.5.14}$$

with $b = \lambda$ and $a/b = \mu(\infty)$ is the necessary and sufficient condition for exponential relaxation of the first moment.

26.6 Assume that an ensemble of diatomic molecules can be modelled by a system of one-dimensional harmonic oscillators whose time dependent distribution function over the semi-infinite energy level system $n = 0, 1, ..., N, ...$ is determined by the master equation of Problem 26.3. Assume that an oscillator dissociates irreversibly whenever its vibrational energy reaches $(N+1)h\nu$. Calculate the mean time to dissociation for an initial delta function distribution of oscillators, $P(n, 0) = \delta_{nr}$, where r is an integer satisfying $0 \leqslant r \leqslant N$.

Solution

Since for the harmonic oscillator employed here transitions can take place only between nearest neighbour levels, $\Delta n = \pm 1$, one need calculate only the mean dissociation time from level 0. To see this, let $\langle T_{r,N+1}\rangle$ be the expected time for the energy to reach the value $(N+1)\theta$ starting from $r\theta$. Then

$$\langle T_{0,N+1}\rangle = \langle T_{0,r}\rangle + \langle T_{r,N+1}\rangle \tag{26.6.1}$$

or

$$\langle T_{r,N+1}\rangle = \langle T_{0,N+1}\rangle - \langle T_{0,r}\rangle\,.$$

Hence we need only discuss the case $r = 0$. The master equation with dissociation at level $N+1$ is (cf Problem 26.3)

$$\frac{\mathrm{d}P(n,\tau)}{\mathrm{d}\tau} = n\mathrm{e}^{-\theta}P(n-1,\tau) - [n+(n+1)\mathrm{e}^{-\theta}]P(n,\tau) + (n+1)P(n+1,\tau)$$

$$n = 0, 1, 2, ..., N-1 \tag{26.6.2}$$

$$\frac{\mathrm{d}P(N,\tau)}{\mathrm{d}\tau} = N\mathrm{e}^{-\theta}P(N-1,\tau) - [N+(N+1)\mathrm{e}^{-\theta}]P(N,\tau)\,.$$

The probability that the molecule is undissociated at dimensionless time τ is

$$\eta(\tau) = P(0,\tau)+P(1,\tau)+\cdots+P(N,\tau) . \tag{26.6.3}$$

The probability that dissociation occurs in the time interval $(\tau, \tau+d\tau)$ will be denoted by $D(\tau)\,d\tau$ and can be calculated from the identity

$$\eta(\tau) = D(\tau)\,d\tau+\eta(\tau+d\tau) . \tag{26.6.4}$$

That is, if the molecule is not dissociated at time τ, it either dissociates in $(\tau, \tau+d\tau)$ or it is still undissociated at $\tau+d\tau$. Passing to the limit $d\tau = 0$ one finds

$$D(\tau) = -\frac{d\eta(\tau)}{d\tau} . \tag{26.6.5}$$

The dimensionless mean time τ_D to dissociation is[2]

$$\tau_D = \int_0^\infty \tau D(\tau)\,d\tau = -\int_0^\infty \tau\frac{d\eta}{d\tau}\,d\tau = \int_0^\infty \eta(\tau)\,d\tau$$

$$= \rho_0+\rho_1+\rho_2+\cdots+\rho_N , \tag{26.6.6}$$

where we have set

$$\rho_j = \int_0^\infty P(j,\tau)\,d\tau . \tag{26.6.7}$$

The ρ_j can be calculated by integrating both sides of Equation (26.6.2) over τ from 0 to ∞. One first notes that

$$\begin{aligned} \int_0^\infty \frac{dP(0,\tau)}{d\tau}\,d\tau &= P(0,\infty)-P(0,0) = -1 , \\ \int_0^\infty \frac{dP(j,\tau)}{d\tau}\,d\tau &= P(j,\infty)-P(j,0) = 0 \qquad j = 1,2,\ldots,N , \end{aligned} \tag{26.6.8}$$

since $P(j, \infty) = 0$ for all $j = 0, 1, \ldots, N$ owing to the dissociation, and $P(j, 0) = 0$ for all $j \neq 0$, owing to the initial condition $P(0, 0) = 1$.

[2] In the integration by parts which leads from the second integral to the third the term $\lim_{\tau\to\infty} \tau\eta(\tau)$ was set equal to zero. This procedure is valid provided that τ_D is finite. To see this we note that

$$\tau\eta(\tau) = \left|\tau\int_\tau^\infty \frac{d\eta(x)}{dx}\,dx\right| \leqslant \left|\int_\tau^\infty x\frac{d\eta(x)}{dx}\,dx\right|$$

Since

$$\int_0^\infty x\frac{d\eta}{dx}\,dx$$

is finite the right side of this equation tends to zero as $\tau \to \infty$.

When the integrations are performed it is found that the ρ_j satisfy

$$\begin{aligned} -1 &= \rho_1 - e^{-\theta}\rho_0 \\ 0 &= e^{-\theta}\rho_0 - (1+2e^{-\theta})\rho_1 + 2\rho_2 \\ 0 &= 2e^{-\theta}\rho_1 - (2+3e^{-\theta})\rho_2 + 3\rho_2 \\ &\vdots \\ 0 &= (N-1)e^{-\theta}\rho_{N-2} - [(N-1)+Ne^{-\theta}]\rho_{N-1} + N\rho_N \\ 0 &= Ne^{-\theta}\rho_{N-1} - [N+(N+1)e^{-\theta}]\rho_N \,. \end{aligned} \tag{26.6.9}$$

These equations can be solved recursively for the ρ_j. This yields

$$\begin{aligned} \rho_1 &= e^{-\theta}\rho_0 - 1 \\ \rho_2 &= \tfrac{1}{2}(1+2e^{-\theta})\rho_1 - \tfrac{1}{2}e^{-\theta}\rho_0 = e^{-2\theta}\rho_0 - (e^{-\theta}+\tfrac{1}{2}) \\ \rho_3 &= \tfrac{1}{3}(2+3e^{-\theta})\rho_2 - \tfrac{2}{3}e^{-\theta}\rho_1 = e^{-3\theta}\rho_0 - (e^{-2\theta}+\tfrac{1}{2}e^{-\theta}+\tfrac{1}{3}) \\ &\vdots \\ \rho_N &= e^{-N\theta}\rho_0 - \left(e^{-(N-1)\theta} + \frac{e^{-(N-2)\theta}}{2} + \frac{e^{-(N-3)\theta}}{3} + \cdots + \frac{1}{N}\right). \end{aligned} \tag{26.6.10}$$

The general term can be written

$$\rho_n = e^{-n\theta}(\rho_0 - s_n)\,, \qquad n = 1, 2, ..., N\,, \tag{26.6.11}$$

where

$$s_n = \sum_{j=1}^{n} \frac{e^{j\theta}}{j}\,. \tag{26.6.12}$$

To obtain ρ_0, one notes that the last line of Equation (26.6.9) can be rewritten as

$$Ne^{-\theta}[e^{-(N-1)\theta}(\rho_0 - s_{N-1})] = [N+(N+1)e^{-\theta}][e^{-N\theta}(\rho_0 - s_N)] \tag{26.6.13}$$

from which ρ_0 is found to be

$$\rho_0 = s_{N+1}\,. \tag{26.6.14}$$

Using Equations (26.6.11), (26.6.12), and (26.6.14), we can now rewrite Equation (26.6.6) as

$$\tau_D = \sum_{n=0}^{n} \rho_n \sum_{n=0}^{N} e^{-n\theta}(s_{N+1} - s_n) = \sum_{n=0}^{N} e^{-n\theta} \sum_{j=n+1}^{N+1} \frac{e^{j\theta}}{j}\,. \tag{26.6.15}$$

Carrying out the indicated summation yields

$$\tau_D = \frac{1}{1-e^{-\theta}} \sum_{j=1}^{N+1} \frac{e^{j\theta}-1}{j}\,. \tag{26.6.16}$$

for the mean dimensionless time to dissociation.

26.7 Calculate the mean chain length at time t of a polymer which can be in one of two states: active, i.e. growing; or inactive, i.e. incapable of further growth. Assume that the concentration of monomer is held constant at its initial value M, that the active polymer can change irreversibly into an inactive state, and that the initial state of the system consists of active monomers. If E_n^* represents the concentration of active n-mer, E_n that of inactive n-mer, the equations describing the reaction are then

$$M + E_n^* \xrightarrow{k} E_{n+1}^* \qquad \text{and} \qquad E_n^* \xrightarrow{\gamma} E_n \,.$$

Solution

Let $P_n^*(t)$ be the probability that a chain chosen at random at time t is an active n-mer, and let $P_n(t)$ be the probability that it is an inactive n-mer. The probability that an active chain will add a monomer in the time interval $(t, t+\mathrm{d}t)$ is $kM\,\mathrm{d}t$ and the probability that an active n-mer will become an inactive n-mer in $(t, t+\mathrm{d}t)$ is $\gamma\,\mathrm{d}t$. A set of equations for the $P_n^*(t)$ and $P_n(t)$ can be derived by writing the various possible events that can occur in $(t, t+\mathrm{d}t)$. These are:

1. $M+E_n^* \to E_{n+1}^*$ with probability $kM\,\mathrm{d}t$;
2. $E_n^* \to E_n$ with probability $\gamma\,\mathrm{d}t$;
3. no transition with probability $1-(kM+\gamma)\,\mathrm{d}t$.

Hence

$$\left.\begin{aligned} P_n^*(t+\mathrm{d}t) &= kMP_{n-1}^*(t)\,\mathrm{d}t + P_n^*(t)[1-(kM+\gamma)\,\mathrm{d}t] \\ P_n(t+\mathrm{d}t) &= P_n^*(t)\gamma\,\mathrm{d}t + P_n(t) \end{aligned}\right\} \quad n \geqslant 1 \qquad (26.7.1)$$

Passing to the limit $\mathrm{d}t \to 0$ one finds the following set of differential equations:

$$\dot{P}_n^*(t) = kMP_{n-1}^*(t) - (kM+\gamma)P_n^*(t) \qquad (26.7.2)$$

$$\dot{P}_n(t) = \gamma P_n^*(t) \qquad n \geqslant 1 \qquad (26.7.3)$$

so that

$$P_n(t) = \gamma \int_0^t P_n^*(t)\,\mathrm{d}t \,. \qquad (26.7.4)$$

Equation (26.7.2) can be solved through the introduction of the generating function

$$G(z, t) = \sum_{n=1}^{\infty} P_n^*(t) z^n \,. \qquad (26.7.5)$$

The function $G(z, t)$ satisfies

$$\frac{\partial G}{\partial t} = [kM(z-1)-\gamma]G \qquad (26.7.6)$$

with the solution

$$G(z, t) = G(z, 0)\exp\{-[\gamma + kM(1-z)]t\} \,. \qquad (26.7.7)$$

For the assumed initial condition

$$P_n^*(0) = \begin{cases} 1 \text{ for } n = 1 \\ 0 \text{ for } n \neq 1 \end{cases} \tag{26.7.8}$$

one finds $G(z, 0) = z$ so that

$$G(z, t) = z\exp\{-[\gamma + kM(1-z)]t\}. \tag{26.7.9}$$

The mean chain length $\mu_1(t)$ is given by

$$\mu_1(t) = \sum_{n=1}^{\infty} n[P_n^*(t) + P_n(t)] = \left[\frac{\partial G(z,t)}{\partial z} + \gamma \int_0^t \frac{\partial G(z,t)}{\partial z} \mathrm{d}t\right]_{z=1}. \tag{26.7.10}$$

The second equality on the right hand side of Equation (26.7.10) follows immediately from the definition of the generating function, Equation (26.7.5). Evaluation of Equation (26.7.10) with $G(z, t)$ given by Equation (26.7.9) yields

$$\mu_1(t) = e^{-\gamma t}(1 + kMt) + \gamma \int_0^t e^{-\gamma t}(1 + kMt)\,\mathrm{d}t = 1 + \frac{kM}{\gamma}(1 - e^{-\gamma t}) \tag{26.7.11}$$

for the mean chain length of the polymer at time t.

The higher moments

$$\mu_r(t) = \sum_{n=1}^{\infty} n^r [P_n^*(t) + P_n(t)] = \left[\left(z\frac{\partial}{\partial z}\right)^r G(z,t) + \gamma \int_0^t \left(z\frac{\partial}{\partial x}\right)^r G(z,t)\,\mathrm{d}t\right]_{z=1} \tag{26.7.12}$$

can also readily be obtained so that it is possible to determine the dispersion of the mean polymer length.

The probabilities $P_n^*(t)$ and $P_n(t)$ can be obtained by expanding the generating function $G(z, t)$ in Equation (26.7.9). This yields

$$P_n^*(t) = \frac{(kMt)^{n-1}}{(n-1)!} \exp[-(\gamma + kM)t] \tag{26.7.13}$$

and from Equation (26.7.4)

$$P_n(t) = \gamma \frac{(kM)^{n-1}}{(n-1)!} \int_0^t \tau^{n-1} \exp[-(\gamma + kM)\tau]\,\mathrm{d}\tau. \tag{26.7.14}$$

26.8 Consider a one-dimensional random walk where a particle can move the distance ϵ either to the right or to the left. The duration of each step is Δt. The probability of moving in either direction depends upon the position of the particle; if the particle is at the point k, the probabilities of moving a distance ϵ to the right or left are, respectively,

$$\tfrac{1}{2}\left(1-\frac{k}{N}\right) \quad \text{and} \quad \tfrac{1}{2}\left(1+\frac{k}{N}\right).$$

The possible positions of the particle are limited by the condition $-N \leqslant k \leqslant N$, where k and N are integers.

(i) By appropriate scaling of the variables and in the limit $N \to \infty$, derive a partial differential equation, the Fokker-Planck equation, in continuous time and space for $p(x, \tau)\,\mathrm{d}x$, the probability of finding the particle between x and $x+\mathrm{d}x$ at time τ.

(ii) The one-dimensional Ornstein-Uhlenbeck equation can be written in the form

$$\frac{\partial p(x,\tau)}{\partial \tau} = -\frac{1}{f}\frac{\partial}{\partial x}[F(x)p(x,\tau)]+D\frac{\partial^2 p(x,\tau)}{\partial x^2}, \qquad (26.8.1)$$

where f is the 'friction coefficient', $F(x)$ is an outside force acting along the x-axis, and D is the diffusion coefficient. Compare your Fokker-Planck equation with the Ornstein-Uhlenbeck equation and from the explicit expression for $F(x)$ discuss the physics of the random walk of part (i).

Solution

(i) Let $P(k\epsilon, s\Delta t)$ be the probability that the particle is at point $k\epsilon$ at time $s\Delta t$. Since the probabilities for moving right or left add up to unity, the probability that the particle remains at point k during the time interval Δt is zero, and the difference equation for the random walk therefore is

$$\begin{aligned} P(k\epsilon, s\Delta t) = \tfrac{1}{2}\Bigg\{&\left[1-\frac{(k-1)\epsilon}{N\epsilon}\right]P[(k-1)\epsilon,(s-1)\Delta t] \\ &+\left[1+\frac{(k+1)\epsilon}{N\epsilon}\right]P[(k+1)\epsilon,(s-1)\Delta t]\Bigg\}. \end{aligned} \qquad (26.8.2)$$

Subtracting $P[k\epsilon, (s-1)\Delta t]$ from both sides and dividing by Δt we obtain, after some algebra,

$$\begin{aligned} &\frac{P[k\epsilon, s\Delta t]-P[k\epsilon,(s-1)\Delta t]}{\Delta t} \\ &= \frac{\epsilon^2}{2\Delta t}\left\{\frac{P[k-1)\epsilon,(s-1)\Delta t]-2P[k\epsilon,(s-1)\Delta t]+P[(k+1)\epsilon,(s-1)\Delta t]}{\epsilon^2}\right\} \\ &\quad+\frac{1}{\Delta t N}\left\{\frac{(k+1)\epsilon P[(k+1)\epsilon,(s-1)\Delta t]-(k-1)\epsilon P[(k-1)\epsilon,(s-1)\Delta t]}{2\epsilon}\right\}. \end{aligned} \qquad (26.8.3)$$

In the limit as

$$\Delta t \to 0, \quad \epsilon \to 0, \quad N \to \infty, \qquad (26.8.4)$$

and subject to the validity of the limiting processes

$$\frac{\epsilon^2}{\Delta t} \to D, \quad \frac{1}{\Delta t N} \to \gamma, \quad s\Delta t \to \tau, \quad k\epsilon \to x, \tag{26.8.5}$$

the above equation reduces to the Fokker-Planck equation

$$\frac{\partial P(x,\tau)}{\partial \tau} = \gamma \frac{\partial}{\partial x}[xP(x,\tau)] + \tfrac{1}{2}D\frac{\partial^2 P(x,\tau)}{\partial x^2}, \tag{26.8.6}$$

where x and τ are the continuous space and time variables. [Note that the passage to the limits as given by Equation (26.8.4) is the crucial step in transforming a discrete difference equation to the corresponding partial differential equation.]

(ii) A comparison of Equation (26.8.6) with the general Ornstein-Uhlenbeck equation shows that

$$F(x) = -\gamma f x = -ax \tag{26.8.7}$$

so that the random walk (i.e. the Brownian motion) is that of an elastically bound particle.

26.9 For a general stochastic process in continuous state space the state probabilities are defined to be $p_r(x_1, t_1; x_2, t_2; \dots x_r, t_r)$ where $p_r \, dx_1 \, dx_2 \dots dx_r$ is the probability that the random function $X(t)$ satisfies $x_1 \leqslant X(t_1) < x_1 + dx_1, x_2 \leqslant X(t_2) < x_2 + dx_2, \dots, x_r \leqslant X(t_r) < x_r + dx_r$. Conditional probabilities $w_r(x_r, t_r \,|\, x_{r-1}, t_{r-1}; \dots x_1, t_1)$ can be defined in terms of the p_r by

$$p_r(x_1, t_1; \dots, x_r, t_r) = w_r(x_r, t_r \,|\, x_{r-1}, t_{r-1}; \dots x_1, t_1) \times p_{r-1}(x_1, t_1; \dots; x_{r-1}, t_{r-1}). \tag{26.9.1}$$

The term $w_r \, dx_r$ represents the probability that $x_r \leqslant X(t_r) < x_r + dx_r$ given the information that $X(t_1) = x_1, X(t_2) = x_2, \dots, X(t_{r-1}) = x_{r-1}$. A Markov process is defined by the requirement that

$$w_r(x_r, t_r \,|\, x_{r-1}, t_{r-1}; \dots x_1, t_1) = w_2(x_r, t_r \,|\, x_{r-1}, t_{r-1}) \tag{26.9.2}$$

for all $r \geqslant 2$. The combination of Equations (26.9.1) and (26.9.2) for a Markov process yield the general relation

$$p_r(x_1, t_1; \dots; x_r, t_r) = p_1(x_1, t_1) w_2(x_2, t_2 \,|\, x_1, t_1) w_2(x_3, t_3 \,|\, x_2, t_2) \times \dots w_2(x_r, t_r \,|\, x_{r-1}, t_{r-1}) \tag{26.9.3}$$

for all $r \geqslant 2$.

A **stationary Gaussian Markov process** is defined by requiring that

$$p_1(x_1, t_1) = \frac{1}{\sigma\sqrt{2\pi}} \exp\left(-\frac{x_1^2}{2\sigma^2}\right),$$

$$w_2(x, t \mid y, s) = \frac{1}{\sigma[2\pi(1-\rho^2/\sigma^4)]^{1/2}} \exp\left\{-\frac{[x-(\rho/\sigma^2)y]^2}{2\sigma^2(1-\rho^2/\sigma^4)}\right\}, \quad (26.9.4)$$

where the variance

$$\sigma^2 = \langle X^2(t)\rangle = \langle X^2(0)\rangle = \text{constant} \quad (26.9.5)$$

and the correlation function

$$\rho = \langle X(t)X(s)\rangle = \rho(|t-s|) = \rho(\tau). \quad (26.9.6)$$

The brackets indicate an ensemble average.

Correlation functions are of great importance in stochastic theory and statistical mechanics. In particular, transport coefficients can be written in terms of time integrals of correlation functions, see Chapters 13, 23, and 24.

Show that for a Gaussian Markov process the correlation function $\rho(\tau)$ of Equation (26.9.6) has the form (β is a constant) of Problem 23.3, i.e.

$$\rho(\tau) = \exp(-\beta|\tau|), \quad \beta \geqslant 0 \quad (26.9.7)$$

or

$$\rho(\tau) \equiv 0. \quad (26.9.8)$$

Solution

We shall derive and solve a functional equation for the correlation function $\rho(\tau)$. One can compute $\rho(t_3-t_1)$ as

$$\rho(t_3-t_1) = \langle X(t_3)X(t_1)\rangle$$

or more explicitly as

$$\rho(t_3-t_1) = \iiint_{-\infty}^{\infty} x_1 x_3 p_3(x_1, t_1;\ x_2, t_2;\ x_3, t_3)\,dx_1\,dx_2\,dx_3$$

$$t_1 < t_2 < t_3. \quad (26.9.9)$$

If use is made of Equations (26.9.3) and (26.9.4) the integrals can be evaluated explicitly as

$$\sigma^2\rho(t_3-t_1) = \rho(t_3-t_2)\rho(t_2-t_1) \quad (26.9.10)$$

or

$$\sigma^2\rho(\tau+\tau') = \rho(\tau)\rho(\tau'), \quad \tau, \tau' \geqslant 0. \quad (26.9.11)$$

It follows from Equation (26.9.11) that either $\rho(\tau) \equiv 0$ for all τ or that $\rho(\tau)$ must have the form

$$\rho(\tau) = \sigma^2 \exp(-\beta\tau), \quad (26.9.12)$$

where β is a constant. Since for a stationary process $\rho(\tau) = \rho(-\tau)$, Equation (26.9.12) is more properly written

$$\rho(\tau) = \sigma^2 \exp(-\beta|\tau|). \tag{26.9.13}$$

The constant β is chosen positive in all cases of physical interest so that the correlation does not go to infinity as $\tau \to \infty$. The correlation function for the important class of stationary Gaussian Markov processes is thus a simple exponentially decreasing function of time.

27
Ergodic theory, H-theorems, recurrence problems[(1)]

D.ter HAAR
(University of Oxford, Oxford)

27.1 In the absence of external forces, the number of atoms of a monatomic gas whose representative points lie in a volume element $du\,dv\,dw \equiv d^3\mathrm{c}$ of velocity space is $f(u,v,w)\,du\,dv\,dw$ [$\equiv f(\mathrm{c})\,d^3\mathrm{c}$], where u, v, w are the Cartesian components of the velocity $\mathbf{c}$ of an atom. Let $A(B)$ be the number of atoms which leave (enter) this volume element per unit time due to collisions. Assuming (i) that the atomic collisions are equivalent to collisions between elastic spheres, and (ii) that the expression $f(\mathrm{c})\,d^3\mathrm{c}$ is correct for *any* point in configuration space, obtain expressions for A and B and show that

$$\frac{\partial f}{\partial t} d^3\mathrm{c} = B - A \,. \tag{27.1.1}$$

The assumption (ii) is the assumption of **molecular chaos** or the so-called **Stosszahlansatz.**

Solution

Consider the collision between two identical atoms. Let $\mathbf{c}_1, \mathbf{c}_2, \mathbf{c}_1', \mathbf{c}_2'$ be the velocities of the two atoms before and after the collision, respectively, and let $\mathbf{w}$ be the centre-of-mass velocity,

$$\mathbf{w} = \tfrac{1}{2}(\mathbf{c}_1 + \mathbf{c}_2) \,. \tag{27.1.2}$$

The velocities $\mathbf{c}_1'$ and $\mathbf{c}_2'$ are not completely determined by $\mathbf{c}_1$ and $\mathbf{c}_2$, since we only have four equations,

$$c_1^2 + c_2^2 = c_1'^2 + c_2'^2 \quad \text{(conservation of energy)}, \tag{27.1.3}$$

$$\mathbf{c}_1 + \mathbf{c}_2 = \mathbf{c}_1' + \mathbf{c}_2' \quad \text{(conservation of momentum)} \tag{27.1.4}$$

for six components.

(1) For general references see: R.Jancel, *The Foundations of Classical and Quantum Statistical Mechanics* (Pergamon Press, Oxford), 1969; D.ter Haar, *Elements of Statistical Mechanics* (Holt, Rinehart, and Winston, New York), 1954, especially Appendix I; D.ter Haar, *Rev.Mod.Phys.*, **27**, 289 (1955); I.E.Farquhar, *Ergodic Theory in Statistical Mechanics* (Interscience, New York), 1964.

Let ω be the unit vector in the direction of the line of centres, that is, the vector connecting the centre of atom 1 with the centre of atom 2 (see Figures 27.1.1 and 27.1.2); we have then

$$\omega = \frac{\mathbf{c}_1 - \mathbf{c}'_1}{|\mathbf{c}_1 - \mathbf{c}'_1|} \,. \tag{27.1.5}$$

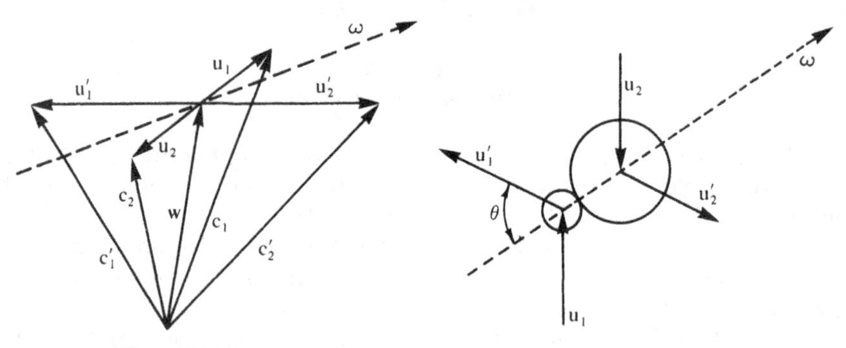

Figure 27.1.1. Figure 27.1.2.

In the centre-of-mass system the description is much simpler. If $\mathbf{u}_1$, $\mathbf{u}_2$, $\mathbf{u}'_1$, $\mathbf{u}'_2$ are the velocities in the centre-of-mass system, we have

$$\mathbf{u}_i = \mathbf{c}_i - \mathbf{w}, \quad \mathbf{u}'_i = \mathbf{c}'_i - \mathbf{w} \,, \tag{27.1.6}$$

$$\mathbf{u}_1 + \mathbf{u}_2 = \mathbf{u}'_1 + \mathbf{u}'_2 = 0 \,, \tag{27.1.7}$$

$$\mathbf{u}_1^2 + \mathbf{u}_2^2 = \mathbf{u}'^2_1 + \mathbf{u}'^2_2 \,, \tag{27.1.8}$$

$$\omega = \frac{\mathbf{u}_1 - \mathbf{u}'_1}{|\mathbf{u}_1 - \mathbf{u}'_1|} \,. \tag{27.1.9}$$

Simple considerations now show (see Figure 27.1.3) that

$$A = f(\mathbf{c}_1)\mathrm{d}^3\mathbf{c}_1 \int f(\mathbf{c}_2)\mathrm{d}^3\mathbf{c}_2 \int a_{12 \to 1'2'}\mathrm{d}^2\omega \,, \tag{27.1.10}$$

where we consider in Equation (27.1.1) a value $\mathbf{c}_1$ of the velocity, and where $a_{12 \to 1'2'}$ is given by

$$a_{12 \to 1'2'} \begin{cases} = D^2 c_{\mathrm{rel}} \cos\theta & (\cos\theta > 0) \,, \\ = 0 & (\cos\theta < 0) \,. \end{cases} \tag{27.1.11}$$

Here D is the diameter of the elastic sphere, $\mathbf{c}_{\mathrm{rel}} = \mathbf{c}_1 - \mathbf{c}_2$, and θ is the angle between ω and $\mathbf{c}_{\mathrm{rel}}$. [Note that in writing down Equation (27.1.10) we have tacitly used the Stosszahlansatz.]

To find B we must consider **inverse** collisions, which are collisions for which the velocities after the collision are $\mathbf{c}_1$ and $\mathbf{c}_2$ and before the collision $\mathbf{c}'_1$ and $\mathbf{c}'_2$. The line of centres will now be $-\omega$ (see Figure 27.1.4 which gives the inverse collision in the centre-of-mass system).

There is a one-to-one correspondence between the original and the inverse collision, since $\mathbf{c}_1$, $\mathbf{c}_2$, and $\boldsymbol{\omega}$ completely determine $\mathbf{c}_1'$, $\mathbf{c}_2'$, and $\boldsymbol{\omega}'$.

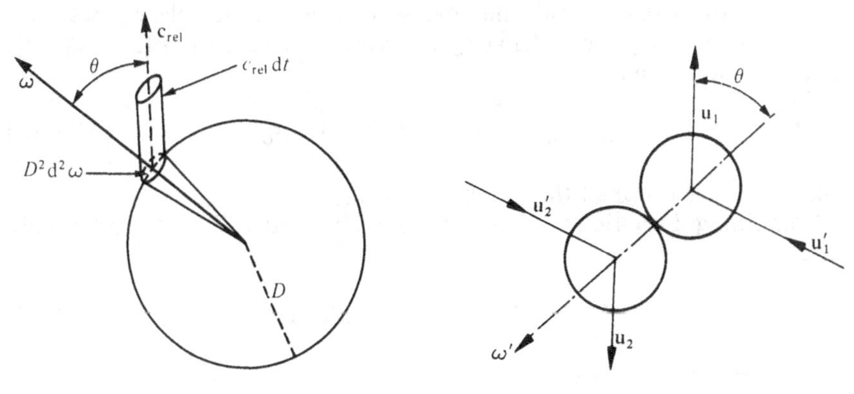

Figure 27.1.3. Figure 27.1.4.

For B we now get

$$B = \int' f(\mathbf{c}_1')\,d^3\mathbf{c}_1' f(\mathbf{c}_2')\,d^3\mathbf{c}_2' \int a_{1'2' \to 12}\,d^2\omega' , \qquad (27.1.12)$$

where the prime on the integration sign indicates that the integration over $\mathbf{c}_1'$, $\mathbf{c}_2'$, and ω' is such that one of the final velocities falls into the previously fixed volume element $d^3\mathbf{c}_1$.

Let J be the Jacobian of the transformation of $\mathbf{c}_1'$, $\mathbf{c}_2'$, ω' to $\mathbf{c}_1$, $\mathbf{c}_2$, ω,

$$J = \frac{\partial(\mathbf{c}_1', \mathbf{c}_2', \omega')}{\partial(\mathbf{c}_1, \mathbf{c}_2, \omega)} = \begin{vmatrix} \frac{\partial u_1'}{\partial u_1} & \frac{\partial v_1'}{\partial u_1} & \cdots\cdots & \frac{\partial w_2'}{\partial u_1} & \frac{\partial \omega_1'}{\partial u_1} & \frac{\partial \omega_2'}{\partial u_1} \\ \cdots & \cdots & \cdots & \cdots & \cdots & \cdots \\ \cdots & \cdots & \cdots & \cdots & \cdots & \cdots \\ \frac{\partial u_1'}{\partial w_2} & \cdots & \cdots & \cdots & \cdots & \frac{\partial \omega_2'}{\partial w_2} \\ \cdots & \cdots & \cdots & \cdots & \cdots & \cdots \\ \frac{\partial u_1'}{\partial \omega_2} & \cdots & \cdots & \cdots & \cdots & \frac{\partial \omega_2'}{\partial \omega_2} \end{vmatrix} , \qquad (27.1.13)$$

where u, v, w are the Cartesian components of $\mathbf{c}$ and where ω_1 and ω_2 are two quantities which determine the unit vector $\boldsymbol{\omega}$ (we could, for instance, use the polar angles ϑ and φ). In the simple case considered here, where we have two identical atoms in collision, we easily find $J = 1$, a result which is generally true and is a consequence of Liouville's theorem (see Problem 27.9).

As $a_{1'2' \to 12}$ and $a_{12 \to 1'2'}$ depend only on c_{rel} and θ, which are the same for the original and the inverse collision, quite clearly we have $a_{1'2' \to 12} = a_{12 \to 1'2'}$, and changing in expression (27.1.12) from $\mathbf{c}_1'$, $\mathbf{c}_2'$, ω' to $\mathbf{c}_1$, $\mathbf{c}_2$, ω, we obtain

$$B = \mathrm{d}^3\mathbf{c}_1 \int f(\mathbf{c}_1')f(\mathbf{c}_2')\,\mathrm{d}^3\mathbf{c}_2 \int a\,\mathrm{d}^2\omega\,, \tag{27.1.14}$$

where we have dropped the index of $a_{1'2' \to 12}$.

Combining Equations (27.1.1), (27.1.10), and (27.1.11), we finally have

$$\frac{\partial f_1}{\partial t} = -\int (f_1 f_2 - f_1' f_2')a\,\mathrm{d}^3\mathbf{c}_2\,\mathrm{d}^2\omega\,, \tag{27.1.15}$$

where $f_i \equiv f(\mathbf{c}_i)$, $f_i' \equiv f(\mathbf{c}_i')$.

27.2 Show that with the assumptions of Problem 27.1 the Maxwell distribution of Problem 3.1 is left unchanged by collisions.

[Note that this result is necessary for the Maxwell distribution to be an equilibrium distribution.]

Solution

If f is the Maxwell distribution, we have ($\beta = 1/kT$, where T is the temperature) from Problem 3.1.

$$f = n\left(\frac{\beta m}{2\pi}\right)^{3/2} \exp(-\tfrac{1}{2}\beta mc^2)\,, \tag{27.2.1}$$

and by virtue of Equation (27.1.3)

$$f_1 f_2 = f_1' f_2'\,. \tag{27.2.2}$$

Hence from Equation (27.1.15)

$$\frac{\partial f}{\partial t} = 0\,, \tag{27.2.3}$$

which proves that the Maxwell distribution is left unchanged by collisions.

27.3 If

$$H \equiv \int f \ln f\,\mathrm{d}^3\mathbf{c}\,, \tag{27.3.1}$$

prove that for a gas of elastic spheres

$$\frac{\mathrm{d}H}{\mathrm{d}t} \leqslant 0\,, \tag{27.3.2}$$

and discuss this result. [H is **Boltzmann's H function.**]

Solution

From Equations (27.3.1) and (27.1.15) we have

$$\frac{dH}{dt} = \int \frac{\partial f}{\partial t}(\ln f + 1)\,d^3c$$

or

$$\frac{dH}{dt} = \int (f'f_1' - ff_1)(\ln f + 1)\,a\,d^3c\,d^3c_1\,d^2\omega\,. \tag{27.3.3}$$

If we interchange c and c_1 and bear in mind that this leaves c_{rel} and thus a unchanged, we can also write

$$\frac{dH}{dt} = \int (f'f_1' - ff_1)(\ln f_1 + 1)\,a\,d^3c\,d^3c_1\,d^2\omega\,. \tag{27.3.4}$$

If we now transform from c, c_1, ω to c$'$, c_1', ω' and use the fact that the Jacobian of this transformation is equal to 1, we can write instead of Equations (27.3.3) and (27.3.4)

$$\frac{dH}{dt} = \int (ff_1 - f'f_1')(\ln f' + 1)\,a\,d^3c\,d^3c_1\,d^2\omega\,, \tag{27.3.5}$$

or

$$\frac{dH}{dt} = \int (ff_1 - f_1'f_1')(\ln f_1' + 1)\,a\,d^3c\,d^3c_1\,d^2\omega\,. \tag{27.3.6}$$

Taking the average of Equations (27.3.3) to (27.3.6) we have

$$\frac{dH}{dt} = -\tfrac{1}{4}\int (f'f_1' - ff_1)\ln\left(\frac{f'f_1'}{ff_1}\right)a\,d^3c\,d^3c_1\,d^2\omega\,. \tag{27.3.7}$$

As for p and q non-negative,

$$(p-q)\ln\frac{p}{q}\begin{cases} > 0 & (p \neq q)\,, \\ = 0 & (p = q)\,, \end{cases}$$

and as a is a positive quantity (compare Problem 27.1), we see that

$$\frac{dH}{dt} < 0\,. \tag{27.3.8}$$

One can prove that H is a bounded function and the result (27.3.8) thus means that H will decrease until the distribution function satisfies Equation (27.2.3). Equation (27.2.3) is thus not only a sufficient, but also a necessary condition for f to be an equilibrium distribution.

27.4 The equilibrium distribution function of a gas of non-interacting particles is Maxwellian. Show that the entropy of such a system is

$$S = -kH + K\,, \tag{27.4.1}$$

where K is some additive constant.

Solution

For a gas of non-interacting particles, the distribution function f is the Maxwell function given by Equation (27.2.1), whence

$$\ln f = \ln n + \tfrac{3}{2}\ln\beta - \tfrac{1}{2}\beta mc^2 + \text{const.} \tag{27.4.2}$$

If v is the volume per unit mass,

$$v = \frac{1}{nm}, \tag{27.4.3}$$

we find from Equations (27.4.2) and (27.3.1)

$$\frac{H}{n} = -\ln v + \tfrac{3}{2}\ln\beta - \tfrac{1}{2}\beta\overline{mc^2} + \text{const}, \tag{27.4.4}$$

where the bar indicates an average,

$$\overline{G} = \int Gf\,\mathrm{d}^3c\,. \tag{27.4.5}$$

Evaluating $\overline{mc^2}$ we find from Equation (27.4.4)

$$\frac{H}{n} = -\ln v + \tfrac{3}{2}\ln\beta + \text{const} \tag{27.4.6}$$

or

$$\frac{H}{n} = -\ln v - \tfrac{3}{2}\ln T + \text{const.} \tag{27.4.7}$$

For the entropy S_v per unit volume of a perfect classical gas, we have (see Problem 1.23 with $A = nk$ and $g = \tfrac{2}{3}$)

$$S_v = nk(\tfrac{3}{2}\ln T + \ln v) + D\,, \tag{27.4.8}$$

where the constant D involves the chemical constant. Thus, apart from a possible addition constant,

$$S_v = -kH\,.$$

27.5 Divide velocity space into non-overlapping cells of size Z_k, each cell being a volume element d^3c. Let the representative points of the N atoms in a gas be distributed over the cells in such a way that there are N_k points in the kth cell. Let $W(N_k)$ be the probability for a given N_k distribution, defined as the fraction of all possible arrangements for which this N_k distribution is realised. Find an expression for $W(N_k)$ assuming that the *a priori* probability for a representative point to fall into the kth cell will be proportional to its size. [Hint: see Problem 2.11.]

Solution

$$W(N_k) = CN!\prod_i \frac{Z_i^{N_i}}{N_i!}, \tag{27.5.1}$$

where C is a normalisation constant.

27.6 Defining the equilibrium distribution as the one which makes $W(N_k)$ of Problem 27.5 a maximum for given values of N and of the total energy E, find the equilibrium distribution and find also a connection between the corresponding probability and Boltzmann's H function for the case of a perfect classical gas. [Hint: see Problem 2.11.]

Solution

From Equation (27.5.1) we have

$$\ln W = \ln N! + \sum_i [N_i \ln Z_i - \ln N_i!] + \text{const.} \tag{27.6.1}$$

If the N_i are sufficiently large that we can use the Stirling formula

$$\ln x! = x \ln x - x , \tag{27.6.2}$$

we have

$$\ln W = \sum_i N_i \ln \frac{N_i}{NZ_i} + \text{const} , \tag{27.6.3}$$

where we have used the fact that the total number of particles is fixed,

$$N = \sum N_i . \tag{27.6.4}$$

The other condition to be satisfied is that of a fixed total energy,

$$E = \sum N_i \epsilon_i , \tag{27.6.5}$$

where ϵ_i is the energy of an atom in the cell Z_i.

Looking for a maximum of $\ln W$ under the conditions (27.6.4) and (27.6.5), we use the method of Lagrangian multipliers. For a variation δN_i in the N_i we have

$$\delta \ln W = 0 = -\sum_i \delta N_i \left(\ln \frac{N_i}{NZ_i} + 1 \right) , \tag{27.6.6}$$

$$\delta N = 0 = \sum_i \delta N_i , \tag{27.6.7}$$

$$\delta E = 0 = \sum_i \epsilon_i \delta N_i . \tag{27.6.8}$$

Taking (27.6.6)+$(\alpha+1)$(27.6.7)$-\beta$(27.6.8) we get

$$\sum_i \delta N_i \left(-\ln \frac{N_i}{NZ_i} + \alpha - \beta \epsilon_i \right) = 0 , \tag{27.6.9}$$

whence

$$N_i = NZ_i \exp(\alpha - \beta \epsilon_i) , \tag{27.6.10}$$

which is the Maxwell distribution.

For the case of a perfect gas we have

$$Z_i = d^3c\,, \tag{27.6.11}$$

$$\frac{N_i}{N} = \frac{1}{n} f(c)\, d^3c\,, \tag{27.6.12}$$

$$\epsilon_i = \tfrac{1}{2}mc^2\,. \tag{27.6.13}$$

From Equations (27.6.3), (27.6.10), (27.6.11), (27.6.12), we then get

$$\ln W = -\int f \ln f\, d^3c + \text{const},$$

and hence, using also Equation (27.4.1),

$$-kH = S_v = k \ln W\,. \tag{27.6.14}$$

27.7 Consider the following simplified model of a gas, due to the Ehrenfests. In the plane of the paper there may be a large number of point particles, N per unit area, which are called P molecules. They do not interact with each other, but they collide elastically with another set of entities which are called Q molecules; these are squares of edge length a, distributed at random over the plane, and fixed in position in such a way that their diagonals are exactly parallel to the x and y axes. Their average surface density is n, and it is assumed that their mean distance apart is large compared to a.

Suppose that at a certain moment all the P molecules have velocities which are of the same absolute magnitude, c, and limited in direction to (1) the positive x axis, (2) the positive y axis, (3) the negative x axis, and (4) the negative y axis (see Figure 27.7.1). As a result the distribution

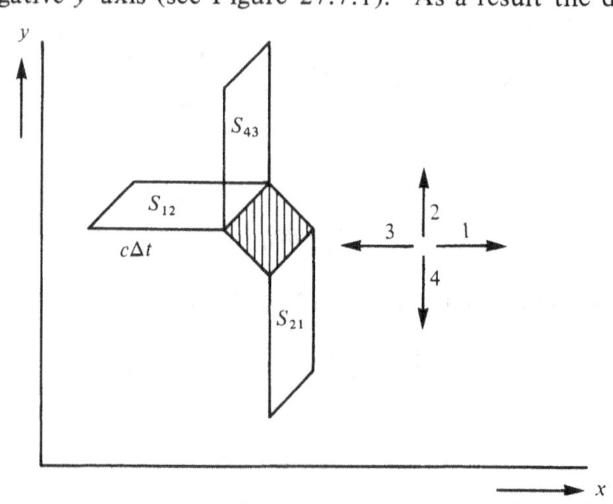

Figure 27.7.1.

function will consist of the four numbers f_1, f_2, f_3, f_4 which are the numbers of P molecules per unit area in the four possible directions.

If we assume that the number $N_{ij}\Delta t$ of P molecules which during a time interval Δt change direction from i to j per unit area is given by the Stosszahlansatz expression

$$N_{ij}\,\Delta t = f_i S_{ij} n\,, \tag{27.7.1}$$

where S_{ij} is the area of a parallelogram of length $c\Delta t$ on that edge of one of the Q molecules which is in the $-i,j$ quadrant (see Figure 27.7.1), prove that the f_i will approach equilibrium values.

Solution

As there is no preferential direction, we expect that the equilibrium distribution is given by the equation

$$f_1^{\mathrm{eq}} = f_2^{\mathrm{eq}} = f_3^{\mathrm{eq}} = f_4^{\mathrm{eq}} = \tfrac{1}{4}N\,. \tag{27.7.2}$$

From Equation (27.7.1) it follows that we have ($A_{ij} = S_{ij}n/\Delta t = A$)

$$\frac{\mathrm{d}f_1}{\mathrm{d}t} = -N_{12} - N_{14} + N_{21} + N_{41}$$

$$= A[f_2 + f_4 - 2f_1]$$

$$\frac{\mathrm{d}f_2}{\mathrm{d}t} = A[f_1 + f_3 - 2f_2]$$

$$\frac{\mathrm{d}f_3}{\mathrm{d}t} = A[f_2 + f_4 - 2f_3]$$

$$\frac{\mathrm{d}f_4}{\mathrm{d}t} = A[f_1 + f_3 - 2f_4]$$

with the solutions

$$f_i(t) = f_i^{\mathrm{eq}} + [f_i(0) - f_i^{\mathrm{eq}}]\mathrm{e}^{-2At}\,.$$

27.8 Consider the Lorentz model of a metal in which the electrons, which are supposed to be non-interacting, are scattered in such a way that the number per unit volume $N_{\omega,\omega}\,\mathrm{d}t\,\mathrm{d}^2\omega\mathrm{d}^2\omega'$ which during a time interval $\mathrm{d}t$ change their directions from within an element of solid angle $\mathrm{d}^2\omega$ to within an element of solid angle $\mathrm{d}^2\omega'$ is given by the equation

$$N_{\omega,\omega'}\mathrm{d}t\,\mathrm{d}^2\omega\mathrm{d}^2\omega' = Af(\vartheta,\varphi)\frac{\mathrm{d}^2\omega}{4\pi}\frac{\mathrm{d}^2\omega'}{4\pi}\mathrm{d}t\,, \tag{27.8.1}$$

where $f(\vartheta,\varphi)\sin\vartheta\,\mathrm{d}\vartheta\,\mathrm{d}\varphi/4\pi$ [$= f(\omega)\mathrm{d}^2\omega/4\pi$] is the number of electrons per unit volume with velocities in a direction within the solid angle $\mathrm{d}^2\omega$. In this model all electrons are moving with the same speed.

Find the equilibrium distribution of the electrons and show that the scattering mechanism will produce an exponential approach to this equilibrium distribution.

Solution

As there is no preferential direction, we expect that the equilibrium distribution is an isotropic one:

$$f^{\mathrm{eq}}(\omega) = N\,, \tag{27.8.2}$$

where N is the number of electrons per unit volume.

From Equation (27.8.1) we find

$$\frac{\mathrm{d}f(\omega)}{\mathrm{d}t} = A\left[\int \frac{f(\omega')}{4\pi}\mathrm{d}^2\omega' - f(\omega)\right] = A(N-f)\,, \tag{27.8.3}$$

whence

$$f(\omega) = N-[N-f_{t=0}(\omega)]\mathrm{e}^{-At}\,. \tag{27.8.4}$$

27.9 A classical system contains N particles, each of s degrees of freedom so that its development can be described by a (representative) point in the $2sN$-dimensional phase space (Γ space) with coordinates $p_1, \ldots, p_{sN}, q_1, \ldots, q_{sN}$. If $D(p_1, \ldots, q_{sN}, t)\mathrm{d}\Omega$ is the number of representative points in a volume element $\mathrm{d}\Omega$ ($\equiv \prod_i \mathrm{d}p_i\mathrm{d}q_i$) of Γ space which represent an ensemble of such systems, prove that

$$\frac{\mathrm{d}D}{\mathrm{d}t} = \frac{\partial D}{\partial t} + \sum_i \left(\frac{\partial D}{\partial p_i}\dot{p}_i + \frac{\partial D}{\partial q_i}\dot{q}_i\right) = 0\,. \tag{27.9.1}$$

This equation shows that the reaction in Γ space is an incompressible flow and is a form of **Liouville's theorem.**

Solution

If a point in Γ space is determined by the values of the $2sN$ variables $q_k^{(j)}$, $p_k^{(j)}$ ($k = 1, \ldots, s$; $j = 1, \ldots, N$) where the $q_k^{(j)}$ and $p_k^{(j)}$ are the generalised coordinates and momenta of the jth particle, the motion in Γ space is governed by the Hamiltonian equations of motion

$$\dot{p}_i = -\frac{\partial H}{\partial q_i}\,, \quad \dot{q}_i = \frac{\partial H}{\partial p_i}\,, \quad i = 1, 2, \ldots, sN\,, \tag{27.9.2}$$

where to simplify matters we have numbered the p's and q's continuously.

If $\mathrm{d}D/\mathrm{d}t$ is the convected rate of change in D, that is, the rate of change when we follow a representative point along its orbit in Γ space and $\partial D/\partial t$ the local rate of change, we have

$$\frac{\mathrm{d}D}{\mathrm{d}t} = \frac{\partial D}{\partial t} + \sum_i \left[\frac{\partial D}{\partial p_i}\dot{p}_i + \frac{\partial D}{\partial q_i}\dot{q}_i\right]\,. \tag{27.9.3}$$

Let N be the number of representative points in $\mathrm{d}\Omega$ at time t so that $N = D\mathrm{d}\Omega$. At $t+\delta t$ we have

$$N+\delta N = \left(D+\frac{\partial D}{\partial t}\delta t\right)\mathrm{d}\Omega\,. \tag{27.9.4}$$

The difference δN arises from the flow of the points in phase space and, by analogy with ordinary flow of a gas or liquid, we find

$$\delta N = -\sum_i \left[D\left(\frac{\partial \dot{q}_i}{\partial q_i} + \frac{\partial \dot{p}_i}{\partial p_i}\right) + \left(\frac{\partial D}{\partial q_i}\dot{q}_i + \frac{\partial D}{\partial p_i}\dot{p}_i\right)\right] \mathrm{d}\Omega\,\delta t\,. \tag{27.9.5}$$

As it follows from Equation (27.9.2) that

$$\frac{\partial \dot{q}_i}{\partial q_i} + \frac{\partial \dot{p}_i}{\partial p_i} = 0\,, \tag{27.9.6}$$

we find from Equations (27.9.4) and (27.9.3) that

$$\frac{\partial D}{\partial t} = -\sum_i \left(\frac{\partial D}{\partial q_i}\dot{q}_i + \frac{\partial D}{\partial p_i}\dot{p}_i\right), \tag{27.9.7}$$

whence

$$\frac{\mathrm{d}D}{\mathrm{d}t} = 0\,.$$

27.10 Let v be the $2sN$-dimensional 'velocity' vector in Γ space with components $\dot{q}_1, \ldots, \dot{q}_{sN}, \dot{p}_1, \ldots, \dot{p}_{sN}$, and let σ be the 'projection' of a cell $\delta\Omega$ in Γ space on a plane at right angles to v. Find an expression for the mean time τ spent by a representative point inside $\delta\Omega$.

Solution

The time τ is clearly given by the equation

$$\tau = \frac{\bar{l}}{v}\,, \tag{27.10.1}$$

where $v = [\dot{q}_1^2 + \cdots + \dot{p}_{sN}^2]^{1/2}$ and $\bar{l}$ the mean segment of the orbit inside $\delta\Omega$. From the definition of σ it follows that

$$\bar{l} = \frac{\delta\Omega}{\sigma}\,, \tag{27.10.2}$$

and hence

$$\tau = \frac{\delta\Omega}{\sigma v}\,. \tag{27.10.3}$$

27.11 Consider a system with a fixed energy E. We shall assume that the energy surface $E(p, q) =$ constant is an invariant indecomposable region Ω of Γ space, which means that (i) for any point P also the total orbit through P lies in Ω and (ii) it cannot be divided into two parts Ω' and Ω'' which are separately invariant. We shall not prove, but only state, the plausible fact that indecomposability of Ω is equivalent to the transitivity of the motion, which means that the orbit from any point P in Ω will come arbitrarily close to any other point P$'$ in Ω. Not only will any point come arbitrarily close to any other point in Ω, but if we look at a cell $\delta\Omega$ in Γ space, the point P will traverse this cell over and over again, provided the total volume of Ω is finite as we shall assume to be the case (**Poincaré's recurrence** theorem).

Give an expression for the mean recurrence time T and estimate this quantity for the case of a gas of 10^{18} atoms of mass 1 g in a volume of 1 cm^3 with an average velocity of 10^4 cm s^{-1} where the cell $\delta\Omega$ has dimensions 10^{-7} cm for its position coordinates and 10^2 g cm s^{-1} for its momentum coordinates.

Solution

The volume swept out per unit time by the orbits passing through $\delta\Omega$ is clearly σv and by Liouville's theorem this will be the volume swept out by those orbits at any time. Moreover, if Ω is indecomposable, all orbits will pass through $\delta\Omega$ and they will fill Ω completely. This means, therefore, that the mean recurrence time will be

$$T = \frac{\Omega}{\sigma v} . \tag{27.11.1}$$

To estimate T for the case considered, we note that

$$\sigma \approx (10^{-7})^{3N}(10^2)^{3N}, \quad v \approx 10^4(3N)^{1/2},$$

and

$$\Omega \approx 1^{3N}\left(\frac{E}{m}\right)^{3N/2} \approx (3N)^{3N/2}(10^4)^{3N} .$$

Hence

$$T \approx (3 \times 10^{18})^{(1\cdot 5 \times 10^{18})} \text{ s} \approx 10^{(10^{19})} \text{ years} .$$

27.12 Using the assumptions of the previous problems, we can now attack a simplified proof of the ergodic theorem, that is, the equality of ensemble and time averages [(2)]. Consider a phase function f(P) which is a function of the representative point P in Γ space. Its time average f^*—which is the average which is measured physically—is given by the equation

$$f^* = \lim_{t \to \infty} \frac{1}{t}\int_0^t \mathrm{f}[P(\tau)]\,\mathrm{d}\tau , \tag{27.12.1}$$

while it is usually only possible to evaluate the phase average $\bar{f}$ given by

$$\bar{f} = \frac{1}{\Omega}\int_\Omega \mathrm{f}(P)\,\mathrm{d}\Omega . \tag{27.12.2}$$

The ergodic theorem now states that

$$f^* = \bar{f} . \tag{27.12.3}$$

By dividing Ω into cells $\delta\Omega_i$ and the integral in Equation (27.12.1) into time intervals corresponding to the passage of P through the different cells prove the theorem (27.12.3).

[(2)] See H.Wergeland, *Acta Chem. Scand.*, **12**, 1117, 1958.

Solution

In Equation (27.12.1) we write

$$\int_0^t \mathrm{f}(t)\,\mathrm{d}t = \lim \sum_n \mathrm{f}(t_n)\delta t_n \qquad (27.12.4)$$

with

$$t = \sum \delta t_n \,. \qquad (27.12.5)$$

Choosing for the δt_n the transit times of P through the cells $\delta\Omega_i$ and using Poincaré's theorem that every cell will be traversed several times, we have

$$\sum \mathrm{f}(t_n)\delta t_n = \sum_{i=0}^{\Omega/\delta\Omega} \sum_{r=0}^{N_i} \mathrm{f}(P_i)\delta t_{ir}\,, \qquad (27.12.6)$$

where r numbers the recurrences of passage of P through $\delta\Omega_i$ and i numbers the cells, and where N_i is the number of recurrences in the time interval $(0, t)$.

We have thus

$$f^* = \lim_{t\to\infty} \lim_{\delta\Omega\to 0} \sum_{i=0}^{\Omega/\delta\Omega} \mathrm{f}(P_i)\frac{\overline{\delta t_i}}{t/N_i}\,. \qquad (27.12.7)$$

We clearly have $\lim_{t\to\infty} \overline{\delta t_i}$ = mean life time of P in $\delta\Omega_i$, which according to Equation (27.10.3) is equal to $\delta\Omega/\sigma v$. We also have $\lim_{t\to\infty} t/N_i$ = mean recurrence time, which according to Equation (27.11.1) is equal to $\Omega/\sigma v$. Hence we have

$$f^* = \lim_{\delta\Omega\to 0} \sum_i \mathrm{f}(P_i)\frac{\delta\Omega_i}{\Omega} = \frac{1}{\Omega}\int_\Omega \mathrm{f}(P)\,\mathrm{d}\Omega = \bar{f}\,.$$

27.13 To study the approach to equilibrium and the recurrence problem, let us again consider the Lorentz model of Problem 27.8. We have already seen that the system in that model has an equilibrium distribution (27.8.2). We now divide phase space into $2m+1$ cells of equal volume, which we number from $-m$ to $+m$. Each cell corresponds to an element of solid angle $\delta\omega$ with

$$\delta\omega = \frac{4\pi}{2m+1}\,. \qquad (27.13.1)$$

The distribution function is now a function of a discrete argument, $f_v(v = -m, ..., +m)$, and its equilibrium value is

$$f_v^{\mathrm{eq}} = \frac{N}{2m+1}\,. \qquad (27.13.2)$$

The $2m+1$ f_v satisfy the condition

$$\sum_{v=-m}^{+m} f_v(t) = N\,. \tag{27.13.3}$$

To measure the departure from equilibrium we introduce a function Δ given by the equation

$$\Delta = \sum_{v=-m}^{+m} (f_v - f_v^{\text{eq}})^2\,. \tag{27.13.4}$$

Assuming that $|f_v - f_v^{\text{eq}}| \ll f_v^{\text{eq}}$, find a relation between H and Δ, where we define H as

$$H = \sum f_v \ln f_v\,. \tag{27.13.5}$$

Solution

$$H = \sum_v f_v \ln f_v = H^{\text{eq}} + \frac{\Delta}{2N}\,,$$

where

$$H^{\text{eq}} = \sum_v f_v^{\text{eq}} \ln f_v^{\text{eq}}\,.$$

27.14 Use the result from probability theory that if

$$\Phi = \sum_k \phi_k(q_i)\,, \tag{27.14.1}$$

where the q_i are random variables such that $\psi(q_i)\prod \mathrm{d}q_i$ is the probability of finding the q_i within ranges q_i to $q_i + \mathrm{d}q_i$. The probability of finding Φ between Φ and $\Phi + \mathrm{d}\Phi$ is given by the equation

$$w(\Phi)\mathrm{d}\Phi = \frac{\mathrm{d}\Phi}{2\pi}\int_{-\infty}^{+\infty} \exp(-\mathrm{i}\rho\Phi)A(\rho)\mathrm{d}\rho\,, \tag{27.14.2}$$

where

$$A(\rho) = \int \dots \int \exp\left(\mathrm{i}\rho \sum \phi_k\right) \psi(q_i) \prod \mathrm{d}q_i\,. \tag{27.14.3}$$

Find the normalised probability $w(\Delta)\mathrm{d}\Delta$ that Δ lies within the interval Δ, $\Delta + \mathrm{d}\Delta$ and discuss the result.

Solution

In the present case, using Equation (27.13.3), we find easily for the probability distribution function $\psi(f_v)$

$$\psi(f_v) = \frac{N!}{\prod f_v!}(2m+1)^{-N}$$

or, using the Stirling formula for the factorial in the form (see Problem 2.11b)

$$\ln x! = x\ln x - x + \tfrac{1}{2}\ln x + \tfrac{1}{2}\ln 2\pi\,,$$

and assuming $|f_v - f_v^{eq}|/f_v \ll 1$,

$$\psi(f_v) = \left(\frac{2m+1}{2\pi N}\right)^m (2m+1)^{1/2} \exp\left[-(2m+1)\sum \frac{(f_v - f_v^{eq})^2}{2N}\right]. \quad (27.14.4)$$

We then find for the function $A(\rho)$

$$A(\rho) = \int \ldots \int \psi(f_v) \exp(-i\rho\Delta) \sum_{v=1}^{m} df_{-v}\, df_v\,, \quad (27.14.5)$$

where the integral is $2m$-fold since the f_v satisfy condition (27.13.3). The integration is straightforward, but tedious, and the result is

$$A(\rho) = \left(1 - 2i\frac{N\rho}{2m+1}\right)^{-m}. \quad (27.14.6)$$

Substituting Equation (27.14.6) into Equation (27.14.2) and evaluating the integral by contour integration, we find finally

$$w(\Delta) = \left(\frac{2m+1}{2N}\right)^m \frac{\Delta^{m-1}}{(m-1)!} \exp\left[-\frac{(2m+1)\Delta}{2N}\right], \quad (27.14.7)$$

which shows that $w(\Delta)$ decreases steeply with increasing Δ, provided N is large.

27.15 Find the average value of Δ and its dispersion (or standard deviation).

Solution

From Equation (27.14.7) we find

$$\Delta_{av} = \int w(\Delta)\Delta\, d\Delta = \frac{2mN}{2m+1}\,,$$

$$(\Delta^2)_{av} = \int \Delta^2 w(\Delta)\, d\Delta = \frac{2m(2m+2)}{(2m+1)^2} N^2\,,$$

and hence, using the fact that $m \gg 1$,

$$[(\Delta - \Delta_{av})^2]_{av} \approx \frac{N^2}{m}\,.$$

27.16 Using the model of Problem 27.8 calculate the probability $w(\Delta, \Delta')$ that Δ changes its value from Δ to Δ' in a time interval τ, assuming that the distribution in space of the scatterers is random.

Solution

The change in Δ is related to the change in the f_v and we have

$$\Delta' - \Delta = 2\sum_v (f_v - f_v^{eq})(f_v' - f_v) = 2\sum_v f_v(f_v' - f_v)\,, \quad (27.16.1)$$

where we have used Equation (27.13.3).

Let $x_{vv'}$ be the number of particles passing from cell v to cell v' during τ. Its average value will be given by the equation [compare Equations (27.8.1) and (27.8.3)]

$$x_{vv'}^{\text{av}} = \frac{Af_v\tau}{2m+1} \equiv af_v \,. \tag{27.16.2}$$

As the distribution in space of the scatterers is random, we can, provided τ is sufficiently small and aN sufficiently large so that

$$A\tau \ll 1 \text{ and } N \gg aN \gg 1 \,, \tag{27.16.3}$$

take for the distribution of the $x_{vv'}$ a Gaussian distribution

$$\psi(x_{vv'}) = (2\pi f_v)^{-½} \exp\left[-\frac{(x_{vv'} - af_v)^2}{2af_v}\right]. \tag{27.16.4}$$

We then have [compare Equation (27.8.3)]

$$f_v' - f_v = \sum_{v'} (x_{v'v} - x_{vv'}) \,. \tag{27.16.5}$$

We now are in a position similar to the one in Problem 27.14 with $w(\Delta)w(\Delta, \Delta')$ for ϕ_k. We then get

$$w(\Delta)w(\Delta, \Delta') = \frac{1}{4\pi^2}\iint \exp[-\mathrm{i}\sigma\Delta - \mathrm{i}\rho(\Delta' - \Delta)]A(\rho, \sigma)\,\mathrm{d}\rho\,\mathrm{d}\sigma \tag{27.16.6}$$

with

$$A(\rho, \sigma) = \int \cdots \int \mathrm{d}f_{-m} \ldots \mathrm{d}f_{-1}\,\mathrm{d}f_1 \ldots \mathrm{d}f_m\,\mathrm{d}x_{-m-m} \ldots \mathrm{d}x_{mm}\,\psi(f_v)\prod \psi(x_{vv'})$$
$$\times \exp\left[\mathrm{i}\sigma\sum(f_v - f_v^{\text{eq}})^2 + 2\mathrm{i}\rho\sum f_v(f_v' - f_v)\right]. \tag{27.16.7}$$

Using Equations (27.16.4) and (27.16.5), the integration over the $(2m+1)^2$ variables $x_{vv'}$ is straightforward. To evaluate the integral over the $2m$ variables $f_v\,(v \neq 0)$ we introduce new variables α_v,

$$\alpha_v = \frac{(f_v - f_v^{\text{eq}})}{f_v^{\text{eq}}} \,, \tag{27.16.8}$$

and neglect cubic terms in the exponent in comparison with quadratic terms. We then obtain

$$A(\rho, \sigma) = \left[\frac{m+\frac{1}{2}}{4Na\rho(2Na\rho + 2m+1) - 2N\mathrm{i}\sigma + 2m+1}\right]^m \tag{27.16.9}$$

and from Equations (27.16.6) and (27.14.7) finally

$$w(\Delta, \Delta') = (32\pi Na\Delta)^{-½} \exp\left\{-\frac{[\Delta' - \Delta + 2(2m+1)a\Delta]^2}{32Na\Delta}\right\}. \tag{27.16.10}$$

27.17 Use the results from the preceding problem to find the average value Δ' of Δ a time τ after its value was Δ, and also to find the average value Δ'' of Δ a time τ before its value becomes Δ.[3]

Solution

We have

$$\Delta'_{\text{av}} = \int w(\Delta, \Delta')\Delta' \, \mathrm{d}\Delta'$$

or

$$\Delta'_{\text{av}} = (1-2a)\Delta ,$$

corresponding to

$$\left(\frac{\mathrm{d}\Delta}{\mathrm{d}t}\right)_{\text{av}} = -2a\frac{\Delta}{\tau} = -\frac{2A}{2m+1}\Delta$$

[compare Equation (27.8.3)].

For Δ''_{av} we have

$$\Delta''_{\text{av}} = \frac{\int \Delta'' w(\Delta'') w(\Delta'', \Delta) \, \mathrm{d}\Delta''}{\int w(\Delta'') w(\Delta'', \Delta) \, \mathrm{d}\Delta''} ,$$

which leads to

$$\Delta''_{\text{av}} = (1-2a)\Delta = \Delta'_{\text{av}}$$

showing the fact that the Δ curve is symmetric in time.

27.18 Evaluate the recurrence time $T(\Delta)$ of a state characterised by Δ.

Solution

To find $T(\Delta)$ we first of all find the mean time of persistence in a state Δ. Let $\phi_\Delta(k\tau)$ be the probability that Δ is observed at times $0, \tau, 2\tau, 3\tau, \ldots, (k-1)\tau$, but not at time $k\tau$. We clearly have

$$\phi_\Delta(k\tau) = w^{k-1}(\Delta, \Delta)[1 - w(\Delta, \Delta)] . \tag{27.18.1}$$

The mean time of persistence $\Theta(\Delta)$ is clearly given by the equation

$$\Theta(\Delta) = \sum_{k=1}^{\infty} k\tau\phi_\Delta(k\tau) = \sum_{k=1}^{\infty} k\tau[1 - w(\Delta, \Delta)]w^{k-1}(\Delta, \Delta)$$

$$= \frac{\tau}{1 - w(\Delta, \Delta)} . \tag{27.18.2}$$

[3] Compare S.Chandrasekhar, *Rev. Mod. Phys.*, **15**, 1, 1943.

Similarly, if $\psi_n(k\tau)$ is the probability that starting from an *arbitrary* state $\not\Delta$ which is not Δ we shall observe states $\not\Delta$ at $0, \tau, 2\tau, \ldots, (k-1)\tau$, but Δ at time $k\tau$, we have

$$T(\Delta) = \sum_{k=1}^{\infty} k\tau\psi_\Delta(k\tau)\,. \tag{27.18.3}$$

If $w(\not\Delta, \not\Delta)$ is the probability that from any $\not\Delta$ there has occurred in the interval τ a transition to some other $\not\Delta$, we have clearly

$$\psi_\Delta(k\tau) = w^{k-1}(\not\Delta, \not\Delta)[1-w(\not\Delta, \not\Delta)]\,, \tag{27.18.4}$$

and hence

$$T(\Delta) = \frac{\tau}{1-w(\not\Delta, \not\Delta)}\,. \tag{27.18.5}$$

We have clearly

$$1-w(\not\Delta, \not\Delta) = w(\not\Delta, \Delta)\,, \tag{27.18.6}$$

and as at equilibrium the number of transitions $\not\Delta \to \Delta$ must be the same as the number of transitions $\Delta \to \not\Delta$, we have

$$[1-w(\Delta)]w(\not\Delta, \Delta) = w(\Delta)[1-w(\Delta, \Delta)]\,. \tag{27.18.7}$$

Combining Equations (27.18.5), (27.18.6), and (27.18.7) we have finally

$$T(\Delta) = \frac{\tau[1-w(\Delta)]}{w(\Delta)[1-w(\Delta, \Delta)]} = \Theta(\Delta)\frac{1-w(\Delta)}{w(\Delta)}\,, \tag{27.18.8}$$

and we see that $T(\Delta)$ decreases steeply with increasing Δ.

28
Variational principles and minimum entropy production[1]

S.SIMONS
(*Queen Mary College, London*)

MACROSCOPIC PRINCIPLES

28.1 Let the rate of entropy production σ in a system be expressed in the form

$$\sigma = \sum_{p=1}^{N} J_p X_p \tag{28.1.1}$$

where J_p and X_p $(1 \leqslant p \leqslant N)$ are respectively the **fluxes** and **forces** for the system. It is known from Onsager's theory of irreversible thermodynamics that if in the steady state the relationship between X_p and J_p is of the form (see Problem 25.7)

$$J_p = \sum_{q=1}^{N} L_{pq} X_q \tag{28.1.2}$$

then, in the absence of a magnetic field and rotation of the system, the **phenomenological coefficient matrix** L **is symmetric**; that is $L_{pq} = L_{qp}$.

Prove that if $X_1, X_2, ..., X_s$ $(s < N)$ are kept constant, then the **minimum rate of entropy production** for variable $X_{s+1}, X_{s+2}, ..., X_N$ corresponds to the steady state of the system in which $J_{s+1}, J_{s+2}, ..., J_N$ are all zero.

Solution

We see from Equations (28.1.1) and (28.1.2) that the rate of entropy production may be expressed in the form

$$\sigma = \sum_{p,q=1}^{N} L_{pq} X_p X_q \,. \tag{28.1.3}$$

In order to minimise this for variable $X_{s+1}, X_{s+2}, ..., X_N$, we equate in turn $\partial\sigma/\partial X_p$ to zero for $s+1 \leqslant p \leqslant N$. This gives

$$\frac{\partial \sigma}{\partial X_p} = \sum_{q=1}^{N} (L_{pq} + L_{qp}) X_q = 0$$

for $s+1 \leqslant p \leqslant N$. Finally, making use of Onsager's symmetry relation

[1] A useful background reference on entropy production is *Thermodynamics of Irreversible Processes* by I.Prigogine (Interscience, New York), 1967.

$L_{pq} = L_{qp}$, we obtain

$$2 \sum_{q=1}^{N} L_{pq} X_q = 0$$

for $s+1 \leqslant p \leqslant N$, which is equivalent to $J_{s+1} = 0 = J_{s+2} = \ldots = J_N$. To confirm that this indeed corresponds to a minimum value of σ (and not to some other stationary value) we need only remark that it follows from the Second Law of Thermodynamics that σ is always positive and thus, since it possesses only a single stationary value, this must be a minimum.

28.2 In the notation of the last problem, the rate of entropy production

$$\sigma = \sum_{p=1}^{N} J_p X_p \ , \tag{28.2.1}$$

and it is clear that in the steady state, when Equation (28.1.2) is applicable, we have

$$X_p = \sum_{q=1}^{N} M_{pq} J_q \ , \tag{28.2.2}$$

where **M** is the inverse of **L**. We then have

$$\sigma = \sum_{p,q=1}^{N} M_{pq} J_p J_q \tag{28.2.3}$$

and, as σ is always $\geqslant 0$, **M** is a symmetric positive semi-definite matrix[(2)]. If the system is not in a steady state so that Equation (28.2.2) does not hold, the two expressions for σ, (28.2.1) and (28.2.3), will in general be different. We shall then refer to these expressions respectively as the **extrinsic** and **intrinsic rates of entropy production** and will denote them by σ_e and σ_i.

(a) If the matrix M_{pq} and the forces F_p $(1 \leqslant p \leqslant N)$ are given, prove that the steady state relationship (28.2.2) corresponds to minimising $\sigma_i - 2\sigma_e$ with respect to varying J_p $(1 \leqslant p \leqslant N)$.

(b) Show also that relationship (28.2.2) is given by maximising σ_i subject to it being kept equal to σ_e.

[Hint: Since **M** is positive definite,

$$\sum_{p,q=1}^{N} M_{pq}(K_p - J_p)(K_q - J_q) \geqslant 0 \ , \tag{28.2.4}$$

where K_p are arbitrary fluxes and J_p satisfies Equation (28.2.2).]

(c) Prove that a further variational principle for Equation (28.2.2) is given by minimising σ_i/σ_e^2.

[Hint: As **M** is positive semi-definite,

$$\sum_{p,q=1}^{N} M_{pq}(\lambda K_p + J_p)(\lambda K_q + J_q) \geqslant 0 \quad \text{for all real } \lambda.$$

[(2)] A matrix **M** is positive semi-definite if $\sum_{p,q} M_{pq} X_p X_q \geqslant 0$ for all X_p.

Express the left hand side as a quadratic in λ and consider the condition for it to be non-negative for real λ.]

Solution

(a) Making use of Equation (28.2.4) and of the symmetry of $\mathbf{M}$ we find

$$\sum_{p,q} M_{pq}K_pK_q - 2\sum_{p,q} M_{pq}K_pJ_q + \sum_{p,q} M_{pq}J_pJ_q \geqslant 0\,. \tag{28.2.5}$$

Since

$$X_p = \sum_q M_{pq}J_q \tag{28.2.6}$$

this gives

$$\sum_{p,q} M_{pq}K_pK_q - 2\sum_p K_pX_p \geqslant -\sum_{p,q} M_{pq}J_pJ_q = \sum_{p,q} M_{pq}J_pJ_q - 2\sum_p J_pX_p\,.$$

Thus $\sigma_\mathrm{i} - 2\sigma_\mathrm{e}$ calculated with general fluxes is greater than or equal to its value when calculated with the steady states fluxes satisfying Equation (28.2.6), and hence these latter fluxes minimise $\sigma_\mathrm{i} - 2\sigma_\mathrm{e}$.

(b) To prove the second variational principle, we note that keeping σ_i equal to σ_e corresponds to the constraint

$$\sum_{p,q} M_{pq}K_pK_q = \sum_p X_pK_p = \sum_{p,q} M_{pq}K_pJ_q\,. \tag{28.2.7}$$

If we make use of this, inequality (28.2.1) becomes

$$\sum_{p,q} M_{pq}J_pJ_q \geqslant \sum_{p,q} M_{pq}K_pK_q\,.$$

This implies that, subject to the constraint $\sigma_\mathrm{i} = \sigma_\mathrm{e}$, σ_i is maximised by the fluxes satisfying Equation (28.2.6).

(c) Proof of the third variational principle proceeds by noting that

$$\sum_{pq} M_{pq}(\lambda K_p + J_p)(\lambda K_q + J_q) \geqslant 0$$

is equivalent to

$$\lambda^2 \sum_{p,q} M_{pq}K_pK_q + 2\lambda \sum_{p,q} M_{pq}K_pJ_q + \sum_{p,q} M_{pq}J_pJ_q \geqslant 0\,.$$

For this to be true for all real λ, the condition

$$\left(\sum_{p,q} M_{pq}K_pJ_q\right)^2 \leqslant \sum_{p,q} M_{pq}K_pK_q \sum_{p,q} M_{pq}J_pJ_q \tag{28.2.8}$$

must hold. As Equation (28.2.6) implies that

$$\sum_{p,q} M_{pq}K_pJ_q = \sum_p X_pK_p\,,$$

it follows from inequality (28.2.8) that

$$\frac{\sum_{p,q} M_{pq}K_pK_q}{\left(\sum_p X_pK_p\right)^2} \geqslant \frac{1}{\sum_{p,q} M_{pq}J_pJ_q} = \frac{\sum_{p,q} M_{pq}J_pJ_q}{\left(\sum_p X_pJ_p\right)^2} .$$

This means that σ_i/σ_e^2 is a minimum when $K_p = J_p$.

Alternative proofs of these results can be formulated using the techniques of calculus. Thus, for the first of the above variational principles, we can find the minimum of $\sigma_i - 2\sigma_e$ (considered as a function of K_p) by equating to zero in turn $\partial(\sigma_i - 2\sigma_e)/\partial K_p$ for $1 \leqslant p \leqslant N$. This gives

$$\sum_q M_{pq}K_q - X_p = 0 \quad (1 \leqslant p \leqslant N)$$

on using the symmetry of **M**, and this corresponds to Equation (28.2.6). To show that this is indeed a minimum let $K_p = J_p + \Delta J_p$, when

$$\begin{aligned}\Delta(\sigma_i - 2\sigma_e) &= \sum_{p,q} M_{pq}(J_p + \Delta J_p)(J_q + \Delta J_q) - 2\sum_p (J_p + \Delta J_p)X_p \\ &\qquad - \sum_{p,q} M_{pq}J_pJ_q + 2\sum_p J_pX_p \\ &= 2\sum_{pq} M_{pq}\Delta J_pJ_q + \sum_{p,q} M_{pq}\Delta J_p\Delta J_q - 2\sum_p X_p\Delta J_p \\ &= \sum_{p,q} M_{pq}\Delta J_p\Delta J_q\end{aligned}$$

on making use of Equation (28.2.6). Since **M** is a positive definite matrix, it follows that $\Delta(\sigma_i - 2\sigma_e) > 0$ for all ΔJ_p and therefore that the stationary value is a true minimum. Similar, though rather more lengthy, proofs can be obtained for the other two variational principles.

28.3 Consider the steady state of three systems characterised in turn by an inverse phenomenological coefficient matrix **M**, **M**′, **M**+**M**′ and suppose the same forces **F** to be applied to each. If in the steady state the rates of entropy production for the systems are respectively $\sigma_{\mathbf{M}}$, $\sigma_{\mathbf{M}'}$, $\sigma_{\mathbf{M}+\mathbf{M}'}$, use the third variational principle of the last problem to prove that

$$\frac{1}{\sigma_{\mathbf{M}+\mathbf{M}'}} \geqslant \frac{1}{\sigma_{\mathbf{M}}} + \frac{1}{\sigma_{\mathbf{M}'}} .$$

[Hint: Consider σ_i/σ_e^2 for the third system, evaluated with the corresponding steady state currents for that system.]

Solution

Let J_p $(1 \leqslant p \leqslant N)$ be the currents in the system characterised by matrix $\mathbf{M}+\mathbf{M}'$ corresponding to the given forces F_p. Then

$$\frac{1}{\sigma_{\mathrm{M+M'}}} = \frac{\sigma_{\mathrm{i}}}{\sigma_{\mathrm{e}}^2} \text{ in the steady state}$$

$$= \frac{\sum\limits_{p,q=1}^{N} (M_{pq}+M'_{pq})J_p J_q}{\sum\limits_{p=1}^{N} J_p X_p}$$

$$= \frac{\sum\limits_{p,q} M_{pq} J_p J_q}{\sum\limits_{p} J_p X_p} + \frac{\sum\limits_{p,q} M'_{pq} J_p J_q}{\sum\limits_{p} J_p X_p} . \tag{28.3.1}$$

Now the first term on the right hand side of Equation (28.3.1) is $\sigma_{\mathrm{i}}/\sigma_{\mathrm{e}}^2$ for the system characterised by matrix $\mathbf{M}$, but calculated with currents which in general are different to the steady state currents for that system. Thus, according to the third variational principle of Problem 28.2, this term is $\geqslant \sigma_{\mathrm{i}}/\sigma_{\mathrm{e}}^2$ for the steady state distribution in the system; that is, this term is $\geqslant 1/\sigma_{\mathrm{M}}$. Similarly the second term on the right hand side of Equation (28.3.1) is $\geqslant 1/\sigma_{\mathrm{M'}}$. Hence it follows from Equation (28.3.1) that

$$\frac{1}{\sigma_{\mathrm{M+M'}}} \geqslant \frac{1}{\sigma_{\mathrm{M}}} + \frac{1}{\sigma_{\mathrm{M'}}} .$$

28.4 In Problem 28.2 we saw that the solution of the steady state equation (28.2.2) is given by certain variational principles (for example, the minimisation of $\sigma_{\mathrm{i}} - 2\sigma_{\mathrm{e}}$ with respect to the fluxes J_p). This offers a technique for obtaining approximate solutions of Equation (28.2.2) for the fluxes (in terms of the given forces) in cases where N is large and the matrix $\mathbf{M}$ cannot therefore be readily inverted. We assume a solution of Equation (28.2.2) containing a set of undetermined parameters (in number less than N) and proceed to find the best values for these parameters by minimising $\sigma_{\mathrm{i}} - 2\sigma_{\mathrm{e}}$ with respect to variation of the parameters. It is convenient to introduce these parameters as undetermined constants in a linear combination of given vectors, and we therefore assume a solution for J_p in the form

$$J_p = \sum_{s=1}^{T} \alpha_s \Gamma_p^{(s)} \quad (1 \leqslant p \leqslant N) . \tag{28.4.1}$$

Here $\Gamma_p^{(s)}$ $(1 \leqslant p \leqslant N, 1 \leqslant s \leqslant T, T < N)$ are a set of T given vectors and α_s are the arbitrary constants. Using this form for J_p, together with the above variational principle, prove that the values of α_s are determined by equations of the form

$$\sum_{t=1}^{T} R_{st}\alpha_t = Q_s \quad (1 \leqslant s \leqslant T)$$

and obtain explicit forms for R_{st} and Q_s in terms of X_p, M_{pq}, and $\Gamma_p^{(s)}$. Show that, with the present approximation, the rate of entropy production in the steady state is given by

$$\sigma = \sum_{s,t=1}^{T} [R^{-1}]_{st} Q_s Q_t \,.$$

Solution

Since

$$J_p = \sum_{s=1}^{T} \alpha_s \Gamma_p^{(s)} \,,$$

$$\sigma_{\rm i} = \sum_{p,q=1}^{N} M_{pq} J_p J_q = \sum_{p,q=1}^{N} M_{pq} \left(\sum_{s=1}^{T} \alpha_s \Gamma_p^{(s)} \right) \left(\sum_{t=1}^{T} \alpha_t \Gamma_q^{(t)} \right)$$

$$= \sum_{s,t=1}^{T} R_{st} \alpha_s \alpha_t \tag{28.4.2}$$

where

$$R_{st} = \sum_{p,q=1}^{N} M_{pq} \Gamma_p^{(s)} \Gamma_q^{(t)} \,. \tag{28.4.3}$$

Also

$$\sigma_{\rm e} = \sum_{p=1}^{N} J_p X_p = \sum_{p=1}^{N} X_p \left(\sum_{s=1}^{T} \alpha_s \Gamma_p^{(s)} \right)$$

$$= \sum_{s=1}^{T} Q_s \alpha_s \tag{28.4.4}$$

where

$$Q_s = \sum_{p=1}^{N} X_p \Gamma_p^{(s)} \,. \tag{28.4.5}$$

Thus

$$\sigma_{\rm i} - 2\sigma_{\rm e} = \sum_{s,t=1}^{T} R_{st} \alpha_s \alpha_t - 2 \sum_{s=1}^{T} Q_s \alpha_s \,. \tag{28.4.6}$$

Now, it readily follows from Equations (28.4.2) and (28.4.3) that R_{st} is a symmetric positive definite matrix, since M_{pq} has these properties. The expression (28.4.6) is therefore of the same form as the expression for $\sigma_{\rm i} - 2\sigma_{\rm e}$ in terms of the original fluxes J_p, if we identify α with J, $\mathbf{R}$ with $\mathbf{M}$, and $\mathbf{Q}$ with $\mathbf{X}$. Thus, by employing the same proof as used in the last problem for showing that $\sigma_i - 2\sigma_e$ is minimised by J_p satisfying Equation (28.2.2), it now follows that $\sigma_i - 2\sigma_e$ is minimised by α_s taking values determined by the equation

$$\sum_{t=1}^{T} R_{st}\alpha_t = Q_s \tag{28.4.7}$$

with **R** and **Q** given by Equations (28.4.2) and (28.4.5).

Alternatively, Equation (28.4.7) can be derived by equating to zero $\partial(\sigma_i - 2\sigma_e)/\partial\alpha_s$ for $1 \leqslant s \leqslant T$, with $\sigma_i - 2\sigma_e$ given by Equation (28.4.6).

In the steady state, the rate of entropy production is given by Equation (28.4.4), where α_s is obtained from Equation (28.4.7). Thus, in the steady state,

$$\sigma = \sum_{s=1}^{T} Q_s\left(\sum_{t=1}^{T} [R^{-1}]_{st}Q_t\right) = \sum_{s,t=1}^{r} [R^{-1}]_{st}Q_sQ_t \,. \tag{28.4.8}$$

We may note that this method of obtaining an approximate solution is of particular value when the ultimate purpose is to calculate the rate of entropy production in the steady state. This is because the quantity being minimised, $\sigma_i - 2\sigma_e$, reduces to $-\sigma$ in the steady state. Hence, if the solution for the currents is in error by a small amount ϵ, the error in the value of σ obtained will be proportional to ϵ^2, which will be much smaller. Also, since the correct value of $-\sigma$ in the steady state is the minimum value of $\sigma_i - 2\sigma_e$, it follows that the value of σ as given by Equation (28.4.8) will be less than the true value of σ; that is, Equation (28.4.8) provides a lower limit for σ.

28.5 In a conducting medium the electric current may be described by a vector J_p $(1 \leqslant p \leqslant 3)$ where J_1, J_2, J_3 are respectively the components of current density along the x, y, z axes of a Cartesian co-ordinate system. Likewise the electric field may be described by a vector $\mathcal{E}_p$ $(1 \leqslant p \leqslant 3)$ where $\mathcal{E}_1, \mathcal{E}_2, \mathcal{E}_3$ are respectively the electric field components along the x, y, z directions. The rate of entropy production per unit volume σ is given by

$$T\sigma = \sum_{p=1}^{3} J_p \mathcal{E}_p \,, \tag{28.5.1}$$

where T is the absolute temperature, and in the steady state $\mathcal{E}$ and **J** are connected by the general relationship

$$\mathcal{E}_p = \sum_{q=1}^{3} r_{pq}J_q \,, \tag{28.5.2}$$

where r_{pq} $(1 \leqslant p, q \leqslant 3)$ is the resistance tensor. In a particular case, with a given set of units, the electric field vector $\mathcal{E}$ is

$$\mathcal{E} = \begin{pmatrix} 1 \\ 2 \\ 3 \end{pmatrix}$$

and the resistance matrix takes the form

$$\mathbf{r} = \begin{pmatrix} 3 & 1 & 1 \\ 1 & 3 & 1 \\ 1 & 1 & 3 \end{pmatrix}.$$

Obtain expressions for σ_i and σ_e in terms of J_1, J_2, J_3. Hence, if an approximate solution of Equation (28.5.2) is given by $J_p = \alpha\Gamma_p$, where

$$\Gamma = \begin{pmatrix} 1 \\ 2 \\ 3 \end{pmatrix}$$

and α is an arbitrary constant, use the method of the last problem to estimate a value for $T\sigma$ in the steady state. Compare this with the true value and comment.

Solution

$$\sigma_i = \frac{1}{T}\sum_{p,q=1}^{3} r_{pq}J_pJ_q = \frac{1}{T}(3J_1^2+3J_2^2+3J_3^2+2J_1J_2+2J_2J_3+2J_3J_1)$$

$$\sigma_e = \frac{1}{T}\sum_{p=1}^{3} \mathcal{E}_pJ_p = \frac{1}{T}(J_1+2J_2+3J_3).$$

Using the given trial solution, we let $J_1 = \alpha, J_2 = 2\alpha, J_3 = 3\alpha$, and obtain $T(\sigma_i - 2\sigma_e) = 64\alpha^2 - 28\alpha$. To find the minimum value of $\sigma_i - 2\sigma_e$, we equate $\partial(\sigma_i - 2\sigma_e)/\partial\alpha$ to zero and obtain $128\alpha - 28 = 0$, whence $\alpha = \frac{7}{32}$. With this value of α,

$$T(2\sigma_e - \sigma_i) = T\sigma = 3{\cdot}06\,. \tag{28.5.3}$$

To obtain the true value of $T\sigma$ we must first solve Equation (28.5.2) for J_p, and this gives

$$J = \tfrac{1}{10}\begin{pmatrix} -1 \\ 4 \\ 9 \end{pmatrix}.$$

On making use of Equation (28.5.3), we then find

$$T\sigma = 3\cdot 40 \tag{28.5.4}$$

and, as expected, the estimated value (28.5.3) is less than the true value. The numerical error in the approximate steady state current is quite large since the latter equals

$$\begin{pmatrix} 0\cdot 22 \\ 0\cdot 44 \\ 0\cdot 66 \end{pmatrix} \quad \text{while the true solution is} \quad \begin{pmatrix} -0\cdot 1 \\ 0\cdot 4 \\ 0\cdot 9 \end{pmatrix}.$$

Nevertheless the error in $T\sigma$ is much less, being about 10%, and this agrees with the discussion at the end of the solution to Problem 28.4.

ELECTRON FLOW PROBLEMS

28.6 We now wish to consider the formulation of variational principles concerning entropy production, based on a description of the system at an atomic level, rather than at the macroscopic level employed hitherto. To be quite definite we shall confine ourselves to particles, such as electrons, which obey Fermi-Dirac statistics, but similar results apply to particles obeying both Maxwell and Bose-Einstein statistics.

In a metal each electron is characterised by a three-dimensional wave vector $\mathbf{k}$ and the number of electrons of given spin with wave number $\mathbf{k}$ is denoted by an occupation number $f(\mathbf{k})$. In thermal equilibrium at temperature T, $f(k)$ is equal to the Fermi-Dirac distribution function [derived in the solutions to Problem 3.12(a)],

$$f^0(\mathbf{k}) = \frac{1}{\exp\left(\dfrac{E-\mu}{\kappa T}\right)+1}, \tag{28.6.1}$$

where $E(\mathbf{k})$ is the energy of an electron with wave number $\mathbf{k}$, μ is a constant—the Fermi potential, and κ is Boltzmann's constant. In the steady state, however, where a net flow of electrons occurs, $f(\mathbf{k})$ differs from the equilibrium value $f^0(\mathbf{k})$. Now it is known that the entropy S of an assembly of electrons is given by

$$S = -\kappa\int\{f(\mathbf{k})\ln f(\mathbf{k})+[1-f(\mathbf{k})]\ln[1-f(\mathbf{k})]\}d\mathbf{k} \tag{28.6.2}$$

where the integration is taken over the whole of $\mathbf{k}$ space [3]. For $f(\mathbf{k})$ close to the Fermi-Dirac distribution we may then let

$$f(\mathbf{k}) = f^0(\mathbf{k})-\kappa^{-1}f^0(1-f^0)\phi(\mathbf{k}), \tag{28.6.3}$$

where $\phi(\mathbf{k})$, which measures the deviation of f from f^0, is small.

[3] P.T.Landsberg, *Thermodynamics with Quantum Statistical Applications* (Interscience, New York), 1961, p.233.

Making use of Equations (28.6.1), (28.6.2), and (28.6.3) show that the rate of entropy production σ (= $\partial S/\partial t$) is given by

$$\sigma = \int \phi \frac{\partial f}{\partial t} \mathrm{d}k + \frac{1}{T}\int (E-\mu)\frac{\partial f}{\partial t}\mathrm{d}k , \qquad (28.6.4)$$

on the assumption that terms in ϕ of degree higher than the first may be neglected[4].

Solution

From the given definition of entropy it follows that

$$\frac{\partial S}{\partial t} = -\kappa \int \left[\frac{\partial f}{\partial t} + \frac{\partial f}{\partial t}\ln f - \frac{\partial f}{\partial t} - \frac{\partial f}{\partial t}\ln(1-f)\right] \mathrm{dk}$$

$$= -\kappa \int \frac{\partial f}{\partial t} \ln\left(\frac{f}{1-f}\right) \mathrm{dk} . \qquad (28.6.5)$$

Now, it follows from the definition of ϕ [Equation (28.6.3)] that

$$\frac{f}{1-f} = \left[\frac{f^0}{1-f^0}\right]\left[\frac{1-\kappa^{-1}(1-f^0)\phi}{1+\kappa^{-1}f^0\phi}\right] . \qquad (28.6.6)$$

On expanding the second factor in Equation (28.6.6) as a power series and retaining only terms linear in ϕ, we obtain $1-\phi/\kappa$. Substituting for f^0 in the first factor from Equation (28.6.1), we then get

$$\frac{f}{1-f} = \exp\left(\frac{\mu - E}{\kappa T}\right)\left(1-\frac{\phi}{\kappa}\right) .$$

This may be substituted into Equation (28.6.5) to give the final result

$$\sigma = \frac{\partial S}{\partial t} = \int \phi \frac{\partial f}{\partial t}\mathrm{dk} + \frac{1}{T}\int (E-\mu)\frac{\partial f}{\partial t}\mathrm{dk} \qquad (28.6.7)$$

on using the first term of the expansion of $\ln(1-\phi/\kappa)$.

28.7 Let us consider the application of Equation (28.6.4) to the flow of electrons through a conductor in the steady state, brought about by an electric field $\mathcal{E}$. The value of f will be changing for two reasons. Firstly the electric field will increase each **k** value at a constant rate, and it may be shown that this produces a 'convective' rate of change of f given by

$$\left.\frac{\partial f}{\partial t}\right)_{\mathrm{conv}} = -e\mathcal{E}\cdot \mathbf{v}\frac{\partial f^0}{\partial E} , \qquad (28.7.1)$$

where e and $\mathbf{v}$ are respectively the charge and velocity of an electron.

[4] A useful reference for this and the next two problems is Chapter 7 of *Electrons and Phonons* by J.M.Ziman (Clarendon Press, Oxford), 1960.

Secondly, the value of f will change due to collisions which the electrons undergo. If we restrict our attention to the important case of collisions with impurities in the conductor, it may be shown that the corresponding 'collision' rate of change of f is given by

$$\left.\frac{\partial f}{\partial t}\right)_{\text{coll}} = \int L(\mathbf{k}, \mathbf{k}')[\phi(\mathbf{k}) - \phi(\mathbf{k}')]\,d\mathbf{k}' \tag{28.7.2}$$

where $\phi(\mathbf{k})$ is defined in Equation (28.6.3) and the integration here is over the whole of $\mathbf{k}'$ space. $L(\mathbf{k}, \mathbf{k}')$ is a kernel describing the type of collision; it generally possesses the following properties

$$L(\mathbf{k}, \mathbf{k}') = L(\mathbf{k}', \mathbf{k}), \qquad L(\mathbf{k}, \mathbf{k}') \geqslant 0 \tag{28.7.3}$$

for all $\mathbf{k}, \mathbf{k}'$. Now, in the steady state of the conductor $\partial f/\partial t = 0$, and this corresponds to

$$\left.\frac{\partial f}{\partial t}\right)_{\text{conv}} + \left.\frac{\partial f}{\partial t}\right)_{\text{coll}} = 0\,.$$

Thus we obtain the steady state condition

$$\int L(\mathbf{k}, \mathbf{k}')[\phi(\mathbf{k}) - \phi(\mathbf{k}')]\,d\mathbf{k}' = e\mathscr{E} \cdot \mathbf{v}\frac{\partial f^0}{\partial E}\,. \tag{28.7.4}$$

This is the celebrated **Boltzmann equation** from whose solution the electric current in the conductor may be calculated. It is closely related to the Boltzmann equation for a gas derived in Problem 17.10.

If we take each of the expressions on the right hand sides of Equations (28.7.1) and (28.7.2) we can, with the aid of Equation (28.6.4), calculate the corresponding forms for $\partial S/\partial t (= \sigma)$. Assuming that $v(-\mathbf{k}) = -v(+\mathbf{k})$ and $E(-\mathbf{k}) = E(+\mathbf{k})$, prove that

$$\sigma_{\text{conv}} = -\frac{1}{T}\sum_{p=1}^{3} \mathscr{E}_p J_p\,, \tag{28.7.5}$$

where $\mathbf{J}$ is the current density. By making use of the first of Equations (28.7.3), prove that

$$\sigma_{\text{coll}} = \tfrac{1}{2}\iint L(\mathbf{k}, \mathbf{k}')[\phi(\mathbf{k}) - \phi(\mathbf{k}')]^2\,d\mathbf{k}\,d\mathbf{k}'\,.$$

It may be shown that in calculating σ_{coll} the contribution of the second term on the right hand side of Equation (28.6.3) is zero; for present purposes assume this to be so.

Solution

From Equations (28.6.7) and (28.7.1) it follows that

$$\sigma_{\text{conv}} = -e\mathscr{E} \cdot \int \mathbf{v}\phi\frac{\partial f^0}{\partial E}\,d\mathbf{k} - \frac{e\mathscr{E}}{T} \cdot \int (E-\mu)\mathbf{v}\frac{\partial f^0}{\partial E}\,d\mathbf{k}\,. \tag{28.7.6}$$

Since $\mathbf{v}(-\mathbf{k}) = -\mathbf{v}(+\mathbf{k})$ and $E(-\mathbf{k}) = +E(+\mathbf{k})$, the integrand in the second term on the right hand side of Equation (28.7.6) is an odd function of k, and thus the value of the integral is zero. As far as the first term is concerned, it follows from the fact that $\partial f^0/\partial E = -f^0(1-f^0)/\kappa T$ (which is readily shown from the definition of f^0) that this term may be expressed as

$$-\frac{1}{T}\mathscr{E}\cdot\int e\mathbf{v}(f-f^0)\,d\mathbf{k}\,, \qquad (28.7.7)$$

making use of Equation (28.6.3). The integral in expression (28.7.7) is just the total charge transported per unit area per second and is thus equal to the current density J. We obtain the required result

$$\left.\frac{\partial S}{\partial t}\right)_{\text{conv}} = -\frac{1}{T}\sum_{p=1}^{3} J_p\mathscr{E}_p\,. \qquad (28.7.8)$$

From Equations (28.6.7) and (28.7.2) we have

$$\sigma_{\text{coll}} = \iint L(\mathbf{k},\mathbf{k}')[\phi(\mathbf{k})-\phi(\mathbf{k}')]\phi(\mathbf{k})\,d\mathbf{k}\,d\mathbf{k}' \qquad (28.7.9)$$

since it may be assumed that the second term in Equation (28.6.7) gives no contribution to $\partial S/\partial t)_{\text{coll}}$. On exchanging k and k′ in Equation (28.7.9) we obtain

$$\left.\frac{\partial S}{\partial t}\right)_{\text{coll}} = \iint L(\mathbf{k},\mathbf{k}')[\phi(\mathbf{k}')-\phi(\mathbf{k})]\phi(\mathbf{k}')\,d\mathbf{k}\,d\mathbf{k}' \qquad (28.7.10)$$

since $L(\mathbf{k},\mathbf{k}') = L(\mathbf{k}',\mathbf{k})$. Finally by adding Equations (28.7.9) and (28.7.10) we find

$$\left.\frac{\partial S}{\partial t}\right)_{\text{coll}} = \tfrac{1}{2}\int L(\mathbf{k},\mathbf{k}')[\phi(\mathbf{k})-\phi(\mathbf{k}')]^2\,d\mathbf{k}\,d\mathbf{k}'\,.$$

28.8 It was shown in Equation (28.7.5) that $-\sigma_{\text{conv}}$ calculated for electron flow is identical with the form for σ_e as given by Equation (28.5.1) for the case of macroscopic variables. This suggests that just as for macroscopic variables it was possible to formulate variational principles for the relationship between forces and fluxes, based on a stationary value of the entropy production, so similarly it may be possible to obtain the Boltzmann flow equation (28.7.4) by a similar device with $\phi(\mathbf{k})$ as the analogue of J_p. Now for the steady state, it follows from Equation (28.7.4) that $-\sigma_{\text{conv}} = \sigma_{\text{coll}}$ and therefore by analogy with the macroscopic approach we suggest the formulation of variational principles by using $-\sigma_{\text{conv}}$ instead of σ_e and σ_{coll} instead of σ_i.

Prove that the Boltzmann Equation (28.7.4) is given by minimising $\sigma_{\text{coll}}+2\sigma_{\text{conv}}$ with respect to varying $\phi(\mathbf{k})$, or by maximising σ_{coll} subject to it equalling $-\sigma_{\text{conv}}$, or by minimising $\sigma_{\text{coll}}/\sigma_{\text{conv}}^2$. [Hint: Follow the same approach as in Problem 28.2 using the continuous variable k instead of the discrete variable p, and replacing summations by integrations.

The analogue of the positive definite character of M_{pq} and its symmetry are the relations (28.7.3) now satisfied by $L(\mathrm{k}, \mathrm{k}')$, so that for all $\theta(\mathrm{k})$ and $\phi(\mathrm{k})$,

$$\iint L(\mathrm{k}, \mathrm{k}')[\theta(\mathrm{k}) - \theta(\mathrm{k}') - \phi(\mathrm{k}) + \phi(\mathrm{k}')]^2 \mathrm{d}\mathrm{k}\,\mathrm{d}\mathrm{k}' \geqslant 0\,.]$$

Solution

Consider a general function $\theta(\mathrm{k})$ and a function $\phi(\mathrm{k})$ satisfying Equation (28.7.4). Then since $L(\mathrm{k}, \mathrm{k}') \geqslant 0$ for all k, k′,

$$\iint L(\mathrm{k}, \mathrm{k}')[\theta(\mathrm{k}) - \theta(\mathrm{k}') - \phi(\mathrm{k}) + \phi(\mathrm{k}')]^2 \mathrm{d}\mathrm{k}\,\mathrm{d}\mathrm{k}' \geqslant 0$$

and thus

$$\int L(\mathrm{k}, \mathrm{k}')[\theta(\mathrm{k}) - \theta(\mathrm{k}')]^2 \mathrm{d}\mathrm{k}\,\mathrm{d}\mathrm{k}' - 2\int L(\mathrm{k}, \mathrm{k}')[\theta(\mathrm{k}) - \theta(\mathrm{k}')][\phi(\mathrm{k}) - \phi(\mathrm{k}')]\,\mathrm{d}\mathrm{k}\,\mathrm{d}\mathrm{k}'$$
$$+ \int L(\mathrm{k}, \mathrm{k}')[\phi(\mathrm{k}) - \phi(\mathrm{k}')]^2 \mathrm{d}\mathrm{k}\,\mathrm{d}\mathrm{k}' \geqslant 0\,. \qquad (28.8.1)$$

Now

$$\int L(\mathrm{k}, \mathrm{k}')[\theta(\mathrm{k}) - \theta(\mathrm{k}')][\phi(\mathrm{k}) - \phi(\mathrm{k}')]\,\mathrm{d}\mathrm{k}\,\mathrm{d}\mathrm{k}'$$
$$= \int L(\mathrm{k}, \mathrm{k}')\theta(\mathrm{k})[\phi(\mathrm{k}) - \phi(\mathrm{k}')]\,\mathrm{d}\mathrm{k}\,\mathrm{d}\mathrm{k}' - \int L(\mathrm{k}, \mathrm{k}')\theta(\mathrm{k}')[\phi(\mathrm{k}) - \phi(\mathrm{k}')]\,\mathrm{d}\mathrm{k}\,\mathrm{d}\mathrm{k}' \qquad (28.8.2)$$

and exchanging k and k′ in the second term, bearing in mind that this leaves L unchanged, we obtain the right hand side of Equation (28.8.2) as

$$2\int \theta(\mathrm{k})\left\{\int L(\mathrm{k}, \mathrm{k}')[\phi(\mathrm{k}) - \phi(\mathrm{k}')]\,\mathrm{d}\mathrm{k}'\right\}\mathrm{d}\mathrm{k}\,.$$

On substituting from Equation (28.7.4) this becomes

$$2e\mathcal{E} \cdot \int \mathrm{v}\frac{\partial f^0}{\partial E}\theta(\mathrm{k})\,\mathrm{d}\mathrm{k}\,.$$

We substitute this expression in the second term of the inequality (28.8.1) and obtain

$$\int L(\mathrm{k}, \mathrm{k}')[\theta(\mathrm{k}) - \theta(\mathrm{k}')]^2 \mathrm{d}\mathrm{k}\,\mathrm{d}\mathrm{k}' - 4e\mathcal{E} \cdot \int \mathrm{v}\frac{\partial f^0}{\partial E}\theta(\mathrm{k})\,\mathrm{d}\mathrm{k}$$
$$\geqslant -\int L(\mathrm{k}, \mathrm{k}')[\phi(\mathrm{k}) - \phi(\mathrm{k}')]^2 \mathrm{d}\mathrm{k}\,\mathrm{d}\mathrm{k}'$$
$$= \int L(\mathrm{k}, \mathrm{k}')[\phi(\mathrm{k}) - \phi(\mathrm{k}')]^2 \mathrm{d}\mathrm{k}\,\mathrm{d}\mathrm{k}' - 4e\mathcal{E} \cdot \int \mathrm{v}\frac{\partial f^0}{\partial E}\phi(\mathrm{k})\,\mathrm{d}\mathrm{k}$$

on applying the above transformation to the case $\theta = \phi$. The left hand side of this inequality is $2(\sigma_{coll} + 2\sigma_{conv})$ calculated with function θ, while the right hand side is the same quantity calculated with function ϕ. It follows that ϕ minimises $\sigma_{coll} + 2\sigma_{conv}$.

By analogy with the methods of Problem 28.2 the other two variational principles may be similarly proved.

The variational principles enunciated here may be used to obtain approximate solutions of the Boltzmann equation following the methods indicated in Problems 28.4 and 28.5. The choice of suitable trial functions is, however, generally rather more difficult.

Author Index

Subject Index

Problem numbers, not pages, are shown. Bold print indicates whole Chapter numbers.